Stochastics, Control and Robotics

Stochastics, Control and Robotics

Harish Parthasarathy
Professor
Electronics & Communication Engineering
Netaji Subhas Institute of Technology (NSIT)
New Delhi, Delhi-110078

CRC Press
Taylor & Francis Group
Boca Raton London New York

CRC Press is an imprint of the
Taylor & Francis Group, an **informa** business

Manakin
PRESS

First published 2021
by CRC Press
2 Park Square, Milton Park, Abingdon, Oxon, OX14 4RN

and by CRC Press
6000 Broken Sound Parkway NW, Suite 300, Boca Raton, FL 33487-2742

CRC Press is an imprint of Informa UK Limited

British Library Cataloguing-in-Publication Data
A catalogue record for this book is available from the British Library

Library of Congress Cataloging-in-Publication Data
A catalog record has been requested

ISBN: 978-1-032-05585-5 (hbk)
ISBN: 978-1-003-19822-2 (ebk)

Manakin
PRESS

Brief Contents

1. Classical Robotics and Quantum Stochastics 1–128

2. Electromagnetus and Related Partial Differential Equation 129–149

3. Radon and Group Theoretic Transforms with Robotics Applications 151–175

4. Stochastic Filtering and Control, Interacting Particles 177–221

5. Classical and Quantum Robotics 223–253

6. Large Deviations, Classical and Quantum General Relativity with GPS Application 255–338

7. Quantum Signal Processing 339–476

Detailed Contents

1. Classical Robotics and Quantum Stochastics **1–128**

[1] A Problem in Robotics with Current Carrying Links 1

[2] Quantum Filtering 4

[3] Evan's Hudson Flow 8

[4] Quantum Version of a Gauss-Markov Process 10

[5] Problems on the δ-Function 13

[6] Conditional Expectation in Quantum Filtering 16

[7] K.R. Parthasarathy Quantum Markov Processes 18

[8] Quantum Stochastic Lyapunov Theory 20

[9] Let P be a Solution to the Martingale Problem $\pi(x, a, b)$, *i.e.* 27

[10] Problem from Revuz and Yor 28

[11] Gough and Kostler Paper on Quantum Filtering Some Remarks 28

[12] Quantization of Master. Slave Robot Motion Using Evans-Hudson Flows 33

[13] Problem From KRP QSC 37

[14] Moment Generating Function for Certain Quantum Random Variables 41

[15] Belavkin Filtering Contd. 42

[16] Modelling Noise in Quantum Mechanical Systems 43

[17] Belavkin has Derived a Quantum Stochastic Differential Equation (*qsde*) for $\widehat{X}(t|t)$ 48

[18] Waveguide Modes 50

[19] Estimating the Initial State from Measurements of an Observable after the State Passes through a Sequence of Quantum Channels 56

[20] Quantum Poisson Random Variable 57

[21] Quantum Relative Entropy Evolution 59
[22] Gauge Group Invariance of a Wave Eqn. 60
[23] Problem 61
[24] Quantum Oscillator Perturbed by an Electromagnetic Field 61
[25] Quantization of Mechanical Systems 64
[26] Quantum Filtering 66
[27] Perspective Projection by Camera Attached to a Robot 71
[28] Proof of The Positivity of Quantum Relative Entropy 72
[29] Entropy of a Quantum System 73
[30] Dirac Eqn. Based Temperature Estimation of Black-Body Temperature Field 76
[31] Quantum Image Processing on a C^∞– Manifold 81
[32] Belavkin Contd. (Quantum Non-Linear Filtering) 82
[33] Remarks on Theorems from V. Kac. Infinite Dimensional Lic Algebra. P.87 83
[34] Quantum Mutual Information 84
[35] Scattering Theory Applied to Quantum Gate Design 86
[36] Quantization of the Simple Exclusion Model 93
[37] Quantum Version of Nearest Neighbour Interactions 96
[38] Remarks and Comments on "Collected Papers of S.R.S Varadhan" Vol. 4, Particle System and Their Large Deviations 101
[39] Quantum String Theory 103
[40] Quantum Shannon Theory Contd. 108
[41] Basics of Cq-Information Theory 111
[42] Typical Sequences and Error Probability in Classical and Quantum Coding 114
[43] Quantum Image Processing Some Basic Problems 125

2. Electromagnetus and Related Partial Differential Equation 129–149

[1] Computation of the Perturbe Characteristics Frequencies in a Cavity Resonator with in Homogeneous Dielectric and Permittivity 129
[2] Numerical Methods for *pde* 133
[3] Tracking of Moving Targets Using a Camera Attached to the Tip of a 2 = Link Robot 137
[4] Problem 140

[5] n-Dimensional Helmholtz Green's Funtion 140

[6] Edge Diffusion 141

[7] Waves in Metamaterials 141

[8] Stroock Partial Differential Equations for Probabilities $\varphi \simeq$ Function, $u \simeq$ Sobolev Distribution 148

3. Radon and Group Theoretic Transforms with Robotics Applications 151–175

[1] Modern Trends in Signal Processing by Harish Parthasarathy, NSIT 151

[2] Radon-Transform Based Image Processing 152

[3] Image Processing Using Invariants of the Permutation Group 155

[4] Radon Transform of Rotated and Translated Image Field 157

[5] Estimating the Rotation Applied to the 3-D Robot Links from Electromagnetic Field Pattern 158

[6] Kinetic Energy of a Two Link Robot 159

[7] Vector Potential Generated by a 2-Link Robot 164

[8] Group Associated with Robot Motion 166

[9] Radon Transform 168

[10] Campbelt Baker-Hausdorff Formula 169

[11] Problem on Rigid Body Motion 171

[12] Remarks on Theorems from V. Kac. Infinite Dimensional Lic Algebra. P.87 171

[13] Theta a Function From Group Theory P 252 (V.Kac) 172

[14] Iwasava Decomposition 175

4. Stochastic Filtering and Control, Interacting Particles 177–221

[1] Bernoulli Filters 177

[2] A Problem in Large Deviation Theory 179

[3] Large Deviations in Image Trajectory on a Screen Taken by a Robot Camera in Motion 180

[4] P406 Revuz and Yor 184

[5] Problem from "Continuous Martingales and Brownian Motion" by Revuz and Yor 188

[6] Problem from Revuz and Yor 189

[7] Collected Papers of Varadhan-Interacting Particle System, Vol. 4. Hamiltonian System and Hydrodynamical Equations 191

[8] (Ph.D. Problem of Rohit) For Rohit and Dr. Vijayant. Skorohod Optional Stochastic Control. Optimal Stochastic Control of Master and Slave Robot 195
[9] English-Chapter-1 198
[10] Markov Processes-Some Application 199
[11] Comments and Proofs on "Particle System" Collected Papers of S.R.S. Varadhan Vol. 4 206
[12] Interacting Diffusions 209
[13] This is Independent of *s*. It is Reasonable to Expert that at Equilibrium 209
[14] Wong Zakai Filtering Equation 211
[15] Stochastic Control of Finance Market 216
[16] Wong-Zokai Equation 218
[17] Filtering Theory with State and Measurement Noises Having Correlation 219

5. Classical and Quantum Robotics **223–253**
[1] Mathematical Pre-Requisites for Robotics 223
[2] Belavkin Filtering Contd. 242
[3] Dirac Eqn., for Rigid Body Robot (Single Link) 243
[4] Quantum Robot Dynamics 249
[5] Hamiltonian Based Quantum Robot Tracking 252
[6] Robot Interacting with a Klein-Gordon Field 253

6. Large Deviations, Classical and Quantum General Relativity with GPS Application **255–338**
[1] Intuitive Derivation of the Gartner Ellis LDP 255
[2] LDP in Robotic Vision 256
[3] LMS Algorithm for Non-Linear Disturbance Observer in a Two-Link Robot System 260
[4] Disturbance Observer Design in Robotics 261
[5] Disturbance Observer Continuation 262
[6] Results in the LMS' Iteration 262
[7] Notation for $\psi(t)$ in the LMS Disturbance Observer 263
[8] Approximate Convergence Analysis 263
[9] Statistical Model for Disturbance Estimation Error Based on LMS 264
[10] Statistics of LMS Coefficients for the Disturbance Observer 264

[11] $\underline{B}(t) \in \mathbb{R}^d$ d-Vector Valued Standard Brownian Motion 265
[12] General Relativistic Correction to GPS 267
[13] Thiemann Modern QGTR 269
[14] Generalization of General Relativistic String Theory 273
[15] Symmetric Space-Times in General Relativity and Cosmology 276
[16] Selected Topics in General Relativity 298
[17] Perturbations in the Tetrad Caused by Metric Perturbation 300
[18] Geodesic Equation in Tetra Formalism 301
[19] Quantum General Relativistic Scattering 302
[20] Derive Dirac Equation in Curved Space-Time Using the Newman Penrose Formalism: Special-Relativity 308
[21] Proofs of Identities from S. Chandra Sekhar, "The Mathematical Theory of Blackholes". Cartain's Second Equation of Structure 313
[22] Mathematical Preliminaries for Cartan's Eqn. of Structure 316
[23] Reflection and Refraction at a Metamaterial Surface 318
[24] Maxwell's Equations in an Unhomogeneous Medium with Background Gravitation Taken into Account 321
[25] Relativistic Kinematics* Motions of a Rigid Body in a Gravitational Field a General Relativistic Calculation 323
[26] Random Fluctuation in the Gravitational Field Produced by Random Fluctuation in the Energy-Momentum Tensor of the Matter Plus Radiation Field 326
[27] Special Relativistic Plasma Physics 330

7. Quantum Signal Processing **339–476**

[1] Quantum Scattering of a Rigid Body 339
[2] Optimal Control of a Field 339
[3] Example from Quantum Mechanics 340
[4] Example from Quantum Field Theory 341
[5] Quantum Signal Processing (Contd.) 342
[6] em Field Theory Based Image Processing 344
[7] Derivation of the Kallianpur-Striebel Formula 346
[8] Slave Dynamics with Parametric Uncertainties Using Adaptive Controllers 348

[9] Belavkin's Theory of Quantum Non-Linear Filtering 367
[10] A Problem Related to the Generator of Brownian Motion 404
[11] A Problem Related to Poisson Random Fields 405
[12] Relativistic Kinematics of Rigid Bodies 412
[13] Quantum Robotics Disturbance Observer 417
[14] Quantum Field Theory in the Presence of Stochastic Disturbance 420
[15] Simultaneous Tracking, Parametric Uncertainties Estimation and Disturbance Observer in a Single Robot 422
[16] Quantum Filtering to Signal Estimation and Image Processing 430
[17] Image Transmission Through Quantum Channels 431
[18] Entanglement Theory in Quantum Mechanics 433
[19] Some Other Remarks Related to The Above Problem 447
[20] Image Processing from the Em Field View Point 454
[21] CAR 455
[22] An Identity Concerning Positive Definite Functions on a Group 455
[23] Example of Construction of a Spherical Function 456
[24] Spherical Functions in Image Processing 458
[25] CAR from CCR (KRP Quantum Stochastic Calculus) 459
[26] Spherical Function (Helgason Vol. II) 461
[27] Clarifications of Problems Related to the I to measure of Brownian motion 463
[28] Problem from Revuz and Yor 464
[29] Group Theory and Robotics 465
[30] Suppose $\nabla^2 f(x) = 0, x \in D$ 469
[31] Entropy Evolution in Sudarshan-Lindblad Equation 472
[32] Problem on Rigid Body Motion 476
[33] Problem from Revuz and Yor 476

Preface

This is a book about electromagnetic, filtering & control of classical and quantum stochastic process including diverse topics such as scattering theory, quantum information, large deviation, classical and quantum general relativity etc. The topics are diverse but I have made some effort to unify them. Some parts of the text also includes classical and quantum robot dynamics with some remarks about how filtering and control can be applied to such systems. Lie group theory has been applied to robotics having 3D links and robotics carrying current interacting with external electromagnetic field having also being studied in this text. Some sections on electromagnetics deal with properties of the Dirac-delta functions and how Green functions for the Laplace and Helmholtz operations can be computed.

Some of the high points of the book are the Hudson Parthasarathy (HP) quantum Ito calculus, Belkavin's theory of quantum filtering point on the HP calculus, techniques for quantizing general relativity and application of general relativity to photon propagation for GPS purposes. I am sure that the material of this book would be useful to research scientists and engineers working on robotics, quantum mechanics, stochastic processes and to physicists working in general relativity. It will also be useful for research students in their fields who are looking for a research problem to solve.

Author

1

Classical Robotics and Quantum Stochastics

[1] A Problem in Robotics with Current Carrying Links:

To determine the configuration of the two link Robot System *i.e.* the 3 Euler angles specifying the configuration of the lower arm and the 3 Euler angles specifying the relative configuration of the upper arm *w.r.t.* the lower arm. The two arms carry current and from the far field radiation pattern, the link configuration arc to be estimated. A point $\underline{r}_0$ in the lower arm at time $t = 0$ moves to

$$\underline{r}(t) = R_1(t)\underline{r}_0$$

After time t. A point $\underline{S}_0$ in the upper arm at time $t = 0$ moves to

$$R_1(t)p_0 + R_2(t)R_1(t)(\underline{S}_0 - p_0)$$

after time t. Here

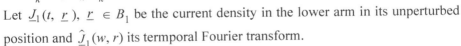

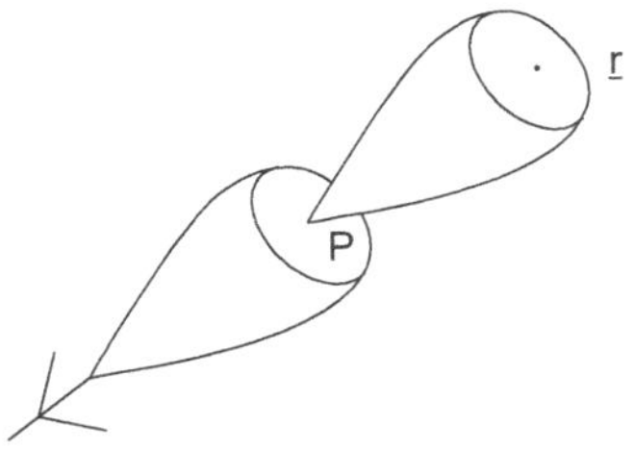

$$R_1(t) = R(\varphi_1(t), \theta_1(t), \psi_1(t)),$$
$$R_2(t) = R(\varphi_2(t), \theta_2(t), \psi_2(t)),$$

Where $\quad R(\varphi, \theta, \psi) = R_z(\varphi)R_x(\theta)R_z(\psi)$

Note that $R_1(t), R_2(t) \in SO$ *(Eq. 3), i.e.*

Det $R_k = 1$, $R_k^T R_k = I$, $k = 1, 2$.

Let $\underline{J}_1(t, \underline{r})$, $\underline{r} \in B_1$ be the current density in the lower arm in its unperturbed position and $\hat{\underline{J}}_1(w, r)$ its termporal Fourier transform.

Let $\underline{J}_2(t, \underline{r})$, $\underline{r} \in B_2$ be the current density in the upper arm in its unperturbed position and $\hat{\underline{J}}_2(w, r)$ is Fourier transform.

Here B_1 is the volume of the lower arm at time $t = 0$ and B_2 the volume of the upper arm at time $t = 0$. Assume t is fixed so we can write R_k for $R_k(t)$, $k = 1, 2$.

The current density in space at time t is then (in the frequency domain)

$$\hat{\underline{J}}(w, \underline{r}) = \hat{\underline{J}}_1(w, R_1^{-1}\underline{r}) + \hat{\underline{J}}_2(wp_0 + (R_2R_1)^{-1}(\underline{r} - R_1p_0))$$

Note that $B_1 \cap B_2 = \phi$ and

$\hat{\underline{J}}_1(w, R_1^{-1}\underline{r})$ is non zero only when

$$\underline{r} \in R_1(B_1) \text{ and } \hat{\underline{J}}_2(w_1 p_0 + (R_2R_1)^{-1}(\underline{r} - R_1 p_0))$$

is non zero only when $\underline{r} \in R_2R_1(B_2 - p_0) + R_1 p_0$

We ignore w and $\wedge, \sim$ and write $R = R_1$, $S = (R_2R_1)^{-1}$. Then

$$\underline{J}(\underline{r}) = \underline{J}_1\left(\underline{\underline{R}}^{-1}\underline{r}\right) + \underline{J}_2\left(p_0 + \underline{\underline{S}}\left(\underline{r} - \underline{\underline{R}}p_0\right)\right) + \underline{W}(\underline{r})$$

R, S need to be estimated.

Here $\underline{W}(\underline{r})$ is the Gaussian noise field.

The far field Magnetic vector potential is proportional to

$$\begin{aligned}
\underline{A}(\hat{\underline{r}}) &= \int \underline{J}(\underline{r}')\exp\left(jk\hat{r}.\underline{r}'\right)d^3r' \\
&= \int \underline{J}_1\left(\underline{\underline{R}}^{-1}\underline{r}'\right)\exp\left(jk\hat{\underline{r}}.\underline{r}'\right)d^3r' \\
&\quad + \int \underline{J}_2\left(p_0 + S\left(\underline{r}' - Rp_0\right)\right)\exp\left(jk\hat{r}.\underline{r}'\right)d^3r' \\
&= \int \underline{J}_1\left(\underline{\underline{R}}^{-1}\underline{r}'\right)\exp\left(jk\hat{\underline{r}}.\underline{r}'\right)d^3r' \\
&\quad + \int J_2(\underline{r}')\exp\left(jk\left(\hat{r}, Rp_0 + S^{-1}\left(\underline{r}' - \underline{p}_0\right)\right)\right)d^3r' \\
&= F_1\left(R^T\hat{r}\right) + \exp\left(jk\left(\hat{r}, \left(R - S^{-1}\right)p_0\right)\right)F_2\left(S\hat{r}\right) \\
&= F_1\left(R^{-1}\hat{r}\right) + \exp\left(jk\left(\left(R^{-1} - S\right)\hat{\underline{r}}, p_0\right)\right)F_2\left(S\hat{r}\right)
\end{aligned}$$

where

$$\begin{aligned}
F_1(\hat{r}) &= \int J_1(\underline{r}')\exp\left(jk\left(\hat{r}, \underline{r}'\right)\right)d^3r', \\
F_2(\hat{r}) &= \int J_2(\underline{r}')\exp\left(jk\left(\hat{r}, \underline{r}'\right)\right)d^3r'.
\end{aligned}$$

The ML estimators of R, S are

$$\left(\hat{R}, \hat{S}\right) = \arg\min \int \| \underline{A}(\hat{\underline{r}}) - F_1(R^{-1}\hat{r})$$

$R, S \in SO(Eq.\ 3)\ S^2 - \exp\left(jk\left(\left(R^{-1} - S\right)\hat{r}, p_0\right)\right)F_2\left(S\hat{r}\right)\|^2\, dS(\hat{r})$

Can those estimates be carried out more efficiently then by search methods using group representation theory?

Suppose S is known and we wish to estimate R.

Note that

$$\begin{aligned}
\underline{A}(\hat{\underline{r}}) &= F_1\left(R_1^{-1}\hat{r}\right) + \exp\left\{jk\left(\left(R_1^{-1} - R_1^{-1}R_2^{-1}\right)\hat{r}, p_0\right)\right\}F_2\left(R_1^{-1}R_2^{-1}\hat{r}\right) \\
&= F_1\left(R_1^{-1}\hat{r}\right) + \exp\left\{jk((R_1^{-1}(I - R_2^{-1})\hat{r}, p_0)\right\}.F_2\left(R_1^{-1}R_2^{-1}\hat{r}\right)
\end{aligned}$$

Let $\left\{Y_{lm}(\vec{r}) \| m \le l, l = 0, 1, 2, \ldots\right\}$ be the spherical harmonic. Thus if $\pi_l(R)$ is the representation of $SO\ (Eq.\ 3)$

in the space $V_l = \text{span}\{Y_{lm} \| m| \le l\}$, then

$$\int_{S^2} F_1\left(R_1^{-1}\hat{r}\right) Y_{lm}\left(\hat{r}\right) dS\left(\hat{r}\right)$$

$$= \int_{S^2} F_1\left(\hat{r}\right) Y_{lm}\left(R_1 \hat{r}\right) dS\left(\hat{r}\right)$$

$$= \int_{S^2} F_1\left(\hat{r}\right) \sum_{|m'| \le l} \left[\pi_l\left(R_1^{-1}\right)\right]_{m'm} Y_{lm'}\left(\hat{r}\right) dS\left(\hat{r}\right)$$

$$= \sum_{|m'| \le l} \overline{\left[\pi_l\left(R_1\right)\right]}_{mm'} \tilde{F}[lm']$$

where $$\tilde{F}_1[lm] = \int_{S^2} F_1\left(\hat{r}\right) Y_{lm}\left(\hat{r}\right) dS\left(\hat{r}\right)$$

Further $$\int_{S^2} \exp\left\{jk\left(R_1^{-1}\left(I - R_2^{-1}\right)\hat{r}, p_0\right)\right\} F_2\left(R_1^{-1}R_2^{-1}\hat{r}\right) Y_{lm}\left(\hat{r}\right) dS\left(\hat{r}\right)$$

$$= \int_{S^2} \exp\left\{jk\left(\left(R_1^{-1}R_2 - I\right)\hat{r}, p_0\right)\right\} F_2\left(R_1^{-1}\hat{r}\right) Y_{lm}\left(R_2 \hat{r}\right) dS\left(\hat{r}\right)$$

$$= \sum_{m'} \overline{\left[\pi_l\left(R_2\right)\right]}_{mm'} \int_{S^2} \exp\left(jk\left(\left(R_1^{-1}R_2 - I\right)\hat{r}, p_0\right)\right) F_2\left(R_1^{-1}\hat{r}\right) Y_{lm'}\left(\hat{r}\right) dS\left(\hat{r}\right)$$

Where $\bar{r}$ is the unit vector in the direction of $\bar{r}$.

Now the function $\hat{r} \to \exp\left(jk\left(\hat{r}, p_0\right)\right)$ on S^2 can be expanded using spherical hormonics, say

$$\exp\left(jk\left(\hat{r}, p_0\right)\right) = \sum_{l,m} C(lm) Y_{lm}\left(\hat{r}\right)$$

Then,

$$\exp\left\{jk\left(R_1^{-1}R_2\hat{r}, p_0\right)\right\} = \sum_{l,m} C(lm) Y_{lm}\left(R_1^{-1}R_2\hat{r}\right)$$

$$= \sum_{l,m} C(lm) \sum_{m'} \left[\pi_l\left(R_2^{-1}R_1\right)\right]_{m'm} Y_{lm'}\left(\hat{r}\right)$$

$$= \sum_{lmm'} C(l,m)\left[\pi_l\left(R_2^{-1}R_1\right)\right]_{m'm} Y_{lm'}\left(\hat{r}\right)$$

Also, $$F_2(\hat{r}) = \sum_{lm} d(lm) Y_{lm}\left(\hat{r}\right)$$

$$(CClm) = \int_{S^2} \exp\left\{jk\left(\hat{r}, p_0\right)\right\} \overline{Y_{lm}\left(\hat{r}\right)} dS\left(\hat{r}\right),$$

$$d(lm) = \int_{S^2} F_2\left(\hat{r}\right) Y_{lm}\left(\hat{r}\right) dS\left(\hat{r}\right)$$

Then,

$$\int \exp\left\{jk\left(\left(R_1^{-1}R_2 - I\right)\hat{r}, P_0\right)\right\} F_2\left(R_1^{-1}\hat{r}\right) Y_{lm}\left(R_2\vec{r}\right) dS\left(\hat{r}\right)$$

$$= \sum_{m'} \overline{\left[\pi_l\left(R_2\right)\right]}_{mm'} \int_{S^2} Y_{lm'}\left(\hat{r}\right) dS\left(\hat{r}\right) \exp\left(-jk\left(\hat{r}, p_0\right)\right)$$

$$\sum_{l_1 m_1 m_2} C\left(l_1, m_1 m_2\right)\left[\pi_{l_1}\left(R_1^{-1}R_1\right)\right]_{m_2 m_1} Y_{l_1 m_2}\left(\vec{r}\right) \sum_{l_3 m_3} d\left(l_3 m_3\right) Y_{l_3 m_3}\left(R_1^{-1}\hat{r}\right)$$

$$= \sum_{l_1 m_1 m_2 l_3 m_3 m'} \overline{\left[\pi_l\left(R_2\right)\right]}_{mm'} \left[\pi_{l_1}\left(R_2^{-1}R_1\right)\right]_{m_2 m_1} \left[\pi_{l_3}\left(R_1\right)\right]_{m_4 m_3}$$

$$C\left(l_1 m_1 m_2\right) d\left(l_3 m_3\right) \int_{S^2} \exp\left(-jk\left(\hat{r}, p_0\right)\right) Y_{l_1 m_2}\left(\hat{r}\right) Y_{l_3 m_4}\left(\hat{r}\right) dS\left(\hat{r}\right)$$

These formulae can be used to derive the least squares estimator for R_1, R_2 (which is the same as the *ML* estimates in the presence of white Gaussian noise.

❖❖❖❖❖

[2] Quantum Filtering:

$$j_t(\chi) = V_t^* \times V_t.\ \underline{X} \varepsilon \mathcal{L}(\hbar)$$

$$dV_t = (-iH\,dt + L_1 dA_t + L_2 dA_t^* + Sd\Lambda_t)V_t$$

where $H, L_1, L_2, S \, \varepsilon \, \mathcal{L}(\hbar)$ are chosen so that

$$V_t^* V_t = I \,\forall\, t \geq 0.\ (X \triangleq X \oplus I)$$

Then $$dj_t(\chi) = j_t(\mathcal{L}_0)dt + j_t(\mathcal{L}_1\chi)dA_t + j_t(\mathcal{L}_2\chi)dA_t^* + j_t(\mathcal{L}_3\chi)d\Lambda_t$$

$\mathcal{L}_0, \mathcal{L}_1, \mathcal{L}_2, \mathcal{L}_3$ are the Evan-Hudson structure map. Computation of $\mathcal{L}_0, \mathcal{L}_2, \mathcal{L}_3$:

$$\begin{aligned} dj_t(X) &= dV_t^* X V_t + V_t^* X dV_t + dV_t^* X dV_t \\ &= V_t^*(i(H^* X - XH^*)dt + (L_2^* X + XL_1)dA_t \\ &\quad + (L_1^* X + XL_2^*)\alpha A_t^* + (S^* X + XS)d\Lambda_t \\ &\quad + L_2^* XL_1 dt + S^* XL_2 dA_t^* + L_1 XS dA_t + S^* XS d\Lambda_t)V_t \\ &= j_t(i(H^* X - XH^*) + L_2^* XL_1)dt + j_t(L_2^* X + XL_1 \\ &\quad + L_1 XS)dA_t + j_t(L_1^* X + XL_2 + S^* XL_2)dAt^* \\ &\quad + j_t(S^* X + XS + S^* XS)d\Lambda_t \end{aligned}$$

Thus, $\mathcal{L}_0(X) = i(H^* X - XH^*) + L_2^* XL_1$,

$$\begin{aligned} \mathcal{L}_1(X) &= L_2^* X + XL_1 + L_1 XS, \\ \mathcal{L}_2(X) &= L_1^* X + XL_2 + S^* XL_2, \\ \mathcal{L}_3(X) &= S^* X + XS + X^* XS. \end{aligned}$$

$\{j_t(X), t \geq 0\}$ in the state process.

Measurement process

$$Y^{out}(t) = V_t^*(I \otimes Y^{in}(t))V_t$$
$$dY^{in}(t) = P_1(t)dA_t + P_1{}^*(t)dA_t^* + P_2(t)d\Lambda_t$$

where $\quad P_1(t), P_2(t) \in \mathcal{L}(\Gamma s(\mathcal{H}_t)), P_2{}^*(t) = P_2(t)$

Clearly $VT^*(I \otimes Y^{in}(t))V_T = Y^{out}(t) \forall T \geq t$

$$Y^{in}(t) = \int_0^t \left(P_1(S)dA_s + \left(P_1^*(S)\right)dA_s^* + \left(P_2(S)\right)d\Lambda_s\right)$$

$$\begin{aligned} dY^{out}(t) &= dY^{in}(t) + dV_t^*(I \otimes dY^{in}(t))V_t + V_t^*(I \otimes dY^{in}(t))dV_t \\ &= P_1(t)dA_t + P_1{}^*(t)dA_t^* + P_2(t)d\Lambda_t \\ &\quad + V_t^*(L_1 \otimes P_1{}^*(t))V_t dt + V_t^*(L_1 \otimes P_2(t))V_t dA_t \\ &\quad + V_t^*(S^* \otimes P_2(t))V_t d\Lambda_t + V_t^*(L_2 \otimes P_1(t))V_t^* dt \\ &\quad + V_t^*(L_2 \otimes P_2(t))V_t^* dA_t^* + V_t^*(S \otimes P_2(t))V_t^* d\Lambda_t \end{aligned}$$

Thus, $\quad Y^{out}(t) \equiv j_t(Y^{in}(t) = j_t(Y^{in}(t)) \ \forall\ T \geq t$

and $\quad dY^{out}(t) = j_t(L_1P_1{}^*(t) + L_2P_1(t))dt + j_t(L_1P_2(t))dA_t$
$\quad\quad + j_t(L_2P_2(t))dA_t^* + j_t(SP_2(t))d\Lambda_t$

For $T \geq t$,

$$\begin{aligned} [j_T(X), Y^{out}(t)] &= [V_T{}^*(\chi \otimes I)V_T, V_T{}^*(I \otimes Y^{in}(t))V_T] \\ &= [V_T{}^*(\chi \otimes I, I \otimes Y^{in}(t))]V_T = 0. \end{aligned}$$

This is the non-demolition property.

Assume that the measurement observables $\{Y^{out}(t) : t \geq 0\}$ are commutative *i.e.*

$$[Y^{out}(t), Y^{out}(S)] = 0 \ \forall\ t, S \geq 0.$$

This is equivalent to saying that

$$[Y^{in}(t), Y^{in}(S)] = 0 \ \forall\ t, S \geq 0$$

Since $\quad Y^{out}(t) = V_{tvs}^* Y^{in}(t) V_{tvs}$

and $\quad Y^{out}(S) = V_{tvs}^* Y^{in}(S) V_{tvs} \ \forall\ t, S \geq 0$

where $tvs = \max(t, S)$

$\{Y^{out}(t) ; t \geq 0\}$ will be commutative if

$$\begin{aligned} [P_1(t), A_S] &= 0, [P_1(t), A_S^*] = 0, \\ [P_1(t), A_S] &= 0, [P_2(t), A_S] = 0, \\ [P_2(t), A_S] &= 0, [P_2(t), A_S^*] = 0, \\ [P_1(t), P_1(S)] &= 0, [P_1(t), P_2(S)] = 0, \\ [P_2(t), P_1(S)] &= 0, \ \forall\ t > S. \end{aligned}$$

These condition hold if for example $P_1(t)$, $P_2(t)$ are scalar valued functions of time *i.e.*

$$P_1(t) = C_1(t)I,$$
$$P_2(t) = C_2(t)I,$$

$C_1(t), C_2(t) \in \mathbb{C}, \bar{C}_2(t) = C_2(t)(i.e.\ C_2(t) \in \mathbb{R})$.

Let $\eta_{t]}^{\text{in}}$ be the Von-Neumann algebra generated by $\{Y^{\text{in}}(S) : S \le t\}$

and $\eta_{t]}^{\text{out}}$ the Von-Neumann algebra generated by $\{Y^{\text{out}}(S) : S \le t\}$.

Then, $$\eta_{t]}^{\text{out}} = V_t^* \eta_{t]}^{\text{in}} V_t$$

Non-Commutative Girsanov Trick

Let $$\pi_t(\chi) = \mathbb{E}[j_t(\chi) \mid \eta_{t]}^{\text{out}}]$$

Then $$\pi_t(\chi) = V_t^* \widetilde{\mathbb{E}}_t [j_t(X) \mid \eta_{t]}^{\text{out}}] V_t$$

Where $$\widetilde{\mathbb{E}}_t[\xi] = \widetilde{\mathbb{E}}_t(V_t^* \xi V_t)$$

Remark:

Let $$\tilde{\mathbb{E}}(X) = \mathbb{E}(U^* X U)$$

and $\tilde{m} = U^* m U$ where U is unitary and m is an Abelian Algebra. Then, for $\tilde{\xi} \in \tilde{m}$, we have $\tilde{\xi} = U^* \xi U$

for some $\xi \in m$ and then

$$\mathbb{E}\left[\left(U^* XU - \mathbb{E}\left[U^* XU \mid \tilde{m}\right]\right)\tilde{\xi}\right] = 0$$

i.e. $$\mathbb{E}\left(U^* X \xi U - \mathbb{E}\left[U^* XU \mid \tilde{m}\right] \xi U^* XU\right) = 0$$

Note that

$$\begin{aligned} \mathbb{E}[U^* XU \mid \tilde{m}]\, U^* \xi U &= \mathbb{E}[U^* XU \mid \tilde{m}]\, \tilde{\xi} \\ &= \mathbb{E}[U^* XU \tilde{\xi} \mid \tilde{m}] \\ &= \mathbb{E}[U^* X \xi U \mid \tilde{m}] \end{aligned}$$

and so $\mathbb{E}\mathbb{E}[U^* X \xi U \mid \tilde{m}] = \mathbb{E}[U^* X \xi U]$

On the other hand, $U^2 \widetilde{\mathbb{E}}[X \mid m] U \in \tilde{m}$ and $\mathbb{E}[(U^* XU - U^* \widetilde{\mathbb{E}}[X \mid m] U)\tilde{\xi}]$

$$\begin{aligned} &= \mathbb{E}[U^* \chi \xi U - U^* \widetilde{\mathbb{E}}[X \mid m] \xi U] \\ &= \mathbb{E}[(U^* X \xi - \widetilde{\mathbb{E}}[X \mid m] \xi) U] \\ &= \widetilde{\mathbb{E}}[X\xi - \widetilde{\mathbb{E}}[X \mid m]\xi] = 0 \end{aligned}$$

This proves that

$$\mathbb{E}[U^* XU \mid \tilde{m}] = U^* \widetilde{\mathbb{E}}[X \mid m] U$$

Suppose that for some adopted process

$F(t) \in \eta_{t]}^{in}$, we have

$$\widetilde{\mathbb{E}}_t[X] = \mathbb{E}[F(t)^* (X \otimes I) F(t)]$$

$\forall X \in \alpha(h)$. This means that

$$\mathbb{E}(V_t^* X V_t) = \mathbb{E}[F(t)^* X F(t)]$$

$\forall X \in \mathcal{L}(h)$

Example: Suppose $\mathbb{E}(\xi) = Tr((\rho_s \otimes \rho_e)\xi)$

where $\rho_s \in \mathcal{L}(\hbar)$ and $\rho_e \in \mathcal{L}\,(\Gamma s(\mathcal{H})$.

Then, $$\mathbb{E}(V_t^* X V_t) = T_r((\rho_s \otimes \rho_e)V_t^* X V_t)$$
$$= T_{r1}[(T_{r2}(V_t(\rho_s \otimes \rho_e)V_t^*))X]$$

Now, $$\rho_s(t) = T_{r2}(V_t(\rho_s \otimes \rho_e)V_t^*) \in \mathcal{L}(\hbar)$$

and it follows that, $\mathbb{E}(V_t^* X V_t) = Tr(\rho_s(t)X)$

We want to write

$$T_r(\,\rho_s(t)X) = T_r((\rho_s \otimes \rho_e)(F(t)^* XF(t))$$

i.e. $$\rho_s(t) = T_{r2}(F(t)\rho_s \otimes \rho_e F^*(t)) \;\forall\; t$$

where $F(t) \in \eta_{t]}^{\text{in}}$.

If we assume that

$F(t) \in \mathcal{L}(\hbar)$ then automatically $F(t) \in \eta_{t]}^{\text{in}}$ and the above condition reduces to

$$\rho_s(t) = F(t)\rho_s F^*(t)$$

Which corresponds to involves evolution.

Let $$\rho_t(X) = V_t^* \mathbb{E}(F(t)^*(X \otimes I)F(t) |\, \eta_{t]}^{\text{in}}] V_t^*$$

Then $$\mathbb{E}(jt(X) |\, \eta_{t]}^{\text{out}}) = \frac{\sigma_t\,(X)}{\sigma_t\,(1)}$$

(Quantum Kalliapur Striebel).

Gough and Kostle have shown that

$$d(F^*(t)XF(t)) = F^*(t)(X\tilde{L}_t + \tilde{L}_t^{\,*}X)F(t)dY^{in}(t)$$
$$+ F^*(t)(\,\tilde{L}_t^{\,*}X\tilde{L}_t + X\tilde{K}_t + \tilde{K}_t X)F(t)dt$$

For $Z \,\varepsilon\, \eta_{t]}^{\text{in}}$,

$$\mathbb{E}\mathbb{E}[F^*(t)\chi F(t) |\, \eta_{t]}^{\text{in}})Z] = \mathbb{E}[F^*(t)XF(t)Z]$$

$$\mathbb{E}[\mathbb{E}(d(F^*(t)XF(t) |\, \eta_{t]}^{\text{in}})Z] = \mathbb{E}[(d(F^*(t)XF(t)|Z]$$

$$\mathbb{E}[F^*(t + dt)XF(t + dt) |\, \eta_{t + dt]}^{\text{in}})]$$
$$= -\,\mathbb{E}[(F^*(t)XF(t)|\eta_{t]}^{\text{in}}]$$
$$= d\mathbb{E}[F^*(t)XF(t)|\eta_{t]}^{\text{in}}] = Z_1 dt + Z_2 dY^{\text{in}}(t) \text{ say}$$

Where $Z_1, Z_2 \in \eta_{t]}^{\text{in}}$

We write

$$d(F^*(t)X(t)) = F^*(t)\psi_1(t)F(t)dY^{\text{in}}(t) + F^*(t)\psi_2(t)F(t)dt$$

Thus from*,

$$\mathbb{E}[F^*(t)\psi_1(t)F(t)dY^{\text{in}}(t) + F^*(t)\psi_2(t)F(t)dt|\eta^{\text{in}}_{t + dt]}]$$
$$+ \mathbb{E}[F^*(t)XF(t)|\eta^{\text{in}}_{t + dt]}] - \mathbb{E}[F^*(t)XF(t)|\eta_{t]}^{\text{in}}]$$
$$= Z_1 dt + Z_2 dY^{\text{in}}(t)$$

$$= \mathbb{E}[F^*(t)\psi_1(t)F(t)|\ \eta^{\text{in}}{}_{t+dt]}]dY^{\text{in}}{}_t$$
$$+ \mathbb{E}[F^*(t)\psi_2(t)F(t)\ |\eta^{in}{}_{t+dt]}]dt$$

(since $F^*(t)XF(t)$ is independent of $dY^{\text{in}}(t)$ and so

$$\mathbb{E}[F^*(t)XF(t)|\eta^{\text{in}}{}_{t+dt]}] = \mathbb{E}[F^*(t)XF(t)|\eta_{t]}{}^{\text{in}}]$$

and likewise $F^*(t)\ \psi_k(t)F(t)$ is independent of $dY^{\text{in}}(t)$

and hence $\mathbb{E}[F^*(t)\psi_k(t)F(t)|\ \eta^{\text{in}}{}_{t+dt]}] = \mathbb{E}[F^*(t)\psi_k(t)F(t)|\eta_{t]}{}^{\text{in}}]$

Thus, $\quad d\tilde{\pi}_t(X) = Z_1 dt + Z_2 dY^{\text{in}}(t)$

where

$$Z_1 = \mathbb{E}[F^*(t)\psi_2(t)F(t)|\eta_{t]}{}^{\text{in}}] = \pi_t(\psi_2(t))$$

and $\quad Z_2 = \mathbb{E}[F^*(t)\psi_1(t)F(t)|\eta_{t]}{}^{\text{in}}] = \pi_t(\psi_1(t)).$

Thus our quantum nonlinear fitting eqn. is

$$d\tilde{\pi}_t(X) = \tilde{\pi}_t(\psi_1(t))dt + \tilde{\pi}_t(\psi_2(t))dY^{\text{in}}(t)$$

Where $\quad \psi_1(t) = \tilde{L}_t{}^* X\,\tilde{L}_t + X\widetilde{K}_t + \widetilde{K}_t X$

$$\psi_2(t) = X\tilde{L}_t + \tilde{L}_t{}^* X$$

Where $\quad \tilde{\pi}_t(X) = \mathbb{E}[F^*(t)XF(t)|\eta_{t]}{}^{\text{in}}].$

Now, $\quad \sigma_t(X) = V_t^*\,\tilde{\pi}_t(X)V_t.$

Now that $\tilde{\pi}_t(X) \in \mathcal{L}(\Gamma_s(\mathcal{H}_{t]})).$

Thus, $\quad d\sigma_t(X) = dV_t^* d\tilde{\pi}_t(X)V_t + V_t^* d\tilde{\pi}_t(X)dV_t + V_t^* d\tilde{\pi}_t(X)V_t$

$$= V_t^*(L_1{}^* dA^* + L_2{}^* dA + S^* d\Lambda)\tilde{\pi}_t(\psi_2)dY^{\text{in}}V_t + V_t^*\,\tilde{\pi}_t$$

$(\psi_2)dY^{\text{in}}(L_1 dA + L_2 dA^* + Sd\Lambda)V_t + V_t^*\,\tilde{\pi}_t(\psi_1)V_t dt + V_t^*\,\tilde{\pi}_t(\psi_2)V_t dY^{\text{in}}(t)$

$$= V_t^* L_2{}^*\,\tilde{\pi}_t(\psi_2)V_t(\bar{C}_1(t)dt + C_2(t)d\Lambda)$$

❖❖❖❖❖

[3] Evan's Hidson Flow:

Classical Markov flow:

$j_t: \mathbb{B}C_0{}^{\infty}(\mathbb{R}^n) \to L^2(\Omega, \mathcal{F}, P)$

$$j_t(f) = f(X_t) \text{ where } \{X_t: t \geq 0\} \text{ t. Then}$$

$$j_t(C_1 f_1 + C_2 f_2) = C_1 j_t(f_1) + C_2 j_t(f_2),$$

$$j_t(f_1 f_2) = j_t(f_1)j_t(f_2).$$

i.e. $\quad j_t(1) = 1, j_t(f^*) = j_t(f)^*$

j_t is a * unital homomorphism.

Let $\quad dX_t = \underline{\mu}(X_t)dt + \underline{\underline{\sigma}}(\underline{X}_t)d\underline{B}_t$

Then $\quad dj_t(f) = df(X_t) = \mathcal{L}f(X_t)dt + d\underline{B}_t^T\sigma^T(\underline{X}_t)\underline{f}'(\underline{X}_t)$

where $\mathcal{L}f(x) = \underline{\mu}(x)^T \nabla_x f(\underline{x}) + \frac{1}{2} T_r \left(a(x)\nabla\nabla^T f(x)\right)$

With $\underline{\underline{a}}(x) = \underline{\underline{\sigma}}(x),\ \underline{\underline{\sigma}}^T(x)$

Define the structure maps

$\theta_D : \mathbb{B}C_0^\infty(\mathbb{R}^n) \to \mathbb{B}C_0^\infty(\mathbb{R}^n),$

$\theta_k : \mathbb{B}C_0^\infty(\mathbb{R}^n) \to \mathbb{B}C_0^\infty(\mathbb{R}^n)$

by
$$\theta_0(f) = \mathcal{L}(f),$$
$$\theta_k(f) = \sigma^T(x) f(x))_k$$
$$= \sum_j \sigma_{jk}(x) \frac{\partial f(x)}{\partial x_j}$$

Thus,
$$\theta_k = (\sigma^T(x)\nabla_x)_k$$
$$= \sum_j \sigma_{jk}(x) \frac{\partial}{\partial x_j}$$

Then
$$dj_t(f) = j_t\left(\theta_0(f)\right)dt + \sum_{k=1}^{n} j_t\left(\theta_k(f)\right)dB_k$$

Gauss-Markov process in the classical case are characterized by the side

$$d\underline{X}(t) = \underline{\underline{A}}(t)\underline{X}(t)\,dt + \underline{\underline{G}}(t)\,d\underline{B}(t)$$

Thus, $df(X(t)) = \left[\underline{X}(t)^T \underline{\underline{A}}(t)^T \nabla_X f(\underline{X}(t)) + \frac{1}{2} T_r \left(G(t)G(t)^T \nabla_X \nabla_X^T f(\underline{X}(t))\right)\right]dt$

$$+ d\underline{B}(t)^T \underline{\underline{G}}(t)^T \nabla_X f(\underline{X}(t))$$

So
$$\theta_{0t}(f)(x) = x^T A(t)^T \nabla_x f(x) + \frac{1}{2} T_r (G(t)G(t)^T \nabla_x \nabla_x^T f(x))$$

and
$$\theta_{k_t}(f)(x) = \sum_j G_{jk}(t) \frac{\partial f(x)}{\partial x_j}$$

Now consider the quantum case. Let $\begin{bmatrix} \underline{q} \\ \underline{p} \end{bmatrix}$ be the position (vector)-momentum (vector) operators with classical Hamiltonian

$$H(\underline{q}, \underline{p}) = \frac{1}{2} \underline{p}^T \underline{\underline{A}}\underline{p} + \frac{1}{2} \underline{q}^T \underline{\underline{K}}\underline{q}$$

$\underline{\underline{A}}, \underline{\underline{K}} > 0$. Then
$$\frac{dq}{dt} = \frac{\partial H}{\partial p} = \underline{\underline{A}}\underline{p},$$
$$\frac{dp}{dt} = -\frac{\partial H}{\partial q} = -\underline{\underline{K}}\underline{q}$$

Adding noise to this system (classical) gives

$$\frac{d}{dt}\begin{bmatrix} \underline{q} \\ \underline{p} \end{bmatrix} = \begin{bmatrix} O & \underline{\underline{A}} \\ -\underline{\underline{K}} & O \end{bmatrix}\begin{bmatrix} \underline{q} \\ \underline{p} \end{bmatrix} dt + \underline{\underline{G}} d\underline{B}(t)$$

Then

$$df(\underline{q},\underline{p}) = \left\{ \left[\underline{q}^T, \underline{p}^T\right]\underline{\underline{C}}^T \begin{bmatrix} \dfrac{\partial}{\partial \underline{q}} \\ \dfrac{\partial}{\partial \underline{p}} \end{bmatrix} f(\underline{q},\underline{p}) + \frac{1}{2} T_r (GG^T \begin{bmatrix} \dfrac{\partial}{\partial \underline{q}} \\ \dfrac{\partial}{\partial \underline{p}} \end{bmatrix}\begin{bmatrix} \dfrac{\partial}{\partial \underline{q}^T} & \dfrac{\partial}{\partial \underline{p}^T} \end{bmatrix} f(\underline{q},\underline{p}) \right\} dt + d\underline{B}^T \underline{\underline{G}}^T \begin{bmatrix} \dfrac{\partial f}{\partial \underline{q}} \\ \dfrac{\partial f}{\partial \underline{p}} \end{bmatrix}.$$

$$= \Lambda \text{ or for } f = f(\underline{q},\underline{p}),$$

$$dj_t(f) = j_t(\theta_0(f))dt + \sum_{k=1}^{2n}\left(\sum_{j=1}^{n} G_{jk}\frac{\partial f}{\partial q_j} + \sum_{j=n+1}^{2n} G_{jk}\frac{\partial f}{\partial p_1}. \right) dB_k$$

$$= j(\theta_0(f))dt + \sum_{k=1}^{2n} j_t(\theta_k(f))dB_k$$

where $\theta_0(f)(\underline{q},\underline{p})$

$$= \left[\underline{q}^T, \underline{p}^T\right]\underline{\underline{C}}^T \begin{bmatrix} \dfrac{\partial}{\partial \underline{q}} \\ \dfrac{\partial}{\partial \underline{p}} \end{bmatrix} f(\underline{q},\underline{p}) + \frac{1}{2} T_r \left(GG^T \begin{bmatrix} \dfrac{\partial}{\partial \underline{q}} \\ \dfrac{\partial}{\partial \underline{p}} \end{bmatrix}\begin{bmatrix} \dfrac{\partial}{\partial \underline{q}^T} & \dfrac{\partial}{\partial \underline{p}^T} \end{bmatrix} f(\underline{q},\underline{p}) \right)$$

and

$$\theta_k(f)(\underline{q},\underline{p}) = \sum_{j=1}^{n}\left(G_{jk}\frac{\partial f}{\partial q_j} + G_{j+nk}\frac{\partial f}{\partial p_j} \right)$$

❖❖❖❖❖

[4] Quantum Version of a Gauss-Markov Process:

$$dj_t(X) = j_t(\theta_0(X))dt + \sum_{k=1}^{d}\left(j_t(\theta_k(X))dA_k(t) + j_t(\theta_k(X^*))dA_k^*(t) \right)$$

where $X \in \mathbb{B}(\hbar)$

$A_k(t)$, $A^*k(t)$

$\theta_k : \mathbb{B}(\hbar) \to \mathbb{B}(\hbar)$

$\in \mathcal{L}(\Gamma s(\mathcal{H}))$

is linear $0 \le k \le d$.

The structure maps θ_0 and θ_k must be chosen to ensure Gaussianity of $\{j_t(X) : 0 \le t \le T\}$ in a given state, $[\varphi]$ *i.e.* if $\{\mathcal{L}t : 0 \le t \le T\}$ is any scalar function, then

$$< \varphi \,|\, \exp\left(\int_0^T \alpha_t(X)\,dt\right) |\, \varphi >$$

Must be a quadratic functional of $\{\alpha_t\} 0 \le t \le T$

Let $$dj_t(X) = j_t(\theta_0(X))dt + j_t(\theta_1(X))dA_t + j_t(\theta_2(X))dA^*_t$$

Then
$$\begin{aligned}
e^{\lambda djt(X)} &= I + \lambda dj_t(X) + \frac{\lambda^2}{2}(dj_t(X))^2 + 0(dt)\\
&= I + \lambda j_t(\theta_0(X))dt + \lambda j_t(\theta_1(X))dA + \lambda j_t(\theta_2(X))dA^*\\
&\quad + \frac{\lambda^2}{2} jt(\theta_1(X))\, jt(\theta_2(X))dt + 0(dt)
\end{aligned}$$

$$\begin{aligned}
< fe(u),\ e^{\lambda dj_t(X)} fe(u) > &= \|f\|^2 \exp(\|u\|^2) + dt\{\lambda < fe(u), j_t(\theta_0(X))fe(u) >\\
&\quad + \frac{\lambda^2}{2} < fe(u), j_t(\theta_1(X))j_t(\theta_2(X))fe(u) >\\
&\quad + \lambda u(t) < fe(u), j_t(\theta_1(X))fe(u) >\\
&\quad + \lambda \bar{u}(t) < fe(u), j_t(\theta_2(X))fe(u) >\}
\end{aligned}$$

$$\begin{aligned}
e^{\lambda j_t(X)}\, e^{\lambda dj_t(X)} &= e(\lambda j_{t+dt}(X) + \alpha_1\lambda^2[j_t(X), dj_t(X)]\\
&\quad + \alpha_2\lambda^3[[j_t(X), dj_t(X)], dj_t(X)] + \alpha_3\lambda^3[j_t(X), [j_t(X), dj_t(X)]\\
&\quad + \ldots]
\end{aligned}$$

We require

$$\begin{aligned}
[[j_t(X), dj_t(X)], j_t(X)] &= 0,\\
[[j_t(X), dj_t(X)], dj_t(X)] &= 0 \qquad \ldots(1)
\end{aligned}$$

upto $O(dt)$. Then Gaussianity of $j_t(X)$ will imply that the $j_{t+dt}(X)$ Moment generating function of is e^{4th} degree polynomial in λ.

$$\begin{aligned}
[j_t(X), dj_t(X)] &= [j_t(X), j_t(\theta_0(X))]dt + [j_t(X), j_t(\theta_1(X))]dA_t\\
&\quad + [j_t(X), j_t(\theta_2(X))]dA_t^*\\
&= j_t([X, \theta_0(X)])dt + j_t([X, \theta_1(X)]dA_t + j_t([X, \theta_2(X)])dA_t^*
\end{aligned}$$

So that condition (1) hold if

$$\begin{aligned}
[[X, \theta_0(X)], X] &= 0,\\
[[X, \theta_0(X)], X] &= 0,\\
[[X, \theta_m(X)], \theta_k(X)] &= 0,\ k = 0, 1\ m = 0, 1.
\end{aligned}$$

$\forall\, X \in \mathbb{B}(\hbar)$

$$dj_t(X) = j_t(\theta_0(X))dt + j_t(\theta_1(X))dA_t + j_t(\theta_2(X))dA_t^*$$

$$\begin{aligned} j_t(XY) &= j_t(X)j_t(Y) \\ dj_t(XY) &= dj_t(X).j_t(Y) + j_t(X)dj_t(Y) + dj_t(X).dj_t(Y) \\ \Rightarrow \quad \theta_0(XY) &= \theta_0(X)Y + X\theta_0(Y) + \theta_1(X)\theta_2(Y), \text{ (Coeff. of } dt) \\ \theta_1(XY) &= \theta_1(X)Y + X\theta_1(Y) \text{ (Coeff of } dA_t) \\ \theta_2(XY) &= \theta_2(X)Y + X\theta_2(Y) \end{aligned}$$

θ_1 and θ_2 are thus derivations of $\mathbb{B}(\hbar)$.

So,

$\theta_1 = ad\, Z_1$ for some $Z_1 \in \mathbb{B}(\hbar)$

$\theta_2 = ad\, Z_2$ for some $Z_2 \in \mathbb{B}(\hbar)$.

$[[X, [Z_k, X]], X] = 0$, $k = 1, 2$.

If $[[X, \theta_k(X)], X] = 0$

and $[[X, \theta_k(X)], \theta_n(X)] = 0$

$k = 0, 1$ then $[j_t(X), dj_t(X)]$ commutes with both $j_t(X)$ and $dj_t(X)$ and hence the above Baker-Campbell Hausd orff formula implies

$$e^{\lambda_1 j_t(X)+\lambda_2 dj_t(X)} = e^{\lambda_1 j_t(X)} e^{\lambda_2 dj_t(X)} e^{-\alpha_1\lambda_1\lambda_2[j_t, djt(X)]}$$

Let $\{e_k\}$ be an ONB for $\hbar \otimes \Gamma s(\mathcal{H})$

Then $< e_k, e^{\lambda_1 j_t(X)+\lambda_2 dj_t(X)} e_m >$

$$= \sum_r < e_k, e^{\lambda_1 jt(X)} e_r >< e_r, e^{\lambda_2 dj_t(X)} e_s >< e_s, e^{-\alpha_1\lambda_1\lambda_2[j_t(X) dj_t(X)]} e_m >$$

$$[j_t(X), dj_t(X)] = j_t([X, \theta_0(X)]dt + j_t([X, \theta_1(X)])dA_t + j_t([X, \theta_2(X)]]dA_t^*$$

$$\begin{aligned} &< fe(u), e^{\lambda[j_t(X), dj_t(X)]} ge(v) > \\ &\quad = < fe(u), ge(v) > + dt\{< fe(u), \{j_t([X, \theta_0(X)]) \\ &\quad\quad + \frac{\lambda^2}{2} j_t\,(X, \theta_1(X)].[X, \theta_2(X)])\} ge(v) > \\ &\quad\quad + \lambda < fe(u), j_t([X, \theta, (X)])\, ge(v) > v(t) \\ &\quad\quad + \lambda < fe(u), j_t([X, \theta_2(X)] ge(v) > \overline{u(t)} \end{aligned}$$

For joint Gaussainty of $j_t(X)$ and $dj_t(X)$, we thus require $[X, \theta_1(X)].[X, \theta_2(X)] = 0$.

Let
$$\psi(t, \lambda) = < fe(u), e^{\lambda j_t(X)} ge(v) >$$

$$\begin{aligned} \psi(t + dt, \lambda) &= < fDe(u), e^{\lambda(j_t(X)+dj_t(X)} ge(v) > \\ &= < fe(u), e^{\lambda j_t(X)} e^{\lambda dj_t(X)} e^{-\alpha\lambda^2[j_t(X), dj_t(X)]} ge(v) > \end{aligned}$$

Instead we define

$$\psi_{km}(t, \lambda) = < e_k, e^{\lambda j_t(X)} e_m >$$

where $\{e_k\}_{k=1}^{\infty}$ is an onb for $\hbar \otimes \Gamma s(\mathcal{H})$.

Then $d\psi_{km}(t, \lambda) + \psi_{km}(t, \lambda)$

$$= \sum_{r,s=1}^{\infty} < e_k, e^{\lambda j_t(X)} e_r >< e_r, e^{\lambda dj_t(X)} e_s >< e_s, e^{-\alpha\lambda^2 [j_t(X), dj_t(X)]} e_m >$$

$$\sum_{r,s=1}^{\infty} \psi_{kr}(t, \lambda) < e_r, e^{\lambda dj_t(X)} e_s >< e_s, e^{-\alpha\lambda^2 [j_t(X), dj_t(X)]} e_m >$$

$$< e_r, e^{\lambda dj_t(\chi)} e_s >$$

$$= \delta_{rs} + \lambda \Big\{ < e_r, j_t(\theta_0(X)) e_s > dt + \sum_p < e_r, j_t(\theta_1(X)) e_p > a(p,s) dt$$

$$\sum_p \overline{a(p,r)} < e_p, j_t(\theta_2(X)) e_s > dt \Big\}$$

$$+ \frac{\lambda^2}{2} < e_r, j_t(\theta_1(X))(\theta_2(X)) e_s > dt + 0(dt)$$

where $\qquad a(p, s)dt = < e_p, dA_t e_s >$

❖❖❖❖❖

[5] Problems on the δ-Function:

[1] X denotes the space of rapidly decreasing function on $\mathbb{R}^n$, *i.e.*

$$\chi = C^{\infty}(\mathbb{R}^n) \text{ and } f \in \chi \Rightarrow \prod_{j=1}^{n} |x_j|^{mj} \left| \frac{\partial^{\sum_1^n m'_j} f(x)}{\partial X_1^{m'_1} \ldots \partial X_n^{m'_n}} \right| \to 0$$

As $||\chi|| \to \infty$

for all m_j, $m'_j \geq 0$ (non-negative integers).

If $\mathcal{F}$ denotes the Fourier transform on X, then show that

$$\mathcal{F}(X) = X.$$

Hint: $\mathcal{F}\left\{ \prod_{j=1}^{n} X_j^{mj} \prod_{j=1}^{n} D_j^{mj'} f \right\} = \sum_i mj \prod_1^n D_j^{mj} \prod_1^n \chi_j^{mj'} \vec{f}(x)$

$(\hat{f} = \mathcal{F}(f))$

Now, $\left| \Pi D_j^{mj} \Pi \chi_j^{mj'} \vec{f}(x) \right| \leq \dfrac{C_p}{||\chi||^{2p}}$

For all sufficiently large p and sufficiently large $||X||$.

Provided that we assume that $\widehat{f}$ is rapidly decreasing. Then

$$\left|\Pi \chi_j^{mj} \Pi D_j^{m'j} f(x)\right| \leq C' \int \left|\Pi D_j^{mj} \Pi Xj_{d^n X}^{m'j} \vec{f}(X)\right| d^n X$$

$$\leq C'' \int_{\|X\|>R} \frac{d^n \chi}{\|\chi\|^{2p}} + \tilde{C}(R) < \infty$$

For sufficiently large R. Thus, if $\widehat{f}$ is rapidly decreasing, then

$$\sup_{X \in \mathbb{R}^n} \left|\Pi X_j^{mj} \Pi D_j^{m'j} f(X)\right| < \infty \forall m_j, m'_j \geq 0.$$

If then readily follows by replacing

m_j by $m_j + 1\ \forall j$ that

$$\left|\Pi X_j^{mj} \Pi D_j^{m'j} f(X)\right| \leq \frac{K}{\prod_{j=1}^{n} |X_j|} \quad \text{for all sufficiently large } |X_j|$$

Proving the claim.

[2] Let

$(D^n + a_1 D^{n-1} + \ldots + a_{n-1} D + a_n) f(X) = \delta(X), X \in \mathbb{R}$

Solve for f without using Laplace transforms.

[3] Let $p(\xi_1, \ldots, \xi_n)$ be a polynomial in n variables. Solve

$p(D_1, \ldots, D_n) f(X) = \delta(X)$.

Hint: $\widehat{f}(\underline{\xi}) = \dfrac{1}{p(\underline{\xi})}$

Hence
$$f(X) = \frac{1}{(2\pi)^n} \int_{\mathbb{R}^n} \frac{\vec{f}(\underline{\xi})}{p(\underline{\xi})} \exp\left(i < \underline{\xi}, \underline{X} >\right) d^n \underline{\xi}$$

[2] Let G be a compact group and $\widehat{G}$ the set of its irreducible (unitary) representation. By the Peter-Weyl theorem.

$$f(g) = \sum_{\pi \in \widehat{G}} d_\pi T_r \left(\widehat{f}(\pi) \pi(g)\right)$$

where
$$\vec{f}(X) = \int_G f(x) \pi^*(x)\, dx$$

Thus,
$$\delta_g(x) = \sum_{\pi \in \widehat{G}} d_\pi X_\pi \left(x^{-1} g\right) = \sum_{\pi \in \widehat{G}} d_\pi \overline{\chi_\pi \left(g^{-1} x\right)}$$

is the Dirac δ-function on G where

$$\int_G f(x) \delta_g(x)\, dx = f(g) \text{ defines } \delta_g(x).$$

[3] Let $(\nabla^2 + k^2)G(\underline{r} \mid \underline{r}') = \delta(\underline{r} - \underline{r}')\underline{r}, \underline{r}' \in v$

$$G(\underline{r} \mid \underline{r}') = 0, \; \underline{r} \in \partial v$$

Assume $(\nabla^2 + k^2)\psi(\underline{r}) = S(\underline{r}), \; \underline{r} \in \partial v,$

$$\psi(\underline{r}) = \psi_0(\underline{r}), \; \underline{r} \in \partial v$$

Then, by Green 's theorem,

$$\int_v \left\{ \psi(\underline{r})(\nabla^2 + k^2)G(\underline{r} \mid \underline{r}') - G(\underline{r} \mid \underline{r}')(\nabla^2 + k^2)\psi(\underline{r}) \right\} d^3 r$$

$$= \int_S \left\{ \psi(\underline{r}) \frac{\partial G(\underline{r} \mid \underline{r}')}{\partial \vec{n}} - G(\underline{r} \mid \underline{r}') \frac{\partial \psi(r)}{\partial \vec{n}} \right\} dS(\underline{r})$$

or equivalently,

$$\int_v \left\{ \psi(\underline{r})\delta(\underline{r} - \underline{r}') - G(\underline{r} \mid \underline{r}')S(\underline{r}) \right\} d^3 r = \int_v \psi_0(\underline{r}) \frac{\partial G(\underline{r} \mid \underline{r}')}{\partial \vec{n}} dS(\underline{r})$$

i.e.
$$\psi(\underline{r}') = \int_v G(\underline{r} - \underline{r}')S(\underline{r}) d^3 r + \int_S \psi_0(\underline{r}) \frac{\partial G(\underline{r} \mid \underline{r}')}{\partial \vec{n}} dS(\underline{r})$$

[4] Generalization to $\mathbb{R}_n$:

$$\nabla^2 = \sum_{\alpha=1}^{n} \frac{\partial^2}{\partial X_\alpha^2}$$

$$\varphi \nabla^2 \psi - \psi \nabla^2 \varphi = \text{div}(\varphi \nabla \psi - \psi \nabla \varphi)$$

Thus it $(\nabla^2 + k^2)G(\underline{X} \mid \underline{X}') = \delta(\underline{X} \mid \underline{X}'), \underline{X}, \underline{X}' \in v \subset \mathbb{R}^n$

$$G(\underline{X} \mid \underline{X}') = 0, X \in \partial v$$

then consider

$$(\nabla^2 + k^2)\psi(\underline{X}) = S(\underline{X}), \underline{X} \in v$$

$$\psi(\underline{X}) = \psi_0(\underline{X}), \underline{X} \in \partial v,$$

Then the same formula is valid:

$$\psi(\underline{X}') = \int_v G(\underline{X} \mid \underline{X}')S(X')d^n X' + \int_S \psi_0(\underline{X}) \frac{\partial G(X \mid X')}{\partial \vec{n}} dS(\chi)$$

Problem: Solve

$$(\nabla^2 + k^2)G(r) = \delta(\underline{r}), \; \underline{r} \in \mathbb{R}^n.$$

Let $d\Omega(\hat{r})$ be the $n-1$ dimensional solid angle measure in $\mathbb{R}^n$. Then

$$d^n r = r^{n-1} dr \; d\Omega(\hat{r}) \; r = \left[\sum_1^n X_\alpha^2 \right]^{1/2}$$

$$\underline{\nabla}G(r) = \sum_{\alpha=1}^{n} \frac{X_\alpha}{r}\hat{e}_\alpha G'(r)$$

$$\nabla^2 G(r) = \sum_{\alpha=1}^{n} \frac{\partial}{\partial X_\alpha}\left(\frac{X_\alpha}{r}G'(r)\right)$$

$$= \sum_{\alpha=1}^{n}\left(\left(\frac{1}{r}-\frac{X_\alpha^2}{r^3}\right)G'(r)+\frac{X_\alpha^2}{r^2}G''(r)\right) = \left(\frac{n-1}{r}\right)G'(r)+G''(r)$$

Let $\int_{S^{n-1}} d\Omega(\hat{r}) = C(n).$

Then $\int_{\hat{r}\in S^{n-1}} d^n r = C(n)r^{n-1}\,dr$ (radial volume measure).

Thus $\int \delta(\underline{r}) - f(r)r^{n-1}dr d\Omega(\hat{r}) = \int f(r)\delta(\underline{r})d^n r = f(0)$

$$\Rightarrow \quad \int_{S^{n-1}} \delta(\underline{r})d\Omega(\hat{r}) = \frac{1}{r^{n-1}}\delta(r)$$

So the green's function $G(r)$ satisfies

$$\left(\frac{n-1}{r}\right)G'(r)+G''(r)+k^2G(r) = \frac{\delta(r)}{C(n)r^{n-1}} \qquad ...(1)$$

To solve this, we assume its validity for $r \in \mathbb{R}$ *i.e.* also admilt –ve values of r.

For $r > 0$,

$$rG''(r)+(n-1)G'(r)+k^2 rG(r) = 0$$

Let $r = \lambda\xi$. Then $\frac{d}{dr} = \frac{1}{\lambda}\frac{d}{d\xi}$, $\frac{d^2}{dr^2} = \frac{1}{\lambda^2}\frac{d^2}{d\xi^2}$.

So $\lambda^{-1}\xi\frac{d^2}{d\xi^2}G(\lambda\xi)+\frac{(n-1)}{\lambda}\frac{dG(\lambda\xi)}{d\xi}+\lambda k_\xi^2 G(\lambda\xi) = 0,\ \xi > 0.$

Let $\lambda = n - 1$. Then

$$\left[\frac{1}{(n-1)}\xi\frac{d^2}{d\xi^2}+\frac{d}{d\xi}+k^2(n-1)\xi\right]G(\lambda\xi) = 0,\ \xi > 0.$$

Solve it by power series.

❖❖❖❖❖

[6] Conditional Expectation in Quantum Filtering:

Let m be an algebra of observables. Assume that m is Abelian. Let $m \subset \infty$. Where ∞ is another algebra (non-commutative in general). Consider the commutant m' of m:

$$m' = \{X \in \infty \mid [X, Y] = 0 \ \forall \ Y \in m\}.$$

Then $m \subset m'$.

Let ρ be a state on ∞. For example, we may take ∞ as the algebra of linear operators in a Hilbert space $\mathcal{H}$ and ρ is a +ve define operator in $\mathcal{H}$ having unit trace. For $X \in \infty$,

$$\mathbb{E}(X) = Tr(pX).$$

Define $\mathbb{E}[.|m] : m' \rightarrow m$

so that (1) $\mathbb{E}[.|m]$ is linear, (2)

$$\mathbb{E}[\mathbb{E}[X|m]Y] = \mathbb{E}[XY]$$

$\forall \ X \in m', Y \in m$.

Suppose for example,

$$m = \{X_\alpha | \alpha \in I\}$$

and let μ_Y be a measure on I, for $Y \in m'$.

For $Y \in m'$, we let

$$\mathbb{E}[Y|m] = \int_I X_\alpha d\mu_Y(\alpha)$$

Then $\mathbb{E}\left[\left(\int_I X_\alpha d\mu_Y(\alpha)\right)Z\right] = \int_I \mathbb{E}(X_\alpha Z)d\mu_Y(\alpha) = \mathbb{E}[YZ] \ \forall \ Z \in m$

Take $Z = X_\beta$ and get

$$\int \mathbb{E}(X_\alpha X_\beta)d\mu_Y(\alpha) = \mathbb{E}[YX_\beta], \ \beta \in I.$$

Solving this equation gives μ_Y.

We can attempt to generalize this as follows. Suppose we can find a set $\{A_k\}_{k=1}^{\infty}$ in m' such that every $X \in m'$ can be expressed as

$$X = \sum_{k=1}^{\infty} A_k X_k \quad \text{for some } Xk \in m, \ k = 1, 2, \ldots$$

Then, for $Y \in m$

$$\mathbb{E}(XY) = \sum_k \mathbb{E}(A_k X_k Y)$$

and

$$\mathbb{E}(X|m) = \sum_k \mathbb{E}(A_k \mid m) X_k$$

so

$$\mathbb{E}\{\mathbb{E}(X|m)Y\} = \sum_k \mathbb{E}(\mathbb{E}(A_k \mid m) X_k Y)$$

In particular taking $X_k = X_0 \delta_{k, k0}$ for some $X_0 \in m$ gives

$$\mathbb{E}(A_{k_0} X_0 Y) = \mathbb{E}(\mathbb{E}(A_{k_0} \mid m) X_0 Y) \forall \in m$$

or equivalently since m is an algebra,

$$\mathbb{E}(A_{k_0} Y) = \mathbb{E}(\mathbb{E}(A_k \mid m) Y) \forall Y \in m.$$

This condition holding $\forall\ k_0 \geq 1$ is a $n.\ s.$

Condition that defines $\mathbb{E}(.|m)$.

Let $F \in m'$ and define assuming $\mathbb{E}(F^*F) = 1$,

$$\mathbb{E}_F(X) = \mathbb{E}(F^*XF) = \mathbb{E}(F^*FX).$$

Then $\quad \mathbb{E}_F(\mathbb{E}_F(X|m)Y) = \mathbb{E}_F(XY),\ Y \in m.$

Gives $\mathbb{E}(F^*F\ \mathbb{E}_F(X|m)Y) = \mathbb{E}(F^*FXY),\ Y \in m, X \in m',$

$\Rightarrow \mathbb{E}(\mathbb{E}(F^*F|m)\mathbb{E}_F(X|m|Y) = \mathbb{E}(\mathbb{E}F^*FX|m)Y)Y \in m, X \in m'$

$\Rightarrow \quad \mathbb{E}[F^*F|m]\mathbb{E}_F(X|m) = \mathbb{E}(F^*FX|m)$

or

$$\mathbb{E}_F(X|m) = \frac{\mathbb{E}\left(F^*FX \mid m\right)}{\mathbb{E}\left(F^*F \mid m\right)}.$$

❖❖❖❖❖

[7] Quantum Markov Processes:

[**Reference:** K.R. Parthasarathy Quantum Markov Processes.]

Quantum probability and Strong quantum Markov processes. Consider first a classical Markov processes $\{X_t : t \geq 0\}$. Let $\{\mathcal{F}_t^X\}_{t \geq 0}$ be the underlying filtration, *i.e.* X_t is $\mathcal{F}_t^X$ measurably.

$$\mathcal{F}_t^X = \sigma\{X_S : S \leq t\}.$$

Let $0 < t_1 < t_2 < ... < t_N$ and let $Y_k \in \mathcal{F}_{t_k}^X$, $1 \leq k \leq N$, *i.e.* Y_k is a functional of $\{X_S : S \leq t_k\}$.

We write

$\lambda(t_N, t_{N-1}, ..., t_1, X_N, Y_{N-1}, ..., Y_1, u)$ for $Y_N Y_{N-1} \cdot\cdot Y_1 u$ where u is $\mathcal{F}_0^X$-measurable.

Then for $t_i < t < t_{i+1}$ for some $i = 1, 2, ..., N-1$ we have if F_t denotes $\mathbb{E}(.|\mathcal{F}_t)$, and $Y \in \mathcal{F}_t$,

$$\begin{aligned}
&F_t Y\lambda(t_N, t_{N-1}, ...,t_1, Y_N, Y_{N-1}, ..., Y_1, u) \\
&= \mathbb{E}[YY_N\,Y_{N-1} \ldots Y_1 u|\mathcal{F}_t] \\
&= \left(\mathbb{E}\left(\mathbb{E}.\mathbb{E}\left(\mathbb{E}\left(Y_N \mid \mathcal{F}_{t_{N-1}}\right)Y_{N-1} \mid \mathcal{F}_{N-2}\right)\ldots Y_{i+1} \mid \mathcal{F}_t\right)Y \mid \mathcal{F}_i\right)Y_i Y_{i-1}\ldots Y_1 u \\
&= \mathbb{E}\left[T_{t,t_i}\left(T_{t_{i+1},t}\left(\ldots T_{t_{N-1},t_{N-2}}\left(T_{t_N,t_{N-1}}\left(Y_N\right)Y_{N-1}\right)Y_{N-2}\right)Y\right)Y_i Y_{i-1\ldots} Y_1 u\right] \\
&= F_0\lambda\Big(t,t_i,\ldots,t_1\Big(T_{t,t_i}\Big(T_{t_{i+1},t}\Big(\ldots\Big(T_{t_{N-1},t_{N-2}} \\
&\quad \left(T_{t_N,t_{N-1}}\left(Y_N\right)Y_{N-1}\right)Y_{N-2}\big)\ldots\Big)Y\Big),Y_i,Y_{i-1},\ldots,Y_1 u\Big)
\end{aligned}$$

We define the operator

$$\tilde{j}_t(Y) = F_t(Y.)$$

i.e.

$$\tilde{j}_t(Y)(Z) = \mathbb{E}[YZ|\mathcal{F}_t]$$

Thus we have

$$\tilde{j}_t(Y)\,\lambda(t_N, t_{N-1}, \ldots, t_1, X_N, X_{N-1}, \ldots, X_1, u) = \lambda(t, t_i, \ldots, t_1, Y_t, Y_i, Y_{i-1} \ldots, Y_1, u)$$

where
$$Y_t = \mathbb{E}[Y_N Y_{N-1} \ldots Y_{i+1} Y | \mathcal{F}_i](t_i < t < t_{i+1})$$
$$= T_{t,t_i}\left(T_{t_{i+1},t}\left(\ldots T_{t_{N-2},t_{N-1}}\left(T_{t_N,t_{N-1}}\left(Y_N\right)Y_{N-1}\right)\right)Y\right)\ldots t \in £_t .$$

Also for

$$t < t_i < t_{i+1}, Y \in \mathcal{F}_t,$$

Define $j_t(Y)$ by its action on $\lambda(t_N, \ldots, t_1, Y_N, \ldots, Y_1, u)$

where $Y_k \in \mathcal{F}_{t_k}$, by

$$j_t(Y)\lambda(t_N, \ldots, t_1, Y_N, \ldots, Y_1, u)$$
$$= \lambda(t_N, \ldots, t_{i+1}, t, t_i, \ldots, t_1, Y_N, \ldots, Y_{i+1}, Y, Y_{i-1}, \ldots, Y_1, u)$$

Then $\quad j_t(cY + Z) = cj_t(Y) + j_t(Z)$

and $\quad j_t(YZ) = j_t(Y) + j_t(Z)$

where $\quad j_t(Y)\lambda(t_N, \ldots, t_{i+1}, t, t_i, \ldots, t_1, Y_N, \ldots, Y_{i+1}, Z, Y_{i-1}, \ldots, Y_1, u)$

$$= \lambda(t_N, \ldots, t_{i+1}, t, t_i, \ldots, t_1, Y_N, \ldots, Y_{i+1}, YZ, Y_{i-1}, Y_1, u)$$

$Y, Z \in \mathcal{F}_t, c \in \mathbb{C}$.

Thus j_t is a unital homomorphism from

$L^2(\Omega, \mathcal{F}_t, P)$ into $(£(L^2(\Omega, \mathcal{L}, P))$

This method of introducing the Markov property for classical stochastic processes can be generalized to quantum stochastic processes.

First note that in the classical case,

$$\mathbb{E}\{\lambda(t_N, \ldots, t_1, Y_N, \ldots, Y_1,)^u\, \lambda(t_N, \ldots, t_1, Z_N, \ldots, Z_1,)^u\}$$
$$= F_0\{\lambda(t_N, \ldots, t_1, Y_N, \ldots, Y_1,)\, \lambda(t_N, \ldots, t_1, Z_N, \ldots, Z_1)\}$$
$$= \mathbb{E}\{Y_N \ldots Y, Z_N \ldots Z_1\}(\text{where } Y_k, Z_k \in \mathcal{F}_{t_k}, 1 \le k \le N,$$
$$\mathbb{E}\{Y_N Z_N Y_{N-1} Z_{N-1} \ldots Y_2 Z_2 . uv Y_1 Z_1\} t_N > t_{N-1} > \ldots > t_1 > a\; u, v \in \mathcal{F}_0)$$
$$= \mathbb{E}\{(\mathbb{E}(\ldots \mathbb{E}(\mathbb{E}(Y_N Z_N | \mathcal{F}_{t_{N-1}})Y_{N-1} Z_{N-1} | \mathcal{F}_{t_{N-2}}) | \mathcal{F}_{t_1})Y_1 Z_1 | \mathcal{F}_0)uv\}$$
$$= F_0\left[\lambda\left(T_{t_2,t_1} \ldots \left(T_{t_{N-1},t_{N-2}}\left(T_{t_N,t_{N-1}}\left(Y_N Z_N\right)Y_{N-1}Z_{N-1}\right)\right)\ldots\right)Y_1 Z_1, uv\right]$$

In the non-commutative case, we cannot couple $(Y_N, Z_N), \ldots, (Y_1, Z_1)$ etc.

We thus get in the non-commutative case,

$$< \lambda(t_N, \ldots, t_1, Y_N, \ldots, Y_1, u), \lambda(t_N, \ldots, t_1, Z_N, \ldots, Z_1, v) >$$
$$= \left\langle T_{u,t_1,0}\left(Y_1 \ldots T_{t_{N-2},t_{N-3}}\left(Y_{N-2} T_{t_{N-1},t_{N-2}}\left(Y_{N-1} T_{t_N,t_{N-1}}\right.\right.\right.\right.$$
$$\left.\left.\left.\left.\left(Y_N Z_N \mid Z_{N-1}\right) Z_{N-2}\right) Z_1\right) v\right\rangle$$

Here $T_{t,s}$ $t \geq s$ are stochastic operators $T_{t,s}(1) = 1$,

$$T_{t_2,t_1} . T_{t_3,t_2} = T_{t_3,t_1}, t_3 \geq t_2 \geq t_1$$

and finally $T_{t,s}$ is completely positive, *i.e.*

if X_α, $\alpha = 1, 2, ..., N$ are operators in the algebra ∞ on which $T_{t,s}$ its, then

$$\left\langle \overset{N}{\underset{\alpha=1}{\oplus}} u_\alpha, \left(\left(T_{t,s}\left(X_\alpha^* X_\beta\right)\right)\right) | \overset{N}{\underset{\alpha=1}{\oplus}} u_\alpha \right\rangle \geq 0$$

Here $\infty = \mathbb{B}(\mathcal{H})$ and $u_\alpha, \in \mathcal{H}$.

Equivalently,

$$\sum_{\alpha,\beta=1}^{N} \left\langle u_\alpha, T_{t,s}\left(X_\alpha X_\beta\right) v_\beta \right\rangle \geq 0.$$

This property is satisfied in the classical case:

$$T_{t,s}(X) = \mathbb{E}(X|\mathcal{F}_s), X \in L^2(\mathcal{F}_t)$$

$T_{t,s} : L^2(\mathcal{F}_t) \to L^2(\mathcal{F}_s)$.

Then, if $X_\alpha \in L^2(\mathcal{F}_t) 1 \leq \alpha \leq N$,

$$T_{t,s}\left(\bar{X}_\alpha X_\beta\right) = \mathbb{E}\left[\bar{X}_\alpha X_\beta \mid \mathcal{F}_s\right]$$

and $\displaystyle\sum_{\alpha,\beta=1}^{N} \bar{u}_\alpha u_\beta \mathbb{E}\left[\bar{X}_\alpha X_\beta \mid \mathcal{F}_s\right] = \mathbb{E}\left[\left|\sum_{\alpha=1}^{N} u_\alpha \bar{X}_\alpha\right|^2 \mathcal{F}_s\right] \geq 0.$

for all $u_\alpha \in \mathbb{C}$, $1 \leq \alpha \leq N$.

❖❖❖❖❖

[8] Quantum Stochastic Lyapunov Theory:

Let $$d\underline{X}(t) = \left[\sum_{j=1}^{p} L_j^{(1)} dA_j(t) + L_j^{(2)} dA_j^*(t) + \sum_{k,j=1}^{p} S_k^j d\Lambda_j^k(t)\right] \underline{X}(t) + M\underline{X}(t)\,dt$$

where $M, L_j^{(1)}, L_j^{(2)}, S_k^j \in \mathcal{L}(\hbar)$ (system space operators) and $A_j(t)$, $A_j^*(t)$, $\Lambda_j^k(t) \in \mathcal{L}(\Gamma_s(\mathcal{H}))$

where $$\mathcal{H} = \mathbb{C}^p \otimes L^2(\mathbb{R}_+)$$

$\{|e_j> : 1 \leq j \leq p\}$ is an ONB for $\mathbb{C}^p$

and $$A_j(t) = a(|e_j> \otimes \chi_{[0,t]}),$$

$$\Lambda_j^k(t) = \lambda(|e_j> < e_k| \otimes \chi_{[0,t]})$$

Thus, $$d\Lambda_j^k d\Lambda_s^r = \delta_s^k d\Lambda_j^k,$$

$$dA_j(t) dA_k^*(t) = \delta_{jk} dt$$

$$dA_j(t) d\Lambda_m^k(t) = \delta_{jm} dA_k(t)$$

$$d\Lambda_m^{\ k}(t)dA_j^{\ *}(t) = \delta_{kj}dA_m^{\ *}(t)$$

Let $V \in \mathcal{L}(\hbar)$, $V > 0$.

Consider

$$f_v(t) = T_r(X^*(t)VX(t)) \equiv T_{r1}(VT_{r2}(X(t)X^*(t)))$$

v≥ 0 and is = 0 iff

$$T_{r2}(X(t)X^*(t)) = 0 \text{ iff } X(t) = 0.$$

$f_v(t)$ is a non-negative real valued function of time. Now for asymptotic stability we required that $f_v'(t) < 0 \ \forall\ t$. But

$d(X^*VX)=dX^*.V.X + X^*.V.dX + dX^*.V.dX$

$$u \equiv ((u_j(t))_{j=1}^{\ p},\ t \geq 0)$$

Now, $< fe(u),\ dX^*.VX.ge(v) >$

$$= \left\langle fe(u), X^*\left(\sum_j\left(L_j^{(1)^*}dA_j^* + L_j^{(2)}dA_j\right)\right.\right.$$

$$\left.\left. + \sum_{j,k} S_k^{j^*}\, d\Lambda_k^j\right) VXge(v)\right\rangle + \left\langle fe(u), X^*M^*VXge(u)\right\rangle dt\Big)$$

$$= \left\{\sum_j \overline{u_j(t)}\left\langle fe(u), X^*L_j^{(1)^*}VXge(v)\right\rangle\right.$$

$$+ \sum_j v_j(t)\left\langle fe(u), X^*L_j^{(2)^*}VXge(v)\right\rangle$$

$$+ \sum_{j,k} \bar{u}_k(t)v_j(t)\left\langle fe(u), X^*S_k^{j^*}VXge(v)\right\rangle$$

$$\left. + \left\langle fe(u), X^*M^*VXge(v)\right\rangle\right\}dt \qquad ...(1)$$

Secondly, $< fe(u),\ X^*VdXge(v) >$ = Conjugate of (1) with $fe(u)$ and $ge(v)$ interchanged

$$= \left\{\sum_j v_j(t)\left\langle fe(u), X^*VL_j^{(1)}Xge(v)\right\rangle\right.$$

$$+ \sum_j \overline{u_j(t)}\left\langle fe(u), X^*VL_j^{(2)}Xge(v)\right\rangle$$

$$+ \sum_{j,k} \overline{u_j(t)}v_k(t)\left\langle fe(u), X^*VS_k^jXge(v)\right\rangle$$

$$\left. + \left\langle fe(u), X^*VMXge(v)\right\rangle\right\}dt \qquad ...(2)$$

$(X = X(t),\ X^* = X^*(t))$

Finally, $< fe(u), dX^* V\, dX\, ge(v) >$

$$= \left\langle fe(u), X^*\left(\sum_j \left(L_j^{(1)} dA_j + L_j^{(2)} dA_j^*\right) + \sum_{j,k} S_k^j d\Lambda_k^j\right) V. \right.$$

$$\left. \sum_j \left(L_j^{(1)} dA_j + L_j^{(2)} dA_j^*\right) + \sum_{j,k} S_k^j d\Lambda_k^j \Big) X ge(u) \right\rangle\Big)$$

$$= \sum_j \left\langle fe(u), X^* L_j^{(2)^*} V L_j^{(2)} X ge(v) \right\rangle dt$$

$$+ \sum_{jkmn} \left\langle fe(u), X^* S_k^{j^*} V S_n^m X ge(v) \right\rangle \delta_n^j v_m(t) \overline{u_k(t)} dt$$

$$+ \sum_{jmn} \left\langle fe(u), X^* L_k^{(2)^*} V S_n^m X ge(v) \right\rangle \delta_{jn} v_m(t)\, dt$$

$$+ \sum_{jkm} \left\langle fe(u), X^* S_k^{j^*} V L_m^{(2)} X ge(v) \right\rangle \delta_{jm} \overline{u_k(t)} dt$$

$$= \sum_j \left\langle fe(u), X^* L_j^{(2)^*} V L_j^{(2)} X ge(v) \right\rangle$$

$$+ \sum_{jkm} \left\langle fe(u), X^* S_k^{j^*} V S_j^m X ge(v) \right\rangle v_m(t) \overline{u_k(t)}$$

$$+ \sum_{km} \left\langle fe(u), X^* L_j^{(2)^*} V S_j^m X ge(v) \right\rangle v_m(t)$$

$$+ \sum_{km} \left\langle fe(u), X^* S_k^{m^*} V L_m^{(2)} X ge(v) \right\rangle \overline{u_k(t)} \Big\}\, dt$$

Using these formulas, find conditions for which

$$\frac{d}{dt} Tr\left[pX^*(t) V X(t)\right] \leq 0 \forall t$$

where $$\rho = \sum_{\alpha,\beta=1}^{N} \left| f_\alpha e(u_\alpha) \right\rangle a_{\alpha\beta} \left\langle f_\beta e(u_\beta) \right|$$

where $$Tr(\rho) = \sum_{\alpha,\beta=1}^{N} a_{\alpha\beta} \left\langle f_\beta \middle| f_\alpha \right\rangle \exp\left(\left\langle u_\beta, u_\alpha \right\rangle\right)$$

and $\left(\left(a_{\alpha\beta}\right)\right)_{\alpha,\beta=1}^{N} \geq 0$, *i.e.* ρ is a state (density matrix) in $h \otimes \Gamma_s(\mathcal{H})$

Combining all these gives

$$\frac{d}{dt}\left\langle fe(u), X^{*}VXge(v)\right\rangle =$$

Stochastic Lyapunov Theorem.

Consider the side

$$d\begin{pmatrix} \underline{X} \\ \underline{Y} \end{pmatrix} = \begin{pmatrix} A & B(X) \\ C(X) & D(X) \end{pmatrix}\begin{pmatrix} \underline{X} \\ \underline{Y} \end{pmatrix} dt + \underline{\underline{G}}d\underline{B}(t)$$

A is a constant square matrix. $\underline{\underline{B}}(\underline{X})$ is a rectangular matrix dependent on $\underline{X}$ but not on $\underline{Y}$ and so are $\underline{\underline{C}}(\underline{X})$ and $\underline{\underline{D}}(\underline{X})$.

B is known and $\underline{\underline{C}}(\underline{X})$ and $\underline{\underline{D}}(\underline{X})$ are to be determined by the condition that

$$\frac{d}{dt}\left\{\left(\underline{X}^T \underline{Y}^T\right)\underline{\underline{Q}}\begin{pmatrix} \underline{X} \\ \underline{Y} \end{pmatrix}\right\} \leq 0$$

When $\underline{\underline{G}} = 0$ (no noise) and $\underline{\underline{Q}}$ is a given +ve definite matrix. We have (assuming $G = 0$)

$$\frac{d}{dt}\left(\underline{X}^T, \underline{Y}^T\right)\underline{\underline{Q}}\begin{pmatrix} \underline{X} \\ \underline{Y} \end{pmatrix} = 2\left(\underline{X}^T, \underline{Y}^T\right)Q\begin{pmatrix} \dot{\underline{X}} \\ \dot{\underline{Y}} \end{pmatrix}$$

$$= 2\left(X^T Y^T\right)Q\begin{pmatrix} A & B \\ C & D \end{pmatrix}\begin{pmatrix} X \\ Y \end{pmatrix}$$

$$= \left(X^T Y^T\right)\left[Q\begin{pmatrix} A & B \\ C & D \end{pmatrix} + \begin{pmatrix} A^T & C^T \\ B^T & D^T \end{pmatrix}Q\right]\begin{pmatrix} X \\ Y \end{pmatrix}$$

Let $Q = \begin{pmatrix} Q_{11} & Q_{12} \\ Q_{21} & Q_{22} \end{pmatrix}$

We required $\frac{dV}{dt} \leq 0$

where $V(\underline{X}, \underline{Y}) = \left(X^T, Y^T\right)Q\begin{pmatrix} X \\ Y \end{pmatrix}$.

and hence we require $Q\begin{pmatrix} A & B \\ C & D \end{pmatrix} + \begin{pmatrix} A^T & C^T \\ B^T & D^T \end{pmatrix}Q \leq 0$

i.e. $$= \begin{pmatrix} Q_{11}A + Q_{12}C & Q_{11}B + Q_{12}D \\ Q_{21}A + Q_{22}C & Q_{21}B + Q_{22}D \end{pmatrix} + (11)^T \leq 0$$

Suppose we take $Q_{12} = 0$ (and hence $Q_{12} = Q_{12}{}^T = 0$). Then

$$\frac{dV}{dt} = \begin{pmatrix} Q_{11}A & Q_{11}B \\ Q_{22}C & Q_{22}D \end{pmatrix} + \begin{pmatrix} A^T Q_{11} & C^T Q_{22} \\ B^T Q_{11} & D^T Q_{22} \end{pmatrix}$$

$$= \begin{pmatrix} Q_{11}A + A^T Q_{11} & Q_{11}B + C^T Q_{22} \\ Q_{22}C + B^T Q_{11} & Q_{22}D + D^T Q_{22} \end{pmatrix} \le 0$$

This can be satisfied by taking Q_{11} so that $Q_{11} > 0$, $Q_{11}A + A^T Q_{11} < 0$,

$C(X) = -Q_{22}^{-1} B(X)^T Q_{11}$,

$D(X) = -Q_{22}^{-1} R$ (or $D(X) = -Q_{22}^{-1} R\ Q_{22}^{1/2}$)

where $R < 0$ is any matrix (even X dependent)

This $D(X) = -Q_{22}^{-1} R$ ($R < 0$ constant)

we get $\dfrac{dV}{dt} = \begin{pmatrix} Q_{11}A + A^T Q_{11} & 0 \\ 0 & -2R \end{pmatrix} < 0$

Note: We can even have $B = B(t, \underline{X})$.

Then $C = C(t, \underline{X}) = -Q_{22}^{-1}\ B(t, \underline{X})\, Q_{11}$,

$D = D(t, \underline{X}) = -Q_{22}^{-1}\ R(t, \underline{X})$

where $R(t, \underline{X}) < 0\ \forall\ t, \underline{X}$,

Now, when $G \neq 0$, we get

$$dV(\underline{X}, \underline{Y}) = d\left(\underline{X}^T \underline{Y}^T\right) \underline{\underline{Q}} \begin{pmatrix} \underline{X} \\ \underline{Y} \end{pmatrix}$$

$$= 2\left(\underline{X}^T \underline{Y}^T\right) \underline{\underline{Q}} \begin{pmatrix} d\underline{X} \\ d\underline{Y} \end{pmatrix} + T_r\left(GG^T Q\right) dt$$

$$= 2\left(X^T Y^T\right) Q \begin{pmatrix} A & B \\ C & D \end{pmatrix} \begin{pmatrix} X \\ Y \end{pmatrix} + T_r\left(GG^T Q\right) dt$$

$$+2\left(\underline{X}^T \underline{Y}^T\right) \underline{\underline{Q}} \underline{G} d\underline{B}(t)$$

Making the same choice of Q, C, D as far the non-random case, we find

$$\frac{d}{dt}\mathbb{E}(V) = \mathbb{E}\left[\left(\underline{X}^T \underline{Y}^T\right) \begin{pmatrix} Q_{11}A + A^T Q_{11} & 0 \\ 0 & -2R \end{pmatrix} \begin{pmatrix} \underline{X} \\ \underline{Y} \end{pmatrix}\right] + T_r\left(GG^T Q\right)$$

The first term is ≤ 0 by our choice of $\underline{\underline{A}}$, $\underline{\underline{R}}$ but the second term is ≥ 0. We need to estimate the first term. We write

$$B(t, \underline{X}) \approx B_0 + \delta . B_1(t, \underline{X})$$

where δ is a small perturbation parameter and B_0 is a constant matrix.

Then $\quad C(t,\underline{X}) = C_0 + \delta.C_1(t,\underline{X})$

where

$$C_0 = -Q_{22}^{-1}B_0Q_{11},$$

$$C_1 = -Q_{22}^{-1}R\,B_1Q_{11},$$

$$D(t,X) = -Q_{22}^{-1}R = D_0 \text{ (a constant matrix)}$$

Thus,
$$d\begin{pmatrix} X \\ Y \end{pmatrix} = \left[\begin{pmatrix} A & B_0 \\ C_0 & D_0 \end{pmatrix} + \delta\begin{pmatrix} 0 & B_1(t,X) \\ C_1(t,\underline{X}) & D_1(t,X) \end{pmatrix}\right]\begin{pmatrix} X \\ Y \end{pmatrix}dt + GdB$$

We write
$$X = X_0 + \delta.X_1 + O(\delta^2),$$
$$Y = Y_0 + \delta.Y_1 + O(\delta^2).$$

Then upto $O(\delta)$, we get

$$d\begin{pmatrix} X_0 \\ Y_0 \end{pmatrix} = \begin{pmatrix} A & B_0 \\ C_0 & D_0 \end{pmatrix}\begin{pmatrix} X_0 \\ Y_0 \end{pmatrix}dt + \underline{\underline{G}}d\underline{B},$$

$$d\begin{pmatrix} X_1 \\ Y_1 \end{pmatrix} = \begin{pmatrix} A & B_0 \\ C_0 & D_0 \end{pmatrix}\begin{pmatrix} X_1 \\ Y1 \end{pmatrix}dt + \underline{f}(t)dt$$

where
$$\underline{f}(t) = \begin{pmatrix} 0 & \underline{\underline{B}}_1(t,X_0(t)) \\ C_1(t,X_0(t)) & D_1(t,X_0(t)) \end{pmatrix}\begin{pmatrix} X_0^{(t)} \\ Y_0^{(t)} \end{pmatrix}$$

Thus,
$$\begin{pmatrix} X_0(t) \\ Y_0(t) \end{pmatrix} = \exp(tF_0)\begin{pmatrix} X(0) \\ Y(0) \end{pmatrix} + \int_0^t \exp\{(t-\tau)F_0\}Gd\underline{B}(\tau)$$

and
$$\begin{pmatrix} X_0(t) \\ Y_1(t) \end{pmatrix} = \int_0^t \exp\{(t-\tau)F_0\}f(\tau)d\tau$$

where
$$F_0 = \begin{pmatrix} A & B_0 \\ C_0 & D_0 \end{pmatrix}.$$

Then
$$\frac{d}{dt}\mathbb{E}(V_t) \equiv \frac{d}{dt}\mathbb{E}\{V(\underline{X}(t),\underline{Y}(t))\}$$
$$= T_r\begin{Bmatrix} Q_{11}A + A^TQ_{11} & 0 \\ 0 & -2R \end{Bmatrix}\mathbb{E}\left[\begin{pmatrix} \underline{X}(t) \\ \underline{Y}(t) \end{pmatrix}\left(X^T(t), Y^T(t)\right)\right]$$
$$+ T_r\left(GG^TQ\right)$$

where
$$\begin{pmatrix} \underline{X}(t) \\ \underline{Y}(t) \end{pmatrix} = \begin{pmatrix} \underline{X}_0(t) \\ \underline{Y}_0(t) \end{pmatrix} + \delta\begin{pmatrix} \underline{X}_1(t) \\ Y_1(t) \end{pmatrix} + O(\delta^2).$$

Now, $\mathbb{E}.\left\{\begin{pmatrix} \underline{X}(t) \\ \underline{Y}(t) \end{pmatrix}\left(X^T(t), Y^T(t)\right)\right\}$

$$= \exp(tF_0)R_0\exp\left(tF_0^T\right) + \int_0^t \exp(\tau F_0)GG^T\exp\left(\tau F_0^T\right)d\tau$$

$$+\delta\left\{\exp(tF_0)\underline{\xi}_0\int_0^t \mathbb{E}\left(\underline{f}^T(\tau)\right)e^{(t-\tau)}F_0^T d\tau + \int_0^t \exp\{(t-\tau)F_0\}\mathbb{E}\{\underline{f}(\tau)\}\right.$$

$$\xi_0^T\exp\left(tF_0^T\right)d\tau + \mathbb{E}\left\{\left(\int_0^t \exp\{(t-\tau)F_0\}Gd\underline{B}(\tau)\right)\int_0^t f^T(\tau')\exp\left((t-\tau')F_0^T\right)d\tau'\right\}$$

$$\left.+\mathbb{E}\left\{\left(\int_0^t f(\tau')\exp\left((t-\tau')F_0\right)d\tau'\right)\left(\int_0^t d\underline{B}^T(\tau)G^T\exp\left((t-\tau)F_0^T\right)d\tau\right)\right\} + O\left(\delta^2\right).$$

Here $\xi_0 = \begin{pmatrix} X(0) \\ Y(0) \end{pmatrix}$, $R_0 = \underline{\xi}_0\underline{\xi}_0^T$

In particular,

$$\frac{d}{dt}\mathbb{E}(V_t) = T_r\left\{\begin{pmatrix} Q_{11}A + A^TQ_{11} & 0 \\ 0 & -2R \end{pmatrix}\exp(tF_0)R_0\exp\left(tF_0^T\right)\right\}$$

$$+T_r\left\{\begin{pmatrix} Q_{11}A + A^TQ_{11} & 0 \\ 0 & -2R \end{pmatrix}\int_0^t \exp(\tau F_0)GG^T\exp\left(\tau F_0^T\right)d\tau\right\}$$

$$+T_r\left(GG^TQ\right) + O(\delta).$$

The first term on the rhs is –ve since $\begin{pmatrix} Q_{11}A + A^TQ_{11} & 0 \\ 0 & -2R \end{pmatrix} < 0.$

Assume that the eigen values of F_0 all have –ve real part. Then we get

$$\lim_{t\to\infty}\frac{d}{dt}\mathbb{E}(V_t)$$

$$= T_r\left(GG^TQ\right) + T_r\left\{\begin{pmatrix} Q_{11}A + A^TQ_{11} & 0 \\ 0 & -2R \end{pmatrix}\int_0^\infty \exp(\tau F_0)GG^T\exp\left(\tau F_0^T\right)d\tau\right\}$$

and we require for asymptotic stability that this quantity be ≤ 0.

Let $$F_0 = \sum_{k=1}^{r}\lambda_k P_k,\ \sum_1^r P_k = I,\ P_kP_j = P_k\delta_{k'j}$$

be the spectral decomposition of F_0.

Then, $$< V_\infty > \triangleq \lim_{t\to\infty} \frac{d}{dt}\mathbb{E}(V_t)$$

$$= T_r\left(GG^TQ\right)+T_r\sum_{k,j=1}^{r} Tr\left\{\begin{pmatrix} Q_{11}A+A^TQ_{11} & 0 \\ 0 & -2R\end{pmatrix}\frac{P_kGG^TP_j}{\left(\lambda_k+\lambda_j\right)}\right\}$$

This must be made ≤ 0 by an appropriate choice of Q_{11}, Q_{22}, $R > 0$.

❖❖❖❖❖

[9] Let *P* be a Solution to the Martingale Problem π(*x*, *a*, *b*), *i.e.*:

$f(X_t)-\int_0^t \alpha f(X_s)ds, t\geq 0$ is a P-Martingale where

$$\alpha f(x) = b(x)^T\frac{\partial f(x)}{\partial x}+\frac{1}{2}T_r\left(a(x)\frac{\partial^2 f(x)}{\partial x\partial x^T}\right)$$

ty

$$\left.\frac{dQ}{dP}\right|_{\mathcal{F}_t} = \exp\left\{\int_0^t < c(X_s), d\bar{X}_s > -\frac{1}{2}\int_0^t < c, ac > (X_s), ds\right\}, \equiv M_t\, t\geq 0.$$

Then Q is a solution to the Martingale problem $\pi(x, a, b+ac)$, where

$$\bar{X}_t = X_t-\int_0^t b(X_s)ds$$

Proof: $$dM_t = M_t < c(X_t), d\bar{X}_t > \qquad (\therefore M \text{ is a } P\text{- Martingale})$$

Hence $$\mathbb{E}_P\left\{d\left(\underline{X}_t M_t\right)\right\} = \mathbb{E}_P\left\{\underline{X}_t dM_t + M_t d\underline{X}_t + d < M, \underline{X} >_t\right\}$$
$$= \mathbb{E}_P\left[M_t\underline{b}(X_t)dt + M_t\underline{\underline{a}}(X_t)\underline{c}(X_t)dt\right]$$

Thus,

$$\mathbb{E}_Q\{dX_t\} = dt\cdot\mathbb{E}_Q\{(b+ac)(X_t)\}$$

More generally,

$$\mathbb{E}_P\left\{d\left(f(X_t)M_t\right)\right\} = \mathbb{E}_P\{(\mathcal{L}f(X_t))M_t\}dt + \mathbb{E}_P\{f(X_t)dM_t\}$$
$$+\mathbb{E}_P\left\{(d < M, X > t)^T f'(X_t)\right\}$$
$$= \mathbb{E}_Q\{\mathcal{L}f(X_t)\}dt + \mathbb{E}_P\left\{M_t(ac)(X_t)^T f'(X_t)\right\}dt$$

i.e. $$\mathbb{E}_Q\{\mathcal{L}f(X_t)\} = \mathbb{E}_Q\{\mathcal{L}f(X_t)\}dt + \mathbb{E}_Q\left\{ac(X_t)^T f'(X_t)\right\}dt$$
$$= \mathbb{E}_Q\left\{\tilde{\mathcal{L}}f(X_t)\right\}dt$$

where $\tilde{\mathcal{L}} = (b(x)+ac(x))^T \frac{\partial}{\partial x}+\frac{1}{2}T_r\left(a\frac{\partial^2}{\partial x\partial x^T}(\cdot)\right) = \mathcal{L}+(ac)^T\frac{\partial}{\partial \underline{\chi}}$

❖❖❖❖❖

[10] Problem from Revuz and Yor:

Still more generally, suppose $Z_t \in \mathcal{F}_t$.

Then

$$\mathbb{E}_P\left\{\left(d\left(f(X_t)M_t\right)\right)\cdot Z_t\right\} = \mathbb{E}_P\left\{\mathcal{L}f(X_t)\cdot M_t Z_t\right\}dt$$
$$+\mathbb{E}_P\left\{M_t\left(ac(X_t)^T f'(X_t)\right)Z_t\right\}dt$$

$$i.e \qquad \mathbb{E}_Q\left[Z_t df(X_t)\right] = \mathbb{E}_Q\left[Z_t\cdot\tilde{\mathcal{L}}f(X_t)\right]dt$$

Taking $Z_t = Z\in\mathcal{F}_t$ for a fixed t, we deduce that for all $T > t$,

$$\mathbb{E}_Q\left[Z\cdot\left(f(X_T)-f(X_t)-\int_t^T\tilde{\mathcal{L}}f(X_s)ds\right)\right] = 0$$

and hence

$$M_t^f = f(X_t)-\int_0^T \tilde{\pounds}f(X_s)ds,\ t\geq 0 \text{ is a } Q\text{-Martingale.}$$

❖❖❖❖❖

[11] Gough and Kostler Paper on Quantum Filtering Some Remarks:

$$\mathbb{E}^{\beta}(\cdot) = <\psi^{\beta},..\psi^{\beta}>$$

$$|\psi^{\beta}> = |\phi>|e(\beta)>\ \exp\left(-\frac{\|\beta\|^2}{2}\right),$$

$$|\varphi>\in \hbar, <\phi|\phi> = 1,\ \beta\in L^2(\mathbb{R}_+),$$

$$<e(\beta_1), e(\beta_2)> = \exp\ (<\beta_1,\beta_2>|.$$

$$dA_t\,|\psi^{\beta}> = \lambda\,|\phi>dA_t\,|e(\beta)>$$

$$= \beta(t)\lambda|\phi>|e(\beta)> = \beta(t)\,|\psi^{\beta}>,\ \lambda = \exp\left(-\frac{\|\beta\|^2}{2}\right).$$

$$<\phi' e(v)|d\Lambda t|\psi> = \lambda<\phi' e(\nu)|d\Lambda_t|\phi e(\beta)>$$

$$= \lambda<\phi'\,|\,\phi>\overline{\nu(t)}\beta(t)<e(\nu)\,|\,e(\beta)>dt$$

$$= \lambda < \phi' | \phi > \beta(t) < e(\nu) | eA_t^* | e(\beta) >$$

Thus, $\quad d\Lambda_t | \psi^\beta > = \beta(t) dA_t^* | \psi^\beta >$

Compute

$$\mathbb{E}^\beta [dj_t(X)] = < \psi^\beta | dj_t(X) | \psi^\beta >$$

$$dj_t(X) = j_t(\mathcal{L}_0 X)dt + j_t(\mathcal{L}_1 X)dA_t + j_t(\mathcal{L}_2 X)dA_t^* + j_t(\mathcal{L}_3 X)d\Lambda_t$$

So,

$$\mathbb{E}^\beta \{j_t(X)\} = \mathbb{E}^\beta [j_t(\mathcal{L}_0 X)]dt + \beta(t)\mathbb{E}^\beta [j_t(\mathcal{L}_1 X)]dt + \overline{\beta(t)}\mathbb{E}^\beta [j_t(\mathcal{L}_2 X)]dt + |\beta(t)|^2 E^\beta [j_t(\mathcal{L}_3 X)]dt$$

Note: If ξ_t is an adapted process *i.e.* $\xi_t \in \mathcal{L}(\hbar_0 \otimes \Gamma_s(\mathcal{H}_t]))$, then

Thus $\quad \mathbb{E}^\beta \{j_t(X)\} = \mathbb{E}^\beta \{j_t(\mathcal{L}^{\beta(t)} X)\} dt$

where $\quad \mathcal{L}^\beta(t) = \mathcal{L}_0 + \beta(t)\mathcal{L}_1 + \bar{\beta}(t)\mathcal{L}_2 + |\beta(t)|^2 \mathcal{L}_3.$

Let

$$Y_t^{\text{in}} = \int_0^t (c_1(\xi)dA_s + \overline{c_1(s)}dA_s^* + c_2(s)d\Lambda_s)$$

Then for $t_1 \geq t_2$

$$\left[Y_{t_1}^{\text{in}}, Y_{t_2}^{\text{in}}\right] = 0$$

Moreover since $V_t^* Y_t^{\text{in}} V_t = V_T^* Y_t^{\text{in}} V_T^*$

$= Y_t^{\text{out}} \ \forall \ T \geq t$, we get if follows that

$$\left[Y_t^{\text{out}}, j_T(X)\right] = 0 \forall T \geq t,\ X \in £(\hbar)$$

where $\quad j_t(X) = V_t^* V_t$

This is because $\left[Y_t^{\text{in}}, X\right] = 0 \ \forall t \geq 0$

since $X \in £(\hbar)$ while $Y_t^{\text{in}} \in £(\Gamma_{s(}\mathcal{H}_t]))$

and $\quad [\hbar, \Gamma s(\mathcal{H})] = 0.$

Let $\quad \psi(t) = V_t.$

$$Y_t^{\text{out}} = V_t^* V_t^{\text{in}} V_t$$

so

$$dY_t^{\text{out}} = dY_t^{\text{in}} + dV_t^* dY_t^{\text{in}} V_t + V_t^* dY_t^{\text{in}} dV_t$$

$$= c_1(t)dA_t + \overline{c_1(t)}dA_t^* + c_2(t)d\Lambda t + V_t^*(L_2^* dAt + S^* d\Lambda_t)$$

$$\left(\overline{c_1(t)}dA_t^* + c_2(t)d\Lambda_t\right)V_t + V_t^*\left(c_1(t)dA_t + c_2(t)d\Lambda_t\right).$$

$$(L_2 dA_t^* + Sd\Lambda_t)V_t$$

$$= c_1(t)dA_t + \overline{c_1(t)}dA_t^* + c_2(t)d\Lambda_t + \overline{c_1}(t)V_t^* L_2^* V_t dt$$

$$+ \overline{c_1(t)}V_t^* S^* V_t dA_t^* + c_2(t)V_t^* S^* V_t d\Lambda_t$$

$$+ c_2(t)V_t^* L_2^* V_t dA_t + c_1(t)V_t^* L_2 V_t dt$$

$$+ c_2(t)V_t^* L_2 V_t dA_t^* + c_1(t)V_t^* SV_t dA_t + c_2(t)V_t^* SV_t d\Lambda_t$$

$$dY_t^{\text{in}} \mid \psi^\beta > \; = c_1(t)dA_t \mid \psi^\beta > + \overline{c}_1(t)dA_t^* \mid \psi^\beta > + c_2(t)d\Lambda_t \mid \psi^\beta >$$

$$= \beta(t)c_1(t)dt + \overline{c}_1(t)dA_t^* + \beta(t)c_2(t)dA_t^* \mid \psi^\beta >$$

So

$$\mathbb{E}^\beta\left\{dY_t^{\text{in}}\right\} = <\psi^\beta \left| dY_t^{\text{in}} \right| \psi^\beta>$$

$$= \beta(t)c_1(t)dt + \overline{c}_1(t)\overline{\beta}(t)dt + \left|\overline{\beta}(t)\right|^2 c_2(t)dt$$

$$= \left\{2\,\mathrm{Re}\{\beta(t)c_1(t)\} + |\beta(t)|^2 c_2(t)\right\}dt$$

Likewise

$$\mathbb{E}^\beta\left\{dY_t^{\text{out}}\right\} = dt\Big\{2Re\{\beta(t)c_1(t)\} + |\beta(t)|^2 c_2(t)$$

$$+ 2Re\left\{c_1(t)\mathbb{E}^R\left\{V_t^* L_2 V_t\right\}\right\}$$

$$+ 2Re\left\{c_1(t)\beta(t)\mathbb{E}^\beta\left\{V_t^* SV_t\right\}\right\}$$

$$+ 2c_2(t)\,\mathrm{Re}\left\{\mathbb{E}^\beta\left\{V_t^* SV_t\right\}|\beta(t)|^2\right.$$

$$\left.+ 2c_2(t)\,\mathrm{Re}\left\{\mathbb{E}^\beta\left\{V_t^* L_2 V_t\right\}\overline{\beta}(t)\right\}\right\}$$

$$= dt\Big\{2\,\mathrm{Re}\{\beta(t)c_1(t)\} + |\beta(t)|^2 c_2(t)$$

$$+ 2\,\mathrm{Re}\,\mathbb{E}^\beta\Big\{j_t\big(c_1(t)L_2 + c_1(t)\beta(t)S$$

$$+ c_2(t)|\beta(t)|^2 S + C_2(t)\overline{\beta}(t)L_2\big)\Big\}\Big\}$$

Recall that

$$dV_t = (-iHdt + L_1 dA_t + L_2 dA_t^* + Sd\Lambda_t)V_t$$

where the system operators ($\in$ £ ($\hbar$)) $= H, L_1, L_2, S$ have been chosen to make V_t unitary, *i.e.*, $d(V_t^* V_t) = 0$ and this has been achieved using the quantum I to formulae:

$$dA_t dA_t^* = dt,\; d\Lambda_t dA_t^* = dA_t^*,$$

$$dA_t d\Lambda_t = dA_t,\; d\Lambda_t d\Lambda_t = d\Lambda_t,$$

all the other products being zero.

$\mathbb{E}^{\beta}\left\{j_t(\chi) \mid \eta_{t]}^{\text{out}}\right\}$

is the filtered estimate of the state $j_t(X)$ given the measurements σ - algebra $\eta_{t]}^{\text{out}}$ upto time t.

We note that

$$j_t(X) = V_t^* X V_t,\ X \in \pounds(\hbar)$$

$$\eta_{t]}^{\text{out}} = V_t^* \eta_{t]}^{\text{in}} V_t,\ \eta_{t]}^{\text{in}} \subset$$

and since $$\eta_{t]}^{\text{out}} = V_T^* \eta_{t]}^{\text{in}} V_T,\ \pounds\left(\Gamma^s\left(H_{t]}\right)\right) T \geq t$$

and it follows using

$$\left[X, \eta_{t]}^{\text{in}}\right] = 0$$

that $$\left[j_T(X), \eta_{t]}^{\text{out}}\right] = 0 \quad \forall T \geq t.$$

Now, $$\mathbb{E}^{\beta}\left[j_t(X) \mid \eta_{t]}^{\text{out}}\right] = V_t^* \tilde{\mathbb{E}}_t\left(X \mid \eta_{t]}^{\text{in}}\right) V_t$$

where $$\tilde{\mathbb{E}}_t(\xi) = \mathbb{E}^{\beta}\left[V_t^* \xi V_t\right]$$

This follows because,

$$\mathbb{E}^{\beta}\left[\left(jt(X) - V_t^* \tilde{\mathbb{E}}_t\left(X \mid \eta_{t]}^{\text{in}}\right) V_t\right) \eta_{t]}^{\text{out}}\right]$$

$$= \mathbb{E}^{\beta}\left\{\left(V_t^* X V_t - V_t^* \tilde{\mathbb{E}}_t\left(X \mid \eta_{t]}^{\text{in}}\right) V_t\right) V_t^* \eta_{t]}^{\text{in}} V_t\right\}$$

$$= \mathbb{E}^{\beta}\left\{V_t^*\left(X \eta_{t]}^{\text{in}} - \tilde{\mathbb{E}}_t\left(X \mid \eta_{t]}^{\text{in}}\right) \eta_{t]}^{\text{in}}\right) V_t\right\}$$

$$= \tilde{\mathbb{E}}_t\left[X \eta_t^{\text{in}} - \tilde{\mathbb{E}}_t\left(X \mid \eta_{t]}^{\text{in}}\right) \eta_{t]}^{\text{in}}\right]$$

$$= \tilde{\mathbb{E}}_t\left[X \eta_{t]}^{\text{in}}\right] - \tilde{\mathbb{E}}_t \tilde{\mathbb{E}}_t\left[X \eta_{t]}^{\text{in}} \mid \eta_{t]}^{\text{in}}\right] = 0.$$

Now,

Suppose $F(t)$ is an adapted process such that

$F(t) \in \eta_{t]}^{\text{in}}$ *i.e.* $[F(t),\ \eta_{t]}^{\text{in}}] = 0\ \forall t$, and $\mathrm{E}^{\beta}\left\{V_t^* X_t V_t\right\}$

$$= \mathbb{E}^{\beta}\left\{F^*(t) X_t F(H)\right\} \forall t$$

(X_t being any adapted process)

Then $$\mathbb{E}^{\beta}\left\{j_t(X) \mid \eta_{t]}^{\text{out}}\right\} = V_t^* \tilde{\mathbb{E}}_t\left\{X \mid \eta_{t]}^{\text{in}}\right\} V_t$$

and $$\tilde{\mathbb{E}} t\left\{X \mid \eta_{t\}}^{\text{in}}\right\} = \frac{\mathbb{E}^{\beta}\left\{F^*(t) X F(t) \mid \eta_{t]}^{\text{in}}\right\}}{\mathbb{E}^{\beta}\left\{F^*(t) F(t) \mid \eta_{t]}^{\text{in}}\right\}}$$

This follows from

$$\mathbb{E}^{\beta}\left\{\tilde{\mathbb{E}}_t\left\{X \mid \eta_{t]}^{\text{in}}\right\} \mathbb{E}^{\beta}\left\{F^*(t) F(t) \mid \eta_{t]}^{\text{in}}\right\} \eta_{t]}^{\text{in}}\right\}$$

$$= \mathbb{E}^{\beta}\left\{\mathbb{E}^{\beta}\left\{V_t^* X V_t \mid \eta_{t]}^{\text{in}}\right\} X \mathbb{E}^{\beta}\left\{F^*(t)F(t) \mid \eta_{t]}^{\text{in}}\right\}\eta_{t]}^{\text{in}}\right\}$$

$$= \mathbb{E}^{\beta}\left[F^*(t)\tilde{\mathbb{E}}_t\left[X\middle|\eta_{t]}^{\text{in}}\right]F(t)\eta_{t]}^{\text{in}}\right]$$

$$= \mathbb{E}^{\beta}\left[V_t^*\tilde{\mathbb{E}}_t\left[X\middle|\eta_{t]}^{\text{in}}\right]\eta_{t]}^{\text{in}}V_t\right]$$

$$= \tilde{\mathbb{E}}t\left[\tilde{\mathbb{E}}\left[X\middle|\eta_t^{\text{in}}\right]\eta_t^{\text{in}}\right] = \tilde{\mathbb{E}}_t\left[X\eta_t^{\text{in}}\right]$$

On the other hand,

$$\mathbb{E}^{\beta}\left\{\mathbb{E}^{\beta}\left\{F^*(t)XF(t)\middle|\eta_{t]}^{\text{in}}\right\}\eta_{t]}^{\text{in}}\right\}$$

$$= \mathbb{E}^{\beta}\left[F^*(t)XF(t)\eta_{t]}^{\text{in}}\right]$$

$$= \mathbb{E}^{\beta}\left[F^*(t)X\eta_{t]}^{\text{in}}F(t)\right]$$

$$= \mathbb{E}^{\beta}\left[V_t^*X\eta_{t]}^{\text{in}}V_t\right]$$

$$= \tilde{\mathbb{E}}_t\left[X\eta_{t]}^{\text{in}}\right]$$

Proving the claim.

Now let $\quad |\psi(t)> = V_t|\psi^{\beta}>$

Then $\quad d\,|\psi(t)> = dV_t|\psi^{\beta}>$

$$= \left(-iHdt + L_1 dA_t + L_2 dA_t^* + Sd\Lambda_t\right)|\psi\beta>$$

$$= \left(-iHdt + \beta(t)L_1 dt + \left(S\bar{\beta}(t) + L_2\right)dA_t^*\right)|\psi^{\beta}>$$

$$= \Bigg[-iHdt + \beta(t)L_1 dt + \left(\frac{S\bar{\beta}(t) + L_2}{\bar{c}_1(t)}\right)$$

$$\times\left(\bar{c}_1(t)dA_t^* + c_1(t)dA_t\right)\Bigg]\Bigg|\psi^{\beta}>$$

$$-\frac{dtc_1(t)}{\bar{c}_1(t)}\left(S\bar{\beta}(t) + L_2\right)\beta(t)\Bigg|\psi^{\beta}>$$

Assume $c_2(t) = 0$, *i.e.*

$$dY_t^{\text{in}} = c_1(t)dA_t + \overline{c_1(t)}dA_t^*$$

Then we get

$$d|\psi(t)> = dV_t|\psi^{\beta}>$$
$$= dF_t|\psi^{\beta}>$$

i.e. $\quad V_t|\psi^{\beta}> = F_t\,|\,\psi^{\beta}>$

and $F_t \in \eta_{t]}^{in/}$.

❖❖❖❖❖

[12] Quantization of Master. Slave Robot Motion Using Evans-Hudson Flows:

$$\begin{aligned}
\underline{\underline{M}}_m(\underline{q}_m)\ddot{\underline{q}}_m + \underline{N}_m(\underline{q}_m, \dot{\underline{q}}_m) &= \underline{\tau}_m(t) + \underline{\underline{F}}_m(\underline{q}_m, \ddot{\underline{q}}_m)\big(K_{mp}(\underline{q}_s - \underline{q}_m) \\
&\quad + K_{md}(\dot{\underline{q}}_s - \dot{\underline{q}}_m)\big) + \underline{d}_m(t) \\
\underline{\underline{M}}_s(\underline{q}_s)\ddot{\underline{q}}_s + \underline{N}_s(\underline{q}_s, \dot{\underline{q}}_s) &= \underline{\tau}_s(t) + \underline{\underline{F}}_s(\underline{q}_s, \ddot{\underline{q}}_s)\big(K_{sp}(\underline{q}_m - \underline{q}_s) \\
&\quad + K_{sd}(\dot{\underline{q}}_m - \dot{\underline{q}}_s)\big) + \underline{d}_s(t)
\end{aligned}$$

$\underline{d}_m(t)$, $\underline{d}_s(t)$ are white noise processes.

Thus,

$$\frac{d}{dt}\begin{bmatrix} \underline{q}_m \\ \dot{\underline{q}}_m \\ \underline{q}_s \\ \dot{\underline{q}}_s \end{bmatrix} = \begin{bmatrix} \dot{\underline{q}}_m \\ \underline{\psi}_1(\underline{q}_m, \underline{q}_s, \dot{\underline{q}}_m, \dot{\underline{q}}_s) \\ \dot{\underline{q}}_s \\ \underline{\psi}_2(\underline{q}_m, \underline{q}_s, \dot{\underline{q}}_m, \dot{\underline{q}}_s) \end{bmatrix} + \psi_3(\underline{q}_m, \underline{q}_s, \dot{\underline{q}}_m, \dot{\underline{q}}_s)\begin{bmatrix} \underline{d}_m(t) \\ \underline{d}_s(t) \end{bmatrix}$$

$$+ \psi_3(\underline{q}_m, \underline{q}_s, \dot{\underline{q}}_m, \dot{\underline{q}}_s)\begin{bmatrix} \underline{\tau}_m(t) \\ \underline{\tau}_s(t) \end{bmatrix}$$

where

$$\begin{aligned}
\psi_1 &= -\underline{\underline{M}}_m(\underline{q}m)^{-1}\underline{N}_m(\underline{q}_m, \dot{\underline{q}}_m) \\
&\quad + K_{mp}\underline{\underline{M}}_m(\underline{q}_m)^{-1}F_m(\underline{q}_m, \dot{\underline{q}}_m)(\underline{q}_s - \underline{q}_m) \\
&\quad + K_{md}\underline{\underline{M}}_m(\underline{q}_m)^{-1}F_m(\underline{q}_m, \dot{\underline{q}}_m)(\dot{\underline{q}}_s - \dot{\underline{q}}_m)
\end{aligned}$$

$$\triangleq \underline{\psi}_{10}(\underline{q}_m, \dot{\underline{q}}_m) + K_{mp}\underline{\psi}_{11}(\underline{q}_m, \underline{q}_s, \dot{\underline{q}}_m, \dot{\underline{q}}_s) + K_{md}\underline{\psi}_{12}(\underline{q}_m, \underline{q}_s, \dot{\underline{q}}_m, \dot{\underline{q}}_s)$$

and likewise for ψ_2.

Here
$$\begin{bmatrix} \underline{d}_m(t) \\ \underline{d}_s(t) \end{bmatrix} = \begin{bmatrix} \underline{\underline{\Sigma}}_m \dfrac{d\underline{B}_m(t)}{dt} \\ \underline{\underline{\Sigma}}_s \dfrac{d\underline{B}_s(t)}{dt} \end{bmatrix} = \underline{\underline{\Sigma}}\frac{d\underline{B}}{dt}$$

$$\underline{\underline{\Sigma}} = \begin{bmatrix} \underline{\underline{\Sigma}}_m & \underline{\underline{O}} \\ \underline{\underline{O}} & \underline{\underline{\Sigma}}_s \end{bmatrix} \quad \underline{B} = \begin{bmatrix} \underline{B}_m \\ \underline{B}_s \end{bmatrix} \in \mathbb{R}^4$$

$\underline{B}$ is a 4-dimensional Brownian motion.

Defining
$$\underline{q} = \begin{bmatrix} \underline{q}_m \\ \underline{q}_s \end{bmatrix} \in \mathbb{R}^4,$$
we get

$$d\begin{bmatrix} \underline{q}(t) \\ \dot{\underline{q}}(t) \end{bmatrix} = \left\{\underline{F}_0\left(\underline{q}(t), \dot{\underline{q}}(t), t\right) + K_{mp}\underline{F}_1\left(\underline{q}(t), \dot{\underline{q}}(t)\right)\right.$$

$$+ K_{md}\underline{F}_2\left(\underline{q}(t), \underline{\dot{q}}(t)\right) + K_{sp}\underline{G}_1\left(\underline{q}(t), \underline{\dot{q}}(t)\right)$$
$$+ K_{sd}\underline{G}_2\left(\underline{q}(t), \underline{\dot{q}}(t)\right)\Big\}dt + \underline{\underline{H}}\left(\underline{q}(t), \underline{\dot{q}}(t)\right)d\underline{B}(t)$$

Let $\begin{bmatrix} \underline{q} \\ \underline{\dot{q}} \end{bmatrix} = \underset{\varepsilon}{\underline{\xi}}\mathbb{R}^8$. Then

$$d\underline{\xi} = \{\underline{F}_0(\underline{\xi}, t) + K_{mp}E_1(\underline{\xi}) + K_{md}\underline{F}_2(\underline{\xi}) + K_{sp}\underline{G}_1(\underline{\xi}) + K_{sd}\underline{G}_2(\underline{\xi})\}dt + \underline{\underline{H}}(\underline{\xi})d\underline{B}$$

Let $f: \mathbb{R}^8 \to \mathbb{R}$ be C^2.

Then,

$$df(\xi_t) = \mathcal{L}_{1t} f(\underline{\xi}_t)dt + \sum_{k=1}^{4} \mathcal{L}_{2k} f(\underline{\xi}_t)dB_k$$

where $$\mathcal{L}_{1t}f(\underline{\xi}) = \underline{P}(\underline{\xi}, t)^T \underline{\nabla}_\xi f(\underline{\xi}) + \frac{1}{2}T_r\left(\underline{\underline{H}}(\underline{\xi})\underline{\underline{H}}(\underline{\xi})^T \underline{\nabla}_\xi \underline{\nabla}_\xi^T f(\underline{\xi})\right)$$

and $$\mathcal{L}_{2k}f(\underline{\xi}) = \sum_{m=1}^{8} H_{mk}(\underline{\xi})\frac{\partial f(\underline{\xi})}{\partial \xi_m}$$
$$= \left[\underline{\underline{H}}(\underline{\xi})^T \underline{\nabla}_\xi f(\underline{\xi})\right]_k$$

Here $P(\underline{\xi}, t) = \underline{F}_0(\underline{\xi}, t) + K_{mp}\underline{F}_1(\underline{\xi}) + K_{md}\underline{F}_2(\underline{\xi}) + K_{sp}\underline{G}_1(\underline{\xi}) + K_{sd}\underline{G}_2(\underline{\xi})$

In the quantum context, $f(\underline{\xi}_t) \to j_t(f)$

j_t being a homomorphism of an "initial algebra" $\mathbb{B}_0$ into $\mathbb{B}_{t]}$ and f is replaced by $X \in \mathbb{B}_0$ where $\mathbb{B}_0$ can be non-Abelian.

We get

$$dj_t(X) = j_t\left(£_0 t(X) + K_{mp}£_1(X) + K_{md}£_2(X)\right.$$
$$\left. + K_{sp}£_3(X) + K_{sd}£_4(X)\right)dt + \sum_{k=1}^{4} j_t\left(£_{5k}(X)\right)dA_k(t)$$
$$+ \sum_{k=1}^{4} j_t\left(£_{6k}(X)\right)dA_k^*(t) + \sum_{j,k=1}^{4} j_t\left(S_j^k(X)\right)d\Lambda_k^j(t)$$

The aim is to choose $\{K_{mp}, K_{md}, K_{sp}, K_{sd}\}$

so that $\int_0^T < fe(u), j_t(X)\ fe(u) > dt$ is a minimum, where $f \in \mathcal{h}$ (initial Hilbert space and χ is a fixed observable in $\mathcal{L}(\mathcal{h})$.

In this classical context, $j_t(f) = f(\underline{\xi}_t)$

$$= \alpha \| \underline{q}_m(t) - \underline{q}_s(t) \|^2 + \beta \| \underline{\dot{q}}_m(t) - \underline{\dot{q}}_s(t) \|^2$$

Or more precisely writing

$$\underline{\xi} = \begin{bmatrix} \underline{q}_m \\ \underline{q}_s \\ \dot{\underline{q}}_m \\ \dot{\underline{q}}_s \end{bmatrix}, f(\xi) = \alpha \| \underline{q}_m - \underline{q}_s \|^2 + \beta \| \dot{\underline{q}}_m - \dot{\underline{q}} s \|^2$$

Note that

$$\begin{aligned} \frac{1}{dt} < fe(u), dj_t(X) fe(u) > &= < fe(u), j_t(\mathcal{L}_{0t} X) fe(u) > \\ &+ K_{mp} < fe(u), j_t(\mathcal{L}_1 X) fe(u) > \\ &+ K_{md} < fe(u), jt(\mathcal{L}_2 X) fe(u) > \\ &+ K_{sp} < fe(u), j_t(\mathcal{L}_3 X) fe(u) > \\ &+ K_{sd} < fe(u), j_t(\mathcal{L}_4 X) fe(u) > \\ &+ 2\,\mathrm{Re}\left[\sum_{k=1}^{4} \langle fe(u), j_t(\mathcal{L}_{sk} X) f_e(u) \rangle u_k(t) \right] \\ &+ \sum_{k,j} \langle fe(u), j_t(S_j^k X) fe(u) \rangle \bar{u}_k(t) u_j(t) \end{aligned}$$

Here, we are assuming

$$S_k^j(X)^* = S_j^k(X)$$

$(X^* = X)$.

Thus

$$\begin{aligned} < fe(u), j_T(X) fe(u) > &= \int_0^T \langle fe(u), j_t(\mathcal{L}_{0t} X) fe(u) \rangle dt \\ &+ \sum_{\alpha=1}^{4} K_\alpha \int_0^T \langle fe(u), j_t(\mathcal{L}_\alpha X) fe(u) \rangle dt \\ &+ 2\,\mathrm{Re} \sum_{k=1}^{4} \int_0^T \langle fe(u), j_t(\mathcal{L}_{5k} X) fe(u) \rangle u_k(t) dt \\ &+ \sum_{k,j} \int_0^T \langle fe(u), j_t(S_j^k X) fe(u) \rangle \bar{u}_k(t) uj(t) dt \end{aligned}$$

Perturbative approximations.

Consider first the following simplified problem

$$dj_t(X) = (j_t(\mathcal{L}_0 X) + K j_t(\mathcal{L}_1 X))dt + j_t(\mathcal{L}_2 X) dA(t) + j_t(\mathcal{L}_3 X) dA^*(t) + j_t(\mathcal{L}_4 X) d\Lambda(t)$$

Where $\mathcal{L}_3(X) = \mathcal{L}_2(X)^* \equiv \mathcal{L}_2^+(X)$ *i.e.* $\mathcal{L}_2^+ = \mathcal{L}_3$,

and $\mathcal{L}_4(X)^* = \mathcal{L}_4(X)$, *i.e.* $\mathcal{L}_4^+ = \mathcal{L}_4$

Where $\chi^* = \chi$.

Also, $\mathcal{L}_0^+ = \mathcal{L}_0, \mathcal{L}_1^+ = \mathcal{L}_1$.

K is a real constant chosen so that

$$\int_0^T < fe(u), j_t(\chi) fe(u) > dt$$

is a minimum.

For example we could choose

$j_t(X) = V_t^* X V_t \equiv V_t^* (X \otimes \mathrm{I}) V_t$

where $\quad dV_t = \left[-iHdt + L_1 dA_t + L_2 dA_t^* + Sd\Lambda_t\right]V_t$

with $\quad H = H_0 + KH_1$

Then,

$$\begin{aligned}
dj_t(X) &= d(V_t^* X V_t) \\
&= dV_t^* X V_t + V_t^* X dV_t + dV_t^* X dV_t \\
&= V_t^* [i(H^* X - XH)dt + (L_2^* X + XL_1)dA_t \\
&\quad + (S^* X + XS)d\Lambda_t + (L_1^* X + XL_2)dA_t^* \\
&\quad + L_2^* X L_2\, dt + L_2^* X SdA_t + S^* XSd\Lambda_t + S^* XL_2 dA_t^*]V_t \\
&= [j_t(\mathcal{L}_0 X) + Kj_1(\mathcal{L}_1 X)]dt + j_t(\mathcal{L}_2 X)dA_t + j_t(\mathcal{L}_3 X)dA_t^* \\
&\quad + jt(\mathcal{L}_4 X)d\Lambda_t
\end{aligned}$$

Where

$$\begin{aligned}
\mathcal{L}_0(X) &= i(H_0^* X - XH_0), \\
\mathcal{L}_1(X) &= i(H_1^* X - XH_1), \\
\mathcal{L}_2(X) &= L_2^* X + XL_1 + L_2^* XS \\
\mathcal{L}_3(X) &= L_1^* X + XL_2 + S^* XL_2 \\
\mathcal{L}_4(X) &= S^* X + XS + S^* XS
\end{aligned}$$

The approximate solution to j_t is obtained from an approximate solution to V_t. We can also incorporate parameters like K into L_1, L_2 and S. That takes care of pd controllers. Feedback $K_p(\underline{q}_m - \underline{q}_s)$ can be achieved by adding a harmonic oscillator potential to H_0. The feedback $K_d(\underline{\dot{q}}_m - \underline{\dot{q}}_s)$ however, is like a velocity damping term and must be taken care of by parameters incorporated in the Sudarshan-Lindblad generator or equivalently in L_1, L_2, S. Thus we write

$$H = H_0 + K_0 H_1, L_2 = L_{20} + K_2 L_{21}, L_1 = L_{10} + K_1 L_{11}$$
$$S = S_0 + K_3 S_1.$$

❖❖❖❖❖

[13] Problem From KRP QSC:

Let A, B be operators such that $[A, B]$ commutes with both A and B.

Let $X(t) = e^{t(A+B)}$. Then

$$X'(t) = (A + B)\, X(t).$$

Put $X(t) = e^{tA} F(t)$. Then,

$$X'(t) = e^{tA}(AF(t) + F'(t)) = (A + B)\, e^{tA}F(t)$$

So,

$$\begin{aligned} F'(t) &= e^{-tA}Be^{tA}F(t) \\ &= e^{-t\, ad\, A}(B)F(t) \\ &= (B - t[A, B])\, F(t) \end{aligned}$$

Since $(ad\, A)^n\, (B) = 0$ for $n \geq 2$. Then,

$$F(t) = e^{tB - \frac{t2}{2}[A, B]} \text{ since } B - t[A, B],\ t \in \mathbb{R}$$

form a commuting family of operators.

It follows that

$$\begin{aligned} X(t) &= e^{tA}e^{tB - \frac{t^2}{2}[A, B]} \\ &= e^{tA}e^{tB}e^{-\frac{t^2}{2}[A, B]} \end{aligned}$$

Now apply this result to the operators $a(u)$, $a^+(v)$.

we get $e^{\lambda a(u)}e^{\mu a^+(v)}e^{-\frac{\lambda\mu}{2}<u, v>} = e^{\lambda a(u) + \mu a^+(v)}$

Since $[a(u), a^+(v)] = <u, v>$ commutes with both $a(u)$ and $a^+(v)$. Likewise

$$e^{\lambda a(u)+\mu a^+(v)} = e^{\mu a^+(v)}e^{\lambda a(u)}e^{-\frac{\lambda\mu}{2}<v, u>}$$

Then $< e(w_1), e^{\lambda a(u)+\mu a^+(v)}e(w_2) >$

$$\begin{aligned} &= e^{-\frac{\lambda\mu}{2}<v, u>} < e(w_1), e^{\mu a^+(v)}e^{\lambda a(u)}e(w_2) > \\ &= e^{-\frac{\lambda\mu}{2}<v, u>} < e^{\bar{\mu} a(v)}e(w_1), e^{\lambda a(u)}e(w_2) > \\ &= e^{-\frac{\lambda\mu}{2}<v, u>} e^{\lambda<u, w_2>}e^{\mu<w_1, v>}\, e^{<w_1, w_2>} \end{aligned}$$

In particular,

$$< e(w_1), e^{\lambda a(u) - \bar{\lambda} a^+(u)}e(w_2) >$$

$$= e^{-\frac{|\lambda|^2}{2}\|u\|^2} e^{\lambda<u, w_2>}e^{-\bar{\lambda}<w_1, u>} \times e^{<w_1, w_2>}$$

On the other hand,

$$< e(w_1), e^{\alpha W(u, I)} e(w_2) > \; = \; < e(w_1), \sum_{n=0}^{\infty} \frac{1}{\lfloor n} £^n W(u, I)^n e(w_2) >$$

Now, $$W(u, I)\, e(w_2) = \exp\left\{-\frac{1}{2}\|u\|^2 - < u, w_2 >\right\} e(w_2 + u).$$

$$W(u, I)^2\, e(w_2) = \exp\left\{-\frac{1}{2}\|u\|^2 - < u, w_2 > - \frac{1}{2}\|u\|^2 - < u, w_2 + u >\right\}$$
$$e(w_2 + 2u)$$
$$= \exp\left\{-2\|u\|^2 - 2 < u, w_2 >\right\} e(w_2 + 2u)$$

$$W(u, I)^3\, e(w_2) = \exp\left\{-\frac{1}{2}\|u\|^2 - < u, w_2 + 2u > -2\|u\|^2 - 2 < u, w_2 >\right\}$$
$$e(w_2 + 3u)$$
$$= \exp\left\{-\frac{9}{2}\|u\|^2 - 3 < u, w_2 >\right\} e(w_2 + 3u)$$

In general let

$$W(u, I)^n\, e(w_2) = \exp\left(-\alpha_n \|u\|^2 - \beta_n < u, w_2 >\right) e(w_2 + nu)$$

Then $$W(u, I)^{n+1}\, e(w_2) = \exp\left(-\alpha_n \|u\|^2 - \beta_n < u, w_2 > - \frac{1}{2}\|u\|^2\right.$$
$$\left. - < u, w_2 + nu >\right) e\left(w_2 + (n+1)u\right)$$
$$= \exp\left(-\left(\alpha_n + \frac{1}{2} + n\right)\|u\|^2 - (\beta + 1) < u, w_2 >\right)$$
$$e\left(w_2 + (n+1)u\right)$$

So $$\alpha_{n+1} = \alpha_n + n + \frac{1}{2},\ \alpha_0 = 0,$$
$$\beta_{n+1} = \beta_n + 1,\ \beta_0 = 0.$$

Thus $$\alpha_n = \sum_{n=0}^{n-1}\left(r + \frac{1}{2}\right) = \frac{n(n-1)}{2} + \frac{1}{2}n = \frac{n^2}{2}$$

$\beta_n = n$. Thus

$$W(u, I)^n\, e(w_2) = \exp\left\{-n^2 \frac{\|u\|^2}{2} - n < u, w_2 >\right\} e(w_2 + nu) < e(w_1),$$
$$W(u, I)^n\, e(w_2) >$$
$$= \exp\left\{-n^2 \frac{\|u\|^2}{2} - n < u, w_2 > + n < w_2, u > + < w_1, w_2 >\right\}$$

In particular,

$$< e(w_1), W(u, I)\, e(w_2) > \; = \exp\left\{-\frac{1}{2}\|u\|^2 - < u, w_2 > + < w_1, u > + < w_1, w_2 >\right\}$$

$$<e(w_1), W(\alpha u, I)\, e(w_2)>$$

$$= \exp\left\{-\frac{1}{2}|\alpha|^2 \|u\|^2 - \bar{\alpha} < u, w_2 > \right.$$

$$\left. + \alpha < w_1, u > + < w_1, w_2 > \right\}$$

$$= \left\langle e(w_1), \exp\left\{\alpha a^+(u) - \bar{\alpha} a(u)\right\} e(w_2)\right\rangle$$

Then,

$$W(\alpha u, I) = \exp\left\{\alpha a^+(u) - \bar{\alpha} a(u)\right\}.$$

KRP QSC.

$$V_\phi(t) = \exp\left(i\,\mathrm{I_m} \int_0^t e^{i\phi(s)} |F(s)|^2\, ds\right) \times W\left((e^{i\phi} - 1) f_t, e^{i\phi_t}\right)$$

$$\phi_t = \phi.\chi[o, t], f_t = f. X_{[O, t]}.$$

$<e(u), V_\phi(t)\, e(v)>$

$$= \exp\left(i\,\mathrm{I_m} \int_0^t e^{i\phi(s)} |f(s)|^2\, ds\right)$$

$$\times \left\langle e(u) \exp\left\{-\frac{1}{2}\|(e^{i\phi} - 1) f_t\|^2 - \left\langle (e^{i\phi} - 1) f_t, e^{i\phi_t} v\right\rangle\right\}\right|$$

$$\left. e\left(e^{i\phi_t} v + (e^{i\phi} - 1) f_t\right)\right\rangle$$

$$= \exp\left(i \int_0^t |f(s)|^2 \sin\{\phi(s)\}\, ds\right)$$

$$\times \exp\left\{-\frac{1}{2}\left(2\|f_t\|^2 - 2\int_0^t |f(s)|^2 \cos(\phi(s))\, ds\right)\right\}$$

$$-\left\langle \left(e^{i\phi} - 1\right) f_t, e^{i\phi_t} v\right\rangle + \left\langle u, e^{i\phi_t} v + \left(e^{i\phi} - 1\right) f_t\right\rangle\Big)$$

$$= \exp\left\{\int_0^t |f(s)|^2 e^{i\phi(s)} ds - \int_0^t |f(s)|^2 ds - \int_0^t (1 - e^{i\phi(s)})\, \overline{f(s)} v(s) ds\right.$$

$$+ \int_0^t (e^{i\phi(s)} - 1) \bar{u}(s) v(s) ds + \int_0^\infty \bar{u}(s) v(s) ds$$

$$\left. + \int_0^t (e^{i\phi(s)} - 1) \bar{u}(s) f(s) ds\right\}$$

Thus, $d < e(u), V_\phi(t)e(v) > = < e(u), dV_\phi(t)e(v) >$

$$= (e^{i\phi(t)} - 1)\left\{|f(t)|^2 + \bar{f}(t)v(t) + \bar{u}(t)v(t) + \bar{u}(t)f(t)\right\}$$

$$\left\langle e(u), V_\phi(t)e(v) \right\rangle$$

Note:

$$\left\langle u, e^{i\phi_t}v\right\rangle = \left\langle u, (e^{i\phi_t} - 1)v\right\rangle + \langle u, v\rangle$$

$$= \int_0^t (e^{i\phi(s)} - 1)\bar{u}(s)v(s)ds + \langle u, v\rangle .$$

$$<u, v> = \int_0^\infty \bar{u}(s)v(s)ds .$$

It follows that

$$dV_\phi(t) = \left(e^{i\phi(t)} - 1\right)\left\{|f(t)|^2 dt + \bar{f}(t)dA^*(t) + f(t)dA^+(t) + d\Lambda(t)\right\}V_\phi(t)$$

$$M(t, \phi, X) = \mathbb{E}_0\{U(t)^* XV_\phi(t)U(t)\}$$

$$d(U^*XV_\phi U) = dU^*XV_\phi U + U^*\chi V_\phi dU + U^*XdV\phi U + dU^*XV\phi dU$$

$$+ dU^*XdV_\phi U + U^*XdV_\phi dU$$

Now,

$$\mathbb{E}_0 [dU^*XV\phi U] = < e(o), dU^*XV_\phi U e(o) >$$

$$= \mathbb{E}_0\left\{U^*\left(iH - \frac{1}{2}L^*L\right)XV_\phi U\right\}dt$$

$$= dt\, M\left[t, \phi, \left(iH - \frac{1}{2}L^*L\right)X\right],$$

$$\mathbb{E}_0 [U^*XV\phi dU] = dt\, M\left(t, \phi, X(-iH - \frac{1}{2}L^*L)\right)$$

$$\mathbb{E}_0 [U^*XdV\phi U] = dt(e^{i\phi} - 1)|f|^2\, \mathbb{E}_0\left[U^*XV_\phi U\right]$$

$$= dt\, (e^{i\phi} - 1)\, |f|^2\, M(t, \phi, \chi)$$

$$\mathbb{E}_0 [dU^{*X}V\phi dU] = \mathbb{E}_0[U^*L^*XV\phi LU]dt$$

$$= \mathbb{E}_0[U^*L^*XLV\phi U]dt$$

($\mathbb{E}_0$ picks only the coefficient of dt)

$$= M(t, \phi, L^*XL)$$

$$(\Gamma V\phi, T) = 0\ \forall T \in £(h) \text{ since}$$

i.e. $V_\phi^{(t)} \in £(\Gamma s(\mathcal{H}))$

$$\mathbb{E}_0[dU^*XdV_\phi U] = \mathbb{E}_0[U^*LXf(e^{i\phi} - 1)V_\phi U]dt$$

$$= (e^{i\phi} - 1)f\, \mathbb{E}_0[U^*LXV_\phi U]dt$$

$$= (e^{i\phi} - 1)FM(t, \phi, LX)dt$$

$$\mathbb{E}_0[U^*XdV_\phi U]dt = \left(e^{i\phi} - 1\right)\bar{f}\, \mathbb{E}_0\left[U^*XV_\phi LU\right]dt$$

$$= \left(e^{i\phi}-1\right)\bar{f}\, dt\, M(t, \phi, \chi L)$$

$$([V_\phi, L] = 0 \because [V_\phi, h] = 0 \because [\Gamma_s(\mathcal{H}), h] = 0)$$

Then,

$$\begin{aligned}\frac{d}{dt}M(t,\phi,X) &= dt^{-1}.\ \mathbb{E}_0\{d(U^* X V_\phi U)]\\ &= M\left(t,\phi,\left(iH-\frac{1}{2}L^* L\right)X - X\left(iH+\frac{1}{2}L^* L\right)\right.\\ &\quad \left. + X|f|^2\left(e^{i\phi}-1\right) + L^* XL + (e^{i\phi}-1)(fLX + \bar{f}XL)\right)\\ &= M\left(t,\phi,\theta(\chi)\right) + (e^{i\phi}-1)M(t,\phi,|f|^2\chi + fL\chi + \bar{f}\chi L)\end{aligned}$$

Note that

$$|f|^2 X + fLX + \bar{f}XL = \bar{f}X(f+L) + fLX.$$

Remark: An additional term also has to be considered in the evaluation of $\dfrac{d\,M(t,\phi,\chi)}{dt}$.

It is $dU^* X\, dV\phi\, dU$

The coefficient of dt in this triple differential product comes from

$$dA\,.\,d\Lambda\,.\,dA^+ = dA\, dA^+ = dt$$

The corresponding coefficient is $(e^{i\phi} - 1)$

$$U^* LX(e^{i\phi} - 1)\, V_\phi LU = (U^* LXLV\phi U)$$

and so, taking into account this correction, we get

$$\begin{aligned}\frac{d}{dt}M(t,\phi,X) &= M\left(t,\phi,\theta(\chi)\right) + (e^{i\phi}-1)M\\ &\quad \left(t,\phi|f|^2 X + fLX + \bar{f}XL + LXL\right)\\ &= M\left(t,\phi,\theta(X)\right) + (e^{i\phi}-1)M\left(t,\phi,(\bar{f}+L)X(f+L)\right).\end{aligned}$$

❖❖❖❖❖

[14] Moment Generating Function for Certain Quantum Random Variables:

$h \simeq$ system space, $\mathcal{H} \simeq$ Bath space for 1 Boson $\Gamma_s(\mathcal{H}) \simeq$ bath Boson Fock space for indefinite number of Bosons.

$L \in \pounds(h)$, $a(u) \in \mathcal{L}(\Gamma_s(\mathcal{H}_1))$

$a(u)$, $a^*(u)$ are the annihilation and creation operators

$$[a(u), a(v)] = \langle u, v\rangle$$

$f \in h.e(u) \in \Gamma_s(\mathcal{H})$ is the exponential vector.

Let $[L, L^*] = 0$. *i.e.* L is a normal operator.

We wish to find

$\left\langle fe(v)\middle|\exp\left(La(u)+L^{*}a^{*}(u)\right)He(v)\right\rangle$

Suppose A, B are operator such that

$$[A, [A. B]] = [B.[A, B]] = 0.$$

Let
$$F(t) = \exp(t(A+B))$$
$$F'(t) = (A+B)F(t).$$

Let
$$F(t) = \exp(tA)G(t). \text{ Then}$$
$$G'(t) = [\exp(-t\, al\, A)(B)]\, G(t)$$
$$= (B - t[A, B])\, G(t)$$

and so,
$$G(t) = \exp\left(tB - \frac{t^2}{2}[A, B]\right)$$
$$= \exp(tB)\cdot\exp\left(-\frac{t^2}{2}[A, B]\right).$$

Thus,
$$\exp(tA + B) = \exp(tA)\cdot\exp(tB)\cdot\exp\left(-\frac{t^2}{2}[A, B]\right)$$

or equivalently

$$\exp(A+B) = \exp(A).\exp(B).\exp\left(-\frac{1}{2}[A, B]\right).$$

Thus since $[L\,a(u), L^*a^*(u)] = \|u\|^2 LL^*$

Commutes with both $La(u)$ and $L^*\,a^*(u)$, it follows that

❖❖❖❖❖

[15] Belavkin Filtering Contd.:

$$d\,\sigma_t(X) = \mathbb{E}\left[F^*(t+dt)XF(t+dt)\,|\,\eta^{in}_{t+dt]}\right]$$
$$-\mathbb{E}\left[F^*(t)XF(t)\,|\,\eta^{in}_{t]}\right]$$
$$= \mathbb{E}\left[F^*(t)XF(t)\,|\,\eta^{in}_{t+dt]}\right] - \mathbb{E}\left[F^*(t)XF(t)\,|\,\eta^{in}_{t]}\right]$$
$$+\mathbb{E}\left[d(F^*(t)XF(t)\,|\,\eta^{in}_{t+dt]}\right]$$

Now
$$F^*(t)XF(t) \in \sigma\{Y^{in}(S), S \le t, \delta(\hbar)\}$$

and since
$$\eta^{in}_{t+dt]} = \sigma\{\eta^{in}_{t],}\, dY^{in}(t)\}$$

it follows two in the state if

$$e(u) > |fe(u)\rangle\exp\left(-\|u\|^2\!\big/_2\right),$$

$$\mathbb{E}\left[F^*(t)XF(t) \mid \eta^{in}_{t+dt]}\right] = \mathbb{E}\left[F^*(t)XF(t) \mid \eta^{in}_{t]}\right]$$

Thus, $$d\sigma_t(X) = \mathbb{E}\left[d\left(F^*(t)XF(t)\right) \mid \eta^{in}_{t+dt]}\right]$$

$$\mathbb{E}\left\{F^*(t)\left[\left(P^*(t)X + XP(t)\right)F(t) \mid \eta^{in}_{t]}\right]dY_{in}(t)\right.$$

$$+\mathbb{E}\left[F^*(t)P^*(t)XP(t)F(t) \middle| \eta^{in}_{t]}\right]dZ_{in}(t)$$

$$+\mathbb{E}\left[F^*(t)\left(Q^*(t)X + XQ(t)\right)F(t) \middle| \eta^{in}_{t]}\right]dt$$

$$= \sigma_t(P^*X + XP)dY_{in} + \sigma_t(P^*XP)dZ_{in} + \sigma_t(Q^*X + XQ)dt$$

$$d\left\{\frac{\sigma_t(X)}{\sigma_t(1)}\right\} = \frac{d\sigma_t(X)}{\sigma_t(1)} - \frac{\sigma_t(X)d\sigma_t(1)}{\sigma_t(1)^2} + \frac{\sigma_t(X)}{\sigma_t(1)^3}(d\sigma_t(1))^2 - \frac{d\sigma_t(X)d\sigma_t(1)}{\sigma_t(1)^2}$$

Let $v_t(X) = \dfrac{\sigma_t(X)}{\sigma_t(1)}$. Then we get

$$\begin{aligned} d\,v_t(X) = {} & v_t(P^*X + XP)\,d\,Y_{in} + v_t(P^*XP)\,d\,Z_{in} + v_t(Q^*X + XQ)\,dt \\ & - v_t(X)\,(v_t(P^* + P)\,d\,Z_{in} + v_t(P^*P)\,d\,Z_{in} + v_t(Q^*Q)\,dt \\ & + v_t(X)\,v_t\,(P^* + P)\,d\,Z_{in} + v_t(P^*X\,P)^2\,d\,V_{in} \\ & + 2\,v_t\,(P^* + P)\,v_t(P^*P)\,dW_{in}) - v_t(P^*X + XP)\,v_t(P^* + P)\,dZ_{in} \\ & - v_t\,(P^*X + XP)\,v_t\,(P^*P) + v_t(P^* + P)\,v_t(P^*XP))\,d\,W_{in} \\ & - v_t(P^*XP)\,d\,V_{in} \end{aligned}$$

where
$$\begin{aligned} (dY_{in})^2 &= dZ_{in} \\ (dY_{in})^3 &= dW_{in} = dY_{in}dZ_{in} \\ (dY_{in})^4 &= dV_{in} = (dZ_{in})^2 \end{aligned}$$

We can express there as

$$\begin{aligned} dv_t(X) = {} & A_t(X)\,dt + B_t(X)\,dY^{in}(t) + C_t(X)\,dZ_{in}(t) \\ & + D_t(X)dW_{in}(t) + E_t(X)\,dV_{in}(t) \end{aligned}$$

where $A_t(X)$, $B_t(X)$, $C_t(X)$, $D_t(X)$ and $E_t(X)$ are linear functions of X with values in the Abelian algebra

$$\eta^{in}_{t]} = \sigma\left\{Y^{in}(s)\; s \le t\right\},$$

❖❖❖❖❖

[16] Modelling Noise in Quantum Mechanical Systems:

[*a*] Dyson series:

Let $\xi(t)$, $t \geq 0$ be a random process

Consider the perturbed quantum Hamiltonian

$$H(t) = H_0 + \varepsilon\xi(t)V$$

Let $U_0(t) = \exp(-itH_0) \simeq$ then unitary evolution operator of the unperturbed system. Them if $U(t, s)$, $t \geq s$ is the perturbed unitary evolution from time s to time t,

$$i\hbar\frac{\partial U(t,s)}{\partial t} = H(t)\,U(t, s),\, t \geq s,$$

$$U(s, s) = I$$

The solution is $\quad U(t, s) = U_0(t)\,W(t, s)$

where $\quad i\hbar\frac{\partial W(t,s)}{\partial t} = \varepsilon\xi(t)\,\bar{V}(t)\,W(t,s),\, t \geq s\,,$

$$W(s, s) = U_0(-s),$$

$$\bar{V}(t) = U_0(-t)VU.(t).$$

Then the solution is

$$U(t, s) = U_0(t-s) + U_0(t)\sum_{n=1}^{\infty}\left(\frac{-i\varepsilon}{h}\right)^n$$

$$\int_{s<t_n<\ldots<t_1<t} \xi(t_1)\ldots\xi(t_n)V(t_1)\ldots\bar{V}(t_n)\,dt_1\ldots.dt_n)\,\underline{U}_0(-s)$$

$$= U_0(t-s)\sum_{n=1}^{\infty}\left(\frac{-i\varepsilon}{\hbar}\right)^n$$

$$\int_{s<t_n<\ldots<t_1<t} \xi(t_1)\ldots\xi(t_n)\,U_0(t-t_1)\,V(t_1)U_0(t_1-t_2)\,V(t_2)\ldots V(t_n)$$

$$U_0(t_n - s)\,dt_1\ldots dt_n$$

Let $|\, n >$, $n = 0, 1, 2, \ldots$ be the stationary states of the unperturbed system.

Thus $\quad H_0|n\rangle = E_n|n\rangle,\, n = 0, 1, 2, \ldots$

Then $\quad U_0(t)\,|n\rangle = \exp\left(-i\frac{E_n t}{\hbar}\right)|n\rangle$

and since $\langle m \mid n\rangle = \delta[m-n]$, then transition probability from $|n\rangle - 1|m\rangle$, $m \neq n$ in time $s \to t$ is given by

$$P(m, t \mid n, s) = \mathbb{E}\left\{|\langle m \mid U(t,s) \mid n\rangle|^2\right\}$$

$$= \mathbb{E}\left\{\left|\sum_{k=1}^{\infty}\left(\frac{-i\,\varepsilon}{\hbar}\right)^k\right.\right.$$

$$\left.\left.\int_{s<t_k<\ldots<t_1<t} \xi(t_1)\ldots\xi(t_k) < m \,|\, \bar{V}(t_1)\ldots\bar{V}(t_k) \,|\, n > dt_1\ldots dt_k\right|^2\right\}$$

$$= \sum_{k,\,r=1}^{\infty}\left(\frac{\varepsilon}{\hbar}\right)^{k+r}$$

$$\operatorname{Re}\left\{i^{r-k} \int_{\substack{S<t_k<\ldots<t_1<t \\ S<S_r<\ldots<S_1<t}} \mathbb{E}\,\{\xi(t_1)\}\ldots\xi(t_k)\xi(s_1)\ldots\xi(s_r)\right\}$$

$$\times\left\langle m \,|\, V(t_1)\ldots\tilde{V}(t_k) \,|\, n\right\rangle\left\langle n \,|\, \bar{v}(s_r)\ldots V(t_r)\ldots\tilde{V}(s_1 \,|\, m)\right\rangle$$

Note that

$$\left\langle m \,|\, \bar{V}(t_1)\ldots\bar{V}(t_k) \,|\, n\right\rangle = \sum_{0\le p_1,\ldots,\, p_{k-1}<\infty} \exp\left(\frac{i(E_m t_1 - E_n t_k)}{\hbar}\right)$$

$$\times\left\langle m \,|\, V \,|\, p_1\right\rangle\left\langle p_1 \,|\, V \,|\, p_2\right\rangle\ldots\left\langle p_{k-1} \,|\, V \,|\, n\right\rangle$$

$$\times \exp\left(\frac{i}{\hbar}\left(E_{p_1}(t_2 - t_1) + E_{p_2}(t_3 - t_2) + \ldots + E_{p_{k-1}}(t_k - t_{k-1})\right)\right)$$

This formula for the average transition probability tells us how to estimate the moments of the random process $\{\xi(t) : t \ge 0\}$ from measurements of the average transition probability.

If the noise is white, *i.e.* derivative of Brownain motion, then an *Ito* correction term has to be included in the Schrödinger eqn. to maintain unitarity of the evolution. Specifically consider the "noisy Schrödinger eqn."

$$d\,|\,\underline{\psi}_t > = \left[-i(H_0 + iP(t)\,dt - i\sum_{j=1}^{k} L_j(t)\,dB_j(t)\right]\left|\,\underline{\psi}_t\right\rangle$$

where H_0 is the System Hamiltonian, $B_1(\cdot), \ldots, B_k(\cdot)$

$P(t)$ is an It_0 correction term. It is determined from the unitarity condition

$$d\left\langle \psi_t \,|\, \psi_t\right\rangle = 0$$

i.e.
$$\left\langle d\psi_t \,|\, \psi_t\right\rangle + \left\langle \psi_t \,|\, d\psi_t\right\rangle + \left\langle d\psi_t \,|\, d\psi_t\right\rangle = 0$$

giving $-i\left\langle \psi_t \,|\, \left(H_o - iP^*\right)dt + \sum L_j^* dB_j \,|\, \psi_t\right\rangle$

$$-i\left\langle \psi_t \,|\, (H_o - iP)\,dt + \sum L_j\, dB_j \,|\, \psi_t\right\rangle$$

$$+\left\langle \psi_t \mid \sum_j L_j^* L_j \mid \psi_t \right\rangle dt = 0$$

so that

$$L_j^* = L_j, \sum_j L_j^* L_j = P + P^*$$

We can absorb $\dfrac{p - p^*}{2i}$ into H_o and hence set

$$P = P^* = \frac{1}{2}\sum_j L_j^* L_j$$

Thus the noisy Schrödinger eqn, is

$$d \mid \psi_t \rangle = \left[\left(-iH(t) + P(t) \right) dt - i \sum_j L_j(t)\, dB_j(t) \right] \Big\rangle \mid \psi_t >$$

where $L_j^* = L_j, H^* - H, P^* = P = \frac{1}{2}\sum L_j^2.$

A more generalized version of the noisy Schrödinger eqn. is based on modelling the noise processes a operators in a Boson Fock space.

$$\Gamma_s(\mathcal{H}) = \mathbb{C} \oplus \bigoplus_{n=1}^{\infty} \mathcal{H}^{(S)n},$$

$$\mathcal{H} = \left(\mathbb{C}^p \oplus L^2 [0, \infty) \right)$$

on this space, we define operations $A_j(t), A_j^*(t), \Lambda_k^j(t), \quad 1 \leq j, k \leq p, t \geq 0$ satisfying the Hudson-Pathasarathy quantum It_o formula:

$$d\, A_j \cdot d\, A_k^* = \delta_{jk}\, dt,$$

$$d\, A_j \cdot d\, \Lambda_s^r = \delta_j^r\, dA_s$$

$$d\, \Lambda_s^r\, d\, A_j^* = \delta_j^r\, d\Lambda_s^*$$

$$d\Lambda_s^r \cdot d\Lambda_k^j = \delta_k^r\, d\Lambda_s^j$$

and all other product differentials are zero.

The noisy Schrödinger evolution eqn. is then

$$dU(t) = \left[-iH(t)\, dt + \sum_{j=1}^{p} (L_j(t)\, dA,(t) - \tilde{L}_j(t)\, dA_j^*(t)) \right.$$

$$\left. + \sum_{j,k=1}^{p} S_k^j(t)\, d\Lambda_j^k(t) \right] U(t)$$

where the operators, $H(t), L_j(t), (\tilde{L}_j(t), S_k^j(t)$ and in the system Hilbert space $\hbar$ and the system plus bath space is $\hbar \otimes \Gamma_s(\mathcal{H})$

The operation $H(t), \left\{L_j(t), \tilde{L}_j(t), S_k^j(t)\right\}$ can be chosen so that $U(t)$ in unitary (by applying the quantum It_o formula t_o

$0 = d(U^*U) = dU^* \cdot U + U^* \cdot dU + dU \cdot dU$

[4] Quantum filtering theory.

Here, a system observable $X \in \mathcal{L}(\hbar)$ evolves as

$$
\begin{aligned}
X(t) &= U^*(t)\, X\, U(t) \\
&= U^*(t)\,(X \otimes I)\, U(t)
\end{aligned}
$$

and the measurement process is of the form

$$
Y_{\text{out}}^t = U^*(t)\left(I \otimes \int_0^t (C_k^j)(t')d\Lambda_j^k(t') + \alpha_j^{(t')}\, dA_j^*(t') + \overline{\alpha_j(t')}dA_j^*(t') \right) U(t)
$$

where $C_k^j(t'), \alpha_j, (t),$ are complex scalar valued functions. It is easy to see that

$$
Y_{\text{out}}(t) = U^*(T)\,(I \otimes Y_{\text{in}}(t))\, U(T),
$$

$\forall^* T \geq t$ where

$$
Y_{\text{in}}(t) = \int_0^t \left(C_k'^j(t')\, d\Lambda_j^k(t') + \alpha_j,(t')\, dA_j(t') + \overline{\alpha,(t)}\, \alpha A_j^*(t') \right)
$$

To make $Y_{\text{out}}(t)$ (or equivalently $Y_{\text{in}}(t)$) Hermitian we require

$$
C_k^j(t) = \overline{C_j^k(t)}
$$

since $\Lambda_j^k(t)^* = \Lambda_k^j(t)$.

It follows that

$$
\left[X, Y_{\text{in}}(t)\right] = 0 \,\forall\, t
$$

and hence, $[X(T),\, Y_{\text{out}}(t)] = 0 \;\forall T \geq t$

In the language of Belavkin, $Y_{\text{out}}(\cdot)$ is a non-demolition measurement process.

Quantum filtering aims at constructing a "conditional expectation".

$$
\widehat{X}(t|t) = \mathbb{E}\left[X(t) \mid Y_{\text{out}}(s), s \leq t\right]
$$

This is possible provided $\left[Y_{\text{out}}(t_1), Y_{\text{out}}(t_2)\right] = 0 \; \forall \; t_1, t_2$ or equivalently

$$\left[Y_{\text{in}}(t_1), Y_{\text{in}}(t_2)\right] = 0 \; \forall t_1, t_2 .$$

The conditional expectation is then defined by the rule

$$\left\langle fe(u) \,|\, (X(t) - \widehat{X}(t\,|\,t)) \,|\, fe(u) \right\rangle = 0$$

For some fixed $f \in \hbar$ and $e(u) = 1 \oplus \overset{\infty}{\underset{n=1}{\otimes}} \dfrac{u^{\otimes n}}{\sqrt{|\underline{n}}}$.

❖❖❖❖❖

[17] Belavkin has Derived a Quantum Stochastic Differential Equation (*qsde*) for $\widehat{X}(t\,|\,t)$:

A block processing approach to quantum conditional expectation is to assume

$$\widehat{X}(t\,|\,t) \approx \int_0^t h(t,s)\, Y_{\text{out}}(s)\,ds$$

where $h(t, s) \in \mathbb{R}$ chosen so that

$$\left\langle fe(u) \,|\, X(t)\, Y_{\text{out}}(s) \,|\, fe(u) \right\rangle$$

$$= \int_0^t h(t,s')\left\langle fe(u) \,|\, Y_{\text{out}}(s')\, Y_{\text{out}}(s) \,|\, fe(u) \right\rangle ds_1' 0 \le s \le t$$

and solve this integral eqn. for $\{h(t, s)\}$.

Note that for $s \le t$,

$$X(t) Y_{\text{out}}(s) = U(t)^* (X \otimes Y_{\text{in}}(s))\, U(t)$$

and hence $\left\langle fe(u) \,|\, X(t) Y_{\text{out}}(s) \,|\, fe(u) \right\rangle$

$$= \left\langle U(t)\,(fe(u)) \,|\, X \otimes Y_{\text{in}}(s) \,|\, U(t) fe(u) \right\rangle .$$

Construction of quantum-Girsanov transformation

Let m be a Von-Neumann algebra and m' its commutator. Assume m' is Abelian. Let $F \in m$ be such that $\mathbb{E}(F^* F) = 1$ where $\mathbb{E}$ is an expectation on m (like for $m = \mathbb{B}(\mathcal{H})$, we can take $\mathbb{E}(X) = T_r(\rho X)$ where $\rho \ge 0, T_r(\rho) = 1$). Put

$$\mathbb{E}_F(X) = \mathbb{E}_F(F^* X\, F),\; X \in m$$

$\mathbb{E}_F(\cdot)$ is another expectation on m.

By definition $\mathbb{E}_F(X \,|\, m') \in m'$, satisfies

$$\mathbb{E}_F(XY) = \mathbb{E}_F\left(\mathbb{E}_F(X \mid m')Y\right), \ \forall Y \in m', X \in m.$$

Thus, $\mathbb{E}(F^* X\, Y\, F) = \mathbb{E}(F^* \mathbb{E}_F(X \mid m')\, YF)Y \in m'$

Equivalently,

$$\mathbb{E}(F^* X\, F Y) = \mathbb{E}(F^* F\, \mathbb{E}_F(X \mid m')Y)$$

so, $$\mathbb{E}_F(X \mid m') = \frac{\mathbb{E}(F^* X\, F \mid m')}{\mathbb{E}(F^*\, F \mid m')}$$

Quantum Kallianpur-Striebel formula:

$X \in \pounds(\hbar)$

$$\infty\{Y_{\text{out}}(s) : s \leq t\} \ \underline{\underline{\Delta}}\ m_t^{\text{out}}$$

$$\infty\{Y_{\text{in}}(s) : s \leq t\} \ \underline{\underline{\Delta}}\ m_t^{\text{in}}$$

Then $$m_t^{\text{out}} = U(t)^* m_t^{\text{in}}\, U(t)$$

since $$Y_{\text{out}}(s) = U(t)^* Y_{\text{in}}(s)\, U(t), t \geq s.$$

$$\text{Let } X(t) = U^*(t)\, X U)(t).$$

Suppose that there exists $F(t) \in m_t^{\text{in}}$ such that

$$\mathbb{E} \mid U^*(t) X\, U(t)\big] = \mathbb{E}\left[F(t)^* X\, F(t)\right], \mathbb{E}\left[F^*(t)\, F(t)\right] = 1.$$

Define $$\tilde{\mathbb{E}}_t(X) = \mathbb{E}\left[U^*(t)\, X\, U(t)\right] \equiv \mathbb{E}_{F(t)}[X]$$

Then $$\mathbb{E}[X(t) \mid m_t^{\text{out}}] = U^*(t)\, \tilde{\mathbb{E}}_t[X \mid m_t^{\text{in}}]\, U(t)$$

$$= \frac{U^*(t)\mathbb{E}[F*(t)X\, F(t) \mid m_t^{\text{in}}]}{\mathbb{E}[F^*(t)F(t) \mid m_t^{\text{in}}]\, U(t)}$$

Remarks: Let m be an Abelian algebra and $X \in m'$. Then if U is a unitary operator, we have

$$\mathbb{E}[U^* X U \mid U^* m U] = U^* 1 E_U[X \mid m]U$$

where $$\mathbb{E}_U[X] = \mathbb{E}[U^* X\, U]$$

To see this, let $Y \in m$. Then

$$\mathbb{E}[\mathbb{E}[U^* X\, U \mid U^* m U]\, U^* Y U] = \mathbb{E}[U^* X U\, U^* Y U] = \mathbb{E}[U^* X\, Y\, U]$$

On the other hand,

$$\mathbb{E}[U^* \mathbb{E}_U[X \mid m] U\, U^* Y U] = \mathbb{E}[U^* \mathbb{E}_U[X \mid m] Y\, U]$$

$$= \mathbb{E}[U^* \mathbb{E}_U[XY \mid m] U]$$

$$= \mathbb{E}_U \mathbb{E}_U [XY \mid m] = \mathbb{E}_U [XY]$$

$$= \mathbb{E}[U^* XYU]$$

❖❖❖❖❖

[18] Waveguide Modes:

$$\underline{E}(x, y, z, w) = e(x, y, w) e^{-\tau(w)z}$$

$$\underline{H}(x, y, z, w) = \underline{H}(x, y, w) e^{-\tau(0)z}$$

$$\underline{\nabla} \perp = \hat{x}\frac{\partial}{\partial x} + \hat{y}\frac{\partial}{\partial y},$$

$$\underline{E} = \underline{E}_{\perp} + E_z \hat{z},$$

$$\underline{H} = \underline{H}_{\perp} + H_z \hat{z}$$

$$\underline{\nabla} \times \underline{E} = -j\omega\mu H$$

$$\underline{\nabla} \times \underline{H} = (\sigma + j\,\omega\varepsilon)\underline{E}$$

Duality: The Maxwell eqns. remain invariant under the transformation $\underline{E} \to \underline{H}$, $\underline{H} \to -\underline{E}$, $j\omega\mu \leftrightarrow \sigma + j\omega\varepsilon$.

or equivalently, $\mu \leftrightarrow \varepsilon' = \varepsilon - \dfrac{j\sigma}{\omega}$

$$\left(\nabla_{\perp}\gamma\hat{z}\right) \times \left(\underline{E}_{\perp} + E_z \hat{z}\right) = -j\omega\mu\left(\underline{H}_{\perp} H_z \hat{z}\right)$$

gives $\nabla_{\perp} \otimes E_{\perp} = -j\,\omega\mu\, H_z \hat{z}$

$$\underline{\nabla}_{\perp} E_z \times \hat{z} - \gamma\hat{z} \times \underline{E}_{\perp} = -j\omega\mu H_{\perp} \tag{1}$$

By duality,

$$\underline{\nabla} \perp Hz \times \hat{z} - \lambda\hat{z} \times \underline{H}_{\perp} = (\sigma + j\omega\varepsilon)\, E_{\perp} = j\omega\varepsilon' H_{\perp} \tag{2}$$

Taking cross product with $\hat{z}$ on both side of (1) gives

$$\underline{\nabla}_{\perp} E_k + \gamma\, \underline{E}_{\perp} = -j\omega\mu\hat{z} \times H_{\perp} \tag{3}$$

Eliminating $\hat{z} \times \underline{H}_{\perp}$ between (2) and (3), we get

$$\underline{\nabla}_{\perp} H_z \times \hat{z} + \frac{\gamma}{j\omega\mu}(\underline{\nabla} E_z + \tau\, \underline{E}_{\perp}) = j\,\omega\,\varepsilon' E_{\perp}$$

or $\left(\gamma^2 + \omega^2 u\,\varepsilon'\right) \underline{E}_{\perp} = -\gamma\underline{\nabla}_{\perp} E_z - j\omega\mu\underline{\nabla}_{\perp} H_z \times \hat{z}$

or
$$\underline{E}_{\perp} = -\frac{\gamma}{h^2}\underline{\nabla}_{\perp} E_z - \frac{j\,\omega\mu}{h^2}\underline{\nabla}_{\perp} H_z \times \hat{z} \tag{4}$$

Again by duality,

$$\underline{H}_\perp = -\frac{\gamma}{h^2}\underline{\nabla}_\perp H_z + \frac{j\omega\varepsilon'}{h^2}\underline{\nabla}_\perp E_z \times \hat{z} \tag{5}$$

Boundary conditions: Sidewalls are perfect electric conductors, so

$$E_z\big|_{\partial D} = 0$$

where D is the cross-section area of the guide and ∂D is the boundary curve of D.

Now assume $\varepsilon' = \varepsilon'(x, y)$, $\mu = \mu\,(x, y)$, *i.e.* the guide is inhomogeneous in the $x - y$ plane at any $z \geq 0$. Then let

$$\varepsilon' = \varepsilon_0\left(1+\delta\cdot\chi_e(x,y)\right)$$

$$\mu = \mu_0\left(1+\delta\cdot\chi_m(x,y)\right)$$

where δ is a small perturbation parameter.

Then $$\omega^2\varepsilon'\mu = \omega^2\,\varepsilon_0\,\mu_0\left(1+\delta\cdot\chi(x,y)\right)+O\left(\delta^2\right)$$

where $$\chi(x, y) = \chi_e(x,y)+\chi_m(x,y)$$

$$\underline{\nabla}_\perp \times \underline{E}_\perp = -j\,\omega\mu\,H_z\hat{z}$$

gives on using (4).

$$\nabla_\perp\left(\frac{\gamma}{h^2}\nabla_\perp E_z\right) = \nabla_\perp\left(\frac{j\omega\mu}{h^2}\underline{\nabla}_\perp H_z\times\hat{z}\right) = j\,\omega\mu\,H_z\hat{z}$$

or equivalently,

$$\underline{\nabla}_\perp\left(\frac{\gamma}{h^2}\right)\times\underline{\nabla}_\perp E_2 + \nabla_\perp\times\left(\frac{j\omega\mu}{h^2}\right)\times\left(\underline{\nabla}_\perp H_z\times\hat{z}\right) - \frac{j\omega\mu}{h^2}\nabla^2{}_\perp H_z\hat{z} - j\omega\mu\;H_z\hat{z} = 0$$

which is equivalent to

$$\nabla_\perp^2 H_z + h^2 H_z - \frac{h^2}{j\omega\mu}\left(\underline{\nabla}_\perp\left(\frac{\gamma}{h^2}\right)\times\underline{\nabla}_\perp E_z, \hat{z}\right.$$
$$\left. - \left(\underline{\nabla}_\perp \log\left(\frac{\mu}{h^2}\right)\times\left(\underline{\nabla}_\perp E_z\times\hat{z}\right), \hat{z}\right)\right) = 0 \tag{6}$$

and by duality,

$$\nabla_\perp^2 E_z + h^2 E_z - \frac{h^2}{j\omega\varepsilon'}\left(\underline{\nabla}_\perp\left(\frac{\gamma}{h^2}\right)\times\underline{\nabla}_\perp H_z, \hat{z}\right.$$
$$\left. = \left(\underline{\nabla}_\perp \log\left(\frac{\varepsilon'}{h^2}\right)\times\left(\underline{\nabla}_\perp E_z\times\hat{z}\right), \hat{z}\right)\right) = 0 \tag{7}$$

(6) and (7) can be simplified to

$$\left(\nabla_\perp^2 + h^2\right)H_z - \frac{h^2}{j\omega\mu}\left(\hat{z}\times\underline{\nabla}_\perp\left(\frac{\gamma}{h^2}\right), \underline{\nabla}_\perp E_z\right) + \left(\underline{\nabla}_\perp \log\left(\frac{\mu}{h^2}\right), \underline{\nabla}_\perp H_z\right) = 0 \quad (8)$$

and

$$\left(\nabla_\perp^2 + h^2\right)E_z - \frac{h^2}{j\omega\varepsilon'}\left(\hat{z}\times\underline{\nabla}_\perp\left(\frac{\gamma}{h^2}\right), \underline{\nabla}_\perp H_z\right) + \left(\underline{\nabla}_\perp \log\left(\frac{\varepsilon'}{h^2}\right), \underline{\nabla}_\perp E_z\right) = 0 \quad (9)$$

This can be expressed in a convenient notation.

$$\left(\nabla_\perp^2 + h^2\right)\begin{pmatrix} E_z \\ H_z \end{pmatrix} + \begin{pmatrix} \left(\underline{\nabla}_\perp \log\left(\frac{\varepsilon'}{h^2}\right), \nabla_\perp\right), & \frac{h^2}{j\omega\varepsilon'}\left(\hat{z}\times\underline{\nabla}_\perp\left(\frac{\gamma}{h^2}\right), \underline{\nabla}_\perp\right) \\ \frac{-h^2}{j\omega\mu}\left(\hat{z}\times\underline{\nabla}_\perp\left(\frac{\gamma}{h^2}\right), \underline{\nabla}_\perp\right), & \left(\underline{\nabla}_\perp \log\left(\frac{\mu}{h^2}\right), \underline{\nabla}_\perp\right) \end{pmatrix}\begin{pmatrix} E_z \\ H_z \end{pmatrix} = 0 \quad (10)$$

$$\begin{aligned} h^2 &= \gamma + \omega^2 \mu_0 \varepsilon_0 (1 + \delta\, \chi(x, y)) + O(\delta^2) \\ &= h_0^2 + \delta k_0^2 \chi(x, y) + O(\delta^2),\ h_0^2 \\ &= \gamma^2 + \omega^2 \mu_0 \varepsilon_0 = \gamma^2 + k_0^2 . \end{aligned}$$

Let $\chi_e(x, y)$, $\chi_m(x, y)$ be random Gaussian fields with zero-mean and correlations

$$\begin{aligned} \left\langle \chi_e(x, y)\, \bar{\chi}_e(x', y')\right\rangle &= R_{ee}(x, y; x', y') \\ \left\langle \chi_e(x, y), \bar{\chi}_m(x', y')\right\rangle &= R_{em}(x, y; x', y') \\ \left\langle \chi_m(x, y), \bar{\chi}_m(x', y')\right\rangle &= R_{mm}(x, y; x', y') . \end{aligned}$$

These are the susceptibility correlations at a given frequency ω.

$$\begin{aligned} \frac{\varepsilon'}{h^2} &= \frac{\epsilon_0\left(1 + \delta\chi_e(x, y)\right)}{h_0^2 + \delta k_0^2 \chi(x, y)} \\ &= \frac{\epsilon_0}{h_0^2} = (1 + \delta\chi_e)\left(1 - \frac{\delta k_0^2}{h_0^2}\chi\right) + O(\delta^2) \\ \frac{\epsilon_0}{h_0^2} &= \frac{\delta\varepsilon_0}{h_0^2}\left(\chi_e - \frac{\delta k_0^2}{h_0^2}\chi\right) + O(\delta^2) \end{aligned}$$

$$\log\left(\frac{\varepsilon'}{h^2}\right) = \log\left(\frac{\varepsilon_0}{h_0^2}\right) + \delta\left(\chi_e - \frac{k_0^2}{h_0^2}\chi\right) + O(\delta^2)$$

$$\underline{\nabla}_\perp \log\left(\frac{\varepsilon'}{h^2}\right) = \delta\underline{\nabla}_\perp\left(\chi_e - \frac{k_0^2}{h_0^2}\chi\right) + O(\delta^2)$$

$$\frac{h^2}{\varepsilon'}\underline{\nabla}\frac{1}{h^2} = -\frac{1}{\varepsilon'}\underline{\nabla}_\perp \log(h^2) = \frac{-1}{\varepsilon_0(1+\delta\cdot\chi_e)}\underline{\nabla}_\perp \log(h_0^2+\delta k_0^2\chi)$$

$$= -\frac{1}{\varepsilon_0}(1-\delta\chi_e)\underline{\nabla}_\perp\left(\log h_0^2+\frac{\delta k_0^2}{h_0^2}\chi\right)+O(\delta^2)$$

$$\frac{\delta k_0^2}{\varepsilon_0 h_0^2}\underline{\nabla}\chi+O(\delta^2)$$

$$\frac{h^2}{\mu}\underline{\nabla}_\perp\frac{1}{h^2} = -\frac{1}{\mu}\underline{\nabla}_\perp \log h^2$$

$$= -\frac{1}{\mu}\underline{\nabla}_\perp \log h^2$$

$$= -\frac{\delta k_0^2}{\mu_0 h_0^2}\underline{\nabla}_\perp\chi+O(\delta^2)$$

$$\underline{\nabla}_\perp \log\left(\frac{\mu}{h^2}\right) = \underline{\nabla}_\perp \log\left\{\frac{\mu_0(1+\delta\cdot\chi_m)}{\left(h_0^2+\delta k^2\chi\right)}\right\}$$

$$= \underline{\nabla}_\perp\left\{(1+\delta\chi_m)\left(1-\frac{\delta k_0^2}{h_0^2}\chi 1\right\}+O(\delta^2)\right.$$

$$= \delta\underline{\nabla}_\perp\left(\chi_m-\frac{k_0^2}{h_0^2}\chi\right)+O(\delta^2)$$

Finally, the 2 × 2 matrix operator (10) can be expressed as

$$\underline{\underline{M}} = \delta\begin{pmatrix}\left(\underline{\nabla}_\perp\left(\chi_e-\frac{k_0^2}{h_0^2}\chi\right),\underline{\nabla}_\perp\right), & \frac{-k_0^2\gamma}{j\omega\,\varepsilon_0 h_0^2}(\underline{\nabla}_\perp X,\underline{\nabla}_\perp) \\ \frac{\gamma k_0^2}{j\omega\mu_0 h_0^2}(\hat{z}\times\underline{\nabla}\chi,\underline{\nabla}_\perp) & \left(\underline{\nabla}_\perp\left(\chi_m-\frac{k_0^2\chi}{h_0^2},\underline{\nabla}_\perp\right),\underline{\nabla}_\perp\right)\end{pmatrix}$$

$$+O(\delta^2) \tag{11}$$

Note that $\gamma^2+k_0^2 = h_0^2,\ k_0^2 = \omega^2\varepsilon_0\mu_0$

Also $h^2 = h_0^2+\delta k_0^2\chi$

Thus defining the operator,

$$\underline{\underline{N}}(h_0^2) = k_0^2\chi\underline{\underline{I}}_2 + \begin{pmatrix} \left(\underline{\nabla}_\perp\left(\chi_e - \frac{k_0^2}{h_0^2}\chi\right), \underline{\nabla}_\perp\right) & \frac{-\gamma k_0^2}{j\omega\varepsilon_0 h_0^2}(\hat{z}\times\underline{\nabla}_\perp, \underline{\nabla}_\perp) \\ \frac{\gamma k_0^2}{j\omega\mu_0 h_0^2}(\hat{z}\times\underline{\nabla}_\perp\chi, \underline{\nabla}_\perp) & \left(\underline{\nabla}_\perp\left(\chi_m - \frac{k_0^2\chi}{h_0^2}\right), \underline{\nabla}_\perp\right) \end{pmatrix}$$

with $\gamma = \sqrt{h_0^2 - k_0^2}$

we get $\left(\nabla_\perp^2 + h_0^2\right)\underline{\psi}(x, y) + \delta.\,\underline{\underline{N}}\left(h_0^2\right)\underline{\psi}(x, y) = 0$

where
$$\underline{\psi} = \begin{pmatrix} E_z \\ H_z \end{pmatrix} \tag{13}$$

Boundary condition: Let D be the cross section of the guide and $\Gamma = \partial D$ its boundary. Then $E_z|_\Gamma = 0,\ H_{\hat{n}}|_\Gamma = 0.$

Now $H_{\hat{n}}|_\Gamma = \left(\underline{H}_\perp \hat{n}\right)|_\Gamma = 0$ ⇔(by (5))

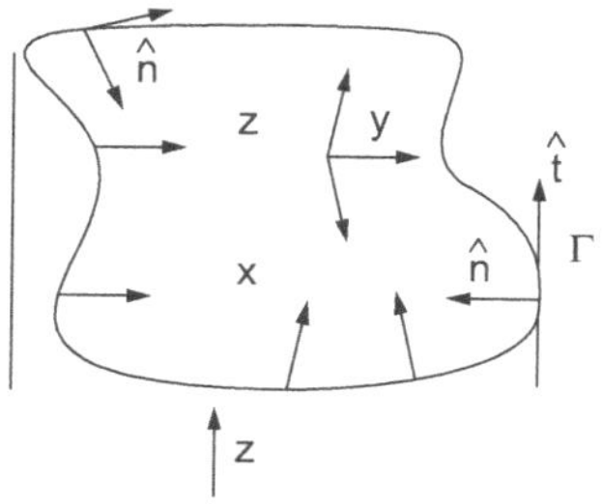

$$\left\{\frac{j\omega\varepsilon'}{h^2}(\underline{\nabla}_\perp E_z \times \hat{z}, \hat{n}) - \frac{\gamma}{h^2}(\underline{\nabla}_\perp h_z \hat{n})\right\}\Bigg|_\Gamma = 0$$

Now
$$\left(\underline{\nabla}_\perp E_z \times \hat{z}, \hat{n}\right)\big|_\Gamma = \left(\underline{\nabla}_\perp E_z, \hat{z}\times\hat{n}\right)\big|_\Gamma = 0$$

since
$$\hat{z}\times\hat{n} = -\hat{t} \text{ and } \left.\frac{\partial E_z}{\partial \hat{t}}\right|_\Gamma = 0 \text{ since } E_z|_\Gamma = 0$$

Thus the required boundary conditions are

$$\left.\frac{\partial H_z}{\partial \hat{n}}\right|_\Gamma = 0,\ E_z|_\Gamma = 0. \tag{14}$$

In other words, E_z satisfies the Dirichlet boundary conditions while H_z satisfies the Newmann boundary condition.

Let $(\nabla_\perp^2 + h_0^2)\,\psi(x, y) = 0$

with the above boundary conditions have eigen-values $h_0^2[n], n = 1, 2, ...$with $\underline{\psi}_n^{10}(x, y), n = 1,2, ...$(– Ve of the eigen-value) as a corresponding set of orthonormal eigen function, *i.e.*

$$\left\langle \psi_n^{(0)}, \psi_m^{(0)} \right\rangle = \int_D \overline{\psi}_n^{(0)}(x, y)^T \, \psi_m^{(0)}(x, y)\, dx\, dy$$
$$= \delta_{n, m}.$$

There may be degeneracy (like in a rectangular waveguide the (n, m^{th}) mode has a degeneracy of two corresponding to the TE and TM modes) so more precisely, let $h_0^2[n], n = 1, 2, ...$ be the **distinct** eigenvalues of $-\nabla_\perp^2$ with the stated boundary conditions and each n, ... let $\left\{\underline{\psi}_{n\beta}^{(0)}, \beta = 1, 2, ...g_n\right\}$ be an orthonormal basis for the corresponding eigen subspace $\mathcal{H}_n$, *i.e.*

$$\mathcal{H}_n = \left\{\underline{\psi} \,\middle|\, \left(\nabla_\perp^2 + h_0^2\,[n]\right)\underline{\psi} = 0, \psi_1\Big|_{\partial D} = 0, \frac{\partial \psi_z}{\partial \hat{n}}\Big|_{\partial D} = 0\right\}$$

with $$\underline{\psi} = \begin{pmatrix} \psi_1 \\ \psi_2 \end{pmatrix}.$$

Let $$\psi = \sum_{\beta=1}^{g_n} C_\beta \psi_{n\beta}^{(0)} + \delta \cdot \psi^{(1)} + O(\delta^2)$$
$$\equiv \underline{\psi}_n^{(0)} + \delta \cdot \underline{\psi}^{(1)} + O(\delta^2)$$

Note that $\underline{\psi}_n^{(0)} \in \mathcal{H}_n$. Then

$$\left[\nabla_\perp^2 + h_0^2 + \delta \cdot N\left(h_0^2\right)\right]\left[\underline{\psi}_n^0 + \delta \cdot \underline{\psi}^{(1)}\right] + \delta \cdot h_1^2 = O + O(\delta^2)$$

i.e. $$\left[\nabla_\perp^2 + h_0^2\right]\psi_n^{(0)} = 0 \qquad (O(\delta^0)\text{ eqn.}),$$

$$\left[\nabla_\perp^2 + h_0^2\right]\underline{\psi}^{(1)} + N\left(h_0^2\right)\underline{\psi}_n^{(0)} + h_1^2\underline{\psi}_n^{(0)} = 0 \qquad (O(\delta')\text{eqn.})$$

Thus $$h_0^2 = h_0^2[n]$$

and $$h_0^2 C_\gamma = -\left\langle \psi_{n\gamma}^{(0)}, N\left(h_0^2\,[n]\right)\underline{\psi}_n^{(0)}\right\rangle$$
$$= -\sum_{\beta=1}^{g_n} \left\langle \psi_{n\gamma}^{(0)}, N\left(h_0^2[n]\right) \underline{\psi}_{n\beta}^{(0)}\right\rangle C_3 \qquad (15)$$

In other words,

$\underline{C} = ((C_\beta))_{\beta=1}^{g_n}$ is an eigenvector of the $g_n \times g_n$ matrix

$$\underset{=n}{A} = -\left(\left(< \psi_{n\gamma}^{(0)}, N(h_0^2[n])\, \psi_{n\beta}^{(0)} >\right)\right)_{1\le\gamma,\beta\le g_n}$$

and the possible values of $h_1^2 \left(\delta \cdot h_1^2\right.$ is the first order perturbation to the mode $\left.h_0^2[n]\right)$ are the eigenvalues of the secular matrix $\underset{=n}{A}$.

❖❖❖❖❖

[19] Estimating the Initial State from Measurements of an Observable after the State Passes through a Sequence of Quantum Channels:

The kth channel transforms a state ρ to

$$T_k(\rho) = \sum_{m=1}^{N} E_{km}\, \rho\, E_{km}^*$$

where $\sum_{m=1}^{N} E_{km}^*\, E_{km}^* = I.$

Let ρ_0 be the initial state. After it passes through T_1 and an observable X measured by noting its outcome as some C_α, $\alpha = 1, 2, \ldots, r$

$$X = \sum_{\alpha=1}^{r} C_\alpha P_\alpha$$

being the spectral decomposition of $X \left(P_\alpha\, P_\beta = P_\alpha\, \delta_{\alpha\beta}, P_\alpha^* = P_\alpha, \sum_1^r P_\alpha = I \right)$

the state is $\dfrac{P_\alpha T_1(\rho_0) P_\alpha}{T_r(T_1(\rho_0) P_\alpha)}$

Then it passer through T_2 and again X is measured and the outcome noted. Let the sequence of outcomes be $C_{\alpha 1}, C_{\alpha 2}, \ldots, C_{\alpha N}$. The probability of this is

$$p(C_{\alpha 1}, C_{\alpha 1}, \ldots, C_{\alpha N} | \rho_0) = T_r(T_1(\rho_0)\, p_{\alpha 1})\; T_r\left(T_2\left(\frac{p_\alpha T_1(\rho_0) P_{\alpha_1}}{T_r(T_1(\rho_0) P_{\alpha_1})} \right) P_{\alpha 2} \right)$$

$$\ldots T_r(T_N(\rho_{N-1}) P_{\alpha N})$$

Where $$\rho_k = \frac{P_{\alpha k}\, T_k(\rho_{k-1}) P_{\alpha k}}{T_r(T_k(\rho_{k-1}) P_{\alpha k})}$$

Making the required cancellations, we get

$$T_r\{(T_N \circ \bar{P}_{\alpha N-1}) \circ (T_{N-1} \circ \bar{P}_{\alpha N-2}) \circ \cdots \circ (T_2 \circ \bar{P}_{\alpha 1}) \circ T_1(\rho_0)\}$$

where $\tilde{P}_{\alpha k}(X) = P_{\alpha k} \times P_{\alpha k}$.

❖❖❖❖❖

[20] Quantum Poisson Random Variable

$\lambda(H)$ is defined by:

$$\lambda(H) = -i\frac{d}{dt}W\left(0, e^{itH}\right)\Big|_t = 0$$

where W is the Weyl operator and H commutes with $\chi_{[a, b]}$. Here $\mathcal{H} = L^2(\mathbb{R})$ and $H : \mathcal{H} \to \mathcal{H}$ is Hermitian.

$$\left\langle e(u), e^{it\lambda(H)} e(u)\right\rangle = \left\langle e(u), W(0, e^{itH})\, e(u)\right\rangle$$

$$= \left\langle e(u), e(e^{itH}u)\right\rangle = \exp\left(\left\langle u, e^{itH}u\right)\right\rangle$$

Let $$|\varphi(u)\rangle = \exp\left(-\frac{\|u\|^2}{2}\right)\left| e(u)\right\rangle.$$

Then $\langle \phi(u)|\exp(it\,\lambda(H))|\varphi(u)\rangle = \exp(\langle u, (e^{itH} - I)\, u\rangle)$

Let $H = \int_{\mathbb{R}} \lambda\, P(d\lambda)$ be the spectral representation of H. Then,

$$\langle \varphi(u)|\exp(t\,\lambda(H))|\varphi(u)\rangle = \exp\left(\left\langle u, \left(e^{iH} - I\right)u\right\rangle\right)$$

$$= \exp\left(\left\langle u, \int_{\mathbb{R}} (e^{t\lambda} - 1)\, P(d\lambda)\, u\right\rangle\right)$$

$$= \exp\left(\|u\|^2\right)\exp\left(\int_{\mathbb{R}} \langle u, p(d\lambda)u\rangle e^{t\lambda}\right)$$

Let $$F_u(d\lambda) = \langle u, P(d\lambda)u\rangle.$$ Then

$$\log\left(\langle \phi(u)\,|\exp(t\,\lambda(H)\,|\,\varphi(u)\rangle\right) = -\|u\|^2 + \int_R e^{t\lambda} d\,F_u(\lambda) \equiv \Lambda(t) \text{ say}$$

Rate function (Legendre transform) of $\lambda(H)$ in the state $|\phi(u)\rangle$:

$$\Lambda^*(x) = \sup_{t \in \mathbb{R}}(tx - \Lambda(t)) = \sup_{t \in \mathbb{R}}\left(tx - \int_{\mathbb{R}} e^{t\lambda} dF_u(\lambda)\right) + \|u\|^2$$

Supremum is attained by setting $\frac{d}{dt}\left(tx - \int_{\mathbb{R}} e^{t\lambda} d\,F_u(\lambda)\right) = 0$ *i.e.*

$$x = \int_{\mathbb{R}} \lambda e^{t\lambda} d\,F_u(\lambda)$$

Let $X = \lambda(H) + a(u) + a^*(u)$

Compute $M_{x, u}(t) = \langle \varphi(v) \mid \exp(tX) \mid \phi(v)\rangle$

where $|\phi(v)> = \exp\left(-\frac{\|v\|^2}{2}\right)\left| e(v)\right\rangle$

$$[a(u), a^*(u)] = \|u\|^2,$$

$$\langle e(v), a(u)\lambda(H)\, e(v)\rangle = \left\langle a^*(u), a(u),\ \lambda(H)\, e(v)\right\rangle$$

$$= \frac{d}{dt}\langle e(v+tu), \lambda(H)e(v)\rangle\Big|_{t=0}$$

$$\frac{d}{dt}\langle v+tu, Hv\rangle \exp\left(\langle v+tu, v\rangle\right|_{t=0}$$

$$= \left[\langle u, H\,v\rangle + \langle v, H\,v\rangle\langle u, v\rangle\right]\langle e(v), e(v)\rangle$$

$$\langle e(v), \lambda(H)\, a(u)\, e(v)\rangle = \langle u, u\rangle\langle v, H\ v\rangle\left(\langle e(v), e(v)\rangle\right.$$

So,

$$\langle e(v), [a(u), \lambda(H)]\rangle_{e(v)} = \langle u, H\,v\rangle\langle e(v), e(v)\rangle$$

$$= \left\langle H^* u, v\right\rangle\langle e(v), e(v)\rangle = \langle Hu, v\rangle\langle e(v), e(v)\rangle$$

$\because$ $H^* = H = \langle e(v), a(Hu)\, e(v)\rangle$

Thus, $[a(u), \lambda(H)] = a(Hu)$

$$[\lambda(H), a^*(u)] = a^*(Hu)$$

* $X = a(u) + a^*(u) + \lambda(H)$

$$[a(u) + \lambda(H), a^*(u)] = \|u\|^2 + a^*(Hu)$$

Let $u \in \mathcal{H}$, $U \in u(H)$ and consider

$$W(u, U) \in u(\Gamma_s(\mathcal{H}))$$

Let $H(u, U)$ be its generator so

$$W(u, U) = \exp(i\,H(u, U)).$$

$$\langle e(v), W(u, U)e(v)\rangle = \exp\left(-\frac{1}{2}\|u\|^2 - \langle u, U\,v\rangle\right)\langle e(v), e(U\,v+u)\rangle$$

So $\langle \varphi(v), W(u, U)\,\varphi(v)\rangle = \exp\left(-\frac{1}{2}\|u\|^2 - \langle u, U\,v\rangle + \langle v, U\ v\rangle + \langle v, u\rangle - \|v\|^2\right)$

$$W(u, U)^x \left\langle e(v), W\left(tu, e^{isH}\right)e(v)\right\rangle$$

$$= \left\langle e(v), \exp\left(-\frac{t^2\|u\|^2}{2} - \left\langle tu, e^{isH}v\right\rangle\right)e\left(e^{isH}v + tu\right)\right\rangle$$

$$= \exp\left(-\frac{t^2\|u\|^2}{2} - t\left\langle u, e^{isH}v\right\rangle + \left\langle v, e^{isH}v\right\rangle + t\langle v, u\rangle\right)$$

$$\log\left\{\left\langle \varphi(v), W\left(tu, e^{isH}\right)\varphi(v)\right\rangle\right\}$$

$$= -\frac{t^2\|u\|^2}{2} - t\left\langle u, e^{isH}v\right\rangle + \left\langle v, e^{isH}v\right\rangle$$

$$+ t\langle v, u\rangle - \|v\|^2$$

$$= -t^2\frac{\|u\|^2}{2} - t\int_{\mathbb{R}} e^{is\lambda}d\,F_{nv}(\lambda)$$

$$+\int_{\mathbb{R}} (e^{is\lambda}-1)\,d\,F_v(\lambda) + t\langle v, u\rangle$$

$$[\lambda(H), a(u) + a^*(u)] = a^*(Hu) - a(Hu)\ (H^* = H)$$

$$[\lambda(H), a^*(Hu) - a(Hu)] = a^*(H^2u) + a(H^2u)\ [a^*(Hu) - a(Hu), a(u) + a^*(u)]$$

❖❖❖❖❖

[21] Quantum Relative Entropy Evolution:

$$*\ \rho'(t) = -i[H, \rho(t)] + \varepsilon\theta_1(\rho(t))$$

$$\sigma'(t) = -i[H, \sigma(t)] + \varepsilon\theta_2(\sigma(t))$$

Some Hamiltonian but different noise operators.

$$\theta_1(\rho) = -\frac{1}{2}\left(L_1^*l_1\rho + \rho L_1^*L_1 - 2L_1\rho L_1^*\right)$$

$$\theta_2(\rho) = -\frac{1}{2}\left(L_2^*l_2\rho + \rho L_2^*L_2 - 2L_2\rho L_2^*\right).$$

$$S(t) \triangleq S(\rho(t), \sigma(t)) = T_r\{\rho(t)(\log\rho(t) - \log\sigma(t))\}$$

Compute $\frac{d\,S(t)}{dt}$ upto $O(\varepsilon)$.

Hint: Let $\log\rho(t) = Z(t)$.

Then $\rho(t) = \exp(Z(t))$.

$$\rho'(t) = \exp(Z(t))_0\frac{\left(I - \exp(-ad\,Z(t))\right)}{ad\,Z(t)}(Z'(t))$$

$$\therefore \quad Z'(t) = \frac{ad\,Z(t)}{1-\exp(-ad\,Z(t))}\left(\rho^{-1}\rho'\right)$$

$$= \frac{ad\log\rho}{(1-\exp(-ad\log\rho))}\left(\rho^{-1}\rho'\right)$$

$$= \left(I+\sum_{n=1}^{\infty} C_n(ad\log\rho)^n\right)\left(\rho^{-1}\rho'\right)$$

Thus, $$T_r(\rho\, Z') = T_r(\rho') + \sum_{n=1}^{\infty} C_n\, T_r(\rho(ad\log\rho)^n(\rho-\rho'))$$

$$\rho\, ad \log \rho\, (X) = \rho[\log \rho, X] = [\log \rho, \rho X]$$

so $$T_r(\rho\, ad \log \rho\, (X)) = 0\ \forall X. \text{ Thus}$$

$$T_r((ad \log \rho)^n\, (\rho^{-1}\rho^1)) = 0 \text{ for } n \geq 1.$$

Also $$T_r(\rho') = 0. \text{ Thus,}$$

$$T_r(\rho\, Z') = 0 \text{ and hence}$$

$$\frac{d}{dt}T_r(\rho\log\rho) = T_r(\rho' \log \rho).$$

$$T_r(\rho\, (\log\sigma)') = T_r\left(\rho\sigma^{-1}\sigma'\right) + \sum_{n\geq 1} C_n\, T_r\left(\rho(ad\log\sigma)^n\left(\sigma^{-1}\sigma'\right)\right)$$

$$T_r(\rho\, (ad \log \sigma)^n\, (\sigma^{-1}\sigma^1)) = T_r(\rho\sigma^{-1}(ad \log \sigma)^n\, (\sigma'))$$

and thus $$\frac{d}{dt}T_r(\rho\log\sigma) = T_r(\rho\sigma^{-1}\sigma') + \sum_{n\geq 1} C_n\, T_r(p\sigma^{-1}(ad\log\sigma)^n(\sigma'))$$

❖❖❖❖❖

[22] Gauge Group Invariance of a Wave Eqn.:

$$\left(\partial_\mu + \alpha A_\mu\right)\psi = 0$$

$A_\mu(x) \simeq$ Lie algebra valued fields.

Group G acts on $\psi(x)$.

$\psi(x) \to g(x)\,\psi(x)\; g = g(x) \in G.$

We require the transformation law

$A_\mu(x) \to A'_\mu(x)$ so that

$$\left(\partial_\mu + \alpha A_\mu\right)g\psi = g\left(\partial_\mu + \alpha A'_\mu\right)\psi$$

Thus, $$\partial_\mu g + \alpha A_\mu g = g\partial_\mu + \alpha g A'_\mu$$

or $$\left(\partial_\mu g\right) = \alpha\left(gA'_\mu - A_\mu g\right)$$

or
$$A'_\mu = g^{-1}A_\mu g + \frac{1}{\alpha}g^{-1}\left(\partial_\mu g\right).$$

❖❖❖❖❖

[23] Problem:

Calculate $\left\langle r\middle|X^m p_X^n\middle|s\right\rangle$

for a one dimensional Harmonioscillator.

Hint:
$$a = \frac{X = ip_x}{\sqrt{2}}, a^* = \frac{X - ip_x}{\sqrt{2}}$$

$$[a, a^*] = 1$$

$$X = \frac{a+a^*}{\sqrt{2}}, p_x = \frac{a-a^*}{i\sqrt{2}}$$

$$\left(a+a^*\right)|S\rangle = (a+a^*)^{m-1}\left(\sqrt{S}\,|\,S-1\rangle+\sqrt{S+1}/S+1\rangle\right)$$

So
$$\left\langle r\middle|X^m\middle|S\right\rangle = \sqrt{S}\left\langle r\middle|\left(a+a^*\right)^{m-1}\middle|S-1\right\rangle$$
$$+\sqrt{S+1}\left\langle r\middle|\left(a+a^*\right)^{m-1}\middle|S+1\right\rangle$$

❖❖❖❖❖

[24] Quantum Oscillator Perturbed by an Electromagnetic Field:

$$\frac{1}{2}\left\{\left(p_x + eB_0^{(t)}\,Y\!\!\left/2\right.\right)^2 + \left(p_Y - eB_0^H\,X\!\!\left/2\right.\right)^2\right\}$$

$$+\frac{1}{2}\left(X^2+Y^2\right)+e\left(E_1(t)X+E_2(t)Y\right)$$
$$= H(t).$$

$$H(t) = H_0 + e\left\{E_1(t)X + E_2(t)Y + \frac{1}{2}B_0(t)\left(Y_{pX} - X_{pY}\right)\right\}$$
$$+\frac{e^2}{8}B_0^2(t)\left(X^2+Y^2\right)$$
$$= H_0 + e\left\{E_1(t)X + E_2(t)Y - \frac{1}{2}B_0(t)L_Z\right\}$$
$$+\frac{e^2}{8}B_0^2(t)\left(X^2+Y^2\right)$$

$$H_0 = \frac{1}{2}\left\{p_X^2 + p_Y^2 + X^2 + Y^2\right\}.$$

Let
$$E_k(t) = \sum_{n=0}^{p} h_k[n]\xi_k(t - n\Delta), k = 1, 2,$$

$$B_0(t) = \sum_{n=0}^{p} h_0[n]\xi_0(t - n\Delta)$$

$$U_0(t) = \exp(-itH_0).$$

$$X(t) = U_0(-t) \times U_0(t), L(t) = U_0(-t)L_z U_0(t)$$

$$Y(t) = U_0(-t)Y\,U_0(t),$$

$$R^2(t) = U_0(-t)\left(X^2 + Y^2\right)U_0(t) = X^2(t) + Y^2(t).$$

$$L(t) = X(t)P_Y(t)PX(t)$$

where
$$P_X(t) = U_0(-t)p_X U_0(t),$$
$$P_Y(t) = U_0(-t)p_Y U_0(t).$$

$$H(t) = H_0 + eV_1(t) + e^2V_2(t).$$

$$iU'(t) = H(t)U(t).\ U(t) = U_0(t)W(t).$$

$$W'(t) = -i\left(e\tilde{V}_1(t) + e^2\tilde{V}_2(t)\right)W(t)$$

$$\tilde{V}_1(t) = E_1(t) + X(t) + E_2(t)Y(t) - \frac{1}{2}B_0(T)L(t)$$

$$\tilde{V}_2(t) = \frac{1}{8}B_0^2(t)R^2(t).$$

$$W(t) = I - i\int_0^t \left(e\tilde{V}_1(t_1) + e^2\tilde{V}_2(t_1)\right)dt - e^2\int_{0<t_2<t_1<t} \tilde{V}_1(t_1)\tilde{V}_1(t_2)dt_1dt_2 + O\left(e^3\right).$$

$$I - W(t) = ie\int_0^t \tilde{V}_1(t_1)dt_1 + e^2\left\{\int_{[0,t]^2} \tilde{V}_1(t_1)\tilde{V}_1(t_2)\theta(t_1 - t_2)dt_1dt_2 + i\int_0^t \tilde{V}_2(t_1)dt_1\right\}$$

$$\int_0^t \tilde{V}_1(t_1)\,dt_1 = \sum_{k=0}^{p} h_1[k]\int_0^t \xi_1(t_1+k\Delta)X(t_1)\,dt_1$$

$$+\sum_{k=0}^{p} h_2[k]\int_0^t \xi_2(t_1+k\Delta)\,Y(t_1)\,dt_1$$

$$-\frac{1}{2}\sum_{k=0}^{p} h_0[k]\int_0^t \xi_0(t_1-k\Delta)\,L(t_1)\,dt_1$$

$$\int_{[0,t]^2} \tilde{V}_1(t_1)\tilde{V}_1(t_2)\,dt_1\,dt_2\,\theta(t_1-t_2)$$

$$= \int_{[0,t]^2} \theta(t_1-t_2)\left\{E_1(t_1)X(t_1)+E_2(t_2)Y(t_1)\frac{1}{2}B0(t_1)L(t_1)\right\}$$

$$\times\left\{E_1(t_2)X(t_2)+E_2(t_2)Y(t_2)-\frac{1}{2}B_0(t_2)L(t_2)\right\}dt_1dt_2$$

$$= \sum_{k,m=0}^{p} h_1[k]h_1[m]$$

$$\int_{[0,t]^2} \theta(t_1-t_2)\xi_1(t_1-k\Delta)\xi_1(t_1-m\Delta)X(t_1)\,X(t_2)\,dt_1\,dt_2$$

$$+\sum_{k,m=0}^{p} h_1[k]h_2[m]$$

$$\int_{[0,t]^2} \theta(t_1-t_2)\xi_1(t_1-k\Delta)\xi_2(t_2-m\Delta)X(t_1)\,X(t_1)\,dt_1\,dt_2$$

$$+\sum_{k,m=0}^{p} h_2[k]h_1[m]\int_{[0,t]^2} \theta(t_1-t_2)\xi_2(t_1-k\Delta)$$

$$\xi_1(t_2-m\Delta)Y(t_1)X(t_2)\,dt_1\,dt_2$$

$$+\sum_{k,m=0}^{p} h_2[k]h_2[m]\int_{[0,t]^2} \theta(t_1-t_2)\xi_2(t_1-k\Delta)$$

$$\xi_2(t_2-m\Delta)Y(t_1)X(t_2)\,dt_1\,dt_2$$

$$+\frac{1}{4}\sum_{k,m=0}^{p} h_0[k]h_0[m]\int_{[0,t]^2} \theta(t_1-t_2)\xi_0(t_1-k\Delta)$$

$$\xi_0(t_2-m\Delta)Y(t_1)L(t_2)\,dt_1\,dt_2$$

$$-\frac{1}{2}\sum_{k,m} h_1[k]h_0[m] \int_{[0,t]^2} \theta(t_1-t_2)\xi_1(t_1-k\Delta)$$
$$\xi_0(t_2-m\Delta)X(t_1)L(t_2)dt_1\, dt_2$$
$$-\frac{1}{2}\sum_{k,m} h_2[k]h_0[m] \int_{[0,t]^2} \theta(t_1-t_2)\xi_2(t_1-k\Delta)$$
$$\xi_0(t_2-m\Delta)Y(t_1)L(t_2)dt_1\, dt_2$$
$$-\frac{1}{2}\sum_{k,m} h_2[k]h_1[m] \int_{[0,t]^2} \theta(t_1-t_2)\xi_0(t_1-k\Delta)$$
$$\xi_1(t_2-m\Delta)X(t_1)L(t_2)dt_1\, dt_2$$
$$-\frac{1}{2}\sum_{k,m} h_0[k]h_2[m] \int_{[0,t]^2} \theta(t_1-t_2)\xi_0(t_1-k\Delta)$$
$$\xi_2(t_2-m\Delta)L(t_1)Y(t_2)dt_1\, dt_2$$

❖❖❖❖❖

[25] Quantization of Mechanical Systems:

Charged string in an em field

$X(t, s)$, $Y(t, s)$, $\left(\frac{\partial X}{\partial s}\right)^s + \left(\frac{\partial Y}{\partial s}\right)^2 = 1.$

$0 \le s \le 1.$

$$A = \frac{1}{2}B_0(t)\hat{Z}\times \underline{r} = \frac{1}{2}B_0(t)(-Y, X, 0)$$
$$\Phi = -(E_{01}(t)X + E_{02}(t)Y)$$
$$E(t, X, Y) = E_{01}(t)\hat{X} + E_{02}(t)\hat{Y}$$
$$+\frac{1}{2}B_0'(t)(Y\hat{X} - X\hat{Y}) \qquad \left(\underline{E} = -\underline{\nabla}\phi - \frac{\partial \underline{A}}{\partial t}\right)$$

Lagrangian for the strings:

$$\mathcal{L} = \mathcal{L}\left(x(t,\cdot), Y(t,\cdot), \overset{\circ}{X}(t,\circ), \overset{\circ}{Y}(t,\circ)\right)$$
$$= \frac{1}{2}\sigma_m\int_0^1\left[\left(\frac{\partial X}{\partial t}\right)^2 + \left(\frac{\partial X}{\partial t}\right)^2\right]ds$$
$$-\frac{1}{2}T\int_0^1\left[\left(\frac{\partial X}{\partial s}\right)^2 + \left(\frac{\partial Y}{\partial s}\right)^2\right]ds$$

$$-\sigma_q\int_0^1 \Phi(t,X,Y)ds+\sigma_q\int_0^1\left(A_X\frac{\partial X}{\partial t}+A_Y\frac{\partial Y}{\partial t}\right)d.$$

Let
$$X(t,s) = \sum_{n\in\mathbb{Z}} X_n(t)\exp(2\pi i\,ns)$$

$$Y(t,s) = \sum_{n\in\mathbb{Z}} Y_n(t)\exp(2\pi i\,ns)$$

$$X_{-n}(t) = \overline{X_n(t)}, Y_{-n(t)} = \overline{Y_{n-1}(t)} \text{ for } X, Y \text{ to be real.}$$

$$\frac{1}{2}\int_0^1\left[\left(\frac{\partial X}{\partial t}\right)^2+\left(\frac{\partial X}{\partial t}\right)^2\right]ds = \frac{1}{2}\sum_n\left(|X_n'(t)|^2+|Y_n'(t)|^2\right)$$

$$\frac{1}{2}\int_0^1\left[\left(\frac{\partial X}{\partial s}\right)^2+\left(\frac{\partial X}{\partial s}\right)^2\right]ds = 2\pi^2\sum_n n^2\left(|X_n(t)|^2+|Y_n(t)|^2\right)$$

$$-\int_0^1\Phi(f,X,Y)\,ds = E_{01}(t)\int_0^1 Xds+E_{02}(t)\int_0^1 X\,ds$$

$$= E_{01}(t)X_0(t)+E_{02}(t)Y_0(t)$$

$$\int_0^1 A_x(t,X,Y)\frac{\partial X}{\partial t}ds = -\frac{1}{2}B_0(t)\int_0^1 Y\frac{\partial X}{\partial t}ds$$

$$\int_0^1 Y\frac{\partial X}{\partial t}ds = \sum_n Y_n(t)\overline{X_n'(t)} = \sum_n X_n'(t)\overline{Y_n(t)}$$

$$\int_0^1 A_Y(t,X,Y)\frac{\partial Y}{\partial t}ds$$

$$\frac{1}{2}B_0(t)\int_0^1 X\frac{\partial Y}{\partial t}ds = \frac{1}{2}B_0(t)\sum_n\overline{X_n(t)}Y_n'(t)$$

$$= \frac{1}{2}B_0(t)\sum_n X_n(t)\overline{Y_n'}(t)$$

Lagrangian:

$$\mathcal{L}(X_n,Y_n,X_n',Y_n') = \frac{1}{2}\sigma_m\sum_n\left(|X_n'(t)|^2+|Y_n'(t)|^2\right)$$

$$-2\pi^2T\left(\sum_n n^2|X_n(t)|^2+|Y_n(t)|^2\right)$$

$$-\sigma_q\left(E_{01}(t)X_0(t)+E_{02}(t)Y_0(t)\right)$$
$$+\frac{\sigma_q B_0(t)}{2}\left(\sum_n \overline{X_n(t)}\,Y_n'(t)-\overline{Y_n(t)}X_n'(t)\right)$$

Classical eqns. of motion:

$$\frac{d}{dt}\frac{\partial\mathcal{L}}{\partial\bar{X}_n'}=\frac{\partial\mathcal{L}}{\partial\bar{X}_n},\ \frac{d}{dt}\frac{\partial\mathcal{L}}{\partial\bar{Y}_n'}=\frac{\partial\mathcal{L}}{\partial\bar{Y}_n}$$

$$\Rightarrow \frac{1}{2}\sigma_m X_n''(t)+2\pi^2Tn^2X_n(t)-\frac{\sigma_q B_0(t)}{2}Y_n'(t)-\sigma_g\left(\frac{B_0(t)}{2}Y_n(t)\right)'=0$$

❖❖❖❖❖

[26] Quantum Filtering:

$$dU(t)=\left[-i\,H(t)\,dt+\sum_{j=1}^{P}\left(L_j^{(t)}dA_j(t)+\tilde{L}_j(t)\,dA_j^*(t)\right)\right.$$
$$\left.+\sum_{j,h=1}^{p}S_k^j(t)d\Lambda_j^k(t)\right]U(t)$$

$$|\psi\rangle=|f\,e(u)\rangle$$

Let

$$|\psi(t)\rangle=U(t)|\psi\rangle$$
$$X(t)=U^*(t)XU(t)=j_t(X)$$
$$dU(t)|\psi\rangle=iH(t)|\psi(t)\rangle\,dt$$
$$+\sum_j L_j(t)u_j(t)|\psi(t)\rangle dt+\sum\tilde{L}_j(t)\,U(t)dA_j^*(t)|\psi\rangle$$
$$+\sum S_k^jU(t)d\Lambda_j^k(t)|\psi\rangle$$

$$\langle g\,e(v)|d\,A_j^*(t)\,|\psi\rangle=\bar{v}_j(t)\,dt\langle ge(v)|\psi\rangle$$

$$\langle g\,e(v)|d\,\Lambda_k^j(t)\,|\psi\rangle=u_j(t)\overline{v_k(t)}dt\langle ge(v)|\psi\rangle$$

$$\langle g\,e(v)|\sum_j\tilde{L}_j(t)d\,A_j^*(t)|\psi\rangle=\sum_j\langle g|\tilde{L}_j(t)\,f\rangle\overline{v_j(t)}\langle e(v)|e(u)\rangle\,dt$$

$$\left\langle g\,e(v)\left|\sum_{j,k}S_k^j(t)\,d\Lambda_j^k(t)\right|\psi\right\rangle=\sum_{j,k}\langle g|S_k^j(t)|\,f\rangle u_j(t)\overline{u_k(t)}\langle e(v)|e(u)\rangle dt$$

Thus, $\left\langle ge(v)\left|\left(\sum\tilde{L}_jdA_j^*+\sum S_k^jd\Lambda_j^k\right)fe(u)\right.\right\rangle$

$$= \left\langle ge(v) \left| \left\{ \sum_j \overline{v_j(t)}\, \tilde{L}_j(t) + \sum_{j,k} S_k^j(t) u_j(t) \overline{v_k(t)} \right\} \right| fe(u) \right\rangle dt$$

$$= \left\langle ge(v) \left| \left\{ \sum_j \overline{v_j(t)} \left(\tilde{L}_k(t) + \sum_j u_j(t) S_k^j(t) \right) \right\} \right| fe(u) \right\rangle dt$$

Let $\quad dY_{\text{in}}(t) = C_j^k(t)\, d\Lambda_k^j(t) + b_j(t)\, dA_j(t) + \overline{b_j(t)}\, d A_j^*(t)$

Then $\quad [dY_{\text{in}}(t), dY_{\text{in}}(s)] = 0 \;\forall\; t, s$

and hence $\quad [Y_{\text{in}}(t), Y_{\text{in}}(s)] = 0 \;\forall\; t, s$

Let $\quad \eta_{t]}^{\text{in}} = \sigma\{Y_{\text{in}}(s) : s \leq t\}$

Let $\quad dF(t) = (P(t)d\,Y_{\text{in}}(t) + Q(t)\,dt)\,F(t)$

where $P(t), Q(t) \in \mathcal{L}(h)$ *i.e.* P, Q are system operators. Then

$$F(t) \in \eta_{t]}^{\text{in}}$$

We want to have P, Q such that

$$U(t)|\psi\rangle = F(t)|\psi\rangle \equiv |\psi(t)\rangle$$

i.e. $\quad dU(t)|\psi\rangle = dF(t)|\psi\rangle$

Now, $\quad \langle ge(v)|dU(t)|\psi\rangle = -i\langle ge(v)|H(t)|\psi(t)\rangle dt + \langle ge(v)|\sum_k u_k(t) L_k(t)$

$$+ \sum_k \overline{v_k(t)}(\tilde{L}_k(t) + \sum_j u_j(t) S_k^j(t)|\psi\rangle dt$$

and $\quad \langle ge(v)|dF(t)|\psi\rangle = \langle ge(v)|P(t)F(t)d\,Y_{\text{in}}(t)|\psi\rangle + \langle ge(v)|Q(t)F(t)|\psi\rangle dt$

$$dY_{\text{in}}(t)|\psi\rangle = C_j^k(t)d\Lambda_k^j(t) + b_j(t)dA_j(t) + \overline{b_j(t)}\, dA_j^*(t)$$

$$\langle g\,e(v)|dy_{\text{in}}(t)|\psi\rangle = C_j^k(t)u_k(t)\overline{v_j(t)}\langle ge(v)|\psi\rangle dt + b_j(t)u_j(t)\langle ge(v)|\psi\rangle dt + \overline{b_j(t)}\overline{v_j(t)}\langle ge(v)|\psi\rangle dt$$

Thus, $\quad \langle ge(v)|dF(t)|\psi\rangle = C_j^k(t)u_k(t)\overline{v_j(t)}\langle ge(v)|\,p(t)|\psi(t)\rangle dt$

$$+ b_j(t)u_j(t)\langle ge(v)|P(t)|\psi(t)\rangle dt$$

$$+ \overline{b_j(t)}\overline{v_j(t)}\langle ge(v)|P(t)|\psi(t)\rangle\, dt$$

$$+ \langle ge(v)|Q(t)|\psi(t)\rangle\, dt$$

So for $U(t)|\psi\rangle = F(t)|\psi\rangle \forall t$, we require that

$$-iH(t) + \sum_k u_k(t) L_k(t) + \sum \overline{v_k(t)} \left(\tilde{L}_k(t) + \sum_j u_j(t) S_k^j(t) \right)$$

$$= C_j^k(t)u_k(t)\,\overline{u_j(t)}P(t)+b_j(t)$$
$$u_j(t)P(t)+\overline{b_j(t)}\,\overline{v_j(t)}P(t)+Q(t)$$

$\forall v$

or equivalently,

$$-iH(t)+\sum_k u_k(t)L_k(t) = Q(t)+\left(\sum_j b_j(t)u_j(t)\right)P(t),$$

$$\tilde{L}_k(t)+\sum_j u_j(t)S_k^j(t) = \left(\sum_j C_k^j(t)u_j(t)+\overline{b_h(t)}\right)P(t)\forall k$$

Assume that there condition hold.

Then $\mathbb{E}\left[j_t(X)\middle|\eta_t^{\text{out}}\right] = \mathbb{E}[U^*(t)\,X\,U(t)|U^*(t)\,\eta_t^{\text{in}}\,U(t)]$

$$= U^*(t)\tilde{\mathbb{E}}_t[X\,|\,\eta_t^{\text{in}}\,]U(t)$$

where $\tilde{\mathbb{E}}_t[X] = \tilde{\mathbb{E}}_t[U^*(t)X\,U\,(t)]$

$$\Big(\text{For } Y\in\eta_{t]}^{\text{in}}, \mathbb{E}\left[U^*(t)\,\tilde{\mathbb{E}}_t[X\,|\,\eta_t^{\text{in}}\,]U(t)\,U^*(t)YU(t)\right]$$

$$= \mathbb{E}\left[U^*(t)\tilde{\mathbb{E}}_t\left[X\,|\,\eta_t^{\text{in}}\right]Y\,U(t)\right]$$

$$= \mathbb{E}\left[U^*(t)\tilde{\mathbb{E}}_t\left[XY\,|\,\eta_t^{\text{in}}\right]U(t)\right]$$

$$\tilde{\mathbb{E}}_t\tilde{\mathbb{E}}_t\left[XY\,|\,\eta_t^{\text{in}}\right] = \tilde{\mathbb{E}}_t[XY] = \tilde{\mathbb{E}}_t[U^*(t)X\,Y\,U(t)]\Big)$$

Thus, $\mathbb{E}\,[J_t(X)\,|\,\eta_t^{\text{out}}] = U^*(t)\dfrac{\mathbb{E}[F^*(t)XF(t)\,|\,\eta_{t]}^{\text{in}}]}{\mathbb{E}[F^*(t)F(t)\,|\,\eta_{t]}^{\text{in}}]}U(t)$

(Quantum Bayes formula). This is called quantum Kallianpur-Striebel formula.

Let $\xi_t(X) = \mathbb{E}\left[F^*(t)\,X\,F(t)\,|\,\eta_{t]}^{\text{in}}\right]$

Let $f(s) \in \mathbb{C}, 0\le s\le t$

Thus we get

$$\mathbb{E}\left[\xi_t(X)\int_0^t f(s)dY^{\text{in}}(s)\right] = \mathbb{E}\left[F^*(t)X\,F(t)\int_0^t f(s)dY^{in}(s)\right]$$

$$d\left[\xi_t(X)\int_0^t f(s)\,dY^{\text{in}}(s)\right] = d\xi_t(X)\int_0^t f(s)\,dY^{\text{in}}(s)+\xi_t(X)f(t)\,dY^{\text{in}}(t)$$
$$+\,d\xi_t(X)\cdot f(t)\,dY^{\text{in}}(t)$$

Here $\mathbb{E}\,(\xi) = \langle \psi \,|\, \xi \,|\, \psi \rangle = \exp\left(-\|u\|^2\right)\langle fe(u)|\xi|\, fe(u)\rangle$

Now $d\left[F^*(t) X\, F(t) \int_0^t f_0(s)\, dY^{\text{in}}(s)\right]$

$$
\begin{aligned}
&= f_0(t)F^*(t)X\, F(t)dY^{\text{in}}(t) \\
&\quad + dF^*(t)X\, F(t)\int_0^t f_0(s)\, dY^{\text{in}}(s) \\
&\quad + F^*(t)X\, F(t)\int_0^t f_0(s)dY^{\text{in}}(s) \\
&\quad + dF^*(t)XdF(t)\int_0^t f_0(s)dY^{\text{in}}(s) \\
&\quad + f_0(t)XF(t)dF^*(t)XF(t)dY^{\text{in}}(t) \\
&\quad + f_0(t)F^*(t)X\, dF(t)\, dY^{in}(t)
\end{aligned}
$$

Recall that $dF\,(t) = (P\,(t)\, dY^{\text{in}}(t) + Q\,(t)\, dt)F(t),\ F(0) = I$

and $dY^{in}\,(t) = C_k^j(t)\, d\Lambda_j^k(t) + b_j(t)dA_j(t) + \overline{b_j(t)}\, dA_j^*(t)$

Where summation over repeated indices is implied.

$$
\begin{aligned}
&\mathbb{E}\left[F_0(t)F^*(t)X\, F(t)\, dY^{\text{in}}(t)\right] \\
&\quad = F_0(t)\langle \psi \,|\, F^*(t)X\, F(t) \,|\, \psi \rangle \\
&\quad \left(C_k^j(t)u_k(t)\, \overline{u,(t)} + b_j(t)u_j(t) + \overline{b_j(t)}\, \overline{u_j(t)}\right)dt\,,
\end{aligned}
$$

$$
\begin{aligned}
&\mathbb{E}\left[dF^*(t)X\, F(t)\int_0^t f_0(s)\, d\,Y^{\text{in}}(s)\right] \\
&\quad = \left\langle \psi \middle| F^*(t)P^*(t)X\, F(t) \middle| \left(\int_0^t f_0(s)\, d\,Y_\psi^{\text{in}}(s)\right)\right\rangle \\
&\quad \left(\overline{C_k^j}(t)u_j(t)\, \overline{u_k(t)} + \overline{b_j(t)}\, \overline{u_j(t)}\right. \\
&\quad \left. + \overline{b_j(t)}\, \overline{u_j(t)} + b_0(t)\, u_j(t)\right)dt \\
&\quad + \left\langle \psi \,|\, F^*(t)\, Q^*(t)X\, F(t)\int_0^t f_0(s)\, dY^{\text{in}}(s) \,|\, \psi \right\rangle
\end{aligned}
$$

$$\mathbb{E}\left[F^*(t)X\,dF(t)\int_0^t f(s)\,dY^{\text{in}}(s)\right]$$

$$= \langle\psi|F^*(t)X\,P(t)\,F(t)\int_0^t f(s)dY^{\text{in}}(s)|\psi\rangle$$

$$\left(C_k^j(t)u_k(t)\,\overline{u_j(t)} + b_j(t)u_j(t) + \overline{b_j(t)}\,\overline{u_j(t)}\right)dt$$

$$+\left\langle\psi\middle|F^*(t)X\,Q(t)\,F(t)\int_0^t f(s)dY^{\text{in}}(s)\middle|\psi\right\rangle dt,$$

$$dF^*(t)X\,dF(t)\int_0^t f(s)\,dY^{\text{in}}(s)$$

$$= \left[F^*(t)P^*(t)\,XP(t)F(t)\int_0^t f_0(s)\,dY^{\text{in}}(s)\right]$$

$$\left(C_k^j d\Lambda_j^k + b_j\,dA_j + \bar{b}_j dA_j^*\right)^2$$

Now, $\left(C_k^j\,d\Lambda_j^k + b_j dA_j + \bar{b}_j\,dA_j^*\right)^2$

$$= C_k^j C_s^r\,d\Lambda_j^k\,d\Lambda_r^s + C_k^j\,\bar{b}_r\,d\Lambda_j^k\,dA_r^*$$

$$+ b_r C_k^j\,dA_r\,d\Lambda_j^* + b_j\,\bar{b}_r\,dA_j\,dA_r^*$$

$$= C_k^j C_s^k\,d\Lambda_j^s + C_k^j\,\bar{b}_k\,dA_j^k + b_j C_k^j\,dA_k + \sum_j |b_j|^2\,dt$$

$$dF(t) = \left(P(t)dY_{\text{in}}(t) + Q(t)dt\right)F(t)$$

$$dF^*(t)XF(t) = dF^*(t)XF(t) + F^*(t)XdF(t) + dF^*(t)XdF(t)$$

$$= F^*(t)P^*(t)XF(t)dY_{\text{in}}(t) + F^*(t)XP(t)F(t)dY_{\text{in}}(t)$$

$$+ F^*(t)P^*(t)XP(t)F(t)\left(dY_{\text{in}}(t)\right)^2$$

$$+ \left(F^*(t)Q^*(t)XF(t) + F^*(t)XQ(t)F(t)\right)dt$$

Now $\left(dY_{in}(t)\right)^2 = C_k^j b_j dA_k + \bar{C}_k^j \bar{b}_j dA_k^* + \sum_j |b_j|^2\,dt + \left(C^2\right)_k^j\,d\Lambda_j^k \equiv dZ_{in}(t)$ (say)

Let $\sigma_t(X) = \mathbb{E}\left\{F^*(t)XF(t)\middle|\eta_{t]}^{\text{in}}\right\}$

Then it follows that

❖❖❖❖❖

[27] Perspective Projection by Camera Attached to a Robot:

The Robot tip is at

$$(x(t), (t)) = \left(x_0(t) + L_1 \cos q_1(t) + L_2 \cos q_2(t),\right.$$
$$\left. y_0(t) + L_1 \sin \dot{q}_1(t) + L_2 \sin \dot{q}_2(t)\right)$$

Point objects are located at (ξ_k, η_k), $k =, 1\ 2, \ldots., d$.

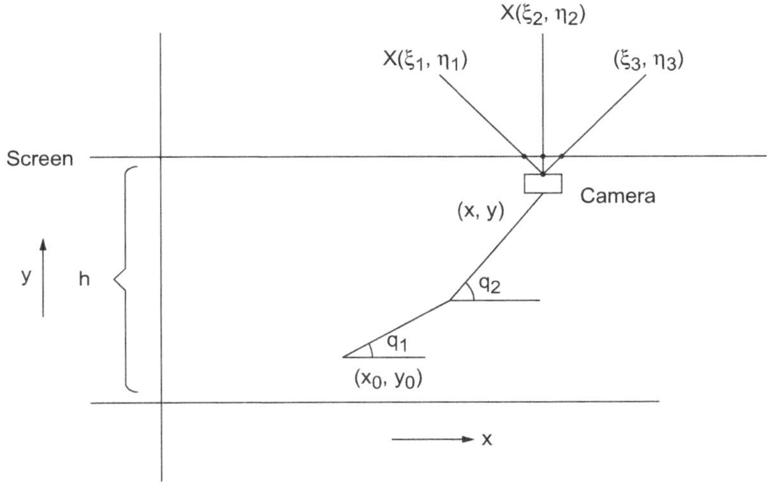

The screen is located at $y = h$.

The line joining the camera at $(x(t), y(t))$ and the kth object at (ξ_k, η_k) has equation

$$\frac{Y - y}{X - x} = \frac{y - \eta_k}{x - \xi_k}$$

When $Y = h$,

$$X = x(t)\frac{(h - y(t))(x(t) - \xi_k)}{(y(t) - \eta_k)} = X_k(t)$$

The camera records the positions $X_k(t)$, $1 \leq k \leq d$ on the screen at different times $t_1, t_,, \ldots., t_N$ and then from this data tries to reconstruct the positions (ξ_k, η_k), $1 \leq k \leq d$ of the objects.

Due to noise in the robot motion, $x(t)$ is replaced by $x(t) + w_1(t)$ and $y(t)$ by $y(t) + w_2(t)$, where $w_1(\cdot)$ and $w_2(\cdot)$ an zero mean Gaussian noise processes. Then, the measurements process is

$$X_k(t) = \frac{\left(h - y(t) - w_2(t)\right)\left(x(t) + w_1(t) - \xi_k\right)}{\left(y(t) + w_2(t) - \eta_k\right)}$$

$$\approx \frac{\left(h - y(t)\right)\left(x(t) - \xi_k\right) + \left(h - y(t)\right)w_1(t) - \left(x(t) - \xi_k\right)w_2(t)}{y(t) - \eta_k + w_2(t)}$$

$$\approx \frac{\left(h - y(t)\right)\left(x(t) - \xi_k\right)}{\left(y(t) - \eta_k\right)} + \frac{\left\{\left(h - y(t)\right)w_1(t) - \left(x(t) - \xi_k\right)w_2(t)\right\}}{\left(Y(t) - \eta_k\right)}$$

$$-\frac{(h-y(t))(x(t)-\xi_k)w_2(t)}{(y(t)-\eta_k)^2}$$

upto linear orders in $w_1(\cdot), w_2(\cdot)$

These measurements can be expressed as

$$X_k(t) =$$
$$F_0(t, \xi_k, \eta_k) + F_1(t, \xi_k, \eta_h)w_1(t)F_2(t, \xi_k, \eta_h)w_2(t), 1 \le k \le d$$

where $$F_0(t,\xi_k,\eta_k) = \frac{(h-y(t))(x(t)-\xi_k)}{(y(t)-\eta_k)},$$

$$F_1(t, \xi_k, \eta_k) = \frac{(h-y(t))}{(y(t)-\eta_k)},$$

$$F_2(t, \xi_k, \eta_k) = \frac{(h-y(t))(x(t)-\xi_k)}{(y(t)-\eta_k)^2} - \frac{(x(t)-\xi_k)}{(y(t)-\eta_k)}$$

$(w_1(\cdot), w_2(\cdot))$ have zero mean and correlations

$$R_{w_1}(\tau) = \mathbb{E}\{w_1(t+\tau)w_1(t)\}$$
$$R_{w_1}(\tau) = \mathbb{E}\{w_2(t+\tau)w_2(t)\}$$
$$R_{w_1}(\tau) = \mathbb{E}\{w_1(t+\tau)w_2(t)\}$$

❖❖❖❖❖

[28] Proof of The Positivity of Quantum Relative Entropy:

$$\rho = \sum_{i=1}^{N} |e_i > p_i < e_i|,$$

$$\sigma = \sum_{i=1}^{N} |f_i > q_i < f_i|$$

$$\langle e_i | f_j \rangle = \delta_{ij}, p_i, q_i > 0, \sum p_i = \sum q_i = 1$$

$$D(\rho|\sigma) = T_r(\rho(\log\rho - \log\sigma))$$
$$= \sum_i p_i \log p_i - \sum_{i,j} p_i |\langle e_i | f_j \rangle|^2 \log q_j$$
$$= \sum_{i,j} p_i S_{ij} \log\left(\frac{pi}{q_j}\right)$$

(where $S_{ij} = |\langle e_i | f_i \rangle i|^2$.

Note that $$\sum_i S_{ij} = \sum_j S_{ij} = 1.)$$

Thus, $$D(\rho|\sigma) = \sum_{i,j} p_i S_{ij} \log\left(\frac{p_i S_{ij}}{q_i S_{ij}}\right) \geq 0$$

since $\{p_i S_{ij}\}_{i,j}$ and $\{q_i S_{ij}\}_{i,j}$ are both probability distributions on $\{1, 2, ..., N\}^2$.

❖❖❖❖❖

[29] Entropy of a Quantum System:

$$\rho'(t) = -i[H, \rho(t)] + \varepsilon\,\theta(\rho(t))$$

$$\theta(\rho) = -\frac{1}{2}\left(L^* L\rho + \rho L^* L - 2L\rho L^*\right)$$

$$S(t) = -T_r(\rho(t) \log \rho(t)).$$

$$S'(t) = T_r\left(\rho'(t)\log\rho(t)\right) = -\varepsilon\, T_r\left(\theta(\rho(t))\log\rho(t)\right)$$

$$= \frac{\varepsilon}{2} T_r\left(2L^* L\,\rho \log\rho - 2L\rho L^* \log\rho\right)$$

$$= \varepsilon T_r\left(L^* L\rho \log\rho - L\rho L^* \log\rho\right)$$

$$= \varepsilon\, T_r\left(\left[L^*, L\rho\right]\log\rho\right)$$

$$\rho(t) = \exp(-it\, ad\, H)(\rho(0)) + \varepsilon\int_0^t \exp\left(-i(t-S)ad\ H\right)\left(\theta(\rho(S)\right)ds$$

$$= \exp(-it\, ad\, H)(\rho(0))$$

$$+\ \varepsilon\int_0^t \exp\left(-i(t-S)ad\ H\right)\theta\left(\exp(-is\ ad\ H)(\rho(0)\right)ds + O(\varepsilon^2)$$

Let $$T_t^{(0)} = \exp(-it\, ad\, H)$$

Then, $$\rho(t) = T_t^{(0)}\left(\rho(0)\right) + \varepsilon\int_0^t T_{t-S}^0\,\theta\left(T_S^0(\rho(0)\right)ds + O(\varepsilon^2)$$

Thus, $$S'(t) = \varepsilon\, T_r\left(\left[L^*, L\rho(t)\right]\log\rho(t)\right)$$

$$= \varepsilon\, T_r\left(\left[L^*, LT_t^{(0)}(\rho(0))\right]\log T_t^{(0)}(\rho(0))\right) + O(\varepsilon^2)$$

Let $$\rho(t) = \sum_k \lambda_k(t)\left|e_k(t)\right\rangle\left\langle e_k(t)\right| \qquad \text{(spectral representation)}$$

Then $$S'(t) = \varepsilon\sum_k \log(\lambda_k)\left\langle e_k\left|\left[L^*, L\rho\right]e_k\right.\right\rangle$$

$$= \varepsilon \sum_k \log \lambda_k \left\langle e_k \left| \left[L^*, L \right] e_k \right. \right\rangle \lambda_k + \varepsilon \sum_k \log(\lambda_k) \left\langle e_k \left| L\left[L^*, \rho \right] e_k \right. \right\rangle$$

$$= \varepsilon \sum_k \lambda_k \log \lambda_k \left\langle e_k \left| L\left[L^*, L \right] \right| e_k \right\rangle$$

$$= \varepsilon \sum_k \log(\lambda_k) \left\langle e_k \left| \left[L, L^* \right] \right| e_k \right\rangle$$

$$+ \varepsilon \sum_k \log(\lambda_k) \left\langle e_k \left| \left[LL^* \right] \right| e_k \right\rangle \lambda_k$$

$$- \varepsilon \sum_k \log \lambda_k \left\langle e_k \left| \left[L^* \rho L \right] \right| e_k \right\rangle$$

$$= \varepsilon \sum_k \lambda_k \log \lambda_k \left\langle e_k \left| L^*, L \right| e_k \right\rangle + \varepsilon \sum_k \lambda_k \log \lambda_k \left\langle e_k \left| LL^* \right| e_k \right\rangle$$

$$+ \varepsilon \sum_{k,m} \lambda_k \log \lambda_k \left| \left\langle e_k \right| L \left| e_k \right\rangle \right|^2 - \varepsilon \sum_k \lambda_k \log \lambda_k \left\langle e_k \left| L^* L \right| e_k \right.$$

$$- \varepsilon \sum_{k,m} \lambda_k \log \lambda_k \left| \left\langle e_k \left| L \right| e_m \right\rangle \right|^2$$

$$= \varepsilon \sum_k \lambda_k \log \lambda_k \left\langle e_k \left| L^* L \right| e_k \right\rangle - \varepsilon \sum_{k,m} \lambda_m \log \lambda_k \left| \left\langle e_k \left| L \right| e_m \right\rangle \right|^2$$

$$= \varepsilon \sum_{k,m} \lambda_k \log \lambda_k \left| \left\langle e_m \left| L \right| e_k \right\rangle \right|^2 - \varepsilon \sum_{k,m} \lambda_m \log \lambda_k \left| \left\langle e_k \left| L \right| e_m \right\rangle \right|^2$$

$$= \varepsilon \sum_{m,k} \left| \left\langle e_k \left| L \right| e_m \right\rangle \right|^2 \lambda_m \log\left(\frac{\lambda_m}{\lambda_k} \right)$$

So $S'(t) \geq 0$ iff

$$\sum_{m,k} \left| \left\langle e_k \left| L \right| e_m \right\rangle \right|^2 \lambda_m \log\left(\frac{\lambda_m}{\lambda_k} \right) \geq 0 \ \forall\, t.$$

Remark:

Let $\log \rho(t) = Z(\text{t})$

$$\rho(t) = e^{Z(t)}$$

$$\rho'(t) = e^{z(t)} \left(\frac{1 - \exp(-ad\ Z(t))}{ad\ Z(t)} \right) (Z'(t))$$

$$Z'(t) = \frac{ad\ Z(t)}{1 - \exp(-ad\ Z(t))} \left(\rho(t)^{-1} \rho'(t) \right)$$

$$= \left(1 + \sum_{m=1}^{\infty} C_m \left(ad\ Z(t)^m\right)\right)\left(\rho^{-1}\ \rho'\right)$$

$$T_r\left[\rho \frac{d}{dt} \log \rho\right] = T_r\left[\rho Z'\right]$$

$$= T_r\left(\rho'\right) + \sum_{m \geq 1} C_m T_r\left[\rho\left(ad \log p\right)^m \left(\rho^{-1}\ \rho'\right)\right]$$

$$T_r\left(\rho'\right) = 0,\ T_r[\rho\ (ad \log \rho)^m(X)]$$

$$\left(X = \rho^{-1} p',\ Y = \left(ad \log \rho\right)^{m-1}(X)\right)$$

$$= T_r[\rho\ ad \log \rho(Y)] = T_r[\rho[\log \rho, Y]] = T_r[[\log \rho,\ \rho Y]] = 0$$

Rate of entropy change of a quantum system

$$\rho'(t) = T(\rho(t)),\ T_r(T(\rho)) = 0$$

$$= i\left[H, e(t)\right] - \frac{1}{2}\left(L^* L\rho(t) + \rho(t) L^* L - 2L\rho(t)L^*\right)$$

$$S(\rho(t)) = -\ T_r(\rho(t) \log e(t))$$

$$\frac{d}{dt} S\left(e(t)\right) = T_r\left(\rho'(t) \log e(t)\right) - T_r\left(\rho(t) \frac{d}{dt} \log \rho(t)\right)$$

Let $\log r(t) = Z(t)$

Then $\rho(t) = \rho^{Z(t)}$

$$\rho'(t) = \rho^{Z(t)} \left(\frac{1 - e^{-ad\ Z(t)}}{ad\left(Z(t)\right)}\right)\left(Z'(t)\right)$$

$$Z'(t) = \left(\frac{ad\ Z(t)}{1 - \exp\left(-ad\ Z(t)\right)}\right)\left(e^{-Z(t)} \rho'(t)\right)$$

$$= \left(\frac{ad \log \rho(t)}{1 - \exp\left(-ad\ \rho(t)\right)}\right)\left(e(t)^{-1} \rho'(t)\right)$$

Let $\frac{x}{1 - e^{-x}} = \sum_{n=0}^{\infty} C_n x^n$. Then $C_0 = 1$

So,

$$\frac{ad \log \rho}{1 - \exp\left(-ad \log \rho\right)} = 1 + \sum_{n=1}^{\infty} C_n \left(ad \log \rho\right)^n$$

$$Z'(t) = \rho^{-1}\ \rho' + \sum_{n=1}^{\infty} C_n \left(ad \log \rho\right)^n \left(\rho^{-1} \rho'\right)$$

$$T_r\left(\rho\frac{d}{dt}\log\rho\right) = T_r(\rho Z')$$

$$= T_r(\rho') + \sum_{n=1}^{\infty} C_n T_r\left(\rho(ad\log\rho)^n(\rho^{-1}\rho')\right)$$

$$= \sum_{n=1}^{\infty} C_n T_r\left(\rho(ad\log\rho)^n(\rho^{-1}\rho')\right)$$

since $\quad T_r(\rho') = \frac{d}{dt}T_r\rho = 0 \qquad (T_r(\rho') = T_r(T(\rho)) = 0)$

Now $\rho\,(ad\log\rho)(X) = \rho[\log\rho, X] = [\log\rho, \rho X] = (ad\log\rho)(\rho X)$

Thus $\rho(ad\log\rho)^n(\rho^{-1}\rho') = \rho\,(ad\log\rho)^n$ and so $T_r(\rho\,(ad\log\rho)^n(\rho^{-1}\rho')) = 0$.

Thus,

$$T_r\left(\rho\frac{d\log\rho}{dt}\right) = 0 \text{ and we get}$$

$$\frac{d}{dt}S(\rho(t)) = -T_r(\rho'(t)\log\rho(t))$$

$$= -T_r(T(\rho(t))\log\rho(t))$$

$$= -T_r[(i[H,\rho] + \theta(\rho)\log\rho]$$

$$= -T_r[(i[H\log\rho, \rho] + \theta(\rho)\log\rho]$$

$$= -T_r(\theta(\rho)\log\rho)$$

where $\quad \theta(\rho) = -\frac{1}{2}(L^*L\rho + \rho L^*L - 2\,L\rho L^*)$

$$\frac{dS(\rho(t))}{dt} = \frac{1}{2}\{T_r(L^*L\,\rho\log\rho) + T_r(\rho\,L^*L\log\rho) - 2\,T_r(L\rho L^*\log\rho)\}$$

$$= T_r(L^*L\,\rho\log\rho) - T_r(L\rho L^*\log\rho)$$

❖❖❖❖❖

[30] Dirac Eqn. Based Temperature Estimation of Black-Body Temperature Field:

Dirac eqn.

$$\left[\gamma^\mu(i\partial_\mu + eA_\mu) - m\right]\Psi = 0.$$

$$\gamma^0 = \begin{pmatrix} I & 0 \\ 0 & -I \end{pmatrix}, \gamma^r = \begin{pmatrix} 0 & i\sigma_r \\ i\sigma_r & 0 \end{pmatrix}, 1 \le r \le 3,$$

$$\sigma_2 = \begin{pmatrix} 0 & -i \\ i & 0 \end{pmatrix}, \sigma_1 = \begin{pmatrix} 0 & 1 \\ 1 & 0 \end{pmatrix}, \sigma_3 = \begin{pmatrix} 1 & 0 \\ 0 & -1 \end{pmatrix},$$

$$\sigma_r\sigma_s + \sigma_s\sigma_r = 2\delta_{rs}, \qquad \gamma^{r^*} = -\gamma^r, \gamma^{0^*} = \gamma^0.$$

Thus, $\gamma^\mu\gamma^\nu + \gamma^\nu\gamma^\mu = 2g^{\mu\nu}$

$((2g^{\mu\nu})) = diag[1,-1,-1,1].$

Other choices:

$$\gamma^0 = \begin{pmatrix} 0 & I \\ I & 0 \end{pmatrix}, \gamma^r = \begin{pmatrix} 0 & \sigma_r \\ -\sigma_r & 0 \end{pmatrix}, 1 \le r \le 3$$

Then $\qquad \gamma^{0^*} = \gamma^0, \gamma^{r^*} = -\gamma^r,$

$$\gamma^\mu\gamma^\nu + \gamma^\nu\gamma^\mu = 2g^{\mu\nu}$$

Let $\qquad \bar{\Psi} = \psi^*\gamma^0 \left(\equiv \psi^{-T}\gamma^0\right).$

Let $\qquad J^\mu = \tilde{\psi}\gamma^\mu\psi = \psi^*\gamma^0\gamma^\mu\psi$

Then $J^0 = \psi^*\psi$ is real,

$$J^r = \psi^*\gamma^0\gamma^\tau\psi$$

$$\Rightarrow \qquad J^{r*} = \psi^*\gamma^{r*}\gamma^{0*}\psi = -\psi^*\gamma^\tau\psi^0\psi$$

$$= \psi^*\gamma^0\gamma^r\,\psi = J^r$$

Hence J^r is also real, $1 \le r \le 3$.

Now $\qquad i\partial_\mu J^\mu = i\partial_\mu\left(\psi^*\gamma^0\gamma^\mu\psi\right)$

$$i\left(\partial_\mu\psi^*\right)\gamma^0\gamma^\mu\psi + i\psi^*\gamma^0\gamma^\mu\partial_\mu\psi$$

Now $\qquad i\gamma^\mu\partial_\mu\psi = -e\gamma^\mu A_{r\mu}\psi + m\psi \qquad (1)$

So $\qquad i\partial_\mu\psi^*\gamma^{\mu*} = -eA_\mu\psi^*\gamma^{\mu*} + m\psi^* \qquad (2)$

Now $\qquad \gamma^0\gamma^r = \begin{pmatrix} 0 & I \\ I & 0 \end{pmatrix}\begin{pmatrix} 0 & \sigma_r \\ -\sigma_r & 0 \end{pmatrix}\begin{pmatrix} -\sigma_r & 0 \\ 0 & -\sigma_r \end{pmatrix}$

and hence $\gamma^0\gamma^\mu$ is Hermitian, *i.e.* $\gamma^{\mu*}\gamma^{0*} = \gamma^{\mu*}\gamma^0 = \gamma^0\gamma^\mu$

So $\qquad -i\partial_\mu\psi^*\gamma^{\mu*}\gamma^o = -eA_\mu\psi^*\gamma^{\mu*}\gamma^o + m\psi^*\gamma^0$

$$\Rightarrow \qquad -i\partial_\mu\psi^*\gamma^0\gamma^\mu = -eA_\mu\psi^*\gamma^0\gamma^\mu + m\psi^*\gamma^0$$

and hence (1)

$$\Rightarrow \qquad i\psi^*\gamma^0\gamma^\mu\partial_\mu\psi = -e\psi^*\gamma^0\gamma^\mu\psi A_\mu + m\psi^*\gamma^0\psi$$

while (2) $\Rightarrow$

$$-i\partial_\mu\psi^*\gamma^0\gamma^\mu\psi = -eA_\mu\psi^*\gamma^0\gamma^\mu\psi + m\psi^*\gamma^0\psi$$

Subtracting, we get

$$\partial_\mu\left(\psi^*\gamma^0\gamma^\mu\psi\right) = 0$$

i.e. $\partial_\mu J^\mu = 0$ which is the charge conservation equation; provided we define the current density as

$$-e\psi^*\gamma^0\gamma^\mu\psi \equiv -e\tilde{\psi}\gamma^\mu\psi$$

Now put $A_\mu(X)$ be a random Gaussian field with zero mean and correlations

$$\left\langle A_\mu\left(t+\tau,\underline{r}_1\right)\cdot A_r\left(t,\underline{r}_2\right)\right\rangle = R_{\mu r}\left(\tau,\underline{r}_1-\underline{r}_2\right)$$

i.e., $$\left\langle A_\mu\left(t+\tau,\underline{r}_0-\underline{r}\right)A_r\left(t,\underline{r}_0\right)\right\rangle = R_{\mu r}\left(\tau,\underline{r}\right)$$

Let $$\psi(X) = \psi^{(0)}(X) + e\psi(X) + O(e^2).$$

Then by first order perturbation theory,

$$\left(\gamma^\mu i\,\partial_\mu - m\right)\psi^{(0)} = 0, \tag{4}$$

$$\left(\gamma^\mu i\,\partial_\mu - m\right)\psi^{(0)} = -A_\mu\gamma^\mu\psi^{(0)} \tag{5}$$

$$\psi_\alpha^{(0)} = \int\left[a_\alpha(\underline{p})\exp(i.p.X) + \overline{b_\alpha(p)}\exp\left(i.\underline{p}-X\right)\right]d^3p \tag{6}$$

where $p = p^\mu = \left(p^0,\underline{p}\right)$

$$p.X = p_\mu X^\mu = p^\mu X_\mu = p^0X^0 - \left(\underline{p},\underline{r}\right) = p^0t - \left(\underline{p},\underline{r}\right)$$

(6) satisfies (4) provided

$$\left(\gamma^\mu p_\mu - m\right)\underline{a}(\underline{p}) = 0$$

$$\left(\gamma^\mu p_\mu + m\right)\overline{\underline{b}(\underline{p})} = 0$$

These eqn. imply $p_\mu p^\mu = m^2$ *i.e.* $P^{02} = \left|\underline{p}\right|^2 + m^2$ or $p^0 = F\left(\underline{p}\right)$

$$= \pm\sqrt{m^2 + \left|\underline{p}\right|^2}.$$

Now $\underline{a}(\underline{p})$ can be determined as follows:

$$\gamma^0 = \begin{pmatrix} 0 & I \\ I & 0 \end{pmatrix}, \gamma^r = \begin{pmatrix} 0 & \sigma_r \\ -\sigma_r & 0 \end{pmatrix}$$

So,
$$\gamma^0 p_\mu = \begin{pmatrix} 0, E(\underline{p}) + (\underline{\underline{\sigma}}, \underline{p}) \\ E(\underline{p}) - (\underline{\underline{\sigma}}\ \underline{p})\ 0 \end{pmatrix}$$

So $\left(\gamma^\mu p_\mu - m\right)\underline{a}(\underline{p}) = 0$

$\Leftrightarrow$

$$m\underline{a}_1(\underline{p}) = \left(E(\underline{p}) + (\underline{\underline{\sigma}}, \underline{p})\right)\underline{a}, (\underline{p})$$

$$\left(\underline{a}(\underline{p}) = \begin{pmatrix} \underline{a}_1(\underline{p}) \\ \underline{a}_2(\underline{p}) \end{pmatrix}\right)$$

Thus
$$\underline{a}(\underline{p}) = \begin{bmatrix} \frac{1}{m}\left(E(\underline{p}) + (\underline{\underline{\sigma}}, \underline{p})\right)\underline{a}_2(\underline{p}) \\ \underline{a}_2(\underline{p}) \end{bmatrix}$$

Normalizing so that $\left\|\underline{a}(\underline{p})\right\|^2 = 1$ gives

$$\left\langle \underline{a}_2(\underline{p}), \frac{1}{m^2}\left(E(\underline{p})^2 + \underline{p}^2 + 2E(\underline{p})(\underline{\underline{\sigma}}, \underline{p})\right)\underline{a}_2(\underline{p})\right\rangle + \left\|\underline{a}(\underline{p})\right\|^2 = 1$$

$$\left\langle \underline{a}_2(\underline{p}), \left(2E(\underline{p})^2 - m^2 + 2E(\underline{p})(\underline{\underline{\sigma}}, \underline{p})\right)\underline{a}_2(\underline{p})\right\rangle$$
$$+ m^2\left\langle \underline{a}_2(\underline{p}), \underline{a}_2(\underline{p})\right\rangle = m^2$$

or
$$2E(\underline{p})\left\langle \underline{a}_2(\underline{p}), \left(E(\underline{p}) + (\underline{\underline{\sigma}}, \underline{p})\right)\underline{a}_2(\underline{p})\right\rangle = m^2$$

$$2E(\underline{p})^2\left\|\underline{a}_2(\underline{p})\right\|^2 + 2E(\underline{p})\left\langle \underline{a}_2(\underline{p}), (\underline{\underline{\sigma}}, \underline{p})\underline{a}_2(\underline{p})\right\rangle = m^2$$

Likewise,
$$2E(\underline{p})^2\left\|\underline{b}_2(\underline{p})\right\|^2 + 2E(\underline{p})\left\langle \underline{b}_2(\underline{p}),\ (\underline{\underline{\sigma}}, \underline{p})\underline{b}_2(\underline{p})\right\rangle = m^2$$

(Simply replace m by $-m$)

Let
$$\underline{a}_2(\underline{p}) = \lambda(\underline{p})\underline{\chi}(\underline{p})$$

where

$$\lambda(\underline{p}) \in \mathbb{C}, \underline{\chi}(\underline{p}) \in \mathbb{C}^2, \left\|\underline{\chi}(\underline{p})\right\|^2 = 1$$

Then $2E(\underline{p})^2\left|\lambda(\underline{p})\right|^2 + 2E(\underline{p})\left|\lambda(\underline{p})\right|^2\left\langle \underline{\chi}(\underline{p}), \left(\underline{\underline{\sigma}}, (\underline{p})\underline{\chi}(\underline{p})\right)\right\rangle = m^2$

i.e. $$|\lambda(\underline{p})|^2 = \frac{m^2}{2E(\underline{p})^2\left(E(\underline{p})+\langle\underline{\chi}(\underline{p}),(\underline{\underline{\sigma}},\underline{p})\underline{\chi}(\underline{p})\rangle\right)}$$

In particular $\underline{a}(\underline{p})$ and $\underline{b}(\underline{p}) \in \mathbb{C}^4$ both have two complex degrees of freedom $\forall\, \underline{p} \in \mathbb{R}^3$.

We now solve the $O(A_\mu)$ eqn *i.e.* (5)

Let $$A_\mu(\underline{X}) = \int\left[\hat{A}_\mu(\underline{p})\exp(-i\, p\cdot X) + \overline{\hat{A}_\mu(\underline{p})}\exp(i\, p\cdot X)\right]d^3p$$

where $p.X = p_\mu X^\mu$ and $\Box A_\mu = \partial_\alpha\partial_\alpha A_\mu = 0$

$$\Rightarrow \qquad p^2 = 0,\ i.e.\ p_\mu p^\mu = 0,\ i.e.\ p^0 = \pm E(\underline{p}) = |\underline{p}|$$

So we have writing $E_0(\underline{p}) = |\underline{p}|$ that

$$A_\mu(\underline{X}) = \int\Big[\hat{A}_\mu(p)\exp\left(-i\left(E_0(\underline{p})t - (\underline{p},\underline{r})\right)\right)$$
$$+\overrightarrow{\hat{A}(\underline{p})}\exp\left(i\left(E_0(\underline{p})t\right)\exp(p,\underline{r})\right)\Big]d^3p$$

Then $A_\mu(X)\psi_\alpha^{(0)}(X)$

$$= \int\Big[\hat{A}_\mu(\underline{p})\exp\left(-iE_0(\underline{p})t\right)\exp\left(i(p,\underline{r})\right)$$
$$+\overrightarrow{\hat{A}_\mu(\underline{p})}\exp\left(i\left(E_0(\underline{p})t\right)\exp\left(-i(p,\underline{r})\right)\right)\Big]d^3p$$
$$\times\int a_\alpha(\underline{p})\exp\left(-iE(\underline{p})t\right)\exp\left(\underline{i}\ \underline{p}\cdot\underline{r}\right)$$
$$+\overline{b_\alpha(\underline{p})}\exp\left(-iE(\underline{p})t\right)\exp\left(-i\ \underline{p}.\underline{r}\right)d^3\underline{p}$$
$$= \int\Big\{\hat{A}_\mu(\underline{p}')\,a_\alpha(\underline{p})\exp\left(-i\left(E(\underline{p})+E_0(\underline{p}')\right)t\right)\exp\left(i\left(\underline{p}+\underline{p}',\underline{r}\right)\right)$$
$$+\ \hat{A}_\mu(\underline{p}')\,\overline{b_\alpha(\underline{p})}\exp\left(i\left(E(\underline{p})-E_0(\underline{p}')\right)t\right)\exp\left(-i\left(\underline{p}+\underline{p}',\underline{r}\right)\right)$$
$$+\ \overline{\hat{A}_\mu(\underline{p}')}\,a_\alpha(\underline{p})\exp\left(-i\left(E(\underline{p})-E_0(p')\right)t\right)\exp\left(i\left(\underline{p}-\underline{p}',\underline{r}\right)\right)$$
$$= +\ \overline{\hat{A}_\mu(\underline{p}')\,b_\alpha(\underline{p})}\exp\left(-i\left(E(\underline{p})+E_0(p')\right)t\right)$$
$$\exp\left(-i\left(\underline{p}-\underline{p}',\underline{r}\right)\right)\Big\}d^3p\,d^3p'$$

Writing $\psi_\alpha^{(1)}(X) = \int\psi_\alpha^{(1)}(k)\exp\left(-i\,\underline{\underline{k}}.\underline{X}\right)d^4k$

where $k.X = k_\mu X^\mu = k^\mu X_\mu = k^0 t - (\underline{k}, \underline{r})$

we get from (5)

$$\int_{\mathbb{R}^4} \left(k_\mu \gamma^\mu - m\right) \underline{\widehat{\psi}}^{(1)}(\underline{k}) \exp(-ik.X) d^4k = -\gamma^\mu \left(A_\mu^{(X)} \underline{\psi}^{(0)}(X)\right)$$

$$= \int -\underline{\underline{\gamma}}^\mu \underline{X}_\mu \left(t, \underline{p}\right) \exp\left(i\, \underline{p}.\underline{r}\right) d^3p$$

where $\chi_\mu\left(t, \underline{p}\right)$

$$= \int \hat{A}_\mu\left(\underline{p}'\right) \underline{a}\left(\underline{p} - \underline{p}'\right) \exp\left(-i\left(E\left(\underline{p} + \underline{p}'\right) + E_0\left(\underline{p}'\right)\right)t\right) d^3p'$$

+ 3 other similar terms.

The correlation of $\left\{\psi_0^{(1)}\right\}$ can be computed from this expression in terms of correlation in $\{A_\mu\}$.

Reference: Rohit Rana, temperature measurement using Klein -Gordon field, M. Tech Thesis, NSIT, 2014, Under supervision of H. Parthasarathy and T. K. Rawat.

❖❖❖❖❖

[31] Quantum Image Processing on a C^∞– Manifold:

The image field on the manifold is diffused using the Diffusion equation.

$$\frac{\partial u(t, \underline{x})}{\partial t} = \frac{1}{2} g^{-\frac{1}{2}} \left(g^{\frac{1}{2}} g^{\alpha\beta} u,_\alpha (t, \underline{X})\right),_\beta$$

$$= \frac{1}{2} g^{\alpha\beta} u,_{\alpha\beta} + \frac{1}{2} g^{-\frac{1}{2}} (g^{\frac{1}{2}} g^{\alpha\beta}),_\beta u,_\alpha$$

We can express thus as

$$u,_t^{(t,x)} = D_1^{\alpha\beta}(X) u,_{\alpha\beta}(t, X) + D_2^\alpha(X) u,\alpha(t, X) \quad (1)$$

where

$$D_1^{\alpha\beta}(X) = \frac{1}{2} g^{\alpha\beta}(X),$$

$$D_2^\alpha(X) = \frac{1}{2} g^{-\frac{1}{2}} \left(g^{\frac{1}{2}} g^{\alpha\beta}\right),_\beta$$

$$= \frac{1}{2} g,_\beta^{\alpha\beta} + \frac{1}{4} g,_\beta g^{\alpha\beta} g^{-1}$$

The image field at time t is then $u(t, \underline{X})$, $\underline{X} \in \mathbb{R}^n$.

To quantize this image field, we must first derive (1) from an action principle.

Diffusion of an edge:

An edge is a line in $\mathbb{R}^2$ having the equation

$$lX + mY = C$$

The corresponding intensity pattern is

$$f(X, Y) = \delta(lX + mY - C)$$

If the image field consists of only edges, then the intensity field in

$$f(X, Y) = \sum_{\alpha} \delta(l_\alpha X + m_\alpha Y - c_\alpha)$$

Under the diffusion equation defined by

$$\frac{\partial u(t, x, y)}{\partial t} = \frac{1}{2} D\left(\frac{\partial^2 u}{\partial x^2} + \frac{\partial^2 u}{\partial y^2}\right),$$

$$u(0, X, Y) = f(X, Y),$$

this field of edges evolves to

$$u(t, X, Y) = (K_t * f)(X, Y)$$

$$\equiv \int K_t(X - X', Y - Y') f(X', Y') dX'dY'$$

❖❖❖❖❖

[32] Belavkin Contd. (Quantum Non-Linear Filtering):

$$d\widehat{X} = \left(\widehat{F}_\nu^\mu - \widehat{X}(t)\delta_\nu^\mu\right) d\widehat{A}_\mu^\nu$$

$$\equiv \left(\widehat{F} - X \otimes \delta\right) d\widehat{A}$$

Let $\widehat{C} = \widehat{F} - X \otimes \delta$. Then

$$d\widehat{X}^* = \widehat{C} d\widehat{A} \equiv \widehat{C}_\nu^\mu \, d\widehat{A}_\mu^\nu$$

$$d\widehat{X}^* = \widehat{C}_\nu^{\mu *} \, d\widehat{A}_\mu^{\nu *} = \widehat{C}_\mu^{b\nu} \, d\widehat{A}_\nu^\mu = \widehat{C}^b \, d\widehat{A}$$

$$d\left(\widehat{X} * \widehat{X}\right) = d\widehat{X} * .\widehat{X} + \widehat{X} * d\widehat{X} + d\widehat{X} * .d\widehat{X}$$

$$= \widehat{C}^b \widehat{X} d\widehat{A} + \widehat{X} * \widehat{C} d\widehat{A} + \widehat{C}^b \widehat{C} d\widehat{A}.d\widehat{A}$$

Note: $dA_-^j dA_k^+ = \delta_k^j dt$

$$dA_j^+ dA_-^k = \delta_-^+ dA_j^k = 0$$

$$dA_k^j dA_m^l = \delta_m^j dA_k^l. \text{ So } dA_\nu^\mu dA_\sigma^\rho = \delta_\sigma^\mu dA_\nu^\rho)$$

$$= d\widehat{A}\left(\widehat{C}^b \widehat{X} + \widehat{X} * \widehat{C} + \widehat{C}^b \widehat{C}, t\right)$$

Now,

$$\widehat{C}^b\widehat{X}+\widehat{X}^*\widehat{C}+\widehat{C}^b\widehat{C} \equiv \left(\widehat{F}^b-\widehat{X}\otimes\delta\right)\left(\widehat{X}\otimes\delta\right)+\widehat{X}^*\otimes\delta\left(\widehat{F}-\widehat{X}\otimes\delta\right)$$

$$+\left(\widehat{F}^b-\widehat{X}^*\otimes\delta\right)\left(\widehat{F}-\widehat{X}\otimes\delta\right)$$

$$= -\left(\widehat{X}^*\widehat{X}\otimes\delta\right)+\widehat{F}^b\widehat{F}$$

So

$$d\left(\widehat{X}^*\widehat{X}\right) = \left(\widehat{F}^b\widehat{F}-\widehat{X}^*\widehat{X}\otimes\delta\right)d\widehat{A}$$

$$\equiv \left(\widehat{F}^b\widehat{F}-\widehat{X}^*\widehat{X}\otimes\delta\right)^{\mu}_{v} d\widehat{A}^{v}_{\mu}$$

$$= \left(\left(\widehat{F}^b\widehat{F}\right)^{\mu}_{v}-\widehat{X}^*\widehat{X}\,\delta^{\mu}_{v}\right)d\widehat{A}^{v}_{\mu}$$

It $\widehat{F}^b\widehat{F} = I$, then $\widehat{X}^*\widehat{X}(t) = I$

$\Rightarrow$ $d\left(\widehat{X}^*\widehat{X}\right) = 0 \Rightarrow \widehat{X}(t)$

is unitary $\forall t$ if it is unitary at $t = 0$.

❖❖❖❖❖

[33] Remarks on Theorems from V. Kac. Infinite Dimensional Lic Algebra. P.87:

$$t_{w(\alpha)}(\lambda) = \lambda+\langle\lambda,k\rangle w(\alpha)-\left((\lambda/w(\alpha))+\frac{1}{2}|\alpha|^2\langle\lambda,k\rangle\right)\delta$$

$$wt_{\alpha}w^{-1}(\lambda) = \lambda+\left\langle w^{-1}(\lambda),k\right\rangle w(\alpha)$$

$$-\left(\left(w^{-1}(\lambda)|\alpha\right)+\frac{1}{2}|\alpha|^2< w^{-1}(\lambda),K>\right| w(\delta)$$

$$w(\delta) = \delta, \left\langle w^{-1}(\lambda),k\right\rangle = \langle\lambda,w(k)\rangle = \langle\lambda,k\rangle$$

$$\left(w^{-1}(\lambda)|\alpha\right) = (\lambda|w(\alpha))$$

so

$$wt_{\alpha}w^{-1}(\lambda) = \lambda+\langle\lambda,k\rangle w(\alpha)$$

$$\left((\lambda|w(\alpha))+\frac{1}{2}|\alpha|^2\langle\lambda,k\rangle\right)\delta = t_w(\alpha)(\lambda).$$

$$\text{P249 } n\in N_{\mathbb{Z}} \Rightarrow {}_{z}n = (\alpha,\beta,u),n,v$$

$$(\alpha,\beta,u)(v) = t_{\beta}(v)+2\pi i\alpha+(u-i\pi(\alpha|\beta))\delta$$

$$
\begin{aligned}
t_\rho(\lambda)(n\cdot v) &= (\lambda + (\lambda\,|\,\delta)\rho + |\,n, v) - \left(\left(\lambda|\rho| + \frac{1}{2}(\rho\,|\,\rho)(\lambda\,|\,\delta)\right)(\delta\,|\,n\cdot v)\right) \\
&= \left(t_\rho(\lambda)\,|\,t_\beta(v) + 2\pi i\alpha + (u - i\pi(\alpha\,|\,\beta))\delta\right) \\
&= \left(t_\rho - \beta(\lambda)\,|\,v\right) + ((\lambda\,|\,\delta) + (\lambda\,|\,\delta)(\rho\,|\,\delta)) \\
&\quad (u - i\pi(\alpha\,|\,\beta)) + 2\pi i(\lambda\,|\,\alpha) \\
(\lambda\,|\,\delta) &= k \in \mathbb{Z}_+ \\
&\quad \rho_2(\rho\,|\,\delta) = 0.
\end{aligned}
$$

So,

$$
\begin{aligned}
t_\rho(\lambda)(n\cdot v) &= \left(t_{\rho-\beta}(\lambda)\,|\,v\right) + k(u - i\pi(\alpha\,|\,\beta)) \\
\delta(n\cdot v) &= (\delta|n\cdot v) = (\delta|t_\beta(v)) = (\delta|v).
\end{aligned}
$$

❖❖❖❖❖

[34] Quantum Mutual Information:

[Reference: Remark from Mark, M. Wilde "Quantum Information Theory. (P. 303.)]

Holevo information: $\{p_x(x), \rho_x\} \simeq$ ensemble prepared by Alice.

$$
\begin{aligned}
\rho^{XA'} &= \sum_x p_x(x)|x\rangle\langle x|^X \otimes \rho_x^{A'} \\
\mathcal{N}^{A'} \to B &= \text{channel.} \\
\rho^{XB} &= \sum_x p_x(x)|x\rangle\langle x|^X \otimes \rho_x^{B}
\end{aligned}
$$

where

$$
\begin{aligned}
\rho_x^B &= \mathcal{N}^{A'\to B}\left(\rho_x^{A'}\right) \\
X(\mathcal{N}) &= \max_{\rho^{XA'}} I(X;B)_\rho \\
I(X;B)_\rho &= -\sum p_x(x) H\left(\rho_x^B\right) + H\left(\sum_x p_x(x)\rho_x^B\right)
\end{aligned}
$$

Note:

$$
\begin{aligned}
\rho^B &= Tr_x\left(\rho^{XB}\right) = \sum_x p_x(x)\rho_x^B \\
H(B\,|\,X) &= \sum_x p_x(x)\rho_x^B
\end{aligned}
$$

Hence $$I(X:B)_\rho = H(B) - H(B \mid X) = H(\rho B) - H(B \mid X)$$

$$= H\left(\sum_X p_x(u)\rho_x^B\right) - \sum_x p_x(x)H\left(\rho_x^B\right)$$

P. 307 Let

Let $$\sum_{x,y} p(y \mid x)\, p(x)|x\rangle\langle x_1^X| \otimes |y\rangle\langle y|^\Upsilon \otimes \rho_{x,y}^A = \sigma$$

Then

$$T_{ry}\sigma = \sum_{x,y} p(y \mid x)\, p(x)|x\rangle\langle x|^X \otimes \rho_{x,y}^A$$

$$= \sum_x p(x)|x\rangle\langle x|^X \otimes \rho_{x,y}^A$$

Where $$\rho_x^A = \sum_y p(y \mid x)\rho_{x,y}^A$$

$\therefore$ $$I(X;A)_\sigma = H\left(\sum p(x)\rho_x^A\right) - \sum p(x)H\left(\rho_x^A\right) = H(X:A)_\rho$$

Data processing inequality; Let $X \to Y \to Z$ be a Markov chain. Then $I(X;Y) \geq I(X;Z)$

$$I(X;Y) = H(Y) - H(Y \mid X),$$

$$I(X;Z) = H(Z) - H(Z \mid X)$$

$$H(Z|X) \geq H(Z \mid Y, X) = H(Z \mid Y)\text{(by Markov property)},$$

So. $$I(X;Z) \leq H(Z) - H(Z \mid Y)$$

If the Markov chain $X \to Y \to Z$ is stationary *i.e,* $p_{Z \mid Y} = p_{Y \mid X}$ and $p_X = p_Y = p_Z$. the $H(Z) - H(Z|Y) = H(Y) - H(Y|X)$ and we are done.

Let $$\rho = \sum_x p(x)|x\rangle\langle x| \otimes \rho_x$$

Let $$\rho_x = \sum_y p(y \mid x)|yx\rangle\langle yx|$$

Then $$\rho = \sum_{xy} p(x)\, p(y \mid x)|x\rangle\langle x| \otimes |y_x\rangle\langle y_x| \tag{1}$$

Since $$\langle x' \mid x\rangle = \delta_{x'x}, \delta_{yx}, \;\; y'_x = \delta_{yx}, y'_x \;\; \forall x$$

it flows that (1) is a spectral representation.

Hence,

$$H(\rho) = -\sum_{x,y} p(x)\, p(y \mid x) \log(p(x)\, p(y \mid x))$$

$$= -\sum_{x} p(x)\log p(x) - \sum_{x,y} p(y,x)\log p(y\,|\,x)$$

$$= H(X) + H(y|X) = H(Y, X)$$

Where $p(y, x) = p(y|x)\,p(x)$ i the joint probability distribution of $r.v$'s (Y, X).

❖❖❖❖❖

[35] Scattering Theory Applied to Quantum Gate Design:

$\psi_i(\underline{r}) = C.\exp\left(jk\hat{n}_i \cdot \underline{r}\right) \simeq$ Incident projectile state. $E = \hbar^2 k^2 / 2m \simeq$ Projectile energy. $\hat{n}_i \simeq$ Incident projectile direction. $\frac{-\hbar^2}{2m}\Delta\psi_i = E\psi_i$ is clearly satisfied. Final state (scattered)

$$\psi_f(\underline{r}) = \underline{\psi}_i(\underline{r}) + \varepsilon\underline{\psi}_{s1}(r) + \varepsilon^2\psi_{s2}(r) + O(\varepsilon^3).$$

Interaction potential energy $\simeq \varepsilon V(\underline{r})$.

$$\left[-\frac{\hbar^2}{2m}\Delta + \varepsilon V\right]\psi_f = E\psi_f$$

or

$$\left[-\frac{\hbar^2}{2m}\Delta + \varepsilon V\right]\left[\psi_f = \varepsilon\psi_{s1} + \varepsilon^2\psi_{s2}\right]$$

$$= E[\psi_i = \varepsilon\,\psi_{s1} + \varepsilon^2\psi_{s2}] + O(\varepsilon^3)$$

Coeffi. of ε^0, ε^1, ε^2, respectively give

$$\frac{-\hbar^2}{2m}\Delta_{s1} + V\psi_i = E\psi_{s1},$$

$$\frac{-\hbar^2}{2m}\Delta\psi_{s2} + V\psi_{s1} = E\psi_{s2}$$

Thus,

$$\left(\Delta + k^2\right)\psi_{s1} = \frac{2m}{\hbar^2}V\psi_i,$$

$$\left(\Delta + k^2\right)\psi_{s2} = \frac{2m}{\hbar^2}V\psi_{s1}$$

$$\psi_{s1}(\underline{r}) = -\frac{m}{2\pi\hbar^2}\int \frac{\exp\left(jk\,|\,\underline{r} - \underline{r}'\,|\right)}{|\,\underline{r} - \underline{r}'\,|}V(\underline{r}')\psi_i(\underline{r}')d^3r'$$

$$= \int G_k(\underline{r} - \underline{r}')V(\underline{r}')\psi_i(\underline{r}')d^3r'$$

$$G_k(\underline{r}) = \frac{-m}{2\pi\hbar^2 r}\exp(jkr)$$

$$\psi_{s2}(\underline{r}) = \int G_k(\underline{r}-\underline{r}')V(\underline{r}')\psi_i(\underline{r}')d^3r'$$

$$= \int G_k(\underline{r}-\underline{r}')V(\underline{r}')G_k(\underline{r}'-\underline{r}'')V(\underline{r}'')\psi_i(\underline{r}'')d^3r'd^3r''$$

Formally, $\psi_{sm}(\underline{r}) = (G_kV)^m\psi_i, m = 1, 2...$

$$\psi_f = \psi_i + \sum_{m=1}^{\infty}\varepsilon^m(G_kV)^m\psi_i$$

exact solution.

Let $\psi \in L^2\left(\mathbb{R}^3\right)$. Then

$$\| G_k V\psi \|^2 = \int d^3r \int G_k(\underline{r}-\underline{r}')V(\underline{r}')\psi(\underline{r}')d^3r'\Big|^2$$

$$\leq \left[\int | G_k(\underline{r}-\underline{r}')V(\underline{r}')|^2 d^3rd^3r'\right]$$

$$\left[\int | \psi(\underline{r}')|^2 d^3r'\right]$$

Hence $\| G_kV \| < \infty$ iff

$$\int \left|G_k(\underline{r}-\underline{r}')\right|^2 V(\underline{r}')^2 d^3rd^3r' < \infty$$

iff $\int | V(\underline{r}')|^2 d^3r' < \infty$ and $\int | G_k(\underline{r})|^2 d^3r < \infty$

The latter in false since

$$\int \frac{d^3r}{r^2} = \int_0^\infty 4\pi dr = \infty$$

with box normalization however, $\|G_kV\|$this can be made finite.

Now formally,

$$\psi_f = \psi_i + \frac{\varepsilon G_kV}{I - \varepsilon G_kV}\psi_i$$

$$= \left[I - \varepsilon G_kV\right]^{-1}\psi_i$$

Now formally,

$$G_k = \frac{2m}{\hbar^2}(\Delta + k^2)^{-1}$$

$$= -\left(-\frac{\hbar^2}{2m}\Delta - \frac{\hbar^2k^2}{2m}\right)^{-1}$$

$$= -(H_0 - E)^{-1}$$

So, $$[I-\varepsilon G_k V]^{-1} = \left[I+\varepsilon(H_0-E)^{-1}V\right]^{-1}$$

$$= [H_0-E+\varepsilon V]^{-1}[H_0-E]$$

$$= [H_1-E]^{-1}[H_0-E]$$

where $$H_1 = H_0+\varepsilon V$$

This formula implies

$$|\psi_f\rangle = [H_1-E]^{-1}[H_0-E]|\psi_i\rangle$$

or

$$[H_1-E]|\psi_f\rangle = [H_0-E]|\psi_i\rangle$$

Both sides one zero and hence no information is gained from then equation. However, we can also write

$$\sum_{m=1}^{\infty}\varepsilon^m (G_k V)^m = G_k\sum_{m=1}^{\infty}\varepsilon^m (V G_k)^{m-1}V = \varepsilon G_k[I-\varepsilon V G_k]^{-1}V$$

and $$I-\varepsilon V G_k = I+\varepsilon V(H_0-E)^{-1}$$

$$= [H_0+\varepsilon V-E][H_0-E]^{-1}$$

$$= [H_1-E][H_0-E]^{-1}$$

so $$\sum_{m=1}^{\infty}\varepsilon^m (G_k V)^m = \varepsilon G_k\left([H_1-E][H_0-E]\right)^{-1}V$$

thus, $$|\psi_f\rangle = \varepsilon G_k(H_0-E)(H_1-E)^{-1}V$$

$$= -\varepsilon(H_1-E)^{-1}V$$

So,

$$|\psi_f\rangle = \left(I-E(H_1-E)^{-1}V\right)|\psi_i\rangle$$

Check: Multiply both sides of this equation by H_1-E to get

$$(H_1-E)|\psi_i\rangle = (H_1-E-\varepsilon V)|\psi_i\rangle$$

$$= (H_0-E)|\psi_i\rangle$$

and both sides evaluate to zero. We write

$$H_1-E = (H_0-E+\varepsilon V)^{-1}$$

$$= \left(I+\varepsilon(H_0-E)^{-1}V\right)^{-1}(H_0-E)^{-1}$$

$$= \left(I+\varepsilon(H_0-E)^{-1}V\right)^{-1}(H_0-E)^{-1}+O(\varepsilon^2)$$

and hence

$$\begin{aligned}|\psi_f\rangle &= |\psi_i\rangle - \varepsilon(H_0 - E)V|\psi_i\rangle \\ &\quad + \varepsilon^2 (H_0 - E)^{-1} V(H_0 - E)^{-1} V|\psi_i\rangle + O(\varepsilon^3). \\ &= \left[I + \varepsilon G_k V + \varepsilon^2 (G_k V)^2\right]|\psi_i\rangle + O(\varepsilon^3).\end{aligned}$$

$|\psi_i\rangle \simeq$ output state evolving according to H_0.

Alternate way of deriving the operator $|\psi_0\rangle \to |\psi_f\rangle |\psi_0\rangle$ evolves according to H_0 white $|\psi_f\rangle$ evolves according to H_1. As $t \to \infty |\psi_f\rangle_t |\psi_0\rangle_t$ since as $t \to \infty$ the effect of the scaterer interaction becomes negligible. Note that $|\psi_0\rangle$, the out put state, is evolving as a force partible while $|\psi_f\rangle$ the scattered state, evolves according to H_1. Thus

$$\lim_{t\to\infty} \left\| e^{-itH,} |\psi_f\rangle - e^{itH_0} |\psi_0\rangle \right\| = 0$$

So $$|\psi_f\rangle = \lim_{t\to\infty} e^{-itH_0,} e^{-itH_0} |\psi_0\rangle$$

Now

$$\frac{d}{dt} e^{itH,} e^{itH_0} = ie^{itH,}(H_1 - H_0)e^{-itH_0} = i\varepsilon\, e^{itH,} V e^{-itH_0}$$

So $$\Omega(t) = \lim_{t\to\infty} e^{itH_1} e^{-itH_0} = I + i\varepsilon \int_0^\infty e^{itH,} V e^{-itH_1} dt$$

Let $$H_0 = \int_{-\infty}^{+\infty} \lambda\, dE(\lambda) \text{ be the spectral rep'n of } H_0.$$

Then $$\begin{aligned}\Omega(t) &= I + i\varepsilon \int_0^\infty dt \int_{\mathbb{R}} e^{it(H_1 - \lambda)} V\, dE(\lambda) \\ &\triangleq I + i\varepsilon \lim_{\delta\to 0} \int_0^\infty dt\, e^{-it^{(H_1 - \lambda + i\delta)}} V\, dE(\lambda) \\ &I - \varepsilon \int_0^\infty (H_1 - \lambda + i\delta)^{-1} V\, dE(\lambda).\end{aligned}$$

i.e. $$|\psi_f\rangle = |\psi_0\rangle - \varepsilon \int_0^\infty (H_1 - \lambda + i\delta)^{-1} V\, dE(\lambda) |\psi_0\rangle$$

If the output state $|\psi_0\rangle$ has energy E, then $dE(\lambda)|\psi_i\rangle = \delta_{E,\lambda} |\psi_i\rangle$ and we get $|\psi_f\rangle = |\psi_0\rangle - \varepsilon(H_1 - E + i\delta)^{-1} V|\psi_0\rangle$

Now consider the input state $|\psi_i\rangle$. At time $t \to -\infty$ $|\psi_i\rangle$ evolves according to H_0 while $|\psi_f\rangle$ evolves according to H_1. In the remote part, these two states must coincide. Thus

$$\lim_{t\to-\infty} \left\| e^{-itH_1}|\psi_f\rangle - e^{-itH_0}|\psi_i\rangle \right\| = 0$$

or
$$|\psi_f\rangle = \Omega(-)|\psi_i\rangle$$

where
$$\Omega(-) = \lim_{t\to-\infty} e^{itH_0}\, e^{-itH_0}$$

$$\Omega(-) = I - \int_{-\infty}^{0} \frac{d}{dt}\left(e^{itH_1} e^{-itH_0}\right) dt$$

$$= I - i\int_{-\infty}^{0} e^{itH_1}(H_1 - H_0)e^{-itH_0}\, dt$$

$$= I - i\varepsilon\int_{-\infty}^{0} e^{itH_1}\, V\, e^{-itH_0}\, dt$$

$$= I - i\varepsilon \int_{\mathbb{R}\times(-\infty,\,0)} e^{it(H_1-\lambda)} V\, dE(\lambda)\, dt$$

$$= I - i\varepsilon \lim_{\delta\to 0+} \int_{\mathbb{R}\times(-\infty,\,0)} e^{it(H_1-\lambda-i\delta)} V\, dE(\lambda)\, dt$$

$$= I - i\varepsilon \int_{\mathbb{R}} (H_1 - \lambda - i\delta)^{-1} V\, dE(\lambda)$$

So

$$|\psi_f\rangle = |\psi_i\rangle - \varepsilon(H_1 - E - i\delta)^{-1} V |\psi_i\rangle$$

$$\Omega_+(E) = I - \varepsilon(H_1 - E + i\delta)^{-1} V$$

$$\Omega_-(E) = I - \varepsilon(H_1 - E - i\delta)^{-1} V$$

$$|\psi_f\rangle = \Omega_+(E)|\psi_0\rangle,$$

$$|\psi_f\rangle = \Omega_-(E)|\psi_i\rangle$$

Born scattering is an first order approximation of

$$|\psi_f\rangle = \Omega_-(E)|\psi_i\rangle$$

Scattering matrix/operator: Since $\Omega_+(E)$ is an isometry, we have

$$\Omega_+(E)^* \Omega_+(E) = I$$

and hence,

$$|\psi_0\rangle = \Omega_+(E)^* |\psi_f\rangle$$

$$= \Omega_+(E)^*\Omega_-(E)|\psi_i\rangle$$

$$S(E) = \Omega_+(E)^*\Omega_-(E)$$

is the scattering matrix at energy E.

$$|\psi_0\rangle = S(E)|\psi_i\rangle$$

$S(E)$ is unitary on the space of definite energy states *i.e.* states defined on the unit sphere *i.e.* $L^2(S^2)$

$$S(E) = \left[I-\varepsilon(H_1-E+i\delta)^{-1}V\right]^*$$

$$\times\left[I-\varepsilon(H_1-E+i\delta)^{-1}V\right] = I-\varepsilon V(H_0-E-i\delta)^{-1}$$

$$-\varepsilon(H_0-E-i\delta)^{-1}V+O(\varepsilon^2)$$

$$= I-\varepsilon\left\{V(H_0-E-i\delta)^{-1}(H_0-E-i\delta)^{-1}V\right\}+O(\varepsilon^2)$$

Thus, $$|\psi_0\rangle = S(E)|\psi_i\rangle = |\psi_i\rangle-\varepsilon(H_0-E)^{-1}V|\psi_i\rangle+O(\varepsilon^2)$$

and $$(H_0-E)|\psi_i\rangle = 0-\frac{i\varepsilon}{\delta}V|\psi_i\rangle$$

Note that

$$(H_0-E-i\delta)^{-1}V|\psi_i\rangle = (-i\delta)^{-1}|\psi_i\rangle = \frac{i}{\delta}|\psi_i\rangle$$

So

$$|\psi_0\rangle = \left(I-i\delta^{-1}\varepsilon V\right)|\psi_i\rangle$$

$$-\varepsilon(H_0-E)^{-1}V|\psi_i\rangle+O(\varepsilon^2)$$

The infinite factor $\frac{i}{\delta}V|\psi_i\rangle$ in $|\psi_0\rangle$ shown that a blind application of perturbation theory *w.r.t.* ε is incorrect. However, if we still persist,then

$$|\psi_0\,|\,\psi_i\rangle = |\psi_0\,|\,\delta(E)\,|\,\psi_i\rangle$$

$$\langle\psi_i\,|\,\psi_i\rangle-\frac{i\varepsilon}{\delta}\langle\psi_i\,|\,V\,|\,\psi_i\rangle$$

$$-\varepsilon\left\langle\psi_i\middle|(H_0-E)^{-1}V\middle|\psi_i\right\rangle+O\left(\varepsilon^2\right)$$

Where the inner product is *w.r.t.* $L^2(S^2)$ in momentum space. More generally,

$$\langle\psi_{i2}\,|\,S(E)\,|\,\psi_{i1}\rangle = \langle\psi_{i2}\,|\,\psi_{i1}\rangle-\frac{i\varepsilon}{\delta}\langle\psi_{i2}\,|\,V\,|\,\psi_{i1}\rangle$$

$$-\varepsilon\langle\psi_{i2}|(H_0-E)^{-1}V|\psi_{i1}\rangle+O(\varepsilon^2)$$

Where $|\psi_{i1}\rangle$ and $|\psi_{i2}\rangle$ are two incident states.

$$\begin{aligned}
\text{Let} \qquad \psi_{i1}(\underline{r}) &= C_i \exp(jk\hat{n}_i\,\underline{r}) \\
\text{and} \qquad \psi_{i2}(\underline{r}) &= C_2 \exp(jk\hat{n}_2\,\underline{r}) \\
\text{Then} \qquad \widehat{\psi}_{i1}(\underline{k}) &= \frac{1}{(2\pi)^{3/2}} \int \psi_{i1}(\underline{r}) \exp(-i\underline{k}'\underline{r})\, d^3\underline{r} \\
&= (2\pi)^{3/2} C_1 \delta(\underline{k}' - k\hat{n}_1) \\
\text{and likewise} \qquad \widehat{\psi}_{i2}(\underline{r}) &= (2\pi)^{3/2} C_2 \delta(\underline{k}' - k\hat{n}_2) \\
\langle \psi_{i2} | \psi_{i1} \rangle &\equiv (2\pi)^3 \overline{C}_2 C_1 \int \delta(\underline{k}' - k\hat{n}_2)\delta(\underline{k}' - k\hat{n}_1)\, d\Omega(k')
\end{aligned}$$

If we instent take (1) over where of $\mathbb{R}^3$, then formally

$$\begin{aligned}
\langle \psi_{i2} | \psi_{i1} \rangle &= \int \overline{\psi_{i2}(\underline{r})}\, \psi_{i1}(\underline{r})\, d^3 r \\
&= \int \overline{\psi_{i2}(\underline{k}')}\, \psi_{i1}(\underline{k}')\, d^3 k' = (2\pi)^3 \overline{C}_2 C_1 \delta\left(k(\hat{n}_2 - \hat{n}_1)\right) \\
&= \frac{(2\pi)^{3/2} \overline{C}_2 C_1}{k} \delta\left(\hat{n}_2 - \hat{n}_1\right), \\
\langle \psi_{i2} | V | \psi_{i1} \rangle &= \int \overline{\psi}_{i2}(\underline{r}) V(\underline{r}) \psi_{i1}(\underline{r})\, d^3 r \\
&= \overline{C}_2 C_1 \int V(\underline{r}) \exp\left(-jk(\hat{n}_2 - \hat{n}_1, \underline{r})\right) d^3 r \\
&= (2\pi)^{3/2} \overline{C}_2 C_1 \widehat{V}\left(k(\hat{n}_2 - \hat{n}_1)\right)
\end{aligned}$$

By the Riemann-Lebesgue Lemma,

$$\lim_{k \to \infty} \widehat{V}\left(k(\hat{n}_2 - \hat{n}_1)\right) = 0$$

So for high energies, $\langle \psi_{i2} | V | \psi_{i1} \rangle$ may be neglected.

Now, the Born term is

$$\begin{aligned}
\left\langle \psi_{i2} | (H_0 - E)^{-1} | \psi_{i1} \right\rangle &= \frac{-2m}{h^2} \left\langle \psi_{i2} | (\Delta + k^2)^{-1} V | \psi_{i1} \right\rangle \\
&= \frac{m}{2\pi\hbar^2} \int \frac{\exp(jk|\underline{r} - \underline{r}'|)}{|\underline{r} - \underline{r}'|} \overline{\psi_{i2}(\underline{r})} V(\underline{r}') \psi_{i1}(\underline{r}')\, d^3 r\, d^3 r'
\end{aligned}$$

Before evaluating this, we make a remark: ψ_{i1} and ψ_{i2} are concentrated at $r \to \infty$ where $V(\underline{r})$ is concentrated around the origin. So ψ_{i1} ψ_{i2} has zero overlap with V which is another justification for taking the overlap integral $\langle \psi_{i2} | V | \psi_{i1} \rangle = 0$. With the assumption,

$$\langle \psi_{i2} | S(E) | \psi_{i1} \rangle \equiv \langle \psi_{i2} | S(k) | \psi_{i1} \rangle$$

$$\approx \frac{(2\pi)^{3/2}\bar{C}_2 C_1}{k}\delta(\hat{n}_2 - \hat{n}_1)$$

$$-\left(\frac{m\varepsilon}{2\pi\hbar^2}\right)\int\frac{\exp(jk|\underline{r}-\underline{r}'|)}{|\underline{r}-\underline{r}|}\bar{C}_2 C_1 V(\underline{r}')$$

$$\times \exp\left(jk\left(\hat{n}_1, \underline{r}'\right)-\left(\hat{n}_2, \underline{r}'\right)\right)d^3r'\,d^3r$$

Now if we assume that $r >> r'$ *i.e.* the final state is concentrated at ∞ while the potential $V(r')$ is concentrated near the origin, then we can make the field approximation to get

$$\frac{\exp(jk|\underline{r}-\underline{r}'|)}{|\underline{r}-\underline{r}'|} \approx \frac{e^{jkr}}{r}\exp(-jk(\underline{r}-\underline{r}'))$$

and then

$$\langle \psi_{i2} | S(E) | \psi_{i1} \rangle \approx \frac{(2\pi)^{3/2}\bar{C}_2 C_1}{k}\delta\left(\hat{n}_2 - \hat{n}_1\right)$$

$$-\frac{m\varepsilon}{2\pi\hbar^2}\bar{C}_2 C_1\int \exp\left\{-jk\left(\hat{r}, \underline{r}'\right)+\left(\widehat{n_2}, \underline{r}\right)-\left(\widehat{n_1}, \underline{r}'\right)\right\}$$

$$\frac{\exp(jkr)}{r}V(\underline{r}')d^3r'\,d^3r = \frac{(2\pi)^{3/2}\bar{C}_2 C_1}{k}\delta\left(\hat{n}_2 - \hat{n}_1\right)$$

$$-\frac{m\varepsilon\bar{C}_2 C_1}{2\pi\hbar^2}(2\pi)^{3/2}\int\hat{V}\left(k\left(\hat{r}-\hat{n}_1\right)\right)$$

$$\frac{\exp\left(jk\left(r-\left(\hat{n}_2, \underline{r}\right)\right)\right)}{r}d^3r$$

If we don't make the far-field approximation, then

$$\langle \psi_{i2} | S(k) | \psi_{i1} \rangle \approx \frac{(2\pi)^{3/2}\bar{C}_2 C_1}{k}\delta\left(\hat{n}_2 - \hat{n}_1\right)$$

$$-\left(\frac{m\varepsilon}{2\pi\hbar^2}\right)\bar{C}_2 C_1\int\frac{\exp(jk|\underline{r}-\underline{r}'|)}{|\underline{r}-\underline{r}'|}$$

❖❖❖❖❖

[36] Quantization of the Simple Exclusion Model:

$$\mathbb{Z}_N = \{0, 1, 2, ..., N-1\}.$$

$$|X\rangle = \left|X_o X_1 ... X_{N-1}\right> \in (\mathbb{C}^2)^{\otimes N} \cong \mathbb{C}^{2^N}$$

$$X_R \in \{0, 1\}.$$

$$a(\underline{X}, r, s),\ \underline{X} = (X_0, X_1 \ldots X_{N-1}) \in \{0,1\}^N 0 \neq r, s \neq N-1$$

Transition probability amplitudes.

$-i\, X_r(1-X_s)a(\underline{X}, r, s) \simeq$ transition probability amplitude per unit time from $|\underline{X}\rangle \to |\underline{X}\rangle^{(r,s)}$

Where $|X\rangle^{(r,s)}$ is obtained from $|\underline{X}\rangle$ by interchanging X_r and X_s.

So

$$|\psi_{t+\delta t}\rangle = |\psi_t\rangle - i\delta t \sum_{\underline{X}, r, s} X_r(1-X_s)a(\underline{X}, r, s)|\underline{X}\rangle^{(r,s)}\langle\underline{X}|\psi_t\rangle$$

or

$$i\frac{d}{dt}|\psi_t\rangle = \sum_{X, r, s} X_r(1-X_s)a(\underline{X}, r, s)|\underline{X}\rangle^{(r,s)}\langle X|\psi_t\rangle$$

$$= H|\psi_t\rangle$$

$$H = \sum_{X, r, s} X_r(1-X_s)a(\underline{X}, r, s)|\underline{X}\rangle^{(r,s)}\langle\underline{X}|$$

$$H^* = \sum_{X\,r\,s} X_r(1-X_s)\,|\underline{X}\rangle\langle\underline{X}|^{(r,s)}\,\overline{a(\underline{X}, r, s)}$$

$$= \sum_{X\,r\,s} X_r(1-X_r)\,|X\rangle^{(r,s)}\langle\underline{X}|\overline{a(\underline{X}^{(r,s)}, r, s)}$$

For $H^* = H$, we requires

$$X_r(1-X_s)a(X, r, s)|X\rangle^{(r,s)}\langle X|$$

$$+X_s(1-X_r)a(X, s, r)|X\rangle^{(s,r)}\langle X|$$

$$= X_s(1-X_r)\overline{a(X^{(r,s)}, r, s)}|X\rangle^{(r,s)}\langle X|$$

$$+X_r(1-X_s)\overline{a(X^{(r,s)}, r, s)}|X\rangle^{(s,r)}\langle X|$$

$\forall\ r \neq s$. Now $|X\rangle^{(r,s)} = |X\rangle^{(s,r)}, \underline{X}^{(r,s)} = \underline{X}^{(s,r)}$ and hence the condition for Hermitian of H reduces to

$$X_r(1-X_s)a(X, r, s) + X_s(1-X_s)a(X, s, r)$$

$$= X_r(1-X_s)\overline{a(X^{(r,s)}, s, r)} + X_s(1-X_r)\overline{a(X^{(r,s)}, r, s)}\forall r \neq s \forall \underline{X}$$

This can be achieved, for example, by taking $a(\underline{X}, r, s) = a(r, s) \in \mathbb{R} = a(s, r)$ independent of $\underline{X}$.

Another way to achieve this is to take

$$a(\underline{X}, r, s) = a(r, s)\overline{a(s, r)}\forall r \neq s$$

i.e. $((a(r, s)))_{0 \neq r, s \neq N-1}$ is a Hermitian. $N \times N$ matrix independent of X.

A still more general way is to choose a such that

$$a(\underline{X}, \underline{r}, \underline{s}) = \overline{a\left(X^{(r,s)}, s, r\right)} \forall X \in \{0,1\}^N, a \le r, s \le N-1.$$

Consider now

$$-\frac{d}{dt}|\psi_t\rangle = H|\psi_t\rangle = \sum_{r,s,X} X_r(1-X_s)a(r,s)|X\rangle^{(r,s)}\langle X|\psi_t\rangle$$

$$a(r,s) = \overline{a(s,r)}.$$

Thus,

$$i\frac{d}{dt}\langle \underline{Y}|\psi_t\rangle = \sum_{r,s,\underline{X}} X_r(1-X_s)a(r,s)\delta\left[\underline{Y}-\underline{X}^{(r,s)}\right] < \underline{X}|\psi_t\rangle$$

$$= \sum_{r,s,\underline{X}} X_r(1-X_s)a(r,s)\delta\left[\underline{Y}^{(r,s)}-\underline{X}\right] < \underline{X}|\psi_t\rangle$$

$$= \sum_{r,s} X_s(1-Y_r)a(r,s) < Y^{(r,s)}|\underline{\psi}_t\rangle$$

Let $\langle \underline{X}|\underline{\psi}_t\rangle = \psi_t(\underline{X})$. Then we can express the above eqn. as

$$i\frac{d}{dt}\psi_t(\underline{X}) = \sum_{r,s}\psi_t\left(X^{(r,s)}\right)X_s(1-X_r)a(r,s) \tag{1}$$

Probability density and probability current:

$$-i\frac{d\overline{\psi_i(\underline{X})}}{dt} = \sum_{r,s}\overline{\psi_t(\underline{X}^{(r,s)})}X_s(1-X_r)a(s,r)(a(s,r) = \overline{a(r,s)}) \tag{2}$$

From (1) and (2),

$$\frac{d}{dt}|\psi_t(\underline{X})|^2 = \overline{\psi_t(\underline{X})}\frac{d\psi_t(\underline{X})}{dt} + \psi_t(\underline{X})\frac{d\overline{\psi_t(\underline{X})}}{dt}$$

or,

$$i\frac{d}{dt}|\psi_t(\underline{X})|^2 = \overline{\psi_t(\underline{X})}\sum_{r,s}\psi_t(\underline{X}^{(r,s)})X_s(1-X_r)a(r,s)$$

$$-\psi_t(X)\sum_{r,s}\overline{\psi_t\left(X^{(r,s)}\right)}X_s(1-X_r)a(r,s)$$

$$= \sum_{r,s} X_s(1-X_r)2iI_m\left(a(r,s)\overline{\psi_t(\underline{X})}\psi_t(\underline{X}^{(r,s)})\right)$$

Suppose $a(r, r+1) = a(r+1, r) = \alpha \in \mathbb{R}$ and $a(r,s) = 0$ for $|r-s| > 1$. Then we get

$$\frac{d}{dt}|\psi_t(X)|^2 = 2\alpha\sum_r X_{r+1}(1-X_r)I_m\left(\overline{\psi_t(X)}\psi_t^{(r,r+1)}(X)\right)$$

$$+2\alpha\sum_r X_{r+1}(1-X_r)I_m\left(\overline{\psi_t(X)}\psi_t^{(r-1,r)}(X)\right)$$

$$= 2\alpha\sum_r \left(X_r + X_{r+1} - 2X_r X_{r+1}\right) I_m\left(\overline{\psi_t(X)}\,\psi_t^{(r,r+1)}(X)\right)$$

$$= 2\alpha\sum_{r:} I_m\left(\overline{\psi_t(X)}\,\psi_t^{(r,r+1)}(X)\right)$$

$$(X_r, X_{r+1}) = (1,0) \text{ or } (0,1)$$

❖❖❖❖❖

[37] Quantum Version of Nearest Neighbour Interactions:

$\psi_t(X_1,\ldots,X_N) \simeq$ state of the N-particle system at time t. The classical equation

$$dX_i(t) = \left(\psi(X_{i-1}(t)) - 2\psi(X_i(t)) + \psi(X_{i+1}(t))\right)dt + dB_{i-1,i}(t) - dB_{i,i+1}(t) \; (*)$$

Needs to be quantized. We write

$$\langle X_1,\ldots,X_N \mid \psi_t\rangle = \psi_t(X_1,\ldots,X_N)$$

One method to start quantity appears to be

$$\psi_{t+dt}(\underline{X}) = \alpha\psi_t(\underline{X} - \varphi(\underline{X})dt - \sigma(X)\,d\underline{B}(t))$$

where $|\alpha| = 1\;\alpha \in \mathbb{C}$. We write $\alpha = 1 + i\beta\, dt$

so $$|\alpha| = \sqrt{1+\beta^2 dt^2} = 1 + O(dt^2),\; \beta = \beta(X).$$

Comments and remarks on Volume 4. of collected papers of *S.R.S.* Varadhan:

$$dX_i(t) = dZ_{i-1,i}(t) - dZ_{i,i+1}(t)$$

$$dZ_{i,i+1}(t) = (\psi(X_i(t)) - \psi(X_{i+1)}(t))\,dt + dB_{i,i+1}(t),\; \mathrm{i} \in \mathbb{Z}.$$

$$X_N(t) = \sum_{i\in\mathbb{Z}_N} J\left(\frac{i}{N}\right)X_i(t)$$

Forward Kolmogorov operator:

$$dX_i(t) = \left(\psi(X_{i-1}(t)) - 2\psi(X_i(t)) + \psi(X_{i+1}(t))dt + dB_{i-1,i}(t) - dB_{i,i+1}(t)\right).$$

$$\mathcal{L} = \sum_i \left(\psi(X_{i-1}) - 2\psi(X_i) + \psi(X_{i+1})\right)\frac{\partial}{\partial X_i} + \frac{1}{2}\sum_i\left(\frac{\partial}{\partial X_i} - \frac{\partial}{\partial X_{i+1}}\right)^2$$

Note: $$d\,f(X_0,\ldots,X_{N-1}) = \sum_i \frac{\partial f}{\partial X_i}dX_i + \frac{1}{2}\sum_{i,j}\frac{\partial^2 f}{\partial X_i \partial X_i}dX_i\,dX_j$$

$$dX_i\,dX_{i+1} = -(d\,B_{i,i+1})^2 = -dt,$$

$$dX_{i-1}\,dX_i = -(dB_{i-1,i})^2 = -dt,$$

$$dX_i^2 = 2dt$$

A stationary density $p(\underline{X})$ should $\mathcal{L}^*p = 0$. Now let

$$p(\underline{X}) = C\prod_i \exp(\lambda X_i - \phi(X_i))$$

$$£^*p = -\sum_i \frac{\partial}{\partial X_i}\left(\left(\psi(X_{i-1}) - 2\psi(X_i) + \psi(X_{i+1})\right)p\right)$$

$$+\frac{1}{2}\sum_i\left(\frac{\partial}{\partial X_i} - \frac{\partial}{\partial X_{i+1}}\right)^2 p$$

$$= 2\sum_i \psi'(X_i)p - \sum_i\left(\psi(X_{i-1}) - 2\tau(X_i) + \psi(X_{i+1})\right)\frac{\partial p}{\partial X_i}$$

$$+\frac{1}{2}\sum_i\left(\frac{\partial}{\partial X_i} - \frac{\partial}{\partial X_{i+1}}\right)^2 p$$

$$\frac{\partial p}{\partial X_i} = (\lambda - \phi'(X_i))p.$$

$$\left(\frac{\partial}{\partial X_i} - \frac{\partial}{\partial X_{i+1}}\right)p = (\phi'(X_{i+1}) - \phi'(X_i))p$$

$$\left(\frac{\partial}{\partial X_i} - \frac{\partial}{\partial X_{i+1}}\right)^2 p = \left(\phi'(X_{i+1}) - \phi'(X_i)\right)^2 p - \left(\phi''(X_{i+1}) + \phi''(X_i)\right)p$$

So $\mathcal{L}^*p/p = 2\sum_i \psi'(X_i) - \sum_i (\lambda - \phi'(X_i))(\psi(X_{i-1}) - 2\psi(X_i) + \psi(X_{i+1}))$

$$+\sum_i \frac{1}{2}\left(\phi'(X_{i+1}) - \phi'(X_i)\right)^2 - \frac{1}{2}\sum_i\left(\phi''(X_{i+1}) + \phi''(X_i)\right)$$

For this to be zero, we require

$$2\psi'(X_i) - \phi'(X_i)\psi(X_i) + \phi_1^2(X_i) - \phi''(X_i) = 0$$

and $\phi'(X_i)\psi(X_{i-1}) + \left(\phi'(X_{i-1})\right)\psi(X_i) - \phi'(X_{i-1})\phi'(X_i) = 0$

Let $\psi(X) = \alpha\phi'(X)$. Then these conditions reduce to

$$2\alpha\phi''(X) + (-2\alpha - \phi'(X))\phi'(X) + \phi'^2(X) - \phi''(X) = 0$$

There are satisfied provided $(2\alpha - 1)\phi''(X) = (2\alpha - 1)\phi'^2(X)$ and $(2\alpha - 1)\phi'(X)\phi'(Y) = 0$

Thus $\alpha = \frac{1}{2}$. It follows that

$$p(X) = C \cdot \prod_i \exp(\lambda X_i - \phi(X_i))$$

is a stationary density provided $\psi(X) = \frac{1}{2}\phi'(X)$,

i.e. $$\phi(X) = 2\int_o^X \psi(\xi)\,d\xi .$$

Now, $$dX_N(t) = \sum_i J\left(\frac{i}{N}\right) dX_i(t)$$

$$= \sum_{i=0}^{N-1} J\left(\frac{i}{N}\right)(\Psi(X_{i-1}(t)) - 2\psi(X_i(t)) + \psi(X_{i+1}(t))\,dt$$

$$+ \sum_{i=0}^{N-1} J\left(\frac{i}{N}\right)(dB_{i-1,i} - dB_{i,i+1}(t))$$

$$= \sum_i \left(J\left(\frac{i+1}{N}\right) - 2J\left(\frac{i}{N}\right) + J\left(\frac{i-1}{N}\right)\right)\psi(X_i(t))\,dt$$

$$+ \sum_i J\left(\frac{i}{N}\right)(dB_{i-1,i}(t) - dB_{i,i+1})$$

$$\approx \frac{1}{N^2}\sum_i J''\left(\frac{i}{N}\right)\psi(X_i(t)) + \sum_i J\left(\frac{i}{N}\right)(dB_{i-1,i} - dB_{i,i+1})$$

$$\mathbb{E}\left(\sum_i J\left(\frac{i}{N}\right)(B_{i-1,i}(t) - B_{i,i+1}(t))\right)^2$$

$$= \mathbb{E}\left(\sum_i \left(J\left(\frac{i}{N}\right) - J\left(\frac{i-1}{N}\right)\right) B_{i-1,i}(t)\right)^2$$

$$= t\sum_i \left(J\left(\frac{i}{N}\right) - J\left(\frac{i-1}{N}\right)\right)^2$$

$$\approx \frac{t}{N^2}\sum_i J'\left(\frac{i}{N}\right)^2 \approx \frac{t}{N}\int_o^t J'(\theta)^2\,d\theta \to 0 \; N \to \infty .$$

So for $N \to \infty$,

$$dX_N(t) \approx \frac{t}{N^2}\sum_i J''\left(\frac{i}{N}\right)\psi(X_i(t))\,dt$$

Now consider r.v's $\xi_1, \xi_2, \ldots$ which are iid and have logarithmic moment generating function

$$\Lambda(\lambda) = \log\left(\mathbb{E}\, e^{\lambda\xi_1}\right). \text{ Then}$$

$$\mathbb{E}\left[\psi(\xi_1)\middle|\frac{\xi_1+\ldots+\xi_n}{n}=a\right] = \int p(\xi_1\,|\,n,a)\psi(\xi_1)\,d\xi_1$$

Where $p(\xi_1\,|n, a)$ is the pdf of ξ_1 given $\dfrac{\xi_1+\ldots+\xi_n}{n} = a$

Now,
$$p(\xi_1\,|\,n,a) = \frac{p\left(\dfrac{\xi_1+\ldots+\xi_n}{n}=a\,|\,\xi_1\right)p(\xi_1)}{p\left(\dfrac{\xi_1+\ldots+\xi_n}{n}=a\right)}$$

$$= \frac{p\left(\dfrac{\xi_1+\ldots+\xi_n}{n-1}=\dfrac{na-\xi_1}{n-1}\middle|\,\xi_1\right)p(\xi_1)}{p\left(\dfrac{\xi_1+\ldots+\xi_n}{n}=a\right)}$$

$$\approx C\frac{\exp\left(-(n-1)I\left(a-\dfrac{\xi_1}{n-1}\right)\right)p(\xi_1)}{\exp(-nI(a))}$$

$$\approx C\cdot\exp(I(a)+I'(a)\xi_1)\,p(\xi_1) \qquad (n\to\infty)$$

Where $I(a) = \sup_\lambda\,(\lambda a - \Lambda(\lambda))$ by large deviation theory. Thus a $n\to\infty$

$$E\left[\psi(\xi_1)\middle|\frac{\xi_1+\ldots+\xi_n}{n}=a\right] \approx C\cdot\int\exp(I(a)+I'(a)\xi)\,p(\xi)\,\psi(\xi)\,d\xi$$

Note

$$\int\exp(I'(a)\xi)\,p(\xi)\,d\xi = \exp(\Lambda(I'(a)))$$

Now
$$I(a) = \sup_\lambda(\lambda a - \Lambda(\lambda)) = \lambda_o a - \Lambda(\lambda_o)$$

where
$$a = \Lambda'(\lambda_o)$$

Also
$$\lambda(\lambda) = \sup_a(\lambda a - I(a)) = \lambda a_o - I(a_o)$$

where

$$\lambda = I'(a_o).$$

Thus,
$$\Lambda(I'(a_o)) = I'(a_o)a_o - I(a_o)$$

So, $\int\exp(I'(a)\xi)\,p(\xi)\,d\xi = \exp\left(I'(a)a - I(a)\right)$

Thus,

$$\int\exp(I(a)+I'(a)(\xi))\,d\xi = \exp(aI'(a)).$$

So, $$C = \exp(-a\,I'(a))$$

and $\lim_{x\to\infty} \mathbb{E}\left[\psi(\xi_1)\left|\frac{\xi_1+\ldots+\xi_n}{n}=a\right.\right]$

$$= \int \frac{\psi(\xi)\exp(I'(a)\xi)\,p(\xi)\,d\xi}{\exp(a\,I'(a)-I(a))}$$

$$= \int \frac{\psi(\xi)\exp(I'(a)\xi)\,p(\xi)\,d\xi}{\exp(\Lambda(I'(a)))}$$

Now if a $N \to \infty$ the $X_i(t)'s$ converge to the stationary Gibbs distribution $C\cdot\prod_i \exp(\sigma X_i - \phi(X_i)) = \prod_i p(X_i)$

Then $$\Lambda(\lambda) = \log\left\{\frac{\int \exp((\lambda+\sigma)X-\phi(X))\,dX}{\int \exp(\sigma X-\phi(X))\,dX}\right\}$$

We define $$F(\sigma) = \log\int \exp(\sigma X-\phi(X))\,dX$$

and so $$\Lambda(\lambda) = F(\lambda+\sigma)-F(\sigma)$$

$$I(a) = \sup_\lambda(\lambda a - F(\lambda+\sigma)+F(\sigma))$$

$$= \sup_\lambda\big((\lambda+\sigma)a - F(\lambda+\sigma)\big)+F(\sigma)-\sigma a)$$

$$= {}^{\prime}{}_F(a)+F(\sigma)-\sigma c.$$

Where $$I_F(a) = \sup_\lambda(\lambda a - F(\lambda)).$$

Now, $$I'(a) = \theta \text{ satisfies}$$

$$\Lambda(\theta) = \sup_\lambda(\theta x - I(x)) = \theta a - I(a)$$

At equilibrium,

$$\lim_{x\to\infty} \mathbb{E}\left[\psi(X_1)\left|\frac{X_1+\ldots+X_N}{N}=a\right.\right]$$

$$= \int \psi(\xi)\exp\big(I'(a)\xi + I(a) - a\,I'(a) + P(\xi)\,d\xi\big)$$

$$= \int \psi(\xi)\exp(I'_F(a)\xi - \sigma\xi + I_F(a) + F(\sigma) - aI'_F(a) + \sigma a)$$
$$\times \exp(\sigma\xi - \phi(\xi))\,d\xi \,/\, \exp(F(\sigma))$$

$$= \frac{\left[\int \psi(\xi)\exp(I'_F(a)\xi + I_F(a) - \phi(\xi))\,d\xi\right]}{\exp(a\,I'_F(a))}$$

$$= \frac{\left[\int \psi(\xi)\exp(I_F'(a)\xi - \phi(\xi))d\xi\right]}{\exp\left(I_F'(a)a - I_F(a)\right)} \equiv \widehat{\Psi}(A) \text{ say.}$$

❖❖❖❖❖

[38] Remarks and Comments on "Collected Papers of S.R.S Varadhan" Vol. 4, Particle System and Their Large Deviations:

Quantization of the simple exclusion process:

$$H = \sum_{\substack{r \neq s \\ X}} a(r, s) X_r (1 - X_s) |X\rangle^{(r, s)} \langle X|$$

$$a(r, s) = \overline{a(s, r)} \cdot \langle X | \psi_t \rangle = \psi_t(\underline{X}).$$

$$i\frac{d\psi_t(X)}{dt} = (H\psi_t)(\underline{X})$$

$$= \sum_{r \neq s} a(r, s) Y_r (1 - Y_s)\, \delta\left[X - Y^{(r, s)}\right] \psi_t(\underline{Y})$$

$$= \sum_{r \neq s} a(r, s) X_s (1 - X_r)\, \psi_t(\underline{X}^{(r, s)})$$

$$\psi_t : \{0, 1\}^N \to \mathbb{C} \quad N \to \infty$$

$$\sum_{X_1, \ldots, X_{N-1} \in \{0, 1\}} |\psi_t(X)|^2 = p_t(X_0)$$

t distribution of particle at 0th site.

$$i\frac{d}{dt}|\psi_t(X)|^2 = i\frac{d\psi_t(X)}{dt}\overline{\psi}_t(X) - \psi_t(X)\left(\overline{\frac{i\, d\psi_t(X)}{dt}}\right)$$

$$\sum_{r,s} a(r,s) X_s (1 - X_r)\psi.\left(\underline{X}^{(r,s)}\right)\overline{\psi}_t(\underline{X})$$

$$\sum_{s} \overline{a(r,s)} X_s (1 - X_r)\psi_t(\underline{X})\overline{\psi_t\left(X^{(r}\right.}$$

More generally,

$$i\frac{d}{dt}\left\{\psi_t(\underline{X})\overline{\psi_t(\underline{Y})}\right\} = \sum_{r,s} a(r,s) X_s (1 - X_r)\psi_t(\underline{X}^{(r,s)})\overline{\psi_t(\underline{Y})}$$

$$-\sum_{r,s} \overline{a(r,s)}\, Y_s (1 - Y_r)\psi_t(\underline{X})\overline{\psi(\underline{Y}^{(r,s)})}$$

So if $\rho_t(X, Y)$ denotes the density matrix of a mixed state, its Von-Neumann equation of evolution is

$$i\frac{\partial \rho_t(X,Y)}{\partial t}$$

$$= \sum_{r,s}\left[a(r,s)X_s(1-X_r)\rho_t\left(\underline{X}^{(r,s)},\underline{Y}\right)-a(s,r)Y_s(1-Y_r)\rho_t\left(X,Y^{(r,s)}\right)\right]$$

Define $$H_{rs} = \sum_{X\in\{0,1\}^N} a(r,s)X_s(1-X_r)|X\rangle^{(r,s)}\langle X| \quad r\neq s.$$

Then $$H_{rs}^{*} = \sum_{\underline{X}\in\{0,1\}^N} \overline{a(r,s)}X_s(1-X_r)|X\rangle\langle X|^{(r,s)}$$

$$= \sum_{\underline{X}\in\{0,1\}^N} a(r,s)X_r(1-X_s)|X\rangle^{(r,s)}\langle X| = H_{sr}$$

This we can define

$$\tilde{H}_{rs} = H_{rs}+H_{rs}^{*} = H_{rs}+H_{sr}\,,\ 0\le r<s\le N-1$$

and then

$$H = \sum_{r\neq s} H_{rs} = \sum_{a\le r<s\le N-1}\tilde{H}_{rs}$$

Let

$$\varphi(Y_o, X) = (Y_o, X_1\, X_2\, \ldots, X_{N-1}) \equiv (Y_o, \underline{\tilde{X}}_o)$$

Then

$$\underline{\tilde{X}}_o = (X_1, X_2, \ldots, X_{N-1}).$$

$$\sum_{\tilde{X}_o}\rho_t\left(X,(Y_o,\tilde{X}_o)\right) = \rho_t^{(1)}(X_o,Y_o)$$

is the density matrix of the zeroth particle. We write the density evolution as

$$i\frac{\partial \rho_t(X,Y)}{\partial t} = \left[\sum_{r<s}\tilde{H}_{rs},\rho_t\right](X,Y)$$

Consider the term $\left[\tilde{H}_{rs},\rho_t\right](X,Y)$ for $0<r<s\le N-1$
This equals

$$a(r,s)X_s(1-X_r)\rho_t\left(\underline{X}^{(r,s)},\underline{Y}\right)+a(s,r)X_r(1-X_s)\rho_t\left(\underline{X}^{(r,s)},\underline{Y}\right)$$

$$-a(s,r)Y_r(1-Y_s)\rho_t\left(\underline{X},\underline{Y}^{(r,s)}\right)-a(r,s)Y_r(1-Y_r)\rho_t\left(\underline{X},\underline{Y}^{(r,s)}\right)$$

$$= a(r,s)\left\{X_s(1-X_r)\rho_t(X^{(r,s)},\underline{Y})-Y_r(1-Y_s)\rho_t(X,Y^{(r,s)})\right\}$$

$$+a(s,r)\left\{X_r(1-X_s)\rho_t(X^{(r,s)},Y)-Y_s(1-Y_r)\rho_t(\underline{X},\underline{Y})^{(r,s)}\right\}$$

We have $(o<r<s)$ $$\sum_{\tilde{X}_o}\left[\tilde{H}_{rs},\rho_t\right]\left(\underline{X}^{(r,s)},\left(Y_o,\underline{\tilde{X}}_o\right)\right)$$

$$= a(r,s)\sum_{\tilde{X}_o}\Big\{X_s(1-X_r)\rho_t(\underline{X}^{(r,s)},(Y_o,\tilde{\underline{X}}_o))$$

$$-X_r(1-X_s)\rho_t(X,(Y_o,\tilde{X}_o^{(r,s)}))\Big\}$$

$$+a(s,r)\sum_{\tilde{X}_o}\Big\{X_s(1-X_s)\rho_t(\underline{X}^{(r,s)},(Y_o,\tilde{X}_o))$$

$$-X_s(1-X_r)\rho_t(\underline{X},(Y_o,\tilde{X}_o^{(r,s)}))\Big\} = 0$$

Consider now the case $r = 0$ $s > r$. We've to evaluate

$$\sum_{\tilde{X}_o}\left[\sum_{s>o}\tilde{H}_{os},\rho_t\right]\left(X,\left(\left(Y,\tilde{\underline{X}}_0\right)\right)\right.$$

For $s > 0$, we get $\sum_{\tilde{X}_o}[\tilde{H}_{0s},\rho_t](X,(Y_0,\tilde{\underline{X}}_{0,}))$

$$= a(0,s)\Bigg[(1-X_0)\sum_{\tilde{X}_{0,}}X_s\,\rho_t(X^{(0,s)},(Y_{0,}\,\tilde{\underline{X}}_0))$$

$$-Y_o\sum_{\tilde{X}_o}(1-x_s)\rho_t\left(X,(X_{s,}(X_{1,..,}Y_0,..,X_{N-1})\right)\Bigg]$$

5th position.

This we can define

❖❖❖❖❖

[39] Quantum String Theory:

[Reference: Lectures for EC-COE-2012 course.]

$$\mathcal{L} = a_1 g_{\mu\nu}(X)X^{\mu}_{,\tau}X^{\nu}_{,\tau} + a_2\, g_{\mu\nu}(X)X^{\mu}_{,\sigma}X^{\nu}_{,\sigma}$$

$$+a_3\, g_{\mu\nu}(X)X^{\mu}_{,\tau}X^{\nu}_{,\sigma}$$

First consider the special relativistic case: The $g_{\mu\nu} = \eta_{\mu\nu}$.

The Euler - Lagrange eqns.-are

$$\partial_r\frac{\partial\mathcal{L}}{\partial X^{\mu}_{,\tau}} + \partial\sigma\frac{\partial\mathcal{L}}{\partial X^{\mu}_{,\sigma}} = \frac{\partial\mathcal{L}}{\partial X^{\mu}}$$

we find that

$$\frac{\partial\mathcal{L}}{\partial X^{\mu}_{,\tau}} = 2a_1\,\eta_{\mu\nu}\,X^{\nu}_{,\tau} + a_3 z_{\mu\nu}X^{\nu}_{,\sigma}$$

$$\frac{\partial\mathcal{L}}{\partial X^{\mu}_{,\sigma}} = 2a_2\eta_{\mu\nu}\,X^{\nu}_{,\sigma} + a_3\, z_{\mu\nu}\,X^{\nu}_{,\tau}$$

$$\frac{\partial \mathcal{L}}{\partial X^{\mu}} = 0 \text{ so the string equations are}$$

$$2a_1 X^{\mu}_{,\tau\tau} + a_3 X^{\mu}_{,\tau\sigma} + 2a_2 X^{\mu}_{,\sigma\sigma} + a_3 X^{\mu}_{,\tau\sigma} = 0$$

$$a_1 X^{\mu}_{,\tau\tau} + a_2 X^{\mu}_{,\sigma\sigma} + a_3 X^{\mu}_{,\tau\sigma} = 0$$

If $a_3 = 0$ and $a_2 = -a_1$, we get the standard form of the string equations:

$$X^{\mu}_{,\tau\tau} - X^{\mu}_{,\sigma\sigma} = 0.$$

Path integral approach to quantum string theory:

$$\mathcal{L}(X^{\mu}_{,\tau}, X^{\mu}_{,\sigma}) = \frac{1}{2}(X_{1\tau}, X_{1\tau}) - \frac{1}{2}(X_{,\sigma}, X_{,\sigma})$$

$$\triangleq \frac{1}{2} z_{\mu\nu}\left\{X^{\mu}_{,\tau} X^{\nu}_{,\tau} - X^{\mu}_{,\sigma} X^{\nu}_{,\sigma}\right\}$$

Let

$$X^{\mu}(\tau,\sigma) = \sum_{n\in\mathbb{Z}} \xi^{\mu}_{n}(\tau)\exp\left(i2\prod n\sigma\right)$$

$$\xi^{\mu}_{n}(\tau) = \overline{\xi^{\mu}_{-n(\tau)}}\,.$$

$$\int_{o}^{1}\mathcal{L}\,d\sigma = \sum_{n_1\, m\in\mathbb{Z}} \frac{1}{2}(\dot{\xi}_m(\tau), \dot{\xi}m(\tau))\int_{o}^{1}\exp\left(i2\pi(n+m)\sigma\right)d\sigma$$

$$+ Z\pi^{2}\sum(\xi_m(\tau), \xi_m(\tau))\,nm\int_{o}^{1}\exp\left(i2\pi(n+m)\sigma\right)d\sigma$$

$$= \sum_{n,\,m}\left\{\frac{1}{2}\left(\dot{\underline{\xi}}_n, \dot{\underline{\xi}}_n\right) - 2\pi^2 n^2(\xi_n, \xi_{-n})\right\}$$

$$= L\left(\xi^{\mu}_{n}, \dot{\xi}^{\mu}_{n}, n\in\mathbb{Z}_{,}\ 0\le\mu\le d\right)$$

$$= \frac{1}{2}(\dot{\xi}^{2}_{o}\dot{\xi}_{o}) + \sum_{n\ge 1}\left\{\left(\dot{\xi}_{n,}\ \bar{\xi}_n\right) - 2\pi^2 n^2(\xi_n, \bar{\xi}_n)\right\}$$

$$= \frac{1}{2}\left(\dot{\xi}_0, \dot{\xi}_0\right) + \sum_{n\ge 1}\left\{\left(\dot{\xi}_{Rn}, \dot{\xi}_{Rn}\right) + (\xi_{In}, \xi_{In})\right.$$

$$\left. -2\pi^2 n^2\left(\xi_{Rn}, \xi_{Rn}\right) + \left(\xi_{In}, \xi_{In}\right)\right\}$$

where $\xi^{\mu}_{n} = \xi^{\mu}_{Rn} + i\xi^{\mu}_{In}$.

To quantize this, we first revise to quantize a 1-D harmonic oscillator

$$\mathcal{L}(q,\dot{q}) = \frac{1}{2}\dot{q}^2 - \frac{1}{2}w_0^2 q^2 \quad 0\le t\le T$$

$$q(t) = \sum_{n=1}^{\infty}\left(q_{Rn}\cos(nwt) + q_{In}\sin(nwt)\right)_1$$

$$+ q_0 \quad w = \frac{2\pi}{T}.$$

$$q_0, q_{Rn}, q_{In} \in \mathbb{R},$$

$$\frac{1}{2}\int_o^T \dot{q}^2\, dt = \sum_{n=1}^{\infty}\frac{n^2 w^2}{4}\left(q_{Rn}^2 + q_{In}^2\right)$$

$$\frac{1}{2}\int_o^T q^2\, dt = \frac{1}{4}\sum_{n=1}^{\infty}\left(q_{Rn}^2 + q_{In}^2\right) + \frac{1}{2}q_o^2 T$$

$$\therefore \qquad \int_o^T \mathcal{L}\, dt = \frac{-1}{2}q_o^2 T + \frac{1}{4}\left\{\left[\sum_{n\geq 1}\left(n^2w^2 - w_0^2\right)\left(q_{Rn}^2 + q_{In}^2\right)\right]\right\}$$

This expansion, however, does not take care of the boundary condition $q(0) = a$, $q(T) = b$. So we use the half wave Fourier expansion:

$$q(t) = \sum_{n=1}^{\infty} q_n \sin\left(\frac{n\pi t}{T}\right) + \alpha t + \beta$$

$$q(0) = a \Rightarrow \beta = a,\ q(T)\, b \Rightarrow \mathcal{L} = \left(\frac{b-\beta}{\alpha T}\right) = \left(\frac{b-a}{T}\right)$$

$$\int_o^T \dot{q}^2\, dt = \sum_{n\geq 1}\left(\frac{n\pi}{T}\right)^2 \times \frac{T}{2}q_n^2 + \alpha^2 T$$

$$\int_o^T q^2 dt = \frac{T}{2}\sum_{n\geq 1} q_n^2 + \int_o^T (\alpha t + \beta)^2 dt$$

$$+ \sum_{n\geq 1} 2\alpha q_n \int_o^T t\sin\left(\frac{n\pi t}{T}\right) dt + 2\beta\sum q_n\left(\frac{T}{n\pi}\right)(1-(-1)^n)$$

$$\int_o^T t\sin(\gamma t)\, dt = \frac{-t}{\gamma}\cos(\gamma t)\Big|_0^T + \frac{1}{y^2}\sin(\gamma t)\Big|_0^T$$

$$= \frac{-T}{\gamma}\cos(\gamma T) \quad \left(\gamma = \frac{nT}{T}\right)$$

$$= \frac{T}{y}(-1)^{n+1} = \frac{T^2}{n\pi}(-1)^{n+1}$$

The Jacobian for the transformation $\{q(t) : 0 \leq t \leq T\}$ to $\{q_n\}_{n\geq 1}$ is not easy to evaluate since the map $\{q_{(t)}\}_{0\leq 0\leq T} \rightarrow \{q_n\}$ is such that

$$\int_o^T q^2(t)\,dt = ||q||^2 = \frac{I}{2}\sum_{n=1}^{\infty} q_n^2 + \int_o^T (\alpha t+\beta)^2\,dt - 2\sum_{n\geq 1} 2\alpha\,\eta_n q_n$$

Its follows that the map $q(.) \rightarrow \sqrt{\frac{T}{2}}\{q_n\}$

i.e. this map is not unitary

We have

$$\approx \frac{-1}{2}q_o^2 + \frac{1}{4}\sum_{h=1}^{N}\lambda_n\, q_R + q_{In}^2$$

$$\int_o^T \mathcal{L}dt = -\frac{T}{4}{w_0}^2\sum_{n\geq 1} q_n^2 - \frac{1}{2}\int_o^T (\alpha t+\beta)^2\,dt$$

$$+\sum_{n\geq 1}\alpha\,\eta_n q_n - \frac{\beta T}{\pi}\sum_{n\geq 1}\left(\frac{1-(-1)^{n)}\,q_n}{n}\right)$$

$$+\left(\pi^2/4T\right)\sum n^2 q_n^2 + \alpha^2 T/2$$

$$\left(\text{Where } \eta_n = \frac{T^2}{n^{\pi}}(-1)^n\right)$$

$$= -\frac{T}{4}w_o^2\sum_{n\geq 1} q_n^2 + \left((\alpha T+\beta)^3 - \beta^3\right)\Big/3\alpha + \alpha\sum_{h\geq 1}\eta_n q_n$$

$$-\beta\sum_{n\geq 1}\left(\frac{T}{n\pi}\right)(1-(-1)^n)q_n + \left(\pi^2\!\big/_{4T}\right)\sum_{n\geq 1} n^2 q_n^2 + \alpha^2 T/2$$

Let
$$\gamma_n = \alpha\eta_n - \frac{T\beta}{n\pi}(1-(-1)^n),$$

and
$$\xi = \left((\alpha T+\beta)^3 - \beta^3\right)\Big/3\alpha$$

$$\lambda_n = \frac{\pi^2 n^2}{\alpha T} - \frac{w_o^2 T}{2}, n \geq 1,$$

Then
$$S_T\left[\{q_n\}\right]_{a,b} = \frac{1}{2}\sum_{n\geq 1}\lambda_n q_n^2 + \sum_{n\geq 1}\gamma_n q_n + \delta$$

Where
$$\delta = \alpha^2 T/2 + \left((\alpha T+\beta)^3 - \beta^3\right)\Big/3\alpha$$

The path integral is then

$$K_T(b|a) = C\int \exp\left(i_{ST}\left[\{q_n\}\right]_{a,b}\right)\prod_{n\geq 1} dq_n$$

$$S_T\left[\{q_n\}\right]_{a,b} = \frac{1}{2}\sum_{n\geq 1}\lambda_n(q_n + \gamma_{n/2\lambda_n})^2 + \delta - \sum_{n\geq 1}\frac{\gamma_n^2}{8\lambda_n}$$

Now the path integral can be evaluated as an elementary Gaussian integral. The Jacobian:

$$q(t) = \sum_{n=1}^{\infty} q_n \sin\left(\frac{n\pi t}{T}\right) + \alpha t + \beta, 0 \leq b \leq T\,.$$

The term $\alpha t + \beta$ is independent of $\{q_n\}$ and hence can be neglected in the evaluation of the Jacobian.
Consider $t = l\Delta^*$. The writing $\tilde{q}(t) = q(t) - \alpha t - \beta$,

we get $$\tilde{q}(l\Delta) \approx \sum_{n=1}^{N} q_n \sin\left(\frac{n\pi l\Delta}{T}\right) N \text{ large.}$$

Choose $\Delta = T/M$. Then

$$\tilde{q}[l] \triangleq \tilde{q}(l\Delta) = \sum_{h=1}^{N} q_n \sin\left(\frac{\pi nl}{M}\right), 1 \leq l \leq N\,.$$

Let $$M = N,$$

Then $$\tilde{q}[l] = \sum_{h=1}^{N} q_n \sin\left(\frac{\pi nl}{N}\right), 1 \leq l \leq N$$

Now, $$\sum_{h=1}^{N} \sin\left(\frac{\pi nl}{N}\right)\sin\left(\frac{\pi ml}{N}\right)$$

$$= \frac{1}{2}\sum_{l=1}^{N}\left\{\cos\left(\frac{\pi(n-m)l}{N}\right) - \cos\left(\frac{\pi(n+m)l}{N}\right)\right\}$$

Let $n \neq 0$.

$$\sum_{l=1}^{N}\cos\left(\frac{\pi nl}{N}\right) = \text{Re}\sum_{l=1}^{N}\exp\left(\frac{i\pi nl}{N}\right) = \text{Re}\exp\left(\frac{i\pi n}{N}\right)\left[\frac{\exp(i\pi n) - 1}{\left(\exp\left(\frac{i\pi n}{N}\right) - 1\right)}\right]$$

$$= \text{Re}\left[\exp\left(\frac{i\pi}{N}\left(n + \frac{nN}{2} - \frac{n}{2}\right)\right)\frac{\sin(\pi n/2)}{\sin(\pi n/2N)}\right]$$

$$= \cos\left[\frac{\pi n}{2N}(N+1)\right]\frac{\sin[\pi n/2]}{\sin[\pi n/2N]}$$

In the Limit $N \to \infty$, this is close to

$$\cos(\pi n/2)\sin(\pi n/2)\frac{2N}{\pi n}$$

$$= \frac{N}{\pi n}\sin(n\pi) = 0.$$

Note that for finite N, we have exactly

$$\sum_{l=1}^{N}\cos\left(\frac{\pi n l}{N}\right) = \frac{\left[\sin\left(\pi n+\frac{\pi n}{2n}\right)-\sin\left(\frac{\pi n}{2N}\right)\right]}{\sin\left(\frac{\pi n}{2n}\right)}$$

$$= (-1)^n - 1$$

Since $(-1)^{n-m} - (-1)^{n+m} = 0$

We set exactly for all $n \neq m$, n, $m \geq 1$,

$$\sum_{l=1}^{N}\sin\left(\frac{\pi n l}{N}\right)\sin\left(\frac{\pi m l}{N}\right) = 0.$$

❖❖❖❖❖

[40] Quantum Shannon Theory Contd.:

Heisenberg-Weyl operators

$\{|j\rangle\}_{j\in\{0,1,2,\ldots,d-1\}}$ is an ONB for ad dimensional Hilbert space $\mathcal{H}$.

$$Z(z)|j\rangle = \exp\left(\frac{2\pi i\, j z}{d}\right)|j\rangle$$

$$X(x)|j\rangle = |j\otimes x\rangle$$

Heisenbrg-Weyl operators are $X(x)\,Z(z) - 0 \leq x, z \leq d-1$ only on $\mathcal{H}$. Consider

$$|\Phi\rangle^{AB} = \frac{1}{\sqrt{d}}\sum_{i=0}^{d-1}|i\rangle^{A}|i\rangle^{B}$$

$$X(x) = \sum_{x'=0}^{d-1}|x\oplus x'\rangle\langle x'|$$

$$Z(z) = \sum_{x'=0}^{d-1}\exp\left(\frac{2\pi i\, x x'}{d}\right)|x'\rangle\langle x'|.$$

$$X(x)|y\rangle = \sum_{x'}|x\oplus x'\rangle\langle x'\,|\,y\rangle = \sum_{x'}|x\oplus x'\rangle\,\delta_{x'y} = |x\oplus y\rangle.$$

$$X(x)Z(z)|y\rangle = \sum_{x',z'}|x\oplus x'\rangle\langle x'|\exp\left(\frac{2\pi i\, z z'}{d}\right)|z'\rangle\langle z'\,|\,y\rangle$$

$$= \sum_{x', z'} \exp\left(\frac{2\pi i z z'}{d}\right)|x \oplus x'\rangle \, \delta_{x', z'}, \delta_{z', y'}$$

$$= \exp\left(\frac{2\pi i z z'}{d}\right)|x \oplus y\rangle$$

$$Z(z) X(x)|y\rangle = Z(z)|y \oplus x\rangle = \exp\left(\frac{2\pi i z(y+x)}{d}\right)|y \oplus x\rangle$$

$$= \exp\left(\frac{2\pi i xz}{d}\right) X(x) Z(z)|y\rangle$$

So we deduce the Weyl Commutation relations

$$Z(z)\, X(x) = \exp\left(\frac{2\pi i xz}{d}\right) X(x) Z(z)\,.$$

$$\left|\Phi_{x,t}\right\rangle^{AB} \equiv \left(X^A(x) Z^A(z) \otimes I\right)|\Phi\rangle^{AB}$$

$$= \frac{1}{\sqrt{d}} \sum_{y=0}^{d-1} X^A(x) Z^A(z)|y\rangle^A |y\rangle^B$$

$$= \frac{1}{\sqrt{d}} \sum_{y} \exp\left(\frac{2\pi i\, yz}{d}\right)|y \oplus x\rangle^A |y\rangle^B$$

$$\left\langle \Phi_{x', z'} \,|\, \Phi_{x, z}\right\rangle = \frac{1}{d} \sum_{y, y'} \exp\left(\frac{2\pi i}{d}(yz - y' z')\right)$$

$$\times^B \langle y'|^A \langle y' \oplus x' \,|\, y \oplus x\rangle^A |y\rangle^B$$

$$= \frac{1}{d} \sum_{y, y'} \exp\left(\frac{2\pi i}{d}(yz - y' z')\right)^B \langle y' \,|\, y\rangle^{BA} \langle y' \oplus x' \,|\, y \oplus x\rangle^A$$

$$= \frac{1}{d} \sum_{y', y} \exp\left(\frac{2\pi i}{d}(yz - y' z_1' \,|\, \delta_{y', y})\right) \delta y' \oplus x', y \oplus x$$

$$= \delta_{x', x} \frac{1}{d} \sum_{y} \exp\left(\frac{2\pi i}{d} y(z - z')\right)$$

$$= \delta_{x', x}\, \delta_{z', z}$$

Hence $\left\{\left|\phi_{x,z}\right\rangle^{AB} \,|\, 0 \le x, z \le d-1\right\}$.

Form an ONB for $\mathcal{H}^A \otimes \mathcal{H}^B$ ($\dim \mathcal{H}^A = \text{din}\, \mathcal{H}^B = d$)

Alice and Bob share the entangled state

$$|\Phi\rangle^{AB} = \frac{1}{\sqrt{d}}\sum_{x=0}^{d-1}|x\rangle^{A}|x\rangle^{B}$$

Alice applies the operator $X(x)\ Z(z)$ to her side giving $\left|\Phi_{x,t}\right\rangle^{A,B}$ for the state shared by herself and Bob. She them Gravimits her system. A over a noisy quantum channel $\mathcal{U}^{A\to B'}$ to Bob. Bob this has an ensuable of state

$$\left(\mathcal{U}^{A\to B'}\otimes I^{B}\right)\left(\left|\Phi_{x,z}^{AB}\right\rangle,\left\langle\Phi_{x,t}^{AB}\right|\right) x,z\in\{0,1,\ldots,d-1\}$$

Bob can now prepare the state

$$\rho XZB'B = \sum_{x,z}\frac{1}{d^2}|x\rangle\langle x|^{N}\otimes|z\rangle\langle z|^{z}$$

$$\otimes\left(\mathcal{N}^{A\to B'}\otimes I^{B}\left(\left|\Phi_{x,z}^{AB}\right\rangle\left\langle\Phi_{x,z}^{AB}\right|\right)\right)$$

Holevo information of the state:

$$I(XZ;BB') = H(BB') - H(BB'\,|\,XZ)$$

$$\rho B'B = T_{rXZ}\left(\rho^{XZB'B}\right) = \left(\mathcal{N}^{A\to B'}\otimes I^{B}\right)\left(\frac{1}{d^2}\sum_{x,z}\left|\Phi_{x,z}^{AB}\right\rangle\Phi_{x,z}^{AB}\right|\right)$$

Now $\left\{\left|\Phi_{x,z}^{AB}\right\rangle|\,0\le x,z\le d-1\right\}$ is an ONB and

hence $\sum_{x,z}\left|\Phi_{x,z}^{AB}\right\rangle\left\langle\Phi_{x,z}^{AB}\right| = I_{d^2}^{AB}$

Thus, with $\pi^{AB} = \frac{1}{d^2}I_{d^2}^{AB}$

We get $\rho^{B'B} = \pi^{AB}$

and $H(B'B) = \log(d^2) = 2\log(d).$

Also, $H(B'B\,|\,XZ) = \sum_{x,z}\frac{1}{d^2}H\left(\mathcal{N}^{A\to B'}\otimes I^{B}\right)\left(\left|\Phi_{x,z}^{AB}\right\rangle\left\langle\Phi_{x,z}^{AB}\right|\right)$

Now $\left|\Phi_{x,z}^{AB}\right\rangle\left\langle\Phi_{x,z}^{AB}\right| = \left(X^{A}(x)Z^{A}(z)\otimes I^{B}\right)|\Phi\rangle\langle\Phi|\left(Z^{A^*}(z)X^{A^*(x)}\otimes I^{B}\right)$

$$= \frac{1}{d^2}\left(X^{A}(x)Z^{A}(z)\otimes I^{B}\right)$$

$$\sum_{y,y'}|y\rangle^{A}|y\rangle^{BA}\langle y'|^{B}\langle y'|\left(Z^{A^*}(z)X^{A^*}(x)\otimes I^{B}\right)$$

$$= \frac{1}{d^2}\sum_{y,y'}\left(X^{A}(x)Z^{A}(z)\right)|y\rangle^{A}|y\rangle^{BA}\langle y'|Z^{A^*}(z)X^{A^*}(x)^{B}\langle y'|$$

$$= \frac{1}{d^2}\sum_{y,y'} \exp\left(\frac{2\pi i z(y-y')}{d}\right)|y\oplus x\rangle^{AA}\langle y'\oplus x|\otimes|y\rangle^{BB}\langle y'|$$

$$= \frac{1}{d^2}\sum_{y,y'} \exp\left(\frac{2\pi i z(y-y')}{d}\right)|y\rangle^{AA}\langle y'|\otimes|y-x\rangle^{BB}\langle y'-x|$$

$$= \frac{1}{d^2}\sum_{y,y'} |y\rangle^{AA}\langle y'|\otimes Z^B(z)X^B(-x)|y\rangle^{BB}\langle y'|X^{B*}(-x)Z^{B*}(z)$$

$$= \left(I^A\otimes Z^B(z)\right)X^B(-x)(|\Phi\rangle\langle\Phi|)\left(I^A\otimes X^B(-x)^*Z^B(x)^*\right)$$

Now, $\quad X(x)|y\rangle = |y\oplus x\rangle.$

$$\langle z|X(x)|y\rangle = \delta_{z,y\oplus x} = \delta_{z,y\oplus x}$$

$$= \langle y|X(-x)|z\rangle$$

Hence relative to the basis $\{|x\rangle\}$, $X(x)^T = X(-x)$. Also

$$Z(z)|y\rangle = \exp\left(\frac{2\pi iyz}{d}\right)|y\rangle$$

So $Z(z)$ is diagonal relative to the basis $\{|x\rangle\}$.

Hence relative to this basis, $Z(z)^T = Z(z)$.

Thus, we can write the above identity as

$$\left(X^A(x)Z^A(z)\otimes I^B\right)(|\Phi\rangle\langle\Phi|)\left(X^A(x)Z^A(z)\otimes I^{B*}\right)$$

$$= \left(I^A\otimes\left(X^B(x)Z^B(z)\right)^T\right)(|\Phi\rangle\langle\Phi|)\left(\overline{I^A\otimes X^B(x)Z^B(z)}\right)$$

Sphere packing Lamma (as a point step towards Shannon's noisy codiny theorem in the quantum setting).

❖❖❖❖❖

[41] Basics of Cq (Classical Quantum) Information Theory

Cq. Codes $x\to\rho_x$

$m \simeq$ message that Alice wishes to transmits to Bob. $Cm \simeq$ Code ward used to transmit m.

m = 1, 2, ..., $|\mu|$. ($|\mu|$ message to be transmitted) $C = C_m|m$ = 1, 2, .., $|\mu|$ are $|\mu|$ independent r.v.s, (random code)

$p(C) = \prod_{m=1}^{|\mu.|} p_X(C_m) \simeq$ probability of random code. When Alice transmits m_1 Bob receives that state σ_{Cm} and he applies that meaurement operate Λ_m. Let M be the message transmitted by Alice using the random code C_m and M' the message received by Bob. Then

$$P\{M' = m| M = m\} = T_r(\Lambda_m \sigma_{Cm})$$

$$\{p_X(x), \sigma_x | x \in X\}$$

Alice's codeward $C_m \in \chi$

$$C_m = (C_m(1), ..., C_m(n_1)), \text{ than}$$

$$\sigma_{Cm} = \underset{j=1}{\otimes} \quad \sigma_{Cm(j)}$$

Packing Lemma:

Assumption

$$T_r\left(\Pi\sigma_x |\geq| -\varepsilon\right.$$

$$T_r\left(\Pi_x \sigma_x |\geq| -\varepsilon\right.$$

$$T_r\left(\Pi_x\right) \leq d$$

$$\Pi\sigma\Pi < \frac{1}{D}\Pi$$

Then $\exists$ *a POVM* $(\Lambda_m)_{m\in\mu}$ such that

$$\mathbb{E}\left\{\frac{1}{|\mu|}\sum_{m\in\mu} T_r(\Lambda_m \sigma_{Cm})\right\} \geq 1 - 2\left(\varepsilon + 2\sqrt{\varepsilon}\right) - 4\left(\frac{D}{d|\mu|}\right)^{-1}$$

$$\sigma = \sum_{x\in\chi} p_X(x)\sigma_x$$

$$T_r(\Pi\,\sigma\,\Pi) = T_r(\sigma\Pi) = \sum_x p_X(x) T_r(\sigma_x\Pi) \geq 1 - \varepsilon$$

Hence $\frac{1}{D}T_r(\Pi) \geq 1-\varepsilon$, or $T_r(\Pi) \geq (1-\varepsilon)D$

The assumption can be interpreted as "there is a large probability $T_r(\Pi_x\ \sigma_x)$ of correct detection using Π_x and the minimum at amount of dimension used $T_r(\Pi)$ for getting such a correct decision must be $\approx D$. To get a correct decision, Π_x should be small when X is transmitted, *i.e.*, it should not "leak" into the other code words.

Gentle operator lemma: Let $0 \leq \Lambda \leq I(POVM)$

Suppose $T_r(\Lambda\rho| \geq 1-\varepsilon$

$$(\|X\|_1 = T_r((X^* X)^{1/2}) \triangleq T_r(|X|)$$

i.e. probability of detecting ρ is high.

Then

$$\|\rho - \sqrt{\Lambda}\rho\sqrt{\Lambda}\|_1 \neq 2\sqrt{\varepsilon}$$

$$\|\rho - \sqrt{\Lambda}\rho\sqrt{\Lambda}\|_1 \leq \|I - \sqrt{\Lambda}\rho\|_1 + \|\sqrt{\Lambda}\rho(I - \sqrt{\Lambda})\|_1$$

$$= T_r\left\{|(I-\sqrt{\Lambda})\sqrt{\rho}\sqrt{\rho}|\right\}$$

$$+T_r\left|\sqrt{\Lambda}\sqrt{\rho}\sqrt{\rho}(I-\sqrt{\Lambda})\right|+T_r\left\{\left|\sqrt{\Lambda}.\sqrt{\rho}\sqrt{\rho}(I-|\sqrt{\Lambda}|)\right|\right\}\Bigg\|$$

Notes: (1)

$$T_r\{|AB|\} = T_r\{(ABB^*A)^{1/2}\}$$

$BB^* = UDU^*$ $D \simeq$ diagonal + ve definity $U \simeq$ unitary. So

$$T_r\{(ABB^*A)^{1/2}\} = T_r\{(A\cup D^2\cup^* A)^{1/2}\}$$

$$= T_r\{U^*(A\cup D^2\cup^* A)^{1/2}U\}$$

$$= T_r\{(U^*A\cup D^2U^*A^*U)^{1/2}\}$$

$$= T_r\left\{\left(XD^2X^*\right)^{1/2}\right\} \quad X = U^*AU$$

Shannon's Capacity Theorem (Clamical). $m \simeq$ message transmitted by Alice through encoder E^n.

Transmitted Code is $E^n(m)$. $\mathcal{N}^n \simeq$ channel between Alice and Bob. Code received by Bob $\simeq \mathcal{N}^n$. $E^n(m)$. Decoder used by Bob $\simeq D^n$. Message decoded by Bob is $D^n \text{ o } \mathcal{N}^n \text{ o } E^n(m)$. Error probability (maximum)

$$p_e = \max_m P\left\{D^n o\, \mathcal{N}^n o\, E^n(m) \neq m\right\} < \varepsilon \text{ is required.}$$

11D channel $P_{Y^n|X^n}\left(y^n|x^n\right) = \prod_{j=1}^{n} p\left(y_j x_j\right)Y|X$

Message slopace $\mathcal{H}$. Transmitted Code alphabet $\simeq \chi$.

Encoder $\qquad E^N : \mathcal{N} \to \chi^n$

Received Code alphabet $\simeq y$. $X \in \chi, y \in Y$. Channel $\mathcal{N}^n \equiv p_{Y^n|X^n}\left(y^n | x^n\right)$ $x^n \in \chi^n \quad y^n \in Y^n$.

Decoder: $\qquad D^n : Y^n \to \mathcal{N}$.

Code rate $\qquad \dfrac{\log|\mu|}{n} \equiv R.$

$(n, R, \in) \simeq$ Channel Code.

How error during decoding can occur:

(a) $y^n \notin T_\delta^{Y^n}$

(b) $y^n \in T_\delta^{Y^n}, y^n \notin T_\delta^{Y^n|x^n(m)}$

(c) $y^n \in T_\delta^{Y^n}, y^n \in T_\delta^{Y^n|x^n(m')x^n(m)=E^n(m)}$ for some $m' \neq m$.

Note: Decoding D^n is performed as follows

$$Y^n = \mathcal{N}^n\left(x^n_{(m)}\right) = \mathcal{N}^n o\, E^n(m) \text{ is what Bob receives.}$$

If $Y^n \in T_\delta^{Y^n}$ error is reported. If $Y^n \in T_\delta^{Y^n|x^n(m)}$ and $Y^n \notin T_\delta^{Y^n|x^n(m')} \forall m' \neq m$ Bob declares that m is the transmitted message. If there are multiple message m' such that $Y^n \in T_\delta^{Y^n|x^n(m')}$ Bob reports error.

Random Code $m \to X^n(m) \in \chi^n$ is a random variable.

$$I_B \simeq \text{ inductor of } B.$$

Error of type (a) $y^n \in 1 - I_{T\delta}^{Y^n}$

Error of type (b) $y^n \in I_{T\delta}^{Y^n}\left(1 - I_{T\delta}^{Y^n|x^n(m)}\right)$

Error of type (c) $y^n \in \sum_{m' \neq m} I_{T\delta}^{Y^n} I_{T\delta}^{Y^n|x^m(m')}$

❖❖❖❖❖

[42] Typical Sequences and Error Probability in Classical and Quantum Coding:

Expected error probability for the random code. $\bar{p}_e = \bar{p}_e^{(a)} + \bar{p}_e^{(b)} + \bar{p}_c^{(c)}$

$$\begin{aligned}
\bar{p}_e^{(a)} &= \mathbb{E}\,[1 - I_{T\delta}Y^n(Y^n)] = \mathbb{E}_{X^n(m)}\mathbb{E}\,[1 - I_{T\delta}Y^n \mid (X^n(m))] \\
&= \mathbb{E}_{X^n(m)}\left[1 - P\left\{Y^n \in T_\delta^{Y^n} \middle| X^n(m)\right\}\right] \\
&= P\left\{Y^n \notin T_\delta^{Y^n}\right\} < \varepsilon
\end{aligned}$$

By definition of δ-typical set. Here

$$\varepsilon = \frac{\text{Var}\log p_Y(g)}{n\delta^2}$$

$$p_Y(9) = \sum_{x \in X} P_{Y|X}(Y \mid X)\, p_C(x)$$

Here the random code C is selected so that for each message m, $x^n(m)$ has the $pdf \prod_{j=1}^{n} p_C(x_j(m))$

$$\begin{aligned}
\bar{p}_e^{(b)} &= \mathbb{E}_{X^n(m)}\mathbb{E}\left[I_\delta^{Y^n}(Y^n)\left(1 - I_{T_\delta}^{Y^n}\middle|X^n(m)(Y^n)\right)\middle|X^n(m)\right] \\
&= p\left\{Y^n \in T_\delta^{Y^n}\right\} - \mathbb{E}_{X^n(m)}\left[P\left\{Y^n \in T_\delta^{Y^n}, Y^n \in T_\delta^{Y^n|X^n(m)}\middle|X^n(m)\right\}\right]
\end{aligned}$$

$$\leq 1-\mathbb{E}_{X^n(m)} P\left\{Y^n \in T_\delta^{Y^n|X^n(m)} \middle| X^n(m)\right\}$$

$$= \mathbb{E}_{X^n(m)} P\left\{Y^n \notin T_\delta^{Y^n|X^n(m)} \middle| X^n(m)\right\} \leq \varepsilon'$$

$$\varepsilon' = \mathbb{E}_{X(m)}\left[\mathrm{Var}(\log p_{Y|X}(Y \mid X(m)))\right] \Big| / n\delta^2 \Big| X(m)$$

$$= \frac{1}{n\delta^2} \sum_{x\in\chi} p_C(x)\left[\mathbb{E}\left(\left(\log p_{Y|X}(Y \mid X)\right)^2 \mid x\right) - \left(\mathbb{E} \log p_{Y|X_{|x}}(Y \mid X)\right)^2\right]$$

where

$$\mathbb{E}\left\{\left(\log p_{Y|X}(Y \mid X)\right)^2 x\right\} = \sum_{y\in Y} p_{Y|X}(y \mid x)(\log p_{Y|X}(y \mid x))^2$$

$$\mathbb{E}\left[\log p_{Y|X}(y \mid x) \mid x\right] = \sum_{y\in Y} p_{Y|X}(y \mid x) \log p_{Y|X}(y \mid x)$$

Finally,

$$\bar{p}_e^{(c)} \leq \sum_{m'\neq m} \mathbb{E}_{(X^n(m))}\left[P\left\{y^n \in T_\delta^{Y^n}, y^n \in T_\delta^{Y^n} \middle| X^n(m')\right\} \middle| X^n(m), X^n(m')\right]$$

(union bound).

$$p_X(x) \equiv p_C(x)$$

$$= \sum_{x^n(m),\, x^n(m'),\, m'\neq m} p_{X^n}(x^n(m))\, p_{X^n}(x^n(m'))$$

$$y^n \in T_\delta Y^n T_\delta^{\cap} Y^n \mid x^n(m')\, p_{Y^n|X^n}(y^n \mid x^n(m))$$

Note: Random coding implies that $X^n(m)$ and $X^n(m')$ are independent for m $\neq m'$ and

$$p_{X^n(m)}(x^n) = \prod_{j=1}^{n} p_C(x_j) = \prod_{j=1}^{n} p_X(x_j) = p_{X^n}(x^n).$$

i.e. $\{X^n(m) \mid m \in \mu\}$ are iid random vectors in χ^n with each $X^n(m)$ having the product density.

Now $$\sum_{x^n(m)} p_{X^n}(x^n(m))\, p_{Y^n|X^n}(y^n \mid x^n(m)) = p_{Y^n}(y^n).$$

Thus, $$\bar{p}_e^{(c)} \leq \sum_{\substack{m\neq m',\, x^n(m'),\, y^n \\ y^n \in T_\delta^{Y^n} \cap T_\delta^{Y^n|x^n(m')}}} p_{Y^n}(y^n)\, p_{X^n}(x^n(m'))$$

Now, $$y^n \in T_\delta^{Y^n}$$

$$\Rightarrow \qquad \left|\frac{1}{n}\sum_{j=1}^{n}\log p_Y(y_j)+H(Y)\right|<\delta$$

$$\Rightarrow \qquad p_{Y^n}(y^n)<e^n(\delta-H(Y))$$

We have $\quad y^n \in T_\delta^{Y^n}\Big|X^n(m')$

$$\Leftrightarrow \qquad \left|\frac{1}{n}\sum_{j=1}^{n}\log p_{Y|X}(y_j\,|\,x_j(m'))+H(Y\,|\,X)\right|<\delta$$

Note that $(x_j(m'), y_j)$, $j = 1, 2, .., n$ are iid r.v's with pdf $P_{Y|X}(y_j \mid x_j(m')) \cdot p_X(x_j(m'))$

Thus,

$$1 \geq \sum_{x^n(m'),\, y^n \in T_\delta^{Y^n|x^n(m')}} p_{X^n}(x^n(m'))\, p_{Y^n|X^n}\left(y^n\big|x^n(m')\right) yn \in T_\delta^{y^n|x^n(m')}$$

$$\geq \sum_{x^n(m'),\, y^n \in T_\delta^{Y^n|x^n(m')}} p_{X^n}(x^n(m'))\exp(-n(\delta+H(Y\,|\,X)))$$

$$= \sum_{x^n(m')} p_{X^n}\left(x^n(m')\,|\,T_\delta^{Y^n|x^n(m')}\right)\exp(-n(\delta+H(Y\,|\,X)))$$

so $$\sum_{x^n(m')} p_{X^n}(x^n(m'))\left|T_\delta^{Y^n|x^n(m')}\right| \leq \exp(n(\delta+H(Y\,|\,X)))$$

Thus, $$\bar{P}_e^{(c)} = \exp(n(\delta-H(Y)))\times \sum_{\substack{m\neq m' \\ x^n(m'), \\ y^n\in T_\delta^{Y^n|x^n(m')}}} p_{X^n}(x^n(m'))$$

$$= \exp(n(\delta-H(Y)))\sum_{\substack{x^n(m') \\ m\neq m'}} p_{X^n}(x^n(m'))\left|T_\delta^{Y^n|x^n(m')}\right|$$

$$\leq \exp(n(2\delta-H(Y)+H(Y\,|\,X)))\,|\,\mu$$

$$= \exp(n(2\delta-I(X,Y)))\,|\,\mu$$

This can be made arbitrary small as $n \to \infty$ iff

$$\frac{1}{n}\log|\mu_1|-I(X,Y)+2\delta<0$$

or $$\frac{1}{n}\log|\mu_1|-I(X,Y)-2\delta$$

i.e. the bit rate $R = \frac{1}{n}\log|\mu_1| < I(X, Y)$ implies that information can be transmitted reliably over the chamber by choosing an appropriate code. This is Shannon's second coding theorem.

Further problems related to quantum Shannon theory.

(Mark M. Widle, Quantum Information theory).

Let
$$\rho^{X^nY^n} = \sum_{\substack{x^n \in X^n \\ y^n \in Y^n}} p_{X^nY^n}(x^n, y^n)\left|x^n \otimes y^n\right\rangle\left\langle x^n \otimes y^n\right|$$

Then,
$$\rho^{X^n} \equiv T_{r_{Y^n}}\left(\rho^{X^nY^n}\right) \equiv \sum_{x^n \in X^n} p_{X^n}(x^n)\left|x^n\right\rangle\left\langle x^n\right|$$

$$\rho^{Y^n} \equiv T_{r_{Y^n}}\left(\rho^{X^nY^n}\right) \equiv \sum_{y^n \in Y^n} p_{Y^n}(y^n)\left|y^n\right\rangle\left\langle y^n\right|$$

Where
$$p_{X^n}(x^n) = \sum_{y^n} p_{X^nY^n}(x^n, y^n),$$

$$p_{Y^n}(y^n) = \sum_{x^n} p_{X^nY^n}(x^n, y^n)$$

$$\prod_\delta^{X^nY^n} = \sum_{(x^n, y^n) \in T_\delta^{X^nY^n}} \left|x^n \otimes y^n\right\rangle\left\langle x^n \otimes y^n\right|$$

$\simeq$ Typical projector (δ) of $\rho^{X^nY^n}$. Here

$$T_\delta^{X^nY^n} = \left\{\left|(x^n, y^n)\frac{1}{n}\log p_{X^nY^n}(x^n, y^n) + H(X, Y) < \delta\right|\right\}$$

Here, we are assuming

$$p_{X^nY^n}^{(x^n, y^n)} = \prod_{j=1}^{n} p_{XY}(x_j, y_j)$$

$$x^n = (x_j)_{j=1}^n, y^n = (y_j)_{j=1}^n$$

Thus,
$$p_{X^n}(x^n) = \prod_{j=1}^{n} p_X(x_j),$$

$$p_{Y^n}(y^n) = \prod_{j=1}^{n} p_Y(y_j)$$

where
$$p_X(x) = \sum_{y \in Y} p_{XY}(x, y), p_Y(y) = \sum_{x \in \chi} p_{XY}(x, y)$$

Then, $$T_\delta^{X^nY^n} = \left\{(x^n, y^n) \left\| \frac{1}{n}, \sum_{j=1}^{n} \log p(x_j, y_j) + H(X,Y) \right| < \delta \right\}$$

Thus,

$$\prod_\delta^{X^nY^n} \left(\rho^{X^n} \otimes \rho^{Y^n}\right) = \sum_{(x^n,\, y^n) \in T_\delta^{X^nY^n}} \left| x^n \otimes y^n \right\rangle \left\langle x^n \otimes y^n \right| \rho^{X^n} \otimes \rho^{Y^n}$$

and $$\xi \underline{\underline{\Delta}}\; T_r\left(\Pi_\delta^{X^nY^n}\left(\rho^{X^n} \otimes \rho^{Y^n}\right)\right)$$

$$= \sum_{(x^n,\, y^n) \in T_\delta^{X^nY^n}} \left\langle x^n \middle| \rho^{X^n} \middle| x^n \right\rangle \left\langle y^n \middle| \rho^{Y^n} \middle| y^n \right\rangle$$

$$= \sum_{(x^n,\, y^n) \in T_\delta^{X^nY^n}} p_{X^n}(x^n)\, p_Y{}^n(y^n) \tag{1}$$

Now $$(x^n, y^n) \in T_\delta^{X^nY^n}$$

$$\Rightarrow \quad \frac{1}{n} \log p_{X^nY^n}(x^n, y^n) < \delta - H(X,Y)$$

$$\Rightarrow \quad \frac{1}{n} \sum_{j=1}^{n} \log p_{XY}(x_j, y_j) < \delta - H(X,Y)$$

$$\Rightarrow \quad p_{X^nY^n}(x^n, y^n) = \prod_{j=1}^{n} p_{XY}(x_j, y_j) < \exp(n(\delta - H(X,Y))) \tag{2}$$

$$\xi = \sum_{(x^n, y^n) \in T_\delta^{X^nY^n}} \exp\left(n \left\{ \frac{1}{n} \sum_{j=1}^{n} \log p_X(x_j) + \frac{1}{n} \sum_{j=1}^{n} \log p_Y(y_j) \right\} \right)$$

Let

$$T_\delta^{X^n} = \left\{ x^n \in \chi^n \left\| \frac{1}{n} \sum_{j=1}^{n} \log p_X(x_j) + H(X) \right| < \delta \right\}$$

$$T_\delta^{Yn} = \left\{ y^n \in Y^n \left\| \frac{1}{n} \sum_{j=1}^{n} \log p_Y(y_j) + H(Y) \right| < \delta \right\}$$

Let $$\tilde{T}_\delta^{X^n} = T_\delta^{X^n} \times y^n, \; \tilde{T}_\delta^{Y^n} = \chi^n \times T_\delta^{Y^n}$$

So $$\eta \equiv \sum_{\substack{(x^n,\, y^n) \in T_\delta^{X^nY^n} \\ \cap \tilde{T}_n^{X^n} \cap \tilde{T}_\delta^{Y^n}}} \left\langle x^n \middle| \rho^{X^n} \middle| x^n \right\rangle \left\langle y^n \middle| \rho^{Y^n} \middle| y^n \right\rangle$$

$$= T_r\left(\prod_\delta^{X^nY^n}\left(\rho^{X^n}\Pi_\delta^{X^n}\otimes\rho_\delta^{Y^n}\Pi_\delta^{Y^n}\right)\right)$$

$$= T_r\left(\prod_\delta^{X^nY^n}\left(\Pi_\delta^{X^n}\otimes\Pi_\delta^{Y^n}\right)\left(\rho_\delta^{X^n}\otimes\rho_\delta^{Y^n}\right)\right)$$

$$\eta = \sum_{\substack{(x^n,y^n)\in T_\delta^{X^nY^n}\\ \cap\tilde{T}_\delta^{X^n}\cap\tilde{T}_\delta^{Y^n}}} \exp\left(n\left\{\frac{1}{n}\sum_1^n \log p_X(x_j)+\frac{1}{n}\sum_1^n \log p_Y(y_j)\right\}\right)$$

$$\leq \#\left(\tilde{T}_\delta^{X^nY^n}\cap\tilde{T}_\delta^{X^n}\cap\tilde{T}_\delta^{Y^n}\right)\exp((2\delta-H(X)-H(Y))^n)$$

(since $\left(x^n,y^n\right)\in\tilde{T}_\delta^{X^n} \Rightarrow x^n\in T_\delta^{X^n} \Rightarrow \frac{1}{n}\sum_1^n \log p_X(x_j)<\delta-H(X)$,

$$\left(x^n,y^n\right)\in\tilde{T}_\delta^{Y^n} \Rightarrow y^n\in T_\delta^{Y^n} \Rightarrow \frac{1}{n}\sum_1^n \log p_Y(x_j)<\delta-H(Y)$$

Thus,

$$n \leq \left(T_\delta^{X^nY^n}\right)\cdot(\exp(2\delta-H(X)-H(Y))n).$$

Now by Chebyshev's inequality,

$$p_{X^nY^n}(T_\delta^{X^nY^n}) \geq 1-\frac{\mathrm{Var}\log(p_{XY}(x,y))}{n\delta^2}$$

i.e.

$$1 \geq \sum_{(x^n,y^n)\in T_\delta^{X^nY^n}} p_{X^nY^n}(x^n,y^n)\geq 1-\frac{\lambda}{n\delta^2}$$

$$\lambda = \mathrm{Var}(\log p_{XY}(x,y))$$

$$\left|T_\delta^{X^nY^n}\right| \min_{\substack{(x^n,y^n)\in\\ T_\delta^{X^nY^n}}} p_{X^nY^n}(x^n,y^n) \leq 1$$

But $\quad p_{X^nY^n}(x^n,y^n) > \exp(-n(\delta+H(X,Y)))\,\forall(x^n,y^n)\in T_\delta^{X^nY^n}$

Thus, $\quad \left|T_\delta^{X^nY^n}\right| < \exp(n(\delta+H(X,Y)))$

and hence

$$\eta < \exp(n(3\delta+H(X,Y)-H(X)))$$

$$= \exp(n(3\delta-I(X,Y)))$$

Also

$$\zeta = \sum_{\substack{(x^n, y^n) \\ \in \cap \tilde{T}_\delta^{X^{nc}} \cap \tilde{T}_\delta^{Y^{nc}}}} \exp\left(n\left(\frac{1}{n}\sum_1^n \log p_X(x_j) + \frac{1}{n}\sum_1^n \log p_Y(y_j)\right)\right)$$

$$= \sum_{\substack{(x^n, y^n) \\ \in \cap \tilde{T}_\delta^{X^{nc}} \cup \tilde{T}_\delta^{Y^{nc}}}} p_{X^n}(x^n)\, p_{Y^n}(y^n)$$

$$\leq \sum_{x^n \in T_\delta^{X^{nc}}} p_{X^n}(x^n) + \sum_{y^n \in T_\delta^{Y^n}} p_{Y^n}(y^n)$$

$$< \frac{\operatorname{Var}\log p_X(x) + \operatorname{Var}\log p_Y(y)}{n\delta^2}$$

$$\equiv \frac{(\lambda_1 + \lambda_2)}{n\delta^2}$$

by Chebyshev's inequality.

Thus $$\xi \leq \eta + \zeta < \exp(n(3\delta - I(X,Y))) + \frac{(\lambda_1 + \lambda_2)}{n\delta^2}$$

and hence for sufficiently large $n, (n > N(\delta))$

$$\xi \neq \exp(n(4\delta - I(X, Y)))$$

Typical sequence

$x^n \in A^n$ $x \in A$ has probability distribution $p(x)$, $x \in A$.

$N(x|x^n)$ is the number of the x appears in x^n.

$$p^n(x^n) = \prod_{x \in A} p(x)^{N(x|x^n)} = \exp\left(\sum_{x \in A} N(x \mid x^n) \log p(x)\right)$$

$$t_{x^n}(x) = \frac{N(x \mid x^n)}{n}$$, t_{x^n} is a probability distribution on A.

So, $$p^n(x^n) = \exp\left(n \sum_{x \in A} t_{x^n}(x) \log p(x)\right) \frac{1}{n}\log p^n(x^n)$$

$$= \sum_{x \in A} t_{x^n}(x) \log p(x) = -D(t_{x^n} \mid p) - H(t_{x^n})$$

or $$-\frac{1}{n}\log p^n(x^n) = D(t_{x^n} \mid p) + H(t_{x^n})$$

where $$D(q \mid p) = \sum_{x \in A} q(x) \log\left(\frac{q(x)}{p(x)}\right)$$

is the Killback distance between the pdf's q and p.

Let t be a type on A^n *i.e.* a pdf on A of the form

$$t(x) = \frac{N(x)}{n}$$

$$\forall x \in A$$

Where $\quad N(x)\{0, 1, 2, ..., n\} \ \forall x \in A.$

Let $$T_t^n = \left\{ x^n \in A^n \left| \frac{N(x \mid x^n)}{n} = t(x) \forall x \in A \right. \right\}$$

Then $$t^n(T_t^n) = \sum_{x^n \in T_t^n} t^n(x^n) = \sum_{x^n \in T_t^n} \left\{ \prod_{x \in A} t(x)^{N(x|x^n)} \right\}$$

$$= \sum_{x^n \in T_t^n} \exp\left(\sum_{x \in A} N(x \mid x^n \mid \log t(x) \right)$$

$$= \sum_{x^n \in T_t^n} \exp\left(n \sum_{x \in A} t(x) \log t(x) \right) = |T_t^n| \exp(-nH(t))$$

Note that t^n is a pdf on A^n and hence we derive the inequality

$$\left| T_T^N \right| \leq \exp(nH(t))$$

or $$\frac{1}{n} \log \left| T_T^N \right| \leq H(t).$$

Let t' be another type. Then

$$t'^n \left(T_t^n \right) = \sum_{x^n \in T_t^n} t^m(x^n) = \sum_{x^n \in T_t^n} \prod_{x \in A} t'(x)^{nt(x)}$$

$$= |T_t^n| \exp\left(n \sum_{x \in A} t(x) \log t'(x) \right)$$

$$= |T_n^t| \exp(n(-D(t \mid t') - H(t)))$$

i.e. $$|T_t^n| = t'^n(T_t^n) \exp(n(D(t \mid t') + H(t)))$$

* Problem: Find a lower bound on $t^n(T_t^n)$ for large n.

Consider $$1 = \sum_t t'(T_t^n)$$

Now the number of types t is the number of ways in which n can be expressed as

$$n = r_1 + r_2 + ... + r_\chi \qquad \chi = |A|.$$

$$r_j \geq 0.$$

This number is $\binom{n+\chi-1}{n}$

Thus, $$1 = \sum_t t'\left(T_t^n\right) \leq \left[\max_t t'\left(T_t^n\right)\right]\binom{n+\chi-1}{n}$$

and hence $$|T_t^n| \geq t'^n(T_t^n)\exp(n(D(t\,|\,t') + H(t)))\,\forall t'$$

implies $$\max_t |T_t^n| \geq \max_t \exp(n\,H(t))\binom{n+\chi-1}{n}^{-1}$$

Channel capacity theorem of Shannon. $X \simeq$ alphabet (finite) $T_\delta^{X^n} \simeq \delta$ typical sequences.

$$T_\delta^{Xn} = \left\{x^n \in X^n \left\| \frac{1}{n}\sum_{j=1}^{n} \log p(x_j) + H(p)\right| < \delta\right\}$$

Here $x^n = (x_j)_{j=1}^n$ $\{p(x), x \in X\}$ is the source alphabet pdf.

Let $$p^n(x^n) = \prod_{j=1}^{n} p(x_j),$$

$$p'^n(x_n) = \frac{p^n(x^n)}{\sum_{x^n \in T_\delta^{X^n}} p^n(x^n)}, \begin{cases} & x^n \in T_\delta^{X^n} \\ 0 & x^n \in X^n \setminus T_\delta^{X^n} \end{cases}$$

$$\sum_{x^n \in X^n} \left|p'^n(x^n) - p^n(x^n)\right| \sum_{x^n \in T_\delta^{X^n}} \left|\frac{p^n(x^n)}{\sum_{\xi^n \in T_\delta^{X^n}} p^n(\xi^n)} - p^n(x^n)\right| + \sum_{x^n \in X^n \setminus T_\delta^{X^n}} p^n(x^n)$$

$$= \frac{\sum_{\xi^n \in T_\delta^{X^{nc}}} p^n(\xi^n)}{\sum_{\xi^n \in T_\delta^{X^n}} p^n(\xi^n)} \sum_{x^n \in T_\delta^{X^n}} p^n(x^n) + \sum_{x^n \in X^n \setminus T_\delta^{X^n}} p^n(x^n)$$

$$= \frac{p^n\left(T_\delta^{X^{nc}}\right)p^n\left(X_\delta^n\right)}{p^n\left(T_\delta^{X^n}\right)} + p^n\left(T_\delta^{X^{nc}}\right)$$

$$= 2p^n\left(X_\delta^{nc}\right) \leq \frac{2\,\mathrm{Var}(\log p(x))}{n\delta^2} \equiv \varepsilon.$$

By Chebyshev's theorem.

Thus, $$\sum_{x^n \in X^n} \left| p'^n(x_n) - p^n(x_n) \right| \xrightarrow[n \to \infty]{} 0$$

Let X'^n be a r.v. (with values in X^n) having pdf p'^n. We've proved that

$$\sum_{x^n} \left| P\left\{X'^n = x^n\right\} - P\left\{X^n = x^n\right\} \right| < \varepsilon\, x^n(m) \varepsilon\, X^n$$

Let Alice selects $|\mu|$ codewords $m = 1, 2, ..., |\mu|$ having pdf

$$q(m) = p'^n(x_n(m)), m = 1, 2, ..., |\mu|$$

so that $$|\mu| = \left| T_\delta^{X^n} \right| + 1$$

For each m are encoded into the quantum states

$$\rho_{x^n(m)} = \overset{n}{\underset{k=1}{\otimes}} \rho_{x_k(m)}, m = 1, 2, ..., |\mu|$$

She transmits this state to Bob along the product channel $\mathcal{N}^{\otimes m}$

so $$\mathcal{N}^{\otimes m}\left(\rho_{x_1(m)} \otimes \ldots \otimes \rho_{x_n(m)}\right) = \overset{n}{\underset{k=1}{\otimes}} \mathcal{N}\left(\rho_{x_k(m)}\right) = \overset{n}{\underset{k=1}{\otimes}} \rho_{x_k(m)} = \rho_{x^n(m)}$$

where $$\sigma_x = \mathcal{N}(\rho_x), x \in \chi$$

we have $$\mathbb{E}[\sigma_{X'^n}] = \sum_{x^n \in \chi^n} p'^n(x^n)\sigma_{x^n}$$

$$= \sum_{x^n \in T_\delta^{X^n}} p^n(x^n)\sigma_{x^n} \Big/ \sum_{x^n \in T_\delta^{X^n}} p^n(x^n)$$

Let $$\sigma = \sum_{x \in \chi} p(x)\sigma_x \left\| \mathbb{E}\left(\sigma_{X'^n}\right) - \sigma^{\otimes n} \right\|_1$$

$$= \left\| \sum_{x^n \in \chi^n} p'^n(x^n)\sigma_{x^n} - \sigma^{\otimes n} \right\|_1$$

$$= \left\| \sum_{x^n \in \chi^n} p'^n(x^n)\sigma_{x^n} - \sum_{x^n \in \chi^n} p^n(x^n)\sigma_{x^n} \right\|_1$$

$$= \left\| \sum_{x^n} \left(p'^n(x^n) - p^n(x^n)\right)\sigma_{x^n} \right\|_1$$

$$\leq \sum_{x^n} \left| p'^n(x^n) - p^n(x^n) \right| < \varepsilon$$

Note: $$\sum_{x^n} p^n(x^n)\sigma_{x^n} = \sum_{x^n}\prod_{j=1}^{n} p(x_j)\bigotimes_{j=1}^{n}\sigma_{x_j}$$

$$= \bigotimes_{j=1}^{n}\left(\sum_{x\in\chi} p(x)\sigma_x\right) = \sigma^{n\otimes n}.$$

Let Bobs Hilbert space by $\mathcal{H}_B$.

$$\sigma \equiv \sigma^B$$

$$\sigma_{x^n}^{B^n} = \bigotimes_{j=1}^{n}\sigma_{x_j}.$$

$$\prod\nolimits_{\delta}^{B^n} = \sum_{x^n\in T_\delta^B} \left|x^n\right\rangle^B \left\langle x^n\right|^B$$

where $$\left|x^n\right\rangle^B = \left|x_1 x_2 \ldots x_n\right\rangle^B = \bigotimes_{k=1}^{n}\left|x_k\right\rangle^B$$

and $$\sigma|x\rangle^B = q_B(x)|x\rangle^B$$

$$|x\rangle^B = |x_1(m)\ldots x_n(m)\rangle$$

$$\sum_{x\in\chi} q_B(x) = 1$$

$$T_\delta^B = \left\{x^n\in\chi^n \left| \left|\frac{1}{n}\sum_{k=1}^{n}\log q_B(x_k) + H(\sigma) < \delta\right|\right.\right\}$$

$$H(\sigma) = -\sum_{x\in\chi} q_B(x)\log(q_B(x))$$

$$\sigma^{\otimes^n}\prod\nolimits_{\delta}^{B^n} = \sigma^{B^n}\prod\nolimits_{\delta}^{B^n} = \sum_{x^n\in T_\delta^B} q_B^n(x^n)\left|x^n\right\rangle^B \left\langle x^n\right|^B$$

$$q_B^n(x^n) = \prod_{i=1}^{n} q_B^n(x_i).$$

So, $$T_r\left\{\sigma^{B^n}\Pi_\delta^{B^n}\right\} = \sum_{x^n\in T_\delta^B} q_B^n(x^n)$$

$$= q_B^n(T_\delta^B) \geq 1 - \frac{\operatorname{Var}\log q_B(x)}{n\delta^2} = 1-\varepsilon$$

where $$\text{Var}\log q_B(x) = \sum_{x\in\chi} q_B(x)(\log q_B(x))^2 - H(q_B)^2$$

$$= T_r(\sigma(\log\sigma)^2) - H(\sigma)^2$$

$$\prod_{\delta}^{B/x^n} = \text{typical projection}$$

for the state $\quad \sigma_{x^n} = \bigotimes_{k=1}^{n} \sigma_{x_k} = \sigma_{x^n} B^n$

❖❖❖❖❖

[43] Quantum Image Processing Some Basic Problems:

[1] $f(X, Y)$ is a classical image; $F(X, Y)$ of a quantum image *i.e.* a quantum field $f(X, Y)$ evolves according to a diffusion equation so that sharp edges get smoothed out:

$$\frac{\partial f(t,X,Y)}{\partial t} = \frac{\partial}{\partial X} D_{XX}(X,Y)\frac{\partial f(t,X,Y)}{\partial X} + \frac{\partial}{\partial X} D_{YY}(X,Y)\frac{\partial f(t,X,Y)}{\partial Y}$$
$$+\frac{\partial}{\partial X} D_{XX}(X,Y)\frac{\partial f(t,X,Y)}{\partial Y} + \frac{\partial}{\partial Y} D_{XX}(X,Y)\frac{\partial f(t,X,Y)}{\partial Y}$$

or $$\frac{\partial f}{\partial t} = \text{div}\left(\underline{\underline{D}}(X,Y)\underline{\nabla} f(t,X,Y)\right)$$

For quantizing this evolution, we use the Lagrangian approach. Introduce an auxiliary field $g(t, X, Y)$ and let.

$$S[f,g] = \int g(f,t - div(D\nabla f))\,dt\,dx\,dy$$

Then $$\frac{\delta S}{\delta g} = 0 \Rightarrow f_{,t} = \text{div}(D\nabla f)$$

Also $$\frac{\delta S}{\delta f} = 0 \Rightarrow -g_{,t} - \text{div}\left(D^T \underline{\nabla} g\right) = 0$$

or $\quad g_{,t} - \text{div}\left(D^T \underline{\nabla} g\right) = 0$

Now for the quantization.

Let $\quad (X,Y)\in[0,1]\,X\,[0,1]$.

$$f(t,X,Y) = \sum_{nm} f_{n_m}(t)e_{nm}(X,Y)$$

$$g(t,X,Y) = \sum_{nm} g_{nm}(t)e_{nm}(X,Y)$$

$$e_{nm}(X,Y) = \exp(2\pi i(nX+mY))$$

$$\underline{D}(X,Y) = \sum_{nm} \underline{\underline{D}}_{nm}\, e_{nm}(X,Y).$$

$$\underline{\underline{D}}\nabla \underline{f} = \sum_{nmn'm'} \underline{\underline{D}}_{nm}\begin{pmatrix}2\pi i n' \\ 2\pi i m'\end{pmatrix} f_{nm}(t)e_{n+n'\,m+m'}(X,Y)$$

$$\operatorname{div}\left(\underline{\underline{D}}\underline{\nabla} f\right) = -4\pi^2 \sum_{n m n' m'} (n+n', m+m') \underline{\underline{D}}_{nm}\begin{pmatrix} n' \\ m' \end{pmatrix} f_{nm}(t) e_{n+n' m+m'}(X, Y)$$

$$\int_{[0,1]^2} g \operatorname{div}\left(D\underline{\nabla} f\right) dX\, dY$$

$$= -4\pi^2 \sum g_{kr}(t)\, f_{nm}(t)(n+n', m+m')\, D_{nm}\begin{pmatrix} n' \\ m' \end{pmatrix}$$

$$\int_{[0,1]^2} e_{n+n'+k, m+m'+r}(X, Y)\, dXdY$$

$$= -4\pi^2 \sum_{n, n'} \overline{g_{n+n', m+m'}(t)}\, f_{nm}(t)(n+n', m+m')\, \underline{\underline{D}}_{nm}\begin{pmatrix} n' \\ m' \end{pmatrix}$$

$$\int_{[0,1]^2} g\, f_{,t}\, dX\, dY = \sum_{nm} \overline{g_{nm}(t)}\, f'_{nm}(t)$$

$$\therefore \quad L(f_{nm}(t), g_{nm}(t), f'_{nm}(t))$$

$$= \sum_{nm} \overline{g_{nm}(t)}\, f'_{nm}(t) + 4\pi^2 \sum_{nmrs} \overline{g_{nm}(t)}\, f_{n-r,m-s}(t)(n, m)\, \underline{\underline{D}}_{n-r, m-s}\begin{pmatrix} r \\ s \end{pmatrix}$$

$$= \sum_{nm} \overline{g_{nm}(t)}\, f'_{nm}(t) + 4\pi^2 \sum_{nmrs} \overline{g_{nm}(t)}\, f_{rs}(t)(n, m)\, \underline{\underline{D}}_{rs}\begin{pmatrix} n-r \\ m-s \end{pmatrix}$$

*****

Appendix

Problems [1]

$$X(t) = X - \int_o^t \left(C(r)\,dr - D^*(r)\,dA_r - D(r)\,dA_r^*\right)$$

Calculate $d(X(t)^* X(t))$. Using the quantum I. to formula

Hint: $d(X(t)^* X(t)) = (dX(t)^*)X(t) + X(t)^* X(t)^* dX(t) + dX(t)^* dX(t)$

$$= -\left(C^*_{(t)}dt - D^*_{(t)}\, dA_t^* - D^*(t)dA_t\right)X(t)$$

$$-X^*(t)\left(C(t)\,dt - D^*(t)\,dA_t - D(t)\,dA_t^*\right) + D^*(t)D(t)dt$$

$$= -(C^*(t)X(t) + X^*(t)C(t) - D^*(t)D(t))\,dt$$

$$+ (D(t)X(t) + X^*(t)D(t))\,dA_t^*$$

$$+\left(D^*(t)X(t) + X^*(t)D^*(t)\right)dA_t$$

$$C(t) = \gamma(t, X)$$

$$X^* = X(o)$$

$$D(t) = \delta(t, X)$$
$$\delta^*(t, X) = D^*(t)$$
$$\delta(t, X^*) = \delta^*(t, X)^*, \gamma(t, X^*) = \gamma(t, X)^*$$

Then, let $\quad i(t, X) = X(t)$

$$\begin{aligned} d(X(t)^*X(t)) &= di(t, X^*X) \\ &= d\left[X^*X - \int_o^t (\gamma(r, X^*X)dr - \delta(r, X^*X))dA_r\right] \\ &= -\gamma(t, X^*X)dt + \delta^*(t, X^*X)dA_t + \delta(r, X^*X)dA_r^*\Big] \\ &\quad +\delta(t, X^*X)dA_t^* \end{aligned}$$

On the other hand and on the other, by quantum Ito's formula,

$$\begin{aligned} d(X(t)^*X(t)) &= dX(t)^*X(t) + X(t)^*dX(t) + dX(t)^*dX(t) \\ &= -[\gamma(t, X^*)\,i(t, X) + i(t, X)^*\gamma(t, X) - \delta^*(t, X)\,\delta(t, X)]dt \\ &\quad +[\delta(t, X)i(t, X) + i(t, X)^*\delta(t, X)]dA_t^* \\ &\quad +[\delta^*(t, X)i(t, X) + i(t, X)^*\delta^*(t, X)]dA_t \end{aligned}$$

So, the condition for $i(t, X^*X) = i(t, X^*)i(t, X)$ is

$$\begin{aligned} \gamma(t, X^*X) &= \gamma(t, X^*)^*i(t, X) + i(t, X)^*\gamma(t, X) - \delta^*(t, X \mid \delta(t, X)) \\ \gamma(t, X^*X) &= \gamma(t, X^*)^*i(t, X) + i(t, X)^*\delta(t, X)^*\delta(t, X), \\ \delta^*(t, X^*X) &= \delta^*(t, X)i(t, X) + i(t, X)^*\delta^*(t, X) \\ \delta^*(t, X^*X)^* &= i(t, X)^*\delta(t, X) + \delta(t, X^*)i(t, X) \end{aligned}$$

Conditions for $X \xrightarrow{i(t,\cdot)} X(t)$ to be a * representation.

$$dX(t) = C(t)dt - D^*(t)dA_t - D(t)dA_t^*$$
$$dY(t) = C(t)dt - F^*(t)dA_t * F(t)dA_t^*$$

Non-demolition conditions:

$$[X(t), Y(s)] = 0, \forall t \geq s\,.$$

Observations do not affect the future values of the state.

$$\begin{aligned} &\equiv [dX(t), Y(s)] = 0\, \forall t \geq s \\ &\equiv [C(t)dt - D^*(t)dA_t - D(t)dA_t^*, Y(s)] = 0\, \forall t \geq s \\ &\equiv [C(t), Y(s)] = [D^*(t), Y(s)] = [D(t), Y(s)] \\ &= 0\, \forall t \geq s\,. \end{aligned}$$

$$[X(t), Y(t)] = 0, [dX(t), Y(t)] = 0$$

$$\Rightarrow \qquad [X(t), dY(t)] + [dX(t), Y(t)] = 0$$

$$\Rightarrow \qquad \left[X(t), G(t)\,dt + F^*(t)dA_t + F(t)\,dA_t^*\right]$$

$$+\left[D^*(t)\,dA_t + D(t)\,dA_t^*, F^*(t)\,dA_t + F(t)\,dA_t^*\right] = 0$$

$$\Rightarrow \qquad [X(t), G(t)] + D^*(t)F(t) - F^*(t)D(t)\,dt$$

$$+\left[X(t), F^*(t)\right]dA_t + [X(t), F(t)]dA_t^* = 0.$$

$$\Rightarrow \qquad [X(t), F^*(t)] = 0$$

$$= [X(t), F(t)], D^*(t)F(t) - F^*(t)D(t) = [G(t), X(t)].$$

❑❑❑

2

Electromagnetus and Related Partial Differential Equation

[1] Computation of the Perturbe Characteristics Frequencies in a Cavity Resonator with in Homogeneous Dielectric and Permittivity:

$$\begin{aligned}
\underline{\nabla} \times \underline{E} &= -jw\mu(w, \underline{r})\underline{H}, \\
\underline{\nabla} \times \underline{H} &= jw\varepsilon(w, \underline{r})\underline{E} \\
\varepsilon(w, \underline{r}) &= \varepsilon_o\left(1 + \delta\chi_e(w, \underline{r})\right) \quad \text{div}\left[(1 + \delta.\chi_e)\underline{E}\right] = 0, \\
\mu(w, \underline{r}) &= \mu_0\left(1 + \delta\chi_m(w, \underline{r})\right) \quad \text{div}\left[(1 + \mu.\chi_m)\underline{H}\right] = 0.
\end{aligned}$$

Side walls $X = 0$, a and $Y = 0$, b are perfect magnetic conductors. Top and bottom surfaces $z = 0$, d are perfect electric conductors.

Boundary conditions

$$\begin{aligned}
H_Z &= 0, Z = 0, d, X = 0, a, Y = 0, b. \\
E_X &= E_Y = 0, Z = 0, d, \\
E_X &= 0, X = 0, a, E_Y = 0, Y = 0, b.
\end{aligned}$$

we get

$$\begin{aligned}
\text{div } \underline{E} &= -\delta(\underline{\nabla}\chi_e, \underline{E}) + O(\delta^2) \\
\text{div } \underline{H} &= -\delta(\underline{\nabla}\chi_m, \underline{H}) + O(\delta^2) \\
\nabla(div\underline{E}) - \nabla^2\underline{E} &= -jw\{\underline{\nabla}\mu \times \underline{H} + \mu\underline{\nabla} \times \underline{H}\} \\
&= -jw\{\underline{\nabla}\mu \times \underline{H} + jw\varepsilon\mu\underline{E}\}
\end{aligned}$$

or

$$\begin{aligned}
&\nabla^2 E + w^2\varepsilon_0\mu_0 E + \delta w^2\varepsilon_0\mu_0(\chi_e + \chi_m)\underline{E} \\
&= \delta jw\underline{\nabla}\chi_m + \underline{H} - \delta\underline{\nabla}(\underline{\nabla}\chi_e, \underline{E}) + O(\delta^2) \qquad \text{...(1)}
\end{aligned}$$

Likewise by duality $\underline{E} \rightarrow \underline{H}$, $\underline{H} \rightarrow -\underline{E}$, $\chi_e \leftrightarrow \chi_{m'}$ $\varepsilon_0 \leftrightarrow \mu_0$, we get

$$\nabla^2\underline{H} + w^2\mu_0\varepsilon_0\underline{H} + \delta w^2\mu_0\varepsilon_0(\chi_e + \chi_m)\underline{H}$$

$$= -\delta jw\underline{\nabla}\chi_e \times \underline{E} - \delta\underline{\nabla}\left(\underline{\nabla}\chi_m, \underline{H}\right) + O\left(\delta^2\right) \qquad ...(2)$$

We write

$$w = w^{(0)} + \delta.w^{(1)} + O\left(\delta^2\right),$$

$$\underline{E} = \underline{E}^{(0)} + \delta.\underline{E}^{(1)} + O\left(\delta^2\right)$$

$$\underline{H} = \underline{H}^{(0)} + \delta\underline{H}^{(1)} + O\left(\delta^2\right)$$

Then the $O(\delta^0)$ equation is

$$\underline{\nabla} \times \underline{E}^{(0)} = -jw^{(0)}\mu_0\underline{H}^{(0)}$$

$$\underline{\nabla} \times \underline{H}^{(0)} = -jw^{(0)}\varepsilon_0\underline{E}^{(0)}$$

The solution modes with the monitor boundary conditions is obtained using standard methods as

$$w^{(0)} = w^{(0)}(mnp) = \left(\frac{m^2}{a^2} + \frac{n^2}{b^2} + \frac{p^2}{c^2}\right)^{1/2} \pi,$$

$$H_z^{(0)} = C(mnp)u_{mnp}\left(\underline{r}\right)$$

i.e. the complete solution in the time domain for $H_z^{(0)}$ is

$$\sum_{mnp} \text{Re}\left\{C(mnp)\exp\left\{jw^{(0)}(mnp)t\right\}u_{mnp}\left(\underline{r}\right)\right\}$$

Where

$$\upsilon_{mnp}\left(\underline{r}\right) = \frac{2\sqrt{2}}{\sqrt{abd}}\sin\left(\frac{m\pi x}{a}\right)\sin\left(\frac{m\pi y}{b}\right)\sin\left(\frac{m\pi z}{c}\right)$$

Likewise

$$E_{-z}^{(0)} = d(\text{mnp})\upsilon_{mnp}\left(\underline{r}\right)$$

Where

$$\upsilon_{mnp}\left(\underline{r}\right) = \frac{2\sqrt{2}}{\sqrt{abd}}\cos\left(\frac{m\pi x}{a}\right)\cos\left(\frac{m\pi y}{b}\right)\cos\left(\frac{m\pi z}{c}\right).$$

Note that the solutions are obtained by the wave guide method in which $\frac{\partial}{\partial t} \to jw$, $\frac{\partial}{\partial z} \to -y$, so the wave-guide equations

$$E_\perp^{(0)} = \frac{-jw\mu_0}{h^2}\underline{\nabla} \perp H_z^{(0)} \times \hat{z} - \frac{Y}{h^2}\underline{\nabla} \perp E_z^{(0)}$$

$$H_\perp^{(0)} = \frac{-jw\varepsilon_0}{h^2}\underline{\nabla} \perp E_z^{(0)} \times \hat{z} - \frac{Y}{h^2}\underline{\nabla} \perp H_z^{(0)}$$

give in the case of a responator

$$E_\perp^{(0)} = -\frac{\mu_0}{h(mn)^2}\frac{\partial}{\partial t}\underline{\nabla}_\perp H_t^{(0)} \times \vec{z} + \frac{1}{h(mn)^2}\frac{\partial}{\partial z}\nabla_\perp E_z^{(0)}$$

and

$$H_\perp^{(0)} = \frac{\varepsilon_0}{h^2}\frac{\partial}{\partial t}\underline{\nabla} \perp E_z^{(0)} \times \hat{z} + \frac{1}{h^2}\frac{\partial}{\partial z}\underline{\nabla} \perp H_z^{(0)}$$

$$h^2 = h(mn)^2 = \pi^2\left(\frac{m^2}{a^2} + \frac{n^2}{b^2}\right).$$

Thus,

$$E_{\perp}^{(0)} = \frac{-jw^{(0)}(mnp)\mu_0}{h(mn)^2}\left(\underline{\nabla} \perp u_{mnp}(\underline{r}) \times \hat{z}\right) C(mnp)$$

$$+\frac{d(mnp)}{h(mn)^2}\frac{\partial}{\partial t}\underline{\nabla} \perp \upsilon_{mnp}(\underline{r})$$

so

$$E_{\perp}^{(0)} = C(mnp)\left\{\frac{-jw^{(0)}(mnp)\mu_0}{h(mn)^2}\left(\underline{\nabla}_{\perp} u_{mnp}(\underline{r}) \times \hat{z}\right)\right\}$$

$$+ d(mnp)\left\{\upsilon_{mnp}(\underline{r})\hat{z} + \frac{1}{h(mn)^2}\frac{\partial}{\partial z}\underline{\nabla}_{\perp}\upsilon_{mnp}^{(r)}\right\}$$

$$= C(mnp)\underline{\psi}_{mnp}^{E}(\underline{r}) + d(mnp)\varphi_{mnp}^{E}(\underline{r})$$

where $\underline{\psi}_{mnp}^{E}(\underline{r})$ is $E(\underline{r})\, a$ 3 × 1 vectors.

Expressible as a linear combination of

$$\upsilon_{mnp}(\underline{r}),\ \frac{\partial^2}{\partial x \partial z}\upsilon_{mnp}(\underline{r}),\ \frac{\partial^2}{\partial y \partial z}\upsilon_{mnp}(\underline{r}),$$

and $\varphi_{mnp}^{E}(\underline{r})$ is a 3 × 1 vector expressible as a $\frac{\partial u_{mnp}}{\partial X}, \frac{\partial u_{mnp}}{\partial Y}$ linear combination of with constant 3 × 1 vector valued coefficients dependent on (m, n, p). Likewise,

$$\underline{H}^{(0)} = C(mnp)\underline{\psi}_{mnp}^{H}(\underline{r}) + d(mnp)\underline{\varphi}_{mnp}^{H}(\underline{r})$$

$$\left\{\underline{\psi}_{mnp}^{E}, \underline{\varphi}_{mnp}^{E}, \underline{\psi}_{mnp}^{H}, \underline{\varphi}_{mnp}^{H}\right\}$$

are orthogonal to $\left\{\underline{\psi}_{m'n'p'}^{E}, \underline{\varphi}_{m'n'p'}^{E}, \underline{\psi}_{m'n'p'}^{H}, \underline{\varphi}_{m'n'p'}^{H}\right\}$

For all $(m'n'p') \neq (mnp)$

as ordered triplets.

Equating coefficient of $O(\delta)$ in (1) and (2) gives

$$\left(\nabla^2 + \frac{w^{(0)^2}}{C^2}\right)\underline{E}^{(1)} + \frac{2w^{(0)}w^{(1)}}{C^2}\underline{E}^{(0)} + \frac{w^{(0)^2}}{C^2}\left(\chi_e^{(0)} + \chi_m^{0}\right)\underline{E}^{(0)}$$

$$= jw^{(0)}\underline{\nabla}\chi_e^{(0)} \times \underline{H}^{(0)} - \underline{\nabla}\left(\underline{\nabla}\chi_e^{(0)}, \underline{E}^{(0)}\right) \quad ...(3)$$

and $$\left(\nabla^2 + \frac{w^{(0)^2}}{C^2}\right)\underline{H}^{(1)} + \frac{2w^{(0)}w^{(1)}}{C^2}\underline{H}^{(0)} + \frac{w^{(0)^2}}{C^2}\left(\chi_e^{(0)} + \chi_m^{0}\right)\underline{H}^{(0)}$$

$$= -jw^{(0)}\underline{\nabla}_e^{(0)} \times \underline{E}^{(0)} - \underline{\nabla}\left(\underline{\nabla}\chi_m^{(0)}, \underline{H}^{(0)}\right) \quad ...(4)$$

where
$$\chi_e^{(0)} = \chi_e\left(w^{(0)}, \underline{r}\right),$$
$$\chi_m^{(0)} = \chi_m\left(w^{(0)}, \underline{r}\right)$$
$$w^{(0)} = w^{(0)}(mnp).$$

We arrange (3) and (4) as a single 6 × 1 single vector (partial differential equation):

$$\left(\nabla^2 + \frac{w^{(0)^2}}{C^2}\right)\begin{pmatrix} \underline{E}^{(1)} \\ \underline{H}^{(1)} \end{pmatrix} + \frac{2w^{(0)}w^{(1)}}{C^2}\begin{pmatrix} \underline{E}^{(0)} \\ \underline{H}^{(0)} \end{pmatrix} + \frac{w^{(0)^2}}{C^2}\chi^{(0)}\begin{pmatrix} E^{(0)} \\ H^{(0)} \end{pmatrix}$$
$$= jw^{(0)}\begin{pmatrix} \underline{\nabla}\chi_m^{(0)} \times \underline{H}^{(0)} \\ -\underline{\nabla}\chi_e^{(0)} \times \underline{E}^{(0)} \end{pmatrix} - \begin{pmatrix} \underline{\nabla}(\underline{\nabla}\chi_e^{(0)} \times \underline{E}^{(0)}) \\ \underline{\nabla}(\underline{\nabla}\chi_m^{(0)} \times \underline{H}^{(0)}) \end{pmatrix} \qquad \text{...(5)}$$

where
$$w^{(0)} = w^{(0)}(mnp) = \pi\left(\frac{m^2}{a^2} + \frac{n^2}{b^2} + \frac{p^2}{c^2}\right)^{1/2},$$
$$\chi_e^{(0)} = \chi_e\left(w^{(0)}, \underline{r}\right)$$
$$\chi_e^{(0)} + \chi_m^{(0)} = \chi^{(0)}$$
$$\chi_m^{(0)} = \chi_m\left(w^{(0)}, \underline{r}\right)$$

Now,
$$\begin{pmatrix} \underline{E}^{(0)} \\ \underline{H}^{(0)} \end{pmatrix} = C(mnp)\begin{pmatrix} \psi_{mnp}^{E} \\ \psi_{mnp}^{H} \end{pmatrix} + d(mnp)\begin{pmatrix} \varphi_{mnp}^{E} \\ \varphi_{mnp}^{H} \end{pmatrix}$$

So (Eq. 5) gives on taking inner products with $\begin{pmatrix} \psi_{mnp}^{E} \\ \psi_{mnp}^{H} \end{pmatrix}$ and $\begin{pmatrix} \varphi_{mnp}^{E} \\ \varphi_{mnp}^{H} \end{pmatrix}$ and note that $\nabla^2 + w^{(0)^2}$ annihilates both of there functions

$$\frac{2w^{(0)}w^{(1)}}{C^2}\left\{C(mnp) + \| \psi_{mnp} \|^2 + d(mnp) < \psi_{mnp}, \varphi_{mnp} >\right\}$$
$$+ \frac{w^{(0)^2}}{C^2}\left\{C(mnp) < \psi_{mnp}, \chi^{(0)}\psi_{mnp} > d(mnp) < \psi_{mnp}, \chi^{(0)}\varphi_{mnp} >\right\}$$
$$= jw^{(0)}\left\{C(mnp) < \psi_{mnp}, \begin{pmatrix} \underline{\nabla}\chi_m^{(0)} \times \underline{\psi}_{mnp}^{H} \\ -\underline{\nabla}\chi_e^{(0)} \times \underline{\psi}_{mnp}^{E} \end{pmatrix} > \right.$$
$$\left. + d(mnp) < \psi_{mnp}, \begin{pmatrix} \underline{\nabla}\chi_m^{(0)} \times \varphi_{mnp}^{H} \\ -\underline{\nabla}\chi_e^{(0)} \times \varphi_{mnp}^{E} \end{pmatrix} > \right\} \qquad \text{...(6)}$$

and

$$\frac{2w^{(0)}w^{(1)}}{C^2}\left\{<\underline{\varphi}_{mnp},\Psi_{mnp}>C(mnp)+d(mnp)\|\underline{\varphi}_{mnp}\|^2\right\}$$

$$+\frac{w^{(0)^2}}{C^2}\left\{C(mnp)<\underline{\varphi}_{mnp},\chi^{(0)}\underline{\Psi}_{mnp}>+d(mnp)<\underline{\varphi}_{mnp},\chi^{(0)}\underline{\Psi}_{mnp}>\right\}$$

$$= jw^{(0)}\left\{C(mnp)<\underline{\varphi}_{mnp},\begin{pmatrix}\underline{\nabla}\chi_m^{(0)}\times\underline{\Psi}_{mnp}^H\\ \underline{\nabla}\chi_e^{(0)}\times\underline{\Psi}_{mnp}^E\end{pmatrix}>\right.$$

$$\left.+d(mnp)<\underline{\varphi}_{mnp},\begin{pmatrix}\underline{\nabla}\chi_m^{(0)}\times\underline{\varphi}_{mnp}^H\\ \underline{\nabla}\chi_e^{(0)}\times\underline{\varphi}_{mnp}^E\end{pmatrix}>\right\} \quad ...(7)$$

where $\Psi_{mnp} = \begin{pmatrix}\underline{\Psi}_{mnp}^E\\ \underline{\Psi}_{mnp}^H\end{pmatrix}, \varphi_{mnp} = \begin{pmatrix}\underline{\varphi}_{mnp}^E\\ \underline{\varphi}_{mnp}^H\end{pmatrix}$

(6) and (7) can be expressed as

$$\left[\begin{pmatrix}A_{11}(mnp)\,A_{12}(mnp)\\ A_{21}(mnp)\,A_{22}(mnp)\end{pmatrix}-w^{(1)}\begin{pmatrix}B_{11}(mnp) & B_{12}(mnp)\\ B_{21}(mnp) & B_{22}(mnp)\end{pmatrix}\begin{bmatrix}C(mnp)\\ d(mnp)\end{bmatrix}\right]=0$$

and have $w^{(1)}$ takes two values for each (*mnp*), namely the roots of

$$\det\left(A(mnp)-w^{(1)}B(mnp)\right) = 0.$$

❖❖❖❖❖

[2] Numerical Methods for *pde*:

(*a*) Variation principles FEM.

Action Integral:

$$S[\phi] = \int_{D^n} \mathcal{L}(\phi_v,\phi_{\psi\mu},x)d^n x$$

$$\phi_v\colon \mathbb{R}^n \to \mathbb{R}^n, v = 1, 2, ..., p.$$

$$\delta S[\phi] = 0 \Rightarrow \int_D\left[\frac{\partial\mathcal{L}}{\partial\phi_v}\delta\phi v+\frac{\partial\mathcal{L}}{\partial\phi_{v,\mu}}\delta\phi_{v,\mu}\right]d^n x = 0$$

$$\Rightarrow \int_D\left[\frac{\partial\mathcal{L}}{\partial\phi_v}-\partial_\mu\frac{\partial\mathcal{L}}{\partial\phi_{v,\mu}}\right]\delta\phi_v d^n x = 0$$

Assuming Direchlet conditions $\delta\phi_v = 0$ on the boundary *i.e.* ϕ_v is prescribed on ∂D.

Thus $\dfrac{\partial\mathcal{L}}{\partial\phi_v}-\partial_\mu\dfrac{\partial\mathcal{L}}{\partial\phi_{v,\mu}} = 0.$

Euler-Lagrange eqns. for fields.

Numerical implementation: Divide D into polyhedron 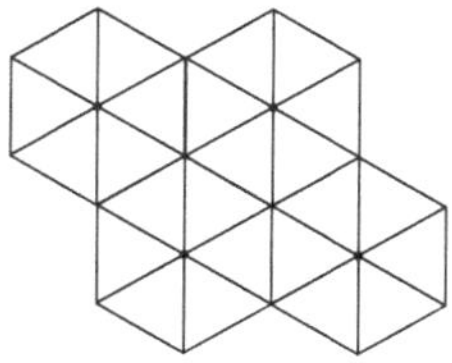 etc.

Express $\phi_\nu(x)$ for $X \in \Delta_k$ (k^{th} polyhedron) as a linear combination of its values at the vertices

$$\phi_\nu(x) = \sum_{m=1}^{q} C_{\nu k}[m]\psi_{km}(x),\ X \in \Delta_k$$

ψ_{km} = basis function chosen so that

$$\psi_{km}(\nu_{k,r}) = \delta_{m,r}, 1 \le m, r \le q$$

where $\nu_{k,r,}$= 1, 2, ..., q are the vertices of Δ_k.

Thus $$\phi_\nu(\nu_{k,r}) = C_{\nu k}[r].$$

then $$S[\phi] \approx \sum_{k=1}^{N} \int_{\Delta_k} \mathcal{L}\left(\sum_{m=1}^{q} C_{\nu k}[m]\psi_{km}(X), \sum_{m=1}^{q} C_{\nu k}[m]\psi_{km,\mu}(X) d^n X\right)$$

$$\equiv S\left[\{C_{\nu k}[m] : 1 \le m \le q, 1 \le k \le N, 1 \le \nu \le p\}\right]$$

Minimize S *w.r.t.* $\{C_{\nu k}[m]\}$.

Example of application (a) Solving the Einstein field equations of gravitation

$$S = \int R\sqrt{-g}\, d^4X$$

R = Curvature scalar = $g^{\mu\nu}R_{\mu\nu}$.

Equivalent action

$$S_1 = \int g^{\mu\nu}\sqrt{-g}\left(\Gamma^{\alpha}_{\mu\nu}\Gamma^{\beta}_{\mu\beta} - \Gamma^{\alpha}_{\mu\beta} - \Gamma^{\beta}_{\nu\alpha}\right)d^4X$$

Equivalent of S and S_1 is proved using the fact that $\delta\Gamma^{\alpha}_{\mu\nu}$ is a tensor and hence

$$\delta R_{\mu\nu} = \delta\left\{\Gamma^{\alpha}_{\mu\alpha,\nu} - \Gamma^{\alpha}_{\mu\nu,\alpha} - \Gamma^{\alpha}_{\mu\nu}\Gamma^{\beta}_{\alpha\beta} + \Gamma^{\alpha}_{\mu\beta}\Gamma^{\beta}_{\nu\alpha}\right\}$$

$$= \left(\delta\Gamma^{\alpha}_{\mu\alpha}\right)_{:\nu} - \left(\delta\Gamma^{\alpha}_{\mu\nu}\right)_{:\alpha}$$

So $$g^{\mu\nu}\sqrt{-g}\,\delta R_{\mu\nu} = \left(g^{\mu\nu}\delta\Gamma^{\beta}_{\mu\alpha}\sqrt{-g}\right)_{,\nu} - \left(g^{\mu\nu}\sqrt{-g}\,\delta\Gamma^{\alpha}_{\mu\nu}\right)_{,\alpha}$$

is a perfect divergence and hence had a vanishing integral.

(b) Solving the Einstein-Maxwell equation in the presence of charged fluid.

$$S = S_1 + S_2 + S_3$$

$$S_1 = C_1\int R\sqrt{-g}\, d^4X,$$

$$S_2 = C_2\int F_{\mu\nu}F^{\mu\nu}\sqrt{-g}\, d^4X,$$

$$S_3 = C_3\int \rho v^{\mu}v^{\nu}g_{\mu\nu}\sqrt{-g}d^4X,$$
$$\equiv C_3\int \rho\sqrt{-g}d^4X$$

Energy-momentum tensor of a fluid with presence taken into account:

$$T^{\mu\nu} = (\rho+p)v^{\mu}v^{\nu} - pg^{\mu\nu}$$

Einstein-Maxwell eqns.

$$R^{\upsilon\nu} - \frac{1}{2}Rg^{\mu\nu} = K_1T^{\mu\nu} + K_2S^{\mu\nu}$$

where $$S^{\mu\nu} = -\frac{1}{4}F_{\alpha\beta}F^{\alpha\beta}g^{\mu\nu} + F^{\mu}_{\alpha}F^{\nu\alpha}$$

is the energy-momentum tensor of the *em* field.

From the Einstein Field equations, we can derive the MHD eqns. in general relativity:

$$K_2S^{\mu\nu}_{:\nu} + K_1((\rho+p)v^{\mu}v^{\nu} - pg^{\mu\nu})_{:\nu} = 0$$

Which can be brought to the generalized Navier-Stokes form used in non relativistic MHD:

$$\rho\left((v,\nabla)v + \frac{\partial v}{\partial t}\right) = -\nabla p + \underline{J}\times\underline{B}$$

where $$\nabla\times H = \underline{J} + \varepsilon\frac{\partial E}{\partial t}$$

The Einstein Field eqns.

$$R^{\mu\nu} - \frac{1}{2}Rg^{\mu\nu} = K_1T^{\mu\nu} + K_2S^{\mu\nu}$$

are a consequence of setting the variation of S *w.r.t.* $g_{\mu\nu}$ to zero. Setting the variation of S *w.r.t.* A_{μ} $(F_{\mu\nu} - A_{\nu,\mu} - A_{\mu\nu})$ to zero gives the Maxwell eqns.

$$F^{\mu\nu}_{:\nu} = K_3J^{\mu}$$

Provided that we add to the action an interaction term $\int J^{\mu}A_{\mu}\sqrt{-g}d^4X$ represents the interaction between the charged fluid and the *em* field.

Here, in MHD we usually assume Ohms law that gives the relation between the current density and the *em* field as

$$\underline{J} = \sigma(\underline{E} + \underline{v}\times\underline{B})$$

In the special relativistic case. In teh general relativistic case, Ohm's law becomes a tensor equation

$$J^{\mu} = \sigma F^{\mu\nu}v_{\nu}$$

and hence the charge – *em* field interaction term in the action amines the form

$$\int j^{\mu}A_{\mu}\sqrt{-g}d^4X = \sigma\int F^{\mu\nu}v_{\nu}A_{\mu}\sqrt{-g}d^4X$$

The finite element method can be applied to MHD in a given metric $g_{\mu\nu}$ by taking as our action.

$$S[v^{\mu}, A^{\mu}] = K_1 \int F_{\mu\nu} F^{\mu\nu} \sqrt{-g} d^4 X + K_2$$
$$\int \rho\, g_{\mu\nu} v^{\mu} v^{\mu} \sqrt{-g} d^4 X + K_3$$
$$\int F^{\mu\nu} v_{\nu} A_{\mu} \sqrt{-g} d^4 X$$

Here we should that $v^{\mu} - \frac{dX^{\mu}}{d\tau}$ and carry out the variation *w.r.t.* X^{μ} rather than v^{μ}.

The complete non-relativistic MHD eqns. are

$$\rho\left((\underline{v}, \underline{\nabla}) \underline{v} + \frac{\partial \underline{v}}{\partial t} \right) = -\nabla p + \eta \nabla^2 v + \sigma(E + v \times B) \times B,$$

$$\text{div } E = 0, \text{ div } B = 0,$$

and
$$\text{Curl } E = -\frac{\partial B}{\partial t}, \text{ Curl } B = \mu_0 J + \frac{1}{c^2}\frac{\partial E}{\partial t}.$$

The last example of a *pde* that can be solved using the FEM is the Klein-Gordon field with Higgs potential.

$$\partial_{\mu}\partial^{\mu}\phi + m^2\phi^2 + \varepsilon V'(\phi) = 0$$

where
$$\partial_{\mu}\partial^{\mu} = \Box = \partial_t^2 - \nabla^2.$$

This *pde* can be derived from the action

$$S[\phi] = \frac{1}{2}\int \left(\partial_{\mu}\phi \partial^{\mu}\phi - m^2\phi^2 - 2\,\varepsilon\, V(\phi) \right) d^4 X$$

and it can be solved using the FEM.

The final example is from quantum field theory where we start with the Lagrangian density $\mathcal{L}\left(\phi_{\mu}, \phi_{\mu,\nu}, X\right)$ with $\phi_{\mu}: \mathbb{R}^4 \to \mathbb{R}$ and construct the Hamiltonian *via* the Lagrangian expressed in terms of function that depend only on time.

$$\phi_{\mu}(X) = \phi_{\mu}(t, \underline{r}) = \sum_{k=1}^{N} \xi_{\mu k}(t) \psi k(\underline{r})$$

where $\{\psi_k\}_{k=1}^{N}$ are basis functions depending only on the spatial coordinates $\underline{r}$:

$$\begin{aligned} S\left[\xi_{\mu k}\right] &= \int \mathcal{L}\left(\phi_{\mu}, \phi_{\mu\nu}, X\right) d^4 X \\ &= \int \mathcal{L}\Bigg(\sum_k \xi_{\mu k}(t) \psi_k(\underline{r}), \sum_k \xi_{\mu k}(t) \underline{\nabla} \psi_k(r), \\ &\qquad \sum_k \xi'_{\mu k}(t) \psi_k(r), t, \underline{r} \Bigg) d^3 r\, dt \\ &= \int L\left(\left\{ \xi_{\mu k}(t), \xi'_{\mu k} \right\}, t \right) dt \end{aligned}$$

with

$$L\left(\left\{\xi_{\mu k}(t), \xi'_{\mu k}(t)\right\}, t\right)$$

$$= \int \mathcal{L}\left(\sum_k \xi_{\mu k}(t)\psi_k(\underline{r}), \sum_k \xi_{\mu k}(t)\underline{\nabla}\psi_k(\underline{r}), \sum_k \xi'_{\mu k}(t)\psi_k(\underline{r}), t, \underline{r}\right) d^3r$$

is the Lagrangian. The Hamiltonian is constructed using the Legendre transformation:

$$p_{\mu k} = \frac{\partial L}{\partial \xi'_{\mu k}},$$

$$H\left(\left\{\xi_{\mu k}, p_{\mu k}\right\}, t\right) = \sum_{\mu k} p_{\mu k}\xi_{\mu k} - L$$

Then the Schrödinger eqn. is formulated:

$$i\hbar \frac{\partial}{\partial t}\psi\left(t, \left\{\xi_{\mu k}\right\}\right) = H\left(\left\{\xi_{\mu k}, -i\hbar \frac{\partial}{\partial \xi_{\mu k}}\right\}, t\right)\psi\left(t, \left\{\xi_{\mu k}\right\}\right)$$

This Schrödinger eqn. is solved using the finite difference scheme. In some cases, we can also derive the Schrödinger eqn. from an action principle and solve it using the FEM. For example, for a single particle, the Schrödinger eqn.

$$i\hbar \frac{\partial \psi(t, \underline{r})}{\partial t} = -\frac{\hbar^2}{2m}\left(\underline{\nabla} + \frac{ic\,A(t, \underline{r})}{\hbar}\right)\psi(t, \underline{r}) + V(t, \underline{r})\,\psi(t, \underline{r})$$

can be derived from the "complex" variational principle

$$\delta S[\psi, \bar{\psi}] = 0$$

where

$$S[\psi, \bar{\psi}] = \int \mathcal{L}(\psi, \bar{\psi}, t, \underline{r})\,dt\, d^3r\,,$$

$$\mathcal{L}(\psi, \bar{\psi}, t, \underline{r}) = \frac{i\hbar}{2}(\bar{\psi}\,\psi_{,t} - \psi\,\bar{\psi}_{,t})$$

$$-\frac{\hbar^2}{2m}\left(\left(\underline{\nabla} + \frac{ieA}{\hbar}\right)\psi, \left(\underline{\nabla} - \frac{ieA}{\hbar}\right)\bar{\psi}\right) - V\psi\bar{\psi}$$

❖❖❖❖❖

[3] Tracking of Moving Targets Using a Camera Attached to the Tip of a 2 = Link Robot:

Let the screen eqn. be $\underline{r} = \underline{R}(u, v) = X(u, v)\hat{x} + y\,(u, v)\,\hat{y} + z\,(u, v)\hat{z}$

The object moves along the trajectory $t \rightarrow \underline{r}_0(t)$

At time t, the link angles are

$$\underline{q}(t) = \underline{q}_o(t) + \varepsilon\, \delta \underline{q}(t),$$

$$\phi(t) = \phi_0(t) + \varepsilon\, \delta\phi(t)$$

where $\underline{F}(\underline{q}, \underline{\dot{q}}, \underline{\ddot{q}}, \dot{\phi}, \ddot{\phi}) = \underline{\tau}_0(t) + \varepsilon\, \omega(t) \in \mathbb{R}^3$

Thus,

$$\underline{F}(\underline{q}_0, \underline{\dot{q}}_0, \underline{\ddot{q}}_0, \dot{\phi}_0, \ddot{\phi}_0) = \underline{\tau}_0(t)$$

$\underline{F}$ can be derived from the Lagrangian

$$\mathcal{L} = \frac{1}{2}\left(q^T, \dot{\phi}\right) J(\underline{q}) \begin{pmatrix} \dot{q} \\ \dot{\phi} \end{pmatrix} - V(\underline{q})$$

where $$\underline{\underline{J}}(\underline{q}) = \begin{pmatrix} \underline{\underline{M}}(\underline{q}) & \underline{0} \\ \underline{0}^T & J_3(\underline{q}) \end{pmatrix} \in \mathbb{R}^{3\times 3}\ \underline{\underline{M}}(\underline{q}) \in \mathbb{R}^{2\times 2}$$

Position of the camera:

$$\underline{r}_2(t) = \underline{r}_1 + \psi(\underline{q}, \phi)$$

$$\underline{r}_2(t) = \underline{r}_{20}(t) + \varepsilon\, \delta \underline{r}_2(t)$$

$$\underline{r}_{20}(t) = \underline{r}_1 + \dot{\psi}(\underline{q}_0(t), \phi_0(t))$$

$$\delta \underline{r}_2(t) = \frac{\partial \psi}{\partial \underline{q}}(\underline{q}_0(t)\, \phi_0(t))\delta q(t) + \frac{\partial \psi}{\partial \psi}(\underline{q}_0(t)\, \phi_0(t), \delta\phi(t))$$

$$= \underline{\underline{A}}(t)\, \underline{\xi}(t)$$

where $$\underline{\underline{A}}(t)\left[\frac{\partial \underline{\underline{\psi}}}{\partial \underline{q}}(q_0(t), \phi_0(\underline{t}))\,\middle|\,\frac{\partial \psi}{\partial \phi}(\underline{q}_0(t), \phi_0(t))\right]$$

Object moves along a trajectory $\underline{r}_0(t)$. At time t, the ray. equation joining the object and the camera is $\lambda \to r_0(t) + \lambda(\underline{r}_2(t) - r_0(t))$.

Assume the parametric equation of the surface as

$$f(x, y, z) = 0,$$

i.e. $$f(\underline{r}) = 0$$

Then the $\lambda = \lambda\ (t)$ corresponding to the image point on the surface is

$$f(r_0 + \lambda(\underline{r}_2 - \underline{r}_0)) = 0.$$

When $\underline{r}^2 \to r_{20} + \delta r_2$, thus $\lambda \to \lambda_0^{10} + \delta\lambda(t)$ and $\underline{r}_0(t) \to \underline{r}_0(t) + \delta r_0(t)$.

Thus,

$$f(\underline{r}_0 + \delta \underline{r}_0 + \delta\lambda(\underline{r}_{20} - \underline{r}_0) + \lambda_0 \delta \underline{r}_2) + \lambda_0(\underline{r}_{20} - \underline{r}_0) = 0$$

or $$\left(\underline{\nabla} f(\underline{r}_0 - \lambda_0(\underline{r}_{20} - \underline{r}_0)), \delta \underline{r}_0 + \delta\lambda(\underline{r}_{20} - \underline{r}_0) + \lambda_0 \delta \underline{r}_2\right) = 0.$$

or, $$(\delta\, \underline{r}(t), \underline{\psi}(t)) + \delta\lambda(t)(\underline{r}_{20} - \underline{r}_0, \underline{\psi}(t)) + \lambda_0(\delta r_2(t), \underline{\psi}(t)) = 0$$

where $\underline{\psi}(t) = \underline{\nabla} f(\underline{r}_0(t) + \lambda_0(t)(r_{20}(t) - r_0(t)))$

Thus, $$\delta\lambda(t) = \frac{-\left[(\delta\underline{r}_0(t), \underline{\psi}(t)) + \lambda_0(t)(\delta\underline{r}_2(t), \underline{\psi}(t))\right]}{\left(r_{20}(t) - \underline{r}_0(t), \underline{\psi}(t)\right)}$$

Here $r_0(t)$ is the non random component of the objects motion and $\delta\underline{r}_0(t)$ is its random component. Likewise $\underline{r}_{20}(t)$ is the non-random component of the camera's motion and $\delta\underline{r}_2(t)$ is the random component

$$\exp(La(u) + L^* a^*(u)) = \exp\left(L^* a^*(u)\right) \cdot \exp(La(u)) \cdot \exp\left(\frac{1}{2}\|u\|^2 LL^*\right)$$

$$= \exp(L^* a^*(u)) \exp\left(\frac{1}{2}\|u\|^2 LL^*\right) \exp(La(u))$$

$$\exp(La(u)) f|e(v)\rangle = \sum_{n=0}^{\infty} \frac{(L^n f)}{\lfloor n} a(u)^n |e(v)\rangle$$

$$= \sum_{n=0}^{\infty} \frac{(L^n f)}{\lfloor n} \langle u,v\rangle^n |e(v)\rangle = \exp(\langle u,v\rangle L) f|e(v)\rangle$$

$$= \left(\exp(\langle u,v\rangle L) f\right)|e(v)\rangle \equiv |\exp\langle u,v\rangle L) f e(v)\rangle$$

and $\langle fe(v)| \exp(La(u) + L^* a(u)) |f e(v)\rangle$

$$= \left\langle f \left| \exp\left(\frac{1}{2}\|u\|^2 LL^* + \langle u,v\rangle L + \langle u,v\rangle L^* \right| f\right\rangle \right).\exp\left(\|v\|^2\right)$$

We have thus proved.

Theorem: Let $|\varphi(v)\rangle = \exp\left(-\frac{\|v\|^2}{2}\right)|e(v)\rangle$ and let $f \in h$ be such that $\|f\|^2 = 1$.

Then the moment generating function of the quantum random variable

$$X = La(u) + L^* a^*(u)$$

In the state $(puv)|f\varphi(v)\rangle$ is given by

$$M_X(t) = \langle f\varphi(v)|\exp(tX)|f\varphi(v)\rangle$$

$$= \left\langle \left| f \exp\left(\frac{1}{2}t^2\|u\|^2 LL^* + t\left(\langle u,v\rangle L + \langle v,u\rangle L^*\right)\right)\right| f\right\rangle$$

$$= \sum_{n=0}^{\infty} \frac{1}{\lfloor n} \left\langle f \left| \left(t\left(\langle u,v\rangle L + \langle v,u\rangle L^*\right) + \frac{1}{2}t^2\|u\|^2 LL^*\right)^n \right| f\right\rangle$$

$$= \sum_{r,s=0}^{\infty} \left\langle f \left| t^r\left(\langle u,v\rangle L + \langle v,u\rangle L^*\right)^r \cdot \frac{t^{2s}}{2^s}\|u\|^{2s}\left(LL^*\right)^s \right| f\right\rangle \cdot \frac{1}{\lfloor r \lfloor s}$$

$$= \sum_{r,\,s=0}^{\infty} \frac{t^{2s+r}}{2^s \lfloor r \lfloor s} \left\langle f \middle| \left(\langle u, v\rangle L + \langle v, u\rangle L^*\right)^r (LL^*)^s \middle| f \right\rangle \|u\|^{2s}$$

Thus, $\langle f\varphi(v) | X^n | f\varphi(v)\rangle$

$$= \sum_{0\le s\le \left[\frac{n}{2}\right]} \frac{\lfloor n}{\lfloor s \lfloor n-2s} \left\langle f \middle| \left(\langle u, v\rangle L + \langle v, u\rangle L^*\right)^{n-2s} \left(LL^*\right)^s \middle| f \right\rangle \|u\|^{2s}$$

These are the moments of the quantum random variable X in the pure state $|f\varphi(v)\rangle$ of $h \otimes \Gamma_s(\mathcal{H})$.

❖❖❖❖❖

[4] Problem:

Design MATLAB programmes for calculating the pressure field, the density field and the external potential field from velocity measurements using the Navier-Stokes equation for a gas:

$$\rho\left((\underline{v}, \underline{\nabla})\underline{v} + \underline{v}, t\right) = -\underline{\nabla} p + \eta \nabla^2 v - \rho \underline{\nabla}\phi,$$

$$p = K\rho^r$$

$$\mathrm{div}(\rho\,\underline{v}) + \frac{\partial \rho}{\partial t} = 0.$$

❖❖❖❖❖

[5] *n*-Dimensional Helmholtz Green's Funtion:

$$\left[\sum_{\alpha=1}^{n} \frac{\partial^2}{\partial x_a^2} + k^2\right] \psi(\underline{x}) = \delta(\underline{x})$$

Let $$\psi(\underline{x}) = \frac{1}{(2\pi)^n} \int_{\mathbb{R}^n} \widehat{\psi}(\underline{\xi}) \exp(i(\underline{\xi}, \underline{x}))\, d^n \underline{\xi}$$

Then $$\left(k^2 - \|\underline{\xi}\|^2\right) \widehat{\psi}(\underline{\xi}) = 1$$

$$\widehat{\psi}(\underline{\xi}) = \frac{1}{k^2 - \|\underline{\xi}\|^2}$$

$$\psi(\underline{x}) = \frac{1}{(2\pi)^n} \int \frac{\exp(i\underline{\xi}, \underline{x})}{\left(k^2 - \|\underline{\xi}\|^2\right)} d^n \underline{\xi}$$

$$d^n \underline{\xi} = r^{n-1} dr\, d\Omega(\hat{\underline{\xi}}), \qquad r = \|\underline{\xi}\|,$$ Let

$$\hat{\xi} = \frac{\underline{\xi}}{r}$$

Then $$\psi(\underline{x}) = \frac{1}{(2\pi)^n}\int \frac{\exp(ir\,\|\underline{x}\|)}{\left(k^2 - r^2\right)} r^{n-1} dr d\Omega(\hat{\xi})$$

Let $\int\limits_{S^n} d\Omega(\hat{\xi}) = S_n$ (a constant)

When

$$S^n = \left\{\hat{\xi} \in \mathbb{R}^n \left\| \hat{\xi} \right\| = 1\right\}$$

Then

$$\psi(\underline{x}) = \frac{S_n}{(2\pi)^n}\int_0^{\infty} \frac{r^{n-1}}{\left(k^2 - r^2\right)} \exp\left(ir\,\|x\|\right) dr$$

❖❖❖❖❖

[6] Edge Diffusion:

$$f_0(x, y) = \delta(px + qy - 1)$$

$$f_t(x, y) = \frac{1}{4\pi Dt}\exp\left(-\frac{\left(x^2 + y^2\right)}{4Dt}\right)^* f_0(x, y)$$

$$= K_t(x, y) * f_0(x, y)$$

$$= \frac{1}{q}\int_{\mathbb{R}^2} \delta\left(\frac{p}{q}\xi + \eta - 1\right) K_t(x - \xi, y - \eta)\, d\xi d\eta$$

$$= \frac{1}{q}\int k_t\left(x - \xi, y - 1 + \frac{p\underline{\xi}}{q}\right) d\xi$$

$\simeq$ Diffused edge.

❖❖❖❖❖

[7] Waves in Metamaterials:

$$\varepsilon_o \Big|_{z=0} - \varepsilon_o(1 - \delta\chi(\underline{r}_1)) = \varepsilon(\underline{r})$$

$$Z < 0 : \left(\nabla^2 + k^2\right)\underline{E}(\underline{r}) = 0$$

$$Z < 0 : \text{div}\left((1 + \delta\chi)\,\underline{E}\right) = 0$$

$$\text{div}\, E = -\delta(\underline{\nabla}\chi, E) + O(\delta^2)\,.$$

$$\nabla \times E = -jw\mu H, \nabla \times H = jw\varepsilon E$$

$$\Rightarrow \quad \nabla(\operatorname{div} E) - \nabla E = (-jw\mu)(jw\varepsilon)\underline{E}$$

$$\Rightarrow \quad \Delta + w^2\mu\varepsilon E - \underline{\nabla}(dw\underline{E}) = 0$$

$$\Rightarrow \quad (\Delta - k^2 - \delta\cdot k^2)_E - \delta\underline{\nabla}(\underline{\nabla}\chi, \underline{E}) = 0$$

$$(\Delta - K^2)\underline{E} = \delta\cdot\left\{k^2\underline{E} + \underline{\nabla}(\underline{\nabla}\chi, \underline{E})\right\}$$

$$E = \underline{E}^{(o)} + \delta\cdot\underline{E}^{(1)}$$

$$(\Delta - k^2)E^{(o)} = 0$$

$$(\Delta - k^2)\underline{E}^{(t)} = k^2\underline{E}^{(o)} + \underline{\nabla}\left(\underline{\nabla}\chi, \underline{E}^{(o)}\right)$$

We can thus write for $z > 0$,

$$\underline{E}(\underline{r}) = \underline{E}^{(o)}(\underline{r}) + \delta\cdot\underline{E}^{(1)}(\underline{r})$$

$$= \int\left[F(\hat{n}) + \delta\cdot\underline{\underline{H}}(r,\hat{n})\underline{F}(\hat{n})\right]\exp(-k\hat{n}.\underline{r})\,d_w(\hat{n}) + O(\delta^2) \qquad (1)$$

where $\underline{\underline{H}}(\underline{r},\hat{n}) = -\frac{1}{4\pi}\int\exp(k\hat{n}\cdot(\underline{r}-\underline{r}'))\exp(-k(\underline{r}-\underline{r}'))$

$$\times\left\{k^2\underline{\underline{I}} + \underline{\nabla}\underline{\nabla}^T\chi(\underline{r}') - k\hat{n}(\underline{\nabla}\chi(\underline{r}'))^T\right\}d^3r' \Big/ |\underline{r}-\underline{r}'|$$

$$\underline{\underline{H}}(\underline{r},\hat{n}) \in \mathbb{C}^{3\times 3},\ F(\hat{n}) \in \mathbb{C}^{3\times 1}$$

Now, $\operatorname{div}(E^{(o)}) = 0 \Rightarrow (\hat{n}, F(\hat{n})) = 0.$

further, $(E_X, E_Y)\big|_{Z=o+}$

$$= (E_X, E_Y)\big|_{Z=o-}$$

$$Z < 0$$

$$\underline{E} = E_{io}\exp(-jk\underline{n}_i\cdot\underline{r})$$

$$(\underline{n}_i, E_{io}) = 0.$$

$$Z < 0$$

$$\underline{E}^{(o)} = \int_{(\hat{n},\,\underline{F}(\hat{n}))=0} \underline{F}(\hat{n})\exp(-k\hat{n}\cdot\underline{r})\,d_w(\hat{n}),$$

$$\underline{E}^{(1)} = -\frac{1}{4\pi}\int\frac{\left\{k^2\underline{F}^{(o)}(\underline{r}') + \underline{\nabla}\left(\underline{\nabla}\chi(\underline{r}'), \underline{E}^{(o)}(\underline{r}')\right)\right\}}{|\underline{r}-\underline{r}'|}\exp(-k|r-r'|)\,d^3r'$$

$$= -\frac{1}{4\pi}\int\exp(-k|\underline{r}-\underline{r}'|)\left\{k^2\underline{F}(\hat{n}) + \underline{\nabla}(\underline{\nabla}\chi(\underline{r}'), E(\hat{n}))\right.$$

$$\left. - k(\nabla\chi(\underline{r}'), F(\hat{n}))\hat{n}\right\}\frac{\exp(-k\hat{n}\cdot\underline{r}')}{|\underline{r}-\underline{r}'|}d^3r'$$

So, $\left(E_{ioX}\widehat{X}+E_{ioY}\widehat{Y}\right)\exp(-jk(n_{iX}X+n_{iY}Y))$

$$= \int\Big[F_X(\hat{n})\widehat{X}+F_Y(\hat{n})\widehat{Y}+\delta(H_{XX}(r_o,\hat{n}))F_x(\hat{n}) + H_{XY}(r_o,\hat{n})F_r(\hat{n})+H_{XZ}(r_o,\hat{n})F_Z(\hat{n}))\widehat{X} + \delta(H_{YX}(\underline{r}_o,\hat{n})F_X(\hat{n})+H_{YY}(\underline{r}_o,\hat{n})F_Y(\hat{n}) + H_{YZ}(\underline{r}_o,\hat{n})F_Z(\hat{n}))\widehat{Y}\Big]\exp(-k(n_XX+n_YY))\,d\Omega(\hat{n})$$

Where $\underline{r}_o = X\widehat{X}+Y\widehat{Y} \equiv \underline{r}_1$.

Let $H_X^T(\underline{r},\hat{n})$ denote the first row of $H(\underline{r},\hat{n})$, $H_Y^T(\underline{r},\hat{n})$ the second row and $H_Z^T(\underline{r},\hat{n})$ the third row. Then we have 3 integral-algebraic equations for $\underline{F}(\hat{n}), n_z > 0\ (i.e.\ \underline{F}(\theta,\varphi), \theta < \pi/2)$. There are

$$E_{ioX}\exp(-jk(n_{iX}X+n_{iY}Y)) = \int\Big[F_X(\hat{n})+\delta H_X^T(\underline{r}_o,\hat{n})\underline{F}(\hat{n})\Big]\exp(-k(n_XX+n_YY))\,d\Omega(\hat{n}) \quad (1)$$

$$F_{-ioY}\exp(-jk(n_{iX}X+n_{iY}Y)) = \int\Big[F_Y(\hat{n})+\delta H_Y^T(\underline{r}_o,\hat{n})F(\hat{n})\Big]\exp(-k(n_XX+n_YY))\,d\Omega(\hat{n}) \quad (2)$$

$$(X, Y) = \underline{r}_o \in \mathbb{R}^2$$

and
$$n_X F_X(\hat{n}) + n_Y F_Y(\hat{n}) + n_Z F_Z(\hat{n}) = 0 \quad (3)$$

More generally, we may assume an arbitrary e-m field on the $Z < 0$ side. We denote this field by $\underline{E}(X, Y, Z)$ and put $\underline{E}(X, Y) = \underline{E}(X, Y, O)$. Then we have instead of the above

$$E_X(\underline{r}_o) = \int\Big[F_X(\hat{n})+\delta H_X^T(\underline{r}_o,\hat{n})\underline{F}(\hat{n})\Big]\exp(-k\hat{n}\cdot\underline{r}_o)\,d\Omega(\hat{n})$$

$$E_Y(\underline{r}_o) = \int\Big[F_Y(\hat{n})+\delta H_Y^T(\underline{r}_o,\hat{n})\underline{F}(\hat{n})\Big]\exp(-k\hat{n}\cdot\underline{r}_o)\,d\Omega(\hat{n})$$

Where $\underline{r}_o = (X, Y, o) \equiv (X, Y)$.

We write $\underline{F}(\hat{n}) = \underline{F}_o(\hat{n}) + \delta\cdot\underline{F}_1(\hat{n}) + O(\delta^2)$

Then equating δ^o, δ^1 turns gives

$$E_X(\underline{r}_o) = \int F_{OX}(\hat{n})\exp(-k\hat{n}\cdot\underline{r}_o)\,d\Omega(\hat{n}),$$

$$\underline{E}_Y(r_o) = \int F_{OY}(\hat{n})\exp(-k\hat{n}\cdot\underline{r}_o)\,d\Omega(\hat{n})$$

$$\int F_{1Y}(\hat{n})\exp(-k\hat{n}\cdot\underline{r}_o)\,d\Omega(\hat{n}) = -\int H_X^T(\underline{r}_o,\hat{n})\underline{E}_o(\hat{n})\exp(-k\hat{n}\cdot\underline{r}_o)\,d\Omega(\hat{n}),$$

$$\int F_{1X}(\hat{n})\exp(-k\,\hat{n}\cdot\underline{r}_o)\,d\Omega(\hat{n})$$

$$= -\int H_Y^T(\underline{r}_o,\hat{n})\,\underline{E}_o(\hat{n})\exp(-k\,\hat{n}\cdot\underline{r}_o)\,d\Omega(\hat{n})$$

and
$$F_{OZ}(\hat{n}) = -(n_X\,F_{OX}(\hat{n}) + n_Y\,F_{OY}(\hat{n}))/n_Z\,,$$
$$F_{1Z}(\hat{n}) = -(n_X\,F_{1X}(\hat{n}) + n_Y\,F_{1Z}(\hat{n}))/n_Z\,.$$

We expand
$$F_{OX}(\hat{n}) = \sum_{l,m} C_{OX}[l,m]Y_{lm}(\hat{n})$$
$$F_{OY}(\hat{n}) = \sum_{l,m} C_{OY}[l,m]Y_{lm}(\hat{n})$$
$$F_{1Y}(\hat{n}) = \sum_{l,m} C_{1Y}[l,m]Y_{lm}(\hat{n})$$
$$F_{1Y}(\hat{n}) = \sum_{l,m} C_{1Y}[l,m]Y_{lm}(\hat{n})$$

$Y_{lm}(\hat{n})$ we are the spherical harmonics.

We consider (1) with

$$\underline{F}(\hat{n}) = \underline{F}_o(\hat{n}) + \delta\underline{F}(\hat{n}) = \sum_{l,m} \underline{C}[l,m]Y_{lm}(\hat{n})$$

The equation

$$(\hat{n}, \underline{F}(\hat{n})) = 0$$

gives
$$F_Z(\hat{n}) = -\left(\frac{n_X}{n_Z}F_X(\hat{n}) + \frac{n_Y}{n_Z}F_Y(\hat{n})\right)$$

and so with
$$\underline{C}[l,m] = \begin{pmatrix} C_X[l,m] \\ C_Y[l,m] \\ C_Z[l,m] \end{pmatrix}$$

it follows that

$$\underline{E}_X(\underline{r}_o) = \int\left(F_X(\hat{n}) + \delta\cdot\underline{H}_X^T(\underline{r}_o,\hat{n})\,\underline{F}(n)\right)\exp(-k\,\hat{n}\cdot r_o)\,d\Omega(\hat{n})$$

$$= \sum_{l,m} C_X[l,m]\int\exp(-k\,\hat{n}\cdot\underline{r}_o)\,Y_{lm}(\hat{n})\,d\Omega(\hat{n})$$

$$+\,\delta\cdot\sum_{l,m}\underline{C}[l,m]^T\int\underline{H}_X(\underline{r}_o,\hat{n})\exp(-k\,\hat{n}\cdot\underline{r}_o)\,Y_{1m}(\hat{n})\,d\Omega(\hat{n})$$

$$E_Y(\underline{r}_o) = \sum_{l,m} C_Y[l,m]\int\exp(-k\,\hat{n}\cdot\underline{r}_o)\,Y_{lm}(\hat{n})\,d\Omega(\hat{n})$$

$$+\,\delta\cdot\sum_{l,m}\underline{C}[l,m]^T\int\underline{H}_Y(\underline{r}_o,\hat{n})\exp(-k\hat{n}\cdot\underline{r}_o)Y_{lm}(\hat{n})\,d\Omega(\hat{n})$$

where $$C_Z[l,m] = -\left\langle Y_{lm}, \frac{n_X}{n_Z} Y_{lm} C_X[l,m] + \frac{n_Y}{n_Z} Y_{lm} C_Y[l,m] \right\rangle$$
$$- C_X[l,m]\left\langle Y_{lm}, \frac{n_X}{n_Z} Y_{lm} \right\rangle - C_Y[l,m]\left\langle Y_{lm}, \frac{n_Y}{n_Z} Y_{lm} \right\rangle$$

Let $$\lambda_X[l,m] = -\left\langle Y_{lm}, \frac{n_X}{n_Z} Y_{lm} \right\rangle,$$
$$\lambda_Y[l,m] = -\left\langle Y_{lm}, \frac{n_Y}{n_Z} Y_{lm} \right\rangle,$$

Then $$C_Z[l,m] = \lambda_X[l,m]C_X[l,m] + \lambda_Y[l,m]C_Y[l,m]$$

Only the coefficient's $\{C_X[l,m], C_Y[l,m]\}$ need to be determined.

We have $$E_X(r_o) = \sum_{l,m} C_X[l,m]\Big(\int \exp(-k\hat{n}\cdot \underline{r}_o) Y_{lm}(\hat{n})\, d\Omega(\hat{n})$$
$$+ \delta\Big(\int H_{XX}(\underline{r}_o, \hat{n}) \exp(-k\hat{n}\cdot \underline{r}_o) Y_{lm}(\hat{n})\, d\Omega(\hat{n})\Big)\Big)$$
$$+ \lambda_X[l,m]\int H_{XZ}(\underline{r}_o, \hat{n}) \exp(-k\hat{n}\cdot \underline{r}_o) Y_{lm}(\hat{n})\, d\Omega(\hat{n})\Big)$$
$$+ \delta \sum_{l,m} C_Y[l,m]\Big(\int H_{XY}(\underline{r}_o, \hat{n}) \exp(-k\hat{n}\cdot \underline{r}_o) Y_{lm}(\hat{n})\, d\Omega(\hat{n})$$
$$+ \lambda_Y[l,m]\int H_{XZ}(r_o, \hat{n}) \cdot \exp(-k\hat{n}\cdot \underline{r}_o) Y_{lm}(\hat{n})\, d\Omega(\hat{n})\Big)$$

$$E_Y(r_o) = \sum_{l,m} C_Y[l,m]\Big(\int \exp(-k\hat{n}\cdot \underline{r}_o) Y_{lm}(\hat{n})\, d\Omega(\hat{n})$$
$$+ \delta\Big(\int H_{YZ}(\underline{r}_o, \hat{n}) \exp(-k\hat{n}\cdot \underline{r}_o) Y_{lm}(\hat{n})\, d\Omega(\hat{n})$$
$$+ \lambda_Y[l,m]\int H_{YZ}(\underline{r}_o, \hat{n}) \exp(-k\hat{n}\cdot \underline{r}_o) Y_{lm}(\hat{n})\, d\Omega(\hat{n})\Big)\Big)$$
$$+ \sum_{l,m} C_X[l,m]\Big(\int H_{YX}(r_o, \hat{n}) \exp(-k\hat{n}\cdot \underline{r}_o) Y_{lm}(\hat{n})\, d\Omega(\hat{n})$$
$$+ \lambda_X[l,m]\int H_{YZ}(\underline{r}_o, \hat{n}) \exp(-k\hat{n}\cdot \underline{r}_o) Y_{lm}(\hat{n})\, d\Omega(\hat{n}) \qquad (3)$$

There equation (2) and (3) can be put in the form

$$E_X(\underline{r}_o) = \sum_{l,m} \psi_{XX}(\underline{r}_o, l, m) C_X[l,m] + \sum_{l,m} \psi_{XY}(\underline{r}_o, l, m) C_X[l,m],$$

$$E_Y(\underline{r}_o) = \sum_{l,m} \psi_{YX}(\underline{r}_o, l, m) C_X[l,m] + \sum_{l,m} \psi_{YY}(\underline{r}_o, l, m) C_Y[l,m]$$

where ψ_{XX}, ψ_{XY}, ψ_{YX}, ψ_{YY}, are known functions.

$$\underline{A}(X,Y,Z) = \underline{A}(X,Y)\exp(-YZ)$$
$$\underline{E} = -\underline{\nabla}\Phi - 1w\underline{A},\; H = \underline{\nabla}\times A/\mu$$

$\underline{\nabla}\cdot(\varepsilon\,\underline{E}) = 0 \Rightarrow \underline{\nabla}\cdot(\varepsilon\,\underline{E}) = 0$ provided that we choose the coulomb gange $\Phi = 0$. Thus,

$$\text{div}\,\underline{A} = -\left(\frac{\nabla\varepsilon}{\varepsilon}, \underline{A}\right) = -\delta(\underline{\nabla}\chi, \underline{A})$$
$$\chi = \chi(x, y),$$
$$\varepsilon = \varepsilon\cdot(1+\delta\cdot\chi(x, y))$$
$$\text{div}\,\underline{A} = -\delta(\underline{\nabla}_\perp\chi, \underline{A}_\perp)$$
$$\nabla\times H = jw\varepsilon\,\underline{E}$$
$$\Rightarrow \quad \nabla\times(\nabla\times A) = -jw\mu\,\varepsilon\,(\underline{\nabla}\Phi + jw\underline{A})$$
$$\Rightarrow \quad \nabla^2\underline{A} + w^2\mu\varepsilon\underline{A} - \underline{\nabla}(div\,\underline{A}) = 0$$
$$\Rightarrow \quad \left(\nabla^2 + k^2(1+\delta\chi(x, y))\underline{A} + \delta\underline{\nabla}(\underline{\nabla}_\perp\chi, \underline{A}_\perp)\right) = 0$$
$$\underline{A} = \underline{A}_o + \delta\underline{A}_1 + O(\delta^2).$$
$$(\nabla^2 + k^2)\underline{A}_o = 0,$$
$$\left(\nabla^2 + h_{oo}^2\right)A_1 + h_{o1}^2\underline{A}_o = -k^2\chi\underline{A}_o - \underline{\nabla}(\underline{\nabla}_\perp\chi, \underline{A}_{o\perp})$$
$$E = -jw\underline{A}.$$
$$E_Z(X, Y) = 0,\; X = 0,\; a \text{ or } Y = 0,\; b.$$
$$\Rightarrow \quad A_Z(X, Y) = 0,\; X = 0,\; a \text{ or } Y = 0,\; b.$$
$$\Rightarrow \quad A_{oZ}(X, Y) = 0,\; X = 0,\; a \text{ or } Y = 0,\; b.$$
$$A_{1Z}(X, Y) = 0,\; X = 0,\; a \text{ or } Y = 0,\; b.$$
$$\left(\nabla_\perp^2 + h_{oo}^2\right)A_o = 0,$$
$$h_o^2 = w^2\mu\varepsilon_o + k^2.$$
$$\Rightarrow \quad A_{oz} = \sum_{n,m=1}^{\infty} C_Z(n, m)u_{nm}(X,Y)\exp(-\gamma_{nm}^{(o)}Z)$$
$$u_{nm}(X,Y) = \frac{2}{\sqrt{ab}}\sin\left(\frac{n\pi x}{a}\right)\sin\left(\frac{m\pi y}{b}\right)$$
$$h_{oo}^2 = \pi^2\left(\frac{n^2}{a^2}+\frac{m^2}{b^2}\right) = h_{oo}^2(n, m).$$
$$\gamma_{nm}^{(o)} = \sqrt{h_{oo}^2(n, m) - k^2},\; k^2 = w^2/\varepsilon_o.$$

$$H = \underline{\nabla} \times \underline{A} = \left(\underline{\nabla}_\perp - \gamma \widehat{Z}\right) \times \left(A_\perp + A_Z \widehat{Z}\right)$$

$$H_Z = \underline{\nabla}_\perp \times \underline{A}_\perp, H_\perp = \underline{\nabla}_\perp \times A_Z \times \widehat{Z} - \gamma \widehat{Z} \times \underline{A}_\perp$$

$$A_Z = 0, X = 0, a \text{ or } Y = a, b$$

$$\Rightarrow \quad \underline{\nabla}_\perp A_Z \times \widehat{Z} = 0, X = 0, a \text{ or } y = 0, b.$$

$$\therefore \quad H_X = 0, X = 0, a \Rightarrow A_Y = 0, X = 0, a$$

$$H_Y = 0, Y = 0, b \Rightarrow A_X = 0, Y = 0, b$$

$$E_X = 0, Y = o, b$$

$$E_Y = 0, X = o, a$$

$$\Rightarrow \quad A_X = 0, Y = o, b$$

$$\Rightarrow \quad A_Y = 0, X = o, a$$

$$A_X = \sum_{n,m \ge 1} C_X(n, m) v_{nm}(X, Y) e^{-\gamma_{nm}^{(o)} Z}$$

$$A_{OX} = \sum_{n,m \ge 1} d_X(n, m) w_{nm}(X, Y) e^{-\gamma_{nm}^{(o)}(Z)}$$

$$\text{div } \underline{A}_o = o \Rightarrow A_{OZ,Z} + A_{OX,X} + A_{OY,Y} = 0$$

So

$$v_{nm}(X, Y) = \frac{2}{\sqrt{ab}} \cos\left(\frac{n\pi X}{a}\right) \cdot \sin\left(\frac{m\pi X}{b}\right)$$

$$w_{nm}(X, Y) = \frac{2}{\sqrt{ab}} \sin\left(\frac{n\pi X}{a}\right) \cos\left(\frac{m\pi X}{b}\right)$$

$$\text{div } \underline{A}_o = 0$$

$$\Rightarrow -\gamma_{nm}^{(o)} C_Z(n, m) - \frac{n\pi}{a} C_X(n, m) - \frac{n\pi}{a} C_Y(n, m) = 0.$$

$$\Rightarrow \quad C_z(n, m) = \gamma_{nm}^{(o)-1} \left(\frac{n\pi}{a} C_X(n, m) - \frac{m\pi}{b} C_Y(n, m)\right)$$

So

$$\underline{A}_o(X, Y, Z) = \sum_{n, m \ge 1} \exp\left(-\gamma_{nm}^{(o)} z\right) \left\{ C_X(n, m) v_{nm}(X, Y) \widehat{X} \right.$$

$$\left. + C_y(n, m) w_{nm}(X, Y) \widehat{Y} - \gamma_{nm}^{(o)-1} \left(\frac{n\pi}{a} C_X(n, m) + \frac{n\pi}{b} C_Y(n, m)\right) u_{nm}(X, Y) \widehat{Z} \right\}$$

$$\left[\nabla_\perp^2 + h_{oo}^2\right] \underline{A}_1 + h_{01}^2 \underline{A}_o = -k^2 \chi \underline{A}_o - \underline{\nabla}\left(\underline{\nabla}_\perp X, \underline{A}_{o\perp}\right)$$

$$\Rightarrow \quad \left[\nabla_\perp^2 + h_{oo}^2\right] A_{1X} + h_{01}^2 A_{oX} = -k^2 \chi A_{oX} - \frac{\partial}{\partial X}(\chi_{,x} A_{ox} + \chi_{,y} A_{oy})$$

$$\left(\nabla_{\perp}^2 + h_{oo}^2\right)A_{1X} + h_{01}^2 A_{oY} = -k^2\chi A_{oY} - \frac{\partial}{\partial Y}(\chi_{,x} A_{ox} + \chi_{,y} A_{oy})$$

$$\left(\nabla_{\perp}^2 + h_{oo}^2\right)A_{1Z} + h_{01}^2 A_{oZ} = -k^2\chi A_{oZ} + Y(\underline{\nabla}_{\perp}\chi A_{o\perp})$$

❖❖❖❖❖

[8] Stroock Partial Differential Equations for Probabilities $\varphi \simeq$ Function, $u \simeq$ Sobolev Distribution:

$$\|\varphi\|_s = \frac{1}{(2\pi)^N}\int(1+|\xi|^2|\varphi(\xi)|^2)\,d\xi$$

$$\|u\|_s = \sup\left\{|\langle\varphi, u\rangle| :\|\varphi\|_{-s} \neq 1\right\}$$

$$H_s = \left\{u \,\|\, u\,\|_s < \infty\right\}.$$

$$|\langle\varphi, u\rangle| = \left|\langle\hat{\varphi}, \hat{u}\rangle\right| = \left|\int(1+|\xi|^2)^{s/2}\overline{\varphi(\xi)}(1+|\xi|^2)^{-s/2}\hat{u}(\xi)\,d\xi\right|$$

$$\leq \left(\int(1+|\xi|^2)^{-s}|\overline{\hat{\varphi}(\xi)}|^2\,d\xi\right)^{1/2}\left(\int(1+|\xi|^2)^s|\hat{u}(\xi)|^2\,d\xi\right)^{1/2}$$

So

$$\sup\left\{|\langle\varphi, u\rangle|\,\|\,\varphi\,\|_{-x} < \infty\right\} = \|u\|_s = \left(\int(1+|\xi|^2)^s\,|\hat{u}(\xi)|^2\,d\xi\right)^{1/2} \equiv \|u\|_s.$$

Let Δ be the Laplacian on $\mathbb{R}^n$. Then

$$\int(1+|\xi|^2)^s\,|\hat{u}(\xi)|^2\,d\xi = \int_{\mathbb{R}^n}\left|(1-\Delta)^{s/2}u(x)\right|^2 dx = \left\|(1-\Delta)^{s/2}u\right\|_2$$

$$H_\infty = \bigcap_{s\in\mathbb{R}} H_s = \left\{u \,|\, \int(1+|\xi|^2)^s\,|\hat{u}(\xi)|^2\,d\xi < \infty \forall s \in \mathbb{R}\right\}$$

$$H_{-\infty} = \left\{u \,|\, \lim_{s\to-\infty}\int(1+|\xi|^2)^s\,|\hat{u}(\xi)|^2\,d\xi < \infty\right\}$$

$$\int(1+|\xi|^2)^s\,|\hat{u}(\xi)|^2\,d\xi < \infty$$

$$\Leftrightarrow \int|\hat{u}(\xi)|^2\,d\xi < \infty \text{ and } \int|\xi|^{2m}|\hat{u}(\xi)|^2\,d\xi < \infty \forall \leq m \leq s$$

its is a positive integer by the binomial theorem.

$$|\langle\varphi, u\rangle|^2 \leq \left(\int(1+|\xi|^2)^s\,|\hat{\varphi}(\xi)|^2\,d\xi\right)\cdot\left(\int(1+|\xi|^2)^{-s}\,|\hat{u}(\xi)|^2\,d\xi\right)$$

$$\leq C(s)\sum_{\|\alpha\|\leq 2s}\left\|\partial^\alpha\varphi\right\|_2^2$$

Provided $u \in H_{-s}$

$u \in H_{-\infty}, \Leftrightarrow u \in H_{-s}$ for some $s \in \mathbb{R}$. If $s \geq 0$ then this implies

$$|\langle\varphi, u|^2 \le C \sum_{\|\alpha\|<2s} \left\|\partial^\alpha \varphi\right\|_2^2$$

If $s < 0$ and $u \in H_{-s}$ then $-s > 0$ and $\int (1+|\xi|^2)^{-s} \hat{u} |\xi|^2 d\xi < \infty$

$$\Rightarrow \qquad \int (1+|\xi|^2)^s |\hat{u}(\xi)|^2 \, d\xi < \infty$$

(since for $s < 0$, $(1+|\xi|^2)^s \le (1+|\xi|^2)^{-2} \forall \xi \in \mathbb{R}^N$)

$$\Rightarrow \qquad |\langle\phi, u\rangle|^2 \le C' \int (1+|\xi|^2)^{-s} |\hat{\varphi}(\xi)|^2 \, d\xi$$

$$\le C \sum_{\|\alpha\| \le 2|s|} \left\|\partial^\alpha \phi\right\|_2^2$$

These discussion imply that

$$H_{-\infty} = \left\{ u \middle\| \langle\varphi, u\rangle \right| \le \sum_{\|\alpha\| \le n} \left\|\partial^\alpha \varphi\right\|_2^2 \forall \varphi \text{ for some } n \ge 0 \Bigg\}$$

$$H_{-\infty} = \bigcup_{s \in \mathbb{R}} H_s = \left\{ u \,\middle|\, \lim_{\delta \to --\infty} \int (1+|\xi|^2)^s |\hat{u}(\xi)|^2 \, d\xi < \infty \right\}.$$

❑❑❑

3

Radon and Group Theoretic Transforms With Robotic Applications

[1] Modern Trends in Signal Processing by Harish Parthasarathy, NSIT:

[**Reference:** Radon Transform-Some Basic Properties.]

[1] Radon transform based computer tomography.

[2] Numerical methods for solving partial differential equation with applications to waveguides and Cavity resonators.

[3] Modelling noise in quantum systems.

[1] Image field $f : \mathbb{R}^n \to \mathbb{C}$ is an image

Hyperplane is $(\widehat{m}, \underline{r}) = p, \widehat{m} \in S^{n-1}, p \in \mathbb{R}$ ·

S^{n-1} is the $n-1$ dimensional sphere in $\mathbb{R}^n$.

$$S^{n-1} = \left\{\underline{r} \in \mathbb{R}^n \mid \|r\| = 1\right\}$$

Projection of f on the hyperplane:

$$(Rf)\ (\widehat{m}, p) = \int_{\mathbb{R}^n} f(\underline{r})\delta\left(p - \left(\widehat{m}, \underline{r}\right)\right) d^n r$$

$Rf \simeq$ Radon transform of f.

Example: $n = 2$

$$(Rf)\ (I, m, p) = \int_{\mathbb{R}^2} f(x, y)\,\delta(p - lx - my)\,dx\,dy$$

Inversion formula:

$$(\mathcal{F}_p Rf)\ (\widehat{m}, w) = \int_{\mathbb{R}} (Rf)(\widehat{m}, p)\exp(-jwp)\,dp$$

$$= \int f(\underline{r})\exp(-jw(\widehat{m}, \underline{r}))d^n r$$

So $$f(\underline{r}) = \frac{1}{(2\pi)^n}\int_{\mathbb{R}^n} (\mathcal{F}_p Rf)\left(\frac{\underline{\xi}}{\|\underline{\xi}\|}, \|\underline{\xi}\|\right)\exp(j(\underline{\xi}, \underline{r}))\, d^n\underline{\xi}$$

Behaviour of Radon transform under rotations and translation

$$SO(n) = \{k \in \mathbb{R}^{n\times n} \mid K^T K = I, \det(K) = 1\}$$

Translates and rotated image field:

$$T_{\underline{k},\underline{a}} f(\underline{r}) = f(\underline{\underline{K}}\underline{r} + \underline{a}), \underline{a} \in \mathbb{R}^n, K \in SO(n).$$

$$(RT_{k,a} f)(\hat{m}, p) = \int_{\mathbb{R}^n} f(\underline{\underline{K}}\underline{r} + \underline{a})\delta(p - (\hat{m}, \underline{r}))d^n r$$

$$\mathcal{F}_p RT_{k,a} f(\hat{m}, w) = \int_{\mathbb{R}} (RT_{k,a} f)(\hat{m}, p)\exp(-j\,wp)\,dp$$

$$= \int f(Kr + a)\exp(-jw(\hat{m}, r))d^n r$$

$$\int f(\underline{x})\exp(-jw(\hat{m}, K^{-1}(x - a))d^n x$$

$(\underline{x} = Kr + a)$ since $\det K = 1$.

Thus $$(\mathcal{F}_p RT_{k,a} f)(\widehat{m}, w) = \exp(jw(K\widehat{m}, a))(\mathcal{F}f)(wk\widehat{m})$$

$$= \exp(jw(K\widehat{m}, a))(\mathcal{F}f\,Rf)(K\widehat{m}, w)$$

$$|(\mathcal{F}_p RT_{k,a} f)(\widehat{m}, w)| = |(\mathcal{F}f)(wK\widehat{m})|$$

K can be recovered from this using Fourier transform on $SO(n)$.

❖❖❖❖❖

[2] Randon-Transform Based Image Processing:

Let $f : \mathbb{R}^2 \to * \ \mathbb{C}$ an image field.

Its projection on the plane

$$lx + my - p = 0$$

where $l^2 + m^2 = 1$ is given by

$$(Rf)(l, m, p) = \int f(x, y)\,\delta(p - lx - my)\,dx\,dy$$

If $f(x, y)$ is a random Gaussian field, then

$$\mathbb{E}\left[(Rf)(l, m, p)\left(\overline{Rf}\right)(l', m', p')\right]$$

$$= \int R_{ff}(x; y; x'y')\,\delta(p - lx - my)\,\delta(p' - l'x' - m'y')\,dx\,dy\,dx'\,dy'$$

Suppose f is a WSS (Wide-sense-stationary) field. Then

$$R_{ff}(x, y; x'\,y') \equiv R_{ff}(x - x', y - y')$$

Then $\mathbb{E}\left[(Rf)(l, m, p)\left(\overline{Rf}\right)(l', m', p')\right]$

$$= \int_{\mathbb{R}^4} R_{ff}(x-x', y-y')\,\delta(p-lx-my)\,\delta(p'-l'x'-m'y')\,dx\,dy\,dx'\,dy'$$

$$\int_{\mathbb{R}^4} R_{ff}(\xi,\eta)\,\delta(p-lx-my)\,\delta(p'-l'(x-\xi)-m'(y-\eta))\,dx\,dy\,d\xi\,d\eta$$

$$= \int R_{ff}(\xi,\eta)d\xi d\eta \int \delta(p-lx-my)\delta(p'-l')(x-\xi)-m'(y-\eta))dx\,dy$$

Let $\phi(p, p', l, m, l', m') = \int \delta(p-lx-my)\,\delta(p'-l'(x-\xi)-m'(y-\eta))\,dx\,dy$

Let $lx + xy = u$

$l'x+m'y = v$

Then $\begin{pmatrix} x \\ y \end{pmatrix} = \begin{pmatrix} l & m \\ l' & m' \end{pmatrix}^{-1} \begin{pmatrix} u \\ v \end{pmatrix}$

$$= \begin{pmatrix} m' & -m \\ -l' & l \end{pmatrix} \begin{pmatrix} u \\ v \end{pmatrix} \Big/ (lm'-l'm)$$

$$= \begin{pmatrix} m'u & -mv \\ lv & -l'u \end{pmatrix} \Big/ (lm'-l'm)$$

So $\Phi(p, p', l, m, l', m') = \int \delta(p-u)\,\delta(p'+l'\xi+m'\eta-v)\,(lm'-l'm)^{-1}\,du\,dv$

$$= (lm'-l'm)^{-1}.$$

Thus, $\mathbb{E}\left[(Rf)(l,m,p)\left(\overline{Rf}\right)(l',m',p')\right]$

$$= (lm'-l'm)^{-1} \int_{\mathbb{R}^2} R_{ff}(\xi,\eta)\,d\xi\,d\eta$$

Inversion formulas:

$$\int_{\mathbb{R}} (Rf)(l,m,p)e^{-j\omega p}dp = \int f(x,y)e^{-j\omega(lX+mY)}dxdy$$

$$= \hat{f}(l\omega, m\omega)$$

where $\hat{f}$ denote 2-D Fourier transform of f.

Thus, writing $\xi = l\omega$, $\eta = m\omega$, we get

$$\hat{f}(\xi,\eta) = \int_{\mathbb{R}} (Rf)\left(\frac{\xi}{\sqrt{\xi^2+\eta^2}}, \frac{\eta}{\sqrt{\xi^2+\eta^2}}, p\right)$$

$$\exp\left(-jp\sqrt{\xi^2+\eta^2}\right)dp$$

and thus $f(x,y) = (2\pi)^{-1}\int (Rf)\left(\frac{\xi}{\sqrt{\xi^2+\eta^2}}, \frac{\eta}{\sqrt{\xi^2+\eta^2}}, p\right)$

$$\exp\left(-jp\sqrt{\xi^2+\eta^2}\right)\exp\left(j(\xi X+\eta Y)\right)dpd\xi d\eta$$

Generalization to n-dimensional images:

$f:\mathbb{R}^n \to \mathbb{C}$.

$$(Rf)\left(\widehat{m},p\right) \triangleq \int_{\mathbb{R}^n} f(x)\delta\left(p-\left(\widehat{m},x\right)\right)d^n x$$

$$(Rf)\left(\widehat{m},-p\right) = (Rf)\left(-\widehat{m},p\right).$$

$$\int_{\mathbb{R}}\left(Rf\left(\widehat{m},p\right)\exp(-)\omega p\right)dp = \int_{\mathbb{R}^n}\left(f(x)\exp\left(-j\left(\omega\widehat{m},x\right)\right)d^n x\right.$$

Equivalently,

$$\int_{\mathbb{R}}(Rf)\left(\frac{\xi}{|\xi|},p\right)\exp(-j\,|\,\xi\,|\,p\,|\,dp = \int_{\mathbb{R}^n} f(x)\exp(-j(\xi,x))\;d^n x, \xi\in\mathbb{R}^n$$

So
$$f(x) = \int_{\mathbb{R}\times\mathbb{R}^n}(2n)^n(Rf)\left(\frac{\xi}{|\xi|},p\right)$$
$$\exp\left\{j\left((\xi,x)-|\xi|p\right)\right\}d\,p\,d^n\xi$$

Radon transform of a moving image:

$$\begin{aligned}
\psi(X,Y,t) &= f\left(X-V_X t, Y-V_Y t\right).\\
&\equiv \psi_t(X,Y).\\
(R\psi_t)(l,m,p) &= \int_{\mathbb{R}^2} f\left(X-V_X t, Y-V_Y t\right)\xi(p-lX-mY)dXdY\\
&= \int f(X',Y')\,\delta\left(p-lX'-mY'-\left(lV_X+mV_Y\right)t\right)dX'dY'\\
&= (Rf)(l,m,p-(lV_X+mV_Y)\,t)
\end{aligned}$$

$$\int_{\mathbb{R}}(R\psi_t)(l,m,p)e^{-j\omega p}dp$$

$$= \exp\left\{-j\omega(lV_x+mV_y)t\right\}\times\int_{\mathbb{R}}(Rf)(l,m,p)\,e^{-j\omega p}dp$$

$$i.e.\left(\mathcal{F}_p R\,\psi_t\right)(l,m,\omega)$$

$$= \exp\left\{-j\omega\left(lV_x+mV_y\right)t\right\}\left(\mathcal{F}_p Rf\right)(l,m,\omega)$$

$$\equiv \left(\mathcal{F}_p R\,\psi_0\right)(l,m,\omega)$$

Thus,

$$\text{Arg}\left\{\left(\mathcal{F}_p R\,\psi_{t1}\right)(l, m, \omega)\right\} - \text{Arg}\left\{\left(\mathcal{F}_p\; R\,\psi_{t2}\right)(l,m,\omega)\right\}$$

$$= \omega\,(t_2 - t_1)\,(lV_x + mV_y)$$

$\psi(X, Y) = 0 \simeq$ eqn. of a curve in the plane. Project the image field $f(x, y)$ on this curve:

$$R_\psi(f) = \int f(x, y)\,\delta(\psi(x, y))dxdy$$

e.g.:

$$y = \varphi(x) \text{ i.e. } \psi(x, y) = y - \varphi(x).$$

Then

$$R_\psi(f) = \int f(x, y)\,\delta\left(y - \varphi(x)\right) dxdy = \int_{\mathbb{R}} f\left(x, \varphi(x)\right) dx$$

$$f(x, y) = s(x, y) + w(x, y)$$

$$R\psi\,(f) = R_\psi(s) + R_\psi(w).$$

$$NSR = \frac{\mathbb{E}\left\{R_\psi(w)^2\right\}}{R_\psi(s)^2} = (SNR)^{-1}$$

$$X \rightarrow X\cos\theta + Y\sin\theta$$

$$Y \rightarrow -X\sin\theta + Y\cos\theta$$

$f(X, Y) \rightarrow f(X\cos\theta + Y\sin\theta - \sin\theta + Y\cos\theta) = \tilde{f}(X, Y) \simeq$ Rotation image.

$$R\tilde{f}(l, m, p) = \int \tilde{f}(X, Y)\,\delta$$

$$\int_{\mathbb{R}} Rf_{k,x}\left(\hat{n}, p\right) e^{-j\omega p} dp = \exp\left(j\omega\left(x, k\hat{n}\right)\right) \int_{\mathbb{R}} \left(R\,f\right)\left(k\hat{n}, p\right) = \exp(-j\omega p)\, dp$$

$$\left(\mathcal{F}_p Rf_{k,x}\right)\left(\hat{n}, \omega\right) = \exp\left(j\omega\left(x, K\hat{n}\right)\right)\left(\mathcal{F}_p R\,f\right)\left(K\hat{n}, \omega\right)$$

If $K \in O\,(n)$ is knows, x can be determined using

$$\omega\left(x, K\hat{n}\right) = \text{Arg}\left\{\left(\mathcal{F}_p R\,f_{k,s}\right)\left(\hat{n}, \omega\right)\right\} - \text{Arg}\left\{\left(\mathcal{F}_p R\,f\right)\left(K\hat{n}, \omega\right)\right\}$$

$$\left|\mathcal{F}_p Rf_{k,x}\left(\hat{n}, \omega\right)\right| = \left|\left(\mathcal{F}_p R\,f\right)\left(K\hat{n}, \omega\right)\right|$$

❖❖❖❖❖

[3] Image Processing Using Invariants of the Permutation Group:

Sources are located at $\underline{r}_1, \underline{r}_2, .., \underline{r}_N$.

Signal field emitted by source at $\underline{r}_k$ is $f_k\left(t - \left|\underline{r} - \underline{r}_k\right|/c\right)$, $1 \le k \le N$. Total signal field generated is

$$f(t, \underline{r}) = \sum_{k=1}^{N} f_k\left(t - |\underline{r} - \underline{r}_k|/c\right)$$

If a permutation σ ($\in S_N$) is applied to the sources, the generated signal field is

$$U_\sigma f(t, \underline{r}) = \sum_{k=1}^{N} f_k\left(t - |\underline{r} - \underline{r}_{\sigma k}|/c\right)$$

$$= \sum_{k=1}^{N} f_{\sigma^{-1}k}\left(t - |\underline{r} - \underline{r}_k|/c\right)$$

$$U_\sigma U_\sigma f(t, \underline{r}) = U_{\sigma\rho} f(t, \underline{r}).$$

Let $$f\left(t, \underline{r}, \underline{r}_1 \ldots, \underline{r}_N\right) = \sum_{k=1}^{N} f_k\left(t - |\underline{r} - \underline{r}_k|/c\right) \equiv f(t, \underline{r})$$

Then $$U_\sigma f(t, \underline{r}) = f\left(t, \underline{r}, \underline{r}_{\sigma 1}, \ldots, \underline{r}_{\sigma N}\right).$$

Group algebra: $\sum_{\sigma \in S_N} a(\sigma) U_\sigma$

Let $\sigma \to \pi(\sigma)$ be a unitary representation on S_N in $\mathbb{C}^p$. Then consider

$$f_\pi(t, \underline{r}) = \sum_{\sigma \in S_N} \pi(\sigma) U_\sigma f(t, \underline{r})$$

We get $$(U_\sigma f)_\pi(t, \underline{r}) = \sum_{\sigma \in S_N} \pi(\sigma) U_{\sigma\rho} f(t, \underline{r})$$

$$= \sum_{\sigma \in S_N} \pi(\sigma\rho^{-1}) U_\sigma f(t, \underline{r}) = f_\pi(t, \underline{r}) \pi^*(\rho)$$

So, $\left(U_\sigma f\right)_\pi(t, \underline{r})\left(\left(U_\sigma f\right)_\pi(t, \underline{r})\right)^* = f_\pi(t, \underline{x})^* f_\pi(t, \underline{r})$

Thus, $f_\pi(t, \underline{r})^* f_\pi(t, \underline{r})$ is an S_N invariant matrix field in space-time.

Young's Tableaux, primitive idempotents and the irreducible rep'm of S_N.

Let $N = 3$, and consider the tableaux

$$T = \begin{array}{|c|c|} \hline 1 & 2 \\ \hline 3 & \\ \cline{1-1} \end{array}$$

$$e_T = \sum_{\substack{p \in R_T, \\ q \in C_T}} Sqn(q) pq$$

in the group algebra. $\sigma_0 = \text{Id}$

Let $$\sigma_1 = \begin{pmatrix} 1 & 2 & 3 \\ 2 & 3 & 1 \end{pmatrix}$$

$$\sigma_2 = \begin{pmatrix} 1 & 2 & 3 \\ 3 & 1 & 2 \end{pmatrix}$$

$$\sigma_3 = \begin{pmatrix} 1 & 2 & 3 \\ 2 & 1 & 3 \end{pmatrix}$$

$$\sigma_4 = \begin{pmatrix} 1 & 2 & 3 \\ 3 & 2 & 1 \end{pmatrix}$$

$$\sigma_5 = \begin{pmatrix} 1 & 2 & 3 \\ 1 & 3 & 2 \end{pmatrix}$$

$$= I_d - (13) - (2\ 3)\ (1\ 3) + (2\ 1)$$

$$= I_d - \sigma_4 - \sigma_5\ \sigma_4 + \sigma_3$$

Now $\sigma_5\sigma_4 = (2\ 3)\ (13) = \sigma_1$

$$e_T = \sigma_0 - \sigma_4 - \sigma_1 + \sigma_3$$

Re_T is an irreducible representation.

$$\sigma_0 e_T = e.e_T = e_T,$$

$$\sigma_1 e_T = \sigma_1$$

❖❖❖❖❖

[4] Radon Transform of Rotated and Translated Image Field:

$f : \mathbb{R}^{\tilde{n}} \to C$

$$O(n) = \left\{K \in \mathbb{R}^{n\times n} \middle| K^T K = I, \det = (K) = 1\right\}.$$

$$Rf_{K,x}\left(\hat{n}, p\right) = \int f(\underline{x} + \underline{K}.y)\delta(p - (y, \hat{n}))d^n y$$

$$\left(\hat{n}, y\right) = p.$$

$n_1 y_1 + \ldots + n_n y_n = p$, $n-1$ dimensional plane in $\mathbb{R}^n$.

$n_1 x + n_2 y + n_3 z = p$ $2-D$ plane in $\mathbb{R}^3$.

Computer Tomography.

$$Rf_{K,x}\left(\hat{n}, p\right) = \int f(\underline{x} + \underline{y})\delta\left(p - \left(y, K\hat{n}\right)\right)d^n y$$

$$= \int f(\underline{y})\delta\left(\underline{p} + \left(x, K\hat{n}\right) - \left(y, K\hat{n}\right)\right)d^n y$$

$Rf\left(\hat{n}, p\right)$ given.

$$Rf_{K,x}\left(\hat{n}, p\right) = Rf\left(K\hat{n}, p + \left(x, K\hat{n}\right)\right)$$

$K \simeq$ rotator $x =$ translation.

❖❖❖❖❖

[5] Estimating the Rotation Applied to the Two 3-D Robot Links from Electromagnetic Field Pattern:

At time t the rotation suffered by the bottom link is $R_1(t)$ and that by the top link is $R_z R_1$. A point $\underline{r}$ in the bottom link moves to $\underline{\underline{R}}_1(t)\,\underline{r} = \underline{\underline{R}}_1(\phi_1(t), \theta_1(t), \psi_1(t))\underline{r}$ and third in the top link to $R_1 p_0 + R_2 R_1(\underline{r} - \underline{p}_0)$.

Where $\quad R_1(t) = \underline{\underline{R}}_1(\phi_2(t), \theta_2(t), \psi_2(t))$

The correct density in the bottom link is $\underline{J}_{=1}\left(R_1^{-1}\,\underline{r}\right)$ and that in the top link is $\underline{J}_{=2}\left(p_0 + (R_2 R_1)^{-1}(\underline{r} - R_1 p_0)\right)$.

Assuming infinite speed of light, the magnetic vector potential at time t is

$$\begin{aligned}\underline{A}(t,\underline{r}) &= \int \frac{\underline{J}_{=1}\left(R_1(t)^{-1}\underline{r}^1\right)}{|\underline{r}-\underline{r}'|} d^3r' \\ &+ \int \frac{\underline{J}_{=1}\left(p_0 + \left(R_2(t)\,R_1(t)^{-1}\right)\left(\underline{r}' - R_1(t)p_0\right)\right)}{|\underline{r}-\underline{r}'|} d^3r' \\ &= \int \frac{\underline{J}_1(\underline{r}')}{\left|\underline{r} - \underline{\underline{R}}_1(t)\,\underline{r}'\right|} d^3r' + \int \frac{\underline{J}_2(\underline{r}')d^3r'}{\left|\underline{r} - R_1^{\theta} p_0 - R_2(t)R_1(t)\left(\underline{r}' - p_0\right)\right|}.\end{aligned}$$

Let $\quad R_2(t)R_1(t) = S(t),\ R_1(t) = R(t)$. Then,

$$\underline{A}(t,\underline{r}) = \int \frac{\underline{J}_1(\underline{r}')}{\left|\underline{r} - \underline{R}(t)\,\underline{r}'\right|} d^3r' + \int \frac{\underline{J}_2(\underline{r}')d^3r'}{\left|\underline{r} - \underline{\underline{R}}(t)p_0 - S(t)\left(\underline{r}' - p_0\right)\right|}.$$

By taking measurements of $\underline{A}(t,\underline{r})$ of different $\underline{r}'s$, $\underline{\underline{R}}(t), \underline{\underline{S}}(t) \in SO(3)$ can be estimated.

For $|\underline{r}| >> |\underline{r}'|$,

$$\begin{aligned}\left|\underline{r} - \underline{\underline{R}}(t)\underline{r}'\right|^{-1} &= \left[r^2 + r'^2 - 2\left(\underline{r}, \underline{\underline{R}}(t)\,\underline{r}'\right)\right]^{-1/2} \\ &\approx r^{-1}\left[1 + \frac{\left(\underline{r}, \underline{\underline{R}}(t)\,\underline{r}'\right)}{r^2}\right] = \frac{1}{r} + \frac{\left(\hat{r}, \underline{\underline{R}}(t)\,\underline{r}'\right)}{r^2}\end{aligned}$$

$$\left|\underline{r} - Rp_0 - S\left(r' - p_0\right)\right|^{-1} \approx \frac{1}{r} + \frac{\left(\hat{r}, Rp_0 + S\left(r' - p_0\right)\right)}{r^2}$$

So the non constant part of the magnetic vector potential at $r >> r'$ is approximates

$$\begin{aligned}\tilde{A}(t,r) &= \int \underline{J}_1(r')\left(\hat{r}, \underline{\underline{R}}(t)\,\underline{r}'\right)d^3r'/r^2 \\ &+ \int \underline{J}_2(r')\left(\hat{r}\right), R_{p_0}^{(t)} + S^{(t)} + (r' - p_0))d^3r'/r^2\end{aligned}$$

Let $\hat{\underline{l}}$ be a constant unit vector. Then

$$\left(\hat{\underline{l}}, \tilde{A}(t, \underline{r})\right) = \left(\underline{\underline{R}}(t)^T \hat{r}, \int \left(\underline{J}_1(\underline{r}'), \hat{\underline{l}}\right) \underline{r}' d^3 r'\right) / r^2$$

$$\left(\underline{\underline{R}}(t)^T \hat{r}, \underline{p}_0 \int \left(\underline{J}_1(\underline{r}'), \hat{\underline{l}}\right) d^3 r' / r^2\right)$$

$$+\left(S(t)^T \hat{\underline{r}}, \int \left(\underline{J}(\underline{r}'), \hat{\underline{l}}\right)\left(\underline{r}' - \underline{p}_0\right) d3r' / r^2\right)$$

This is a linear equation for $\underline{\underline{R}}(t), \underline{\underline{S}}(t)$ and

The rotations $\underline{\underline{R}}(t), \underline{\underline{S}}(t)$ can be estimated by least squares techniques.

❖❖❖❖❖

[6] Kinetic Energy of a Two Link Robot:

$\dot{\underline{r}} \to R_1(t)\underline{r}$ ($\underline{r} \in$ second link).

$\dot{\underline{r}} \to R_1(t) p_0 + R_2(t) R_1(t)(\underline{r} - p_0)$ ($r \in$ second link).

Let B_1 denote the volume of the bottom link at time $t = 0$ and B_2 that of two second link. Then the kinetic energy is

$$T = \frac{1}{2}\rho \int_{B_1} \left\| \underline{\underline{R_1'(t)\underline{r}}} \right\|^2 d^3 r + \frac{1}{2}\rho_2 \int_{B_2} \left\| R_1'(t) p_0 + \left(R_2 R_1\right)'(t)\left(\underline{r} - \underline{p}_0\right)\right\|^2 d^3 \underline{r}$$

Let $R_1(t) = R(t), R_2(t) R_1(t) = S(t)$. Then

$$T = \frac{1}{2}\rho_1 \int_{B_1} \|R'(t)\underline{r}\|^2 d^3 r + \frac{1}{2}\rho_2$$

$$\int_{B_2} \left\|R'(t) p_0 + S'(t)\left(\underline{r} - \underline{p}_0\right)\right\|^2 d^3 r$$

Let $\rho_1 \int_{B_1} \underline{r}\,\underline{r}^T d^3 r = \underline{\underline{I}}_1$,

Then $T_1 = \frac{1}{2} \int_{B_1} \|R'(t)\underline{r}\|^2 d^3 \underline{r} = \frac{1}{2} T_r \left\{\underline{\underline{I}}_1 R'^T(t) R'(t)\right\}$

$$\rho_2 \int_{B_2} \left\|R'(t) p_0 + S'(t)\left(\underline{r} - p_0\right)\right\|^2 d^3 r$$

$$= \mu\left(B_2\right)\|R'(t) p_0\|^2 + T_r\left\{\underline{\underline{I}}_2 S'^T(t) S'(t)\right\}$$

$$+2T_r\left\{R'(t) I_3 S'(t)\right\}$$

where $\mu(B_2) \equiv \mu_2 = \rho_2 \int_{B_2} d^3 r$,

$$\underline{\underline{I}}_2 = \rho_2 \int_{B_2} \left(\underline{r} - \underline{p}_0\right)\left(\underline{r} - \underline{p}_0\right)^T d^3r,$$

$$\underline{\underline{I}}_3 = \rho_2 \int_{B_2} \underline{p}_0 \left(\underline{r} - \underline{p}_0\right)^T d^3\underline{r} = \mu_2\left(\underline{p}_0\left(\underline{R}_2 - \underline{p}_0\right)^T\right)$$

where
$$\underline{R}_2 = \int_{B_2} \underline{r} d^3\underline{r}$$

Group theoretic analysis of the motion of a two-link Robot with the link being 3-D rigid bodies:

Each link of the Robot is a 3-D rigid Top.

The first link has its base point attached to the origin and the second link has its base point attached to some fixed point p on the top surface of the first link. The group of motions is shown to be a representation of $SO(3) \times SO(3)$. We assume that each link carries a d.c. current density an then computer the magnetic field produced by this current field after both links have undergone rotations around their respective base points.

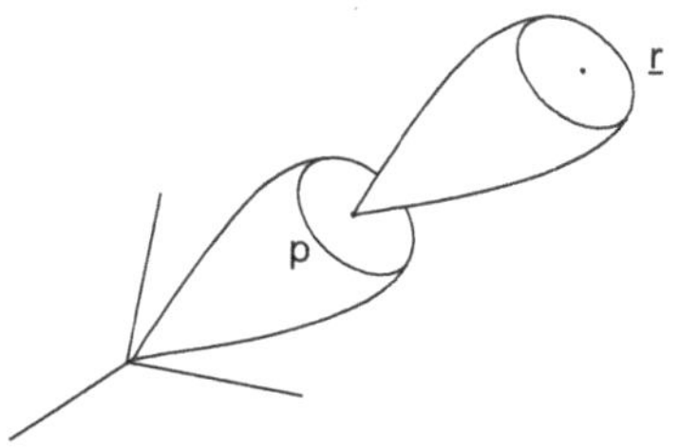

At time $t = 0$, the point p at the joint of the two links is at p_0 and the point r at a fixed location in the second link is $\underline{r}_0$. $\underline{q}_0 = \underline{r}_0 - \underline{p}_0$ is the position of $\underline{r}$ relative to $\underline{p}$ at time $t = 0$. The final configuration is arrived at by first applying a rotation R_1 to the entire system about the origin O, taking $\underline{p}_0 \to \underline{\underline{R}}_1 \underline{p}_0, \underline{r}_0 \to \underline{\underline{R}}_1 r_0$

So that $q_0 \to R_1 q_0$. We then apply a rotation R_2 to the second link around $p = R_1 p_0$ taking $R_1 q_0 \to R_2 R_1 q_0$.

The magnetic field produced by the Robot after the application of the two rotations is computed using the Biot-Savart law (non- relative) calculation. The final position of $\underline{r}$ is thus

$$\begin{aligned}\underline{r} &= R_1 p_0 + R_2 R_1 q_0 \\ &= R_1 p_0 + R_2 R_1 (r_0 - p_0) \\ &= R_2 R_1 r_0 + (R_1 - R_2 R_1) p_0\end{aligned}$$

and
$$\underline{p} = R_1 p_0$$

The group action is then

$$\begin{bmatrix} \underline{r}_0 \\ \underline{p}_0 \end{bmatrix} \to \begin{bmatrix} R_2 R_1 \underline{r}_0 + (R_1 - R_2 R_1)\underline{p}_0 \\ R_1 \underline{p}_0 \end{bmatrix} = \begin{bmatrix} R_2 R_1 & R_1 - R_2 R_1 \\ 0 & R_1 \end{bmatrix} \begin{bmatrix} \underline{r}_0 \\ \underline{p}_0 \end{bmatrix}$$

The group G is this given by

$$G = \left\{ \begin{bmatrix} R_2 R_1 & R_1 - R_2 R_1 \\ 0 & R_1 \end{bmatrix} : R_1, R_2 \in SO(3) \right\}$$

or equivalently,

$$G = \left\{ \begin{bmatrix} R & S - R \\ 0 & S \end{bmatrix} : R, S \in SO(3) \right\}$$

Composition law:

Let

$$T(R, S) = \begin{bmatrix} R & S - R \\ 0 & S \end{bmatrix}$$

Then

$$T(R_2, S_2)\cdot T(R_1, S_1) = \begin{bmatrix} R_2 & S_2 - R_2 \\ 0 & S_2 \end{bmatrix} \cdot \begin{bmatrix} R_1 & S_1 - R_1 \\ 0 & S_1 \end{bmatrix}$$

$$\begin{bmatrix} R_2 R_1 & S_2 R_1 - R_2 R_1 \\ 0 & S_2 S_1 \end{bmatrix} = T(R_2 R_1, S_2 S_1).$$

This shows that $(R, S) \to T(R, S)$ is a representation of $SO(3) \times SO(3)$ in $\mathbb{R}^6$. Now suppose that lower link carries an initial current density $(t = 0)$ $\underline{\underline{J}}_1(\underline{r})$ and the top link $\underline{\underline{J}}_2(\underline{r})$. Then after time t, the current density in space is given by

$$\underline{J}(t, \underline{r}) = \underline{J}_1\left(R_1^{-1}(t)\underline{r}\right) + \underline{J}_2\left(R_1^{-1}(t) R_2^{-1}(t)\underline{r} - \tilde{\underline{\underline{R}}}(t)\underline{p}_0\right)$$

Writing $R(t) = R_1(t)$ and $S(t) = R_2(t)\, R_1(t)$, we get

$$\underline{J}(t, \underline{r}) = J_1\left(R^{-1}(t)\underline{r}\right) + J_2\left(S(t)^{-1}\left(\underline{r} - \tilde{R}(t) p_0\right)\right)$$

where

$$\tilde{R}(t) = R_1(t) - R_2\, R_1(t)$$

Note that $J_1(\underline{r})$ is zero for $\underline{r} \in B_1$ and $J_2(\underline{r})$ is zero for $\underline{r} \in B_2$ where B_1 and B_2 are respectively the volumes of the first and the second link. Note that $B_1 \cap B_2 = \phi$. The magnetic field produced by this current density in space

$$B(t, \underline{r}) = \int_{\mathbb{R}^3} \frac{J(t, \underline{r}') \times (\underline{r} - \underline{r}')}{|\underline{r} - \underline{r}|^3} d^3 r'$$

(By Biot Savart's law): Assuming that $\hat{x}, \hat{y}, \hat{z}$ are the column vectors

$$\begin{pmatrix} 1 \\ 0 \\ 0 \end{pmatrix}, \begin{pmatrix} 0 \\ 1 \\ 0 \end{pmatrix} \text{ and } \begin{pmatrix} 0 \\ 0 \\ 1 \end{pmatrix}$$

respectively and $\underline{r} = \begin{pmatrix} X \\ Y \\ Z \end{pmatrix}, \underline{r}' = \begin{pmatrix} X' \\ Y' \\ Z' \end{pmatrix}$,

we have $B(t, \underline{r}) = \int \underline{\underline{G}}(\underline{r} - \underline{r}') \underline{J}(t, \underline{r}') d^3 r'$

where
$$\underline{\underline{G}}(\underline{r}) = \left[\frac{\hat{x} \times \underline{r}}{r^3}, \frac{\hat{y} \times \underline{r}}{r^3}, \frac{\hat{z} \times \underline{r}}{r^3} \right] = \frac{1}{r^3} \begin{bmatrix} 0 & -z & -y \\ -z & 0 & x \\ y & -x & 0 \end{bmatrix}$$

Here, $\underline{J}(t, \underline{r})$ is a 3 × 1 column vector and $\underline{\underline{G}}(\underline{r})$ is a 3 × 3 matrix $\underline{B}(t, \underline{r})$ is a 3 × 1 column vector. We can simplify the above to

$$B(t, \underline{r}) = \int_{B_1} \underline{\underline{G}}(\underline{r} - R(t)\underline{r}') \underline{J}_1(r')\, d^3 r'$$
$$+ \int_{B_2} \underline{\underline{G}}(\underline{r} - \tilde{R}(t)\, p_0 - S(t)\, \underline{r}') \underline{J}_2(r')\, d^3 r'$$

Note that
$$\widetilde{R} = R - S \tag{1}$$

The problems are (1) estimate the matrices $R(t)$, $S(t) \in SO\,(3)$ for a fixed t, from measurements, of $\underline{B}(t, \underline{r})$ at different $\underline{r} = \underline{r}_1\, \underline{r}_2, ..., \underline{r}_N$. (2) Estimate the square $\underline{\underline{R}}(t_k)$, $\underline{\underline{S}}(t_k)$, $k = 1, 2, ..., N$ at a square of time points $t_1\, t_2, ..., t_N$ from measurements of $\underline{B}(t, \underline{r})$ at times $t_1, ..., t_N$, $\underline{r} \in \{\underline{r}_1, ..., \underline{r}_M\}$. Here relativistic effects are not being taken into account *i.e.* we are not using related potentials .

Taking the spatial Fourier Transform of (1) gives

$$\hat{\underline{B}}(t, \underline{k}) = \int_{\mathbb{R}_3} \underline{B}(t, \underline{r}) \exp(i\, \underline{k} \cdot \underline{r}) d^3 r$$
$$= \int_{B_1} \exp\left(i\left(k, \underline{\underline{R}}(t)\, \underline{r}'\right)\right) \hat{\underline{\underline{G}}}(\underline{k})\, \underline{J}_1(\underline{r}') d^3 r'$$
$$+ \int_{B_2} \exp\left(i\left(\underline{k}, \tilde{\underline{\underline{R}}}(t)\, \underline{p}_0 + \underline{\underline{S}}(t)\, \underline{r}'\right)\right) \hat{\underline{\underline{G}}}(\underline{k})\, \underline{J}_2(\underline{r}') d^3 r'$$

Or equivalently,

$$\hat{G}(k)^{-1} \hat{\underline{B}}(\underline{k}) = \int_{B_1} \exp\left(i\left(\underline{k}, \underline{\underline{R}}(t)\, \underline{r}'\right)\right) \underline{J}_1(\underline{r})\, d^3 r$$
$$\int_{B_2} \exp\left(i\left(\underline{k}, \tilde{\underline{\underline{R}}}(t)\, p_0 + \underline{\underline{S}}(t) \underline{r}\right)\right) \underline{J}_2(\underline{r})\, d^3 r$$

Consider the function

$$\exp\left(i\left(\underline{k}, \underline{r}\right)\right)$$

We can express it as

$$\exp\left(i\left(\underline{k}, \underline{r}\right)\right) = \sum_{l,\, m} C(l, m, \underline{k}, r)\, Y_{lm}(\hat{r})$$

where $\left\{Y_{lm}\left(\hat{r}\right)\right\}_{|m| \le l,\, l \ge 0}$ are the spherical harmonics. They satisfy the transformation property

$$Y_{lm}\left(R^{-1}\hat{r}\right) = \sum_{|m'| \le l} \left[\pi_l(R)\right]_{m'm} Y_{lm'}(\hat{r}),\; R \in SO(3)$$

Where π_l is the lth irreducible representation of SO (3). They coefficients (l, m, k, r) are given by

$$\begin{aligned} C(l, m, \underline{k}, r) &= \int_{S^2} \exp\left(i\left(\underline{k}, \underline{r}\right)\right) \overline{Y_{lm}(\hat{r})}\, dS(\hat{r}) \\ &= \int_0^{\pi} \int_0^{2\pi} \exp\left(ikr\left(\sin\alpha \sin\theta \cos(\phi - \beta) + \cos\alpha \cos\theta\right)\right) \\ &\quad \overline{Y_{lm}(\theta, \phi)} \sin\theta\, d\theta\, d\phi \end{aligned}$$

Where $\underline{k} = k\,(\cos\beta \sin\alpha, \sin\beta \sin\alpha, \cos\alpha)$

Then
$$\begin{aligned} \exp\left(i\left(\underline{k}, R\underline{r}\right)\right) &= \sum_{l,\, m} C(l, m, \underline{k}, r)\, Y_{lm}\left(R\hat{r}\right) \\ &= \sum_{l,\, m,\, m'} C(l, m, \underline{k}, r) \left[\pi_l\left(R^{-1}\right)\right]_{m'm} Y_{lm'}(\hat{r}) \end{aligned}$$

and hence we get

$$\begin{aligned} \underline{\underline{\widehat{G}}}(\underline{k})^{-1} \underline{\widehat{B}}(\underline{k}) &= \sum_{lmm'} \left(\int C(l, m, \underline{k}, r) Y_{lm'}(\hat{r})\, \underline{J}_1(\underline{r})\, d^3r\right) \left[\pi_l\left(R^{-1}\right)\right]_{m'm} \\ &\quad + \exp\left(i\left(\underline{k}, \underline{\underline{\widetilde{R}}}\underline{p}_0\right)\right) \sum_{lmm'} \left(\int C(l, m, \underline{k}, r)\, Y_{lm'}(\hat{r}) \underline{J}_2(\underline{r}) d^3r\right) \end{aligned}$$

$$\begin{aligned} \left[\pi_l\left(S^{-1}\right)\right]_{m'm} &= \sum_{lmm'} \psi_{\underline{k}}\,[lmm'] \left[\overline{\pi_l(R)}\right]_{mm'} + \exp\left(i\left(\underline{k}, \underline{\underline{\widetilde{R}}}\, \underline{p}_0\right)\right) \\ &\quad \sum_{lmm'} \chi_{\underline{k}}\,[lmm'] \left[\overline{\pi_l(R)}\right]_{mm'} \end{aligned}$$

where $\psi_{\underline{k}}\,[lmm'] = \int_{B_1} C(l\; m\; \underline{k}, r)\, Y_{lm}(\hat{r}) \underline{J}_1(\underline{r}) d^3r$ and $\chi_{\underline{k}}\,[lmm']$

$$= \int_{B_2} C(l\; m\; \underline{k}, r)\, Y_{lm}(\hat{r})\, \underline{J}_2(\hat{r})\, d^3r$$

Now,
$$\exp\left(i\left(\underline{k}, \underline{\underline{R}}\, \underline{p}_0\right)\right) = \sum_{lm} C(l, m, \underline{k}, p_0) Y_{lm}\left(R\hat{p}_0\right)$$

$$= \sum_{lmm'} C(l, m, \underline{k}, p_0)\left[\overline{\pi_l(R)}\right]_{mm'} Y_{lm}\left(\hat{p}_0\right)$$

So $$\underline{\underline{\hat{G}}}(\underline{k})^{-1}\underline{\hat{B}}(\underline{k}) = \sum_{lmm'} \psi_k [lmm']\left[\overline{\pi_l(R)}\right]_{mm'} + \sum_{\substack{l_1 m_1 m_1' \\ l_2 m_2 m_2' \\ l_3 m_3 m_3'}} C(l_1, m_1, \underline{k}, p_0)$$

$$\chi_{\underline{k}}[l_2\, m_2\, m_2']\left[\overline{\pi_{l_1}(R)}\right]_{m_1 m'_1}\left[\overline{\pi_{l_2}(S)}\right]_{m_2 m'_2} - \left(\hat{p}_0\right)$$

$$\left(Y_{l_1 m_1'}\, Y_{l_3 m_3'}\right) \times C(l_3, m_3, \underline{k}, p_0)\left[\overline{\pi_{l_3}(S)}\right]_{m_3 m_3'}$$

❖❖❖❖❖

[7] Vector Potential Generated by a 2-Link Robot:

$\underline{r}_0 \to \underline{R}_1\, \underline{r}_0$, $\underline{r}_0 \in \underline{B}_1$ (lower link)

$$\underline{r}_1 \to R_1\, p_0 + R_2 R_1\left(\underline{r}_1 - \underline{p}_0\right),\ \underline{r}_1 \in B_2$$

$R_1 = R, R_2 R_1 = S$ (upper link)

$\underline{J}_1(\underline{r}) \simeq$ current density in lower link.

$\underline{J}_2(\underline{r}) \simeq$ current density in upper link at time $t = 0$

Current density at time t,

$$\underline{J}(\underline{r}) = \underline{J}_1(\underline{\underline{R}}^{-1}\underline{r}) + \underline{J}_2\left(p_0 + S^{-1}\left(\underline{r} - R\underline{p}_0\right)\right)$$

For field magnetic vector potential

$$\underline{\tilde{A}}(\hat{r}) = \int \underline{J}(\underline{r}')\,|\exp\left(jk\hat{r}\cdot\underline{r}'\right) d^3r'$$

$$= \int \underline{J}(\underline{r}')\,|\exp\left(jk\left(R^{-1}\hat{r}\cdot\underline{r}'\right)\right) d^3r'$$

$$+\int J_2(\underline{r}')\exp\left(jk\left(\hat{r}, Rp_0 + S(\underline{r}' - p_0)\right)\right) d^3r'$$

$$= \underline{A}_1(R^{-1}\hat{r}) + \underline{A}_2(S^{-1}\hat{r}) \times \exp(jk(R^{-1} - S^{-1})\hat{r}, p_0)$$

Let $$\underline{A}_1(\hat{r}) = \sum_{l,m} C_1(lm) Y_{lm}(\hat{r}),$$

$$\exp\left(-jk\left(\hat{r}, p_0\right)\right)\underline{A}_2(\hat{r}) = \sum_{l,m} C_2(lm) Y_{lm}(\hat{r}).$$

Then, $$\underline{A}_1(R^{-1}\hat{r}) = \sum_{lmm'} \underline{C}_1(lm)[\pi_l(R)]_{m'm}\, Y_{lm'(\hat{r})}$$

$$\underline{\tilde{A}}_2(\hat{r}) = \exp\left(jk\left(R^{-1}\hat{r}, p_0\right)\right) \sum_{l,mm'} \underline{C}_2(lm)[\pi_l(S)]_{m'm}\, Y_{lm'(\hat{r})}$$

$$\equiv \underline{A}_2(S^{-1}\hat{r})\exp(jk(R^{-1}-S^{-1})\hat{r},p_0)$$

Let $$\exp(jk(\hat{r},p_0)) = \sum_{l,m} d(lm)Y_{lm}(\hat{r})$$

So, $$\int_{s^2} \tilde{\underline{A}}_2(\hat{r})\overline{Y}_{l_o m_o}(\hat{r})dS(\hat{r})$$

$$= \int \sum_{\substack{lmm' \\ l'',m''}} C_2(lm)d(l''m'')Y_{l''m''}(R^{-1}\hat{r})[\pi_l(S)]_{m'm}Y_{lm'}(\hat{r})\overline{Y}_{l_o m_o}(\hat{r})dS(\hat{r})$$

$$\tilde{\underline{A}}(\hat{r}) = \underline{A}_1(R^{-1}\hat{r})$$

$$= \sum_{lm} C_1(lm)Y_{lm}(R^{-1}\hat{r}) = \sum_{lmm'} C_1(lm)[\pi_1(R)]_{m'm}Y_{lm'}(\hat{r})$$

$$\int \tilde{A}_1(\hat{r})\overline{Y_{lm}(\hat{r})}\, dS(\hat{r}) = \sum_{m'} C_1(lm')[\pi_l(R)]_{mm'}$$

So $$\int_{S^2} \tilde{\underline{A}}_1(\hat{r})\overline{Y_{l_o m_o}(\hat{r})}dS(\hat{r}) = \int_{S^2} (\tilde{A}_1(\hat{r})+\tilde{A}_2(\hat{r}))\overline{Y_{l_o m_o}(\hat{r})}dS(\hat{r})$$

$$= \sum_m \left[\pi_{l_0}(R)\right]_{m_0 m} C_1(lm)$$

$$+ \sum_{l_1 l_2 m_1 m_2 m_3 m_4} C_2(l_1 m_1)d(l_2 m_2)[\pi_{l_2}(R)]_{m_3 m_2}[\pi_{l_1}(S)]_{m_4 m_1}$$

$$\int Y_{l_1 m_4}(\hat{r})\overline{Y}_{l_0 m_0}(\hat{r})\dot{Y}_{l_2 m_3}(\hat{r})dS(\hat{r})$$

***n*-link Robot**

First link is attached to ground at its base point. Second link is attached to the first link at p_1.

⋮

k^{th} link is attached to the $(k-1)^{\text{th}}$ link at p_{k-1}.

⋮

n^{th} link is attached to the $(n-1)^{\text{th}}$ link at p_{n-1}.

All links first suffer a rotation R_1 around O.

2nd to n^{th} links suffer a rotation R_2 around $R_1 p_1$.

⋮

n^{th} link suffers a rotation R_n around

$R_1p_1+R_2R_1(p_2-p_1)+R_3R_2R_1(p_3-p_2)+\ldots+R_{n-1}\ldots R_1(p_{n-1}-p_{n-2})$

So let $\underline{r}_k$ be an initial point in the kth link, $k = 1,2,\ldots,n$. Then after the sequence of rotations $R_1, \ldots, R_n$,

$\underline{r}_k \to R_1p_1 + R_2R_1(p_2-p_1) + \ldots + R_{k-1}\ldots R_1(p_{k-1}-p_{k-2}) + R_kR_{k-1}\ldots R_1(r_k-p_{k-1})$
$k = 1, 2, \ldots, n.$

Writing $R_1 = S_1$, $R_2R_1 = S_2, \ldots R_k\ldots R_1 = S_k$, $k = 1, 2, \ldots, n$, we get

$\underline{r}_k \to S_1p_1 + S_2(p_2-p_1) + \ldots + S_{k-1}(p_{k-1}-p_{k-2}) + S_k(\underline{r}_k - \underline{p}_{k-1}), 1 \le k \le n.$

If $J_k(\underline{r})$ is the initial current density in the kth link, then after the sequence of rotation, its current density is

$$\tilde{J}_k(\underline{r}) = J_k(T_r^{-1}\underline{r})$$

where $T_k(\underline{r}) = S_1p_1 + S_2(p_2 - p_1) + \ldots + S_{k-1}(p_{k-1}-p_{k-2}) + S_k(\underline{r} - p_{k-1})$.

$$\text{Thus, } T_k^{-1}(\underline{r}) = \underline{p}_{k-1} + S_k^{-1}\left(\underline{r} - \sum_{j=1}^{k-1} S_j(p_j - p_{j-1}\right) p_0 = 0$$

and the total current density in the find state is given by

$$\tilde{J}(\underline{r}) = \sum_{k=1}^{n} \tilde{J}_k(\underline{r})$$

$$= \sum_{k=1}^{n} J_k\left(p_{k-1} + S_k^{-1}\left(\underline{r} - \sum_{j=1}^{k-1} S_j(\underline{p}_j - \underline{p}_{j-1})\right)\right)$$

❖❖❖❖❖

[8] Group Associated with Robot Motion:

$$(x, y, \theta_1, \theta_2) \to (x+a, y+b, \theta_1+\alpha_1, \theta_2+\alpha_2)$$

A belian group of transformation.
When the rigid rods have breadth and width dimension, *i.e.* they are like tops.
Euler angles for the first top $(\varphi_1, \theta_1, \psi_1)$.

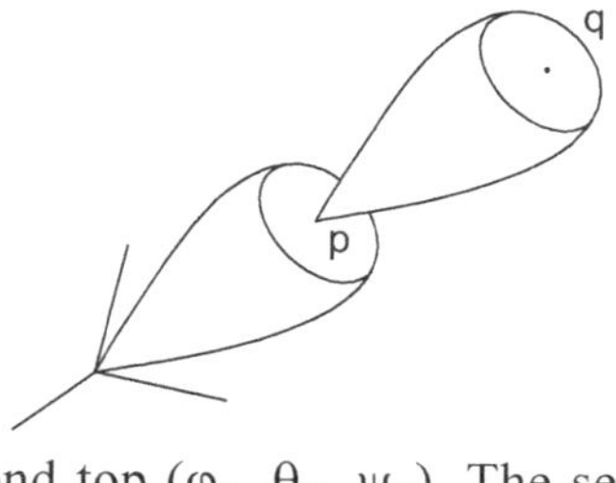

Eular angles for the second top $(\varphi_2, \theta_2, \psi_2)$. The second top is attached to a point $\underline{p}_0 = (x_0, y_0, l)$ on the first top at time $t = 0$. At time t, this point moves to

$$R(\varphi_1,\theta_1,\psi_1)\underline{p}_0 = \underline{p}_0 \equiv \underline{p}_t .$$

where $\varphi_1 = \theta_1(t)$, $\theta_1 = \theta_1(t)$, $\psi_1 = \psi_1(t)$.

The point $\underline{Q}$ on the second top was at time $t = 0$ located at

$$\begin{aligned}
\underline{p}_0 + \underline{q}_0 &= \underline{r}_0 \\
&= \int_0^\infty P_0^t dt \int_0^\infty n(Y_0 k_u \cap \{u < R\})\, du \\
&= \int_0^\infty \left\{ \left(\int_0^R Y_0 z k_u \, du \right) \right\} dt \\
&= \int_0^\infty p_k(r)\, dr \int_r^\infty \\
&= \int_0^R \int_R p(r)\, dr \int_r^\infty \frac{Q_u du}{\sqrt{2\pi u}} \int_0^r Y_0 k_u du \\
&= \int_r^\infty \frac{du}{\sqrt{2\pi u}} \mathbb{E} \int_0^r \{Y_0 k_u du \mid R = v\}
\end{aligned}$$

and at time, it is located at

$$\underline{p}_t + \underline{q}_t = \underline{r}_t = \underline{\underline{R}}(\varphi_1,\theta_1,\psi_1)\,\underline{p}_0 + \underline{\underline{R}}(\varphi_2,\theta_2,\psi_2)\,\underline{q}_0$$

Thus defines a transformation

$$\underline{r}_0 \to \underline{r}_t$$

by

$$\begin{aligned}
\underline{r}_t &= \underline{\underline{R}}(\varphi_1,\theta_1,\psi_1)\, p_0 + \underline{\underline{R}}(\varphi_2,\theta_2,\psi_2)(\underline{r}_0 - \underline{p}_0) \\
&= \underline{\underline{R}}_2 \underline{r}_0 + \left(\underline{\underline{R}}_1 - \underline{\underline{R}}_2\right)\underline{p}_0
\end{aligned}$$

With $\underline{p}_0$ fixed, the transformation group action is

$$\begin{aligned}
&T(\varphi_1,\theta_1,\psi_1,\varphi_2,\theta_2,\psi_2)\underline{r}_0 \\
&= \underline{\underline{R}}_2 \underline{r}_0 + \left(R_1 - \underline{\underline{R}}_2\right) p_0
\end{aligned}$$

Or equivalently

$$\begin{bmatrix} \underline{r}_0 \\ p_0 \end{bmatrix} \to \begin{bmatrix} R_2 & R_2 - R_1 \\ 0 & I_3 \end{bmatrix} \begin{bmatrix} \underline{r}_0 \\ p_0 \end{bmatrix}$$

Find generators for lie algebra of the lie group generated by all 6 × 6 matrix of the form

$$\begin{bmatrix} R_2 & R_2 - R_1 \\ 0 & I_3 \end{bmatrix}, \qquad R_1, R_2 \in SO(3) .$$

❖❖❖❖❖

[9] Radon Transform:

$$\widehat{f}(\underline{\alpha}, p) = \int_{\mathbb{R}^n} f(x)\delta(\langle\alpha, x\rangle - p)dx$$

$\|\underline{\alpha}\| -1, \alpha \in \mathbb{R}^n, p \in \mathbb{R}$. Then

$$\widehat{f}(\underline{\alpha}, -p) = \widehat{f}(-\underline{\alpha}, p),$$

$$\int_{\mathbb{R}} \widehat{f}(\underline{\alpha}, p)\exp(-iwp)dp = \int_{\mathbb{R}^n} f(x)\exp(-iw\langle\alpha, x\rangle)dx$$

$$\equiv \widetilde{f}(\underline{\alpha}, w) \text{ say.}$$

Thus,

$$f(x) = \int_{\substack{w \geq 0 \\ \alpha \in S^{n-1}}} \widehat{f}(\underline{\alpha}, w)\exp\Big(iw\langle\alpha, x\rangle w^{n-1} d\Omega(\alpha)dw\Big)(2\pi)^n$$

where $w^{n-1}d\Omega(\underline{\alpha})dw \equiv d^n(w\underline{x})$ (Lebesgue measure on $\mathbb{R}^n$)

Thus,

$$f(x) = \int \widehat{f}(\underline{\alpha}, p)\exp(-iwp)\exp\big(iw\langle\alpha, x\rangle\big)$$

$$\alpha \in S^{n-1}, w \geq 0, p \in \mathbb{R} \quad w^{n-1} d\Omega(\alpha)dwdp / (2\pi)^n$$

$$= (2\pi)^{-n}\int \widehat{f}\big(\hat{\underline{\xi}}, p\big)\exp\big(-i\|\underline{\xi}\|p\big)$$

$$\exp\big(i\langle\underline{\xi}, \underline{x}\rangle\big)$$

$$\|\underline{\xi}\|^{n-1}$$

$$d\Omega\big(\hat{\xi}\big)d\|\xi\|dp$$

$$= (2\pi)^{-n}\int i^{n-1}\left(\frac{d^{n-1}}{dp^{n-1}}\exp(-i\|\xi\|p)\right)$$

$$\widehat{f}\big(\hat{\underline{\xi}}, p\big)\exp\big(i\langle\underline{\xi}, \underline{x}\rangle\big)$$

$$d\Omega\big(\hat{\xi}\big)d\|\xi\|dp$$

$$= (2\pi)^{-n}(-i)^{n-1}\int \exp(-i\|\xi\|p)$$

$$\left(\frac{\partial^{n-1}}{\partial p^{n-1}}\widehat{f}(\hat{\xi}, p)\right)$$

$$\exp\big(i\langle\underline{\xi}, \underline{x}\rangle\big)$$

$$d\Omega\big(\hat{\xi}\big)d\|\xi\|dp$$

$$\left(\alpha = \hat{\xi} = \frac{\xi}{\|\underline{\xi}\|}\right)$$

Let $\psi_n\left(\underline{x}, p, \hat{\xi}\right) = \int_0^\infty \exp\left\{iw\left(\left\langle\hat{\xi}, \underline{x}\right\rangle - p\right)\right\} dw$

Then, $f(\underline{x}) = C_n \int \psi_n(\underline{x}, p, \hat{\xi}) S^{n-1} \times \mathbb{R} \frac{\partial^n}{\partial p^{n-1}} \widehat{f}\left(\hat{\xi}, p\right)$

where $C_n = (2\pi)^{-n}(-2)^{n-1}, d\Omega\left(\hat{\xi}\right)dp$

Equivalently,

$$\begin{aligned}
f(x) &= (2\pi)^{-n}(-2)^{n-1}\Box_x^{\frac{n-1}{2}}\left\{\int \widehat{f}\left(\hat{\xi}, p\right)\right. \\
&\quad \times \exp\left\{i\left(\left\langle\underline{\xi}, \underline{x}\right\rangle - \|\xi\| p\right)\right\} \\
&\quad d\Omega\left(\hat{\xi}\right) d\|\xi\| dp \\
&= C_n \Box_x^{\frac{n-1}{2}} \int \widehat{f}(\hat{\xi}, p) \\
&\quad S^{n-1} \times \mathbb{R}\, \psi_n\left(x, p, \hat{\xi}\right) \\
&\quad d\Omega\left(\hat{\xi}\right) dp \\
&= C_n \int \widehat{f}\left(\hat{\xi}, p\right)\left\{\Box_x^{\frac{n-1}{2}} \psi_n\left(x, p, \hat{\xi}\right)\right\} \\
&\quad S^{n-1} \times \mathbb{R} \quad d\Omega\left(\hat{\xi}\right) dp
\end{aligned}$$

❖❖❖❖❖

[10] Campbelt Baker-Hausdorff Formula:

$$ez(t) = e^X e^{tY} \cdot z(t) = \log\left(e^X \cdot e^{tY}\right)$$

$$\frac{d}{dt} e^{z(t)} = e^{z(t)} \frac{(I - \exp(-ad\ z(t)))}{ad\,(z(t))}\left(\frac{dz(t)}{dt}\right)$$

$$e^{-z(t)} \frac{d}{dt} e^{z(t)} = \frac{(I - \exp\,(-ad\, z(t)))}{ad\,(z(t))}\left(\frac{dz(t)}{dt}\right)$$

But $\frac{d}{dt} e^{z(t)} = e^{z(t)} Y.$

So $e^{-z(t)} \frac{d}{dt} e^{z(t)} = Y = \left(\frac{I - \exp(-odz(t))}{ad\,(z(t))}\right)\left(\frac{dz(t)}{dt}\right)$

or $$\frac{d}{dt}z(t) = \left(\frac{adz(t)}{I-\exp(-adz(t))}\right)(Y) = adz(t) = \log\left(e^{adX}e^{tadY}\right)$$

So, $$\frac{dz(t)}{dt} = \frac{\log\left(e^{adX}e^{tadY}\right)}{1-e^{adX}e^{tadY}}(Y)$$

$$= g\left(e^{adX}e^{tadY}\right)(Y)$$

where $$g(\xi) = \frac{\log(\xi)}{1-\xi}$$

Thus, $$z(1)-z(0) = \log\left(e^{X}e^{Y}\right)-X = \left[\int_0^1 g\left(e^{adX}e^{tadY}\right)dt\right](Y)$$

Suppose $$g(\xi) = \sum_{r=0}^{\infty} C_r(1-\xi)^r g(\xi) = \frac{\log(1-(1-\xi))}{(1-\xi)}$$

$$= -\sum_{r=1}^{\infty}\frac{(1-\xi)^{r-1}}{r}$$

$$= -\sum_{r=0}^{\infty}\frac{(1-\xi)^{r}}{(r+1)}$$

Thus, $$C_r = -\frac{1}{(1+r)}, r \geq 0.$$

Then $$g(\xi) = \sum_{r=0}^{r} C_r \sum_{k=0}^{r}\binom{r}{k}(-1)^k\xi^k$$

$$= \sum_{k=0}^{\infty} d_k\xi^k \text{ where } d_k = \sum_{r=k}^{\infty} C_r\binom{r}{k}(-1)^k$$

Not a convergent services. So

$$g(\xi) = \sum_{0\leq k\leq r\leq\infty} C_r\binom{r}{k}(-1)^k\xi^k$$

$$g\left(e^{adX}\cdot e^{tadY}\right) = \sum_{0\leq k\leq r\leq\infty} C_r\binom{r}{k}(-1)^k\left(e^{adX}\cdot e^{tadY}\right)^k$$

$$\left(e^{adX}\cdot e^{tadY}\right)^k(W) = e^{adX}\cdot e^{tadY}e^{adX}e^{tadY}\cdot e^{adX}e^{tadY}(W)$$

$$= \left(e^{adX}\sum_{m=0}^{\infty}\frac{t^m}{\lfloor m}(adY)^m\right)^k(W)$$

❖❖❖❖❖

[11] Problem on Rigid Body Motion:

[1] Consider the two link Robot with each link being a 3-D top. We've seen that the group of motion is defined by

$$G = \left\{\begin{bmatrix} R_2R_1 & R_1 - R_2R_1 \\ 0 & R_1 \end{bmatrix} : R_1, R_2 \in SO(3)\right\}$$

$$= \left\{\begin{bmatrix} R & S-R \\ 0 & S \end{bmatrix} : R_1, S \in SO(3)\right\}$$

Show that if we define

$$T(R, S) = \begin{bmatrix} R & S-R \\ 0 & S \end{bmatrix}$$

Then

$$T(R_1, S_2). T(R_1, S_1) = \begin{bmatrix} R_2 & S_2 - R_2 \\ 0 & S_2 \end{bmatrix} . \begin{bmatrix} R_1 & S_1 - R_1 \\ 0 & S_1 \end{bmatrix}$$

$$= \begin{bmatrix} R_2R_1 & S_2S_1 - R_2R_1 \\ 0 & S_2S_1 \end{bmatrix}$$

$$= T(R_2R_1, S_2S_1)$$

and hence

$(R, S) \to T(R, S)$

is a representation of $SO(3) \times SO(3)$ in $\mathbb{R}^6$.

❖❖❖❖❖

[12] Remarks on Theorems from V. Kac. Infinite Dimensional Lic Algebra. P.87

$$t_{w(\alpha)}(\lambda) = \lambda + \langle \lambda, k\rangle w(\alpha) - \left((\lambda / w(\alpha)) + \frac{1}{2}|\alpha|^2 \langle \lambda, k\rangle\right)\delta$$

$$wt_\alpha w^{-1}(\lambda) = \lambda + \left\langle w^{-1}(\lambda), k\right\rangle w(\alpha)$$

$$-\left(\left(w^{-1}(\lambda) | \alpha\right) + \frac{1}{2}|\alpha|^2 < w^{-1}(\lambda), K >| w(\delta)\right.$$

$$w(\delta) = \delta, \left\langle w^{-1}(\lambda), k\right\rangle = \langle \lambda, w(k)\rangle = \langle \lambda, k\rangle$$

$$\left(w^{-1}(\lambda) | \alpha\right) = (\lambda | w(\alpha))$$

so $$wt_\alpha w^{-1}(\lambda) = \lambda + \langle \lambda, k\rangle w(\alpha)$$

$$\left((\lambda | w(\alpha)) + \frac{1}{2}|\alpha|^2 \langle \lambda, k\rangle\right)\delta = t_w(\alpha)(\lambda).$$

$$\text{P249 } n \in N_{\mathbb{Z}} \Rightarrow {}_z n = (\alpha, \beta, u), n, v$$

$$(\alpha, \beta, u)(v) = t_\beta(v) + 2\pi i\alpha + (u - i\pi(\alpha \mid \beta))\delta$$

$$t_\rho(\lambda)(n \cdot v) = (\lambda + (\lambda \mid \delta)\rho + |n, v) - \left(\left(\lambda|\rho| + \frac{1}{2}(\rho \mid \rho)(\lambda \mid \delta)\right)(\delta \mid n \cdot v)\right.$$

$$= \left(t_\rho(\lambda) \mid t_\beta(v) + 2\pi i\alpha + (u - i\pi(\alpha \mid \beta))\delta\right)$$

$$= \left(t_\rho - \beta(\lambda) \mid v\right) + ((\lambda \mid \delta) + (\lambda \mid \delta)(\rho \mid \delta))$$

$$(u - i\pi(\alpha \mid \beta)) + 2\pi i(\lambda \mid \alpha)$$

$$(\lambda \mid \delta) = k \in \mathbb{Z}_+$$

$$\rho_2(\rho \mid \delta) = 0.$$

So,

$$t_\rho(\lambda)(n \cdot v) = \left(t_{\rho-\beta}(\lambda) \mid v\right) + k(u - i\pi(\alpha \mid \beta))$$

$$\delta(n \cdot v) = (\delta|n \cdot v) = (\delta|t_\beta(v)) = (\delta|v).$$

❖❖❖❖❖

[13] Theta a Function From Group Theory P 252 (V.Kac):

$$F(n.v) = F(v)\forall n \in N_{\mathbb{Z}}\ .$$

Let $n = (\alpha, \beta, u)$

$$n.u = t_\beta(u) + 2\pi i\alpha + (u - i\pi(\alpha \mid \beta))\delta$$

$$= v + (v \mid \delta)\beta - \left((v \mid \beta) + \frac{1}{2}(\beta \mid \beta)(v \mid \delta)\right)\delta$$

$$+ 2\pi i\alpha + (u - i\pi(\alpha \mid \beta))\delta$$

$$u \in 2\pi i\mathbb{Z}, u + i\pi(\alpha \mid \beta) \in 2\pi i\mathbb{Z}$$

$$\Rightarrow \quad u \in i\pi(\alpha \mid \beta) \in 2u - 2\pi i\mathbb{Z} \in 2\pi i\mathbb{Z}$$

$$p_\alpha(v) = v + 2\pi i\alpha.$$

$$F(n.v) = F(u)\forall n \in N_{\mathbb{Z}}$$

$\Rightarrow$ (taking $\beta = 0$)

$$F(v + 2\pi i\alpha + (u - i\pi(\alpha \mid \beta)\delta)) = F(v)$$

$$\Rightarrow \quad F(v + 2\pi i\alpha)\exp\{k(u - i\pi(\alpha \mid \beta))\delta\} = F(v)$$

$\Rightarrow$ $F(v + 2\pi i\alpha) = F(v)$ since $u - i\pi(\alpha|\beta) \leftarrow 2\pi i\mathbb{Z}$.

So

$$F(v + a\delta) = e^{ka}F(v),$$

$$F(v + 2\pi i\alpha) = F(v)a \in M, a \in \mathbb{C}$$

In term of co-ordination

$$v = 2\pi i\left[\sum_{s=1}^{l} z_s u_s - \tau\Lambda_0 + u\delta\right] \qquad \text{(sec (13.24))}$$

Consider $\quad F = e^{k\Lambda_0} \sum_{\gamma\in M^*} a_\tau(\gamma)e^{\gamma}$

Then

$$F = (v + a\delta) = e^{k}(\Lambda_0(v) + a\Lambda_0(\delta))$$

$$\sum_{\gamma\in M^*} a_\tau(\gamma)e^{\gamma(v)+a(\gamma|\delta)}$$

$$= e^{ka}F(v)$$

Since

$$\Lambda_0(\delta| = (\Lambda_0|\delta) = 1,$$

$$(\gamma|\delta) = 0\,(M\perp\delta)$$

Also,

$$F(v + 2\pi i\alpha) = e^{k\Lambda_0(v)} \sum_{\gamma\in M^*} a_\tau(\gamma)e^{(\gamma(v)+2\pi i(\gamma|\alpha))}$$

$$= F(v), \alpha\in M$$

Since $(\gamma\,|\,\alpha)e\mathbb{Z}$ for $\gamma\in M^*, \alpha\in M$ by definition of M^*

$$F(n.v) = F(v)$$

$$\Rightarrow \quad F(t_B(v)) = F(v) \text{ (take } \alpha = 0,\ u = 0)$$

$$\forall\beta\in M$$

$$\Rightarrow \quad \exp\big(k(\Lambda_0\,|\,t\beta(v))\big) \sum_{\gamma\in M^*} a_\tau(\gamma)\exp\big(\gamma\,|\,t_\beta(v)\big)\big)$$

$$= \exp(k(\Lambda_0(v))) \sum_{\gamma\in M^*} a_\tau(\gamma)\exp((\gamma\,|\,v))$$

$$\rho(\Lambda_0\,|\,\tau_\beta(v)) = (t_{-\beta}(\Lambda_0\,|\,v))$$

$$= \left(\Lambda_0 - (\Lambda_0\,|\,\delta)\beta + \left((\Lambda_0\,|\,\beta) - \frac{1}{2}(\beta\,|\,\beta)(\Lambda_0\,|\,\delta)\right)\delta\,|\,v\right)$$

$$= (\Lambda_0\,|\,u) - (\beta\,|\,v) + -\frac{1}{2}(\beta\,|\,\beta)(v\,|\,\delta)$$

$(\therefore (\Lambda_0|\delta) = 1, (\Lambda_0|\beta) = 0)$

$$(\gamma\,|\,t_\beta(v)) = (t_{-\beta}(\gamma)\,|\,v)$$

$$= \left(\gamma + (\gamma\,|\,\beta) - \frac{1}{2}(\beta\,|\,\beta)(\gamma\,|\,\delta)\delta\,|\,v\right)$$

$$= (\gamma\,|\,v) + (\gamma\,|\,\beta)(\delta\,|\,v)$$

$$\therefore \qquad (\gamma|\delta) = 0(M \perp \delta)$$

$$\therefore \quad k(\Lambda_0 \mid t_\beta(v)) + (\gamma \mid t_\beta(v)) = k(\Lambda_0 \mid v) + (\gamma \mid v) + \left((\gamma \mid \beta) - \frac{k}{2}(\beta \mid \beta)\right)\delta - k(\beta \mid v)$$

So,
$$F_\circ t_\beta = \exp(e\Lambda_0) \sum_{\gamma \in M^*} a_\tau(\tau) \exp\left\{\left(\gamma + (\gamma \mid \beta) - \frac{k}{2}(\beta \mid \beta)\right)\delta\right\}$$
$$\exp(-k\beta)$$
$$= \exp(k\Lambda_0) \sum_{\gamma \in M^*} a_\tau(\gamma).\exp(\gamma - k\beta)$$
$$\exp\left\{\left((\gamma \mid \beta) - \frac{k}{2}(\beta \mid \beta)\right)\delta\right\}$$

(Note that $\gamma - k\beta \in M^*$) (Let $\gamma - k\beta = \xi$)

$$= \exp(k\Lambda_0) \sum_{\xi \in M^*} a_\tau(k\beta + \xi)\exp(\xi)$$
$$\times \exp\left\{\left(\frac{k}{2}(\beta \mid \beta) + (\xi \mid \beta)\right)\delta\right\}$$

So
$$F_\circ t\beta = F$$

$$\Rightarrow \sum_{\xi \in M^*}\left[a\tau(k\beta + \xi)\exp\left\{\left(\frac{k}{2}(\beta \mid \beta) + (\xi \mid \beta)\right)\delta\right\} - a_\tau(\xi)\right].\exp(\xi)$$
$$= 0$$

Now
$$(\delta|v) = \left(\delta \mid 2\pi i\left(\sum_1^l z_s v_s - \tau\Lambda_0 + \delta\right)\right) = -2\pi i\tau$$

Hence
$$F_\circ t\beta = F \Rightarrow \sum_{\xi \in M^*}\left[a_\tau(k_\beta + \xi)\exp\left\{-\left(\frac{k}{2}(\beta \mid \beta) + (\xi \mid \beta)\right)2\pi i\tau\right\}\right.$$
$$\left. - a_\tau(\xi)\right]\exp((\xi \mid v)) = 0$$

$\forall \beta \in M$ Thus,

$$a_\tau(k\beta + \xi).\exp\{-ik\pi(\beta \mid \beta)\tau\}.\exp\{-2\pi i(\xi \mid \beta)\tau\}$$
$$= a_\tau(\xi) \ \forall \xi \in M^* \ \forall \beta \in M.$$

Now, it follows that

$$a_\tau(k\beta + \xi)\exp\left\{i\pi k^{-1}(k\beta + \xi \mid k\beta + \xi)\tau\right\}$$
$$= a_\tau(k\beta + \xi).\exp\{i\pi k(i\pi k(\beta \mid \beta)\tau\}.\exp\{-2\pi i(\beta \mid \xi)\tau\}$$
$$\times \exp\left\{-i\pi k^{-1}(\xi \mid \xi)\tau\right\}$$

$$= a_\tau(\xi)\exp\left\{-i\pi k^{-1}(\xi \mid \xi)\tau\right\}$$

i.e., $$a_\tau(\gamma)\exp\left\{-i\pi k^{-1}(\gamma \mid \gamma)\tau\right\}$$

depends only on γ mod kM.

❖❖❖❖❖

[14] Remark about Iwasawa decomposition

[Remark from Helgason's Book]

$$G = KMAN \; M \text{ centralizes } A,\ M \subset K$$

$$H \in a,\ X \in \mathcal{H},\ [H, X] = \alpha(H)\, X.$$

$$Z \in \mathcal{H} \Rightarrow [H, [Z, X]] = -\{[Z, [X, H]] + [X, [H, Z]]\}$$

$$= [Z, \alpha(H) X] = \alpha(H)\,[Z, X]$$

So $[Z, X] \in N$. Actually, we've proved that

$$[M, g_\alpha] \subset g_\alpha\ g(g_\alpha) \subset g - \alpha$$

Since $$[H, X] = \alpha(H) X$$

$$[\theta(H), \theta(X)] = \alpha(H)\,\theta(X) \Rightarrow -[H, \theta(X)] = \alpha(H)\,\theta(X)$$

Since

$$H \in a \subset p \Rightarrow \theta(H) = -H.$$

$$m * \bar{n}_1 = \bar{n}(m*\bar{n}_1)m(m*\bar{n}_1)\exp(B(m*\bar{n}_1)n_B(m*\bar{n}_1))$$

$$\in \bar{N}MAN$$

$$\therefore \quad m^{*2}\,\bar{n}_1 = \bar{n}_1 = m*\bar{n}(m*\bar{n}_1)m(m*\bar{n}_1)\exp(B(m*\bar{n}_1))n_B(m*\bar{n}_2)$$

$$= \bar{n}(m*\underline{n}(m*\bar{n}_1))m(m*\bar{n}(m*\bar{n}_1))$$

$$\exp(B(m*\bar{n}(m*\bar{n}_1))n_B(m*\bar{n}(m*\bar{n}_1))$$

$$\times m(m*\bar{n}_1)\exp(B(m*\bar{n}_1))n_B(m*\bar{n}_2)$$

Since $m(m*\bar{n}_1)$ commutes with $\exp(B(m*\bar{n}(m*\bar{n}_1)))$ it follows that

$$\exp(B(m*\bar{n}(m*\bar{n}_1))) + B(m*\bar{n}_1)$$

$$= 1$$

Hence

$$B(m*\bar{n}(m*\bar{n}_1)) = B(m*\bar{n}(m*\bar{n}_1)) - B(m*\bar{n}_1)$$

❑❑❑

4

Stochastic Filtering and Control, Interacting Partials

[1] Bernoulli Filters:

Generalization of Bernoulli filters to K state filter. At each time n, one of {0, 1, 2, ..., k–1} particles (targets) is present.

$\phi_{n+1|n}\left(X_r, r \mid X_s, S\right)$

is the probability density at time $n+1$ of the state specified by $X_r \in \mathbb{R}^r$ and $r \in \{0, 1, 2, ... k-1\}$ given that at time n, there were S particles located at

$X_S \in \mathbb{R}^S, s \in \{0,1,2, ... k-1\}$

Note that $X_0 = \phi$ (empty) and so

$$\phi_{n+1|n}\left(X_0, 0 \mid X_s, s\right) \equiv q_s(X_s)$$
$$\phi_{n+1|n}\left(X_r, r \mid X_0, 0\right) \equiv p_r(X_r)$$

where $q_0(X_0) = q_0$ are + *ve* constants

$$= p_0(X_0) = p_0$$

$\phi_{n+1|n}$ is the transition probability density of a K-particle filter. if $K = 2$, we get the Bernoulli filter. $\phi_{n+1|n}$ describes a Markov process whose state space is random being either one of $\mathbb{R}^r, 0 \le r \le K-1$ at the each time instant.

The state at any time n is (X_r, r)

$\in \mathbb{R}^r \times \{0,1, ... k-1\}$.

We have $\sum_{r=1}^{k-1} \int_{\mathbb{R}^r} \phi_{n+1|n}(X_r\ r \mid X_s, s)\, d^r X_r + q_s(X_s) = 1.$

$\forall\, s = 0, 1, 2, ..., k-1.$

Note that for $s = 0$, this condition becomes

$$\sum_{R=1}^{k-1} \int_{\mathbb{R}^r} p_r(X_r) d^r X_r + q_o \ = 1$$

The measurement model is

$$\underline{Z}_n \ = \ \underline{h}(X_n, n) + \underline{V}_n, n = 0, 1, 2, \ldots \tag{1}$$

where $\{\underline{V}_n\}$ is iid with pdf $p_v(\underline{v})$.

where $h(X_n, n) \ = \ h(X_{rn}, r_n, n) \ = \ h_n(X_{rn}, r_n)$

where h_n: $\bigcup_{r=0}^{k-1} (\mathbb{R}^r \times \{r\}) \to \mathbb{R}^6$

(X_{r_n}, n, r_n) is the state of the system at time n, $X_{r_n}, n \in \mathbb{R}^{r_n}$ conditional pdf given observations.

Let $\xi_n = \left(X_{r_n}, n, r_n\right)$ be the state at time n.

Let $Y_n = \{z_k : k \le n\}$ be the collection of observation upto time n. Then

$$p\left(\xi_{n+1} \middle| Y_n + 1\right)$$

$$= \frac{p(\underline{Z}_{n+1}, \underline{Y}_n, \xi_{n+1})}{p(\underline{Z}_{n+1}, \underline{Y}_n)}$$

$$= \frac{p\left(\underline{Z}_{n+1} \middle| \underline{\xi}_{n+1}\right) p\left(\underline{\xi}_{n+1} \middle| \underline{Y}_n\right)}{\int p\left(\underline{Z}_{n+1} \middle| \xi_{n+1}\right) p\left(\underline{\xi}_{n+1} \middle| \underline{Y}_n\right) d\underline{\xi}_{n+1}}$$

$$= \frac{\int p_v\left(\underline{Z}_{n+1} - h_{n+1}(\xi_{n+1})\right) p(\xi_{n+1} \mid \xi_n) d(\xi_n \mid Y_n) d\xi_n}{\int p_v\left(Z_{n+1} - h_{n+1}(\xi_{n+1})\right) p(\xi_{n+1} \mid \xi_n) p(\xi_n \mid Y_n) d\xi_n d\xi_{n+1}} \tag{2}$$

By this we mean that

$$p\left(X_{r_{n+1}, n+1} r_{n+1} \mid Y_{n+1}\right)$$

$$= \frac{\displaystyle\sum_{r_n=0}^{k-1} \int p_v\left(\underline{Z}_{n+1} - h_{n+1}\left(X_{r_{n+1}, n+1, r_{n+1}}\right)\right) p\left(X_{r_{n+1}, n+1}, r_{n+1} \mid X_{r_n}, n, r_n\right) p\left(X_{r_n}, n \mid Y_n\right) d^{r_n} X_{r_n}, n}{\displaystyle\sum_{r_n, r_{n+1}=0}^{k-1} \int p_v\left(Z_{n+1} - h_{n+1}\left(X_{r_{n+1}, n+1, r_{n+1}}\right)\right) p\left(X_{r_{n+1}, n+1}, r_{n+1} \mid X_{r_n}, n, r_n\right) p\left(X_{r_n, n}, n \mid y_n\right) d^{r_n} X_{n_{r,n}} \, d^{r_{n+1}} X_{r_{n+1}, n+1}}$$

Or equivalently,

$$p\left(X_{r_{n+1},n+1} r_{n+1} \mid Y_{n+1}\right)$$

$$= \frac{\sum_{r_n} \int \phi_{n+1|n}\left(X_{r_{n+1},\,n+1}, r_{n+1} \mid Xr_{n,n},\, r_n\right) \times p_v\left(\underline{Z}_{n+1} - h_{n+1}\left(X_{r_{n+1},\,n+1},\, r_{n+1}\right)\right) p\left(X_{r_{n,n}}, r_n \mid Y_n\right) d^{r_n} X}{\sum_{r_n,\, r_{n+1}} \int \phi_{n+1|n}\left(X_{r_{n+1},\,n+1}, r_{n+1} \mid Xr_{n,n},\, r_n\right) \times p_v\left(\underline{Z}_{n+1} - h_{n+1}\left(X_{r_{n+1},\,n+1},\, r_{n+1}\right)\right) p\left(X_{r_{n,n}}, r_n \mid Y_n\right) d^{r_n} X\, d^{r_{n+1}} X}$$

or equivalently,

$$p_{n+1}\left(X_r, r \mid Y_{n+1}\right)$$

$$= \frac{\sum_s \int \phi_{n+1}\,|_n\left(X_r, r \mid X_s, s\right) p_\text{v}\left(\underline{Z}_{n+1} - h_{n+1}\left(X_r, r\right)\right) p_n\left(X_s, s \mid Y_n\right) d^s X_s}{\sum_{r,\,s} \int n\, d^r X_r\, d^s X_s}$$

$$= p_\text{v}\left(\underline{Z}_{n+1} - \underline{h}_{n+1}\left(X_r, r\right)\right) \times \frac{\sum_s \int \phi_{n+1|n}\left(X_r, r \mid X_s, s\right) p_\text{v}\left(X_s, s \mid Y_n\right) d^s X_s}{\sum_{r,s} \int p_\text{v}\left(\underline{Z}_{n+1} - h_{n+1}\left(X_r, r\right)\right) \phi_{n+1|n}\left(X_r, r \mid X_s, s\right) p_n\left(X_s, s \mid Y_n\right) d^r X_r d^s X_s}$$

❖❖❖❖❖

[2] A Problem in Large Deviation Theory:

$$L'_n = n^{-1} \sum_{i=i}^{m} X_i \delta_{yi} \{X_i\}_{i=1}^{\infty} \text{ iid } r.v.\text{'s}.$$

$$n = n(m),\ n/m\ \beta \to \in (0, \infty).$$

i.e. $$\lim_{m \to \infty} \frac{n(m)}{m} = \beta\, L^y_m \to \mu$$

where $$L^y_m = \frac{1}{m} \sum_{i=1}^{m} \delta_{y_i}$$

$$\int f dL'_n = n^{-1} \sum_{i=1}^{m} X_i f(y_i)$$

$$\mathbb{E} \exp\left(n \int f\, d\, L'_n\right) = \mathbb{E} \exp\left(\sum_{i=1}^{m} X_i\, f(y_i)\right) = \prod_{i=1}^{m} \exp\left(\Lambda_X\left(f\left(y_i\right)\right)\right)$$

$$\frac{1}{n}\log \mathbb{E}\exp\left(n\int f dL_n'\right) = \frac{1}{n}\sum_{i=1}^{m} \Lambda_X\left(f\left(y_i\right)\right)$$

$$\underset{m\to\infty}{\to} \frac{1}{\beta}\int \Lambda_X\left(f\left(y\right)\right)d\mu\left(y\right)$$

(Gartner-Ellis limiting logarithmic moment generating function)

Rate function of $\{L_n'\}$: $I_X(v) =$

Verification of the formulas.

❖❖❖❖❖

[3] Large Deviations in Image Trajectory on a Screen Taken by a Robot Camera in Motion:

Abstract: We consider a camera attached to the end of single and double link Robot arms. The Robot link moves by virtue of motors attached to its base-point. There is a flat screen separating the object from the Robot. Noise in the motors (assumed to be weak) causes random fluctuation in the Robot motion result resulting in random motion of the image of the object taken by the Robot Camera on the screen. We compute the large-deviation rate function of the image trajectory and use it to determine the (small) probability that the image escapes away from the field of vision on the screen.

Camera is located at

$$\underline{r}_2(t) = \underline{r}_1^{(t)} + l_1\hat{n}_1(t) + l_2\hat{n}_2(t)$$

where $\underline{r}_1(t)$ is the base-point of the robot and l_1 is the length of the first link while l_2 is the length of the second link. $\hat{n}_1(t)$ and $\hat{n}_2(t)$ are respectively unit vectors along the first and second links.

If $\underline{w}_1(t)$ is the angular velocity of the first link relative to its base point at $\underline{r}_1(t)$ and $\underline{w}_2(t)$ is the angular velocity of the second link relatives to its base point at $\underline{r}_1(t) + l_1\hat{n}_1(t)$.

Then $$\dot{\underline{r}}_2(t) = \dot{\underline{r}}_1(t) + l_1\underline{w}_1(t)\times\hat{n}_1(t) + l_2\underline{w}_2(t)\times\hat{n}_2(t)$$

In spherical polar co-ordinates $\hat{n}_1(t) = (\theta_1(t), \phi_1(t))$ and $\hat{n}_2(t) = (\theta_2(t), \phi_2(t))$, we have

$$\hat{n}_1(t) = \hat{X}\cos\phi_1\sin\theta_1 + \hat{Y}\sin\phi_1\sin\theta_1 + \hat{Z}\cos\theta_1,$$

$$\hat{n}_2(t) = \hat{X}\cos\phi_2\sin\theta_2 + \hat{Y}\sin\phi_2\sin\theta_2 + \hat{Z}\cos\theta_2$$

Thus, $$\frac{d\hat{n}_1}{dt} = \dot{\theta}_1\hat{\theta}_1 + \sin\theta_1\dot{\phi}_1\hat{\phi}_1$$

$$\frac{d\hat{n}_2}{dt} = \dot{\theta}_2\,\hat{\theta}_2 + \sin\theta_2\,\dot{\phi}_2\hat{\phi}_2$$

Let
$$\underline{w}_1 = w_{1\theta}\hat{\theta}_1 + w_{1\phi}\hat{\phi}_1$$
$$\underline{w}_2 = w_{2\theta}\hat{\theta}_2 + w_{2\phi}\hat{\phi}_2$$

Then,
$$\underline{w}_1 \times \hat{n} = -w_{1\theta}\hat{\phi}_1 + w_{1\phi}\hat{\theta}_1$$
$$\underline{w}_2 \times \hat{n} = -w_{2\theta}\hat{\phi}_2 + w_{2\phi}\hat{\theta}_2$$

Thus,
$$w_{1\theta} = -\sin\theta_1\dot{\phi}_1,\ w_{2\theta} = -\sin\dot{\theta}_2\dot{\phi}_2$$
$$w_{1\theta} = \dot{\theta}_1, w_{2\phi} = \dot{\theta}_2$$

The Camera at $\underline{r}_2$ (t) takes a picture of the object at $\underline{r}_0$ = and the image is recorded on the screen placed on the X-Z plane. Let $(\xi_x(t),\xi_z(t))$ denote the coordinates (X and Z) of the image on the screen. Then writing $\underline{r}_2(t) = (X_2(t), Y_2(t), Z_2(t))$ and $\underline{r}_0$ $= (X_0, Y_0, Z_0)$, we have for some $\lambda \in \mathbb{R}$, $\underline{r}_0 + \lambda(\underline{r}_2 - \underline{r}_0) = (\xi_x, 0, \xi_z)$

Thus equality the Y-component, we get

$$Y_0 + \lambda\,(Y_2 - Y_0) = 0, \text{ or } \lambda = \frac{Y_0}{Y_0 - Y_2}$$

Then,
$$\xi_X = X_0 + \lambda\,(X_2 - X_0)$$

$$= X_0 - \frac{Y_0(X_2 - X_0)}{(Y_2 - Y_0)} = \frac{(X_0Y_2 - X_2Y_0)}{Y_2 - Y_0}$$

$$\xi_X = Z_0 + \lambda\,(Z_2 - Z_0) = Z_0 - \frac{Y_0(Z_2 - Z_0)}{(Y_0 - Y_2)} = \frac{(Z_0Y_2 - Z_2Y_0)}{Y_2 - Y_0}$$

Assuming $\underline{r}_2(t) = (X_2(t), Y_2(t), Z_2(t))$ to be differentiable, we have

$$\dot{\xi}_z = \frac{\left((Y_2 - Y_0(Z_0\dot{Y}_2 - Y_0\dot{Z}_2) - (Z_0Y_2 - Z_2Y_0)\dot{Y}_2\right)}{(Y_2 - Y_0)^2}$$

$$= \frac{\left\{\left(Y_2Z_0\,\dot{Y}_2 - Y_2Y_0\dot{Z}_2 - Y_0Z_0\dot{Y}_2 + Y_0^2\dot{Z}_2\right)\left(Z_0Y_2\dot{Y}_2 - Z_0Y_2\dot{Y}_2\right)\right\}}{(Y_2 - Y_0)^2}$$

$$= \left\{\dot{Y}_2(Z_2Y_0 - Y_0Z_0) + \dot{Z}_2\left(Y_0^2 - Y_2Y_0\right)\right\}/(Y_2 - Y_0)^2$$

$$= \left\{(Z_2 - Z_0)Y_0\dot{Y}_2 + Y_0\dot{Z}_2(Y_0Y_2)\right\}/(Y_2 - Y_0)^2$$

and likewise $\left\{(X_2 - X_0)Y_0\dot{Y}_2 + Y_2\dot{X}(Y_2 - Y_2)\right\}/(Y_2 - Y_0)^2$.

Large deviations for the image of an object on a screen taken by a Robot Camera in the presence of machine noise

Problem Formulation

The Camera attached to the tip of a single link Robot arm is located at

$$= \underline{r}_1 + l\hat{n}(t) \equiv \underline{r}_2(t)$$

where $\underline{r}_1 \in \mathbb{R}^3$ is fixed and $\hat{n}(t)$ is a unit vector

$$\hat{n}(t) = \cos\phi(t)\sin\theta(t)\hat{x} + \sin\phi(t)\sin\theta(t)\hat{y} + \cos\theta(t)\hat{z}$$

We have $\frac{d\underline{r}_2(t)}{dt} = l\frac{d\hat{n}(t)}{dt} = l\underline{w}(t)\times\hat{n}(t) = \underline{w}(t)\times\left(\underline{r}_2(t)-\underline{r}_1\right)$ (1)

$w(t)$ can be found using

$$\frac{d\hat{n}(t)}{dt} = \dot{\theta}(t)\hat{\theta}(t) + \dot{\phi}(t)\sin\theta(t)\hat{\phi}(t) \tag{2a}$$

On the one hand while on the other.

Writing $\underline{\omega}(t) = \underline{\omega}_\theta(t)\hat{\theta}(t) + \underline{\omega}_\phi(t)\hat{\theta}_\phi(t)$

we get $\underline{\omega}(t)\times\hat{n}(t) = \omega_\theta\hat{\theta}\times\hat{n} + \omega_\phi\hat{\phi}\times\hat{n} = -\omega_\phi\hat{\phi} + \omega_\phi\hat{\theta}$

Thus,

$$\omega_\theta = -\dot{\phi}\sin\dot{\theta},$$

$$\omega_\phi = \dot{\theta}$$

Lagragian for the link:

$$\mathcal{L}\left(\theta, \varphi, \dot{\theta}, \dot{\varphi}\right) = \frac{1}{2}mL^2\left(\dot{\theta}^2 + \sin^2\theta\dot{\phi}^2\right) - \frac{mgL}{2}\cos\theta$$

Two motors are attached at the base providing $\tau_\theta(t)$ and $\tau_\phi(t)$ torques. Thus the equations of motion of the links are

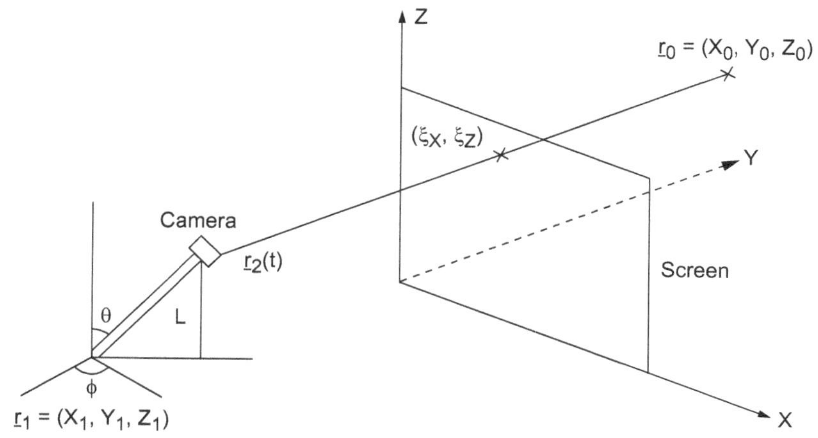

Eqns. of motion

$$\frac{d}{dt}\frac{\partial\alpha}{\partial\dot{\theta}} - \frac{\partial\alpha}{\partial\theta} = \tau_\theta(t),$$

$$\frac{d}{dt}\frac{\partial\alpha}{\partial\dot{\phi}} - \frac{\partial\alpha}{\partial\phi} = \tau_\phi(t)$$

These give

$$mL^2\ddot{\theta} - mL^2\sin\dot{\theta}\cos\theta\dot{\phi}^2 - mgL\sin\dot{\theta} = \tau_\theta(t),$$

$$mL^2\frac{d}{dt}\left(\sin^2\theta\dot{\phi}\right) = \tau_\phi(t)$$

$$i.e.\ \ddot{\theta} - \sin\theta\cos\theta\dot{\phi}^2 - \frac{g}{L}\sin\dot{\theta} = \frac{\tau_\phi(t)}{mL^2},$$

$$\sin^2\theta\ddot{\phi} + 2\sin\dot{\theta}\cos\theta\dot{\theta}\dot{\phi} = \frac{\tau_\phi(t)}{mL^2}$$

Steady state (equilibrium position of the link):

$$\dot{\theta} = 0,\ \ddot{\theta} = 0,\ \ddot{\phi} = 0,\ \ddot{\phi} = 0.$$

Thus $\quad \tau_\phi(t) = 0,\ \theta = \theta_0 = -\sin^{-1}\left(\frac{\tau_{\phi^0}}{mgL}\right)$

Assuming $\underline{\tau}_\theta(t) = \text{Constant } \underline{\tau}_{\theta^0}$

Let $\phi = \phi_0$.

Small deviations from equilibrium:

$$\theta(t) = \theta_0 + \varepsilon\delta\,\underline{\theta}\,(t),\ \phi(t) = \phi_0 + \varepsilon\delta\theta(t)$$

Let

$$\tau_\theta(t) = \underline{\tau}_{\theta_0} + \varepsilon\, w_1(t)$$

$$\tau\phi(t) = \varepsilon w_2(t)$$

where $w_1(t)$ and $w_2(t)$ are independent zero mean Gaussian random processes. Then the $O(\varepsilon)$ terms in the dynamical eqn. give

$$\delta\ddot{\theta} - \frac{g}{L}\cos\theta_0\delta\theta = \frac{w_1(t)}{mL^2},$$

$$\sin^2\theta_0\delta\ddot{\phi} = \frac{w_2(t)}{mL^2}.$$

So, $\quad \delta\theta(t) = \left(\frac{L}{g}\sec\theta_0\right)^{1/2}\left(mL^2\right)^{-1}\int_0^t \sinh\left(\sqrt{\frac{g}{L}\cos\theta_0}\,(t-\tau)\right)w_1(\tau)\,d\tau$

$$\delta\phi(t) = \frac{1}{mL^2\sin^2\theta_0}\int_0^t (t-\tau)\,w_2(\tau)\,d\tau$$

Thus, $\{(\delta\theta(t), \delta\phi(t))\}_{t\geq 0}$ is zero mean Gaussian random process with correlations

$$\mathbb{E}\{(\delta\theta(t+\tau)\,\delta\theta(t))\} = \frac{1}{mL^2 g\cos\theta_0}\int_0^{t+\tau}\int_0^t \sinh\left\{\sqrt{\frac{g}{L}\cos\theta_0}\,(t+\tau-t_1)\right\}$$

$$\times \sinh\left\{\sqrt{\frac{g}{L}\cos\theta_0}\,(t-t_2)\right\} R_{w_1 w_1}(t_1,t_2)\,dt_1\,dt_2$$

$$\equiv R_\theta(t+\tau, t)$$

$$\mathbb{E}\{\delta\theta(t+\tau)\cdot\delta\phi(t)\} = 0,$$

$$\mathbb{E}\{\delta\phi(t+\tau)\cdot\delta\phi(t)\} = \frac{1}{(mL^2\sin^2\theta_0)^2}\int_0^{t+\tau}\int_0^{t}(t+\tau-t_1)(t-t_2)R_{w_2 w_2}(t,t_2)\,dt_1 dt_2$$

$$= R_{\phi\phi}(t+\tau, t)$$

The rate function for $(\delta\theta(\cdot), \delta\phi(\cdot))$ over the interval $[0, T]$ is then

$$I_T(\delta\theta(\cdot), \delta\phi(\cdot)) = \frac{1}{2}\int_0^T\int_0^T \Big(R_{\theta\theta}(t_1,t_2)\,\delta\theta(t_1)\,\delta\theta(t_2)$$
$$+R_{\phi\phi}(t_1,t_2)\,\delta\phi(t_1)\,\delta\phi(t_2)\Big)\,dt_1\,dt_2$$

❖❖❖❖❖

[4] P406 Revuz and Yor:

$$a^{-2}C_t = \inf\left\{u : a2\int_0^u \exp\left(2a\beta_s^{(a)}\right)ds > t\right\}$$

is to be proved

$$\beta_s^{(a)} = a^{-1}\beta_{a^2 t}$$

$$\sigma_r = \inf\{t \ge 0 \mid \log\rho_s = r\}$$

$$C_t = \langle \log\rho\rangle_t = \int_0^t \frac{ds}{\rho_s^2}$$

$$B_t = \log\rho_{\tau_t} \qquad \tau = C^{-1}$$

$$\int_0^{\tau_t}\frac{ds}{\rho_s^2} = t\frac{d\tau_t}{\rho_{\tau_t}^2} = dt$$

$$\beta_t^{(a)} = a^{-1}\beta_{a^2 t} = a^{-1}\log\rho_{\tau a^2 t}$$

$$\rho_{\tau_t} = \exp(\beta_t)$$

$$\exp\left(2a\beta_t^{(a)}\right) = \exp\left(2\beta_{a^2 t}\right)$$

$$= \rho^2_{\tau_{a^2 t}}$$

$$\int_0^{u_t} \exp\left(2a\beta_s^{(a)}\right)ds = \frac{t}{a^2}$$

$$\Rightarrow \quad \int_0^{u_t} \rho^2_{\tau a^2 s}\, ds = \frac{t}{a^2}$$

$$\Rightarrow \quad \int_0^{a^2 u_t} \rho^2_{\tau_s} \frac{ds}{a^2} = \frac{t}{a^2}$$

$$\Rightarrow \quad \int_0^{a^2 ut} \rho^2_{\tau_s}\, ds = t$$

$$\Rightarrow \quad \int_0^{\tau u_t a^2} \rho^2_{\tau C_s}\, dC_s = t \qquad \left(\tau = C^{-1}\right)$$

$$\Rightarrow \quad \int_0^{\tau u_t a^2} \rho^2_s dC_s = t$$

$$\Rightarrow \quad \rho^2_{\tau a^2 u_t} dC_{\tau a^2 u_t} = dt$$

$$\Rightarrow \quad \rho^2_{\tau a^2 u_t} d\left(a^2 u_t\right) = dt$$

$$\Rightarrow \quad d\left(a^2 u_t\right) = \frac{dt}{\rho^2_{\tau a^2 u_t}}$$

Comparing this with

$$dC_t = \frac{dt}{\rho_t^2} = \frac{dt}{\rho^2_{\tau C_t}}$$

gives $\quad a^2 u_t = C_t$

i.e. $\quad u_t = a^{-2} C_t$ **Q.E.D.**

Problems from Revuz and Yor

(Continuous Marketing and BM).

Let $\underline{B}^{(t)} = B_1(t) + iB_2(t) \approx$ Conformal planer BM.

$\log \underline{B} = \log|\underline{B}| + i\, Arg\, B$ is a Martingale (local)

since its Laplacian Vanishes.

Let $\quad |\underline{B}| = \rho_t \equiv \sqrt{B_1^2 + B_2^2}\,,$

$$\theta_t = \text{Arg}\, \underline{B} = tan^{-1}\left(\frac{B_2(t)}{B_1(t)}\right)$$

$\log \rho$ and θ are Martingales (local).

$$(d \log \rho_t)^2 = \left(\frac{d\rho_t}{\rho_t}\right)^2 = \left(\frac{B_1 dB_1 + B_2 dB_2}{\rho_t^2}\right)^2$$

$$= \frac{dt}{\rho_t^2} = (d\theta_t)^2. \qquad \text{(property of conformal Martingales)}$$

$$\langle \log \rho \rangle_t = \int_0^t \frac{ds}{\rho_s^2} = C_t \qquad \text{say.}$$

Let $\tau_t = (C^{-1})_t$

i.e. $\displaystyle\int_0^{\tau_t} \frac{ds}{\rho_s^2} = C\tau_t = t$

Then $\dfrac{d\tau_t}{\rho_t^2} = dt,$

i.e. $d\tau_t = \rho_t^2 dt$

$$\tau_t = \int_0^t \rho_s^2 ds$$

$(\log \rho)_{\tau_t} = \log \rho_{\tau_t}$ is a B.M. and θ_{τ_t} is also a BM and there two B.M.'s are independent since $d \log \rho . d\theta = 0$.(property of conformal Martingales.

Let $\beta_t = \log \rho_{\tau_t}$, $Y_t = \theta_{\tau_t}$

Let $\sigma_r = \min\{t \geq 0 | \rho_t = r\}$

Then $\rho_{\tau_t} = e^{\beta_t}$

$$\rho_t = e^{\beta_{C_t}}$$

and hence
$$\sigma_r = \min\{t \geq 0 | e^{\beta_{C_t}} = r\}$$
$$= \min\{t \geq 0 | \beta_{C_t} = \log r\}$$
$$= \tau_{T_{\log r}}$$

or equivalently $T_{\log r} = C_{\sigma_r}$

where $T_a = \min\{t \geq 0 | \beta_t = a\}$

By optional stopping theorem

$$\mathbb{E}\left[\exp\left(\lambda \beta_{C_t} - \frac{\lambda^2 C_t}{2}\right)\right] = e^{\lambda \beta_0} = \rho_0^{\lambda} \qquad (\rho_0 \approx \text{non randomly})$$

i.e. $$\mathbb{E}\left[\rho_t^\lambda \exp\left(-\frac{\lambda^2 C_t}{2}\right)\right] = \rho_0^\lambda$$

Now let $r_1 < r_2$. Then assuming $\rho(0) \in (r_1, r_2)$,

$$\sigma_{r_1} \Lambda \sigma_{r_2} = \min\{t \geq 0 | \rho_t(r_1, r_2)\}.$$

and since $$\rho_t^\lambda \exp\left(-\frac{\lambda^2 C_t}{2}\right) = \exp\left(\lambda \beta_{C_t} - \frac{\lambda^2 C_t}{2}\right)$$

is a Martingale, to follows that

$$\mathbb{E}\left[\rho_{\sigma_{r_1}\Lambda\sigma_{r_2}}^\lambda \exp\left(-\frac{\lambda^2}{2} C\sigma_{r_1}\Lambda\sigma_{r_2}\right)\right] = \rho_0^\lambda$$

Thus $$\mathbb{E}\left[r_1^\lambda \exp\left(-\tfrac{\lambda^2}{2} C\sigma_{r_1}\right)\chi_{\sigma_{r_1} < \sigma_{r_2}} + r_2^\lambda \exp\left(-\frac{\lambda^2}{2} C\sigma_{r_2}\right)\chi_{\sigma_{r_2} < \sigma_{r_1}}\right]$$
$$= \rho_0^\lambda$$

Equivalently

$$\mathbb{E}\left[\exp\left(-\frac{\lambda^2}{2}\left(T_{\log r_1} \Lambda T_{\log r_2}\right)\right)\left(r_1^\lambda - r_2^\lambda\right)\chi_{\sigma_{r_1} < \sigma_{r_2}}\right]$$
$$= \rho_0^\lambda - r_2^\lambda \mathbb{E}\left[\exp\left(-\frac{\lambda^2}{2}\left(T_{\log r_2} \Lambda T_{\log r_1}\right)\right)\right]$$

Now $$\sigma_{r_1} < \sigma_{r_2} \Rightarrow C_{\sigma_{r_1}} \leq C_{\sigma_{r_2}}$$

$\Rightarrow$ $$T_{\log r_1} \leq T_{\log r_2}.$$

Thus,

$$\mathbb{E}\left[\exp\left(-\frac{\lambda^2}{2} T_{\log r_1}\right)\chi_{\{\sigma_{r_1} < \sigma_{r_2}\}}\right]$$
$$= \frac{\left\{\rho_0^\lambda - r_2^\lambda \mathbb{E}\left[\exp\left(-\frac{\lambda^2}{2}\left(T_{\log r_2} \Lambda T_{\log r_1}\right)\right)\right]\right\}}{r_1^\lambda - r_2^\lambda}$$

Now $$\mathbb{E}\left[\exp\left(\lambda \beta_{Ta} - \frac{\lambda^2 T_a}{2}\right)\right] = \rho_0^\lambda = e^{\lambda \beta_0}$$

Thus $$\mathbb{E}\left[\exp\left(\frac{-\lambda^2 T_a}{2}\right)\right] = \rho_0^\lambda e^{-\lambda a}$$

$$\mathbb{E}\left[\exp\left(-\frac{\lambda^2 T_{\log r_1}}{2}\right)\chi_{\{\sigma_{r_1} < \sigma_{r_2}\}}\right] = \mathbb{E}\left[\exp\left(-\frac{\lambda^2}{2} C_{\sigma_{r_1}}\right)\chi_{\{\sigma_{r_1} < \sigma_{r_2}\}}\right]$$

$$= \frac{\rho_0^{\lambda} - r_2^{\lambda}\mathbb{E}\left[\exp\left(-\frac{\lambda^2}{2}T_{\log r_1}\Lambda T_{\log r_2}\right)\right]}{r_1^{\lambda} - r_2^{\lambda}}$$

Now $$\mathbb{E}\left[\exp\left(\lambda\beta_{T_a\Lambda T_b} - \frac{\lambda^2}{2}T_a\Lambda T_b\right)\right] = \exp\left(\lambda_0^{\beta}\right)$$

where $a < \beta_0 < b$, β_0 non random.

Thus,

$$\mathbb{E}\left[\exp\left(\lambda\left(\beta_{T_a\Lambda T_b} - \frac{a+b}{2}\right) - \frac{\lambda^2}{2}T_a\Lambda T_b\right)\right] = \exp\left(\lambda\left(\beta_0 - \left(\frac{a+b}{2}\right)\right)\right)$$

Changing λ to $-\lambda$, adding and using

$$\cosh\left\{\lambda\left(\beta_{T_a\Lambda T_b} - \frac{a+b}{2}\right)\right\} = \cosh\left(\frac{\lambda(b-a)}{2}\right)$$

$\left(\text{since } \beta_{T_a\Lambda T_b} \in \{a,b\}\right)$, it follows that

$$\mathbb{E}\left[\exp\left(-\frac{\lambda^2}{2}T_a\Lambda T_b\right)\right] = \frac{\cosh\left\{\lambda\left(\beta_0 - \frac{a+b}{2}\right)\right\}}{\cosh\left\{\frac{\lambda(b-a)}{2}\right\}}$$

Thus, $$\mathbb{E}\left[\exp\left(-\frac{\lambda^2}{2}C_{\sigma_{r_1}}\right)\chi_{\{\sigma_{r_1} < \sigma_{r_2}\}}\right]$$

$$= \frac{\left\{\rho_0^{\lambda} - r_2^{\lambda}\frac{\cosh\left\{\lambda\left(\log\rho_0 - \log\sqrt{r_1 r_2}\right)\right\}}{\cosh\left\{\frac{\lambda}{2}\log\left(\frac{r_2}{r_1}\right)\right\}}\right\}}{\left(r_1^{\lambda} - r_2^{\lambda}\right)}$$

❖❖❖❖❖

[5] Problem from "Continuous Martingales and Brownian Motion" by Revuz and Yor:

Consider the filteration

$$\begin{aligned}\tilde{\mathcal{F}}_S^t &= \sigma\left\{\mathcal{F}_S^B, B_t\right\}\\ &\equiv \sigma\left\{\mathcal{F}_S^B U\{B_t\}\right\}, S \geq 0\end{aligned}$$

A process of the form $S \to f(B_S, B_t)$, $S \geq 0$ is adapted to $\left\{\tilde{\mathcal{F}}_S^t\right\}_{S\geq 0}$ (t is fixed)

With respect fo this "augmented filteration", we defite the process

$$Z_S = Z_S^t = (B_s - B_t)\,\theta(B_s - B_t).$$

Then $\{Z_S^t\}_{S\geq 0}$ is adapted to $\{\tilde{\mathcal{F}}_S^t\}_{S\geq 0}$. We can define stochastic integrals *w.r.t.* dB_S for process $s \to f(B_s, B_t)$ relative to this augmented filteration. Then

$$dZ_S = \theta(B_S - B_t)dB_S + \frac{1}{2}\delta(B_S - B_t)dS$$

and hence
$$Z_t - Z_0 = \int_0^t \theta(B_S - B_t)dB_S + \frac{1}{2}\int_0^t \delta(B_S - B_t)dS$$

or equivalently,

$$Bt\,\theta(-B_t) = \int_0^t \theta(B_S - B_t)dB_S + \frac{1}{2}L_t^{B_t}$$

i.e.
$$\frac{1}{2}L_t^{B_t} = B_t\theta(-B_t) - \int_0^t \theta(B_S - B_t)dB_S\,.$$

Note that
$$\theta(x) = \begin{cases} 1 & x \geq 0 \\ 0 & x < 0 \end{cases}$$

❖❖❖❖❖

[6] Problem from Revuz and Yor:

$$\sqrt{n}\left(\int_0^t\left(\delta\left(B_s - \frac{x+a}{n}\right) - \delta\left(B_s - \frac{x}{n}\right)\right)ds\right)$$
$$= X_n(t, x)$$

$$\int_{\mathbb{R}} f(x)X_n(t,x)\,dx = n\sqrt{n}\int_0^t\big(f(nB_s - a) - f(nB_s)\big)ds$$
$$= Mn(t) \text{ say.}$$

Let
$$g_\alpha^\mu(x) = f(x-a) - f(x)\,.$$

$$dg_a(nB_t) = ng_a'(nB_t)dB_t + \frac{n^2}{2}g_a^u(nB_t)dt$$

Thus,
$$g_a(nB_t) - g_a(O) = n\int_0^t g_a'(nB_t)dB_t + \frac{\sqrt{n}}{2}M_n(t)$$
$$= \frac{1}{\sqrt{n}}(g_a(nB_t) - g_a(0))$$
$$= \sqrt{n}\int_0^t g_a'(nB_t)dB_t + \frac{1}{2}M_n(t)$$

It follows that

$$\lim_{x\to\infty}\frac{1}{2}M_n(t) = \lim_{x\to\infty}\sqrt{n}\int_0^t g'_a(nB_t)\,dB_t$$

Consider the Martingale

$$M^n_{(t)} = \sqrt{n}\int_0^t \psi(nB_s)\,dB_s$$

We have

$$\begin{aligned}
\langle M^n\rangle_t &= n\int_0^t \psi^2(nB_s)\,ds \\
&= n\int_{[0,\,t]\times\mathbb{R}} \psi^2(x)\,\delta(x-nB_s)\,ds\,dx \\
&= \int_{[0,\,t]\times\mathbb{R}} \psi^2(x)\,\delta\left(\frac{x}{n}-B_s\right)ds\,dx \\
&\to \left(\int_{\mathbb{R}}\psi^2(x)\,dx\right)\int_0^t \delta(B_s)\,ds \\
&= \|\psi\|^2 L_t
\end{aligned}$$

It follows that these exists a *BM* $Y(.)$ independent of $B(.)$ such that

$$\lim_{n\to\infty}\frac{1}{2}M^n(t) = Y(\|\psi\|L_t)\forall t$$

where L_t is the local time of $B(.)$ at zero

Thus,

$$\lim_{n\to\infty}\frac{1}{2}M_n(t) = Y(\|g'_a\|L_t)$$

Now take $f(x) = \delta(x-x_0)\ x_0 \in \mathbb{R}$ fixed.

Then

$$\begin{aligned}
M_n(t) &= X_n(t,\,x_0) \\
&= \sqrt{n}\int_0^t\left(\delta\left(B_s-\frac{x+a}{n}\right)-\delta\left(B_s-\frac{x}{n}\right)\right)ds
\end{aligned}$$

Then

$$\begin{aligned}
g''_a(x) &= f(x-a)-f(x) \\
&= \delta(x-a-x_0)-\delta(x-x_0)
\end{aligned}$$

so

$$g''_a(x) = \theta(x-a-x_0)-\theta(x-x_0)$$

Thus,

$$\|g'_a\| = \int_{x_0}^{a+x_0} dx = a$$

and we decline that

$$\lim_{n\to\infty}\frac{1}{2}M_n(t) = \lim_{n\to\infty}\frac{\sqrt{n}}{2}\int_0^t\left(\delta\left(B_s-\frac{x+a}{n}\right)-\delta\left(B_s-\frac{x}{n}\right)\right)ds$$

❖❖❖❖❖

[7] Collected Papers of Varadhan-Interacting Particle System, Vol. 4. Hamiltonian System and Hydrodynamical Equations:

$$\mathcal{H}\left(\underline{X},\underline{p}\right) = \sum_{\alpha=1}^{N} \underline{p}^2\alpha/2m + \sum_{1\le\alpha<\beta\le N} V\left(\underline{X}_\alpha-\underline{X}_\beta\right)$$

$$\underline{p}_\alpha;\underline{X}_\alpha\in\mathbb{R}^3,\underline{X} = \left(\underline{X}_\alpha\right)_{\alpha=1}^{N}\in\mathbb{R}^{3N},\underline{p} = \left(\underline{p}_\alpha\right)_{\alpha=1}^{N}\in\mathbb{R}^{3N}$$

Initial configuration is random (random configuration envolving under determination dynamics). Let $f_t(\underline{X},\underline{p})$ be the *pdf* of $\left(\underline{X}(t),\underline{p}(t)\right)$. Liowvilles equation implies

$$\frac{\partial f_t}{\partial t}+\sum_\alpha\left(\frac{d\underline{X}_\alpha}{dt},\frac{\partial f_t}{\partial \underline{X}\alpha}\right)+\sum_\alpha\left(\frac{d\underline{p}_\alpha}{dt},\frac{\partial f_t}{\partial \underline{p}_\alpha}\right) = 0$$

or
$$\frac{\partial f_t}{\partial t}+\frac{1}{m}\sum_\alpha\left(\underline{p}_\alpha,\frac{\partial f_t}{\partial \underline{X}_\alpha}\right)-\sum_{\alpha\ne\beta}\left(V'\left(\underline{X}_\alpha-\underline{X}_\beta\right),\frac{\partial f_t}{\partial \underline{p}_\alpha}\right) = 0$$

(1) Assuming $V(\vec{X}) = V(-\vec{X}),\vec{X}\in\mathbb{R}^3$.

More generally for any Hamiltonian $\mathcal{H}(\underline{X},\underline{p})$, we have

$$\frac{d\underline{X}_\alpha}{dt} = \frac{\partial\mathcal{H}}{\partial\underline{p}_\alpha},\frac{d\underline{p}_\alpha}{dt} = -\frac{\partial\mathcal{H}}{\partial\underline{X}_\alpha}$$

and therefore

$$\frac{\partial f_t}{\partial t}\sum_\alpha\left\{\left(\frac{\partial\mathcal{H}}{\partial\underline{p}_\alpha},\frac{\partial f_t}{\partial\underline{X}_\alpha}\right)-\left(\frac{\partial\mathcal{H}}{\partial\underline{X}_\alpha},\frac{\partial f_t}{\partial\underline{p}_\alpha}\right)\right\} = 0$$

Empirical density

$$\rho\left(\underline{t},x\right) = \sum_{\alpha=1}^{N}\delta\left(\underline{X}-\underline{X}_\alpha(t)\right)\in\mathbb{R}$$

Emprical momentum density

$$\underline{\pi}(t,\underline{\mu}) = \sum_{\alpha=1}^{N}\underline{p}_\alpha(t)\delta\left(\underline{x}-\underline{x}_\alpha(t)\right)\in\mathbb{R}^3$$

Empirical momentum flux

$$\underline{\pi}(t,\underline{x}) = \frac{1}{m}\sum_{\alpha=1}^{N}\underline{p}_\alpha(t)\,\underline{p}_\beta(t)^T\delta\left(\underline{x}-\underline{x}_\alpha(t)\right)$$

Empirical energy density

$$e(t, \underline{x}) = \sum_{\alpha=1}^{N} e_\alpha(t)\delta(\underline{x} - \underline{x}_\alpha(t))$$

When

$$e_\alpha(t) = \frac{\underline{p}_\alpha^2(t)}{\alpha m} + \sum_{\beta:\beta\neq\alpha} V\left(\underline{x}_\alpha^{(t)} - \underline{x}_\beta^{(t)}\right)$$

is the energy of the αth particle. $\left(\underline{X} \in \mathbb{R}^3, \underline{p} \in \mathbb{R}^3\right)$.

In general for any of observable $\psi(\underline{x}, \underline{p})$ we define its empirical density as

$$\rho_\psi(t, \underline{x}) = \sum_{\alpha-1}^{N} \psi\left(\underline{x}_\alpha(t), \underline{p}_\alpha(t)\right)\delta\left(x - \underline{x}_\alpha(t)\right)$$

Its average value is

$$\langle\rho\psi\rangle\langle t, \underline{x}\rangle = \int f_t\left(\{\underline{X}_\alpha\}, \{\underline{p}_\alpha\}\right)$$

$$= \sum_{\beta=1}^{N} \psi\left(\underline{x}_\beta, \underline{p}_\beta\right) \delta(\underline{x} - \underline{x}_\beta)$$

$$\prod_{\alpha=1}^{N} d^3\underline{x}_\alpha d^3\underline{p}_\alpha$$

$$= \sum_{\beta=1}^{N} \int \psi\left(\vec{\underline{x}}, \vec{\underline{p}}\right) f_{t\alpha}\left(\vec{x}, \vec{p}\right) d^3\vec{p}$$

where $f_{t\alpha}\left(\vec{x}, \vec{x}\right)$ is the Martingale density of $\left(x_\alpha(t), \underline{p}_\alpha(t)\right)$:

$$f_{t\alpha}\left(\underline{x}_\alpha, \underline{p}_\alpha\right) = \int f_t\left(\left\{\underline{x}_\beta, \underline{p}_\beta\right\}_1^N\right)$$

$$\prod_{\beta\neq\alpha} d^3\underline{x}_\beta d^3\underline{p}_\beta$$

We get from (1), by integrating over $\underline{x}_\beta, \underline{p}_\beta, \beta \neq \alpha$

$$= \frac{\partial f_{t\alpha}}{\partial t}(\underline{x}_\alpha, \underline{p}_\alpha) + \frac{1}{m}\left(\underline{p}_\alpha, \frac{\partial f_{t\alpha}}{\partial x}(\underline{x}_\alpha, \underline{p}_\alpha)\right)$$

$$- \sum_{\beta:\beta\neq\alpha} \int \left(V'(\underline{x}_\alpha - \underline{x}_\beta), \frac{\partial f_{t\alpha\beta}}{\partial \underline{p}_\alpha}\left(\underline{x}_\alpha, \underline{x}_\beta, \underline{p}_\alpha, \underline{p}_\beta\right)\right)$$

$$d^3\underline{x}_\beta d^3\underline{p}_\beta = 0$$

where $f_{t\alpha\beta}$ is the joint (marginal) density of $\left(\underline{x}_\alpha, \underline{x}_\beta, \underline{p}_\alpha, \underline{p}_\beta\right)$:

$$f_{t\alpha\beta}\left(\underline{x}_\alpha, \underline{x}_\beta, \underline{p}_\alpha, \underline{p}_\beta\right) = \int f_t(\underline{x}, \underline{p}) \prod_{\gamma\neq\alpha,\beta} d^3\underline{x}_\gamma d^3\underline{p}_\gamma$$

If we assume that all first and second order marginal densities are the same and we denote there marginals by $f_t^{(1)}$ and $f_t^{(2)}$ respectively, then

$$\frac{\partial f_t^{(1)}}{\partial t}(\underline{x}_1, \underline{p}_1) + \frac{1}{m}\left(\underline{p}, \frac{\partial f_t^{(1)}}{\partial x}(\underline{x}_1, \underline{p}_1)\right)$$

$$-(N-1)\int\left(V'(\underline{x}_1 - \underline{x}_2).\frac{\partial f_t^{(2)}}{\partial \underline{p}_1}(\underline{x}_1, \underline{x}_2, \underline{p}_1, \underline{p}_2)\right)$$

$$d^3\underline{x}_2 d^3\underline{p}_2 = 0$$

The BBGKY can be continued further. ($p[q]$ Vol. 4. S.R.S Varadhan collected papers) We've to prove $\int\sum_{\mu,j} A_j^\mu \lambda_j^\mu\, dx = 0$. From (2.25) (assuming $\lambda^4 = \frac{1}{4} =$ constant)

$$\lambda^4.A_j^\mu\lambda_j^\mu = -\left(\lambda^j q^0\right)\lambda_j^0 - \left(\lambda^j q^i + P\delta_{ij}\right)\lambda^i_{,j}$$

$$+\left(-\lambda^j q^4 + \lambda^j P\right)\lambda^4_{,j}$$

$$= -\lambda^j q^0 \lambda^0_{,j} - \lambda^j q^i \lambda^i_{,j} - \psi\lambda^i_{,i}$$

$$(\psi = P) = -\left\{\lambda^j \frac{\partial\psi}{\lambda^0}\lambda^0_{,j} + \lambda^j\lambda^i_{,j}\frac{\partial\psi}{\partial\lambda^i} + \psi\lambda^i_{,i}\right\}$$

$$\left(\lambda q^\mu = \partial\psi/\partial\lambda^\mu\right)$$

$$= -\lambda^j\left\{\frac{\partial\psi}{\partial\lambda^0}\lambda^0_{,j} + \frac{\partial\psi}{\partial\lambda^i}\lambda^i_{,j}\right\} + \lambda^i\psi_{,i}$$

(after neglecting a total divergence $(\psi\lambda^i)_{,i}$)

Using $\psi_{,i} = \frac{\partial\psi}{\partial\lambda^0}\lambda^0_{,i} + \frac{\partial\psi}{\partial\lambda^j}\lambda^i_{,i}$

this becomes

$$\lambda^4 A_j^\mu \lambda_j^\mu = -\lambda^j\psi_{,j} + \lambda^j\psi_{,j} = 0.$$

Current in a simple exclusion model on $\mathbb{Z}^3 N$.

Each particle has a charge $q.X, Y \in \mathbb{Z}_N^3, \frac{X}{N}, \frac{Y}{N} \in [0,1]^3$.

If $\eta_t(X) = 1, \eta_t(Y) = 0$ and in time dt, the particle of X jumps to the site Y, then its velocity is $\frac{Y-X}{dt}$ or after scaling $\frac{(Y-X)}{N\,dt}$. The outward current density at $X\left(\equiv\frac{X}{N}\right)$ is times taken to be proportional to

$$\sum_{Y:Y\neq x} \eta_t(X)(1-\eta_{t.}(Y))\frac{dN_t^{(X,Y)}}{dt}(Y-X)$$

We write this as (after time scaling)

$$\underline{J}_N^{(1)}(t,X) = \frac{\alpha(N)}{N^2}\sum_{Y:Y\neq X}\eta_{N^2 t}(X)\left(1-\eta_{N^2 t}(Y)\right)$$
$$\frac{d}{dt} - N_{N^2 t}^{(X,Y)}(Y-X)$$

and so

$$\mathbb{E}\left\{J_N^{(1)}(t,x)\right\} = \frac{\alpha(N)}{N^2}\sum_{Y:Y\neq X}\mathbb{E}\left\{\eta_{N^2 t}(X)\left(1-\eta_{N^2 t}(Y)\right)\right\}$$
$$\frac{d}{dt}\lambda_{(Y-X)}^{N^2}tp(Y-X)$$
$$= \lambda\alpha(N)\sum_{X,Y}\left\{\mathbb{E}\left\{\eta_{N^2 t}(X)(1-\eta_{N^2 t}(X+Z))\right\}zp(Z)\right\}$$

Likewise the **and current into X is after rescaling

$$J_N^{(2)}(t,X) = \frac{\alpha(N)}{N^2}\sum_{Y:Y\neq X}\eta_{N^2 t}(Y)\left(1-\eta_{N^2 t}(X)\right)$$
$$\frac{d}{dt}N_{N^2 t}^{(Y,X)}(X-Y)$$

The total current density it $\underline{X}$ is

$$J_N(t,X) = J_N^{(1)}(t,X)+J_N^{(2)}(t,X).$$

We get $$\mathbb{E}\left\{J_N(t,X)\right\} = \lambda\alpha(N)\sum_{Y:Y\neq X}\left\{\mathbb{E}\left[\eta_{N^2 t}(X)\left(1-\eta_{N^2 t}(Y)\right)\right]p(Y-X)\right.$$
$$\left.-\mathbb{E}\left[\eta_{N^2 t}(Y)\left(1-\eta_{N^2 t}(X)\right)\right]p(X-Y)\right\}$$
$$\approx \alpha(N)\sum_{Y:Y\neq X}\left[p(Y-X)\rho_N\left(t,\frac{X}{N}\right)\left(1-\rho_N\left(t,\frac{Y}{N}\right)\right)\right.$$
$$\left.-p(X-Y)\rho_N\left(t,\frac{Y}{N}\right)\left(1-\rho_N\left(t,\frac{X}{N}\right)\right)\right](Y-X)$$

$$= \lambda\alpha(N)\sum_{z\neq 0} z\left\{p(z)\rho_N\left(t,\frac{X}{N}\right)\left(1-\rho_N\left(t,\frac{X+Z}{N}\right)\right)\right.$$

$$\left. - p(-z)\rho_N\left(t,\frac{x+z}{N}\right)\left(1-\rho_N\left(t,\frac{x}{N}\right)\right)\right\}$$

Note that $X, Y, Z \in \mathbb{Z}_N^3$.

❖❖❖❖❖

[8] (Ph.D. Problem of Rohit) For Rohit and Dr. Vijayant. Skorohod Optional Stochastic Control. Optional Stochastic Control of Master and Slave Robot:

$$\underline{\underline{M}}_m(\underline{\theta}_m)\ddot{\underline{\theta}}_m + \underline{N}_m(\underline{\theta}_m, \dot{\underline{\theta}}_m) = \underline{\underline{F}}_m(\underline{\theta}_m, \dot{\underline{\theta}})\left(\underline{f}_n(t) + \underline{w}_m(t)\right)$$

$$\underline{\underline{M}}_s(\underline{\theta}_s)\ddot{\underline{\theta}}_s + \underline{N}_s(\underline{\theta}_s, \dot{\underline{\theta}}_s)$$

$$= \underline{F}_s(\underline{\theta}_s, \dot{\underline{\theta}}_s)\left(\underline{f}_e(t) + \underline{w}_e(t)\right) + \underline{\underline{K}}(t)\underline{\psi}(\underline{\theta}_m, \dot{\underline{\theta}}_m, \underline{\theta}_s, \dot{\underline{\theta}}_s)$$

Aim: $\{\underline{f}_n(t)\}$ choose the feedback controller matrix $\{\underline{\underline{K}}(t)\}$ so that $\mathbb{E}\left\{\int_0^T L(\underline{\theta}_m, \underline{\theta}_m, \underline{\theta}_s, \underline{\theta}_s)dt\right\}$ is a minimum.

Let $$\underline{w}_m(t) = \sigma_m \frac{d\underline{B}_m}{dt}, \underline{w}_e(t) = \sigma_e \frac{d\underline{B}_s}{dt}$$

When $\underline{B}_m(.)$ and $\underline{B}_s(.)$ are independent 3-D Brownian motion processes. In stochastic differential form, the above equation. Assume the form $d\underline{\theta}_m = \underline{w}_m dt$

$$d\underline{\theta}_s = \underline{w}_s dt$$

$$d\underline{w}_m = \underline{G}_m(\underline{\theta}_m, \underline{w}_m)dt$$

$$+\underline{\underline{H}}_m = (\underline{\theta}_m, \underline{w}_m)\underline{f}_n(t)dt$$

$$+\sigma\underline{\underline{H}}_m = (\underline{\theta}_m, \underline{w}_m)dB_m(t)$$

$$d\underline{w}_s = \underline{G}_s(\underline{\theta}_s, \underline{w}_s)dt + \underline{\underline{H}}_s(\underline{\theta}_s, \underline{w}_s)f_e(t)dt$$

$$+\sigma_e \underline{H}_s(\underline{\theta}_s, \underline{w}_s)d + \underline{\underline{B}}_s(t)$$

$$+\underline{\underline{K}}(t)\psi(\underline{\theta}_m, \underline{w}_m, \underline{\theta}_s \underline{w}_s)dt$$

We assume $+\underline{\underline{K}}(t)$ is a function of *t* and

$$\left(\underline{\theta}_m(t),^T \underline{w}_m(t),^T \underline{\theta}_s(t)^T \underline{w}_s(t)^T\right) \in \mathbb{R}^8$$

(two link robot arm). Here,

$$\underline{G}_m = -\underline{\underline{M}}_m^{-1}\underline{N}_m, \underline{G}_s = -\underline{\underline{M}}_s^{-1}\underline{N}_s,$$

$$\underline{\underline{H}}_{sm} = \underline{\underline{M}}_m^{-1}\underline{F}_m, \underline{\underline{H}}_s = \underline{\underline{M}}_s^{-1}\underline{F}_s,$$

Let

$$\min_{\substack{k(s)\\ t<s\leq T}} \mathbb{E}\left\{\int_t^T L\left(\underline{\theta}_m(\tau), \underline{w}_m(\tau), \theta_s(\tau), \underline{w}_s(\tau)\right)d\tau \mid \underline{\theta}_m(t), \underline{w}_m(t), \underline{\theta}_s(t), \underline{w}_s(t)\right.$$

$$= V\left(t, \underline{\theta}_m(t), \underline{w}_m(t), \underline{\theta}_s(t), \underline{w}_s(t)\right)$$

Then,

$$V\left(t, \underline{\theta}_{ms}, \underline{w}_{m0}, \underline{\theta}_{s0}, \underline{w}_{s0}\right) = L\left(\underline{\theta}_{ms}, \underline{w}_{m0}, \underline{\theta}_{s0}, \underline{w}_{s0}\right)dt$$

$$+\min_{\underline{\underline{K}}(t)} \mathbb{E}\left\{V\left(t+dt, \underline{\theta}_m(t+dt), \underline{w}_m(t+dt),\right.\right.$$

$$\left.\underline{\theta}_s(t+dt), \underline{w}_s(t+dt)\right) \mid \underline{\theta}_m(t) = \underline{\theta}_{m0},$$

$$\left.\underline{w}_m(t) = \underline{w}_{m0}, \underline{\theta}_s(t) = \underline{\theta}_{s0}, \underline{w}_s(t) = \underline{w}_{w0}\right\}$$

Or equivalently,

$$-\frac{\partial V}{\partial(t)}\left(t, \underline{\theta}_{m0}, \underline{w}_{m0}, \underline{\theta}_{s0}, \underline{w}_{s0}\right)$$

$$= L\left(\underline{\theta}_{m0}, \underline{w}_{m0}, \underline{\theta}_{s0}, \underline{w}_{s0}\right)$$

$$+\min_{K(t)}\left\{\left[\frac{\partial V}{\partial \underline{\theta}_m}\left(t, \underline{\theta}_{m0}, \underline{w}_{m0}, \underline{\theta}_{s0}, \underline{w}_{s0}\right)\right]^T \frac{1}{dt}\mathbb{E}\left\{d\underline{\theta}_m(t) \mid \underline{\theta}_m(t)\right.\right.$$

$$\left.= \underline{\theta}_{ms} \ldots \underline{w}_s = \underline{w}_{s0}\right\}$$

$$+\left[\frac{\partial V}{\partial \underline{w}_m}\left(t, \underline{\theta}_{m0}, \underline{w}_{m0}, \underline{\theta}_{s0}, \underline{w}_{s0}\right)\right]^T \frac{1}{dt}\mathbb{E}\left\{d\underline{w}_m(t) \mid \underline{\theta}_m(t)\right.$$

$$\left.= \underline{\theta}_{m0} \ldots w_s(t) = w_{s0}\right\}$$

$$+\frac{1}{2}T_r\left[\frac{\partial^2 V\left(t, \underline{\theta}_{m0}, \underline{w}_{m0}, \underline{\theta}_{s0}, \underline{w}_{s0}\right)^T}{\partial \underline{w}_m \partial \underline{w}_m^T}\frac{1}{dt}\mathbb{E}\left\{d\underline{w}_m(t) \mid .d\underline{w}_m(t)^T\right.\right.$$

$$\left.\mid \underline{\theta}_m(t) = \underline{\theta}_{m0} \ldots w_s(t) = w_{s0}\right\}$$

$$+\left[\frac{\partial^2 V}{\partial \underline{w}_s}\left(t, \underline{\theta}_{m0}, \underline{w}_{m0}, \underline{\theta}_{s0}, \underline{w}_{s0}\right)\right]^T \frac{1}{dt}\mathbb{E}\left\{d\underline{w}_s(t) \mid\right.$$

$$\left.\underline{\theta}_m(t) = \underline{\theta}_{m0}, \ldots w_s(t) = w_{s0}\right\}$$

$$+\left[\frac{\partial V}{\partial \underline{ws}}\left(t, \underline{\theta}_{m0}, \underline{w}_{m0}, \underline{\theta}_{s0}\underline{w}_{s0}\right)\right]^T$$

$$\frac{1}{dt}\mathbb{E}\left\{d\underline{w}_s(t)\middle|\underline{\theta}m(t)=\underline{\theta}_{m0},\ldots,\underline{w}s(t)=\underline{w}_{s0}\right\}$$

$$+\frac{1}{2}T_r\left\{\frac{\partial^2 V}{\partial \underline{w}_s \partial \underline{w}_s^T}\left(t, \underline{\theta}_{m0}, \underline{w}_{m0}, \underline{\theta}_{s0}, \underline{w}_{s0}\right)\right.$$

$$\left.\frac{\cdot}{t}\mathbb{E}\left\{d\underline{\underline{w}}_s(t)d\underline{w}_s(t)^T \mid \underline{\theta}_m(t= \underline{\theta}_{m0}, \ldots w_s(t)=w_{s0}\right\}\right.$$

Now,

$$\mathbb{E}\left\{d\underline{w}_m(t)\mid \underline{\theta}_{m0},\ldots,\underline{w}_{s0}\right\} = \underline{w}_{m0}\,dt$$

$$\mathbb{E}\left\{d\underline{\theta}_s(t)\middle|\underline{\theta}_{m0},\ldots,\underline{w}_{s0}\right\} = \underline{w}_{s0}\,dt$$

$$\mathbb{E}\left\{d\underline{w}_m(t)\middle|\underline{\theta}_{m0},\ldots,\underline{w}_{s0}\right\} = \left[\underline{G}_m\left(\underline{\theta}_{m0}, \underline{w}_{m0}\right)+\underline{\underline{H}}_m\left(\underline{\theta}_{m0}, \underline{w}_{m0}\right)\underline{f}_n(t)\right]dt,$$

$$\mathbb{E}\left\{d\underline{w}_s(t)\mid \underline{\theta}_{m0},\ldots,\underline{w}_{s0}\right\} = \left[\underline{G}_m\left(\underline{\theta}_{s0}, \underline{w}_{s0}\right)+\underline{\underline{H}}_s\left(\underline{\theta}_{s0}, \underline{w}_{s0}\right)\underline{f}_e(t)\right.$$

$$\left.+\underline{K}(t)\underline{\psi}\left(\underline{\theta}_{m0}, \underline{w}_{m0}, \underline{\theta}_{s0}, \underline{w}_{s0}\right)\right]dt$$

$$\mathbb{E}\left\{d\underline{w}_m(t)d\underline{w}_m(t)^T = \mid \underline{\theta}_{m0},\ldots,\underline{w}_{s0}\right\}$$

$$\sigma^2\underline{\underline{H}}_m\left(\underline{\theta}_{m0},\underline{w}_{m0}\right)\cdot\underline{\underline{H}}_m\left(\underline{\theta}_{m0},\underline{w}_{m0}\right)^T dt$$

$$\mathbb{E}\left\{d\underline{w}_s(t)d\underline{w}_s(t)^T\middle|\underline{\theta}_{m0},\ldots,\underline{w}_{s0}\right\}$$

$$= \sigma^2\underline{\underline{H}}_s\left(\underline{\theta}_{s0}, \underline{w}_{s0}\right)\underline{\underline{H}}_s\left(\underline{\theta}_{s0}, \underline{w}_{s0}\right)^T dt,$$

So V satisfies the *pde*

$$\frac{\partial V}{\partial t}\left(t, \underline{\theta}_m, \underline{w}_m, \underline{\theta}_s, \underline{w}_s\right) = L\left(\underline{\theta}_m, \underline{w}_m, \underline{\theta}_s, \underline{w}_s\right)$$

$$+\min_{\underline{K}(t)}\left\{\left[\frac{\partial V}{\partial \underline{\theta}_m}\right]^T \underline{w}_m + \left[\frac{\partial V}{\partial \underline{w}_m}\right]^T\left(\underline{\underline{G}}_m + \underline{\underline{H}}_m\underline{f}_n(t)\right)\right.$$

$$\left.+\frac{\sigma^2}{2}T_r\left\{\frac{\partial^2 V}{\partial \underline{w}_m \partial \underline{w}_m^T}\underline{\underline{H}}_m\underline{\underline{H}}_m^T\right\}\right.$$

❖❖❖❖❖

[9] English-Chapter-1:

$$+a(s,0)\left[x_0\sum_{\tilde{x}_0}(1-x_s)\rho_t((x_s,x_1,-,x_0,-,x_{n-1}),(y_0,\tilde{\underline{x}}_0))\right.$$

$$\left.-(1-y_0)\sum_{\tilde{x}_0}x_s\,\rho_t(\underline{x},(x_s,(x_1,..,y_0,..,x_{N-1})))\right]$$

5th position molecular (choose)

$$\sum_{\tilde{x}_0}x_s\,\rho_t\left(\underline{x}^{(0,s)},(y_0,\widetilde{x_0})\right) = \sum_{\tilde{x}_0}x_s\,\rho_t\left((x_s,x_1.,x_0,..,x_{N-1})(y_0.x_1,..,x_{N-1})\right)$$

$$\sum_{x_s}(x_s)\rho_t^{(0,s)}\left((x_s,x_0),(y_0,x_s)\right)$$

$$\sum_{\tilde{x}_0}(1-x_s)\rho_t(\underline{x},(x_s,x_1,..,x_0,.,x_{N-1}))$$

$$= \sum_{\tilde{x}_0}(1-x_s)\rho_t((x_0,.,x_{N-1}),(x_s,x_1,..,y_0,.,x_{N-1}))$$

$$= \sum_{\tilde{x}_0}(1-x_s)\rho_t^{(0,s)}((x_0,x_s),(x_s,y_0))$$

$$\sum_{\tilde{x}_0}(1-x_s)\rho_t((x_s,x_1,.,x_0,.,x_{N-1}),(y_0,x_1,.,x_{N-1}))$$

$$= \sum_{x_s}(1-x_s)\rho_t^{(0,s)}((x_s,x_0),(y_0,x_s))$$

And finally,

$$\sum_{\tilde{x}_0}x_s\,\rho_t(\underline{x},(x_s,x_1,.,y_0,.,x_{N-1}))$$

$$= -\sum_{x_s}x_s\,\rho_t^{(0,s)}((x_0,x_s),(x_s,y_0))$$

Where $\rho_t^{(\alpha,\beta)}$ is the joint marginal density of the αth and β^{th} particle obtained by tracing out ρ_t over the other particles. We thus get

$$i\frac{\partial}{\partial t}\rho_t^{(0)}(x_0,y_0) = \sum_{s\geq 1}a(0,s)\left\{(1-x_0)\sum_{x_s}\rho_t^{(0,s)}((x_s,x_0),(y_0,x_s))x_s\right.$$

$$\left.-y_0\sum_{x_s}(1-x_s)\rho_t^{(0,s)}((x_0,x_s),(x_s,x_0))\right\}$$

$$+\sum_{s\geq 1} a(s,0)\left\{x_0 \sum_{x_s} \rho_t^{(0,s)}((x_s,x_0),(y_0,x_s))(1-x_s)\right.$$

$$\left.-(1-y_0)\sum_{x_s} x_s \rho_t^{(0,s)}((x_0,x_s),(x_s,y_0))\right\}$$

or

$$i\frac{\partial}{\partial t}\rho_t^{(0)}(x_0,y_0) = \sum_{s\geq 1, x_s} \{(a(0,s)(1-x_0)x_s + a(s,0)x_0(1-x_s))$$

$$\rho_t^{(0,s)}((x_0,x_s),(x_s,y_0))\}$$

Note that this is a special case of the quantum Botzmann equation.

Continuation of remarks on volume iv collated papers of S.R.S. Varadhan.

"particle system their large deviation."

Linear response theory in quantum system

$$H(t) = H_0 + \varepsilon V(t)$$

Equilibrium state is $\rho_0 = \frac{1}{z(B)} e^{-\beta H_0}$

$$Z(\beta) = T_r\left(e^{-\beta H_0}\right)$$

$$\rho(t) = \rho_0 + \varepsilon\rho_1(t) + O(\varepsilon^2)$$

Note that $[H_0, \rho_0] = 0$, so ρ_0 satisfies the stationary Von. Neumann equation.

$$i\rho_1'(t) = [H_0, \rho_1(t)] + [V(t), \rho_0]$$

$$\rho_1(t) = -i\int_0^t \exp(-i(t-\tau)adH_0)([V(\tau), \rho_0])d\tau$$

X any system observable

$$\langle X\rangle(t) = T_r((\rho_0 + \varepsilon\rho_1(t)))(X)$$

$$= \langle X\rangle_0 + \varepsilon T_r(\rho_1(t)X)$$

$$\langle X\rangle(t) - \langle X\rangle_0 = -i\varepsilon\int_0^t tr\left(\exp\left(-i\left(t-\tau adH_0\right)\right)\right)\left([V(\tau), \rho_0]X\right)d\tau$$

❖❖❖❖❖

[10] Markov Processes-Some Application:

(1) Simple exclusion models in Markov process theory.

(2) Transmission lines, waveguides and Antennas excited by Markov chains.

(1) Consider $\mathbb{Z}_N = \{0, 1, 2, ..., N-1\}$. $\eta(x) = 1$

If site $x \in \mathbb{Z}_N$ is occupied by a particle and $\eta(x) = 0$ otherwise $p(x)$, $x = 0, 1, 2,..,$ $N-1$ is a probability distribution on

$$\mathbb{Z}_N.\ p(y-x) \equiv p(y-x \bmod N), \quad \sum_{x=0}^{N-1} p(x) = 1.$$

If site x is occupied and site $y \neq x$ is unoccupied, then with a probability $\lambda p(y-x)dt$, the particle at x jumps to the site y in time $[t, t+dt]$. The different jumps processes are independent and here up to $O(dt)$, at must only one jump takes place in $\mathbb{Z}_N$ in time dt. Let $X = \{\eta : \mathbb{Z}_N \rightarrow \{0, 1\}\}$

Thus X is the state space of the above Markov Process $\{\eta_t(x)\}$ and η_t satisfies the side

$$d\eta_t(x) = \sum_{\substack{y \in \mathbb{Z}\ N \\ y \neq x}} \left\{\eta_t(y)(1-\eta_t(x))dN_t(y,x) - \eta_t(x)(1-\eta_t(y))dN_t^{(x,y)}\right\}$$

Where $\left\{dN_t^{(x,y)} : x, y \in \mathbb{Z}^N, x \neq y\right\}$ are independent Poison processes with the rate of $N_t^{(x,y)}$ being $\lambda p(y-x)$. More generally for $f : X \rightarrow \mathbb{R}$,

$$\mathrm{df}(\eta_t) = \sum_{\substack{x, y \in \mathbb{Z}^N, \\ x \neq y}} \left(f\left(\eta_t^{(x,y)}\right) - f(\eta_t)\right)\eta_t(x)(1-\eta_t(y))dN_t^{(x,y)}$$

Generator

$$\begin{aligned}\mathcal{L}f(\eta) &= \mathbb{E}\left[df(\eta_t) \mid \eta_t = \eta\right]/dt \\ &= \lambda \sum_{\substack{x, y \in \mathbb{Z}^N, \\ x \neq y}} \left(f(\eta(x,y)) - f(\eta)\right)\eta(x)(1-\eta(y))p(y-x) \\ &= \lambda \sum_{\substack{x, z \in \mathbb{Z}^N, \\ z \neq 0}} \left(f(\eta(x,x+z)) - f(\eta)\right)\eta(x)(1-\eta(x+z))p(z)\end{aligned}$$

Here,

$$\eta^{(x,y)}(z) = \begin{cases} \eta(x) & z = y \\ \eta(y) & z = x \\ \eta(z), & z \neq x, y \end{cases}$$

Exponential Martingale: Let $N(t, dx)$ be a space-time Poisson random field with $\mathbb{E}(N(t, dx)) = t\,.\,dF(x)$, $x \in Y$.

Then

$$d\exp\left(\int_Y f(x)N(t, dx)\right) = \int_{x \in Y} (\exp(f(x)) - 1)N(dt, dx)\exp\left(\int_Y f(z)N(t, dz)\right)$$

i.e if

$$\xi_f(t) = \exp\left(\int_Y f(x)N(t, dx)\right),$$

then $$d\xi_f(t) = \xi_f(t) \int_{x \in Y} (\exp(f(x)) - 1) N(dt, dx)$$

$$\therefore \quad \mathbb{E}\left[d\xi_t(t) \mid \xi_t(t)\right] = \xi_t(t) \int_Y (\exp(f(x)) - 1) dF(x) dt$$

Let $$M_f(t) = \exp\left(\int_Y F(x) N(t, dx)\right) \varphi(t)$$

When $Q(t)$ is a differentiable function. Then,

$$dM_f(t) = \xi_t(t) \left\{ dt \int_Y (\exp(f(x)) - 1) dF(x) + \frac{\varphi'(t)}{\varphi(t)} dt \right\}$$

Thus choosing

$$\log \varphi(t) = t \int_Y (\exp(f(x)) - 1) dF(x)$$

gives a Martingale $M_f(t)$:

$$M_f(t) = \exp\left(\int_Y f(x) N(t, dx) - t \int_Y (\exp(f(x)) - 1) dF(x) \right)$$

This is the exponential Martingale associated to f for the Possion measure $\{N(t, dx)\}$.

Note that it

$$f(\eta_t) - \int_0^t \mathcal{L} f(\eta_s) ds = M_t, \text{ then}$$

$$dM_t = \sum_{x \neq y} \left(f\left(\eta_t^{(x,y)}\right) - f(\eta_t) \right) \eta_t(x)(1 - \eta_t(y))$$

$$\left(dN_t^{(x,y)} - \lambda p(y - x) dt\right)$$

and M_t is a Martingale.

Let $J : [0, 1] \to \mathbb{R}$. Then define the process

$$X_t = \frac{1}{N} \sum_{x \in \mathbb{Z}^N} J\left(\frac{x}{N}\right) \eta_t(x)$$

Suppose $\eta_t[N\theta] \overrightarrow{N \to \infty} \rho(t, \theta), \theta \in [0, 1]$

Then

$$X_t \overrightarrow{N \to \infty} \int_0^1 J(\theta) \rho(t, \theta) d\theta$$

Now, $$\mathbb{E}\left[d\eta_t(x) \mid \eta_t\right] = \lambda \sum_{y: y \neq x} \left[\eta_t(y)(1 - \eta_t(x)) p(x - y)\right.$$

$$\left. - \eta_t(x)(1 - \eta_t(y)) p(y - x)\right] dt$$

so $$\mathbb{E}\left[dx_t \mid \eta_t\right] = \frac{\lambda}{N}\sum_{x\neq y} J\left(\frac{x}{N}\right)(\eta_t(y)(1-\eta_t(x))\,p(x-y)$$

$$-\eta_t(x)\left(1-\eta_t(y)\right)p\left(y-x\right)\right)dt$$

So $$dx_1 - \mathbb{E}\left[dx_t \mid \eta_t\right] = \frac{1}{N}\sum_{x\neq y} J\left(\frac{x}{N}\right)\left\{\eta_t(y)(1-\eta_t(x))\,dM_t^{(x,y)}\right.$$

$$\left.-\eta_t(x)(1-\eta_t(y))\,dM_t^{(x,y)}\right\}$$

where $$M_t^{(x,y)} = N_t^{(x,y)} - \lambda p(y-x)t$$

is a Martingale.

This can also be expressed as

$$dX_t = \mathbb{E}\left[dX_t \mid \eta_t\right]$$

$$= \frac{1}{N}\sum_{x\neq y}\left(J\left(\frac{x}{N}\right)-J\left(\frac{y}{N}\right)\right)\eta_t(y)(1-\eta_t(x))\,dM_t^{(y,x)}$$

Thus,

Assume that p has finite rage, *i.e.* $p(z) = 0$ if $|z| > q$ when is q a finite positive integer.

Then $$Var\{dx_t \mid \eta_t\} = \frac{\lambda dt}{N^2}\sum_{x,z}\left(J'\left(\frac{x}{N}\right)\frac{z}{N}\right)^2 \eta_t(x)^2(1-\eta_t(x+z))^2\,p(z)$$

$$+\,dt.O\left(\frac{1}{N}4\right).$$

$$\xrightarrow[N\to\infty]{} 0$$

In fact, we have a stronger convergence result:

$$Var\left\{dX_{Nt}^2 \middle| \eta_{Nt}^2\right\} \text{ (time scaling)}$$

$$= \frac{\lambda dt}{N^2}\sum_{x,z}\left\{J'\left(\frac{x}{N}\right)^2 z^2 p(z)\,\eta_N^t(x)^2\left(1-\eta_N^t(x+z)\right)^2\right\}$$

$$\leq \frac{\lambda dt}{N^2}\sum_{x} J'\left(\frac{x}{N}\right)^2 \sum_{z} z^2 p(z) \xrightarrow[N\to\infty]{} 0$$

Since $$\frac{1}{N}\sum_{x} J'\left(\frac{x}{N}\right)^2 \overrightarrow{N\to\infty}\int_0 J'(0)^2\,d\theta < \infty$$

and $\sum_x z^2 p(z) < \infty$ since $p(.)$ has finite range. We note that $Var\{dX_{Nt} \mid \eta_{Nt}\}$

$$= \mathbb{E}\left\{(dX_{Nt})^2 \mid \eta_{Nt}\right\} - \left(\mathbb{E}\left\{dX_{Nt} \mid \eta_{Nt}\right\}\right)^2$$

and hence we have proved that

$$\mathbb{E}\left\{(dX_{Nt})^2 \mid \eta_{Nt}\right\} - \mathbb{E}\left(\mathbb{E}\left\{dX_{Nt} \mid \eta_{Nt}\right\}\right)^2 \xrightarrow[N\to\infty]{} 0$$

Or equivalently,

$$\mathbb{E}\left\{\left(dX_{Nt} - \mathbb{E}\left\{dX_{Nt} \mid \eta_{Nt}\right\}\right)^2\right\} \xrightarrow[N\to\infty]{} 0$$

So $$dX_{Nt} - \mathbb{E}\left\{dX_{Nt} \mid \eta_{Nt}\right\} \xrightarrow[L^2]{N\to\infty} 0$$

Now,

$$X_t = \frac{1}{N}\sum_{x\in\mathbb{Z}_N} J\left(\frac{x}{N}\right)\eta_t(x)$$

$$= \int_0^1 J(\theta)\, d\mu_{N,t}(0)$$

where $$\mu_{N,t} = \frac{1}{N}\sum_{x\in\mathbb{Z}_N} \eta_t(x)\delta_{x/N}$$

Thus, $$X_{Nt} = \int_0^1 J(\theta)\, d\mu_N, N_t(\theta)$$

where $$\mu_{N,Nt} \equiv V_{N,\, t} = \frac{1}{N}\sum_{x\in\mathbb{Z}_N} \eta_{Nt}(x)\delta_{x/N}$$

We have

$$X_t = \frac{1}{N}\sum_{x,z}\left(J\left(\frac{x+z}{N_t}\right) - J\left(\frac{x}{N}\right)\right)$$

$$\int_0^t \eta_s(x)(1-\eta_s(x+z))\, dN_x^{(x,x+z)}$$

So, $$\mathbb{E}\, X_t = \frac{1}{N}\sum_x J\left(\frac{x}{N}\right)\mathbb{E}\{\eta_t(x)\}$$

We define $$\rho_N(t, 0) = \mathbb{E}\left\{\eta_{N^2 t}([N\theta])\right\} - \theta \in [0, 1]$$

So $$\mathbb{E}\left\{x_{Nt}^2\right\} = \frac{1}{N}\sum_x J\left(\frac{x}{N}\right)\rho_N\left(t, \frac{x}{N}\right)$$

Thus, $$\frac{d}{dt}\mathbb{E}\left\{X_{Nt}^2\right\} = \frac{1}{N}\sum_x J\left(\frac{x}{N}\right)\frac{\partial\rho_N}{\partial t}(t,\theta)\frac{x}{N}$$

On the other hand, $$\approx \int_0^1 J(\theta)\frac{\partial\rho_N}{\partial t}(t,\theta)\, d\theta$$

$$Var\left(X_{Nt}^2\right) = \sum_{x,z}\left(J\left(\frac{x+z}{N}\right) - J\left(\frac{x}{N}\right)\right)^2 p(z)$$

$$\int_0^t \mathbb{E}\left\{\eta^2_{N^2s}(x)\left(1-\eta^2_{Ns}(x+z)\right)^2 ds\right\}$$

$$= \sum_{x,z} J'\left(\frac{x}{N}\right)^2 \frac{z^2}{N^2} p(z) \int_0^t \mathbb{E}\left\{\eta^2_{N^2s}(x)\left(1-\eta_{N^2s}(x+z)\right)^2 ds\right\}$$

$$+O\left(\frac{1}{N^2}\right)$$

$$\leq \frac{\lambda t}{N}\left(\sum_z z^2 p(z)\right)\int_0^1 J'^2(\theta)\,d\theta + O\left(\frac{1}{N^2}\right) \xrightarrow[N\to\infty]{} 0$$

It follows that

$$X_{N^2t} - \mathbb{E}\left\{X_{N^2t}\right\} \xrightarrow[N\to\infty]{} 0$$

i.e.

$$\frac{1}{N}\sum_{x\in Z_N} J\left(\frac{x}{N}\right)\eta_{N^2t}(x) - \frac{1}{N}\sum_{x\in\mathbb{Z}_N} J\left(\frac{x}{N}\right)\rho_N\left(t,\frac{x}{N}\right)$$

$$\xrightarrow[N\to\infty]{} 0$$

Talking differential *w.r.t. t*, it is natural to expect that

$$\frac{1}{N}\sum_{x\in Z_N} J\left(\frac{x}{N}\right)d\eta_{N^2t}(x) - \frac{dt}{N}\sum_{x\in\mathbb{Z}_N} J\left(\frac{x}{N}\right)\frac{\partial\rho_N}{\partial t}\left(t,\frac{x}{N}\right)$$

$$\xrightarrow[N\to\infty]{} 0$$

Since the " Martingale part" of

$$\frac{1}{N}\sum_x J\left(\frac{x}{N}\right)\eta_{N^2t}(x)$$

tends to zero, it follows that

$$N^2\frac{\lambda dt}{N}\sum_{x,y\in\mathbb{Z}_N}\left\{J\left(\frac{x}{N}\right)-J\left(\frac{y}{N}\right)\right\}\eta_{N^2t}(y)\left(1-\eta_{N^2t}(x)\right)$$

$$-\frac{dt}{N}\sum_{x\in Z_N} J\left(\frac{x}{N}\right)\frac{\partial\rho_N}{\partial t}\left(t,\frac{x}{N}\right) \xrightarrow[N\to\infty]{} 0$$

Thus,

$$\lambda N\sum_{x,z}\left(J\left(\frac{x+z}{N}\right)-J\left(\frac{x}{N}\right)\right)\eta_{N^2t}(x)\left(1-\eta_{N^2t}(x+z)\right)p(z)$$

$$-\frac{1}{N}\sum_x J\left(\frac{x}{N}\right)\frac{\partial\rho_N}{\partial t}\left(t,\frac{x}{N}\right) \xrightarrow[N\to\infty]{} 0$$

Assume that $\sum_z zp(z) = 0$. Then we get from the above, by replacing $\eta_{N^2t}(x)$

with $\rho_N\left(t,\frac{x}{N}\right)$,

$$\lambda N\sum_{x,z}\left\{\left(J'\left(\frac{x}{N}\right)+\frac{1}{2}J''\left(\frac{x}{N}\right)\frac{z2}{N^2}\right)\right.$$

$$\left.\rho_N\left(t,\frac{x}{N}\right)\left(1-\rho_N\left(t,\frac{x+z}{N}\right)\right)p(z)\right\}$$

$$-\int_0^1 J(\theta)\frac{\partial\rho_N}{\partial t}(t,\theta)\,d\theta\to 0$$

or

$$\lambda\sum_{x,t}\left\{\left(J'\left(\frac{x}{N}\right)z+\frac{1}{\partial N}J''\left(\frac{x}{N}\right)z^2\right)\right.$$

$$\left.\rho_N\left(t,\frac{x}{N}\right)\left(1-\rho_N\left(t,\frac{x}{N}\right)-\frac{\partial\rho_N}{\partial\theta}\left(t,\frac{x}{N}\right)\frac{z}{N}\right)p(z)\right\}$$

$$-\int_0^1 J(\theta)\frac{\partial\rho_N}{\partial t}(t,\theta)\,d\theta\to 0$$

or equivalently with $D=\sum_z z^2 p(z)$,

$$\frac{\lambda D}{\partial N}\sum_x J''\left(\frac{x}{N}\right)\rho_N\left(t,\frac{x}{N}\right)\left(1-\rho_N\left(t,\frac{x}{N}\right)\right)$$

$$-\frac{\lambda D}{N}\sum_x J''\left(\frac{x}{N}\right)\frac{\partial\rho_N}{\partial\theta}\left(t,\frac{x}{N}\right)\rho_N\left(t,\frac{x}{N}\right)$$

$$-\int_0^1 J(\theta)\frac{\partial\rho_N}{\partial\theta}(t,\theta)\,d\theta\xrightarrow[N\to\infty]{}0$$

or

$$\frac{\lambda D}{2}\int_0^1 J''(\theta)\rho_N(t,\theta)(1-\rho_N(t,\theta))\,d\theta$$

$$-\lambda D\int_0^1 J'(\theta)\rho_N(t,\theta)\frac{\partial\rho_N}{\partial\theta}(t,\theta)\,d\theta$$

$$-\int_0^1 J(\theta)\frac{\partial\rho_N}{\partial\theta}(t,\theta)\,d\theta\xrightarrow[N\to\infty]{}0$$

Integrated by parts gives assuming $\rho_N(t,\theta)\to\rho(t,\theta)$, $N\to\infty$

$$\frac{\lambda D}{2}\frac{\partial^2}{\partial\theta^2}(\rho(1-\rho))+\frac{\lambda D}{2}\frac{\partial^2(\rho^2)}{\partial\theta^2}-\frac{\partial\rho}{\partial t}=0$$

We are justified in replacing $\eta_{N^2t}(x)$ by $\rho_N\left(t,\frac{x}{N}\right)\equiv\mathbb{E}\left\{\eta_{N^2t}(x)\right\}$ because taking J.

❖❖❖❖❖

[11] Comments and Proofs on "Particle System" Collected Papers of S.R.S. Varadhan Vol. 4:

(1) Simple exclusion:

$$\begin{aligned} d\eta_t(x) &= \sum_y \Big\{\eta_t(x)(1-\eta_t(x))\,dN_t^{(y,x)} \\ &\quad - \eta_t(x)(1-\eta_t(y))\,dN_t^{(x,y)}\Big\} \\ N_t^{(x,y)} &= \text{Poisson process with mean } p(y-x)\,dt. \\ df(\eta t) &= \sum_{x,y}\Big(f\big(\eta_t^{(x,y)}\big) - f(\eta_t)\Big) \\ &\quad \eta_t(y)(1-\eta_t x) - dN_t^{(y,x)} \\ &= (\mathcal{L}f)(\eta_t)\,dt + dM_t \end{aligned}$$

where

$$\begin{aligned} \mathcal{L}f(\eta) &= \sum_{x,y}\Big(f\big(\eta^{(x,y)}\big) - f(\eta)\Big) \\ &\quad (\eta(y)(1-\eta(x))\,p(x-y)) \end{aligned}$$

and

$$\begin{aligned} dM_t &= \sum_{x,y}\Big(f\big(\eta_t^{(x,y)}\big) - f(\eta)\Big) \\ &\quad \eta_t(y)(1-\eta_t(x))\,dM_t^{(x,y)} \end{aligned}$$

Where $M_t^{(x,y)} = N_t^{(x,y)} - p^{(y-x)}t$

is a Martingale

$$\begin{aligned} d\big(M_t^2\big) &= 2M_t dM_t + d\langle M, M\rangle_t \\ d\,\mathbb{E}\big(M_t^2\big) &= d\,E\big[\langle M, M\rangle_t\big] \\ d\langle M, M\rangle_t &= \sum_{x,y}\Big(F\big(\eta_t^{(x,y)}\big) - f(\eta_t)^2\eta_t(y)^2(1-\eta_t(x)\Big)^2 \\ &\quad d\Big\langle M_t^{(x,y)}, M^{(x,y)}\Big\rangle_t \end{aligned}$$

$$\begin{aligned} \mathbb{E}\Big[d\Big\langle M^{(x,y)}, M^{(x,y)}\Big\rangle_t\Big] &= \mathbb{E}\Big[\Big(d_N^{(x,y)} - p(y-x)dt\Big)^2\Big] \\ &= p(y-x)\,dt \end{aligned}$$

So,

$$\begin{aligned} \frac{d}{dt}\mathbb{E}\big[M_t^2\big] &= \sum_{x,y} E\Big\{\eta_t^{(y)^2}(1-\eta_t(x))^2\Big(f\big(\eta_t^{(x,y)}\big) - f(\eta_t)\Big)^2\Big\} \\ &\quad p(y-x) \end{aligned}$$

Now, But latence by $\mathbb{Z}$ and consider

$$X_N(t) = \frac{1}{N}\sum_{x_t \mathbb{Z}_N} J\left(\frac{x}{N}\right)\eta_t(x)$$

We define the empirical measures on [0.1]

$$\mu_{t, N} = \frac{1}{N}\sum_{x\in\mathbb{Z}_N} \eta_t(x)\delta_{x/N}$$

Then $$X_N(t) = \int_{[0,1]} J(x)\,d\mu_{t,N}(x).$$

We have

$$dX_N(t) = \frac{1}{N}\sum_{x\in\mathbb{Z}_N} J\left(\frac{x}{N}\right)d\eta_t(x)$$

$$= \frac{1}{N}\sum_{x,y\in\mathbb{Z}_N} J\left(\frac{x}{N}\right)\Big[\eta_t(y)(1-\eta_t(x))\,dN_t^{(x,y)}$$

$$-\eta_t(x)(1-\eta_t(y))\,dN_t^{(x,y)}\Big]$$

$$= \frac{1}{N}\sum_{x,y\in\mathbb{Z}_N}\left(J\left(\frac{x}{N}\right)-J\left(\frac{y}{N}\right)\right)\eta_t(y)(1-\eta_t(x))\,dN_t^{(y,x)}$$

$$= \frac{1}{N}\sum_{x,y}\left(J\left(\frac{x}{N}\right)-J\left(\frac{y}{N}\right)\right)\eta_t(y)(1-\eta_t(x))\,p(x-y)\,dt$$

$$+\frac{1}{N}\sum_{x,y}\left(J\left(\frac{x}{N}\right)-J\left(\frac{y}{N}\right)\right)\eta_t(y)(1-\eta_t(x))\,dM_t^{(y,x)}$$

We estimate the Martingale part $M_t^{(N)}$ where

$$M_t^{(N)} = \frac{1}{N}\sum_{x,y\in\mathbb{Z}_N}\left(J\left(\frac{x}{N}\right)-J\left(\frac{y}{N}\right)\right)^2 \eta_t(y)(1-\eta_t(x))\,dM_t^{(y,x)}$$

as $N \to \infty$. We have

$$\frac{d}{dt}\mathbb{E}\left[M_t^{(N)^z}\right] = \frac{1}{N^2}\sum_{x,y\in\mathbb{Z}_N}\left(J\left(\frac{x}{N}\right)-J\left(\frac{y}{N}\right)\right)^2$$

$$\mathbb{E}\,(\eta_t(y))(1-\eta_t(x))\,p(x-y)$$

Consider finite range interaction only *i.e.*

$$p(z) = 0 \text{ for } |z| > \mathbb{R}, \text{ where } R < \infty.$$

Then, $$\frac{d}{dt}\mathbb{E}\left(M_t^{(N)^2}\right) = \frac{1}{N^2}\sum_{\substack{|z|\neq R\\ x\in\mathbb{Z}_N}}\left\{\left(J\left(\frac{x+z}{N}\right)-J\left(\frac{x}{N}\right)\right)^2\right.$$

$$\left.\times E\big(\eta_t(x+z)(1-\eta_t(x))\big)\,p(-z)\right\}$$

We have since $|z| < R$.

$$J\left(\frac{x+z}{N}\right)-J\left(\frac{x}{N}\right) = J'\left(\frac{x}{N}\right)\frac{z}{N}+\frac{1}{2}J''\left(\frac{x}{N}\right)\frac{z^2}{N^2}+O\left(\frac{1}{N^3}\right)$$

assuming J to be bounded on [0, 1]. Thus,

$$\left(J\left(\frac{X+Z}{N}\right)-J\left(\frac{X}{N}\right)\right)^2 = J'\left(\frac{X}{N}\right)^2\frac{Z^2}{N^2}+O\left(\frac{1}{N^3}\right)$$

and hence $$\overline{\lim_{N\to\infty}}\frac{d}{dt}\mathbb{E}\left(M_t^{(N)^2}\right)$$

$$\leq \overline{\lim_{N\to\infty}}\sum_{\substack{x\in\mathbb{Z}_N\\|Z|\leq R}} J'\left(\frac{X}{N}\right)^2\left(Z^2p(-Z)/N^2\right)\mathbb{E}\left(\eta_t(X+Z)(1-\eta_t(X))\right)$$

We also have

$$\overline{\lim_{N\to\infty}}\frac{d}{dt}\mathbb{E}\left(M_t^{(N)^2}\right) \leq \lim_{x\to\infty}\frac{2D}{N}\int_0^1 J'^2(\theta)\,d\theta = 0$$

Here

$$D = \sum_{|Z|\leq R} Z^2p(-Z) = \sum_{|Z|\leq R} Z^2p(Z)$$

We have,

$$\frac{d}{dt}\mathbb{E}\left(M_t^{(N)^2}\right) = \frac{1}{N^2}\sum_{X\in\mathbb{Z}_N}\left(J\left(\frac{X+Z}{N}\right)-J\left(\frac{X}{N}\right)\right)^2$$

$$|Z| \leq R\mathbb{E}\left(\eta_t(X+Z)(1-\eta_t(X))\right)p(-Z)$$

Exponential Martingale:

$$X_N(t) = \sum_X \eta_t(X)J\left(\frac{X}{N}\right)$$

$$dX_N(t) = \sum_{X,Y} J\left(\frac{X}{N}\right)\Big(\eta_t(Y)(1-\eta_t(X))\,dN_t^{(Y,X)}$$

$$-\eta_t(X)\left(1-\eta_t(Y)\right)dN_t^{(X,Y)}\Big)$$

$$= \sum_{X,Y}\left(J\left(\frac{X}{N}\right)-J\left(\frac{Y}{N}\right)\eta_t(Y)(1-\eta_t(X))\right)dN_t^{(Y,X)}$$

$$d\exp\left(X_N(t)\right) = \sum_{X,Y}\left(\exp\left(\left(J\left(\frac{X}{N}\right)-J\left(\frac{Y}{N}\right)\right)\right)-1\right)dN_t^{(Y,X)}\eta_t(Y)(1-\eta_t(X))$$

$$d\exp\left(X_N(t)+\int_o^t f(s)\,ds\right) = \exp\left(X_N(t)+\int_0^t f(s)\,ds\right)$$

$$\times\left(\exp\left(dX_N(t)+f(t)\,dt\right)-1\right)$$

$$= \exp\left(X_N(t) + \int_o^t f(s)\,ds \right)\left(\exp dX_N(t)\right)\cdot(1 + f(t)\,dt - 1)$$

$$= \xi(t)\left(\left(1 + \sum_{X,Y}\left(\exp\left(J\left(\frac{X}{N}\right) - J\left(\frac{Y}{N}\right)\right) - 1\right) dN_t^{(Y,X)}\right)\right.$$

$$\left.\times (1 + f(t)\,dt)\eta_t(Y)(1 - \eta_t(X)) - 1\right)$$

$$= \xi(t)\left\{ f(t)\,dt + \sum_{X,Y}\left(\exp\left(J\left(\frac{X}{N}\right) - J\left(\frac{Y}{N}\right)\right) - 1\right)\right.$$

$$\left.\times\, \eta_t(Y)(1 - \eta_t(X))\,dN_t^{(Y,X)}\right\}$$

where
$$X_N(t) = \sum_X J\left(\frac{X}{N}\right)\eta_t(X),$$

$$\xi(t) = \exp\left(X_N(t) + \int_o^t f(s)\,ds \right)$$

Taking

$$f(t) = -\sum_{X,Y}\left(\exp\left(J\left(\frac{X}{N}\right) - J\left(\frac{Y}{N}\right)\right) - 1\right)\eta_t(Y)(1 - \eta_t(X))\,p(X - Y)$$

we get

$$d\xi(t) = \xi(t)\cdot \sum_{X,Y}\left(\exp\left(J\left(\frac{X}{N}\right) - J\left(\frac{Y}{N}\right)\right) - 1\right)\eta_t(Y)(1 - \eta_t(X))\,dM_t^{(Y,X)}$$

Where $M_t^{(Y,X)} = N_t^{(Y-X)} - p(Y-X)t$ is a Martingale. Thus, $\{\xi(t)\}_{t \geq 0}$ is a Martingale

❖❖❖❖❖

[12] Interacting Diffusions:

$$dX_i(t) = -\,dZ_{i,\,i+1}(t) + dZ_{i-1,\,i}(t)$$

$$dZ_{i,\,i+1}(t) = \left(\psi(X_i(t)) - \psi(X_{i+1}(t))\right)dt + dB_{i,\,i+1}(t)$$

$B_{i,\,i+1}(\cdot), i \in \mathbb{Z}$ are independent Brownian motion processes

❖❖❖❖❖

[13] This is Independent of *s*. It is Reasonable to Expert that at Equilibrium:

$$\mathbb{E}\left|\psi(X_i) - \widehat{\psi}\left(\sum_{m=-N\varepsilon}^{N\varepsilon} X_{i+m}/2N\varepsilon + 1\right)\right| \to 0, N \to \infty, \forall\, \varepsilon > 0$$

Note that

$$\lim_{x\to\infty}\frac{1}{(2N\varepsilon+1)}\sum_{m=-N\varepsilon}^{N\varepsilon}X_{i+m} = \int X_p(X)\,dX$$

$$= \frac{\int X\exp(\sigma X-\phi(X))\,dX}{\int \exp(\sigma X-\phi(X))\,dX} = F'(\sigma)$$

Thus, writing $\qquad f_{N,t} = N\cdot\sum X_i(t)\chi_{\left[\frac{i}{N},\frac{i+1}{N}\right]}$

$f_{N,t}$ is a random function on [0, 1]. We have $\dfrac{(i+2)}{N}$

$$\int_0^1 J(\theta)f_{N,t}(\theta)d\theta = \sum_i X_i(t)N\int_{i/N} J(\theta)d\theta \approx \sum_i J\left(\frac{i}{N}\right)X_i(t) = X_N(t)$$

We get $\int J(\theta)f_{N,t}(\theta)d\theta \approx \sum_I J\left(\frac{i}{N}\right)dX_i(t) \approx \frac{1}{N^2}\sum_i J''\left(\frac{i}{N}\right)\psi(X_i(t))\,dt$

$$\approx \frac{1}{N^2}\sum_i J''\left(\frac{i}{N}\right)\psi\left(\sum_{[m]\neq N\in}\frac{X_{i+m}}{(2N\varepsilon+1)}\right)dt$$

Writing

$$\sum_{[m]\neq N\in}\frac{X_{i+m}}{(2N\varepsilon+1)} \approx \sum_{[m]\neq N\in}\bar{f}_{N,t}\left(\frac{i+m}{N}\right)\Big/(2N\varepsilon+1) \approx \rho_t\left(\frac{i}{N}\right)$$

$$\varepsilon \to o$$
$$N \to \infty\,.$$

Now, $\displaystyle\int_{i/N}^{(i+N)/N} f_{N,t}(\theta)d\theta = X_i(t)$

We this get

$$\sum_{|k-i|\neq N\varepsilon} X_k(t) = \sum_{k=i-N\in}^{i+N\in}\int_{k/N\in}^{(k+1)/N} f_{N,t}(\theta)d\theta = \int_{i/N-\varepsilon}^{i/N+\varepsilon} f_{N,t}(\theta)d\theta$$

$$\mathbb{E}\left[X_i(t)\,\middle|\,\sum_{|k-i|\leq N\varepsilon} X_k(t)/(2N\varepsilon+V)=a\right] \approx \frac{\int\xi\exp(I_F'(a)\xi-\phi(\xi))d\xi}{\exp(I_F'(a)a-I_f(a))}$$

$$F(\sigma) = \sup_x(\sigma_x - I_F(x))$$
$$= \sigma a - I_F(a)$$

Where $\qquad I_F'(a) = \sigma$

So $$\mathbb{E}\left[X_i(t)\middle|\sum_{|k-i|\le N\varepsilon} X_k(t)/(2N\varepsilon+1)=a\right]\approx F'(\sigma) = F'(I_F'(a))$$

So,

$$\frac{d}{dt}\mathbb{E}\left(X_N(t)\middle|\frac{1}{(2N\varepsilon+1)}\sum_{|k-i|\le N\varepsilon} X_k(t)=a\right)$$

$$\approx \frac{1}{N^2}\sum_i J''\left(\frac{i}{N}\right)\mathbb{E}\left[\psi(X_i(t))\middle|\frac{1}{(2N\varepsilon+1)}\sum_{|k-i|\le N\varepsilon} X_k(t)=a\right]$$

$$\approx \frac{1}{N^2}\sum_i J''\left(\frac{i}{N}\right)\widehat{\psi}(a)$$

Writing $a = \rho_t\left(\frac{i}{N}\right)$, we get

$$\frac{d}{dt}\cdot\sum_i J\left(\frac{i}{N}\right)F'\left(I_F'\,\rho_t\left(\frac{i}{N}\right)\right) \approx \frac{1}{N^2}\sum_i J''\left(\frac{i}{N}\right)\widehat{\psi}\left(\rho_t\left(\frac{i}{N}\right)\right)$$

Now $F'(\sigma) = a,\ I'_F(a) = \sigma$ so

$$F'(I_F'(a)) = F'(\sigma) = a = \rho_t\left(\frac{i}{N}\right).$$

and hence,

$$\frac{d}{dt}\sum_i J\left(\frac{i}{N}\right)\rho_t\left(\frac{i}{N}\right) \approx \frac{1}{N^2}\sum_i J''\left(\frac{i}{N}\right)\widehat{\psi}\left(\rho_\tau\left(\frac{i}{N}\right)\right)$$

Now change the time scale:

$N^2\tau = t$. Then let

$$\tilde{\rho}_t(\theta) = \frac{\rho_t}{N^2}(\theta).$$

We get

$$\frac{d}{d\tau}\sum_i J\left(\frac{i}{N}\right)\tilde{\rho}_\tau\left(\frac{i}{N}\right) \approx \sum_i J''\left(\frac{i}{N}\right)\widehat{\psi}\left(\tilde{\rho}_\tau\left(\frac{i}{N}\right)\right)$$

❖❖❖❖❖

[14] Wong Zakai Filtering Equation:

$$dz = -C_t(z)\,dt + a_t(z)\,dv_t$$

$$dy = g_t(z)\,dt - dv_t$$

$$p\left(\underset{+dt}{t},\ T \mid \underset{+dt}{Y_1}\right)$$

$$p(dy_t \mid z_t = z)$$

$$\int p(z(t+dt) \mid z(t), Y_t\ dy(t))\ p(z(t) \mid Y_t, dy(t))\ dz(t)$$

$$= \int \frac{p(z(t)dz(t), Y_t, dy(t))}{p(z(t), Y_t, dy(t))}\ p(dy(t) \mid z(t), y(t))$$

$$\frac{p(z(t), Y_t)}{p(Y_t, dy(t))}\ dt(z)$$

$$= \int \frac{p(dy(t), dz(t) \mid y(t), z(t))}{p(dy(t) \mid z(t), Y_t)}\ p(dy(t) \mid z(t), y(t))$$

$$\frac{dz(t)\, p(z(t) \mid Y_t)}{p(dy(t) \mid Y_t)}$$

$$= \frac{\int p(dy(t), dz(t) \mid y(t), z(t))\, p(z(t) \mid Y_t)\, dz(t)}{\int p(dy(t), dz(t) \mid y(t)\ z(t))\, p(z(t) \mid Y_t)\, dz(t)\, d\, dz(t)},$$

$$\begin{pmatrix} dz \\ dy \end{pmatrix} = \begin{pmatrix} -C_t(z) \\ g_t(z) \end{pmatrix} dt + \begin{pmatrix} a_t(z) \\ -1 \end{pmatrix} dv_t$$

$$p(dy, dz, y \mid, z) = \mathbb{E}\left[\varphi(z(t+dt)) \mid Y_{t+dt}\right] = \mathbb{E}\left[\varphi(z(t), dy(t)) \mid Y_{t+dt}\right]$$

$$\begin{pmatrix} a_t(z) \\ -1 \end{pmatrix} (a_b(z) - 1) + \varepsilon I_2 = B_t(z):$$

$$p(dy, dz \mid y, z) = (2\pi)^{-1} \mid B_t(z) \mid^{-1/2} \exp\left\{-\frac{1}{2} dt\, (dz + C_t(z)\, dt, dy\right.$$

$$-g_t(t)\, dt) B_t(z)^{-1}$$

$$\left.(dz + C_t(z)\, dt\, dy - g_t(z)\, dt)\right\}$$

$$B_t(z) = \begin{pmatrix} \varepsilon + a_t^2(z) & -a_t(z) \\ -a_t(z) & 1+t \end{pmatrix}$$

$$|B_t(z)| = (1+\varepsilon)(\varepsilon + a_t^2(z)) - a_t^2(z)$$

$$= \varepsilon + \varepsilon^2 + \varepsilon a_t^2(z) = \varepsilon(1 + a_t^2(z)) + O(\varepsilon^2)$$

$$\mid B_t(z) \mid^{-1/2} \approx \varepsilon^{-1/2} \Big/ \left(1 + a_t^2(z)\right)^{1/2}$$

$$B_t(z)^{-1} = \begin{pmatrix} 1+\varepsilon & a_t(z) \\ a_t(z) & \varepsilon + a_t^2(z) \end{pmatrix} \Big/ \varepsilon\left(1 + a_t^2(z)\right)$$

$$= \begin{pmatrix} 1 & a_t(z) \\ a_t(z) & a_t^2(z) \end{pmatrix} \Big/ \varepsilon\left(1 + a_t^2(z)\right)$$

$$p(dy, dz \mid y, z) = (2\pi)^{-1}\varepsilon^{-1/2}\left(1+a_t^2(z)\right)^{-1/2}$$

$$\cdot \exp\left\{-\frac{1}{d_t^2\,\varepsilon(1+a_t^2(z))}(dz + C_t(t)\,dt, dy - g_t(z)\,dt)\right.$$

$$\left.\begin{pmatrix} 1 & a_t(z) \\ a_t(z) & a_t^2(z) \end{pmatrix}\begin{pmatrix} dz + C_t\,dt \\ dy - g_t dt \end{pmatrix}\right\}$$

$$\mathcal{L}\left(t + a_t^2\right)^{-1/2} \exp\left\{\frac{-1}{2\varepsilon\left(1+a_t^2\right)dt}((dz + C_t dt)^2 + a_t^2(dy - g_t\,dt)^2\right.$$

$$\left. + 2\,a_t(dz + C_t dt)(dy - g_t\,dt)\right)$$

$$\left(1+a_t^2\right)^{-1/2} \exp\left\{1 - \frac{1}{\varepsilon\left(1+a_t^2\right)}\left(8C_t\,dz - a_t^2\,g_t\,dy - a_t g_t dz + C_t a_t dy\right)\right\}$$

$$= \left(1+a_t^2\right)^{-1/2} \exp\left\{\frac{-1}{\varepsilon\left(1+a_t^2\right)}\left((C_t - a_t g_t)\,dz + \left(C_t a_t - a_t^2 g_t\right)dy\right)\right\}$$

$$= \left(1+a_t^2\right)^{-1/2} \exp\left\{-\frac{(C_t - a_t g_t)}{\varepsilon\left(1+a_t^2\right)}(dz + a_t\,dy)\right\}$$

$$= \left(1+a_t^2\right)^{-1/2} \exp\left\{1 - \frac{(C_t - a_t g_t)}{\varepsilon\left(1+a_t^2\right)}(dz + a_t\,dy)\right\}$$

$$\left. + \frac{(C_t - a_t g_t)}{2\varepsilon^2(1+a_t^2)^2}(1+a_t^2)\,dt\right\}$$

$$p(dy \mid z, y) = p(dy \mid z)$$

$$= \frac{1}{\sqrt{2\pi}\sqrt{dt}} \exp\left(-\frac{1}{2dt}(dy - g_t\,dt)^2\right)$$

$$\alpha \exp\left(g_t\,dy - \frac{g_t^2}{2}\,dt\right)$$

$$= 1 + g_t dy - \frac{g_t^2}{2}\,dt + \frac{g_t^2}{2}\,dt$$

$$= 1 + g_t(z)\,dy.$$

$$\text{Numerator} = \int\left(1+a_t(z)^2\right)^{-1/2}\left\{1-\frac{(C_t-a_t\,g_t)(dz+a_t\,dy)}{\varepsilon(1+a_t^2)}\right.$$

$$\left.+\frac{\left(C_t-a_t\,g_t\right)^2}{2\varepsilon^2\left(1+a_t^2\right)}dt\right\}\{1+g_t(z)dy\}\,p_t(z\,|\,Y_t)\,dz$$

Let $\varepsilon\,(1+a_t^2)\to\varepsilon$

The numerator is

$$\int\left[1-\frac{(C_t-a_tg_t)(dz+a_tdy)}{\varepsilon}\right.$$

$$\left.+\left(1+a_t^2\right)\frac{\left(C_t-a_tg_t\right)^2}{2\varepsilon^2}dt+g_t\,dy\right]p_t(z\,|\,Y_t)\,dz$$

Numerator of

$$p_{t+dt}(z'\,|\,Y_{t+dt}) = 1-\frac{1}{\varepsilon}\int\Psi_t(z)(z'-z)\,p_t(z\,|\,Y_t)\,dz$$

$$-\frac{dy}{\varepsilon}\int a_t(z)\Psi_t(z)\,p_t(z\,|\,Y_t)\,dz$$

$$+\frac{dt}{2\varepsilon^2}\int\Psi_t^2(z)\,p_t(z\,|\,Y_t)\,dz\left(1+a_t^2(z)\right)$$

$$\Psi_t(z) = C_t(z)-a_t(z)\,g_t(z).$$

$$\int\Psi_t(z)(z'-z)\,p_t(z\,|\,Y_t)\,dz$$

$$= \mathbb{E}\left[\Psi_t\left(z(t)\right)\left(z'-z(t)\right)|\,Y_t\right]$$

$$= \mathbb{E}\left[\Psi_t d\left(z'-dz(t)\right)dz(t)\,|\,Y_t\right]$$

Doemo Work.

Alternately Writing

$$\text{Numerator} = \int\left[1-\frac{\Psi_t(z)}{\varepsilon}\left(z'-z+a_t(z)dy\right)\right.$$

$$\left.+\frac{\Psi_t^2(z)}{2\varepsilon^2}\left(z'-z+a_t\left(z\right)dy\right)^2\right]p_t(z|Y_t)dz$$

$$= \int\left[1-\frac{\Psi_t\left(z'-\delta z\right)}{\varepsilon}\left(\delta z+a_t\left(z'-\delta z\right)dy\right)\right.$$

$$\left.+\frac{\Psi_t^2\left(z'-dz\right)}{2\varepsilon^2}(\delta z+a_t(z'-\delta z)dy)^2\right]p_t(z'-\delta z|Y_t)d\delta z$$

$$= \int \left\{ p_t\left(z'|Y_t\right) - \delta z p_t'\left(z' - Y_t\right) + \frac{(\delta z)^2}{2} p_t''\left(z'|Y_t\right) \right\}$$

$$\times \left\{ 1 - \frac{1}{\varepsilon}(\psi_t(z') - \delta z\, \psi_t'(z'))(\delta z + a_t(z')dy - a_t'(z')\delta z\, dy) \right.$$

$$\left. + \frac{\Psi_t^2(z')}{2\varepsilon^2} dt\left(1 + a_t^2(z')\right) \right\} d\,\delta z$$

Let the range of δz be $\left[-\frac{\Delta}{2}, \frac{\Delta}{2}\right]$. Then

$$\text{Numerator} = p_t(z'|Y_t)\Delta + dt\frac{a_t^2(z')}{2} p''(z'|Y_t)\Delta + \frac{\Delta}{\varepsilon}\Psi_t(z')p'(z'|Y_t)$$

$$- \frac{\Delta}{\varepsilon}\Psi_t(z')p_t(z'|Y_t)a_t(z')dy\, a_t^2(z')dt - a_t^2(z')dt$$

$$+ p_t(z'|Y_t)\frac{\Psi_t^2(z')}{2\varepsilon^2} dt(1 + a_t^2(z'))\Delta$$

$$- \frac{\Delta}{\varepsilon}\Psi_t(z')p_t(z'|Y_t)a_t'(z')a_t(z')dt$$

Cancelling out the factor D gives the numerator as

$$p_t(z'|Y_t) + \left\{ \frac{a_t^2(z')}{2} p_t''(z'|Y_t) \right.$$

$$- \frac{1}{\varepsilon} a_1'(z')a_t(z')\Psi_t(z')p_t(z'|Y_t)$$

$$\left. + p_t(z'|Y_t)\frac{\Psi_t^2(z')}{2\varepsilon^2}(1 + a_t^2(z')) \right\} dt$$

$$- \frac{1}{\varepsilon}\Psi_t(z')p_t(z'|Y_t)a_t(z')dy(t)$$

The denominator is the integral of the numerator *w.r.t.* z. Hence

$$p_{t+dt}(z'|Y_{t+dt}) = \frac{p_t(z'|Y_t) + \Lambda_t^\varepsilon(z'|Y_t)dt + M_t^\varepsilon(z'|Y_t)dy}{\left(1 + dt\int \Lambda_t^\varepsilon(z'|Y_t)dz' + dy\int M_t^\varepsilon(z'|Y_t)dz'\right)}$$

$$= p_t(z'|Y_t) + dt\,\Lambda_t^\varepsilon(z'/Y_t) - dt\, p_t(z'|Y_t)\int \Lambda_t^\varepsilon(z'|Y_t)dz'$$

$$- M_t^\varepsilon(z'|Y_t)\left(\int M_t^\varepsilon(z'|Y_t)dz'\right)dt$$

$$- p_t(z'|Y_t)\left(\int M_t^\varepsilon(z'|Y_t)dz'\right)dy(t)$$

Thus,

$$dp_t(z'\,|\,Y_t) = \left\{\frac{a_t^2(z')}{2}p_t''(z'\,|\,Y_t) - \frac{1}{\varepsilon}a_t'(z')a_t(z')\Psi_t(z')\,p_t(z'\,|\,Y_t)\right.$$
$$+\frac{\Psi_t^2(z')}{2\varepsilon^2}\left(1+a_t^2(z')\right)p(z'\,|\,Y_t) - p_t(z'\,|\,Y_t)$$

❖❖❖❖❖

[15] Stochastic Control of Finance Market:

$P_k(t)$, $k = 1, 2, ..., d$ are the share prices. They evolve according to random rates:

$$dP_k(t) = \left(r_k(t)\,dt + \sigma_k\, dB_k(t)\right)P_k, 1 \le k \le d$$

where $r_k(t)'^s$ are non random rates.

At time t, the share holds has $N_k(t)$ shares of type k, $k = 1, 2, ..., d$. His gain in Wealth in time $[t, t + dt]$ is $\sum_{k=1}^{d} N_k(t)\,dP_k(t)$ and his consumption in time $[t, t + dt]$ is $C(t)\,dt$. His total Wealth at time T is then

$$X(T) = \int_0^T\left(\sum_{k=1}^{d} N_k(t)\,dP_k(t)C(t)\,dt\right)$$

The quantity $N_k(.)$ and $C(.)$ satisfy certain constants like N'^s_k must not be too large etc. We write there constrants as

$$\mathbb{E}\,f_r(N_1(t), .., N_d(t), C(t), -\lambda o, r = 1, 2, ..., M.$$
$$P_1(t), ..., P_d(t)$$

Let $$D_r = \text{dias}\left[\{r_k(t)\}_{k=1}^{d}\right].$$

$\mathbb{E}\,[X(t)]$ is to be maximized subject to these constaints.

We consider maximizing

$$\mathbb{E}\;X(T) - \sum_{r=1}^{M}\int_0^T \lambda_r(t)\,\mathbb{E}\;f_r(\mathbb{N}\,(t), \underline{P}(t), C(t))\,dt$$

$$= \mathbb{E}\int_o^T\left\{\underline{N}(t)^T \underline{\underline{D}}_r\,\underline{P}(t) - C(t) - \underline{\lambda}(t)^T\,\underline{f}(\underline{N}(t), \underline{P}(t), C(t))\right\}dt \qquad (1)$$

The optimum $\underline{N}(t)$, $C(t)$ are obtained as functions of $\underline{P}(t)$ that maximize (1)

Let

$$\max_{x\neq \underline{N}(t),\, C(t)\neq T}\int_S^T \mathbb{E}\left[\underline{N}(t)^T \underline{\underline{D}}_r\,\underline{P}(t) - C(t) - \underline{\lambda}(t)^T\,\underline{f}(\underline{N}(t), \underline{P}(t), C(t))\right]$$

$$= V_*(P(s), \lambda_s) \,|\, P(s)]dt$$

Where $\underline{\lambda}_s = \{\underline{\lambda}(t) : s \le t \le T\}$

Then, $V_*(\underline{P}(s), \underline{\lambda}_s)$

$$= \max_{N(s),\, C(s)} \Big\{\big(\underline{N}(s)^T \underline{\underline{D}}_r\, \underline{P}(s) - C(s) - \underline{\lambda}^T(s)\, \underline{f}(\underline{N}(s), \underline{P}(s), C(s)\big) ds$$

$$+ E\big[V_*\big(\underline{P}(s+ds), \lambda_{s+ds}\big) \,|\, \underline{P}(s)\big]\Big\}$$

we write $V_*\big(\underline{P}(s), \underline{\lambda}_s\big) = V_*\big(\underline{P}(s), s\big)$

Note that λ is non-random. Then

$$V_*(\underline{P}(s), s) = \max_{\underline{N}(s),\, C(s)} \Big\{\big(\underline{N}(s)^T \underline{\underline{D}}_r\, \underline{P}(s) - C(s) - \underline{\lambda}(s)^T\, \underline{f}(\underline{N}(s), \underline{P}(s), C(s)\big) ds$$

$$+ V_*(\underline{P}(s), s) + \left[\frac{\partial V_*(\underline{P}(s), s)}{\partial \underline{P}}\right]^T \underline{\underline{D}}_r\, \underline{P}(s)\, ds$$

$$+ \frac{1}{2} T_r \left\{\left[\frac{\partial^2 V_*(\underline{P}(s), s)}{\partial \underline{P}\, \partial \underline{P}^T}\right]^T \underline{\underline{D}}_\sigma(s)\right\} ds + \frac{\partial V_*(\underline{P}(s), s)}{\partial s}\, ds\Big\}$$

or

$$\max_{\underline{N}(s),\, C(s)} \left\{\underline{N}(s)^T \underline{\underline{D}}_r\, \underline{P}(s) - C(s) - \underline{\lambda}(s)^T\, \underline{f}(\underline{N}(s), P(s), C(s))\right\}$$

$$+ \frac{\partial V_*(P(s), s)}{\partial s} + \left(D_r P_T^{(s)}\right)^T \frac{\partial V_*(P(s), s)}{\partial \underline{P}}$$

$$+ \frac{1}{2} T_r \left\{\underline{\underline{D}}_\sigma(s) \frac{\partial^2 V_*(P(s), s)}{\partial \underline{P}\, \partial \underline{P}^T}\right\} = 0$$

We thus get the following partial differential equation for $V_*(\underline{P}(s))$:

$$\max_{\underline{N},\, \underline{P}} \left\{\underline{N}^T \underline{\underline{D}}_r\, \underline{P} - C - \underline{\lambda}(s)^T\, \underline{f}(\underline{N}, \underline{P}, C)\right\}$$

$$+ \frac{\partial V_*(\underline{P}, s)}{\partial s} + r(s)^T \frac{\partial V_*(\underline{P}, s)}{\partial \underline{P}} + \frac{1}{2} T_r \left(\underline{\underline{D}}_\sigma(s) \frac{\partial^2 V_*(\underline{P}, s)}{\partial \underline{P}\, \partial \underline{P}^T}\right) = 0$$

Where $D_\sigma(s) = \text{diag}\left[\left\{\sigma_k^2\, P_k^2(s)\right\}_{k=1}^d\right]$

The optimal scheme of choosing $\underline{N}$, C at each stage is thus given by

$$\left(\widehat{N}(t), \widehat{C}(t)\right) = \underset{N,\, P}{\arg\max} \left\{\underline{N}^T \underline{\underline{D}}_r \underline{P} - C - \lambda(t)^T\, \underline{f}(\underline{N}, \underline{P}, C)\right\}$$

$$= \left\{\widehat{N}(\underline{P}, \underline{\lambda}(t)), \widehat{C}(\underline{P}, \underline{\lambda}(t))\right\}$$

Finally $\{\underline{\lambda}(t)\}_{t \in [o, T]}$ is obtained using this equation combined with

$$\mathbb{E}\left\{f_r(\widehat{N}(\underline{P}(t),\underline{\lambda}(t)),\underline{P}(t),\widehat{C}(\underline{P}(t),\underline{\lambda}(t)))\right\}$$
$$=0,\ 1\le r\le d.$$

The expectation is to be taken *w.r.t.* the random vector $\underline{P}(t)$.

❖❖❖❖❖

[16] Wong-Zokai Equation:

$$dz = -C_t(z)dt + a_t(z)dv_t \qquad \text{(State model)}$$

$$dy = g_t(z)dt - dv_t \qquad \text{(Measurement model)}$$

$$\begin{aligned}
p(Z(t+dt)\,|\,Y_{t+dt}) &= p(Z(t+dt)\,|\,Y_t, dy(t))\\
&= \frac{p(Z(t+dt)\,|\,Y_t, dy(t))}{p(Y_t, dy(t))}\\
&= \frac{p(Z(t+dt), dy(t)\,|\,Y_t)}{p(dy(t)\,|\,Y_t)}\\
&= \frac{\int p(Z(t+dt), dy(t)\,|\,Y_t\,Z(t))\,\dot{p}(Z(t)\,|\,Y_t)\,dZ(t)}{p(dy(t)\,|\,Y_t)}\\
&= \frac{\int p(Z(t+dt), dy(t)\,|\,Z(t), y(t))\,p(Z(t)\,|\,Y_t)\,dZ(t)}{p(dy(t)\,|\,Y_t)}\\
&\equiv \frac{\chi(Z(t+dt), dy(t)\,|\,Y_t)}{\int \chi(Z(t+dt, dy(t)\,|\,Y_t)\,dZ(t+dt)}
\end{aligned}$$

Let $$\begin{bmatrix} dz \\ dy \end{bmatrix} = \begin{bmatrix} -c_t(z) \\ g_t(z) \end{bmatrix} dt + B_t(Z)\begin{bmatrix} dv_1(t) \\ dv_2(t) \end{bmatrix}$$

$$\begin{aligned}
&\mathbb{E}\left[\psi(Z(t+dt), dy(t)\,|\,Z(t), y(t))\right]\\
&= \psi(Z(t),0) - \frac{\partial\psi}{\partial Z}(Z(t),0)\,c_t(Z(t))\,dt\\
&\quad + \frac{1}{2}\frac{\partial^2\psi}{\partial Z^2}(Z(t),0)\,at^2(Z)dt + \frac{\partial\psi}{\partial y}(Z(t),0)\,g_t(Z(t))dt\\
&\quad + \frac{1}{2}\frac{\partial^2\psi}{\partial y^2}(Z(t),0)\,dt\\
&= \int \psi(Z(t+dt), dy(t))\,\chi(Z(t+dt), dy(t)\,|\,Y_t)\,dz(t+dt)\,d\,dy(t)\\
&= \int \psi(Z,o) - \psi_{,z}(Z,o)\,c_t(Z)\,dt + \frac{1}{2}\psi_{,zz}(Z,o)\,a_t^2(Z)\,dt\\
&\quad \left. + \psi_{,Y}(Z,o)\,g_t(z)dt + \frac{1}{2}\psi_{,YY}(Z,o)\,dt\right] p_t(Z\,|\,Y_t)\,dZ
\end{aligned}$$

$$= \int \psi(Z, dy)\chi(Z, dy \mid Y_t)\, dz\, d\, dy$$

Hence, $\chi(Z, dy \mid Y_t)$

$$= \delta(dy)\, p_t(Z \mid Y_t)$$

$$+\left[\frac{\partial}{\partial Z}(C_t(Z)\, p_t(Z \mid Y_t))\delta(dy) + \frac{1}{2}\frac{\partial^2}{\partial Z^2}\left(a_t^2(Z)\, p_t(Z \mid Y_t)\right)\delta(dy)\right.$$

$$\left. - \delta'(dy)\, g_t(Z)\, p_t(Z \mid Y_t) + \frac{1}{2}\delta''(dy)\, p_t(Z \mid Y_t)\right] dt$$

❖❖❖❖❖

[17] Filtering Theory with State and Measurement Noises Having Correlation:

State mode:

$$d\underline{X}(t) = \underline{f}(\underline{X}(t))\,dt + \underline{\underline{g}}(\underline{X}(t))\,d\underline{B}(t)$$

$$d\underline{Z}(t) = \underline{h}(X(t))\,dt + \underline{\underline{K}}(\underline{X}(t))\,d\underline{B}(t)$$

where $\underline{X}(t) \in \mathbb{R}^n, Z(t) \in \mathbb{R}^n, \underline{\underline{g}}(\underline{X}(t)) \in \mathbb{R}^{n\times q}, \underline{\underline{K}}(\underline{X}(t)) \in \mathbb{R}^{p\times q}, \underline{B}(t) \simeq$ standard q dimensional Brownain motion:

$$\underline{Z}_t = \{d\underline{Z}(s), s \neq t\}$$

We need an equation for $p(\underline{X}(t) \mid \underline{Z}_t)$

$$p(\underline{X}(t+dt) \mid \underline{Z}_{t+dt})$$

$$= \int p(\underline{X}(t+dt) \mid \underline{X}(t), \underline{Z}_t, d\underline{z}(t))\, p(\underline{X}(t) \mid \underline{Z}_t, d\underline{z}(t))\, d\underline{X}(t)$$

$$= \int p(\underline{X}(t+dt) \mid \underline{X}(t), Z_t, dZ(t)) \frac{p(\underline{X}(t), Z_t)\, p(\underline{X}(t) \mid \underline{Z}_t, dZ(t))}{p(X(t), \underline{Z}_t, dZ(t)\, d\underline{X}(t))}$$

$$= \int [p(\underline{X}(t+dt), \underline{Z}(t+dt) \mid \underline{X}(t), \underline{Z}(t))$$
$$\times p(\underline{X}(t), \underline{Z}_t) / p(\underline{Z}_t, dZ(t))]\, d\underline{X}(t)$$

$$= \frac{\int [p(\underline{X}(t+dt), \underline{Z}(t+dt) \mid \underline{X}(t), \underline{Z}(t)) \times p(\underline{X}(t), \underline{Z}_t)\, d\underline{X}(t)]}{\int dX(t+dt)}$$

We write $\underline{X}(t+dt) = X'$

$\underline{Z}(t+dt) = Z'$,

$X(t) = X$,

$Z(t) = Z$

and $p(\underline{X}(t+dt), \underline{Z}(t+dt) \mid \underline{X}(t), \underline{Z}(t))$

$$= p(t+dt, X', Z \mid t, X, Z)$$

$$= \delta(X'-X)\delta(Z'-Z) + dt\,\mathcal{L}^{*\prime}(\delta(X'-X)\delta(Z'-Z))$$

Where $\mathcal{L}^*$ is the formed Kolmogorov operator for the state plus observable process:

$$\mathcal{L}\psi^{*\prime} = -\nabla_{X'}^T(\underline{f}(\underline{X}')\psi(X',Z')) = -\nabla_{Z'}^T(\underline{h}(\underline{X}')\psi(X',Z'))$$

$$+\frac{1}{2}T_r\left(\begin{pmatrix}\nabla_{X'}\\ \nabla_{Z'}\end{pmatrix}\left(\nabla_{X'}^T, \nabla_{Z'}^T\right)\begin{pmatrix}g(X')\\ K(X')\end{pmatrix} g(X')_1^T\,K(X')^T\,\psi(X',Z')\right)$$

$$= -\operatorname{div}'_X(\underline{f}(\underline{X}')\psi(X',Z')) - \operatorname{div}'_Z(\underline{h}(X')\psi(X',Z'))$$

$$+\frac{1}{2}T_r\left(\nabla_{X'}\nabla_{X'}^T\,g(X')g(X')^T\,\psi(X',Z')\right)$$

$$+\frac{1}{2}T_r\left(\nabla_{X'}\,\nabla_{X'}^T\,K(X')g(X')^T\,\psi(X',Z')\right)$$

$$+\frac{1}{2}T_r\left(\nabla_{Z'}\,\nabla_{X'}^T\,g(X')K(X')^T\,\psi(X',Z')\right)$$

$$+\frac{1}{2}T_r\left(\nabla_{Z'}\,\nabla_{Z'}^T\,K(X')K(X')^T\,\psi(X',Z')\right)$$

We than get $\quad p(t+dt, X' \mid Z_{t+dt})$

$$= \frac{\int\left(\delta(X'-X)\delta(Z'-Z) + dt\,\mathcal{L}^{*\prime}(\delta(X'-X)\delta(Z'-Z))\,p(t,X \mid Z_t)\,dX\right)}{\int\left(\delta(X'-X)\delta(Z'-Z) + dt\,\mathcal{L}^{*\prime}(\delta(X'-X)\delta(Z'-Z))\,p(t,X \mid Z_t)\,dX\,dX'\right)}$$

$$= \frac{\left[p(t,X' \mid Z_t)\delta(Z'-Z) + dt\,\mathcal{L}^{*\prime}(p(t,X' \mid Z_t)\delta(Z'-Z))\right]}{\delta(Z'-Z) + dt\int\mathcal{L}^{*\prime}(p(t,X' \mid Z_t)\delta(Z'-Z))\,dX'}$$

Now, $\quad \int\mathcal{L}^{*\prime}(p(t,X' \mid Z_t)\delta(Z'-Z))\,dX$

$$= -\int\operatorname{div}_{Z'}(\underline{h}(X')\,p(t,X' \mid Z_t)\delta(Z'-Z))\,dX'$$

$$+\frac{1}{2}\int T_r\left(\nabla_{Z'}\,\nabla_{Z'}^T\left(K(X')K(X')^T\,p(t,X' \mid Z_t)\delta(Z'-Z)\right)\right)dX'$$

$$= -\operatorname{div}_{Z'}\left(\underline{\hat{h}}_t\nabla_{Z'}\,\delta(Z'-Z)\right) + \frac{1}{2}T_r\left(\nabla_{Z'}\,\nabla_{Z'}^T\,\widehat{K_tK_t^T}\,\delta(Z'-Z)\right)$$

$$-\underline{\hat{h}}_t^T\,\nabla_{Z'}\delta(Z'-Z) + \frac{1}{2}T_r\left(\widehat{K_tK_t^T}\,\nabla_{Z'}\nabla^T{}_{,}\delta(Z'-Z)\right)$$

Where $\quad \underline{\hat{h}}_t = \mathbb{E}\,[\underline{h}(X(t)) \mid Z_t],$

$$\widehat{K_tK_t^T} = \mathbb{E}\left[K(\underline{X}(t))K(\underline{X}(t))^T \mid Z_t\right]$$

This we get

$$p(t+dt, X' \mid Z_{t+dt})$$

$$= \frac{\left[\delta(Z'-Z)\cdot p(t, X' \mid Z_t) + dt\,\mathcal{L}^{*\prime}(p(t, X' \mid Z_t)\delta(Z'-Z))\right]}{\delta(Z'-Z) + dt\left(-\underline{\hat{h}}\nabla_{Z'}\delta(Z'-Z) + \frac{1}{2}T_r\left(\widehat{K_t K_t^T}\right)\nabla_{Z'}\nabla_{Z'}^T\delta(Z'-Z)\right)}$$

where

$$Z = Z(t), Z' = Z(t+dt) = Z(t) + dZ(t).$$

$$\mathcal{L}^{*\prime}(p(t, X' \mid Z_t)\delta(Z'-Z))$$

$$= \mathcal{L}o^{*\prime}\left(p(t, X' \mid Z_t)\right)\delta(Z'-Z) - \underline{h}(\underline{X}')^T\nabla_{Z'}\left(\delta(Z'-Z)\right)p(t, X' \mid Z_t)$$

References:

[1] Research Papers of V. Belavkin on quantum filter.

[2] Technical Report on Wave propagation in Metamaterial by Karishma Sharma, D.K. Upadhyay and H. Parthasarthy, NSIT, 2014.

[3] Technical Reports on Robotics by Vijayant Agrawal and Harish Parthasarathy, NSIT, 2014.

❑❑❑

5

Classical and Quantum Robotics

[1] Mathematical Pre-Requisites for Robotics:

[1] Kinetic and potential energies of an n-link system.

[2] Relativistic kinetic energies.

[3] Variational calculus.

[4] Perturbation theory for algebraic and differential equations.

[5] Brownian motion, Poinon process, Levy process (processes with independent increments.

[6] Ito stochastic calculus for Brownian motion and Poinon processes.

[7] Stochastic differential equations driven by Levy processes.

[8] Forward and Backward Kolmogorov equations.

[9] Markov chains.

[10] Approximate mean and variance propagation in stochastic differential equations.

[11] The maximum likelihood method for parameter estimation.

[12] The Cramer-Rao lower bound (CRLB).

[13] Optimal control theory.

[14] Stochastic optimal control.

[15] The n-link Robot differential equations.

[16] Optimal control applied to master-slave Robot tracking.

[17] Leaso mean square (LMS) algorithm applied to tracking problem in Robotics. (adaptive)

[18] Recursive leaso squares (RLS) algorithm applied to tracking problem in Robotics (adaptive).

[19] Maximum likelihood estimation of Pd controller for master-slave Robot system.

[20] Design of feedback and operator force process for optimal control of LTI systems with feedback noise – an application of algebraic perturbation theory.

[21] The rotation group SO(3).

[22] The kinetic energy of a rigid body.

[23] Quantization of rigid body motion.

[24] Feynman path integral approach to quantum Robotics.

[1] The lengths of the links are $L_1, L_2, .., L_n$. The first link has its free end at $(X(t), Y(t))$ and the angles of the ith link relative to the X-axis is $\theta_i(t)$. Let σ_i be the mass pen unit length of the ith link. Then the position of a point P on the ith link at a distance ξ from its joint with the $(i-1)^{th}$ links is Y given by

* Maximum likelihood estimation of parameters in dynamical systems.

Let $\{\underline{W}[n]\}_{n=0}^{\infty}$ be iid random vectors with values in $\mathbb{R}^p$ having pdf $\Phi_w(\underline{W})$. Consider the dynamical system.

$$\underline{Z}[n+1] = \underline{f}(\underline{Z}[n]) + \underline{\underline{G}}(\underline{Z}[n])\underline{W}[n+1],\ n \geq 0$$

Then $\{\underline{Z}[n]\}_{n\geq 0}$ is a Markov process provided $\underline{Z}[0]$ is independent of $\{\underline{W}[n]\}_{n\geq 1}$. The conditional pdf of $\underline{Z}[n+1]$ gives $\underline{Z}[n]$ is

$$p(Z[n+1]|Z[n]) = \left|\underline{\underline{G}}(\underline{Z}[n])\right|^{-1}\Phi_W\left(\underline{\underline{G}}(\underline{Z}[n])^{-1}\left(\underline{Z}[n+1]-\underline{f}(\underline{Z}[n])\right)\right)$$

where $$\left|\underline{\underline{G}}(\underline{Z})\right| = \left|\det(\underline{\underline{G}}(\underline{Z}))\right|$$

Hence if f and $\underline{\underline{G}}$ depend on an unknown parameter vector $\underline{\theta}$, *i.e.* $f(\underline{Z}, \underline{\theta})$, $G(Z, \underline{\theta})$ then the joint pdf of $\{Z[n]\}_{0\leq n\leq N}$ given $\underline{\theta}$ is

$p(\{\underline{Z}[n]\}_{p\leq n\leq N}|\underline{Z}[0], \underline{\theta})$

$$= \prod_{n=0}^{N}\left\{\left|\underline{\underline{G}}(\underline{Z}[n], \underline{\theta}\right|^{-1}\right.$$

$$\left.\Phi_W\left(\underline{\underline{G}}(\underline{Z}[n],\theta)^{-1}\left(\underline{Z}[n+1]-\underline{f}(\underline{Z}[n],\underline{\theta})\right)\right)\right\}$$

or equivalently, the log-likelihood function is

$L(\{\underline{Z}[n]\}_{1\leq n\leq N} \mid \underline{Z}(0), \underline{\theta}) = \log p(\underline{Z} \mid \underline{Z}(0), \underline{\theta})$

$$-\sum_{n=0}^{N}\log\left\{\left|G(\underline{Z}[n],\underline{\theta})\right|\right\}$$

$$+\sum_{n=0}^{N}\log\Phi_W\left(\underline{\underline{G}}(\underline{\underline{Z}}[n],\underline{\theta})^{-1}\left(\underline{Z}[n+1]-\underline{f}(\underline{z}[n],\underline{\theta})\right)\right)$$

The ml of $\underline{\theta}$ based on $\underline{Z} = \{\underline{Z}[n]\}_{P\leq n\leq N}$ is

$$\hat{\underline{\theta}}(N) = \underset{\underline{\theta}}{\arg\max}\, L\left(\underline{Z}|\underline{Z}(0),\underline{\theta}\right)$$

Example:

$$\underline{\underline{Z}}[n+1] = \sum_{k=1}^{p} \theta_k \underline{f}_k(\underline{Z}[n]) + \underline{\underline{G}}(\underline{Z}[n])\underline{W}[n+1]$$

We wish to estimate $\underline{\theta} = (\theta_k)_{k=1}^{p}$ when $\{\underline{W}(n)\}_{n\geq 1}$ is iid $N(0, \underline{\underline{R}}_W)$.

$$\underline{P} = \left(X(t) + \sum_{k=1}^{i-1} L_k \cos\theta_k + \xi \cos\theta_i \,,\; Y(t) + \sum_{k=1}^{i-1} L_k \sin\theta_k + \xi \sin\theta_i \right), \; 1 \leq i \leq n$$

and its velocity is therefore given by

$$\frac{d\underline{P}}{dt} = \left(X'(t) - \sum_{k=1}^{i-1} L_k\theta_k' \sin\theta_k - \xi\theta_i' \sin\theta_i \,,\; Y'(t) + \sum_{k=1}^{i-1} L_k\theta_k' \cos\theta_k + \xi\theta_i' \cos\theta_i \right)$$

The kinetic energy of the ith link is then for $i > 1$,

$$\begin{aligned}
T_i &= \frac{1}{2}\sigma_i \int_0^{L_i} \left|\frac{dp}{dt}\right|^2 d\xi \\
&= \frac{1}{2}\sigma_i \int_0^{L_i} \Bigg(X'^2 + Y'^2 + \xi^2\theta_i'^2 + \sum_{k=1}^{i-1} L^2{}_k \theta_k'^2 \\
&\quad -2X' \sum_{1}^{i-1} L_k\theta_k' \sin\theta_k + 2Y' \sum_{k=1}^{i-1} L_k\theta_k' \cos\theta_k \\
&\quad + 2 \sum_{1\leq k<m\leq i-1} L_k L_m \theta_k'\theta_m' \cos(\theta_k - \theta_m) \\
&\quad + 2\xi \sum_{1\leq k\leq i-1} \theta_i'\theta_k' L_k \cos(\theta_k - \theta_i) \\
&\quad + 2\theta_i'\xi(Y'\cos\theta_i - X'\sin\theta_i) \Bigg) d\xi \\
&= \frac{1}{2}\sigma_i \Bigg(X'^2 + Y'^2 + \sum_{k=1}^{i-1} L_k \theta_k'^2 \\
&\quad - 2X' \sum_{1}^{i-1} L_k\theta_k' \sin\theta_k + 2Y' \sum_{1}^{i-1} L_k\theta_k' \cos\theta_k \\
&\quad + 2 \sum_{l\leq k<m\leq i-1} L_k L_m \theta_k'\theta_m' \cos|\theta_k - \theta_m| \Bigg) L_i \\
&\quad + \sigma_i L_i^3 \frac{\theta_i'^2}{6}
\end{aligned}$$

$$\frac{\sigma_i}{2} \sum_{l\leq k\leq i-1} L_i^2 L_k \theta_i'\theta_k' \cos(\theta_k - \theta_i) + \frac{\sigma_i}{2} L_i^2 \theta_i' (Y'\cos\theta_i - X'\sin\theta_i)$$

For $i = 1$, we have

$$T_1 = \frac{1}{2}\sigma_1(X'^2 + Y'^2) + \sigma_1 L_1^3 \frac{\theta_i^2}{6} + \frac{1}{2}\sigma_1 L_1^2 \theta_1'(Y'\cos\theta_1 - X'\sin\theta_1)$$

The relationship between the kinetic energy T and velocity v of a point mars having rest mars m_0 is

$$T = \frac{m_0 c^2}{\sqrt{1 - \frac{v^2}{c^2}}}.$$

Hence the kinetic energy of the ith link is

$$T_i = \sigma_i c^2 \int_0^{L_i} (F_i X'(t), Y'(t), \theta_j'(t), \theta_j(t), 1 \le j \le i, \xi) d\xi$$

where $F_i(X', Y', \theta'_j, \theta_j, 1 \le j \le i, \xi) = \left(1 - \frac{v^2}{c^2}\right)^{-1/2}$ with

$$\begin{aligned} v^2 = {} & X'^2 + Y'^2 + \sum_{k=1}^{i-1} L_k^2 \theta_k'^2 \\ & + 2 \sum_{l \le k < m \le i-1} L_k L_m \theta_k' \theta_m' \cos(\theta_k - \theta_m) \\ & - 2X' \sum_{k=1}^{i-1} L_k \theta_k' \sin\theta_k + 2Y' \sum_{k=1}^{i-1} L_k \theta_k' \cos\theta_k \\ & + 2\xi \left\{ \theta_i' \sum_{k=1}^{i-1} L_k \theta_k' \cos(\theta_k - \theta_i) \right. \\ & \left. + \theta_i'(Y'\cos\theta_i - X'\sin\theta_i) \right\} + \xi^2 \theta_i'^2 = v_i^2(\xi) \end{aligned}$$

[3] Variational calculus:

We wish to find a path $\underline{q}(t) \in \mathbb{R}^n$, $0 \le t \le T$ (*i.e.* q: $[0, T] \to \mathbb{R}^n$) so that $\underline{q}(0) = \underline{a}$, $\underline{q}(T) = \underline{b}$ are fixed and

$$S[\underline{q}] = \int_0^T L(\underline{q}(t), \underline{q}'(t), t) dt$$

is an extremum (*i.e.* local minimum or local maximum) (over all differentiable paths from $(0, \underline{a})$ to $(T, \underline{b})$).

If $S[\underline{q}]$ is an extremum at $\underline{q}(.) = \underline{q}_*(.)$, then

$$\left.\frac{\delta S[\underline{q}]}{\delta \underline{q}}\right|_{\underline{q}=\underline{q}_*} = 0$$

i.e. under an arbitrary small change $\underline{q}_*(t) \to \underline{q}_*(t) + \delta\underline{q}(t)$, $0 \le t \le T$ with $\underline{q}_*(0) = a$, $\underline{q}_*(T) = b$, $\delta\underline{q}(0) = 0$, $\delta\underline{q}(T) = 0$, and $\{\delta\underline{q}(t) \mid 0 < t < T\}$ arbitrary, we must have

$$\int_0^T L\left(\underline{q}_*(t) + \delta\underline{q}(t), \underline{q}_*'(t) + \delta\underline{q}'(t), t\right)dt - \int_0^T L\left(\underline{q}_*(t) + \delta\underline{q}(t), \underline{q}_*'(t) + \delta\underline{q}'(t), t\right)dt \ge 0$$

or $$\int_0^T\left[\frac{\partial L}{\partial q_\alpha}(\underline{q}_*(t), \underline{q}_*'(t), t)\, \delta q_\alpha(t) + \frac{\partial L}{\partial q_\alpha'}(\underline{q}_*(t), \underline{q}_*'(t), t)\, \delta q_\alpha'(t)\right]dt$$

$$+\frac{1}{2}\int_0^T\left\{\frac{\partial^2 L}{\partial q_\alpha \partial q_\beta}(\underline{q}_*(t), \underline{q}_*'(t),t)\, \delta q_\alpha(t)\, \delta q_\beta(t)\right.$$

$$+\frac{\partial^2 L}{\partial q_\alpha' \partial q_\beta'}(\underline{q}_*(t), \underline{q}_*'(t), t)\, \delta q_\alpha'(t)\, \delta q_\beta'(t)$$

$$\left.+\frac{2\partial^2 L}{\partial q_\alpha \partial q_\beta'}(\underline{q}_*(t), \underline{q}_*'(t), t)\, \delta q_\alpha(t)\, \delta q_\beta'(t)\right\}dt + O\left(\|\delta q\|^3\right) \ge 0.$$

if $\underline{q}_*(.)$ is a local minimum.

These equation simply on integration by parts and using $\delta\underline{q}'(0) = \delta\underline{q}'(T) = 0$, that

$$\frac{\partial L}{\partial q_\alpha}\left(\underline{q}_*(t), \underline{q}_*'(t), t) - \frac{d}{dt}\frac{\partial L}{\partial q_\alpha'}(\underline{q}_*(t), \underline{q}_*'(t), t\right) = 0$$

and $$\begin{bmatrix} \frac{\partial^2 L}{\partial \underline{q}\partial \underline{q}^T}(\underline{q}_*(t), \underline{q}_*'(t), t) & \frac{\partial^2 L}{\partial \underline{q}\partial \underline{q}'^T}(\underline{q}_*(t), \underline{q}_*'(t), t) \\ \frac{\partial^2 L}{\partial \underline{q}'\partial \underline{q}^T}(\underline{q}_*(t), \underline{q}_*'(t), t) & \frac{\partial^2 L}{\partial \underline{q}'\partial \underline{q}'^T}(\underline{q}_*(t), \underline{q}_*'(t), t) \end{bmatrix}^{\forall\alpha} \ge 0$$

$\forall t$

Note: If F: $\mathbb{R}^N \to \mathbb{R}$ is twice differently, then F has a local minimum at $\underline{X}_* \in \mathbb{R}^N$ iff $F(\underline{X}_* + \delta\underline{X}) - F(\underline{X}_*) \ge 0$ upto $O(\|\delta\underline{X}\|^2)$ $\forall \delta\underline{X}$ small, *i.e.*

iff

$$\underline{\nabla}F(\underline{X}_*)^T\,\delta\underline{X} + \frac{1}{2}\delta\underline{X}^T(\underline{\nabla}\underline{\nabla}^T F(\underline{X}_*))\delta\underline{X} + O(\|\delta\underline{X}\|^3) \ge 0\ \forall \delta\underline{X} \in \mathbb{R}^N$$

Putting $\delta\underline{X} = \varepsilon\xi$, dividing by ε and letting $\varepsilon \to 0$

gives $\quad \underline{\nabla} F(\underline{X}_*)^T \underline{\xi} \;\geq\; 0 \forall \underline{\xi} \in \mathbb{R}^N$

which is possible iff $\underline{\nabla} F(\underline{X}_*) = \underline{0}$.

Hence $\delta\underline{X}^T \underline{\nabla}\underline{\nabla}^T F(\underline{X}_*)\, \delta\underline{X} + O(\|\delta\underline{X}\|^3) \geq 0 \;\forall \delta\underline{X} \in \mathbb{R}^N$.

Putting $\delta\underline{X} = \sqrt{\varepsilon}\underline{\xi}$, dividing by $\in$ and letting $\varepsilon \to 0$

gives $\underline{\xi}^T(\underline{\nabla}\underline{\nabla}^T F(\underline{X}_*))\,\underline{\xi} \;\geq\; 0 \forall \underline{\xi} \in \mathbb{R}^N$,

i.e. $\quad \underline{\nabla}\underline{\nabla}^T F(\underline{X}_*) \;\geq 0$ (positive semidefinite matrix)

This variational calculus can be extended to fields *i.e.* functions of several variables $\varphi(X_1, ..., X_N)$ instead of function $q(t)$ of one variable t: Let
$\varphi_k: \mathbb{R}^N \to \mathbb{R}$, $k = 1, 2, ..., p$ be smooth functions.
We wish to extremize

$$S[\underline{\varphi}] = \int_B L(\varphi_k^{(X)}, \varphi_{k,r}^{(X)} : 1 \leq k \leq p, 1 \leq r \leq N, \underline{X})\, d^N X$$

where $BC\mathbb{R}^N$ is a connected Bond set with non-zero Lebesgue measure and boundary ∂B.

The extremization is subject to the boundary condition $\varphi_k|_{\partial B} = 0 \;\forall k$.

Then

$$\begin{aligned}\delta S[\varphi] = \int_B \Bigg\{ & \frac{\partial L}{\partial \varphi_k}\delta\varphi_k + \frac{\partial L}{\partial \varphi_{k,r}}\delta\varphi_{k,r} \\ & + \frac{1}{2}\Bigg(\frac{\partial^2 L}{\partial\varphi_k \partial\varphi_m}\delta\varphi_n\delta\varphi_m + \frac{\partial^2 L}{\partial\varphi_{k,r}\partial\varphi_{m,s}}\delta\varphi_{k,r}\delta\varphi_{m,s} \\ & + 2\frac{\partial^2 L}{\partial\varphi_k \partial\varphi_{m,s}}\delta\varphi_k\delta\varphi_{m,s}\Bigg)\Bigg\} d^N X \geq 0\end{aligned}$$

for a local minimum at $\{\varphi_k\}$.

Integration by parts gives in the usual way

$$\frac{\partial L}{\partial \varphi_k} - \partial r \frac{\partial L}{\partial \varphi_{k,r}} = 0 \quad 1 \leq k \leq p, \; (\alpha)$$

$$\begin{bmatrix} \left(\left(\dfrac{\partial^2 L}{\partial\varphi_k\partial\varphi_m}\right)\right) & \left(\left(\dfrac{\partial^2 L}{\partial\varphi_k\partial\varphi_{m,s}}\right)\right)_{(k,(m,s))} \\ \left(\left(\dfrac{\partial^2 L}{\partial\varphi_{m,s}\partial\varphi_k}\right)\right)_{((m,s),k)} & \left(\left(\dfrac{\partial^2 L}{\partial\varphi_{k,r}\partial\varphi_{m,s}}\right)\right)_{((k,r)(m,s))} \end{bmatrix} \geq 0$$

at the solution to (α).

Summation over the repeated indices is assumed.

[4] Perturbation theory for algebriac equations.

Let f_0: $\mathbb{R}^N \to \mathbb{R}^N$, g_0: $\mathbb{R}^N \to \mathbb{R}^N, \tau > 0$

a small parameter. Assume that we've solved

$$f_0(\underline{X}) = 0$$

getting the solution $\underline{X} = \underline{X}_0$. We wish then to approximately solve

$$f_0(\underline{X}) + \varepsilon g_0(\underline{X}) = 0 \qquad (\alpha)$$

Let $$\underline{X} = \underline{X}_0 + \varepsilon\underline{X}_1 + \varepsilon^2\underline{X}_2 + \ldots = \underline{X}_0 + \sum_{m=1}^{\infty} \varepsilon^m \underline{X}_m \qquad (\beta)$$

Equating substitutively (β) into (α) and equality coeffs of ε^m, $m = 0, 1, 2, \ldots$ gives successively.

$$f_0(X_0) = 0, O(\varepsilon^0)$$

$$\sum_{j=1}^{N} \underline{f}_0, j(\underline{X}_0)\underline{X}_{1j} + \underline{g}_0(\underline{X}_0) = 0, O(\in^0),$$

$$\frac{1}{2}\sum_{jk} \underline{f}_0, jk(\underline{X}_0)X_{1j}\, X_{1k}$$

$$+\sum_{j} f_{0,j}(\underline{X}_0)\underline{X}_{2j} + \sum_{j} \underline{g}_{0,j}(\underline{X}_0)X_{1j} = 0, O(\varepsilon^2)$$

etc. Writing $\underline{\underline{D}}f_0(\underline{X}_0) = \left[\underline{f}_{0,1}(\underline{X}_0), \underline{f}_{0,N}(\underline{X}_0)\right]$

we get on solving the above,

$$\underline{X}_1 = -\left[\underline{\underline{D}}f_0(X_0)\right]^{-1} \underline{g}_0(\underline{X}_0),$$

$$\underline{X}_2 = -\left[\underline{\underline{D}}f_0(\underline{X}_0)\right]^{-1} \left\{\frac{1}{2}\sum_{j,k} \underline{f}_{0,\,jk}(\underline{X}_0)X_{1j}\, X_{1k} + \sum_{j} \underline{g}_{0,\,j}(\underline{X}_0)X_{1j}\right\}$$

etc.

Thus in general, $\underline{X}_m$ can be solved in terms of $\underline{X}_{m-1}, \ldots, \underline{X}_0$ by inverting the matrix $\underline{\underline{D}}f_0(\underline{X}_0)$.

This complete the problem of algebraic perturbation theory:

For differential equations:

$$\frac{d\underline{X}(t)}{dt} = \underline{f}_0\left(t, \underline{X}(t)\right) + \varepsilon f_1\left(t, \underline{X}(t)\right) \qquad (\beta)$$

Let $\underline{X}_0(t)$ be a solution to the unperturbed problem:

$$\frac{d\underline{X}_0(t)}{dt} = \underline{f}_0(t, \underline{X}_0(\mathrm{t}))$$

Let $$\underline{X}(t) = \underline{X}_0(t) + \sum_{m=1}^{\infty} \varepsilon^m \underline{X}_m(t)$$

Equating coeffs. of ε, $\underline{\varepsilon}^2$, ... in (β) given

O (ε): $$\frac{d\underline{X}_1}{dt} = \sum_k \underline{f}_{0,k}(t, \underline{X}_0(t))\, X_{1k}(t) + \underline{f}_1(t, \underline{X}_0(t))$$

O (ε²): $$\frac{d\underline{X}_2}{dt} = \sum_k \underline{f}_{0,k}(t, \underline{X}_0(t)) X_{2k}(t) + \frac{1}{2}\sum_{k,m} \underline{f}_{0,\,km}(t, \underline{X}_0(t)) X_{1k}(t) X_{1m}(t) + \sum_k \underline{f}_{1,\,k}(t, \underline{X}_0(t)) X_{1k}(t)$$

Let $$\underline{\underline{D}}f(t, \underline{X}_0(t)) = \left[\underline{f}_{0,1}(t, \underline{X}_0(t)), .., \underline{f}_{0,N}(t, \underline{X}_0(t))\right] = \underline{\underline{A}}(t).$$

Then if $$\frac{\partial \Phi(t,\tau)}{\partial t} = A(t)\Phi(t,\tau),\ t \ge \tau,$$
$$\Phi(\tau, \tau) = I$$

i.e. $$\Phi(t, \tau) = I + \sum_{n=1}^{\infty} \int_{\tau < t_n < \ldots < t_1 < t} A(t_1) \ldots A(t_n) dt_1, \ldots dt_n$$

is the state transition matrix corresponding to the "force" $\underline{\underline{A}}(t)$.

Then,

$$\underline{X}_1(t) = \int_0^t \Phi(t,\tau) \underline{f}_1(\tau, \underline{X}_0(\tau)) d\tau,$$

$$\underline{X}_2(t) = \int_0^t \Phi(t,\tau) \left\{ \sum_{k,m} \underline{f}_{0,km}(\tau, \underline{X}_0(\tau)) X_{1k}(\tau) X_{1m}(\tau) + \sum_k \underline{f}_{1,k}(\tau, \underline{X}_0(\tau)) X_{1k}(\tau) \right\} d\tau$$

[3] Scattering amplitude computation using FPI (Feynman Path Integral)

$$S[\varphi] = \int \left(\frac{1}{2} \left(\partial_\mu \varphi \partial^\mu \varphi \right) - \frac{1}{2} m^2 \varphi^2 \right) d^4 X$$
$$= -\int \frac{1}{2} (p^2 + m^2) \left| \hat{\varphi}(p) \right|^2 \frac{d^4 p}{(2\pi)^4}$$

where $$\hat{\varphi}(p) = \int \varphi(X) \exp(ip.X) d^4 X \quad p : X = p_\mu X^\mu$$

$$p^2 = p_\mu p^\mu = p^{02} - \sum_{r=1}^{3} p^{r2} = E^2 - P^2;$$

$$P = (p_r)_{r=1}^{3} = \text{3-momentum.}$$

Propagator

$$\Delta(X_1, X_2) = \left\langle Vac \middle| T\{\phi(t_1, \underline{r}_1)\phi(t_2, \underline{r}_2)\} \middle| Vac \right\rangle$$

$$= \int \phi(t_1, \underline{r}_1)\phi(t_2, \underline{r}_2)\exp(iS[\varphi]) \prod_{\substack{\underline{r}\in\mathbb{R}_3 \\ t\in\mathbb{R}}} d\varphi(t, \underline{r})$$

$$(t, \underline{r}) = X, (t_1, \underline{r}_1) = X_1, (t_2, \underline{r}_2) = X_2.$$

After normalization,

$$\Delta(X, Y) = \frac{\displaystyle\int \exp\left(\frac{i}{\hbar}\int \varphi(X_1)\varphi(X_2)K(X_1, X_2)d^4X_1d^4X_2\right) \varphi(X)\varphi(Y) \prod_{\xi\in\mathbb{R}^4} d\varphi(\xi)}{\displaystyle\int \exp\left(\frac{i}{\hbar}\int \varphi(X_1)\varphi(X_2)K(X_1, X_2)d^4X_1d^4X_2\right) \prod_{\xi\in\mathbb{R}^4} d\varphi(\xi)}.$$

Now,

$$K(X_1, X_2) = -\frac{1}{2}(\partial_\mu\partial^\mu\delta)(X_1 - X_2) - \frac{1}{2}m^2\delta(X_1 - X_2)$$

$$= -\frac{1}{2}\Box\delta(X_1 - X_2) - \frac{1}{2}m^2\delta(X_1 - X_2)$$

where $\Box = \partial_\mu\partial^\mu.$

so from standard multivariate Gaussian integral theory,

$$\Delta(X, Y) = \frac{\hbar}{i}K^{-1}(X, Y) \equiv \frac{\hbar}{i}K^{-1}(X, Y)$$

$$= \Delta(X - Y).$$

Now,

$$\widehat{K}(p) = -\frac{1}{2}(p^2 + m^2)$$

$$\widehat{K}^{-1}(p) = (K(p))^{-1} = \frac{-2}{p^2 + m^2}$$

So $\Delta(X, Y) \propto \int \frac{e^{-ip(X-Y)}}{p^2 + m^2} d^4p$

Propagator for the KG field.

QFT (Quantum field theory) *via* creation and annihilation operators and analogy with path integrals.

Current density in the rigid body when its angular configuration is $\underline{\theta}$: $J(t, \underline{\xi}|\, \underline{\theta})$.

KG field $\varphi(t, \underline{\xi})$.

Action for the KG field: $\frac{1}{2}\int\left(\partial_\mu\varphi\partial^\mu\varphi - m^2\frac{\varphi^2}{2}\right)d^3\xi dt$.

Action for rigid body is $\int\frac{1}{2}\underline{\theta}'(t)\underline{\underline{J}}(\theta(t))\underline{\theta}'(t)dt$

Action for interaction between KG field $\varphi(t, \xi)$ and rigid body:

$\int J\left(t, \underline{\xi}|\underline{\theta}(t)\right)\varphi(t, \underline{\xi})d^3\xi\, dt$.

Total action for KG field interacting with the rigid body is

$$
\begin{aligned}
S[\varphi, \underline{\theta}] &= \frac{1}{2}\int\left(\left(\partial_t\varphi(t,\underline{\xi})\right)^2 - \left|\underline{\nabla}\varphi(t,\underline{\xi})\right|^2 - m^2\varphi^2(t,\underline{\xi})\right)d^3\xi dt \\
&\quad + \frac{1}{2}\int\underline{\theta}'(t)\,\underline{\underline{J}}\left(\theta(t)\right)\underline{\theta}'(t)dt \\
&\quad + \int J\left(t,\underline{\xi}|\underline{\theta}(t)\right)\varphi(t,\underline{\xi})d^3\xi dt
\end{aligned}
$$

and this has to be quantized using the P.I. approach.

Exercise: Write down the classical equation for $\underline{\theta}(t)$, $\varphi(\underline{X})$:

[16] Optimal control applied to master-slave Robot tracking:

Slave system is described by the differential eqns:

$$
\underline{\theta}_s''(t) = F_1\left(\underline{\theta}_s(t), \underline{\theta}_s'(t)\right) + F_2\left(\underline{\theta}_s(t), \underline{\theta}_s'(t)\right)\underline{\tau}_s(t)
$$

where the slave torque is generated by passing the position error between master and slave tips through a matrix PID system:

$$
\begin{aligned}
\underline{\tau}_s(t) &= \underline{h}_s'\left(\underline{\theta}_s(t)\right)\underline{F}_s(t) \\
\underline{X}_s(t) &= \underline{h}_s\left(\underline{\theta}_s(t)\right) \\
&= \left(\sum_{k=1}^{d} L_{sk}\cos\theta_{sk}, \sum_{k=1}^{d} L_{sk}\sin\theta_{sk}\right)
\end{aligned}
$$

is the position of the slave tip in the special case of planar motion and if circular rotation around the Z-axis is taken into account, then

$$
\begin{aligned}
\underline{X}_s(t) &= \underline{h}_s\left(\underline{\theta}_s(t), \phi_s(t)\right) \\
&= \left(\sum_k L_{sk}\cos\theta_{sk}\cos\phi_s, \sum_k L_{sk}\cos\theta_{sk}\sin\phi_s, \sum_k L_{sk}\sin\theta_{sk}\right).
\end{aligned}
$$

We shall assume planer motion.

$F_s(t)$ = Slave force (at its tip) = o/p of pid controller with input as $\underline{X}_m(t) - \underline{X}_s(t)$

$\underline{X}_m(t) = \underline{h}_m(\underline{\theta}_m(t))$, $\underline{X}_s(t) = \underline{h}_s(\underline{\theta}_s(t))$.

$$F_s(t) = \underline{\underline{K}}_0 \left(\underline{X}_m(t) - \underline{X}_s(t) \right) + \underline{\underline{K}}_1 \left(\underline{X}'_m(t) - \underline{X}'_s(t) \right)$$

$$K_2 \int_0^t (\underline{X}_m(\tau) - \underline{X}_s(\tau)) d\tau$$

$$\equiv \left(K_0 + K_1 S + \frac{K_2}{S} \right) \left(\underline{h}_m(\underline{\theta}_m(t)) - \underline{h}_s(\theta_s(t)) \right)$$

$S = \frac{d}{dt}$, $\frac{1}{S} = \int_0^t (.) dt$. So after absorbing $\underline{h}'_s(\underline{\theta}_s)$ into F_2, *i.e.* denoting F_2 $(\underline{\theta}_s, \underline{\theta}'_s)$ $\underline{h}'_s(\theta_s)$ by $+\ F_2\ (\underline{\theta}_s, \underline{\theta}'_s)\ (K_0 + K_1 S + K_2 S^{-1})\ (\underline{h}_m(\underline{\theta}_m) - \underline{h}_s(\underline{\theta}_s))$

on operating on both sides by S,

$$\underline{\theta}'''_s = \psi_1(\underline{\theta}_s, \underline{\theta}'_s, \underline{\theta}''_s) + \frac{d}{dt} F_2(\underline{\theta}_s, \underline{\theta}'_s)$$

$$\left(K_0 \frac{d}{dt} + K_1 \frac{d^2}{dt^2} + K_2 \right) \left(\underline{h}_m(\underline{\theta}_m) - \underline{h}_s(\underline{\theta}_s) \right)$$

By constructing a parameter vector

$$\underset{\sim}{\xi} = \begin{bmatrix} Vec(K_0) \\ Vec(K_1) \\ Vec(K_2) \end{bmatrix}$$

These differential equations can be expressed as

$$y(t) = \underline{\underline{F}}(t) \underline{\xi}$$

where $\underline{y}(t)$ is a function of $\underline{\theta}'''_s(t)$, $\underline{\theta}''_s(t)$, $\underline{\theta}'_s(t)$, $\underline{\theta}_s(t)$, $\underline{\theta}''_m(t)$, $\underline{\theta}'_m(t)$, $\underline{\theta}_m(t)$ and so also is $\underline{\underline{F}}(t)$.

From master-slave angular position measurements at discrete time points $n\Delta$, n = 0, 1, 2, ... and replacing $\underline{\theta}_s(t)$, $\underline{\theta}_m(t)$ by $\underline{\theta}_s(n\Delta)$, $\underline{\theta}_m(m\Delta)$, $\underline{\theta}'_s(t)$, $\underline{\theta}'_m(t)$ by

$$\left(\frac{1 - Z^{-1}}{\Delta} \right) \underline{\theta}_s(n\Delta) = \frac{\left(\underline{\theta}_s(n\Delta) - \underline{\theta}_s((n-1)\Delta) \right)}{\Delta},$$

$$\left(\frac{1 - Z^{-1}}{\Delta} \right) \underline{\theta}_m(n\Delta) = \frac{\left(\underline{\theta}_m(n\Delta) - \underline{\theta}_m((n-1)\Delta) \right)}{\Delta}$$

$\underline{\theta}''_s(t)$, $\underline{\theta}''_m(t)$ likewise by

$$\left(\frac{1 - Z^{-1}}{\Delta} \right)^2 \underline{\theta}_s(n\Delta) = \frac{\left(1 + Z^{-2} - 2z^{-1} \right)}{\Delta} \underline{\theta}_s(n\Delta)$$

$$= \frac{\left(\underline{\theta}_s(n\Delta) + \underline{\theta}_s((n-2)\Delta) - 2\underline{\theta}_s((n-1)\Delta)\right)}{\Delta^2}$$

$$\left(\frac{1-Z^{-1}}{\Delta}\right)^3 \underline{\theta}_s(n\Delta) = \frac{\left(1 - Z^{-3} - 3z^{-1} + 3Z^{-2}\right)}{\Delta^3} \underline{\theta}_s(n\Delta)$$

$$= \Delta^{-1}[\underline{\theta}_s(n\Delta) - \underline{\theta}_s((n-3)\Delta) - 3\underline{\theta}_s((n-1)\Delta) + 3\underline{\theta}_s((n-1)\Delta)]$$

$$\frac{\left(1-Z^{-1}\right)^3}{\Delta^3} \underline{\theta}m(n\Delta) = \frac{\left(1 - Z^{-3} - 3Z^{-1} + 3Z^{-2}\right)}{\Delta^3} \underline{\theta}_m(n\Delta)$$

$$= \frac{\left[\underline{\theta}_m(n\Delta) - \underline{\theta}_m((n-3)\Delta) - 3\underline{\theta}_m((n-1)\Delta) + 3\underline{\theta}_m((n-2)\Delta)\right]}{\Delta^3}$$

We can write the dynamical system model as

$$y(t) = \underline{\underline{F}}(t)\underline{\xi}$$

or in the discrete domain as

$$\underline{y}[n] = \underline{\underline{F}}[n]\underline{\xi}, \; n = 0, 1, 2, \ldots$$

Suppose coloured Gaussain noise $\underline{W}[n]$ of mean zero corrupts this measurement model:

$$\underline{y}[n] = \underline{\underline{F}}[n]\underline{\xi} + \underline{W}[n]$$

(If the noise $\underline{W}[n]$ is non-Gaussian but with independent samples having pdf $p_W(\underline{w})$, then the joint pdf f_N of the measured angular data given the unknown parameter vector $\underline{\xi}$ is

$$\log f_N^{(\xi)} = \sum_{n=0}^{N} \log p_W(\underline{y}[n] - \underline{\underline{F}}[n]\underline{\xi}) \equiv L_N(\xi)$$

ξ can be estimated by maximizing this.

Approximate iteration: Let $\hat{\underline{\xi}}[N]$ maximize

$L_N(\xi)$. Let $\quad \hat{\underline{\xi}}[n+1] = \hat{\underline{\xi}}[N] + \Delta\hat{\underline{\xi}}[N+1]$

Then since $\quad L(\xi) = L_N(\xi) + l_{N+1}(\xi)$

where $\quad l_n(\xi) = \log P_W\left(\underline{y}[n] - \underline{\underline{F}}[n]\underline{\xi}\right)$

We get assuming $l_{N+1}(\xi)$ of the same order of smallness as $\Delta\hat{\xi}(N+1)$,

$$L'_N\left(\hat{\underline{\xi}}[N] + \Delta\hat{\underline{\xi}}(N+1)\right)$$

$$+ l'_{N+1}\left(\hat{\underline{\xi}}[N] + \Delta\hat{\underline{\xi}}[N+1]\right) = 0$$

$$\Rightarrow \qquad L''_N\left(\underline{\hat{\xi}}[N]\right)\Delta\underline{\hat{\xi}}[N+1] + l'_{N+1}(\underline{\hat{\xi}}[N]) = 0$$

$$\Rightarrow \qquad \Delta\underline{\hat{\xi}}[N+1] = -L''_N\left(\underline{\hat{\xi}}[N]\right)^{-1} \underline{l}'_{N+1}\left(\underline{\hat{\xi}}[N]\right).$$

Note: If f: $\mathbb{R}^p \to \mathbb{R}$, f' denotes the gradient ($\mathbb{R}^p$-column vector) of f and f'' the Herrian ($\mathbb{R}^{p\times p}$ matrix) of f:

$$f'(\xi) = \left(\left(\frac{\partial f(\underline{\xi})}{\partial \xi_\alpha}\right)\right)_{\alpha=1}^{p}, f''(\xi) = \left(\left(\frac{\partial^2 f(\underline{\xi})}{\partial \xi_\alpha \partial \xi_\beta}\right)\right)_{1\le\alpha,\,\beta\le p},$$

RLS: If $\underline{W}[n]$ is iid $N(0, \underline{\underline{R}}_W)$ then the maximum likelihood estimate of $\underline{\xi}$ based on $\{\underline{y}[n]\}_{n=0}^{N}$ and $\{F[n]\}_0^N$ is given by

$$\begin{aligned}\underline{\hat{\xi}}[N] &= \arg\min_{\underline{\xi}} \exp\left(-\frac{1}{2}\sum_{n=0}^{N}\left(\underline{y}[n]-\underline{\underline{F}}[n]\underline{\xi}\right)^T \underline{\underline{R}}_W^{-1}\left(\underline{y}[n]-\underline{\underline{F}}[n]\underline{\xi}\right)\right) \\ &= \arg\min_{\underline{\xi}} \sum_{n=0}^{N}\left(\underline{y}[n]-\underline{\underline{F}}[n]\underline{\xi}\right)^T \cdot \underline{\underline{R}}_W^{-1}\left(\underline{y}[n]-\underline{\underline{F}}[n]\underline{\xi}\right) \\ &= \left(\sum_{n=0}^{N}\underline{\underline{F}}[n]^T \underline{\underline{R}}_W^{-1}\underline{\underline{F}}[n]\right)^{-1}\left(\sum_{n=0}^{N}\underline{\underline{F}}[n]^T \underline{\underline{R}}_W^{-1}\underline{y}[n]\right)\end{aligned}$$

This can be implemented recursively:

Let

$$\underline{\underline{P}}[N] = \left(\sum_{n=0}^{N}\underline{\underline{F}}[n]^T \underline{\underline{R}}_W^{-1}\underline{\underline{F}}[n]\right)^{-1}$$

$$\underline{q}[N] = \sum_{n=0}^{N}\underline{\underline{F}}[n]^T \underline{\underline{R}}_W^{-1}\underline{y}[n]$$

Then

$$\begin{aligned}\underline{q}[N+1] &= \underline{q}[N] + \underline{\underline{F}}[N+1]^T \underline{\underline{R}}_W^{-1}\underline{y}[N+1], \\ \underline{\underline{P}}[N+1] &= \left(\underline{\underline{P}}[N]^{-1} + \underline{\underline{F}}[N+1]^T \underline{\underline{R}}_W^{-1}\underline{\underline{F}}[N+1]\right)^{-1} \\ &= \underline{\underline{P}}[N] - \underline{\underline{P}}[N]\underline{\underline{F}}[N+1]^T \\ &\quad \left(\underline{\underline{R}}_W + \underline{\underline{F}}[N+1]\underline{\underline{P}}[N]\underline{\underline{F}}[N+1]^T\right)^{-1}\underline{\underline{F}}[N+1]\underline{\underline{P}}[N]\end{aligned}$$

Thus,

$$\begin{aligned}\underline{\hat{\xi}}[N+1] &= \underline{\underline{P}}[N+1]\underline{q}[N+1] \\ &= \left(\underline{\underline{P}}[N] - \underline{\underline{P}}[N]\underline{\underline{F}}[N+1]^T\right. \\ &\quad \left.\left(R_W + F[N+1]P[N]F[N+1]^T\right)^{-1} F[N+1]P[N]\right) \\ &\quad \left(\underline{q}[N] + \underline{\underline{F}}[N+1]^T \underline{\underline{R}}_W^{-1}\underline{y}[N+1]\right)\end{aligned}$$

$$= \underline{\underline{P}}[N]\underline{q}[N] + P[N]F[N+1]^T R_W^{-1} y[N+1]$$
$$-P[N]F[N+1]^T \left(R_W + F(N+1)P[N]F(N+1)^T\right)^{-1}$$
$$F[N+1]P[N]\underline{q}[N]$$
$$-P[N]F[N+1]^T \left(R_W + F(N+1)P[N]F(N+1)^T\right)^{-1}$$
$$F[N+1]P[N]F[N+1]^T R_W^{-1} y[N+1]$$
$$= \underline{\hat{\xi}}(N) + P[N]F[N+1]^T \left(R_W + F[N+1]P[N]F[N+1]^T\right)^{-1}$$
$$\left(\underline{y}[N+1] - F[N+1]\underline{\hat{\xi}}(N)\right)$$

This is the recursive least squares algorithm.

* Continuation of [23] Interaction between KG field and Robot and its quantization:

$$S[\varphi, \underline{\theta}] = \int \left\{ \frac{1}{2}\partial_\mu \varphi(X)\partial^\mu \varphi(X) - \frac{1}{2}m^2\varphi^2(X) \right.$$
$$+\frac{1}{2}\underline{\theta}'(t)\underline{\underline{J}}_0(\underline{\theta}(t))\underline{\theta}'(t)\delta^3(\underline{r}) - V(\underline{\theta}(t))\delta^3(\underline{r})$$
$$\left. +J(X \mid \theta(t))\varphi(X) \right\} d^4X$$

Classical equation of motion of the field and Robot:

$$\delta_\varphi S[\varphi, \underline{\theta}] = 0 \Rightarrow$$
$$\partial_\mu \partial^\mu \varphi(X) + m^2\varphi(X) + J(X|\underline{\theta}(t)) = 0,$$
$$\delta_\theta S[\varphi, \underline{\theta}] = 0 \Rightarrow$$
$$\left(J_0\left(\underline{\theta}(t)\underline{\theta}'(t)\right)' - \frac{1}{2}\theta'(t)\underline{\underline{J}}_0'\left(\theta(t)\right)\underline{\theta}'(t) \right.$$
$$+V'\left(\underline{\theta}(t)\right) - \int \frac{\partial J}{\partial \underline{\theta}}(X \mid \underline{\theta}(t))\varphi(X)d^3X = 0$$

Stochastic optimal control applied to Master-Slave Robot tracking

Master Robot follows the dynamics

$$\ddot{\underline{q}}_m = \underline{f}_m(\underline{q}_m, \dot{\underline{q}}_m) + \underline{\underline{g}}_m(\underline{q}_m, \dot{\underline{q}}_m)\left(\underline{W}_m(t) + \underline{\tau}_m(t)\right)$$
$$+ \underline{\underline{h}}_m(\underline{q}_m, \dot{\underline{q}}_m, \underline{q}_s, \dot{\underline{q}}_s)\underline{f}_m(t)$$

and slave Robot follows the dynamics

$$\ddot{\underline{q}}_s = \underline{f}_s(\underline{q}_s, \dot{\underline{q}}_s) + \underline{\underline{g}}_s(\underline{q}_s, \dot{\underline{q}}_s)\left(\underline{W}_s(t) + \underline{\tau}_s(t)\right)$$
$$+ \underline{\underline{h}}_s(\underline{q}_s, \dot{\underline{q}}_s, \underline{q}_m, \dot{\underline{q}}_m)\underline{f}_s(t)$$

where $\underline{f}_m(t), \underline{f}_s(t)$ are control inputs like pid coefficients which vary with time, $\underline{\tau}_m(t)$ is the master torque, $\underline{\tau}_s(t)$ is the slave torque. The function $\underline{\underline{h}}_m(\underline{q}_m, \dot{\underline{q}}_m, \underline{q}_s, \dot{\underline{q}}_s)$ and $\underline{\underline{h}}_s(\underline{q}_s, \dot{\underline{q}}_s, \underline{q}_m, \dot{\underline{q}}_m)$ measure the error between master and slave dynamics (position and angular velocity). The processes $\underline{W}_m(.)$ and $\underline{W}_m(.)$ are independent white noise processes:

$$\underline{W}_m(t) = d\underline{B}_m(t)/dt,$$
$$\underline{W}_s(t) = d\underline{B}_s(t)/dt$$

where $\underline{B}_m(.)$ and $\underline{B}_s(.)$ are independent standard vector valued Brownian motion processes. This dynamics can be expressed in stochastic differential form as

$$d\begin{pmatrix} \underline{q}_m \\ \underline{q}_s \\ \underline{W}_m \\ \underline{W}_s \end{pmatrix} = \begin{bmatrix} \underline{W}_m \\ \underline{W}_s \\ -\underline{f}_m(\underline{q}_m, \dot{\underline{q}}_m) \\ -\underline{\underline{g}}_m(\underline{q}_m, \dot{\underline{q}}_m)\underline{\tau}_m(t) \\ -\underline{f}_s(\underline{q}_s, \dot{\underline{q}}_s) \\ -\underline{\underline{g}}_s(\underline{q}_s, \dot{\underline{q}}_s)\underline{\tau}_s(t) \end{bmatrix} dt$$

$$+\begin{bmatrix} \underline{\underline{h}}_m(\underline{q}_m, \dot{\underline{q}}_m, \underline{q}_s, \dot{\underline{q}}_s) & \underline{\underline{O}} \\ \underline{\underline{O}} & \underline{\underline{h}}_s(\underline{q}_s, \dot{\underline{q}}_s, \underline{q}_m, \dot{\underline{q}}_m) \end{bmatrix} \begin{bmatrix} \underline{f}_m(t) \\ \underline{f}_s(t) \end{bmatrix} dt$$

$$+\begin{bmatrix} \underline{\underline{g}}_m(\underline{q}_m, \dot{\underline{q}}_m) & \underline{\underline{O}} \\ \underline{\underline{O}} & \underline{\underline{g}}_s(\underline{q}_s, \dot{\underline{q}}_s) \end{bmatrix} d\begin{bmatrix} \underline{B}_m(t) \\ \underline{B}_s(t) \end{bmatrix}$$

or equivalently since $\dot{\underline{q}}_m = \underline{W}_m$, $\dot{\underline{q}}_s = \underline{W}_s$,

$$d\begin{pmatrix} \underline{q}_m \\ \underline{q}_s \\ \underline{W}_m \\ \underline{W}_s \end{pmatrix} = \begin{pmatrix} \underline{W}_m \\ \underline{W}_s \\ -\underline{f}_m(\underline{q}_m, \underline{W}_m) \\ -\underline{\underline{g}}_m(\underline{q}_m, \underline{W}_m)\underline{\tau}_m(t) \\ -\underline{f}_s(\underline{q}_s, \underline{W}_s) \\ -\underline{\underline{g}}_s(\underline{q}_s, \underline{W}_s)\underline{\tau}_s(t) \end{pmatrix} dt$$

$$+\begin{bmatrix} \underline{\underline{h}}_m(\underline{q}_m, \underline{W}_m, \underline{q}_s, \underline{W}_s) & \underline{\underline{O}} \\ \underline{\underline{O}} & \underline{\underline{h}}_s(\underline{q}_s, \underline{W}_s, \underline{q}_m, \underline{W}_m) \end{bmatrix} \underline{f}(t)$$

$$+\begin{bmatrix} \underline{\underline{g}}_m(\underline{q}_m, \underline{W}_m) & \underline{\underline{O}} \\ \underline{\underline{O}} & \underline{\underline{g}}_s(\underline{q}_s, \underline{W}_s) \end{bmatrix} d\underline{B}(t)$$

where $\underline{f}(t) = \begin{bmatrix} \underline{f}_m(t) \\ \underline{f}_s(t) \end{bmatrix}, \underline{B}(t) = \begin{bmatrix} B_m(t) \\ B_s(t) \end{bmatrix}$

$\underline{f}(t)$, $0 \le t \le T$ is to be choosen so that $\underline{f}(t)$ is of the form $\underline{\psi}$ $(t, \underline{q}_m(t), \underline{q}_s(t), \underline{W}_m(t), \underline{W}_s(t))$ (state feedback controller) and such that

$$\mathbb{E}\left\{\int_0^T \mathcal{L}\left(\underline{q}_m(t), \underline{W}_m(t), \underline{q}_s(t), \underline{W}_s(t)\right) dt\right\}$$

is a minimum.

The optimal equations are easily formulated using the Bellman dynamic programming approach.

[23] Contd. (quantum Robotics based on Schrödenger and Dirac equation).

Dirac's equation for a rigid body.

The single link case: Let $(X_0(t), Y_0(t))$ denote the position of the centre of mass of the rigid rod

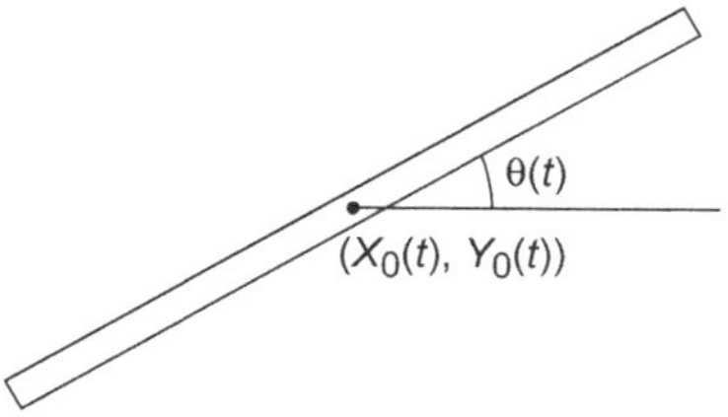

The Dirac Hamiltonian of a single point man m moving freely is

$$H_D = C(\alpha, \underline{p}) + \beta mc^2$$

where

$$\alpha_r = \begin{pmatrix} 0 & \sigma_r \\ \sigma_r & 0 \end{pmatrix}, r = 1, 2, 3,$$

$$\beta = \begin{pmatrix} I & 0 \\ 0 & -I \end{pmatrix}.$$

$$\sigma_1 = \begin{pmatrix} 0 & 1 \\ 1 & 0 \end{pmatrix}, \sigma_2 = \begin{pmatrix} 0 & -i \\ i & 0 \end{pmatrix}, \sigma_3 = \begin{pmatrix} 1 & 0 \\ 0 & -1 \end{pmatrix}$$

We take for the rigid rod

$\underline{p}(\xi)d\xi = \sigma_0 d\xi\ (X'_0 - \xi\theta' \sin\theta, Y'_0 + \xi\theta' \cos\theta)$

σ_0 being the man per unit length of the rod.

$d\xi$ being the length clement, ξ is measured from the C.M. total momentum of the rigid rod is then

$$\underline{P} = \int_{-\frac{L}{2}}^{\frac{L}{2}} \underline{p}(\xi)d\xi = \sigma_0 L(X'_0, Y'_0)$$

i.e. the momentum of the CM. So if we were to blindly integrate the Dirac Hamiltonian density

$$H_{D0} = \int_{-\frac{L}{2}}^{\frac{L}{2}} H_D(\xi)d\xi = \int_{-\frac{L}{2}}^{\frac{L}{2}} \left(c(\underline{\alpha}, p(\xi)) + \beta\sigma_0 c^2\right)d\xi$$

we would get

$$\begin{aligned} H_{D0} &= \sigma_0 Lc(\underline{\alpha}, (X_0', Y_0', 0)) + \beta\sigma_0 Lc^2 \\ &= c(\underline{\alpha}, (P_X, P_Y, 0)) + \beta mc^2 \\ &= c(\alpha_1 P_X + \alpha_2 P_Y) + \beta mc^2 \end{aligned}$$

where $P_X = \sigma L X_0',\ P_Y = \sigma L Y_0'$.

Taking into account the gravitational potential and external torque, the total Dirac Hamiltonian of the rod is

$$\begin{aligned} H_D &= H_{D0} + \mathrm{mgY} - \tau(t)\theta \\ &= c(\alpha_1 P_X + \alpha_2 P_Y) + \beta mc^2 + mgY - \tau(t)\theta \\ &= -i\hbar c\left(\alpha_1 \frac{\partial}{\partial X} + \alpha_2 \frac{\partial}{\partial Y}\right) + \beta mc^2 + mgY - \tau(t)\theta \end{aligned}$$

This Hamiltonian does not account for the rotational kinetic energy.

To take into account the rotational K.E., we first note then the Lagragian of the free rod (relativistic) is

$$\begin{aligned} \mathcal{L} &= -\sigma_0 c^2 \int_{-\frac{L}{2}}^{\frac{L}{2}} \left(1 - \frac{v^2}{c^2}\right)^{\frac{1}{2}} d\xi \\ &= -\sigma_0 c^2 \int_{-\frac{L}{2}}^{\frac{L}{2}} \Big[1 - \frac{1}{c^2}\big(X_0'^2 + Y_0'^2 + \xi^2\theta'^2 \\ &\quad + 2\xi\theta'(Y_0'\cos\theta - \chi_0'\sin\theta)\big)\Big]^{1/2} d\xi \\ &= \mathcal{L}(X_0', Y_0', \theta, \theta'). \end{aligned}$$

We approximate this as

$$\begin{aligned} \mathcal{L} = -\sigma_0 c^2 \int_{|\xi| \le \frac{L}{2}} \Big[1 &- \frac{1}{2c^2} X_0'^2 + Y_0'^2 + \xi^2\theta'^2 \\ &+ 2\xi\theta'(Y_0'\cos\theta - X_0'\sin\theta)\big] - \frac{1}{8c^4} X_0'^2 + Y_0'^2 + \xi^2\theta'^2 \\ &+ 2\xi\theta'(Y_0'\cos\theta - X_0'\sin\theta)\big)^2\Big] d\xi \end{aligned}$$

$$= -\sigma_0 c^2 \left\{ L - \frac{L}{2c^2}\left(X_0'^2 + Y_0'^2\right) - \frac{L^3\theta'^2}{24c^2} \right.$$

$$- \frac{\left(X_0'^2 + Y_0'^2\right)^2 L}{8c^4} - \frac{\theta'^4}{8c^4}\frac{2.L^5}{2^5.5}$$

$$- \frac{1}{8c^4}\left(X_0'^2 + Y_0'^2\right)\theta'^2 \frac{4.L^3}{2^3.3}$$

$$\left. - \frac{1}{8c^4} 4\theta'^2 \left(Y_0' \cos\theta - X_0' \sin\theta\right)^2 \frac{2.L^3}{2^3.3} \right\}$$

$$= -\sigma_0 c^2 + \frac{\sigma_0 L}{2}\left(X_0'^2 + Y_0'^2\right) + \frac{\sigma_0 L^3}{24}\theta'^2$$

$$+ \frac{\sigma_0 L (X_0'^2 + Y_0'^2)^2}{8c^2} + \frac{\sigma_0 L^5 \theta'^4}{640c^2} + \frac{\sigma_0 L^3}{48c^2}(X_0'^2 + Y_0'^2)\theta'^2$$

$$+ \frac{\sigma_0 L^3}{24c^2}\theta'^2 (Y_0' \cos\theta - X_0' \sin\theta)^2$$

$$\equiv \frac{1}{2}m(X_0'^2 + Y_0'^2) + \frac{I\theta'^2}{2} + \varepsilon_1 (X_0'^2 + Y_0'^2)^4$$

$$+ \varepsilon_2 \theta'^4 + \varepsilon_3 (X_0'^2 + Y_0'^2)\theta'^2$$

$$+ \varepsilon_4 \theta'^2 (Y_0' \cos\theta - X_0' \sin\theta)^2$$

The canonical moments are

$$p_X = \frac{\partial \mathcal{L}}{\partial X_0'}$$

$$= mX_0' + 4\varepsilon_1 X_0' (X_0'^2 + Y_0'^2)^3 + 2\varepsilon_3 X_0' \theta'^2$$

$$-2\varepsilon_4 \theta'^2 \sin\theta (Y_0' \cos\theta - X_0' \sin\theta)$$

$$\equiv mX_0' + \delta. f_X(X_0', Y_0', \theta', \theta)$$

$$p_Y = \frac{\partial \mathcal{L}}{\partial Y_0'} = mY_0' + \delta. f_Y(X_0', Y_0', \theta', \theta),$$

$$p_\theta = \frac{\partial \mathcal{L}}{\partial \theta^1} = I\theta' + \delta. f_\theta(X_0', Y_0', \theta', \theta)$$

where δ is a small perturbation parameter. The Hamiltonian is

$$H = p_X X_0' + p_Y Y_0' + p_\theta \theta' - \mathcal{L}$$

we write

$$\mathcal{L} = \frac{1}{2}m(X_0'^2 + Y_0'^2) + \frac{I\theta'^2}{2} + \delta. \psi(X_0', Y_0', \theta', \theta)$$

where $\delta \,.\, \psi$ consists of the 4 ε-term. Thus,

$$f_X = \frac{\partial\psi}{\partial X_0'}, f_Y = \frac{\partial\psi}{\partial Y_0'}, p_\theta = \frac{\partial \mathcal{L}}{\partial\theta'}.$$

Then,

$$H = \frac{1}{2}m(X_0'^2 + Y_0'^2) + \frac{I\theta'^2}{2} + \delta(f_X + f_r + f_\theta - \psi)$$

We can insert the equation for p_X, p_Y, p_θ and write

$$X_0' = \frac{p_X}{m} - \frac{\delta}{m} f_X + O(\delta^2)$$

$$Y_0' = \frac{p_Y}{m} - \frac{\delta}{m} f_Y + O(\delta^2)$$

$$\theta' = \frac{p_\theta}{I} - \frac{\delta}{I} f_\theta + O(\delta^2)$$

where

$$f_X = f_X\left(\frac{p_X}{m}, \frac{p_Y}{m}, \frac{p_\theta}{I}, \theta\right),$$

$$f_Y = f_Y\left(\frac{p_X}{m}, \frac{p_Y}{m}, \frac{p_\theta}{I}, \theta\right),$$

$$f_\theta = f_\theta\left(\frac{p_X}{m}, \frac{p_Y}{m}, \frac{p_\theta}{I}, \theta\right)$$

and than

$$H = \frac{1}{2m}\left(p_X^2 + p_Y^2\right) + \frac{p_\theta^2}{2I} + \delta.\chi(p_X, p_Y, p_\theta, \theta) + O(\delta^2)$$

where $\chi(p_X, p_Y, p_\theta, \theta) = (f_X + f_Y + f_\theta - \psi)\left(\frac{p_X}{m}, \frac{p_Y}{m}, \frac{p_\theta}{I}, \theta\right).$

χ is a 4th degree homogeneous polynomial in (p_X, p_Y, $p\theta$) with coefficients depending on θ. We can thus write

$\chi(p_X, p_Y, p_\theta, \theta) = \chi_1(\theta)\,(p_X^2 + p_Y^2)^2 + \chi_2(\theta)\, p_X p_Y p_\theta^2 + \chi_3(\theta)\, p_X^2 p_\theta^2 + \chi_4(\theta)\, p_Y^2 p_\theta^2 + \chi_5(\theta)\, p_\theta^4$

To pass over to the quantum theory, we make χ Hamilton by replacing it by

$$\chi = \chi_1(\theta)(p_X^2 + p_Y^2) + \frac{p_X p_Y}{2}\left(\chi_2(\theta)P_\theta^2 + p_\theta^2\chi_2(\theta)\right)$$

$$+\frac{p_X^2}{2}\left(\chi_3(\theta)p_\theta^2 + p_\theta^2\chi_3(\theta)\right)$$

$$+\frac{p_Y^2}{2}\left(\chi_4(\theta)p_\theta^2 + p_\theta^2\chi_4(\theta)\right)$$

$$+\frac{1}{2}\left(\chi_5(\theta)p_\theta^4 + p_\theta^4\chi_5(\theta)\right)$$

and use it to calculate the transition probability amplitudes.

We can also use it in scattering theory based on Born's method.

The incident projectile is in the free particle state

$$\left[-\frac{\hbar^2}{2m}\left(\frac{\partial^2}{\partial X^2}+\frac{\partial^2}{\partial Y^2}\right)-\frac{\hbar^2}{2I}\frac{\partial^2}{\partial\theta^2}\right]\psi_2.(X,Y,\theta)$$

$$= E\,\psi_i(X, Y, \theta)$$

This equation is satisfied by

$$\psi_i(X, Y, \theta) = C.\exp(i\,(k_X X + k_Y Y + K_\theta \theta))$$

where $\frac{\hbar^2}{2m}\left(k_X^2 + k_Y^2\right) + \frac{\hbar^2 k_\theta^2}{2I} = E.$

k_θ is quantized by the single-valued condition

$\psi_i(X, Y, \theta + 2\pi) = \psi_i(X, Y, \theta)$. Thus,

$k_\theta = n,\ n \in \mathbb{Z}.$

The final state after Born scattering is

$\Psi_i\,(X, Y, \theta) + \delta\,.\,\psi_s\,(X, Y, \theta) + O(\delta^2)$

where by perturbation theory,

$$\left[\frac{-\hbar^2}{2m}\left(\frac{\partial^2}{\partial X^2}+\frac{\partial^2}{\partial Y^2}\right)-\frac{\hbar^2}{2I}\frac{\partial^2}{\partial\theta^2}\right]\psi_s(X,Y,\theta)+\chi\psi_i(X,Y,\theta)$$

$$= E\,\psi_s(X, Y, \theta)$$

❖❖❖❖❖

[2] Belavkin Filtering Contd.:

$$\begin{aligned} d\,\sigma_t(X) &= \mathbb{E}\left[F^*(t+dt)XF(t+dt)\,|\,\eta^{in}_{t+dt]}\right] \\ &\quad -\mathbb{E}\left[F^*(t)XF(t)\,|\,\eta^{in}_{t]}\right] \\ &= \mathbb{E}\left[F^*(t)XF(t)\,|\,\eta^{in}_{t+dt]}\right]-\mathbb{E}\left[F^*(t)XF(t)\,|\,\eta^{in}_{t]}\right] \\ &\quad +\mathbb{E}\left[d(F^*(t)XF(t)\,|\,\eta^{in}_{t+dt]}\right] \end{aligned}$$

Now $\quad F^*(t)XF(t) \in \sigma\{Y^{in}(S), S \le t, \delta(\hbar)\}$

and since $\quad \eta^{in}_{t+dt]} = \sigma\{\eta^{in}_{t]}, dY^{in}(t)\}$

it follows two in the state if

$$e(u) > |fe(u)\rangle \exp\left(-\|u\|^2\!/_2\right),$$

$$\mathbb{E}\left[F^*(t)XF(t)\,|\,\eta^{in}_{t+dt]}\right] = \mathbb{E}\left[F^*(t)XF(t)\,|\,\eta^{in}_{t]}\right]$$

Thus, $\quad d\sigma_t(X) = \mathbb{E}\left[d\left(F^*(t)XF(t)\right)|\,\eta^{in}_{t+dt]}\right]$

$$\mathbb{E}\left\{F^*(t)\left[\left(P^*(t)X + XP(t)\right)F(t) \mid \eta_{t]}^{in}\right]dY_{in}(t)\right.$$

$$+\mathbb{E}\left[F^*(t)P^*(t)XP(t)F(t)\middle|\eta_{t]}^{in}\right]dZ_{in}(t)$$

$$+\mathbb{E}\left[F^*(t)\left(Q^*(t)X + XQ(t)\right)F(t)\middle|\eta_{t]}^{in}\right]dt$$

$$= \sigma_t(P^*X + XP)dY_{in} + \sigma_t(P^*XP)dZ_{in} + \sigma_t(Q^*X + XQ)dt$$

$$d\left\{\frac{\sigma_t(X)}{\sigma_t(1)}\right\} = \frac{d\sigma_t(X)}{\sigma_t(1)} - \frac{\sigma_t(X)d\sigma_t(1)}{\sigma_t(1)^2} + \frac{\sigma_t(X)}{\sigma_t(1)^3}(d\sigma_t(1))^2 - \frac{d\sigma_t(X)d\sigma_t(1)}{\sigma_t(1)^2}$$

Let $v_t(X) = \dfrac{\sigma_t(X)}{\sigma_t(1)}$. Then we get

$$\begin{aligned} d\, v_t(X) = {} & v_t(P^*X + XP)\, d\, Y_{in} + v_t(P^*XP)\, d\, Z_{in} + v_t(Q^*X + XQ)\, dt \\ & - v_t(X)\,(v_t(P^* + P)\, d\, Z_{in} + v_t(P^*P)\, d\, Z_{in} + v_t(Q^*Q)\, dt \\ & + v_t(X)\, v_t(P^* + P)\, d\, Z_{in} + v_t(P^*X\,P)^2\, d\, V_{in} \\ & + 2\, v_t(P^* + P)\, v_t(P^*P)\, dW_{in}) - v_t(P^*X + XP)\, v_t(P^* + P)\, dZ_{in} \\ & - v_t(P^*X + XP)\, v_t(P^*P) + v_t(P^* + P)\, v_t(P^*XP))\, d\, W_{in} \\ & - v_t(P^*XP)\, d\, V_{in} \end{aligned}$$

where

$$\begin{aligned} (dY_{in})^2 &= dZ_{in} \\ (dY_{in})^3 &= dW_{in} = dY_{in}dZ_{in} \\ (dY_{in})^4 &= dV_{in} = (dZ_{in})^2 \end{aligned}$$

We can express there as

$$\begin{aligned} dv_t(X) = {} & A_t(X)\, dt + B_t(X)\, dY^{in}(t) + C_t(X)\, dZ_{in}(t) \\ & + D_t(X)dW_{in}(t) + E_t(X)\, dV_{in}(t) \end{aligned}$$

where $A_t(X)$, $B_t(X)$, $C_t(X)$, $D_t(X)$ and $E_t(X)$ are linear functions of X with values in the Abelian algebra

$$\eta_{t]}^{in} = \sigma\left\{Y^{in}(s)\, s \le t\right\},$$

❖❖❖❖❖

[3] Dirac Eqn., for Rigid Body Robot (Single Link):

C. M. of the body is $r_0(t) = (x(t), y(t), z(t))$.

Euler angles of rotation about C. M. an $(\varphi(t), \theta(t), \psi(t))$,

T = Kinetic energy of the robot = K. E. of translation of C. M. + K. E. or rotation about C. M. Non-relativistic case:

$$T = \frac{1}{2}m(\dot{x}^2 + \dot{y}^2 + \dot{z}^2) + \int_B \frac{1}{2}\rho\left\|\frac{d}{dt}R(\varphi(t), \theta(t), \psi(t))\xi\right\|^2 d^3\xi$$

$$= \frac{1}{2}m\,|\,\dot{\underline{r}}_0\,|^2 + \frac{1}{2}\rho\int_B \left\| \left(\dot{\varphi}\frac{\partial R}{\partial \varphi} + \dot{\theta}\frac{\partial R}{\partial \theta} + \dot{\psi}\frac{\partial R}{\partial \psi}\xi \right) \right\|^2 d^3\xi$$

$$= \frac{1}{2}m\,|\,\dot{r}_0\,|^2 + \frac{1}{2}T_r\left[\frac{dR}{dt}J\left(\frac{dR}{dt}\right)^T\right]$$

where $$J = \rho\int_B \underline{\xi\xi}^T d^3\xi$$

and $$\frac{dR}{dt} = \varphi\frac{\partial R}{\partial \varphi} + \dot{\theta}\frac{\partial R}{\partial \theta} + \dot{\psi}\frac{\partial R}{\partial \psi}$$

Relativistic Lagrangian with gravity accounted for

$$\mathcal{L} = -\rho_0 c^2 \int_B \left(1 - |\,\dot{r}_0 + \dot{R}(t)\xi\,|^2 / c^2\right)^2 d^3\xi$$

$$= -\rho_0 c^2 \int_B \left[1 - (\dot{r}_{0(0)}^2 + \underline{\xi}^T \dot{R}(t)^T \dot{R}(t)\xi + 2\dot{r}_{0(t)}^T R(t)\underline{\xi})/c^2\right]^{1/2} d^{3\xi}$$

$$\approx -\rho_0 c^2 \int_B \left(1 - \frac{\dot{r}_0^2}{c^2}\right)^{1/2},$$

$$\left[\frac{\left[1 - \frac{1}{2C^2}\left(\xi^T \dot{R}^T \dot{R}\xi + 2\dot{r}_0^T \dot{R}\underline{\xi}\right)\right]}{(1 - \dot{r}_0^2/c^2)} + \frac{3}{8c^4}\frac{(2\dot{r}_0^T \dot{R}\underline{\xi})^2}{(1 - \dot{r}_0^2/c^2)^2}\right] d^3\xi$$

Now $\int_B \underline{\xi} d^3\xi = 0$, by definition of C. M. Hence,

This writing $J = \rho\int_B \underline{\xi\xi}^T d^3\xi$, we get

$$\mathcal{L} \approx -m_0 c^2 \sqrt{1 - \frac{\dot{r}_0^2}{c^2}} + \frac{1}{2}\left(1 - \frac{\dot{r}_0^2}{c^2}\right)^{-1/2} T_r(\dot{R}^T \dot{R}\,J)$$

$$-\frac{3}{2c^2}\left(1 - \frac{\dot{r}^2}{c^2}\right)^{-3/2} \dot{r}_0^2 \dot{R}J\dot{R}^T \dot{r}_0.$$

$$\equiv -m_0 c^2 \left(1 - \frac{\dot{r}_0^2}{c^2}\right)^{1/2} + \varepsilon V_1(\dot{r}_0, \dot{\varphi}, \dot{\theta}, \dot{\psi}, \varphi, \theta, \psi)$$

$$\underline{p}_0 = \frac{\partial \pounds}{\partial \dot{\underline{r}}_0} = \frac{m_0 \dot{\underline{r}}_0}{\sqrt{1 - \dot{r}_0^2/c^2}} + \varepsilon\frac{\partial V_1}{\partial \dot{\underline{r}}_0}$$

$$p_{\varphi} = \frac{\partial \pounds}{\partial \dot{\varphi}} = \varepsilon \frac{\partial V_1}{\partial \dot{\pounds}},$$

$$p_{\theta} = \in \frac{\partial \pounds_1}{\partial \dot{\theta}} = \varepsilon \frac{\partial V_1}{\partial \dot{\theta}},$$

$$p_{\psi} = \frac{\partial \pounds}{\partial \dot{\psi}} = \varepsilon \frac{\partial V_1}{\partial \dot{\psi}}$$

$$H = (\underline{p}_0, \dot{\underline{r}}_0) + p_{\varphi}\dot{\varphi} + p_0\dot{\theta} + p_{\psi}\dot{\psi} - \pounds$$

$$= \frac{m_0 \dot{r}_0^2}{\sqrt{1-\dot{r}_0^2/c^2}} + \varepsilon\left(\dot{r}_0 \frac{\partial V_1}{\partial \dot{r}_0}\right) + \varepsilon\left(\dot{\varphi}\frac{\partial V_1}{\partial \dot{\varphi}} + \dot{\theta}\frac{\partial V_1}{\partial \dot{\theta}} + \dot{\psi}\frac{\partial V_1}{\partial \dot{\psi}}\right)$$

$$+ m_0 c^2\left(1 - \frac{\dot{r}_0^2}{c^2}\right)^{1/2} - \varepsilon V_1$$

$$\equiv \frac{m_0 c^2}{\sqrt{1 - \dot{r}_0^2 / c^2}} + \varepsilon H_1$$

Now,
$$\frac{\dot{r}_0^2}{1 - \dot{r}_0^2/c^2} = \left(p_0 - \varepsilon \frac{\partial V_1}{\partial \dot{r}_0}\right)^2 \Big/ m_0^2$$

so
$$\dot{r}_0^2 = \left[\left(p_0 - \varepsilon \frac{\partial V_1}{\partial \dot{r}_0}\right)^2 \Big/ m_0^2 c^{2+1}\right]^{-1} \left(p_0 - \varepsilon \frac{\partial V_1}{\partial \dot{r}_0}\right)^2 \Big/ m_0^2$$

$$m_0 c^2\left(1 - \frac{\dot{r}_0^2}{c^2}\right)^{-1/2} = c^2 \dot{r}_0^{-1}\left(p_0 - \varepsilon \frac{\partial V_1}{\partial \dot{r}_0}\right)$$

$$= \left[1 + \left(p_0 - \varepsilon \frac{\partial V_1}{\partial \dot{r}_0}\right)^2 \Big/ m_0^2 c^2\right]^{1/2} \times m_0 c^2$$

$$= \left[m_0^2 c^4 + c^2\left(p_0 - \varepsilon \frac{\partial V_1}{\partial \dot{r}_0}\right)^2\right]^{1/2}$$

Note that
$$-V_1 = \frac{1}{2}\left(1 - \frac{\dot{r}_0^2}{c^2}\right)^{-1/2} T_r(\dot{R}J\dot{R}^T) - \frac{3}{2c^2}\left(1 - \frac{\dot{r}_0^2}{c^2}\right)^{-3/2} \dot{\underline{r}}_0^T \dot{R}J\dot{R}^T \dot{\underline{r}}_0$$

is a quadratic function of $(\dot{\varphi}, \dot{\theta}, \dot{\psi})$.

Hence

$$\frac{\partial V_1}{\partial \dot{\varphi}} = V_{11}\dot{\varphi} + V_{12}\dot{\theta} + V_{13}\dot{\psi},$$

$$\frac{\partial V_1}{\partial \dot{\theta}} = V_{21}\dot{\varphi} + V_{22}\dot{\theta} + V_{23}\dot{\psi},$$

$$\frac{\partial V_1}{\partial \dot{\psi}} = V_{31}\dot{\varphi} + V_{32}\dot{\theta} + V_{33}\dot{\psi}$$

where $\{V_{\alpha\beta}\}$ are function of $\{\dot{r}_0, \varphi, \dot{\theta}, \psi\}$ only.

Solving gives $$\begin{pmatrix} \dot{\varphi} \\ \dot{\theta} \\ \dot{\psi} \end{pmatrix} = \varepsilon^{-1} \underline{\underline{V}}^{-1} \begin{pmatrix} p_\varphi \\ p_\theta \\ p_\psi \end{pmatrix}$$

and so,

$$p_\varphi\dot{\varphi} + p_\theta\dot{\theta} + p_\psi\dot{\psi} = \varepsilon^{-1}(p_\varphi, p_\theta, p_\psi)\underline{\underline{V}}^{-1} \begin{pmatrix} p_\varphi \\ p_\theta \\ p_\psi \end{pmatrix}$$

$$\therefore \qquad H = \left[m_0^2 c^4 + c^2 \left(p_0 - \varepsilon \frac{\partial V_1}{\partial \dot{r}_0} \right)^2 \right]^{1/2} + \varepsilon H_1$$

where $$\varepsilon H_1 = -\varepsilon V_1 + \varepsilon\left(\underline{\dot{r}}, \frac{\partial V_1}{\partial \dot{r}_0} \right) + \varepsilon\left(\dot{\varphi}\frac{\partial V_1}{\partial \dot{\varphi}} + \frac{\partial V_1}{\partial \dot{\theta}} + \frac{\partial V_1}{\partial \dot{\psi}} \right)$$

$$= -\varepsilon V_1 + \varepsilon\left(\underline{\dot{r}_0}, \frac{\partial V_1}{\partial \dot{r}_0} \right) + \frac{1}{\varepsilon}(p_\varphi p_\theta p_\psi)\underline{\underline{V}}^{-1} \begin{pmatrix} p_\varphi \\ p_\theta \\ p_\psi \end{pmatrix}$$

We can write

$$V_1 = \frac{1}{2}(\dot{\varphi}, \dot{\theta}, \dot{\psi})\,\underline{\underline{V}} \begin{pmatrix} \dot{\varphi} \\ \dot{\theta} \\ \dot{\psi} \end{pmatrix}$$

where $\underline{\underline{V}} = ((V_{\alpha\beta}))$ are function of $\{\dot{r}_0, \varphi, \theta, \psi\}$ only.

We write $$V_{\alpha\beta} = V_{\alpha\beta}(\underline{\dot{r}_0}, \varphi, \theta, \psi).$$

Now, $$\varepsilon V_1 = \frac{1\varepsilon}{2\varepsilon}(p_\varphi, p_\theta, p_\psi)\underline{\underline{V}}^{-1} \begin{pmatrix} p_\varphi \\ p_\theta \\ P_\psi \end{pmatrix}$$

and hence,

$$\varepsilon H_1 = \frac{1}{2\varepsilon}(p_\varphi, p_\theta, p_\psi)\underline{\underline{V}}^{-1}\begin{pmatrix} p_\varphi \\ p_\theta \\ P_\psi \end{pmatrix}$$

$$+\frac{1}{2\varepsilon}(p_\varphi, p_\theta, \dot{p}_\psi)\sum_{k=1}^{3}\dot{X}_k V^{-1}\frac{\partial V}{\partial X_k}V^{-1}\begin{pmatrix} p_\varphi \\ p_\theta \\ p_\psi \end{pmatrix}$$

Note that $p_{\varphi,} p_{\theta,} p_\psi$ must be $O(\varepsilon)$.

This suggests that the relativistic Hamiltonian for the rigid body should be taken as

$$H = c\left(\underline{\alpha}, \underline{p}_0 - \varepsilon\frac{\partial V_1}{\partial \dot{r}_0}\right) + \beta m_0 c^2 + \varepsilon H_1$$

where $$\dot{r}_0^2 \to \left(\frac{p_0^2}{m_0^2 c^2}\right)^{-1}\frac{p_0^2}{m_0^2}$$

and

$$\dot{r}_0 \to \frac{p_0}{m_0}\left(\frac{p_0^2}{m_0^2 c^2} + 1\right)^{-1/2}$$

$$= \frac{c^2 \underline{p}_0}{\left(p_0^2 + m_0^2 c^2\right)^{1/2}}$$

i.e. $$H = c(\underline{\alpha}, \underline{p}_0) + \beta m_0 c^2 + \varepsilon H_1(p_0, p_\varphi, p_\theta, p_\psi, \varphi, \theta, \psi).$$

Dirac robot with torque and gravity

$$£ = -\rho c^2 \int_D (1 - \left|\dot{\underline{r}}_0 + R'(t)\xi^2\right|^2 / c^2)^{1/2} d^3\xi$$

$$-mg[R(t)]_3 l + (\tau_1(t)\dot{\varphi} + \tau_2(t)\dot{\theta} + \tau_3(t)\dot{\psi})$$

$$R(t) = R\left(\varphi(t), \theta(t), \psi(t)\right)$$

$$\left(1 - \left|\dot{r}_0 + R'(t)\underline{\xi}\right|^2 / C^2\right)^{1/2}$$

$$= \left(1 - \left(\dot{r}_0^2 + \left\|R'(t)\xi\right\|^2 + 2\dot{\underline{r}}_0^T R'(t)\xi\right) / C^2\right)^{1/2}$$

$$\approx \left(1 - \frac{\dot{r}_0^2}{c^2}\right)^{1/2}\left\{1 - \frac{1}{2c^2\left(1 - \frac{\dot{r}_0^2}{c^2}\right)}\left(\left\|R'(t)\xi\right\|^2 + 2\dot{r}_0^T R'(t)\xi\right)\right.$$

$$+\frac{3}{8c^4\left(1-\frac{\dot{r}_0^2}{c^2}\right)^2}4(\dot{r}_0^T R'(t)\xi)^2\Bigg\}\Bigg]$$

$$£ = -m_0c^2\left(1-\frac{\dot{r}_0^2}{C^2}\right)^{1/2} + \frac{1}{2}(\dot{\varphi},\dot{\theta},\dot{\psi})\,\underline{\underline{V}}(\dot{r}_0,\varphi,\theta,\psi)\begin{pmatrix}\dot{\varphi}\\ \dot{\theta}\\ \dot{\psi}\end{pmatrix}$$

$$-\,mgl\,R_{33}(\varphi,\theta,\psi)+\tau_1\dot{\varphi}+\tau_2\dot{\theta}+\tau_3\dot{\psi}$$

where $$\underline{\underline{V}} = -\frac{1}{c^2}\left(1-\frac{\dot{r}_0^2}{c^2}\right)^{-1/2}$$

$$\begin{matrix} T_r\left(\frac{\partial R}{\partial\varphi}J\frac{\partial R^T}{\partial\varphi}\right) & T_r\left(\frac{\partial R}{\partial\varphi}J\frac{\partial R^T}{\partial\theta}\right) & T_r\left(\frac{\partial R}{\partial\varphi}J\frac{\partial R^T}{\partial\psi}\right)\\ T_r\left(\frac{\partial R}{\partial\varphi}J\frac{\partial R^T}{\partial\theta}\right) & T_r\left(\frac{\partial R}{\partial\varphi}J\frac{\partial R^T}{\partial\theta}\right) & T_r\left(\frac{\partial R}{\partial\varphi}J\frac{\partial R^T}{\partial\psi}\right)\\ T_r\left(\frac{\partial R}{\partial\varphi}J\frac{\partial R^T}{\partial\psi}\right) & T_r\left(\frac{\partial R}{\partial\varphi}J\frac{\partial R^T}{\partial\psi}\right) & T_r\left(\frac{\partial R}{\partial\psi}J\frac{\partial R^T}{\partial\psi}\right) \end{matrix}$$

$$+\frac{3}{C4\left(1-\frac{\dot{r}_0^2}{C^2}\right)^{3/2}}Y\cdot$$

$$\begin{bmatrix} \dot{\underline{r}}_0^T\frac{\partial R}{\partial\varphi}J\frac{\partial R^T}{\partial\varphi}\dot{r}_0 & \dot{r}_0^T\frac{\partial R}{\partial\varphi}J\frac{\partial R^T}{\partial\theta}\dot{r}_0 & \dot{r}_0^T\frac{\partial R}{\partial\varphi}J\frac{\partial R^T}{\partial\psi}\dot{r}_0\\ \dot{r}_0^T\frac{\partial R}{\partial\varphi}J\frac{\partial R^T}{\partial\theta}\dot{r}_0 & \dot{r}_0^T\frac{\partial R}{\partial\theta}J\frac{\partial R^T}{\partial\theta}\dot{r}_0 & \dot{r}_0^T\frac{\partial R}{\partial\theta}J\frac{\partial R^T}{\partial\psi}\dot{r}_0\\ \dot{r}_0^T\frac{\partial R}{\partial\varphi}J\frac{\partial R^T}{\partial\psi}\dot{r}_0 & \dot{r}_0^T\frac{\partial R}{\partial\theta}J\frac{\partial R^T}{\partial\psi}\dot{r}_0 & \dot{r}_0^T\frac{\partial R}{\partial\psi}J\frac{\partial R^T}{\partial\psi}\dot{r}_0 \end{bmatrix}$$

We can thus write $$\underline{\underline{V}} = -\frac{1}{c^2}\left(1-\frac{\dot{r}_0^2}{c^2}\right)^{-1/2}\underline{\underline{A}}(\varphi,\theta,\psi)+\frac{3}{c^4}\left(1-\frac{\dot{r}_0^2}{c^2}\right)^{-3/2}$$

$$(I_3\otimes\dot{\underline{r}}_0^T)\underline{\underline{B}}(\varphi,\theta,\psi)(I_3\otimes\dot{\underline{r}}_0)$$

❖❖❖❖❖

[4] Quantum Robot Dynamics:

$$\underline{\underline{M}}(\underline{q})\ddot{\underline{q}} + \underline{N}\left(\underline{q}, \dot{\underline{q}}\right) = \underline{\underline{F}}(\underline{q}, \dot{\underline{q}})\left(K_p\left(\underline{q}_d - \underline{q}\right) + K_d\left(\dot{\underline{q}} - \dot{\underline{q}}\right)\right) + \underline{\underline{G}}\left(\underline{q}, \dot{\underline{q}}\right)\underline{W}_t$$

This equation describes the motion of a classical Robot used to track a desired trajectory $\underline{q}_d$.

$\underline{W}_t = \frac{d\underline{B}_t}{dt}$, $\underline{B}_t \in \mathbb{R}^3$ is 2-D standard Brownian motion. Thus

$$d\begin{pmatrix}\underline{q}\\ \dot{\underline{q}}\end{pmatrix} = \begin{pmatrix}\dot{\underline{q}}\\ -\underline{\underline{M}}(\underline{q})^{-1}\underline{N}(\underline{q}, \dot{\underline{q}})\end{pmatrix}dt + K_P\begin{pmatrix}0\\ \underline{\underline{M}}(\underline{q})^{-1}\underline{\underline{F}}(\underline{q}, \dot{\underline{q}})\left(\underline{q}_d - \underline{q}\right)\end{pmatrix}dt$$

$$+ K_d\begin{pmatrix}0\\ \underline{\underline{M}}(\underline{q})^{-1}\underline{\underline{F}}(\underline{q}, \dot{\underline{q}})\left(\dot{\underline{q}}_d - \dot{\underline{q}}\right)\end{pmatrix}dt + H\left(\underline{q}, \dot{\underline{q}}\right)d\,\underline{B}_t$$

where $H\left(\underline{q}, \dot{\underline{q}}\right) = \begin{pmatrix}\underline{\underline{0}}\\ \underline{\underline{G}}\left(\underline{q}, \dot{\underline{q}}\right)\end{pmatrix}$

We write the eqn. as

$$d\underline{\xi} = \underline{F}_0(\underline{\xi})\,dt + K_p\underline{F}_1\left(t, \underline{\xi}\right)dt + K_d\underline{F}_2\left(t, \underline{\xi}\right)dt + H(\underline{\xi})d\underline{B}_t$$

So
$$df\left(\underline{\xi}\right) = \tilde{\mathcal{L}}_0 f(\underline{\xi})\,dt + k_p\mathcal{L}_{1t}f(\underline{\xi})\,dt + K_d\mathcal{L}_{2t}f(\underline{\xi})\,dt$$

$$+\frac{1}{2}T_r\left(H(\underline{\xi})\underline{H}(\underline{\xi})^T\underline{\nabla}\underline{\nabla}^T f(\underline{\xi})\right)dt + d\underline{B}^T\underline{\underline{H}}^T(\underline{\xi})\underline{\nabla}f(\underline{\xi}) \qquad (1)$$

for $f: \mathbb{R}^4 \to \mathbb{R}$. Let

$$\mathcal{L}_0 f = \tilde{\mathcal{L}}_0 f + \frac{1}{2}T_r\left(HH^T\nabla\nabla^T f\right)$$

Then the quantum version of (1) reads

$$dj_t(f) = j_t(\mathcal{L}_0 f)\,dt + K_p j_t(\mathcal{L}_{1t}f)\,dt + K_d j_t(\mathcal{L}_{2t}f)\,dt + \sum_{k=1,2} j_t\left(\mathcal{L}_{3k}f\right)dB_k$$

where j_t: $f \to f(\underline{\xi}_t)$

Is a * homomorphism from $C^\infty(\mathbb{R})$ into $L^2(\Omega\,\mathcal{F}_t, P)$

We consider a "simplified" quantum model

$$dj_t(X) = (j_t(\mathcal{L}_0 X) + K_1 j_t(\mathcal{L}_1 X))dt + j_t(\mathcal{L}_3 X)dA_t + j_t(\mathcal{L}_3^t X)dA_t^* + j_t(\mathcal{L}_4 X)d\Lambda_t$$

where
$$\mathcal{L}_3 = \mathcal{L}_{30} + K_{2\mathcal{L}31}$$

$$\mathcal{L}_4 = \mathcal{L}_{40} + K_{3\mathcal{L}41}$$

K_1, K_2, K_3 are to be chosen so that

$$\xi = \int_0^T \langle fe(u), j_t(X)\, fe(u)\rangle dt$$

is a minimum. Now, $\xi = \int_{0<S<t<T} dt\, d_s \langle fe(u), j_s(X) fe(u)\rangle$

$$= \int_0^T (T-S)\, d_s \langle fe(u), j_s(X)\, fe(u)\rangle$$

$$= \int_0^T w(S) \langle fe(u), dj_s(X)\, fe(u)\rangle$$

where $w(S) = T - S.$

Now, $\langle fe(u), dj_t(X) fe(u)\rangle = dt\{\langle fe(u), j_t(\mathcal{L}_0 X) fe(u)\rangle$

$$+ 2\,\text{Re}\{u(t)\langle fe(u), j_t(\mathcal{L}_{30} X)\, fe(u)\rangle\}$$

$$+ |u(t)|^2 \langle fe(u), j_t(\mathcal{L}_{40} X)\, fe(u)\rangle\}$$

$$dt\{K_1 \langle fe(u), j_t(\mathcal{L}, X)\, fe(u)\rangle$$

$$+ 2K_2\, \text{Re}\{u(t)\langle fe(u), j_t(\mathcal{L}_{31} X)\, fe(u)\rangle$$

$$+ K_3 |u(t)|^2 \langle fe(u), j_t(\mathcal{L}_{41} X)\, fe(u)\rangle\}$$

Suppose we make K_1, K_2, K_3 non-random *i.e.* (C-number) functions of time. We choose $K_j(s)$, $s \le t$, $j = 1, 2, 3$ so that

$$\int_0^t \langle fe(u), j_s(X) fe(u)\rangle\, ds$$

is a minimum. Then we choose

$\underline{K}(t + dt)$ so that

$$\int_0^{t+dt} \langle fe(u), j_s(X) fe(u)\rangle\, ds = \int_0^{t+dt} (t + dt - S)\langle fe(u), dj_s(X) fe(u)\rangle$$

is a minimum. Since $j_s(X)$, $S \le t$ is known, that is equivalent to minimizing

$$\langle fe(u), dj_s(X) fe(u)\rangle$$

or equivalently

$$K_1 \langle fe(u), j_t(\mathcal{L}_1 X) fe(u)\rangle$$

$$+2K_2\,\text{Re}\{u(t)\langle fe(u),\, j_t(\mathcal{L}_{31}X)fe(u)\rangle\}$$

$$+K_3\,|u(t)|^2\langle fe(u),\, j_t(\mathcal{L}_4X)fe(u)\rangle$$

w.r.t. K_1, K_2, K_3.

The minimizing values are $\underline{K}(t+dt)$

$$\left(\underline{K}(t) = (K_1(t), K_2(t), K_3(t))\right)$$

We will have to incorporate energy constraint on $\underline{K}(t)$ like for example

$$\|\underline{K}(t+dt) - \underline{K}(t)\|^2 \le \varepsilon$$

Here, we are treaty *dt* as a finite small position number. More generally, we could replace this condition by

$$\underline{K}'(t)^T Q(t)\underline{K}'(t) \le \varepsilon,\ \underline{\underline{Q}} \ge 0$$

So letting

$$\alpha(t) = \langle fe(u),\, j_t(\mathcal{L}_1X)fe(u)\rangle,$$

$$\beta(t) = 2\,\text{Re}\{\langle fe(u),\, j_t(\mathcal{L}_{31}X)fe(u) > u(t)\rangle\},$$

$$\gamma(t) = |u(t)|^2\langle fe(u),\, j_t(\mathcal{L}_{41}X)fe(u)\rangle$$

The optimization problem is to minimize

$$\alpha(t)K_1'(t) + \beta(t)K_2'(t) + \gamma(t)K_3'(t) - \lambda(t)\left(\underline{K}'(t)^T\underline{\underline{Q}}(t)\underline{K}(t) - \varepsilon\right)$$

Let

$$\underline{l}(t) = (\alpha(t), \beta(t), \gamma(t))^T.$$

Then the function to be minimized is

$$\underline{l}(t)^T\underline{K}'(t) - \lambda(t)\left(\underline{K}'(t)^T\underline{\underline{Q}}(t)\underline{K}'(t) - \varepsilon\right)$$

Setting the gradient of the *w.r.t.* $\underline{K}'(t)$ to zero gives

$$\underline{l}(t) - 2\lambda(t)\underline{\underline{Q}}(t)\underline{K}'(t) = 0$$

or

$$\underline{K}'(t) = \frac{1}{2\lambda(t)}\underline{\underline{Q}}(t)^{-1}\underline{l}(t) \text{ writing } \lambda(t) \text{ given by}$$

$$\frac{1}{4\lambda^2(t)}\underline{l}(t)^T\underline{\underline{Q}}(t)^{-1}\underline{l}(t) = \varepsilon$$

i.e.

$$\lambda(t) = \frac{1}{2\sqrt{\varepsilon}}\left(\underline{l}(t)^T\underline{\underline{Q}}(t)^{-1}\underline{l}(t)\right)^{1/2}$$

Note that $\underline{l}(t)$ is a function of $\underline{K}(S)$, $S \le t$, and hence we get a functional differential equation for $\underline{K}(\cdot)$.

❖❖❖❖❖

[5] Hamiltonian Based Quantum Robot Tracking:

$$dV_t = \left[-i\left(H_0 + K_0H_1\right)dt + \left(L_{10} + K_1L_{11}\right)dA_t + \left(L_{20} + K_2L_{21}\right)dA_t^* + \left(S_0 + K_3S_1\right)d\Lambda_t\right]V_t$$

$\{K_0, K_1, K_2, K_3\}$ are control parameters that can very subject to the unitarity constant on V_t.

Writing $H = H_0 + K_0H_1, L_1 = L_{10} + K_1 L_{11}$

$L_2 = L_{20} + K_2 L_{21}, S = S_0 + K_3 S_1$, we find using the quantum. Its formula that for $\underline{d}\left(V_t^* V_t\right) = 0$, we require

$$i\left(H^* - H\right) + L_2^* L_2 = 0,$$

$$L_1 + L_2^* + L_2^* S = 0,$$

$$S^*S + S + S^* = 0$$

Thus $H = P + i\,Q, P^* = P, Q^* = Q,$

$$2Q + L_2^* L_2 = 0,$$

$$S = -I + F, \text{ where } F^* F = I.$$

So $H = P_0 + K_0 P_1 \pm\frac{i}{2}\left(L_{20} + K_1 L_{21}\right)^*\left(L_{20} + K_1L_{21}\right)$

where $P_k^* = P_k, k = 0, 1,$

$$L_1 = \left(L_{20} + K_1L_{21}\right)^* F$$

where $F^*F = I$ and F is fixed. The *qsde* satisfied by V_t is thus

$$dV_t = \left[-\left(i\left(P_0 + K_0P_1\right) + \frac{1}{2}\left(L_{20} + K_1L_{21}\right)^*\left(L_{20} + K_1L_{21}\right)\right)^* dt - \left(L_{20} + K_1L_{21}\right)^* F\,d\,A_t + \left(L_{20} + K_1L_{21}\right)dA_t^* + \left(F - I\right)d\Lambda\right]V_t$$

Quantum Stochastic Calculus some remarks

Let $d\Lambda_t = \dfrac{dA_t^* dA_t}{dt}$ *i.e* $\Lambda_t = \int_0^t \dfrac{dA_t^*}{dt}\dfrac{dA_t}{dt}\,dt$ (formally)

Thus $(d\Lambda_t)^2 = \dfrac{1}{dt^2}\, dA_t^*\, dA_t dA_t^*\, dA_t$

$= \dfrac{dA_t^* dA_t}{dt} = d\Lambda_t$ since $dA_t\; dA_t^* = dt$

So formally, we can derive the conservation process Λ from the creation and annihilation process dA_t^*, A_t

Note that $\left\langle e(u), \frac{dA_t^* dA_t}{dt} e(v) \right\rangle = \langle dA_t e(u), dA_t(u) \rangle / dt$

$$= \bar{u}(t)\, v(t) \frac{dt^2}{dt} = \bar{u}(t)\, v(t)\, dt$$

In agreement with the desired properties of Λ_t.

❖❖❖❖❖

[6] Robot Interacting with a Klein-Gordon Field:

Application of QFT to Robotics. The current through a Robot $J(x) = J(t, \underline{r})$ interacts with the quantized KG field $\varphi(x)$ with interaction Lagrangian $\int J(x)\varphi(\underline{x})d^2x$ and interaction action $\int J(x)\varphi(\underline{x})d^4x$. The current through the Robot links is a function of the position vector $\underline{\xi}$ in the link relative to the origin and also a function of the angles $\theta_1, \ldots \theta_d$ of the links. We can express this current densities as

$$J = J\left(t, \underline{\xi} \mid \theta_1(t), \ldots \theta_d(t)\right)$$

The kinetic and potential energies of the Robot system have the form

$$T = \frac{1}{2}\underline{\theta}'(t)^T J_0\left(\underline{\theta}(t)\right)\underline{\theta}'(t)$$

and $$V(\underline{\theta}(t)) = \sum_{k=1}^{d} \lambda_k \delta_m(\theta_k(t)) \text{ (gravitational potential energy)}$$

The action of the system consisting of the quantum Klein-Gordon field, the Robot link system and the interaction of the two is thus

$$S[\varphi, \underline{\theta}] = \int\left(\frac{1}{2}\partial_\mu \varphi . \partial^\mu \varphi - \frac{1}{2}m^2\varphi^2\right)d^3x\, dt$$

$$\cdot \int\left(1/2\underline{\theta}' J_0(\underline{\theta})\underline{\theta}' - V(\underline{\theta})\right)dt + \int J(t, x \mid \underline{\theta}(t))\varphi(t, x)d^3x a$$

Reference (1) Rohit Singla, Vijayant Agrawal and Harish Parthasarathy, "Two links planar Robot Manipulator," Technical. Report, NSIT.

(2) Belavkin "Nonlinear filtering" Journals of Multivariate Analysis.

(3) S. Chandrasekhar, 'The mathematics theory of Blackholes' Oxford.

❑❑❑

6

Large Deviations, Classical and Quantum General Relativity with GPS Application

[1] Intuitive Derivation of the Gartner Ellis LDP:

$Z_n \to 0$. Then put

$$M_n(\lambda) = \mathbb{E}\{e^{\lambda Z_n}\}, \Lambda_n(\lambda) = \log M_n(\lambda),$$

$$\bar{\Lambda}(\lambda) = \lim_{n\to\infty} \frac{1}{n}\Lambda_n(n\lambda) \qquad \text{(Assume this limit exists)}$$

We have for large n,

$$\mathbb{E}\left[e^{n(\lambda Z_n - \bar{\Lambda}(\lambda))}\right] \approx 1 \qquad \left(\text{Since } \mathbb{E}[e^{n(\lambda Z_n - \frac{1}{n}\Lambda_n(n\lambda))}] = 1\right)$$

Assume for simplicity that Z_n takes discrete values $\{x \in I\}$ I countable set with $P_r\{Z_n = x\} = P_n(x)$.

Then,

$$\begin{aligned} 1 &\approx \mathbb{E}\left[e^{n(\lambda Z_n - \bar{\Lambda}(\lambda))}\right] \\ &= \sum_{x\in I} e^{n(\lambda x - \bar{\Lambda}_n(\lambda))} p_n(x) \\ &\approx \geq e^{n(\lambda x - \bar{\Lambda}(\lambda))} p_{n(x)} \forall x, \lambda \end{aligned}$$

Thus,

$$\frac{1}{n}\log p_n(x) \ \tilde{\leq}\ -(\lambda x - \bar{\Lambda}(\lambda)) \forall x, \lambda$$

and in particular,

$$\begin{aligned} \frac{1}{n}\log p_n(x) &\ \tilde{\leq}\ -\sup(\lambda x - \bar{\Lambda}(\lambda)) \\ &= \frac{1}{n}\log p_n(x) \end{aligned}$$

On the other hand, let B be a n hood of 0. Then for large n_1 $Z_n \in B$ with high probability.

Assume $B = (-a, a), a > 0$.

Then for $\lambda > 0$ $e^{n\lambda x}$ is large for $x > a$ and thus

$1 \tilde{<} p_n(B^c)\ e^{n(\lambda x - \bar{\Lambda}(\lambda))}, \lambda > 0, x > a$

$\leq p_n(B^c)e^{nI_1(x)}$

where $$I_1(x) = \sup_{\lambda>0}\left(\lambda x - \bar{\Lambda}(\lambda)\right)$$

Thus,

$$\frac{1}{n}\log p_n\left(B^c\right) \geq -I_1(x), x > a$$

So, $$\frac{1}{n}\log p_n\left(B^c\right) \geq -\inf_{x>a} I_1(x)$$

Likewise for $x < -a$, $e^{n\lambda x}$ is large for $\lambda < 0$ and therefore

$$1 \tilde{<} p_n(B^c)e^{n\left(\lambda x - \bar{\Lambda}(\lambda)\right)}$$

$$x < -a, \lambda < 0 \leq p_n(B^c)\ e^{nI_2(x)},$$

$$x < -a$$

where $$I_2(x) = \sup_{\lambda<0}\left(\lambda x - \bar{\Lambda}(\lambda)\right)$$

Thus ,

$$\frac{1}{n}\log p_n\left(B^c\right) \geq -I_2(x)\ x < -a$$

and hence $$\frac{1}{n}\log p_n\left(B^c\right) \geq -\inf_{x<-a} I_2(x)$$

❖❖❖❖❖

[2] LDP in Robotic Vision:

$R_1(t) \simeq$ rotation applied to first link $= R(\varphi_1(t), \theta_1(t), \psi_1(t))$.

$R_2(t) \simeq$ rotation applied to second link $= R(\varphi_2(t), \theta_2(t), \psi_2(t))$.

General points on the first link:

$$\underline{\underline{R}}_1(t)\underline{r},\ \underline{r} \in B_1. \tag{1}$$

General point on the second link:

$$\underline{\underline{R}}_1(t)\underline{\underline{p}} + \underline{\underline{R}}_2(t)\underline{\underline{R}}_1(t)(\underline{r} - \underline{p}), \underline{r} \in B_2.$$

Let $$R_1(t) \equiv R(t),\ R_2(t)\ R_1(t) \equiv S(t).$$

Let $(\varphi_1(t), \theta_1(t), \psi_1(t)) \equiv (\varphi(t), \theta(t), \psi(t))$,

$$S(t) = R\left(\tilde{\varphi}(t), \tilde{\theta}(t), \tilde{\psi}(t)\right)$$

Note: $R(\varphi, \theta, \psi) = R_z(\varphi)\, R_x(\theta)\, R_z(\psi)$ (2)

Then kinetic energy of the system is

$$T\left(\varphi,\theta,\psi,\tilde{\varphi},\tilde{\theta},\tilde{\psi},\dot{\varphi},\dot{\theta},\dot{\psi},\dot{\tilde{\varphi}},\dot{\tilde{\theta}},\dot{\tilde{\psi}}\right)$$

$$= \frac{1}{2}P\int_{B_1} \|\underline{\underline{R}}'_F(t)\underline{r}\|^2\, d^3r + \frac{1}{2}\rho\int_{B_2} \|\underline{\underline{R}}'(t)\underline{p}$$

$$+\underline{\underline{S}}'(t)(\underline{r}-\underline{p})\|^2\, d^3r$$

$$= \frac{p}{2}T_r(\underline{\underline{R}}'(t)^T\underline{\underline{R}}'(t)\int_{B_1}\underline{r}\,\underline{r}^T d^3r)$$

$$+\frac{p}{2}\left(T_r\left[(R'(t)-S'(t))^T(R'(t)-S'(t)pp^T\right]v(B_2)\right.$$

$$+T_r\left[S'(t)^T S'(t)\int_{B_2}\underline{r}\,\underline{r}^T d^3r\right]$$

$$\left.+\,2T_r\left[(R'(t)-S'(t))^T S'(t)\int_{B_2}\underline{r}\underline{p}^T d^3r\right]\right) \quad (3)$$

Let

$$\rho\int_{B_1}\underline{r}\,\underline{r}^T\, d^3r = \underline{\underline{J}}_1,$$

$$\rho\int_{B_2}\underline{r}\,\underline{r}^T\, d^3r = \underline{\underline{J}}_2,$$

$$\rho\int_{B_2}\underline{r}\, d^3r = m_2\,\underline{\xi}$$

$$\rho v\,(B_2) = m_2,\ m_2\underline{p}\,\underline{p}^T = \underline{\underline{Q}}_{1+n},\ m_2\underline{\xi}\,\underline{p}^T = \underline{\underline{Q}}_2 \quad (4)$$

Then

$$T = \frac{1}{2}T_r\left(R'(t)^T R'(t)\underline{\underline{J}}_1\right)$$

$$+\frac{1}{2}T_r\left((R'-S')^T(R'-S')\underline{\underline{Q}}_1\right)$$

$$+\frac{1}{2}T_r\left[S'^T S'\underline{\underline{J}}_2\right]+T_r\left[(R'-S')^T S'Q_2\right]$$

Note that

$$R' = R'(t) = \frac{\partial}{\partial\varphi}R(\varphi,\theta,\psi,)\varphi'$$

$$+\frac{\partial}{\partial\theta}R(\varphi,\theta,\psi,)\theta' + \frac{\partial R}{\partial\psi}(\varphi,\theta,\psi)\psi'$$

$$\frac{\partial}{\partial\varphi}R(\varphi,\theta,\psi,) = R'_Z(\varphi)R_X(\theta)\,R_Z(\psi)$$

$$\frac{\partial R}{\partial\theta}(\varphi,\theta,\psi) = R_Z(\varphi)R'_X(\theta)\,R_Z(\psi),$$

$$\frac{\partial R}{\partial\psi}(\varphi,\theta,\psi) = R_Z(\varphi)R_X(\theta)\,R'_Z(\psi) \tag{5}$$

Thus $R'^T R'$ is a quadratic function of $(\varphi, \theta, \theta)$, $(R' - S')^T (R' - S')$ is a quadratic function of $\left(\dot{\varphi},\dot{\theta},\dot{\psi},\dot{\tilde{\varphi}},\dot{\tilde{\theta}},\dot{\tilde{\psi}}\right)$ and $S'^T S'$ is a quadratic function of $\left(\dot{\tilde{\varphi}},\dot{\tilde{\theta}},\dot{\tilde{\psi}}\right)$. It follows therefore that T is a quadratic function of $\left(\varphi,\theta,\psi,(\tilde{\varphi},\tilde{\theta},\tilde{\psi}\right)$. with coefficients being trigonometric functions of $\left(\varphi,\theta,\psi,(\tilde{\varphi},\tilde{\theta},\tilde{\psi}\right)$. The Lagrangian of the Robot taking into account gravity is then

$$T - V = \mathcal{L} = \frac{1}{2}\dot{\underline{\xi}}\underline{\underline{J}}(\underline{\xi})\underline{\xi} - V(\underline{\xi}) \tag{6}$$

where $\underline{\xi} = \left(\varphi,\theta,\psi,\tilde{\varphi},\tilde{\theta},\tilde{\psi}\right)^T$ and

Reference:

[1] Lectures on Robotic vision, "Naman Garg, Vijayant Agrawal, H. Parthasarathy and D.K. Upadhayay," Technical Report, NSIT, 2014.

$$\begin{aligned} V(\xi) &= \rho g \int_{B_1} \left(\hat{z}, R_1(\varphi_1,\theta_1,\psi_1)\underline{r}\right) d^3r \\ &\quad + \rho g\{(\hat{z},p) + \int_{B_2} (z, R_2(\varphi_2,\theta_2,\psi_2)\ (\underline{r} - \underline{p}))d^3r\} \\ &= \rho g p_z + m_1 g(\hat{z}, R_1(\varphi_1,\theta_1,\psi_1)\underline{\xi}) \\ &\quad + m_2 g(\hat{z}, R_2(\varphi_2,\theta_2,\psi_2)(\underline{\xi} - \underline{p})) \end{aligned}$$

where
$$\underline{\xi_1} = \int_{B_1} r\frac{d^3r}{v(B_1)},\ m_1 = \rho v(B_1),\ v(B_1) = \int_{B_1} d^3r \tag{7}$$

The Robot eqns. in the presence of noisy torque are

$$\frac{d}{dt}\frac{\partial\mathcal{L}}{\partial\dot{\varphi}} = \frac{\partial\mathcal{L}}{\partial\varphi},\ \frac{d}{dt}\frac{\partial\mathcal{L}}{\partial\dot{\theta}} = \frac{\partial\mathcal{L}}{\partial\theta} + \sqrt{\varepsilon}w_\theta(t), + \sqrt{\varepsilon}W_\theta(t),$$

$$\frac{d}{dt}\frac{\partial\mathcal{L}}{\partial\dot{\psi}} = \frac{\partial\mathcal{L}}{\partial\psi} + \sqrt{\varepsilon}W_\psi(t), \tag{8}$$

$$\frac{d}{dt}\frac{\partial \mathcal{L}}{\partial \dot{\tilde{\varphi}}} = \frac{\partial \mathcal{L}}{\partial \tilde{\varphi}} + \sqrt{\varepsilon}\, W_{\dot{\tilde{\varphi}}}(t), \quad \frac{d}{dt}\frac{\partial \mathcal{L}}{\partial \dot{\tilde{\psi}}} = \frac{\partial \mathcal{L}}{\partial \tilde{\psi}} + \sqrt{\varepsilon} W_{\dot{\tilde{\psi}}}(t)$$

$\frac{d}{dt}\frac{\partial \mathcal{L}}{\partial \dot{\tilde{\theta}}} = \frac{\partial \mathcal{L}}{\partial \tilde{\theta}} + \sqrt{\varepsilon} W_{\tilde{\theta}}(t)$ provided through the motives.

We add a dc torque $\times \left\{\tau_{\varphi}, \tau_{\theta,} \tau_{\psi,} \tilde{\tau}_{\varphi}, \tilde{\tau}_{\theta,} \tilde{\tau}_{\psi,}\right\}$ to maintain the Robot at a fixed point.

$\left(\varphi_0, \theta_0, \psi_0, \tilde{\varphi}_0, \tilde{\theta}_0, \tilde{\psi}_0\right) = \underline{\xi}_0, \; \dot{\underline{\xi}}_0 = 0$ in phase space in the absence of noise. The eqns. of motion in vector form are then

$$\left(\underline{\underline{J}}(\underline{\xi})\underline{\xi}'\right)' - V'(\underline{\xi}) = \underline{\tau} + \sqrt{\varepsilon}\underline{W}(t) \tag{9}$$

where $\underline{\tau} \in \mathbb{R}^6$ is a constants (dc torque) vector.

Assuming $W(t) \in \mathbb{R}^6$ to be white Gaussian noise with autocorrelation $\underline{\underline{I}}_6\delta(\tau)$, the rate function for ξ over the interval $[0, T]$ is

$$I_T(\underline{\xi}) = \frac{1}{2}\int_0^T \| \underline{\underline{J}}(\underline{\xi})\underline{\xi}'' + \underline{N}(\underline{\xi}, \underline{\xi}') - V'(\underline{\xi}) - \underline{\tau} \|^2 dt$$

where $$N(\underline{\xi}, \underline{\xi}') = \underline{\underline{J}}'(\underline{\xi})(\underline{\xi}' \otimes \underline{\xi}') - V'(\underline{\xi}) \tag{10}$$

No assume a camera to be attached at

$$\underline{\underline{R}}_1(t)p + \underline{\underline{R}}_2(t)\underline{\underline{R}}_1(t)(\underline{q} - \underline{p}) = \underline{\underline{R}}(t)\underline{p} + \underline{\underline{S}}(t)(\underline{q} - \underline{p})$$

i.e. at a point on the tip of the top link that way at the position $\underline{q}$ at time $t = 0$.

Denote this position vector by $\chi(t)$:

$$\underline{\chi}(t) = \underline{\underline{R}}(t)\underline{p} + \underline{\underline{S}}(t)(\underline{q} - \underline{p}) \tag{11}$$

The object is located at a fixed point $\underline{r}_0$ and the screen on which the image is formed has equation.

$$\left(\hat{l}, \underline{r}\right) = d \tag{12}$$

i.e. $\quad l_x X + l_y Y + l_z Z = d$

Then join the camera at $\underline{\chi}(t)$ to the object at $\underline{r}_0$ *via* the line.

$$\underline{\chi}(t) + \lambda\,(\underline{r}_0 - \underline{\chi}(t)),\ \lambda \in \mathbb{R} \tag{13}$$

This line will intersect to the screen at $\underline{\eta}(t)$ where

$$\left(\hat{l}, \underline{\eta}(t)\right) = \underline{d}, \; \underline{\eta}(t) = \underline{\chi}(t) + \lambda(r_0 - \underline{\chi}(t)).$$

Thus $\left(\hat{\underline{l}}, \underline{\chi} + \lambda(\underline{r}_0 - \underline{\chi})\right) = d$

or $$\lambda = \frac{\left(d - \left(\hat{l}, \underline{\chi}(t)\right)\right)}{\left(\hat{\underline{l}}, \underline{r}_0\underline{\chi}(t)\right)} \tag{14}$$

Thus, $$\underline{\eta}(t) = \underline{\chi}(t) + \frac{\left(d - \left(\hat{l}, \underline{\chi}(t)\right)\right)\left(\underline{r}_0 - \underline{\chi}(t)\right)}{\left(\hat{\underline{l}}, \underline{r}_0 - \underline{\chi}(t)\right)} \quad (15)$$

Linearizing (9): $$\underline{\xi}(t) = \underline{\xi}^{(0)}(t) + \sqrt{\varepsilon}\,\underline{\xi}^{(1)}(t) \quad (16)$$

given

$$\left(\underline{\underline{J}}\left(\underline{\xi}^{(0)}(t)\right)\xi^{(0)\prime}(t)\right)' - V'\left(\underline{\xi}^{(0)}(t)\right) = \underline{\tau} \quad (17)$$

If we assume $\xi^{(0)\prime} = 0 \forall t$, then $\underline{\xi}^{(0)}$ = constant must satisfy

$$V'(\underline{\xi}^{(0)}) = -\underline{\tau} \quad (18)$$

Equating $0\left(\sqrt{\varepsilon}\right)$ terms gives

$$\underline{\underline{J}}(\underline{\xi}^{(0)}) + \underline{\xi}^{(1)\prime\prime} + \underline{\underline{J}}'(\underline{\xi}^{(0)})\left(\underline{\xi}^{(1)} \otimes \underline{\xi}^{(0)\prime\prime}\right)$$
$$+\underline{\underline{N}}_{,1}(\underline{\xi}^{(0)}, \underline{0})\underline{\xi}^{(1)} + \underline{\underline{N}}_{,2}(\underline{\xi}^{(0)}, 0)\xi(1)' = \underline{W}(t) \quad (19)$$

or since $\xi^{(0)\prime\prime} = 0$, we get

$$\underline{\underline{F}}_0\underline{\xi}^{(1)\prime\prime} + \underline{\underline{F}}_1\underline{\xi}^{(1)\prime} + \underline{\underline{F}}_2\underline{\xi}^{(1)} = \underline{W}(t) \quad (20)$$

where $\underline{\underline{F}}_0$, $\underline{\underline{F}}_1$ and $\underline{\underline{F}}_2$ are the constant matrices defined by

$$\underline{\underline{F}}_0 = \underline{\underline{J}}(\underline{\xi}^{(0)}), \quad \underline{\underline{F}}_1 = N_{,2}\left(\underline{\xi}^{(0)}, 0\right),$$
$$\underline{\underline{F}}_2 = N_{,1}\left(\underline{\xi}^{(0)}, 0\right) \quad (21)$$

❖❖❖❖❖

[3] LMS Algorithm for Non-Linear Disturbance Observer in a Two-Link Robot System:

Abstract: This work deals with a generalized non-linear disturbance observer for a two-link Robot. The observer is based on constructing a linear combination of velocity dependent test functions with the coefficients of the linear combination being adaptive with the update being based on the LMS algorithm. A statistical analysis of the convergence of this algorithm is based on assuming that the samples of the disturbance are white, *i.e.* independent.

Problem formulation

The Robot dynamics is described by

$$\underline{\underline{M}}(\underline{q})\ddot{\underline{q}} + \underline{N}(\underline{q}, \dot{\underline{q}}) = \underline{\tau}(t) + \underline{d}(t)$$

where $\underline{\tau}$ is the applied torque (for following the denied trajectory $\underline{qd}$, $\underline{\tau}$ is taken as the computed torque with PD error feedback:

$$\underline{\tau}(t) = \underline{\underline{M}}(\underline{q}_d)\left(\ddot{\underline{q}}_d + K_p(\underline{q}_d - \underline{q}) + K_d(\dot{\underline{q}}_d - \dot{\underline{q}})\right) + \underline{N}(\underline{q}_d, \dot{\underline{q}}_d)$$

$d(t)$ is the disturbance. As in [.], we construct an estimate $\hat{\underline{d}}(t)$ of $\underline{d}(t)$ based on $(\underline{q}^{(t)}, \dot{\underline{q}}^{(t)})$, $\underline{\tau}(t)$, using the algorithm

$$\hat{\underline{d}}(t) = \underline{Z}(t) + \underline{p}(\dot{\underline{q}}(t)),$$

$$\dot{\underline{Z}} = L\left(\underline{q}(t), \dot{\underline{q}}(t)\right)\left(\underline{N}(\underline{q}(t), \dot{\underline{q}}(t)) - \underline{\tau}(t) - p\left(\dot{\underline{q}}(t) - \underline{Z}(t)\right)\right)$$

From these two equations, we derive

$$\dot{\hat{\underline{d}}} = \dot{\underline{Z}} + \underline{p}'(\dot{\underline{q}})\ddot{\underline{q}} = L\left(N - \tau - p - Z + L^{-1}p^{1}\ddot{q}\right)$$

so that if we choose $L = p'(\dot{\underline{q}})M^{-1}(\underline{q})$, then $L^{-1}p'(\dot{q}) = M(\underline{q})$ and then by virtue of the eqns. of motion,

$$\dot{\hat{\underline{d}}} = L(\underline{d} - \hat{\underline{d}})$$

which implies that if $\dot{\underline{d}} \to 0$, then $\underline{d} - \hat{\underline{d}} \to 0$ provides that the eigen values of $L\left(\underline{q}(t), \dot{\underline{q}}(t)\right)$ all have positive real part for all $t \geq 0$.

❖❖❖❖❖

[4] Disturbance Observer Design in Robotics:

Suppose we design the disturbance observer $\underline{p}(\dot{q})$ as follow: Choose two functions $\underline{p_k}(\dot{\underline{q}}), k = 1, 2, \ldots, p$ and put

$$\underline{p}(\dot{\underline{q}}) = \sum_{h=1}^{p} c_k p_k(\dot{\underline{q}})$$

where now the $C_k's$ may be function's θ time. In the discrete time scenario with discretization step size Δ, the eqn. $\dot{\hat{\underline{d}}} = \underline{\underline{L}}(\underline{q}(t), \dot{q}(t))(\underline{d}(t) - \hat{\underline{d}}(t))$ may be expressed as

$$\hat{\underline{d}}(t+1) = \hat{\underline{d}}(t) + \Delta.\underline{\underline{L}}(t)(\underline{d}(t) - \hat{\underline{d}}(t)) = (I - \Delta\,.\underline{\underline{L}}(t))\hat{\underline{d}}(t) + \Delta.\underline{\underline{L}}(t).\underline{d}(t).$$

We note that

$$\underline{\underline{L}}(t) \underline{\underline{\Delta}}\, L(\underline{q})(t), \dot{\underline{q}}(t)) = \sum_{k=1}^{p} C_k(t)\underline{\underline{p}}'_k(\dot{\underline{q}}(t))\underline{\underline{M}}^{-1}(\underline{q}(t)) = \sum_{k=1}^{p} C_k(t) L_k(\underline{q}(t), \dot{\underline{q}}(t)),$$

where $\underline{\underline{L}}_k(q, \dot{\underline{q}}) = \underline{\underline{p}}'_k(\dot{\underline{q}})\underline{\underline{M}}^{-1}(\underline{q}), 1 \leq k \leq p,$ are known functions of $(\underline{q}, \dot{\underline{q}})$.

❖❖❖❖❖

[5] Disturbance Observer Continuation:

Assuming that $\underline{C}(t) = ((C_k(t)))^p{}_{k=1}$ is known, in accord with the LMS elgorithm, we above.

$$\underline{C}(t+1) = \underline{C}(t) - \mu(t)\frac{\partial}{\partial \underline{C}(t)}\|\underline{d}(t+1) - \hat{\underline{d}}(t+1)\|^2$$

simplying that if the disturbance estimation error energy $\left\|\underline{d}(t+1) - \hat{\underline{d}}(t+1)\right\|^2$ increases with increasing $C_k(t)$ then at the next time step, decrease $C_k(t)$ and *vice versa* in order to lessen the error energy.

Now, $$\left\|\underline{d}(t+1) - \hat{\underline{d}}(t+1)\right\|^2 = \|\underline{e}(t+1)\|^2$$

$$= \left\|\underline{d}(t+1) - \hat{\underline{d}}(t) - \Delta.\sum_{k=1}^{p} C_k(t)\underline{\underline{L}}_k(t)\underline{e}(t)\right\|^2$$

And thus, $\frac{\partial}{\partial C_k(t)}\|\underline{e}(t+1)\|^2$

$$= -2\Delta\left\langle \underline{d}(t+1) - \hat{\underline{d}}(t) - \Delta\cdot\sum_{m=1}^{p} C_k(t)\underline{\underline{L}}_m(t)e(t), \underline{\underline{L}}_k(t)e(t)\right\rangle$$

$$= -2\Delta(\underline{d}(t+1) - \hat{\underline{d}}(t))^T \underline{\underline{L}}_k(t)\underline{e}(t)$$

$$+2\Delta^2\sum_{m=1}^{p} C_m(t)\underline{e}(t)^T \underline{\underline{L}}_k{}^T(t)\underline{\underline{L}}_m(t)e(t)$$

❖❖❖❖❖

[6] Results in the LMS' Iteration:

$$\underline{C}(t+1) = \left[\underline{\underline{I}}_p - 2\Delta^2\mu\underline{\underline{F}}(\underline{q}(t), \underline{e}(t)\right]\underline{C}(t)$$

$$+2\mu\Delta\underline{\psi}\left(\underline{q}(t), \dot{\underline{q}}(t), \underline{e}(t), \underline{d}(t+1) - \underline{d}(t)\right)$$

where

$$\underline{\underline{F}}(\underline{q}(t), \dot{\underline{q}}(t),)^{e(t)} \equiv \underline{\underline{F}}(t) = \left(\underline{e}(t)^T \underline{\underline{L}}_k^T(t)\, \underline{\underline{L}}_m(t)\underline{e}(t)\right), 1 \le k, m \le p$$

Let $$\tilde{\underline{\underline{L}}}(t) = \tilde{\underline{\underline{L}}}\left(\underline{q}(t), \dot{\underline{q}}(t)\right) = \begin{pmatrix} \underline{\underline{L}}_1^T(t) \\ \underline{\underline{L}}_2^T(t) \\ \vdots \\ \underline{\underline{L}}_p^T(t) \end{pmatrix}^T$$

$$\underline{\underline{F}}(t) = (\underline{\underline{L}}_1(t), \underline{\underline{L}}_2(t), \ldots, \underline{\underline{L}}_p(t)). \in \mathbb{R}^{2\times 2p}$$

Then $\quad \underline{\underline{\tilde{L}}}(t)^T \underline{\underline{\tilde{L}}}(t) = ((\underline{\underline{L}}_k^T(t)\underline{\underline{L}}_m(t)))\ 1 \le k, m \le p \in \mathbb{R}^{2p\times 2p}$

and $\left(\underline{\underline{I}}_p \otimes \underline{e}(t)^T\right) (\underline{\underline{\tilde{L}}}(t)^T \underline{\underline{\tilde{L}}}(t)) \left(\underline{\underline{I}}_p \otimes \underline{e}(t)\right)$

$$= \underline{\underline{F}}(t).$$

❖❖❖❖❖

[7] Notation for ψ(*t*) in the LMS Disturbance Observer:

In the above formula,

$$\underline{\psi}(\underline{q}(t), \underline{\dot{q}}(t), \underline{e}(t), \underline{d}(t+1) - \underline{d}(t)) \equiv \underline{\psi}(t)$$

$$= \begin{pmatrix} \left(\underline{d}(t+1) - \underline{\hat{d}}(t)\right)^T & \underline{\underline{L}}_1(t)\underline{e}(t) \\ \vdots & \\ \left(\underline{d}(t+1) - \underline{\hat{d}}(t)\right)^T & \underline{\underline{L}}_p(t)\underline{e}(t) \end{pmatrix}$$

$$= \begin{pmatrix} \left(\underline{d}(t+1) - \underline{d}(t) + \underline{e}(t)\right)^T & \underline{\underline{L}}_1(t)\underline{e}(t) \\ \vdots & \\ \left(\underline{d}(t+1) - \underline{d}(t) + \underline{e}(t)\right)^T & \underline{\underline{L}}_p(t)\underline{e}(t) \end{pmatrix} \in \mathbb{R}^p$$

Note that

$$\underline{\psi}(t) = \begin{pmatrix} \underline{e}(t)^T \underline{\underline{L}}_1(t)^T & \left(\underline{d}(t+1) - \underline{d}(t)\underline{e}(t)\right) \\ \vdots & \\ \underline{e}(t)^T \underline{\underline{L}}_p(t)^T & \left(\underline{d}(t+1) - \underline{d}(t)\underline{e}(t)\right) \end{pmatrix}$$

$$= (I_p \otimes \underline{e}(t)^T)\underline{\underline{\tilde{L}}}(t)(\underline{d}(t+1) - \underline{d}(t) + \underline{e}(t))$$

❖❖❖❖❖

[8] Approximate Convergence Analysis:

Assume that the $\underline{\underline{L}}_k{}'s$ are constant 2 × 2 matrices.

Then so is $\tilde{L}(t)^T \tilde{L}(t) = \left\| \underline{\underline{L}}_K^T \underline{\underline{L}}_m \right\| \underline{\underline{\Delta G}}.$

We have $\underline{\underline{F}}(t) = (\underline{\underline{I}}_p \otimes \underline{e}(t)^T)\underline{\underline{G}}(\underline{\underline{I}}_p \otimes \underline{e}(t))$ and the LMS iteration simplifies to

$$\underline{C}(t+1) = \left[\underline{\underline{I}}_p - 2\mu\Delta^2(\underline{\underline{I}}_p \otimes \underline{e}(t)^T).\underline{\underline{G}}.(\underline{\underline{I}}_p \otimes \underline{e}(t))\right]\underline{C}(t)$$

$$+ 2\mu\Delta(\underline{\underline{I}}_p \otimes \underline{e}(t)^T)\underline{\underline{\tilde{L}}}(\underline{d}(t+1) - \underline{d}(t) + \underline{e}(t) \qquad (\alpha_1)$$

This must be combined with

$$\hat{\underline{d}}(t+1) = \hat{\underline{d}}(t)+\Delta.\left(\sum_{k=1}^{p} C_k(t)\underline{\underline{L}}_k\right)(\underline{d}(t)-\hat{\underline{d}}(t))$$

which is the same as

$$\underline{e}(t+1) = \underline{d}(t+1)-\underline{d}(t)+\underline{e}(t)-\Delta.\left(\sum_{k=1}^{p} C_k(t)\underline{\underline{L}}_k\right)\underline{e}(t)$$

$$= \left[\underline{\underline{I}}_2-\Delta\tilde{\underline{\underline{L}}}(\underline{C}(t)\otimes\underline{\underline{I}}_2)\right]\underline{e}(t)+\underline{d}(t+1)-\underline{d}(t) \qquad (\alpha_2)$$

❖❖❖❖❖

[9] Statistical Model for Disturbance Estimation Error Based on LMS:

Note that $\tilde{\underline{\underline{L}}}$ is now under our assumptions a constant $2\times 2p$ matrix.

The find set of coupled equations to be analyzed statistically are (α_1) and (α_2), which can be expressed as

$$\underline{e}(t+1) = \left[\underline{\underline{I}}_2-\Delta\tilde{\underline{\underline{L}}}\left(\underline{C}(t)\otimes\underline{\underline{I}}_2\right)\right]\underline{e}(t)+\varepsilon\underline{W}(t+1) \qquad (\alpha_3)$$

and

$$\underline{C}(t+1) = \left[\underline{\underline{I}}_p-2\mu\Delta^2(\underline{\underline{I}}_p\otimes\underline{e}(t)^T)\underline{\underline{G}}(\underline{\underline{I}}_p\otimes\underline{e}(t))\right]\underline{C}(t)$$

$$+2\mu\Delta(\underline{\underline{I}}_p\otimes\underline{e}(t)^T)\,\tilde{\underline{\underline{L}}}(\varepsilon\underline{W}(t+1)+\underline{e}(t)) \qquad (\alpha_4)$$

where

$$\varepsilon\underline{W}(t+1) = \underline{d}(t+1)-\underline{d}(t), \qquad (\alpha_5)$$

with the "perturbation parameter" ε being introduced to show that $\underline{d}(t+1)-\underline{d}(t)$, is a small disturbance.

❖❖❖❖❖

[10] Statistics of LMS Coefficients for the Disturbance Observer:

Note that $\underline{e}(t) = \underline{d}(t)-\hat{\underline{d}}(t)$.

Assuming that $\{\underline{e}(t)\}$ is an independent sequence it follows from (α_4) by taking expectation and assuming $\underline{e}(t)$ to be uncorrelation with $\underline{W}(t+1)$ that

$$\mathbb{E}\{\underline{C}(t+1)\} = \left[\underline{\underline{I}}_p-2\mu\Delta^2\mathbb{E}\left\{\left(\underline{\underline{I}}_p\otimes\underline{e}(t)^T\,\underline{\underline{G}}\left(\underline{\underline{I}}_p\otimes\underline{e}(t)\right)\right)\right\}\right]\mathbb{E}\{\underline{C}(t)\} \qquad (\alpha_6)$$

Note that $\mathbb{E}\{(\underline{\underline{I}}_p\otimes\underline{e}(t)^T).\underline{\underline{G}}.(\underline{\underline{I}}_p\otimes\underline{e}(t))\}$

$$= \left(\left(\mathbb{E}\left\{\underline{e}(t)^T\underline{\underline{L}}_k^T\underline{\underline{L}}_m\underline{e}(t)\right\}\right)\right)_{1\le k,m\le p}$$

$$= \left(\left(T_r(\underline{\underline{L}}_k^T \underline{\underline{L}}_m \cdot \underline{\underline{R}}_{ee}(t))\right)\right)_{1\le k,m\le p}$$

$$= \left(\left(T_r(\underline{\underline{L}}_m \cdot \underline{\underline{R}}_{ee}(t).\underline{\underline{L}}_k^T)\right)\right)_{1\le k,m\le p}$$

where $\underline{\underline{R}}_{ee}(t) = \mathbb{E}\left\{\underline{e}(t)\underline{e}^T(t)\right\} \equiv \underline{\underline{H}}_e(t)$ say.

Thus, $$\mathbb{E}\{\underline{C}(t+1)\} = \left[\underline{\underline{I}}_p - 2\mu\Delta^2 \underline{\underline{H}}_e(t)\right].\mathbb{E}\{\underline{C}(t+1)\} + 2\mu\Delta.\underline{f}_e(t) \quad (\alpha_7)$$

where $$\underline{f}_e(t) = \mathbb{E}\left\{\left(\underline{\underline{I}}_p \otimes \underline{e}(t)^T\right)\cdot \underline{\underline{\tilde{L}}}\cdot \underline{e}(t)\right\}$$

$$= \mathbb{E}\left(\left(\underline{e}(t)^T \underline{\underline{L}}_k\, \underline{e}(t)\right)\right)_{k=1}^{p} = \left(\left(T_r(\underline{\underline{L}}_k\, \underline{\underline{R}}_{ee}(t))\right)\right)_{k=1}^{p} \quad (\alpha_8)$$

We now approximate (α_3) by replacing $\underline{C}(t)$ with $\mathbb{E}\{\underline{C}(t+1)\}$. Then taking covariances gives

$$\underline{\underline{R}}_{ee}(t+1) = \left(\underline{\underline{I}}_2 - \Delta\underline{\underline{\tilde{L}}}\left(\mathbb{E}\{\underline{C}(t)\}\otimes \underline{\underline{I}}_2\right)\right)\cdot \underline{\underline{R}}_{ee}(t)\cdot$$
$$\left(\underline{\underline{I}}_2 - \Delta\left(\mathbb{E}\left\{\underline{C}^T(t)\right\}\otimes \underline{\underline{I}}_2\right)\underline{\underline{\tilde{L}}}^T\right) + \varepsilon^2 \underline{\underline{R}}_{ww}(t+1) \quad (\alpha_9)$$

where $$\underline{\underline{R}}_{ww}(t) = \mathbb{E}\left\{\underline{W}(t).\underline{W}^T(t)\right\}$$

$$= \mathbb{E}\left\{(\underline{d}(t) - \underline{d}(t-1)\cdot(\underline{d}(t) - \underline{d}(t-1)^T\right\} \approx 2\underline{\underline{R}}_{dd} \quad (\alpha_{10})$$

If we assume $\{\underline{d}(t)\}$ to be a stationary white sequence.

Reference:

[1] Vijayant Agrawal and H.Parthasarathy, "Lectures on Robotics" Technical Report, NSIT, 2014.

❖❖❖❖❖

[11] $\underline{B}(t) \in \mathbb{R}^d$ *d*-Vector Valued Standard Brownian Motion:

$G \simeq$ connected open set in $\mathbb{R}^d$.

$X \in G$, $Y \in \partial G$. By LDP

$$P\{\sqrt{\varepsilon}\underline{B}(\cdot) = \underline{f}(\cdot)\} \approx \exp\left(-\frac{1}{2\varepsilon}\int_0^T \left|\underline{\dot{f}}(t)\right|^2 dt\right)$$

where $$\underline{B}(\cdot) = \{\underline{B}(t) : 0 \le t \le T\}$$

$$\underline{f}(\cdot) = \left\{\underline{f}(t) : 0 \le t \le T\right\}$$

Let $$V_T(x,y) = \inf \frac{1}{2}\int_0^T \left|\underline{\dot{f}}(t)\right|^2 dt$$

$$\underline{f}(O) = x$$

$$\underline{f}(T) = y \qquad \text{for } x \in G`v,\ y \in \partial G$$

Let $$\tau_\varepsilon(x) = \inf\left\{t \geq 0 \middle| \sqrt{\varepsilon}\underline{B}(t) \in \partial G, \sqrt{\varepsilon}\underline{B}(0) = x\right\}$$

Let $$V_T(x) = \inf V_T(x,y),\ x \in G,\ y \in \partial \text{G}.$$

Then, $$\mathbb{E}\left[e^{\langle \underline{\lambda}.\underline{B}(\tau_\varepsilon(u))\rangle - \frac{1}{2}|\underline{\lambda}|^2 \tau_\varepsilon(x)}\right] = 1$$

(By Doob's optional stopping theorem for Martingale).

Also, by the same theorem since $\left|\underline{B}(t)\right|^2 - d \cdot t$ is a Martingale,

$$\mathbb{E}\left[\left|\underline{B}(\tau_\varepsilon(x))\right|^2 - d \cdot \tau_\varepsilon(x)\right] = x$$

Thus, $$\mathbb{E}\{\tau_\varepsilon(x)\} = \frac{1}{d}\mathbb{E}\left[\left|\underset{-x/d}{\underline{B}}(\tau_\in(x))^2\right|\right]$$

$$\lim_{\varepsilon \to 0} \varepsilon \log \mathbb{E}\{\tau_\varepsilon(x)\} = \lim_{\varepsilon \to 0} \varepsilon \log \mathbb{E}\left[\left|\underline{B}(\tau_\varepsilon(x))\right|^2\right]$$

Now, $$\mathbb{E}\left\{\varepsilon\left|\underline{B})(\tau_\varepsilon)\right|^2\right\} \underset{\varepsilon \to 0}{\approx} \sum_f \left|\underline{f}(\tau_\varepsilon)\right|^2 \cdot \exp\left(-\frac{1}{2\varepsilon}\int_0^{\tau_\varepsilon}\left|\underline{\dot{f}}(t)^2\right| dt\right)$$

Thus $$\varepsilon \log \mathbb{E}\left[\left|\underline{B}(\tau_\varepsilon)\right|^2\right] \underset{\varepsilon \to 0}{\approx} \varepsilon \log \sum_{\underline{f}} \left|\underline{f}(\tau_\varepsilon)\right|^2 \cdot \exp\left(-\frac{1}{2\varepsilon}\int_0^{\tau_\varepsilon}\left|\underline{\dot{f}}(t)\right|^2 dt\right)$$

$$\underset{\varepsilon \to 0}{\approx} \varepsilon \sup_{\underline{f}}\left(\log\left|\underline{f}(\tau_\varepsilon)\right|^2 - \frac{1}{2\varepsilon}\int_0^{\tau_\varepsilon}\left|\underline{\dot{f}}(t)\right|^2 dt\right)$$

$$\underset{\varepsilon \to 0}{\approx} \varepsilon \sup_{\underline{f}}\left[\varepsilon \log\left|\underline{f}(\tau_\varepsilon)\right|^2 - \frac{1}{2}\int_0^{\tau_\varepsilon}\left|\underline{\dot{f}}(t)\right|^2 dt\right]$$

$$= \sup_{y \varepsilon \partial G} \sup_{\substack{\underline{f}: \\ f(0)=x \\ f(T)=y, \\ T>0}} \left[\varepsilon \log\left[f\left|(T)\right|^2\right] - \frac{1}{2}\int_0^T \left|\underline{\dot{f}}(t)\right|^2 dt\right]$$

We first compute $X(T, \varepsilon) = \sup_{\substack{f \\ f(0)=x, \\ f(T)=y,}} \left[\varepsilon \log \left[\underline{f} |(T)|^2 \right] - \frac{1}{2} \int_0^T |\dot{\underline{f}}(t)|^2 \, dt \right]$

Using variational principles for fixed T:

$$
\begin{aligned}
0 &= \delta_f \left\{ \varepsilon \log |\underline{f}(T)|^2 - \frac{1}{2} \int_0^T |\dot{\underline{f}}(t)|^2 \, dt \right\} \\
&= 2\varepsilon \left(\underline{f}(T), \delta \underline{f}(T) \right) - \frac{1}{2} \int_0^T \left(\dot{\underline{f}}(t), \delta \dot{\underline{f}}(t) \right) dt \\
&= -\frac{1}{2} \int_0^T \left(\dot{\underline{f}}(t), \delta \dot{\underline{f}}(t) \right) dt
\end{aligned}
$$

(since $f(T) = y$ fixed $\Rightarrow$ $\delta \underline{f}(T) = 0$)

$$
= \frac{1}{2} \int_0^T \left(\ddot{\underline{f}}(t), \delta \underline{f}(t) \right) dt
$$

Thus, $\ddot{\underline{f}}(t) = 0, \; 0 \le t \le T, \; \underline{f}(0) = x, \underline{f}(T) = y$

So $\underline{f}(t) = \alpha t + \beta, \; \beta = x, \; \alpha T + \beta = y$

$\Rightarrow \quad \alpha = \dfrac{y - x}{T} \Rightarrow \dot{\underline{f}}(t) = \alpha = \dfrac{y - x}{T}.$

Thus $X(T, \varepsilon) = 2\varepsilon \log |\underline{y}| - \dfrac{1}{2T} |\underline{y} - \underline{x}|^2.$

❖❖❖❖❖

[12] General Relativistic Correction to GPS:

Gravitational effects of earth's gravity on photon propagation studies.

$(r_o, \phi_0) \simeq$ location of photon transmitter.

$(r_f, \phi_f) \simeq$ target location.

as measured by an observer a $r \longrightarrow \infty$.

Photon path:

$$
d\tau^2 = 0 = \alpha(r)\, dt^2 - \alpha^{-1}(r)\, dr^2 - r^2 d\phi^2
$$

$$
\mathcal{L} = \frac{d\tau}{d\lambda} = \tau' = \left(\alpha(r) t'^2 - \alpha^{-1}(r) r'^2 - r^2 \phi'^2 \right)^{1/2}
$$

Euler-Lagrange eqns: $$\alpha(r)\frac{dt}{d\tau} = E,$$

$$r^2\frac{d\phi}{d\tau} = h$$

$E, h \simeq$ infinite constants

$$\frac{r^2}{\alpha(r)}\frac{d\phi}{dt} = K = \frac{h}{E} \qquad \text{(finite constant)}$$

Null Condition

$$\alpha(r) - \alpha^{-1}(r)\left(\frac{dr}{dt}\right)^2 - r^2\left(\frac{d\phi}{dt}\right)^2 = 0$$

Trajectory eqn.

$$\frac{\alpha(r)}{K^2}\frac{r^4}{\alpha^2(r)} - \alpha^{-1}(r)\left(\frac{dr}{d\phi}\right)^2 - r^2 = 0$$

or $$\left(\frac{dr}{d\phi}\right)^2 + r^2\alpha(r) = \frac{r^4}{k^2}$$

$$r = \frac{1}{u},$$

$$\left(\frac{du}{d\phi}\right)^2 + u^2\alpha\left(\frac{1}{u}\right) = \frac{1}{k^2}$$

$$i.e.\left(\frac{du}{d\phi}\right)^2 + u^2(1-2mu) = \frac{1}{k^2} m = \frac{GM}{C^2}$$

Photon trajectory eqn.

$$\left.\frac{du}{d\phi}\right|_{\phi=\phi_0,\, r=r_0} = \eta \text{ recorded when photon arrives at target.}$$

Then $$\frac{1}{k^2} = \eta + \frac{1}{r_0^2}\left(1 - \frac{2m}{r_0}\right)$$

K calculate from this

$$\int_{u_0}^{u}\left(\frac{1}{k^2} - u^2(1-2mu)\right)^{-1/2} du = \phi - \phi_0$$

Gives u-$u(\phi)$ trajectory eqn. Equivalently,

$$r = r(\phi) = \frac{1}{u(\phi)}$$

Time (coordinate) to reach the target $\simeq t_f$.

$$\frac{r^2(\phi)}{\alpha(r(\phi))} = \frac{d\phi}{dt} = k = (\eta).$$

$$\int_{\phi_0}^{\phi_f} \frac{r^2(\phi)}{\alpha(r(\phi))} d\phi = k(\eta)t_f$$

ϕ_f calculated using t_f from this expression. Then $r_f = r(\phi_f)$ obtained from trajectory eqn.

Remark: Trajectory eqn. is

$$\left(\frac{du}{d\phi}\right)^2 + u^2 - 2mu^3 = \frac{1}{k^2}$$

In the absence of general relativistic correction term $2mu^3$, this is

$$\frac{du}{d\phi} = \pm\sqrt{a^2 - u^2},\ a = \frac{1}{k}$$

so
$$\sin^{-1}\left(\frac{u}{a}\right)\Bigg|_{uo}^{u} = \pm\phi\,\Big|_{\phi_0}^{\phi}$$

$$\sin^{-1}\left(\frac{u}{a}\right) - \sin^{-1}\left(\frac{u}{a}\right) = \pm(\phi - \phi_0)$$

$$\lambda\frac{u}{a} = \pm\sin\left(\phi - \phi_0 \pm \sin^{-1}\frac{u_0}{a}\right)$$

eqn. of a straight line.

❖❖❖❖❖

[13] Thiemann Modern QGTR:

$$H_a = -2q_{ac}D_b p^{bc}$$

$$H = \int[\lambda C + \lambda^a C_a + N^a H_a + |N|H]d^D X$$

$$= C(\lambda) + \vec{C}(\vec{\lambda}) + \vec{H}(\vec{N}) + H(|N|)$$

Poisson brackets: $\{C(f), H\}$

$$\frac{\delta S}{\delta \dot{N}^a} = \Pi_a\,\Pi_a = C_a$$

$$\frac{\delta S}{\delta \dot{N}} = \Pi,\ \Pi = C.\ \{N, \Pi\} = 1.$$

$$\{N_a, \Pi_b\} = \delta_{ab}.$$

$$\{C(f), H\} = \left\{\int f(x)C(x)\,dx, \int |N(x)|H(x)\,dx\right\}$$

$$= \int f(x)H(y)\{C(x),|N(y)|\}\,dx\,dy$$

$$= -\int f(x)H(x)\operatorname{sgn}(N(x))\,dx$$

$$= -H(sgn\,(N)\,f) = -H\left(\frac{N}{|N|}f\right)$$

$$\left\{\vec{H}(\vec{f}), \vec{H}\left(\vec{f}'\right)\right\}$$

$$= \int f_a(x)f_b'(y)\{H_a(x), H_b(y)\}\,dx\,dy$$

$$= 4\int f_a(x)f_b'(y)\left\{q_{ac}(x)\,P_e^{ec}(x), q_{bh}(y)P_{,k}^{kh}(y)\right\}dx\,dy$$

$$= 4\int f_a(x)f_b'(y)q_{bh}(y)\left\{q_{ac}(x)P_{,k}^{kh}(y)\right\}p_{,e}^{ec}(x)dx\,dy$$

$$+ 4\int f_a(x)f_b'(y)\,q_{ac}(x)\left\{P_{,e}^{ec}(x), q_{bh}(y)\right\}P_{,k}^{kh}(y)dx\,dy$$

$$= -4\int f_a(x)f_b'(y)\delta_a^k\,\delta_c^h\,\delta_{,k}(x-y)q_{bh}(y)p_{,e}^{ec}(x)\,dx\,dy$$

$$-4\int f_a(x)f_b'(y)q_{ac}^{(x)}\,\delta_b^e\,\delta_h^c\,\delta_{,e}(x-y)\,p_{,k}^{kh}(y)\,dx\,dy$$

$$= 4\int f'_{\;b}(x)q_{bc}^{(x)}\left(f_a(x)P_{,e}^{ec}(x)\right), a$$

$$+4\int\left(f_a(x)q_{ac}(x)\right), bf_b'(x)P_{,k}^{kc}(x)dx$$

$$= 4\int f_b'q_{bc}\left(f_{a,a}P_{,e}^{ec} + f_aP_{,ea}^{ec}\right)dx$$

$$+4\int(f_{a,b}\;q_{ac} + f_a\;q_{ac}, b)f_b'(x)P_{,k}^{kc}\;dx$$

$$= -4\int\left(f_b'\;q_{bc}\right)_{,a}^{fa}\;P_{,e}^{ec}dx + 4\int p_{,e}^{ec}\left(f_b'f_{a,b}q_{cc} + f_b'f_a\;q_{ac,\,b}\right)dx$$

$1-D$ quantum gravity gate.

$$d\tau^2 = A(t,x)\,dt^2 - B(t,x)\,dx^2$$

(Example radial component of Schwarz child matrix).

$$g_{oo} = A,\; g_n = -B.$$

$$R_{00} = \Gamma_{0\alpha,0}^{\alpha} - \Gamma_{00,\alpha}^{\alpha} - \Gamma_{00}^{\alpha}\Gamma_{\alpha\beta}^{\alpha} + \Gamma_{0\beta}^{\alpha}\Gamma_{0\alpha}^{\beta}$$

$$= \cancel{\Gamma_{00,0}^{0}} - \Gamma_{01,0}^{1} - \cancel{\Gamma_{00,0}^{0}} - \Gamma_{00,1}^{1} - \Gamma_{00}^{02} - \Gamma_{00}^{0}\,\Gamma_{01}^{1} - \cancel{\Gamma_{00}^{1}\Gamma_{10}^{0}}$$

$$\Gamma_{00}^{1}\,\Gamma_{11}^{1} + \cancel{\Gamma_{00}^{02}} + \cancel{\Gamma_{01}^{0}\Gamma_{00}^{1}} + \Gamma_{00}^{1}\Gamma_{01}^{0} + \Gamma_{01}^{12}$$

$$= \Gamma_{01,0}^{1} - \Gamma_{00,1}^{1} - \cancel{\Gamma_{00}^{02}} - \Gamma_{00}^{0}\,\Gamma_{01}^{1} - \Gamma_{00}^{1}\,\Gamma_{11}^{1} + \Gamma_{00}^{1}\,\Gamma_{01}^{0}$$

$$\delta\left(g^{\mu\nu}\sqrt{-g}\,\Gamma_{\mu a,\nu}^{\alpha}\right) \approx -\delta\left(\left(g^{\mu\nu}\sqrt{-g}\right)_{,\nu}\Gamma_{\mu\alpha}^{\alpha}\right)$$

$$\mathcal{L}_0 = -\Gamma^0_{00}\Gamma^1_{01} - \Gamma^1_{00}\Gamma^1_{11} + \Gamma^1_{00}\Gamma^0_{01}$$

$$\Gamma^0_{00} = \frac{1}{2}g^{00}g_{00,0} = \frac{1}{2}(\log A),_0$$

$$\Gamma^1_{01} = g''\Gamma_{101} = \frac{1}{2}g''\left(\cancel{g_{10,1}} + g_{11,0} - \cancel{g_{10,1}}\right) = \frac{1}{2}(\log B)_{,0},$$

$$\Gamma^1_{00} = g''\cancel{\Gamma_{100}} = \frac{1}{2}g''\left(g_{10,0} + g_{10,0} - g_{00,1}\right)$$

$$= \frac{A_{,1}}{2B}, \cancel{\Gamma^1_{00}} = g''\cancel{\Gamma_{100}} = \frac{1}{2}g''\left(g_{10,0} + g_{10,0} - g_{00,1}\right)$$

$$\Gamma^1_{11} = \frac{1}{2}g''g_{11,1}\ \frac{1}{2}(\log B)_{,1}$$

$$\Gamma^0_{01} = g^{00}\Gamma_{001} = \frac{1}{2}g^{00}g_{00,1} = \frac{1}{2}(\log A)_{,1}$$

$$\mathcal{L} = -\frac{1}{4}(\log A)_{,0}(\log B)_{,0} - \frac{A_{,1}}{4B}(\log B)_{,1} + \frac{A_{,1}}{4B}(\log A)_{,1}$$

$$4\mathcal{L}_0 = \frac{A^2_{,1}}{AB} - \frac{A_{,0}B_{,0}}{AB} - \frac{A_{,1}B_{,1}}{B^2}$$

$$R_{11} = \Gamma^\alpha_{1\alpha,1} - \Gamma^\alpha_{11,\alpha} - \Gamma^\alpha_{11}\Gamma^\beta_{1\alpha} + \Gamma^\alpha_{1\beta}\Gamma^\beta_{1\alpha} \approx \mathcal{L}_1$$

$$= -\Gamma^1_{11}\Gamma^0_{10} - \Gamma^0_{11}\Gamma^0_{00} + \Gamma^0_{11}\Gamma^1_{10}$$

$$A \leftrightarrow -B\ X \leftrightarrow t$$

$$4\mathcal{L}_1 = \frac{B^2_{,0}}{AB} - \frac{A_{,1}B_{,1}}{AB} - \frac{A_{,0}B_{,0}}{B^2}$$

$4\mathcal{L} \simeq 4(g^{00}\mathcal{L}_0 + g''\mathcal{L}_1)$ Lagrangian for the 1-D gravitational field.

$$\frac{\partial X^\mu(t,x)}{\partial t} = Nn^\mu + N^\mu$$

$$g_{\mu\nu}n^\mu n^\nu = s,$$

$$g_{\mu\varpi}n^\mu X^n_{,a} = 0,\ i.e.\ (n^\mu) \perp \textstyle\sum_t$$

$$(N^\mu) \parallel \textstyle\sum_t$$

Here $X^\mu_{,a}\, a = 1, 2, 3$ are tangent vectors to the spatial surface Σ_t. Spatial matrix on Σ_t:

$$q_{\mu\nu} = g_{\mu\nu} - s\,n_\mu n_\nu$$

Covariant derivative on Σ_t: D_μ

$$\text{Let } D_\mu u_\nu = q^\rho_\mu q^\sigma_\nu \nabla_\rho \mu_\sigma \text{ if } u_\mu n^\mu = 0.$$

Then $n^{\mu} D_{\mu} u_{v} = 0$ since $n^{\mu} q_{\mu}^{\rho} = 0.$

Ranging and lowering of indices are carried out using $g^{\mu v}$, $g_{\mu v.}$

Claim: $D_{\mu} q_{v\rho} = 0.$

Let $T_{\alpha\beta}$ spatial

i.e. $n^{\alpha} T_{\alpha\beta} = 0$

$n^{\beta} T_{\alpha\beta} = 0$

For $T_{\alpha\beta} = u_{\alpha} v_{\rho}$,

Then, $D_{\mu} T_{\alpha\beta} = D_{\mu}(u_{\alpha} v_{\beta})$

$$\rho = (D_{\mu} v_{\alpha})\, v_{\beta} + u_{\alpha} D_{\mu} v_{\beta}$$

$$= \left(q_{\mu}^{\rho}\, q_{\alpha}^{\sigma}\, \nabla_{\rho}\, u_{\sigma}\right) v_{\beta} + u_{\alpha} q_{\mu}^{\rho} \nabla_{\rho} v_{\sigma}$$

$$= q_{\mu}^{\rho} \left\{ q_{\alpha}^{\sigma} v_{\beta} \nabla_{\rho} u_{\sigma} + q_{\beta}^{\sigma} u_{\alpha} \nabla_{\rho} v_{\sigma} \right\}$$

$$= q_{\mu}^{\rho} \left\{ q_{\alpha}^{\sigma} v_{\beta} \left(u_{\sigma,\rho} - \Gamma_{e\sigma}^{v} u_{v}\right) + q_{\beta}^{\sigma} u_{\alpha} \left(v_{\sigma,\rho} - \Gamma_{e\sigma}^{v} u_{v}\right) \right\}$$

$$= q_{\mu}^{\rho}\, q_{\alpha}^{\sigma}\, u_{\sigma,\rho} v_{\beta} + q_{\mu}^{\rho}\, q_{\beta}^{\sigma} u_{\alpha} v_{\sigma,\rho} - q_{\mu}^{\rho}\, q_{\alpha}^{\sigma}\, \Gamma_{\rho\sigma}^{v}\, T_{v\beta} - q_{\mu}^{\rho}\, q_{\beta}^{\sigma}\, \Gamma_{\rho\sigma}^{\rho}\, T_{\alpha v}$$

Now, $q_{\mu}^{\rho}\, q_{\alpha}^{\sigma}\, u_{\sigma,\rho} v_{\beta} + q_{\mu}^{\rho}\, q_{\beta}^{\sigma} u_{\alpha} v_{\sigma,\rho}$

$$= q_{\mu}^{\rho}\, v_{\beta} ((q_{\alpha}^{\rho} u_{\sigma}), \rho - u_{\sigma}\, q_{\alpha,\rho}^{\sigma}) + q_{\mu}^{\rho} v_{\alpha} ((u_{\beta}^{\rho}\, u_{\sigma}) \rho - u_{\sigma} q_{\beta,\rho}^{\sigma})$$

$$= q_{\mu}^{\rho} v_{\beta} (u_{\alpha,\sigma} u_{\sigma} q_{\alpha,\rho}^{\sigma}) + q_{\mu}^{\rho} v_{\alpha} ((v_{\beta,\rho} - u_{\sigma}\, q_{\beta,\rho}^{\sigma})$$

$$= q_{\mu}^{\rho}\, T_{\alpha\beta,\rho} - q_{\mu}^{\rho} q_{\alpha,\rho}^{\rho}\, T_{\sigma\beta} - q_{\mu}^{\rho} q_{\beta,\rho}^{\sigma} T_{\alpha\sigma}$$

Equivalently,

$$D_{\mu}(T_{\alpha\beta}) = q_{\mu}^{\rho}\, \rho_{\alpha}^{\sigma} (\nabla_{\rho}\, u_{\sigma}) v_{\beta} + q_{\mu}^{\rho}\, q_{\beta}^{\sigma} (\nabla_{\rho} v_{\sigma}) u_{\alpha}$$

$$= q_{\mu}^{\rho}\, \rho_{\alpha}^{\sigma} (u_{\sigma,\rho} - \Gamma_{\rho\sigma}^{v}\, u_{v} \mid v_{\beta} + q_{\mu}^{\rho} q_{\beta}^{\sigma})(v_{\sigma,\rho} - \Gamma_{\sigma\rho}^{v} v_{v}) u_{\alpha}$$

$$= q_{\mu}^{\rho} ((u_{\alpha} v_{\beta}),_{\rho} - u_{\sigma} v_{\beta} q_{\alpha,\rho}^{\sigma} - u_{\alpha} v_{\sigma}\, q_{\beta,\rho}^{\sigma})$$

$$= -q_{\mu}^{\rho}\, q_{\alpha}^{\sigma}\, \Gamma_{\rho\sigma}^{v}\, u_{v} v_{\beta} - q_{\mu}^{\rho}\, q_{\beta}^{\sigma}\, \Gamma_{\rho\sigma}^{v}\, u_{\alpha} v_{v}$$

$$= \left(q_{\mu}^{v}\, u_{v} = u_{\mu} q_{\mu}^{v} v_{v} = v_{\mu} \because n^{v} u_{v} = 0, n^{v} v_{v} = 0\right)$$

or for spatial temous $T_{\alpha\beta}$,

$$D_{\mu} T_{\alpha\beta} = q_{\mu}^{\rho} \left(T_{\alpha\beta,\rho} - q_{\alpha,\rho}^{\sigma} T_{\sigma\beta} - q_{\beta,\rho}^{\sigma} T_{\alpha\sigma}\right)$$

$$-q_{\mu}^{\rho} \left(q_{\alpha}^{\rho}\, \Gamma_{\rho\sigma}^{v}\, T_{v\beta} + q_{\beta}^{\sigma}\, \Gamma_{\rho\sigma}^{\sigma}\, T_{\alpha v}\right)$$

$$g_{\mu\nu} = q_{\mu\nu} + s\, n_\mu n_\nu \qquad \textbf{Take } s = 1$$

$$\begin{aligned}
q_\mu^\rho q_\alpha^\sigma \Gamma_{\nu\rho\alpha} &= \frac{1}{2} q_\alpha^\rho q_\alpha^\sigma \left(g_{\nu\rho,\sigma} + g_{\nu\sigma,\rho} - g_{\rho\sigma,\nu}\right) \\
&= q_\mu^\rho q_\alpha^\sigma g_{\nu\rho,\sigma} - \frac{1}{2} q_\mu^\rho q_\alpha^\sigma g_{\rho\sigma,\nu} \\
&= q_\mu^\rho q_\alpha^\sigma q_{\nu\rho,\sigma} - \frac{1}{2} q_\mu^\rho q_\alpha^\sigma q_{\rho\sigma,\nu} + q_\mu^\rho q_\alpha^\sigma (n_\nu n_\rho)_{,\sigma} \\
&\quad - \frac{1}{2} q_\mu^\rho q_\alpha^\sigma \left(n_\rho n_\sigma\right)_{,\nu} q_\mu^\rho q_\alpha^\sigma n_\nu n_{\rho,\sigma} \qquad \left(\because q_\mu^\rho n_\rho = 0\right) \\
&= q_\alpha^\sigma n_\nu ((q_\mu^\rho n_\rho)_{,\sigma} - n_\rho q_{\mu,\sigma}^\rho) \\
&= -q_\alpha^\sigma q_{\mu,\sigma}^\rho n_\rho n_\nu
\end{aligned}$$

$$q_\mu^\sigma q_\alpha^\sigma \Gamma_{\rho\sigma}^\nu q_{\nu\beta} = q_\mu^\rho q_\alpha^\sigma \Gamma_{\nu\rho\sigma} \rho_\beta^\nu = 0$$

$$q_\beta^\nu n_\nu = 0.$$

Consider

$$\xi = \left(q_{\alpha,\rho}^\sigma q_{\sigma\beta} + \rho_{\beta,\rho}^\sigma q_{\alpha\sigma}\right) q_\mu^\rho$$

$$q_{\alpha,\rho}^\sigma = \left(\delta_\alpha^\sigma - n^\sigma n_\alpha\right)_{,\rho} = -\left(n^\sigma n_\alpha\right)_{,\rho}$$

$$= q_{\alpha,\rho}^\sigma q_{\sigma\beta} = -\left(n^\sigma n_\alpha\right)_{,\rho} q_{\sigma\beta} = -n_{,\rho}^\sigma n_\alpha q_{\sigma\beta}$$

So,

$$\begin{aligned}
\xi &= -q_\mu^\rho \left(q_{\sigma\beta} n_\alpha n_{,\rho}^\sigma + q_{\sigma\alpha} n_\beta n_{,\rho}^\sigma\right) \\
&= -q_\mu^\rho n_{,\rho}^\sigma \left(q_{\sigma\beta} n_\alpha + q_{\sigma\alpha} b_\beta\right)
\end{aligned}$$

❖❖❖❖❖

[14] Generalization of General Relativistic String Theory:

A p-dimension surface embedded in $\mathbb{R}^n$ has the equation $y^n(t, x', .., x^p)$, $1 \le n \le N$ at time t. Here $0 \le x', .., x^p \le 1$. The p-dimensional surface element is

$$dS = \sqrt{g}\, dx' \ldots dx^p$$

where

$$g_{\mu\nu} = \sum_{n=1}^{N} \frac{\partial y^n}{\partial x^\mu} \frac{\partial y^n}{\partial x^\nu}$$

and $g = det\,((g_{\mu\nu})) = g\,(t, \underline{x})$, $\underline{x} = ((x^\mu))_{\mu=1}^p$. The surface tension is F and so the potential energy of the surface due to stretchy is

$$V_t = \int F \sqrt{g(t,x)}\, dx' \ldots dx^p$$

and the kinetic energy is

$$T_t = \frac{\sigma}{2}\sum_{n=1}^{N}\int\left|\frac{\partial y^n}{\partial t}\right|^2 \sqrt{g(t,x)}\,dx'\ldots dx^p$$

Write down the eqns. of motion.

Problem: A sheet has the parametric equation

$(u, v) \rightarrow \underline{r}(t, u, v) \in \mathbb{R}^3$, $t \geq 0$, $0 \leq u, v \leq 1$.

Let σ be the surface man density of the sheet. The kinetic energy of the sheet is

$$T_t = \frac{1}{2}\sigma \int_{[0,1]^2} \left|\frac{\partial r}{\partial t}\right|^2 \left|\frac{\partial r}{\partial u} \times \frac{\partial r}{\partial v}\right| du\, dv$$

Let F denote the surface tension of the sheet. The potential energy of the sheet due to stretchy is

$$V_t = \int_{[0,1]^2} F\left(\left|\frac{\partial r}{\partial u} \times \frac{\partial r}{\partial v}\right| - 1\right) du\, dv$$

write the down the Euler-Lagrange eqns. of the sheet for the Lagragian

$$L = T_t - V_t$$

Hint: If $\mathcal{L} = \frac{1}{2}\sigma\left|\frac{\partial r}{\partial t}\right|^2 \left|\frac{\partial r}{\partial u} \times \frac{\partial r}{\partial v}\right| - F\left|\frac{\partial r}{\partial u} \times \frac{\partial r}{\partial v}\right|$ (Lagrangain density)

Then the Euler-Lagrange eqns. are

$$\frac{\partial}{\partial t}\frac{\partial \mathcal{L}}{d\underline{r}_{,t}} + \frac{\partial}{\partial u}\frac{\partial \mathcal{L}}{\partial \underline{r}_{,u}} + \frac{\partial}{\partial v}\frac{\partial \mathcal{L}}{\partial \underline{r}_{,v}} = 0.$$

Design of large sized quantum gates using charged strings perturbed by *em* fields.

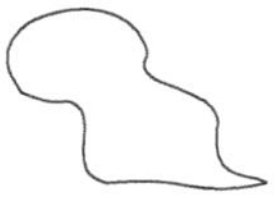

The string moves in the XY plane. Its parametric equation is $S \rightarrow (X(t, s), Y(t, s))$ at times t when $0 \leq s \leq 1$, $(X(t, 1), Y(t, 1)) = (X(t, 0), Y(t, 0))$.

Let σ be the linear change density on the string and $(A_x(x, y, t), (A_y(x, y, t), 0)$ be the magnetic vector potential, $\Phi(x, y, t)$ the electron scalar potential. The Lagragian is

$$\mathcal{L} = \frac{1}{2}\sigma_m \int_0^1 \left[\left(\frac{\partial X}{\partial t}\right)^2 + \left(\frac{\partial Y}{\partial t}\right)^2\right] ds$$

$$- \sigma_q \int_0^1 \Phi(x,y,t)\,ds + \sigma_q \int_0^1 \left[A_x \frac{\partial X}{\partial t} + A_y \frac{\partial Y}{\partial t}\right] ds$$

Note that ds is the length parameter:

$$\left(\frac{\partial X}{\partial s}\right)^2 + \left(\frac{\partial Y}{\partial s}\right)^2 = 1.$$

The classical eqns. of string motion are

$$\frac{\partial}{\partial t}\frac{\partial \mathcal{L}}{\partial X_{,t}} = \frac{\partial \mathcal{L}}{\partial X}, \frac{\partial}{\partial t}\frac{\partial \mathcal{L}}{\partial Y_{,t}} = \frac{\partial \mathcal{L}}{\partial Y}.$$

It easily follows that these Euler Lagrange eqns. decouple for different ρ. However, suppose we now include a string tension term. Then the Lagrangian is

$$\mathcal{L} = \frac{1}{2}\sigma_m \int_0^1 \left[\left(\frac{\partial X}{\partial t}\right)^2 + \left(\frac{\partial Y}{\partial t}\right)^2\right] ds$$

$$-\sigma_q \int_0^1 \Phi(x,y,t)\,ds + \sigma_q \int_0^1 \left(A_x \frac{\partial X}{\partial t} + A_y \frac{\partial Y}{\partial t}\right) ds$$

$$-\frac{1}{2}T\int_0^1 \left[\left(\frac{\partial X}{\partial s}\right)^2 + \left(\frac{\partial Y}{\partial s}\right)^2\right] ds$$

Let
$$X(t,s) = \sum_{n\in\mathbb{Z}} X_t[n]\exp(2\pi i\, ns)$$

$$Y(t,s) = \sum_{n\in\mathbb{Z}} Y_t[n]\exp(2\pi\, i\, n\, s)$$

Let the magnetic field $B_z(t,x,y) = A_{y,x} - A_{x,y}$ be constant in space, say $B_0(t)$. Then we can take

$$\underline{A}(t,x,v) = \frac{1}{2}B_0(t)(\hat{Z}\times \underline{r})$$

i.e.
$$A_x = \left(\frac{X}{2}B_0(t), \frac{Y}{2}B_0(t), 0\right)$$

Likewise assume that the electric field is also constant in space. Then we have approximately

$$\Phi(t,x,y) = (-E_0(t)X, -E_0(t)Y, 0).$$

$$\int_0^1 \left|\frac{\partial X}{\partial t}\right|^2 ds = \frac{1}{2}\sum_{n\in\mathbb{Z}} \left|\dot{X}_t[n]\right|^2$$

$$\int_0^1 \left|\frac{\partial Y}{\partial t}\right|^2 ds = \frac{1}{2}\sum_{n\in\mathbb{Z}} \left|\dot{Y}_t[n]\right|^2$$

$$\int_0^1 \left|\frac{\partial X}{\partial s}\right|^2 ds = \sum_n (2\pi n)^2 \left|X_t[n]\right|^2$$

$$\int_0^1 \left|\frac{\partial Y}{\partial s}\right|^2 ds = \sum_n (2\pi n)^2 \left|Y_t[n]\right|^2$$

$$\int_0^1 \left(A_x\frac{\partial X}{\partial t} + A_y\frac{\partial Y}{\partial t}\right) ds = \frac{B_o(t)}{2}\int_0^1 \left(-X\frac{\partial X}{\partial t} + Y\frac{\partial Y}{\partial t}\right) ds$$

$$\int_0^1 X\frac{\partial X}{\partial t}ds = \frac{1}{2}\frac{\partial}{\partial t}\int_0^1 X^2 ds = \frac{1}{2}\frac{\partial}{\partial t}\sum_{n\in\mathbb{Z}}\left|X_t[n]\right|^2$$

$$\int_0^1 Y\frac{\partial Y}{\partial t}ds = \frac{1}{2}\frac{\partial}{\partial t}\int_0^1 Y^2 ds$$

$$= \frac{1}{2}\frac{\partial}{\partial t}\sum_{n\in\mathbb{Z}}\left|Y_t[n]\right|^2$$

❖❖❖❖❖

[15] Symmetric Space-Times in General Relativity and Cosmology:

If a metric $g_{\mu\nu}(\bar{x})$ is invariant under a diffeomorphisn $X^\mu \xrightarrow{T} \bar{x}^\mu$, then the space-time metric is said to be symmetric *w.r.t.* *T*, *i.e.*

$$\bar{g}_{\mu\nu}(\bar{x}) = g_{\mu\nu}(\bar{x})$$

i.e. $g_{\mu\nu}(\bar{x})\dfrac{\partial\bar{x}^\mu}{\partial x^\alpha}\dfrac{\partial\bar{x}^\nu}{\partial x^\beta}g_{\alpha\beta}(x)$

Suppose $x^\mu \to \underline{x}^\mu + \xi^\mu(x)$ is an infinitesimal transformation *w.r.t.* which the metric $g_{\mu\nu}$ is symmetric. Then

$$g_{\mu\nu}(x+\xi(x))\left(\delta^\mu_\alpha + \xi^\mu_\alpha(x)\right)\left(\delta^\nu_\beta + \xi^\nu_\beta(x)\right) = g_{\alpha\beta}(x)$$

gives upto $O(\xi^\mu)$.

$$\left(g_{\mu\nu}(x) + g_{\mu\nu},_\rho(x)\xi^\rho(x)\right)\left(\delta^\mu_\alpha + \xi^\mu_{,\alpha}(x)\right)$$

$$\left(\delta^\nu_\beta + \xi,^\nu_\beta(x)\right) = g_{\alpha\beta}(x)$$

Or equivalently

$$g_{\mu\beta}\xi,^\mu_\alpha + g_{\nu\alpha}\xi,^\nu_\beta + g_{\alpha\beta,\rho}\xi^\rho = 0 \tag{1}$$

Now, **Theorem**:

$$\xi_{\mu:\nu} + \xi_{\nu:\mu} = \xi_{\mu:\nu} + \xi_{\nu:\mu} - 2\overline{\left|\mu\nu\right.}\xi_\alpha$$

$$= \xi_{\mu,\nu} + \xi_{\nu,\mu} - \left(g_{\alpha\mu,\nu} + g_{\alpha\nu,\mu} - g_{\mu\nu,\alpha}\right)\xi^\alpha = 0 \tag{2}$$

Iff (1) holds.

Proof: Assume (1). Then

$$\xi_{\mu:\nu} + \xi_{\nu:\mu} = g_{\alpha,\mu}\xi^\alpha_{:\nu} + g_{\nu\alpha}\ \xi^\alpha_{:\mu}$$

$$= g_{\mu\alpha}\left(\xi^\alpha_{,\nu} + \Gamma^\alpha_{\sigma\nu}\xi^\sigma\right) + g_{\nu\alpha}\left(\xi^\alpha_{,\mu} + \Gamma^\alpha_{\sigma\nu}\xi^\sigma\right)$$

$$= g_{\mu\alpha}\xi^\alpha_{,\nu} + g_{\nu\alpha}\xi^\alpha_{,\mu} + \left(\Gamma_{\mu\sigma\nu} + \Gamma_{\nu\sigma\mu}\right)\xi^\sigma$$

$$= g_{\mu\alpha}\xi^{\alpha}_{,\nu} + g_{\nu\alpha}\xi^{\alpha}_{,\mu} + g_{\mu\nu,\sigma}\xi^{\sigma} = 0$$

By (1). Hence (1) $\Rightarrow$ (2).

Conversely by reversing the above argument, we get (2) $\Rightarrow$ 1.

Definition: A vector field $\xi^{\mu}(x)$ such that $g_{\mu\nu}(x)$ is symmetric *w.r.t.* the group of diffeomorphisms (one parametric) generatric by ξ^{μ} is called a killing vector field. Thus the condition for $\xi^{\mu}_{(x)}$ to be a killing vector field is that

$$\xi_{\mu:\nu} + \xi_{\nu:\mu} = 0.$$

Quantum Signal Processing

Show that fidelity is invariant under an isometry.

Let ρ and σ be states of system A and $|\phi_\rho\rangle^{RA}$ and $|\phi_\sigma\rangle^{RA}$ purification of ρ and σ respectively.

By definition

$$F(\rho, \sigma) = \max_{|\phi_\rho\rangle^{RA}, |\phi_\sigma\rangle^{RA}} \left|\langle\phi_\rho|\phi_\sigma\rangle^{RA}\right|^2$$

The maximum is over all $(|\phi_\rho\rangle, |\phi_\sigma\rangle)$ such that

$$Tr_R(|\phi_\rho\rangle\langle\phi_\rho|) = \rho,\ Tr_R(|\phi_\sigma\rangle\langle\phi_\sigma|) = \sigma.$$

Let $U: \mathcal{H}_A \to \mathcal{H}_A$ be an isometry. Then

$$F(U\rho U^*, U\sigma U^*) = \max_{|\phi_\rho\rangle^{RA}, |\phi_\sigma\rangle^{RA}} \left|\langle\phi_\rho|(I\otimes U^*)(I\otimes U)|\phi_\sigma\rangle^{RA}\right|^2$$

since $((I\otimes U)|\phi_\rho\rangle, (I\otimes U)|\phi_\sigma\rangle)$ ranges over all joint purification of $(U\rho U^*, U\sigma U^*)$ as $(|\phi_\rho\rangle, |\phi_\sigma\rangle)$ ranges over all joint purifications of (ρ, σ)

Since, $(I\otimes U^*)(I\otimes U) = I\otimes U^*U = I$, we get

$$F(U\rho U^*, U\sigma U^*) = F(\rho, \sigma).$$

* Show that Fidelity is monotone under a noisy quantum operation $\mathcal{U}$:

$$F(\rho, \sigma) \le F(\mathcal{U}(\rho), \mathcal{U}(\sigma)).$$

Let $\quad \rho \xrightarrow{\mathcal{U}} \sum_{k=1}^{d} E_k\rho E_k^*,\ \sum_{k=1}^{d} E_k^* E_k = I$

be the noisy quantum operation. We've to show that

$$F(\rho, \sigma) \le F\left(\sum_k E_k\rho F_{-k}^*, \sum_k E_k\sigma E_k^*\right)$$

Now let be $\left(|\phi_\rho\rangle^{RA}, |\phi_\sigma\rangle^{RA}\right)$ a joint purification of (ρ, σ). Consider the pure state

$$|\psi_\rho\rangle^{SRA} = \sum_k |e_k\rangle^S \otimes (I_R \otimes E_k)|\phi_\rho\rangle^{RA}$$

where $\{|e_k\rangle\}_{k=1}^d$ is an ONB for $\mathcal{H}_S = \mathbb{C}^d$.

We have $Tr_{SR}\left\{|\psi_\rho\rangle^{SRA}\langle\psi_\rho|^{SRA}\right\}$

$$= \sum_{k,k'} Tr_{SR}\left\{|e_k\rangle^S (I_R \otimes E_k)|\phi_\rho\rangle^{RA}\langle e'_k|^S\langle\phi_\rho|^{RA}(I_R \otimes E^*_{k'})\right\}$$

$$= \sum_{k,k'} \langle e_{k'} | e_k\rangle^S Tr_R\left\{(I_R \otimes E_k)|\phi_\rho\rangle^{RA}\langle\phi_\rho|^{RA}(I_R \otimes E^*_{k'})\right\}$$

$$= \sum_k Tr_R\left\{(I_R \otimes E_k)|\phi_\rho\rangle^{RA}\langle\phi_\rho|^{RA}(I_R \otimes E^*_k)\right\}$$

$$= \sum_k E_k\rho E^*_k = \mu(\rho)$$

Likewise if

$$|\psi_\sigma\rangle^{SRA} = \sum_k |e_k\rangle^S \otimes (I_R \otimes E_k)|\phi_\sigma\rangle^{RA}$$

then $\quad Tr_{SR}\left(|\psi_\sigma\rangle^{SRA}\langle\psi_\sigma|^{SRA}\right) = \mu(\sigma).$

i.e. $|\psi_\rho\rangle^{SRA}$ and $|\psi_\sigma\rangle^{SRA}$ are respectively purification of $\mu(\rho)$ and $\mu(\sigma)$. Then,

$$F(\mu(\rho), \mu(\sigma)) \geq \left|\langle\psi_\rho||\psi_\sigma\rangle^{SRA}\right|^2$$

$$= \left|\sum_k \langle\phi_\rho|(I_R \otimes E^*_k)(I_R \otimes E_k)|\phi_\sigma\rangle^{RA}\right|^2$$

$$= \left|\langle\phi_\rho|I_R \otimes \sum_k E^*_k E_k|\phi_\sigma\rangle^{RA}\right|^2$$

$$= \left|\langle\phi_\rho||\phi_\sigma\rangle\right|^2$$

Taking maximum over all $\left(|\phi_\rho\rangle^{RA}, |\phi_\sigma\rangle^{RA}\right)$

gives $F(\mu(\rho), \mu(\sigma)) \geq F(\rho, \sigma)$.

We explain here some of the details of Belavkin's original paper on quantum filtering. Belavkin quantum filtering (contd.)

$$X(t) = \int_0^t (Z^b FZ - X \otimes \delta)^\nu_\lambda d\hat{A}^\lambda_\nu \qquad \text{(state processes)}$$

$$Y_i(t) = \int_0^t (Z^b D_i Z)^\lambda_\nu d\hat{A}^\nu_\lambda \qquad \text{(output processes)}$$

$1 \leq i \leq n$

$$\left[Y_i(t), Z^\lambda_\nu(s)\right] = 0, \ \left[Y_i(t), X(s)\right] = 0, \ t \leq s$$

$$\mathbb{D}_0^t = \left\{ D \in \mathcal{F}_t' \middle| D_-^- = 0 = D_+^+, \left(DZ_+ \middle| DZ_+ \right) = 0 \right\}$$

$$\mathcal{F}_t = \left\{ F \in \infty_t \middle| [D_i, F] = 0 \right\}$$

$$\left(DZ_+ \middle| DZ_+ \right) = \left\langle \xi \middle| (Z^b D^b DZ)_+^- \middle| \xi \right\rangle$$

$$Z = \begin{pmatrix} I & Z_0^- & Z_+^- \\ 0 & Z_0^0 & Z_+^0 \\ 0 & 0 & I \end{pmatrix} g = \begin{pmatrix} 0 & 0 & I \\ 0 & I & 0 \\ I & 0 & 0 \end{pmatrix}$$

$$Z^b = gZ^* g = \begin{pmatrix} 0 & 0 & I \\ 0 & I & 0 \\ I & 0 & 0 \end{pmatrix} \begin{pmatrix} I & 0 & 0 \\ Z_0^{-*} & Z_0^{0*} & 0 \\ Z_+^{-*} & Z_+^{0*} & I \end{pmatrix} \begin{pmatrix} 0 & 0 & I \\ 0 & I & 0 \\ I & 0 & 0 \end{pmatrix}$$

$$= \begin{pmatrix} Z_+^{-*} & Z_+^{0*} & I \\ Z_0^{-*} & Z_0^{0*} & 0 \\ I & 0 & 0 \end{pmatrix} \begin{pmatrix} 0 & 0 & I \\ 0 & I & 0 \\ I & 0 & 0 \end{pmatrix}$$

$$= \begin{pmatrix} I & Z_+^{0+} & Z_+^{-*} \\ 0 & Z_0^{0*} & Z_0^{-*} \\ 0 & 0 & I \end{pmatrix}$$

$$D = \begin{pmatrix} 0 & D_0^- & D_+^- \\ 0 & D_0^0 & D_+^0 \\ 0 & 0 & 0 \end{pmatrix}$$

$$DZ = \begin{pmatrix} 0 & D_0^- & D_+^- \\ 0 & D_0^0 & D_+^0 \\ 0 & 0 & 0 \end{pmatrix} \begin{pmatrix} I & Z_0^- & Z_+^- \\ 0 & Z_0^0 & Z_+^0 \\ 0 & 0 & I \end{pmatrix}$$

$$= \begin{pmatrix} 0 & D_0^- Z_0^0 & D_0^- Z_+^0 + D_+^- \\ 0 & D_0^0 Z_0^0 & D_0^0 Z_+^0 + D_+^0 \\ 0 & 0 & 0 \end{pmatrix}$$

$$Z^b D^b = gZ^* g g D^* g = g(DZ)^* g = (DZ)^b$$

$$= \begin{pmatrix} 0 & \left(D_0^0 Z_+^0 + D_+^0 \right)^* & \left(D_0^- Z_+^0 + D_+^- \right)^* \\ 0 & \left(D_0^0 Z_0^0 \right)^* & \left(D_0^- Z_0^0 \right)^* \\ 0 & 0 & 0 \end{pmatrix}$$

$$\left(Z^b D^b DZ \right)_+^- = (1, 3) \text{ element of } \left(Z^b D^b DZ \right)_+^-$$

$$= \left(D_0^0 Z_+^0 + D_+^0\right)^* \left(D_0^0 Z_+^0 + D_+^0\right)$$

Hence $D \in \mathbb{D}_0^t$

$$\Rightarrow \quad \left(D_0^0 Z_+^0 + D_+^0\right)\xi = 0$$

$$X(t) = \int_0^t \left(Z^b FZ - X \otimes \delta\right)_\lambda^\nu d\hat{A}_\nu^\lambda$$

Define M_t on ξ by

$$M_t\xi = \int_0^t (E_t - E_s)\left(Z^b FZ\right)_\nu^\lambda dA_\lambda^\nu \xi$$

Then for $0 < \tau < t$,

$$E_\tau M_t \xi = \int_0^\tau (E_\tau - E_s)\left(Z^b FZ\right)_\nu^\lambda dA_\lambda^\nu \xi$$

$$= M_\tau \xi$$

since $\quad E_\tau (E_t - E_s) = 0$ for $t > \delta > \tau$

$(E_\tau E_t = E_\tau,\ E_\tau E_s = E_\tau$ for $t > \tau,\ s < \tau)$

$$\int_0^t E_s \left(Z^b FZ\right)_+^- \xi ds + M_t\xi = E_t \int_0^t (Z^b FZ)_\nu^\lambda dA_\lambda^\nu \xi$$

since $E_s \left(Z^b FZ\right)_\nu^\lambda dA_\lambda^\nu \xi = E_s \left(Z^b FZ\right)_+^- \xi ds$

Observation processes

$D_i(t) \in \mathbb{D}^t$

$$dB(t) = \left(Z^b DZ\right)_\nu^\lambda dA_\lambda^\nu$$

$$D = D_0 + \lambda_s^i \tilde{D}_i$$

$$\varepsilon_t(X(t)) = \int_0^t \varepsilon s \left(Z^b FZ\right)_+^- ds + M_t$$

To get the filtering eqn, we need M_t in term of the observation processes $B(t)$. Let

$$M_t = \int_0^t \left(Z^b \tilde{D}_i Z\right)_\nu^\lambda (s) K_s^i dA_\lambda^\nu (s) K_t^i \in \infty_t' Y_i(t)$$

$$- \int_0^t \varepsilon_s \left(Z^b D_i Z\right)_+^- (s) ds$$

$$= \int_0^t \left[(Z^b D_i Z)_\lambda^\nu (s) d\hat{A}_\nu^\lambda (s) - \varepsilon_s \left(Z^b D_i Z\right)_+^- (s) ds\right]$$

$$= \int_0^t \left[(Z^b D_i Z)_+^- - \varepsilon_s (Z^b D_i Z)_+^-\right] ds$$

$$+\int_0^t \sum_{(v,\lambda)\neq(-+)} \left(Z^b D_i Z\right)_\lambda^v (s) d\hat{A}_v^\lambda(s)$$

$$Z^b DZ = \begin{pmatrix} I & Z_+^{0*} & Z_+^{-*} \\ 0 & Z_0^{0*} & Z_0^{-*} \\ 0 & 0 & I \end{pmatrix} \begin{pmatrix} 0 & D_0^- Z_0^0 & D_0^- Z_+^0 + D_+^- \\ 0 & D_0^0 Z_0^0 & D_0^0 Z_+^0 + D_+^0 \\ 0 & 0 & 0 \end{pmatrix}$$

$$= \begin{pmatrix} 0 & D_0^- Z_0^0 & D_0^- Z_+^0 + D_+^- + Z_+^{0*}\left(D_0^0 Z_+^0 + D_+^0\right) \\ 0 & Z_0^{0*} D_0^0 Z_0^0 & Z_0^{0*}\left(D_0^0 Z_+^0 + D_+^0\right) \\ 0 & 0 & 0 \end{pmatrix}$$

$$\begin{aligned} \varepsilon_s\left(Z^b DZ\right)_+^- &= D_0^- \varepsilon_s(Z_+^0) + D_+^- + \varepsilon_s\left(Z_+^{0*} D_0^0 Z_+^0\right) \\ &\quad + \varepsilon_s\left(Z_0^{-*}\right) D_+^0 \left(Z^b D_i Z\right)_+^- - \varepsilon_s\left(Z^b D_i Z\right)_+^- \\ &= D_0^-\left(Z_+^0 - \varepsilon_s\left(Z_+^0\right)\right) + \left(Z_0^{-*} - \varepsilon_s\left(Z_0^{-*}\right)\right) D_+^0 \\ &\quad + \left(Z_+^{0*} D_0^0 Z_+^0 - \varepsilon_s\left(Z_+^{0*} D_0^0 Z_+^0\right)\right) \end{aligned}$$

Compute
$$Z^b \tilde{D} Z = \begin{pmatrix} I & Z_+^{0*} & Z_+^{-*} \\ 0 & Z_0^{0*} & Z_0^{-*} \\ 0 & 0 & I \end{pmatrix} \begin{pmatrix} 0 & \tilde{D}_0^- Z_0^0 & \tilde{D}_0^- Z_+^0 + \tilde{D}_+^- \\ 0 & \tilde{D}_0^0 Z_0^0 & \tilde{D}_0^0 Z_+^0 + \tilde{D}_+^0 \\ 0 & 0 & 0 \end{pmatrix}$$

$$\begin{aligned} \hat{A}\left(Z^b \tilde{D} Z, ds\right) &= \left(\tilde{D}_0^- Z_0^0\right) d\hat{A}_-^0 + \left(\tilde{D}_0^- Z_+^0 + \tilde{D}_+^- + Z_+^{0*} \tilde{D}_0^0 Z_+^0 + Z_+^{0*} D_+^0\right) ds \\ &\quad + \left(Z_0^{0*} \tilde{D}_0^0 Z_0^0\right) d\hat{A}_0^0 + \left(Z_0^{0*} \tilde{D}_0^0 Z_+^0 + Z_0^{0*} \tilde{D}_+^0\right) d\hat{A}_0^+ \end{aligned}$$

Coeff. of ds in this is

$$\tilde{D}_+^- + \tilde{D}_0^- Z_+^0 + Z_+^{0*} \tilde{D}_0^0 Z_+^0 + Z_+^{0*} \tilde{D}_+^0$$

$$= D_+^- - \varepsilon_s\left(Z^b DZ\right)_+^- + D_0^- Z_+^0 + Z_+^{0*} D_0^0 Z_+^0 + Z_+^{0*} D_+^0$$

If we operate on this by ε_s, we get

$$D_+^- + \varepsilon_s\left(Z_+^{0*} D_0^0 Z_+^0\right) + \varepsilon_s\left(Z_+^{0*}\right) D_+^0 + D_0^- \varepsilon_s\left(Z_+^0\right) - \varepsilon_s(Z^b DZ)$$

$$= \varepsilon_s\left(Z^b DZ\right)_+^- - \varepsilon_s\left(Z^b DZ\right)_+^- = 0.$$

Hence $\int_0^t \hat{A}\left(Z^b \tilde{D} Z, ds\right)$ is a Martingale.

Also $Y(t) - \int_0^t \varepsilon_s\left(Z^b DZ\right)_+^- ds$

$$= \int_0^t \left(\left(Z^b DZ\right)_\mu^v (s) d\hat{A}_v^\mu(s) - \varepsilon_s\left(Z^b DZ\right)_+^-(s) ds\right) = \xi(t) \text{ say}$$

$$\begin{aligned}
d\xi(t) &= dY(t) - \varepsilon_t\left(Z^b DZ\right)_+^- dt \\
&= D_0^- Z_0^0 d\hat{A}_-^0 + \left(D_0^- Z_+^0 + D_+^- + Z_+^{0*} D_0^0 Z_+^0 \right. \\
&\quad \left. + Z_+^{0*} D_+^0 - \varepsilon_t\left(Z^b DZ\right)_+^-\right) dt \\
&\quad + Z_0^{0*} D_0^0 Z_0^0 d\hat{A}_0^0 + \left(Z_0^{0*} D_0^0 Z_+^0 + Z_0^{0*} D_+^0\right) d\hat{A}_0^+ \\
&= \tilde{D}_0^- Z_0^0 d\hat{A}_-^0 + \left(\tilde{D}_0^- Z_+^0 + \tilde{D}_+^- + Z_+^{0*}\tilde{D}_0^0 Z_+^0 + Z_+^{0*} D_+^0\right) dt \\
&\quad + Z_0^{0*}\tilde{D}_0^0 Z_0^0 d\hat{A}_0^0 + \left(Z_0^{0*}\tilde{D}_0^0 Z_+^0 + Z_0^{0*}\tilde{D}_+^0\right) d\hat{A}_0^0
\end{aligned}$$

since
$$\tilde{D}_\nu^\mu = D_\nu^\mu \text{ if } (\mu, \nu) \neq (-+),$$
$$\tilde{D}_+^- = D_+^- - \varepsilon_t\left(Z^b DZ\right)_+^-$$

Quantum Filtering

$$dX = C_\nu^\mu \, dA_\mu^\nu \quad A_+^- = t$$
$$A_+^{j*} = A_j^-, \; A_j^{-*} = A_+^j, \; A_k^{j*} \simeq \text{creation operators}$$
$$A_k^{j*} = A_j^k \;\; A_j^- \simeq \text{annilutation operators}$$
$$dA_+^j dA_k^- = \delta_k^j dt = \delta_k^j dA_+^- \;\; A_k^j \simeq \text{conservation operators}$$
$$dA_k^j dA_l^- = \delta_l^j dA_k^-$$
$$dA_+^j dA_l^k = \delta_l^j dA_+^k$$

$\{\infty_t, t \geq 0\}$ increasing family of von-Neumann algebras. $\mathbb{B}_t \subset \infty_t \cap \bigcap_{s \geq t} \infty_s'$

i.e. $Y \in \mathbb{B}_t \Rightarrow Y \in \infty_t$ and $[Y, X] = 0 \;\; \forall X \in \infty s > t$.

Conditional expectation: Let $X \in \infty_s, s > t$.

Assume $\{\infty_s\}_{t \geq 0}$ operate on $\mathcal{H}$ a Hilbert space.

$\varepsilon_t(X) \in \mathbb{B}_t$ is such that

$\varepsilon_t(X)\xi = E_t(X\xi)$ where E_t is the orthogonal projection of ∞_∞ onto $\overline{\mathbb{B}_t\xi}$.

$$(X - \varepsilon_t(X))\xi = X\xi - E_t(X\xi) \perp \overline{\mathbb{B}_t\xi}.$$

Let $A, B \in \mathbb{B}_t, X \in \infty_s, s > t$

$$\varepsilon_t(AX)\xi = E_t(AX\xi)$$
$$AX\xi - E_t(AX\xi) \perp \mathbb{B}_t\xi$$
$$X\xi - E_t(X\xi) \perp \mathbb{B}_t\xi$$

Second one implies

$$UX\xi - UE_t(X\xi) \perp U\mathbb{B}_t\xi = \mathbb{B}_t\xi$$

$\forall$ Unitary $U \in \mathbb{B}_t$

$\Rightarrow$ $AX\xi - A\,E_t(X\xi) \perp \mathbb{B}_t\xi$

Thus $E_t(AX\xi) - A\,E_t(X\xi) \perp \mathbb{B}_t\xi$

But $E_t(AX\xi) - A\,E_t(X\xi) \subset \mathbb{B}_t\xi$

Hence $E_t(AX\xi) = A\,E_t(X\xi)$

or $\varepsilon_t(AX)\xi = A\,\varepsilon_t(X)\xi$

It fixed ε_t exists $\forall\xi$, then it follows that

$$\varepsilon_t(AX) = A\varepsilon_t(X)$$

Let the measurement model be

$$dY = D_\nu^\mu dA_\mu^\nu$$

i.e.

$$Y(t) = \int_0^t D_\nu^\mu(s)dA_\mu^\nu(s),$$

$$X(t) = \int_0^t C_\nu^\mu(s)dA_\mu^\nu(s)$$

For unitary evolution of X, we require

$$0 = d(X^*X) = dX^*.X + X^*.dX + dX^*.dX$$

$$= C_\nu^{\mu*}C_\beta^\alpha dA_\mu^{\nu*}dA_\alpha^\beta + C_\nu^{\mu*}XdA_\mu^{\nu*} + X^*C_\nu^\mu dA_\mu^\nu$$

$$C_\nu^{\mu*}dA_\mu^{\nu*} = C_\nu^{b\mu}dA_\mu^\nu$$ (Belavkin-Quantum non-linear filtering)

So for unitarity, we require

$$0 = C_\nu^{b\mu}C_\beta^\alpha\delta_\alpha^\nu dA_\mu^\beta + C_\nu^{b\mu}XdA_\mu^\nu + X^*C_\nu^\mu dA_\mu^\nu$$

i.e. $C_\beta^{b\mu}C_\nu^\beta + C_\nu^{b\mu}X + X^*C_\nu^\mu = 0.$

Let $\xi = f\otimes e(0)\, f\in h$ (System Hilbert space and $e(0)$ = Vacuum exponential vector in $\Gamma_s(\mathcal{H})$.

$$X(t) = \int_0^t C_\nu^\mu(s)dA_\mu^\nu(s)$$

$$dA_\mu^\nu(t)\xi = C_+^-(t)\xi dt$$

Non-demolition condition:

$$[Y(t), X(s)] = 0\ \forall s \geq t.$$

Thus $$\left[\int_0^t D_\nu^\mu(t_1)dA_\mu^\nu(t_1), \int_0^s C_\nu^\mu(t_2)dA_\mu^\nu(t_2)\right] = 0\ \forall s \geq t$$

This condition is satisfied if

$$(a)\qquad \left[D_{\nu}^{\mu}(t)dA_{\mu}^{\nu}(t),\, C_{\beta}^{\alpha}(t)dA_{\alpha}^{\beta}(t)\right] = 0$$

$$\text{or}\quad D_{\nu}^{\mu}(t)C_{\beta}^{\alpha}(t)\,\delta_{\alpha}^{\nu}dA_{\mu}^{\beta}(t) - C_{\beta}^{\alpha}(t)D_{\nu}^{\mu}(t)\,\delta_{\mu}^{\beta}dA_{\alpha}^{\nu}(t) = 0$$

$$\text{or}\quad D_{\nu}^{\mu}(t)C_{\beta}^{\nu}(t)\,dA_{\mu}^{\beta}(t) - C_{\mu}^{\alpha}(t)D_{\nu}^{\mu}(t)\,dA_{\alpha}^{\nu}(t) = 0$$

$$\text{or}\quad D_{\nu}^{\mu}(t)C_{\beta}^{\nu}(t) - C_{\alpha}^{\mu}(t)D_{\beta}^{\alpha}(t) = 0$$

$$\text{or}\quad D_{\nu}^{\mu}C_{\beta}^{\nu} - C_{\nu}^{\mu}D_{\beta}^{\nu} = 0$$

$$(b)\qquad \left[D_{\nu}^{\mu}(t)dA_{\mu}^{\nu}(t),\, C_{\beta}^{\alpha}(s)dA_{\alpha}^{\beta}(s)\right] = 0,\ s > t.$$

$$\text{or}\quad \left\langle e(u), \left[D_{\nu}^{\mu}(t), dA_{\mu}^{\nu}(t), C_{\beta}^{\alpha}(s)dA_{\alpha}^{\beta}(s)\right]e(v)\right\rangle = 0\ \forall s > t,\ u, v\in\mathcal{H}.$$

$$\text{or}\quad \left\langle dA_{\mu}^{\nu *}(t)e(u)\middle|D_{\nu}^{\mu}(t)C_{\beta}^{\alpha}(s)\middle|dA_{\alpha}^{\beta}(s)e(v)\right\rangle$$
$$-\left\langle dA_{\alpha}^{\beta}(s)^{*}e(u)\middle|C_{\beta}^{\alpha}(s)D_{\nu}^{\mu}(t)\middle|dA_{\mu}^{\nu}(t)e(v)\right\rangle = 0,\ s > t,\ u, v\in\mathcal{H}.$$

(b) requires

$$D_{\nu}^{\mu}(t)dA_{\mu}^{\nu}(t)C_{\beta}^{\alpha}(s)dA_{\alpha}^{\beta}(s)$$
$$- C_{\beta}^{\alpha}(s)dA_{\alpha}^{\beta}(s)D_{\nu}^{\mu}(t)dA_{\mu}^{\nu}(t) = 0\ s > t$$

This is satisfied if

$$\left[dA_{\mu}^{\nu}(t), C_{\beta}^{\alpha}(s)\right] = 0\ \forall s > t$$

$$\text{and}\quad \left[D_{\nu}^{\mu}(t), C_{\beta}^{\alpha}(s)\right] = 0.$$

Assume that the non-demolition condition is satisfied. Consider the algebra $\mathcal{B}_t$ generated by the operators $\{Y(s): s \le t\}$.

$$X(t) = X(s) + \int_s^t C_{\nu}^{\mu}(\tau)dA_{\mu}^{\nu}(\tau)$$

$$Y(t) = Y(s) + \int_s^t D_{\nu}^{\mu}(\tau)dA_{\mu}^{\nu}(\tau),\ t \ge s$$

Then

$$Y(t) = Y(s) + \int_s^{s} D_{+}^{-}(\tau)d\tau + \int_s^t \Big(D_{j}^{+}(\tau)dA_{+}^{j}(\tau)\,.$$
$$+ D_{-}^{j}(\tau)dA_{j}^{-}(\tau) + D_{k}^{j}(\tau)dA_{j}^{k}(\tau)\Big)$$

Now,

$$dA_{j}^{-}(\tau)\xi = dA_{j}^{-}(\tau)f\otimes|e(0)\rangle$$
$$= f\otimes dA_{j}^{-}(\tau)|e(0)\rangle = 0$$

$$dA_{j}^{k}(\tau)\xi = dA_{j}^{k}(\tau)f\otimes|e(0)\rangle$$
$$= f\otimes dA_{j}^{k}(\tau)|e(0)\rangle = 0$$

$$\left\langle e(u)\middle|dA_{+}^{j}(\tau)\middle|e(0)\right\rangle = \left\langle dA_{j}^{-}(\tau)e(u)\middle|e(0)\right\rangle$$

$$= u_j(\tau)d\tau\langle e(u)|e(0)\rangle = u_j(\tau)\mathrm{d}\tau$$

Now, consider $\quad M_1(t) = \int_0^t D_-^j(\tau)dA_j^-(\tau)$

$$\varepsilon_s\left(D_-^j(s)dA_j^-(s)\right)\xi = E_s\left(D_-^j(s)dA_j^-(s)\xi\right) = 0$$

Since $\xi = f\otimes|e(0)\rangle$. Thus

$$\varepsilon_s\left(D_-^j(s)dA_j^-(s)\right) = 0$$

when acting on vectors of the form $f\otimes|e(0)\rangle, f\in\hbar$.

Thus,

$$\varepsilon_t(M_1(t+dt)) = M_1(t)$$

when acting on $\mathcal{B}_t\xi$ since $D_-^j(t)dA_j^-(t)\mathbb{B}_t\xi$

$$= D_-^j(t)\mathbb{B}_t dA_j^-(t)\xi = 0 \qquad \because dA_j^-(t)$$

Commutes with $\mathbb{B}_t$.

Likewise define

$$M_2(t) = \int_0^t D_k^j(\tau)dA_j^k(\tau)$$

$$\varepsilon_t(M_2(t+dt))\xi = M_2(t)\xi + \varepsilon_t(D_k^j(t)dA_j^k(t))\xi$$
$$= M_2(t)\xi$$

More generally,

$$\varepsilon_t(M_2(t+dt))B\xi = M_2(t)B\xi + \varepsilon_t(D_k^j(t)dA_j^k(t)B)\xi$$

$$= M_2(t)B\xi + E_t(D_k^j(t)BdA_j^k(t)\xi)$$

$$= M_2(t)B\xi,\ B\in\mathbb{B}t$$

Note that

$$\mathbb{B}_t = Clspan\left\{D_\nu^\mu(s)dA_\mu^\nu(s) : s < t\right\}$$

$$= Clspan\left\{\int_0^s D_\nu^\mu(\tau)dA_\mu^\nu(\tau) : s \le t\right\}$$

Thus $M_2^{1.1}$ is also a Martingale.

Belavkin contd. (Nonlinear filtering)

$$d\widehat{X} = \left(\widehat{F} - \widehat{X}\otimes\delta\right)d\widehat{A}$$

$$\Rightarrow \qquad d\left(\widehat{X}^*\widehat{X}\right) = \left(\widehat{F}^b\widehat{F} - \widehat{X}^*\widehat{X}\otimes\delta\right)d\widehat{A}$$

$\widehat{X}$ = Unitary iff $\widehat{F}^b = \widehat{F}^{-1}$

Measurement (output) process

$$dY = \left(Z^b GZ - Y \otimes \delta\right) d\hat{A}$$

Let $\hat{X} = U,\ F_\nu^\mu = UZ_\nu^\mu$. Then,

$$dU = \left(UZ_\nu^\mu - U\delta_\nu^\mu\right) d\hat{A}_\mu^\nu$$

$$= U\left(Z_\nu^\mu - I\delta_\nu^\mu\right) d\hat{A}_\mu^\nu$$

$U =$ Unitary iff $\left(U(Z \otimes \delta)\right)^b = \left(U(Z \otimes \delta)\right)^{-1}$

$$\left(U(Z \otimes \delta)\right)_-^{b-} = Z_-^- U^*$$

$$Z \oplus \delta = \begin{pmatrix} Z_-^- & Z_0^- & Z_+^- \\ Z_-^0 & Z_0^0 & Z_+^0 \\ Z_-^+ & Z_0^+ & Z_+^+ \end{pmatrix}$$

$$U(Z \oplus \delta) = \begin{pmatrix} UZ_-^- & UZ_0^- & UZ_+^- \\ UZ_-^0 & UZ_0^0 & UZ_+^0 \\ UZ_-^+ & UZ_0^+ & UZ_+^+ \end{pmatrix}$$

$$(U(Z \otimes \delta))^b = \begin{pmatrix} Z_+^{+*}U^* & Z_+^{0*}U^* & Z_+^{-*}U^* \\ Z_0^{+*}U^* & Z_0^{0*}U^* & Z_0^{-*}U^* \\ Z_-^{+*}U^* & Z_-^{0*}U^* & Z_-^{-*}U^* \end{pmatrix}$$

$$= \begin{pmatrix} U^* & Z_+^{0*}U^* & Z_+^{-*}U^* \\ 0 & Z_0^{0*}U^* & Z_0^{-*}U^* \\ 0 & 0 & U^* \end{pmatrix}$$

$$U^* = U^{-1}\,(U(Z \otimes \delta))^{-1} = (Z \otimes \delta)^{-1} U^*$$

$$= \begin{pmatrix} I & Z_0^- & Z_+^- \\ 0 & Z_0^0 & Z_+^0 \\ 0 & 0 & U^* \end{pmatrix}^{-1} U^*$$

For U to be unitary, we require $\hat{F}^b = \hat{F}^{-1}$, *i.e.*,

$$\begin{pmatrix} I & Z_+^{0*} & Z_+^{-*} \\ 0 & Z_0^{0*} & Z_0^{-*} \\ 0 & 0 & I \end{pmatrix} = \begin{pmatrix} I & Z_0^- & Z_+^- \\ 0 & Z_0^0 & Z_+^0 \\ 0 & 0 & I \end{pmatrix}^{-1}$$

or $$\begin{pmatrix} I & Z_0^- & Z_+^- \\ 0 & Z_0^0 & Z_+^0 \\ 0 & 0 & I \end{pmatrix}\begin{pmatrix} I & Z_+^{0*} & Z_+^{-*} \\ 0 & Z_0^{0*} & Z_+^{0*} \\ 0 & 0 & I \end{pmatrix} = \begin{pmatrix} I & 0 & 0 \\ 0 & I & 0 \\ 0 & 0 & I \end{pmatrix}$$

or $$\begin{aligned} Z_+^{0*} + Z_0^- Z_0^{0*} &= 0, \\ Z_+^{-*} + Z_0^- Z_0^{-*} + Z_+^- &= 0, \\ Z_0^0 Z_0^{0*} &= I, \\ Z_0^0 Z_0^{-*} + Z_+^0 &= 0, \end{aligned}$$

These are equivalent to

$$\begin{aligned} Z_0^{0*} &= Z_0^{0-1},\ Z_+^{0*} = -Z_0^- Z_0^{0*} \\ Z_+^0 &= -Z_0^0 Z_0^{-*}, \\ Z_0^{0*} &= Z_0^{0-1},\ Z_0^{-*} = -Z_0^{0-1} Z_+^0 = Z_0^{0-1} Z_0^0 Z_0^{-*} \text{ consistent}, \\ Z_+^- + Z_+^{-*} &= -Z_0^- Z_0^{-*} \end{aligned}$$

Let $$\hat{Y} = UYU^*$$

Then $$\begin{aligned} d\hat{Y} &= dU.Y.U^* + U.dY.U^* + U.Y.dU^* \\ &\quad + dUdY.U^* + UdYdU^* + dUYdU^*. \end{aligned}$$

$$\begin{aligned} dU.Y.U^* &= U\left(Z_\nu^\mu - I\delta_\nu^\mu\right)YU^* d\hat{A}_\mu^\nu \\ U.dY.U^* &= U\left(Z^b GZ - Y\otimes\delta\right)_\nu^\mu U^* d\hat{A}_\mu^\nu \\ &= U\left(\left(Z^b GZ\right)_\nu^\mu - Y\delta_\nu^\mu\right)U^* d\hat{A}_\mu^\nu \\ U.Y.dU^* &= U.Y.\left(Z_\nu^{\mu*}U^* - U^*\delta_\nu^\mu\right)d\hat{A}_\nu^\mu \\ dU.Y.U^* &= \left(UZ_\nu^\mu - U\delta_\nu^\mu\right)\left(\left(Z^b GZ\right)_\sigma^\rho - Y\delta_\sigma^\rho\right)U^* d\hat{A}_\mu^\nu d\hat{A}_\rho^\sigma \\ &= \left(UZ_\nu^\mu - U\delta_\nu^\mu\right)\left(\left(Z^b GZ\right)_\sigma^\rho - Y\delta_\sigma^\rho\right)\delta_\rho^\nu U^* d\hat{A}_\mu^\sigma \\ &= \left(UZ_\nu^\mu - U\delta_\nu^\mu\right)\left(\left(Z^b GZ\right)_\sigma^\nu - Y\delta_\sigma^\nu\right)U^* d\hat{A}_\mu^\sigma \\ &= \Big[U\left(ZZ^b GZ\right)_\sigma^\mu U^* - UZ_\sigma^\mu YU^* \\ &\quad - U\left(Z^b GZ\right)_\sigma^\mu U^* + UYU^*\delta_\sigma^\mu\Big]d\hat{A}_\mu^\sigma. \end{aligned}$$

Remarks from "Quantum Stochastic Calculus and Quantum Nonlinear Filtering" by V.P. Belavkin (1992).

$$C = \begin{pmatrix} 0 & C_0^- & C_+^- \\ 0 & C_0^0 & C_+^0 \\ 0 & 0 & 0 \end{pmatrix}, C^b = \begin{pmatrix} 0 & C_+^{0*} & C_+^{-*} \\ 0 & C_0^{0*} & C_0^{-*} \\ 0 & 0 & 0 \end{pmatrix}$$

$$g = \begin{pmatrix} 0 & 0 & 1 \\ 0 & 1 & 0 \\ 1 & 0 & 0 \end{pmatrix}$$

$$C^+ = \begin{pmatrix} 0 & 0 & 0 \\ C_0^{-*} & C_0^{0*} & 0 \\ C_+^{-*} & C_+^{0*} & 0 \end{pmatrix}$$

$$gC^+g = \begin{pmatrix} 0 & 0 & 1 \\ 0 & 1 & 0 \\ 1 & 0 & 0 \end{pmatrix}\begin{pmatrix} 0 & 0 & 0 \\ C_0^{-*} & C_0^{0*} & 0 \\ C_0^{-*} & C_0^{0*} & 0 \end{pmatrix}\begin{pmatrix} 0 & 0 & 1 \\ 0 & 1 & 0 \\ 1 & 0 & 0 \end{pmatrix}$$

$$= \begin{pmatrix} C_+^{-*} & C_+^{0*} & 0 \\ C_0^{-*} & C_0^{0*} & 0 \\ 0 & 0 & 0 \end{pmatrix}\begin{pmatrix} 0 & 0 & 1 \\ 0 & 1 & 0 \\ 1 & 0 & 0 \end{pmatrix}$$

$$= \begin{pmatrix} 0 & C_+^{0*} & C_+^{-*} \\ 0 & C_0^{0*} & C_0^{-*} \\ 0 & 0 & 0 \end{pmatrix} = C^b$$

$$A(C, t) = C_+^- t + C_0^- A\_(t) + C_+^0 A^+(t) + C_0^0 N(t)$$

$$dA\_(t) dA^+(t) = dt$$

So $$[A\_(t), A^+(t')] = t \Lambda t'$$

It $a(u)$, $a^+(u)$ are the annihilation and creation operator fields, then we know that

$$[a(u), a^+(v)] = <u, v> \qquad (1)$$

$e(u)$ is the exponential vector, so

$$a(u)e(v) = <u, v>\, e(v).$$

$$a^+(u)e(v) = \frac{d}{dt} e(v + tu)\Big|_{t=0}$$

$$\langle e(w_1), a(u)a^+(v)e(w_2)\rangle = \frac{\partial^2}{\partial t_1 \partial t_2}\langle e(w_1 + t_1 u), e(w_2 + t_2 v)\rangle\Big|_{t_1 = t_2 = 0}$$

$$= \frac{\partial^2}{\partial t_1 \partial t_2} \exp(\langle w_1 + t_1 u, w_2 + t_2 v\rangle)\Big|_{t_1 = t_2 = 0}$$

$$= \frac{\partial}{\partial t_2}\langle u, w_2 + t_2 v\rangle \exp(\langle w_1, w_2 + t_2 v\rangle)\Big|_{t_2 = 0}$$

$$= (\langle u, v\rangle + \langle u, w_2\rangle\langle w_1, v\rangle)\langle e(w_1), e(w_2)\rangle$$

$$\langle e(w_1), a^+(v)a(u)e(w_2)\rangle = \langle u, w_2\rangle\langle w_1, v\rangle\langle e(w_1), e(w_2)\rangle$$

So $\langle e(w_1), [a(u), a(v)]e(w_2)\rangle = \langle u, v\rangle\langle e(w_1), e(w_2)\rangle$

This proves (1).

Let $u,v \in L^2(\mathbb{R}_+)$.

$$A\_(t) = a(\chi_{[0,t]}),\ A^+(t) = a^+(\chi_{[0,t]}),$$

Then $[A\_(t), A^+(t')] = \langle \chi_{[0,t]}, \chi_{[0,t']}\rangle$

$$= t\Lambda t'$$

Conservation process $H \in \mathcal{L}(L^2(\mathbb{R}_+))$

$\Lambda_H(t)$ ($\mathcal{L}(X) \simeq$ linear operator in the vector space X).

$$W(0, \exp(i\lambda H\chi_{[0,t]})) = \exp(i\lambda\Lambda_H(t)).$$

$[\chi_{[0,t]}, H\} = 0\ \forall t$ is assumed.

$$\langle e(w_1), \Lambda_H(t)e(w_2)\rangle = \langle w_1, H\chi[0,t\}w_2\rangle\langle e(w_1), e(w_2)\rangle$$

$$\langle e(w_1), [\Lambda_H(t), a(u)]e(w_2)\rangle$$

$$= \langle e(w_1), \Lambda_H(t)a(u)e(w_2)\rangle - \langle e(w_1), a(u)\Lambda_H(t)e(w_2)\rangle$$

$$= \langle w_1, H\chi_{[0,t]}w_2\rangle\langle u, w_2\rangle\langle e(w_1), e(w_2)\rangle$$

$$-\frac{d}{d\varepsilon}\langle e(w_1+\varepsilon u), \Lambda_H(t)e(w_2)\rangle\Big|_{t=0}$$

$$= \langle w_1, H_t w_2\rangle\langle u, w_2\rangle\langle e(w_1), e(w_2)\rangle$$

$$-\frac{d}{d\varepsilon}(\langle w_1+\varepsilon u, H_t w_2\rangle\langle e(w_1+\varepsilon u), e(w_2)\rangle)\Big|_{\varepsilon=0}$$

$$= [\langle w_1, H_t w_2\rangle\langle u, w_2\rangle - (\langle u, H_t w_2\rangle$$

$$+\langle w_1, H_t w_2\rangle\langle u, w_2\rangle]\langle e(w_1), e(w_2)\rangle$$

$$= -\langle u, H_t w_2\rangle\langle e(w_1), e(w_2)\rangle$$

Hence $\langle e(w_1), [\Lambda_H(t), A\_(t')]e(w_2)\rangle$

$$= \langle e(w_1), [\Lambda_H(t), a(\chi_{[0,t']})]e(w_2)\rangle$$

$$= -\langle \chi_{[0,t']}, H_t w_2\rangle\langle e(w_1), e(w_2)\rangle$$

$$= -\langle \chi_{[0,t\Lambda t']}, Hw_2\rangle\langle e(w_1), e(w_2)\rangle$$

$$= \left(-\int_0^{t\Lambda t'} (Hw_2)(\tau)d\tau\right)\langle e(w_1), e(w_2)\rangle$$

On the other hand,

$$\begin{aligned}\langle e(w_1), a(\chi_{[0,t\Lambda t']})e(w_2)\rangle &= \langle \chi_{[0,t\Lambda t']}, w_2\rangle\langle e(w_1), e(w_2)\rangle \\ &= \left(\int_0^{t\Lambda t'} w_2(\tau)d\tau\right)\langle e(w_1), e(w_2)\rangle\end{aligned}$$

So if $H = I$, the identity operator, then we get

$$\langle e(w_1), [\Lambda_I(t), A\_(t')]e(w_2)\rangle = -\langle e(w_1), A\_(t\Lambda t')e(w_2)\rangle$$

i.e. $\qquad [A\_(t), \Lambda_I(t')] = A\_(t\Lambda t')$

We define $\qquad N(t) = \Lambda I(t).$

$$\begin{aligned}&\langle e(w_1), [a^+(u), \Lambda_H(t)]e(w_2)\rangle \\ &= \langle a(u)e(w_1), \Lambda_H(t)e(w_2)\rangle - \langle e(w_1), \Lambda_H(t)a^+(u)e(w_2)\rangle \\ &= \langle w_1, u\rangle\langle w_1, H_t w_2\rangle\langle e(w_1), e(w_2)\rangle \\ &\quad -\frac{d}{d\varepsilon}\langle e(w_1), \Lambda_H(t)e(w_2+\varepsilon u\rangle\Big|_{\varepsilon=0} \\ &= \langle w_1, u\rangle\langle w_1, H_t w_2\rangle\langle e(w_1), e(w_2)\rangle \\ &\quad -\frac{d}{d\varepsilon}\left(\langle w_1, H_t(w_2+\varepsilon u)\rangle \exp(\langle w_1, w_2+\varepsilon u\rangle)\right)\Big|_{\varepsilon=0} \\ &= \left[\langle w_1, u\rangle\langle w_1, H_t w_2\rangle - (\langle w_1 H_t u\rangle\right. \\ &\quad \left.+\langle w_1, H_t w_2\rangle\langle w_1, u\rangle)\right]\langle e(w_1), e(w_2)\rangle \\ &= -\langle w_1, H_t u\rangle\langle e(w_1), e(w_2)\rangle\end{aligned}$$

Thus, putting $u = \chi_{[0,t']}$,

$$\begin{aligned}\langle e(w_1), [A^+(t'), \Lambda_H(t)]e(w_2)\rangle &= \left(-\int (H^*_{t\Lambda t'}\bar{w}_1)(\tau)d\tau\right)\langle e(w_1), e(w_2)\rangle \\ &= \left(-\int_0^{t\Lambda t'} (H^*\bar{w}_1)(\tau)d\tau\right)\langle e(w_1), e(w_2)\rangle \\ \langle e(w_1), A^+(t)e(w_2)\rangle &= \langle A(t)e(w_1), e(w_2)\rangle \\ &= \overline{\langle \chi_{[0,t]}, w_1\rangle}\langle e(w_1), e(w_2)\rangle \\ &= \left(\int_0^t \bar{w}_1(\tau)d\tau\right)\langle e(w_1), e(w_2)\rangle \\ [N(t), A^+(t')] &= [\Lambda_I(t), A^+(t')] = A^+(t\Lambda t')\end{aligned}$$

To summarize

$$\left[A_-(t), A^+(t')\right] = t\Lambda t', \left[A_-(t), N(t')\right] = A_-(t\Lambda t'),$$

$$\left[N(t), A^+(t')\right] = A^+(t\Lambda t').$$

Here, the noise Hilbert space is $\Gamma_s\left(L^2(\mathbb{R}_+)\right)$

Now $L^2(\mathbb{R}_+) = L^2(\mathbb{R}_+ \to \mathbb{C})$. More generally, we can consider $L^2(\mathbb{R}_+ \to K)$ where $K = \mathbb{C}^m$.

Regard c_0^- as a row vector row ($c_0^-(j)$: $j = 1, 2, .., m$), c_+^0 = column ($c_+^0(j)$: $j = 1, 2, .., m$) as a column vector, $c_0^0 = \left(\left(c_0^0(i, j)\right)\right)_{1\le i,\, j\le m}$ as an $m \times m$ complex matrix, $c_+^- \in \mathbb{C}$ is a scalar (complex). Then

$$c = \begin{pmatrix} 0 & c_0^- & c_+^- \\ 0 & c_0^0 & c_+^0 \\ 0 & 0 & 0 \end{pmatrix} \in \mathbb{C}^{(m+2)\times(m+2)}$$

Let
$$A(c,t) \triangleq Ic_+^- t + \sum_{j=1}^m A_-^j(t)c_0^-(j) + \sum_{j=1}^m A_j^+(t)c_+^0(j) + \sum_{k,\, j=1}^m c_0^0(k,t)N_{kj}(t)$$

Here
$$A_-^j(t) = a(\varphi_j\chi_{[0,t]}),$$

$$A_j^+(t) = a^+(\varphi_j\chi_{[0,\,t]})$$

$$N_{kj}(t) = \Lambda_{|\varphi_k\rangle\langle P_j|}(t)$$

where $\left\{|\varphi_j\rangle\right\}_1^m$ is an ONB for $\mathbb{C}^m$.

Belavkin $\simeq$ quantum filtering P. 187 (Remarks).

$E_t \simeq$ orthoprojector on $\overline{\mathbb{B}^t\xi}$, $\mathbb{B}^t = \infty_t'$.

E_t commutes with ∞_t'.

Proof: Let $\mathbb{B} \in \infty_t'$, $\xi \in \mathcal{H}$. fixed. $\overline{\mathbb{B}^t\xi} = \overline{\infty_t'\xi}$.

$\varepsilon_t = \overline{\infty_t'\xi}$ is ∞_t'-invariant.

$$B\eta - E_t B\eta \perp \varepsilon_t\, \eta - E_t\eta \perp \varepsilon_t\ \forall \eta \in \mathcal{H}.$$

$$\frac{B+B^*}{2}, \frac{B-B^*}{2i} \in \infty_t'.$$

Let
$$U_s = \exp\left(\frac{is(B+B^*)}{2}\right),$$

$$V_s = \exp\left(\frac{is(B-B^*)}{2i}\right).$$

Then $U_s, V_s \in \infty_t'$,

U_s, V_s are unitary. Hence

$$\eta - E_t\eta \perp \varepsilon_t \Rightarrow U_s\eta - U_sE_t\eta \perp \varepsilon_t, \qquad (\because \varepsilon_t \text{ is } \infty_t' \text{ invariant})$$

$$V_s\eta - V_sE_t\eta \perp \xi_t \ \forall s\in\mathbb{R}.$$

Taking $\frac{d}{ds}$ at $s = 0 \Rightarrow B\eta - BE_t\eta \perp \xi_t$

Thus $E_tB\eta - BE_t\eta \perp \xi_t$

But $E_tB\eta\in\xi_t$, $BE_t\eta\in\xi_t$.

Hence $E_tB\eta - BE_t\eta\in\xi_t$

$\Rightarrow$ $[E_t, B]\eta = 0 \ \forall\eta\in\mathcal{H}$

$\Rightarrow$ $[E_t, B] = 0.$

$$(Z^bDZ) = \begin{pmatrix} 0 & Z_+^{0*} & Z_+^{-*} \\ 0 & Z_0^{0*} & Z_+^{0*} \\ 0 & 0 & 0 \end{pmatrix}\begin{pmatrix} 0 & D_0^- & D_+^- \\ 0 & D_0^0 & D_+^0 \\ 0 & 0 & 0 \end{pmatrix} \times \begin{pmatrix} 0 & Z_0^- & Z_+^- \\ 0 & Z_0^0 & Z_+^0 \\ 0 & 0 & 0 \end{pmatrix}$$

$$= \begin{pmatrix} 0 & Z_+^{0*}D_0^0 & Z_+^{0*}D_+^0 \\ 0 & Z_0^{0*}D_0^0 & Z_0^{0*}D_+^0 \\ 0 & 0 & 0 \end{pmatrix}\begin{pmatrix} 0 & Z_0^- & Z_+^- \\ 0 & Z_0^0 & Z_+^0 \\ 0 & 0 & 0 \end{pmatrix}$$

$$= \begin{pmatrix} 0 & Z_+^{0*}D_0^0Z_0^0 & Z_+^{0*}D_0^0Z_+^0 \\ 0 & Z_0^{0*}D_0^0Z_0^0 & Z_0^{0*}D_0^0Z_+^0 \\ 0 & 0 & 0 \end{pmatrix}$$

$$\left(Z^bD_0^bFZ\right)_+^- = \left(Z^bD_0^b(X\otimes\delta + C'Z)\right)_+^-$$

$$Z^bD_0^b(X\otimes\delta)Z = \begin{pmatrix} 0 & Z_+^{0*} & Z_+^{-*} \\ 0 & Z_0^{0*} & Z_+^{0*} \\ 0 & 0 & 0 \end{pmatrix}\begin{pmatrix} 0 & (D_+^{0*})_0 & (D_+^{-*})_0 \\ 0 & (D_0^{0*})_0 & (D_+^{0*})_0 \\ 0 & 0 & 0 \end{pmatrix}$$

$$\begin{pmatrix} X & 0 & 0 \\ 0 & X & 0 \\ 0 & 0 & X \end{pmatrix}\begin{pmatrix} 0 & Z_0^- & Z_+^- \\ 0 & Z_0^0 & Z_+^0 \\ 0 & 0 & 0 \end{pmatrix}$$

$$\begin{pmatrix} 0 & Z_+^{0*}(D_0^{0*})_0 & Z_+^{0*}(D_+^{0*})_0 \\ 0 & Z_0^{0*}(D_0^{0*})_0 & Z_0^{0*}(D_+^{0*})_0 \\ 0 & 0 & 0 \end{pmatrix} \times \begin{pmatrix} 0 & XZ_0^- & XZ_+^- \\ 0 & XZ_0^0 & XZ_+^0 \\ 0 & 0 & 0 \end{pmatrix}$$

$$= \begin{pmatrix} 0 & Z_+^{0*}(D_0^0)_0^*XZ_0^0 & Z_+^{0*}(D_0^{0*})_0XZ_+^0 \\ 0 & Z_0^{0*}(D_0^{0*})_0XZ_0^0 & Z_0^{0*}(D_0^{0*})_0XZ_+^0 \\ 0 & 0 & 0 \end{pmatrix}$$

$$\begin{pmatrix} 0 & D_0^- & D_+^- \\ 0 & D_0^0 & D_+^0 \\ 0 & 0 & 0 \end{pmatrix}\begin{pmatrix} 1 & Z_0^- & Z_+^- \\ 0 & Z_0^0 & Z_+^0 \\ 0 & 0 & 1 \end{pmatrix}$$

$$= \begin{pmatrix} 0 & D_0^- Z_0^0 & D_0^- Z_+^0 + D_+^- \\ 0 & D_0^0 Z_0^0 & D_0^0 Z_+^0 + D_+^0 \\ 0 & 0 & 1 \end{pmatrix}$$

$$Z^b DZ \equiv \begin{pmatrix} 1 & Z_+^{0*} & Z_+^{-*} \\ 0 & Z_0^{0*} & Z_0^{-*} \\ 0 & 0 & 1 \end{pmatrix}\begin{pmatrix} 0 & D_0^- & D_+^- \\ 0 & D_0^0 & D_+^0 \\ 0 & 0 & 0 \end{pmatrix}\begin{pmatrix} 1 & Z_0^- & Z_+^- \\ 0 & Z_0^0 & Z_+^0 \\ 0 & 0 & 1 \end{pmatrix}$$

$$= \begin{pmatrix} 0 & D_0^- + Z_+^{0*} D_0^0 & D_+^- + Z_+^{0*} D_+^0 \\ 0 & Z_0^{0*} D_0^0 & Z_0^{0*} D_+^0 \\ 0 & 0 & 0 \end{pmatrix} \times \begin{pmatrix} 1 & Z_0^- & Z_+^- \\ 0 & Z_0^0 & Z_+^0 \\ 0 & 0 & 1 \end{pmatrix}$$

$$\therefore \quad \left(Z^b DZ\right)_+^- = \left(D_0^- + Z_+^{0*} D_0^0\right) Z_+^0 + D_+^- + Z_+^{0*} D_+^0$$

$$= D_+^- + D_0^- Z_+^0 + Z_+^{0*} D_+^0 + Z_+^{0*} D_0^0 Z_+^0$$

$$C = F - X \otimes \delta = \begin{pmatrix} 0 & C_0^- & C_+^- \\ 0 & C_0^0 & C_+^0 \\ 0 & 0 & 0 \end{pmatrix}$$

$$F = X \otimes \delta + C = \begin{pmatrix} X & C_0^- & C_+^- \\ 0 & X + C_0^0 & C_+^0 \\ 0 & 0 & X \end{pmatrix}$$

$$Z^b D_0^b CZ = (D \equiv D_0)$$

$$\begin{pmatrix} 1 & Z_+^{0*} & Z_+^{-*} \\ 0 & Z_0^{0*} & Z_+^{0*} \\ 0 & 0 & 1 \end{pmatrix}\begin{pmatrix} 0 & D_+^{0*} & D_+^{-*} \\ 0 & D_0^{0*} & D_+^{0*} \\ 0 & 0 & 0 \end{pmatrix}$$

$$\begin{pmatrix} 0 & C_0^- & C_+^- \\ 0 & C_0^0 & C_+^0 \\ 0 & 0 & 0 \end{pmatrix}\begin{pmatrix} 1 & Z_0^- & Z_+^- \\ 0 & Z_0^0 & Z_+^0 \\ 0 & 0 & 1 \end{pmatrix}$$

$$= \begin{pmatrix} 0 & D_+^{0*} + Z_+^{0*} D_0^{0*} & D_+^{-*} + Z_+^{0*} D_+^{0*} \\ 0 & Z_0^{0*} D_0^{0*} & Z_0^{0*} D_+^{0*} \\ 0 & 0 & 0 \end{pmatrix}$$

$$\times \begin{pmatrix} 0 & C_0^- Z_0^0 & C_0^- Z_+^0 + C_+^- \\ 0 & C_0^0 Z_0^0 & C_0^0 Z_+^0 + C_+^0 \\ 0 & 0 & 0 \end{pmatrix}$$

Hence $$(Z^b D_0^b C Z)_+^- = \left(D_+^{0*} + Z_+^{0*} D_0^{0*}\right) \times \left(C_0^0 Z_+^0 + C_+^0\right)$$
$$= \left(D_0^0 Z_+^0 D_+^0\right)^* \left(C_0^0 Z_+^0 + C_+^0\right)$$

* eqn. (2.2)

$$d\widehat{X} = \left(\widehat{F} - \widehat{X} \otimes \delta\right) d\widehat{A}$$
$$= \left(\widehat{F}_\nu^\lambda(t) - \widehat{X}(t)\delta_\nu^\lambda\right) d\widehat{A}_\lambda^\nu(t) = \widehat{C}_\nu^\lambda(t) d\widehat{A}_\lambda^\nu(t)$$
$$d\widehat{A}(c,t) = \widehat{C}_+^- dt + \widehat{C}_0^-(j)\widehat{A}_-^j(t) + \widehat{C}_+^0(j) d\widehat{A}_j^+(t)$$
$$\widehat{C}_0^0(j,m) d\widehat{N}_m^j(t)$$

$\widehat{A}_-^+(t) = t$, $\widehat{C}_+^-(t)$ is the coeff. of $d\widehat{A}_-^+(t) = dt$ in $d\widehat{A}(c,t)$.

$$\widehat{C}_j^-(t) = \widehat{C}_0^-(j),\ \widehat{C}_+^j(t) = \widehat{C}_+^0(j),$$
$$\widehat{A}_m^j(t) = \widehat{N}_m^j(t),\ \widehat{C}_j^m(t) = \widehat{C}_0^0(j,m).$$
$$d\widehat{A}_-^j(t) d\widehat{A}_k^+(t) = \delta_k^j dt = \delta_k^j d\widehat{A}_-^+(t),$$
$$d\widehat{A}_j^+(t) d\widehat{A}_-^k(t) = 0 = \delta_-^+ d\widehat{A}_j^k(t)$$

$\left(\delta_-^+ = 0\right)$

$$d\widehat{A}_-^j(t) dN_m^k(t) = \delta_m^j d\widehat{A}_-^k(t)$$
$$d\widehat{A}_j^+(t) dN_m^k(t) = 0 = \delta_m^+ d\widehat{N}_j^k(t)$$
$$= \delta_m^+ d\widehat{A}_j^k(t)$$

$\left(\delta_m^+ = 0\right)$

$$dN_m^k(t) dA_j^+(t) = \delta_j^k dA_m^+(t)$$

These quantum Ito formula of Hudson and Parthasarthy can be summarized as

$$d\widehat{A}_\nu^\mu(t) d\widehat{A}_\sigma^\rho(t) = \delta_\sigma^\mu d\widehat{A}_\nu^\rho(t)$$

where μ, ν ρ, $\sigma = \{+, -, 1, 2, 3, \ldots\}$

$$d\widehat{X} = \widehat{C}_\nu^\lambda d\widehat{A}_\lambda^\nu = dA(C, t)$$
$$d\widehat{X}^* = \widehat{C}_\nu^{\lambda *} d\widehat{A}_\lambda^{\nu *}$$

$$d\widehat{A}_j^{+*} = d\widehat{A}_-^j,\ d\widehat{A}_-^{j*} = d\widehat{A}_j^+,$$

$$d\widehat{N}_j^{k*} = d\widehat{N}_k^j\ (i.e.\ d\widehat{A}_j^{k*} = d\widehat{A}_k^j)$$

so,
$$d\widehat{X}^* = \widehat{C}_k^{j*}\, d\widehat{A}_k^j + \widehat{C}_+^{-*}\, dt + \widehat{C}_j^{-*}\, d\widehat{A}_-^{j*} + \widehat{C}_+^{j*}\, d\widehat{A}_j^{+*}$$
$$= \widehat{C}_k^j d\widehat{A}_k^j + \widehat{C}_j^{-*}\, dt + \widehat{C}_j^{-*}\, d\widehat{A}_j^+ + \widehat{C}_+^{j*}\, d\widehat{A}_-^j$$

Now
$$C^b = \begin{pmatrix} 0 & C_+^{0*} & C_+^{-*} \\ 0 & C_0^{0*} & C_0^{-*} \\ 0 & 0 & 0 \end{pmatrix}$$

So
$$d\widehat{X}^* = \widehat{C}(k,j)^* d\widehat{A}_j^k + \widehat{C}_+^{-*}\, dt + \widehat{C}_0^-(j)^* d\widehat{A}_-^{j*} + \widehat{C}_+^0(j)^* d\widehat{A}_j^{+*}$$
$$= \widehat{C}^b(k,j) dA_k^j + \widehat{C}_+^{-*}\, dt + \widehat{C}_0^-(j)^* d\widehat{A}_j^+ + \widehat{C}_+^0(j)^* d\widehat{A}_-^j$$
$$= \widehat{C}^b(j,k) dA_k^j + C_+^{b-} dt + C_+^{b0}(j) d\widehat{A}_j^+ + C_0^-(j) d\widehat{A}_-^j$$
$$= d\widehat{A}(C^b, t)$$

Thus if
$$\widehat{X}(t) = A(C,t), \text{ then } \widehat{X}^*(t) = d\widehat{A}(C^b, t).$$

Now
$$dX = \widehat{C}_\nu^\lambda(t) d\widehat{A}_\lambda^\nu(t) = d\widehat{A}(\widehat{C}, t),$$
$$dX^* = \widehat{C}_\nu^{b\lambda}(t) d\widehat{A}_\lambda^\nu(t) = d\widehat{A}(\widehat{C}^b, t)$$

So using the quantum Ito formula,

$$dX^*.dX = \widehat{C}_\nu^{b\lambda}\widehat{C}_{\nu'}^{\lambda'}\, d\widehat{A}_\lambda^\nu d\widehat{A}_{\lambda'}^{\nu'}$$
$$= \widehat{C}_\nu^{b\lambda}\widehat{C}_{\nu'}^{\lambda'}\, \delta_{\lambda'}^\nu . d\widehat{A}_\lambda^{\nu'}$$
$$= \widehat{C}_\nu^{b\lambda}\widehat{C}_{\nu'}^\nu d\widehat{A}_\lambda^{\nu'}$$
$$= \left(\widehat{C}^b\widehat{C}\right)_{\nu'}^\lambda d\widehat{A}_\lambda^{\nu'}$$
$$= d\widehat{A}\left(\widehat{C}^b\widehat{C}, t\right)$$

We have
$$\left[A_-^j(t), A_k^+(t')\right] = \left\langle \varphi_j\chi_{[0,t]}, \varphi_k\chi_{[0,t']}\right\rangle$$
$$= \delta_{jk} t\Lambda t'$$

$$\left[A_-^j(t), N_{km}(t')\right] = \left[a\left(\varphi_j\chi_{[0,t]}\right), \Lambda_{|\varphi_k\rangle\langle\varphi_m|}(t')\right]$$
$$= a\left(|\varphi_m\rangle\langle\varphi_k|\varphi_j\rangle\chi_{[0,t\Lambda t']}\right)$$
$$= \delta_{kj} a\left(|\varphi_m\rangle\chi_{[0,t\Lambda t']}\right)$$

$$= \delta_{kj} A_-^m (t \Lambda t')$$

It is better to denote $N_{km}(t)$ by

$N_k^m(t)$:

$$N_k^m(t) = \Lambda_{|\varphi_k\rangle\langle\varphi_m|}(t)$$

Then, $\left[A_-^j(t), N_k^m(t')\right] = \delta_k^j A_-^m(t\Lambda t')$

Rough Calculation:

$$\langle e(w_1), [\Lambda_H(t), a(u)] e(w_2)\rangle$$

$$= -\langle u, H_t w_2\rangle \langle e(w_1), e(w_2)\rangle \langle e(w_1), a(H_t^* u) e(w_2)\rangle$$

$$= \langle H_t^* u, w_2\rangle \langle e(w_1), e(w_2)\rangle$$

$$= \langle u, H_t w_2\rangle \langle e(w_1), e(w_2)\rangle$$

so $\quad [a(u), \Lambda_H(t)] = a(H_t^* u)$

Also, $\left[N_k^m(t), A_j^+(t')\right] = \left[\Lambda_{|\varphi_k\rangle\langle\varphi_m|}(t), a^*\left(\varphi_j \chi_{[0,t']}\right)\right]$

$$= a^*\left(|\varphi_k\rangle\langle\varphi_m|\varphi_j\rangle \chi_{[0, t\Lambda t']}\right)$$

$$= \delta_j^m a^*\left(|\varphi_k\rangle \chi_{[0, t\Lambda t']}\right)$$

$$= \delta_j^m A_k^+(t\Lambda t')$$

Finally,

$$\left\langle e(w_1), \left[\Lambda_{H_1}(t), \Lambda_{H_2}(t')\right] e(w_2)\right\rangle = \left\langle e(w_1), \Lambda_{H_1}(t)\Lambda_{H_2}(t') e(w_2)\right\rangle - \left\langle e(w_1), \Lambda_{H_2}(t')\Lambda_{H_1}(t) e(w_2)\right\rangle$$

$$\left\langle e(w_1), \Lambda_{H_1}(t)\Lambda_{H_2}(t') e(w_2)\right\rangle$$

$$= -\frac{\partial^2}{\partial\varepsilon_1\partial\varepsilon_2} \left\langle e(w_1), e\left(\exp(i\varepsilon_1 H_{1t}).\exp(i\varepsilon_2 H_{2t'}) w_2\right)\right\rangle \Big| \varepsilon_1 = \varepsilon_2 = 0$$

$$-\frac{\partial^2}{\partial\varepsilon_1\partial\varepsilon_2} \exp\left(\langle w_1, w_2 + i(\varepsilon_1 H_{1t} + \varepsilon_2 H_{2t'}) w_2 - \varepsilon_1\varepsilon_2 H_{1t} H_{2t'} w_2\rangle\right)\Big|_{\varepsilon_1 = \varepsilon_2 = 0}$$

Rough Calculation:

$$[a(u), \Lambda_H(t)] = a\left(H_t^* u\right)$$

$$\Rightarrow \quad \left[\Lambda_H^*(t), a^*(u)\right] = a^*\left(H_t^* u\right)$$

$$\Rightarrow \quad \left[\Lambda_{H^*}(t), a^*(u)\right] = a^*\left(H_t^* u\right)$$

$\forall H$ commuting with $\chi_{[0,\,t]}$

$$\Rightarrow \quad [\Lambda_H(t), a^*(u)] = a^*(H_t u)$$

$$= -\frac{\partial}{\partial \varepsilon_2}\left\{\left(i\langle w_1, H_{1t} w_2\rangle - \langle w_1, \varepsilon_2 H_{1t} H_{2t'} w_2\rangle\right)\right.$$

$$\left.\exp\left(\langle w_1, w_2 + i\varepsilon_2 H_{2t'} w_2\rangle\right)\right\}\Big|_{\varepsilon_2=0}$$

$$= \left\{\langle w_1, H_{1t} w_2\rangle\langle w_1, H_{2t'} w_2\rangle + \langle w_1, H_{1t} H_{2t'} w_2\rangle\right\}$$
$$\langle e(w_1), e(w_2)\rangle$$

Thus, $\left\langle e(w_1), \left[\Lambda_{H_1}(t), \Lambda_{H_2}(t')\right] e(w_2)\right\rangle$

$$= \left\langle w_1, [H_1, H_2]_{t\Lambda t'} w_2\right\rangle\langle e(w_1), e(w_2)\rangle$$

and hence $\left[\Lambda_{H_1}(t), \Lambda_{H_2}(t')\right] = \Lambda_{[H_1, H_2]}(t\Lambda t')$

$$\text{Thus, } \left[N_k^m(t), N_{k'}^{m'}(t')\right] = \left[\Lambda_{|\varphi_k\rangle\langle\varphi_m|}(t), \Lambda_{|\varphi_{k'}\rangle\langle\varphi_{m'}|}(t')\right]$$
$$= \left(\delta_{k'}^m \Lambda|\varphi_k\rangle\langle\varphi_{m'}| - \delta_k^{m'}\Lambda|\varphi_{k'}\rangle\langle\varphi_m|\right)(t\Lambda t')$$
$$= \left(\delta_{k'}^m N_k^{m'} - \delta_k^{m'} N_{k'}^m\right)(t\Lambda t')$$

$$\text{Thus } [A(c,t), A(d,t')] = \sum_{j,k} C_0^-(j) d_+^0(k)\left[A_-^j(t), A_k^+(t')\right]$$
$$+\sum_{j,k} C_+^0(j) d_0^-(k)\left[A_j^+(t), A_-^k(t')\right]$$
$$+\sum_{jkr} C_0^-(j) d_0^0(k,r)\left[A_-^j(t), N_k^r(t')\right]$$
$$+\sum_{jkr} C_0^0(k,r) d_0^-(j)\left[N_k^r(t), A_-^j(t')\right]$$
$$+\sum_{krpq} C_0^0(k,r) d_0^0(p,q)\left[N_k^r(t), N_p^q(t')\right]$$
$$+\sum_{jkr} C_+^0(j) d_0^0(k,r)\left[A_j^+(t), N_k^r(t')\right]$$

$$+\sum_{jkr} C_0^0(k,r)d_+^0(j)\left[N_k^r(t), A_j^+(t')\right]$$

$$= \sum_{j,r} C_0^-(j)d_+^0(k)\delta_k^j t\Lambda t'$$

$$-\sum_{j,k} C_+^0(j)d_0^-(k)\delta_j^k t\Lambda t'$$

$$+\sum_{jkr} C_0^-(j)d_0^0(k,r)\delta_k^j A_-^r(t\Lambda t')$$

$$-\sum_{jkr} C_0^0(k,r)d_0^-(j)\delta_k^j A_-^r(t\Lambda t')$$

$$-\sum_{jkr} C_+^0(j)d_0^0(k,r)\delta_j^r A_k^+(t\Lambda t')$$

$$+\sum_{jkr} C_0^0(k,r)d_+^0(j)\delta_j^r A_k^+(t\Lambda t')$$

$$+\sum_{krpq} C_0^0(k,r)d_0^0(p,q)\delta_p^r N_k^q(t\Lambda t') - \delta_k^q N_p^r(t\Lambda t')$$

$$= t\Lambda t' \sum_j \left(C_0^-(j)d_+^0(j) - C_+^0(j)d_0^-(j)\right)$$

$$+\sum_{jr}\left(C_0^-(j)d_0^0(j,r) - C_0^0(j,r)d_0^-(j)\right)A_-^r(t\Lambda t')$$

$$+\sum_{jr}\left(C_0^0(k,j)d_+^0(j) - d_0^0(k,j)C_+^0(j)\right)A_k^+(t\Lambda t')$$

$$+\sum_{kpq}\left[C_0^0(k,p)d_0^0(p,q) - d_0^0(k,p)C_0^0(p,q)\right]$$

$$N_k^q(t\Lambda t').$$

The above calculation may be referral to while reading Belavkin's original paper to make the reading easier.

❖❖❖❖❖

[16] Selected Topics in General Relativity:

1. Tetrad formulation of field equations.

$g_{mn}(x)$ is the metric; $e_a^m(x)$, $a = 0, 1, 2, 3$ is a tetrad basis in the sense that for each a, $(e_a^m)_{m=0}^3$. is a contravarient vector field and

$$\eta_{ab} = g_{\mu\nu}e_a^\mu e_b^\nu \text{ are scalar constraints.}$$

We have

$$d\tau^2 = g_{\mu\nu}dx^\mu dx^\nu = \eta_{ab}e_\mu^a e_\nu^b dx^\mu dx^\nu.$$

$$= \eta_{ab}(e_\mu^a dx^\mu)(e_\nu^b dx^\nu)$$

$e_\mu^a dx^\mu$ is a one form. Here $((e_\mu^a))$ is the matrix of $((e_\mu^a))$: $e_a^\mu e_\nu^a = \delta_\nu^\mu$, $e_a^\mu e_\mu^b = \delta_a^b$.

$e_a = e_a^\mu \dfrac{\partial}{\partial x^\mu}, a = 0, 1, 2, 3$ form a local basic for the tangent space at x. e_a is a vector field. Likewise, $e_{a\mu}\, dx^\mu$, $a = 0, 1, 2, 3$ form a local basis for the dual to the tangent space at x. Here, $e_{a\mu} = \eta_{ab} e_\mu^b = g_{\mu\nu} e_a^\nu$.

Example: The perturbed Kerr metric has the form

$$d\tau^2 = e^{2\nu} dt^2 - e^{2\mu} dr^2 - e^{2\mu_2} d\theta^2$$
$$- e^{2\psi}\left(d\varphi - w dt - q_1 dr - q_2 d\theta - q_3 d\varphi\right)^2$$

Where $\nu, \mu_1, \mu_2, \psi, w, q_1, q_2, q_3$ are functions of t, r, θ, φ.

We define

$$e^0 = e^\nu dt, e^1 = e^{\mu_1} dr, e^2 = e^{\mu_2} d\theta,$$
$$e^3 = e^\psi \left(d\varphi - w dt - q_1 dr - q_2 d\theta - q_3 d\varphi\right)$$
$$\eta_{00} = 1, \eta_{11} = -1, \eta_{22} = -1, \eta_{33} = -1.$$

Then $$\eta_{ab} e^a e^b = d\tau^2.$$

Note $e_\mu^a\, dx_a^\mu = 0, 1, 2, 3$ are also one for and the from a basis for the co-tangent space at x.

Co-varient derivative:

$$A^\mu_{:\nu} = \left(A^a e_a^\mu\right)_{:\nu} \quad (A^a = e_\mu^a A^\mu \text{ are scalar fields})$$
$$= A^a_{,\nu} e_a^\mu + A^a e^\mu_{a:\nu}.$$
$$= A^\mu_{:\nu} e_\mu^b e_c^\nu \underline{\underline{\Delta}} A^b / c = A^a_{,\nu} e_a^\mu e_\mu^b e_c^\nu + A^a e^\mu_{a:\nu} e_\mu^b e_c^\nu$$
$$= A^b_{,\nu} e_c^\nu + A^a \tau^b_{ac} = A^b_{,c} + \tau^b_{ac} A^a$$

where $A^b_{,c} = e_a(A^b) = e_c^\nu \partial\nu\, A^b = e_c^\nu A^b_{,\nu}$ (directional derivatives along the e_c direction),

$$\tau^b_{ac} = e_\mu^b e^\mu_{a;\nu} e_c^\nu \text{ (Spin coefficients. They are scalar).}$$

Maxwell equation in tetrad notation.

$$F^{\mu\nu}_{;\nu} = KJ^\mu \text{ (Four vector notation)}$$
$$F^{\mu\nu} = F^{ab} e_a^\mu e_b^\nu \qquad F^{ab} = \text{scalars:}$$
$$F^{\mu\nu}_{;\nu} = F^{ab}_{;\nu} e_a^\mu e_b^\nu\ F_{ab} = F^{\mu\nu} e_\mu^a e_\nu^b + F^{ab}\left(e^\mu_{a;\nu} e_b^\nu + e_\nu^\mu e^\nu_{b;\nu}\right)$$
$$= F^{ab}_{,b} e_a^\mu + F^{ab}\left(e^\mu_{a:\nu} e_b^\nu + e_\mu^a e^\nu_{b:\mu}\right)$$
$$e_\mu^a F^{\mu\nu}_{:\nu} = F^{ab}_{,b} + F^{cb}\left(e_\mu^a e^\mu_{c:\nu} e_b^\nu + \delta_e^a e^\nu_{b:\nu}\right)$$
$$= F^{ab}_{,b} + F^{cb}\tau^a_{cb} + F^{ab} e^\nu_{b:\nu}$$

Now, $$e^{\nu}_{b:\mu} = e^{a}_{\mu} e^{\mu}_{b:\nu} e^{\nu}_{a} = \tau^{a}_{ba}.$$

So the Maxwell equation, can be expressed as

$$F^{ab}_{,b} + F^{cb}\tau^{a}_{cb} + F^{ab}\tau^{c}_{bc} = kJ^{a}$$

❖❖❖❖❖

[17] Perturbations in the Tetrad Caused by Metric Perturbation:

$$g_{\mu\nu} = \eta_{ab} e^{a}_{\mu} e^{b}_{\nu}$$

$$\delta g_{\mu\nu} = \eta_{ab}\left(e^{a}_{\mu}\,\delta e^{b}_{\nu} + e^{b}_{\nu}\,\delta e^{a}_{\mu}\right)$$

$$= e_{b\mu}\delta e^{b}_{\nu} + e_{a\nu}\delta e^{a}_{\mu}$$

So

$$e^{b\mu}\delta g_{\mu\nu} = \delta e^{b}_{\nu} + e^{b\mu} e_{a\nu}\delta e^{a}_{\mu}$$

We note that

$$e^{b\mu}\delta e^{a}_{\mu} = \delta\left(e^{b\mu} e^{a}_{\mu}\right) - e^{a}_{\mu}\delta e^{b\mu}$$

$$= -e^{a}_{\mu} + \delta e^{b\mu}$$ .Thus,

$$e^{b\mu} e_{a\nu}\delta e^{a}_{\mu} = -\,e_{a\nu} e^{a}_{\mu}\delta e^{b\mu} = -\,g_{\mu\nu}\delta e^{b\mu}$$

So,

$$\delta e^{b}_{\nu} - g_{\mu\nu}\delta e^{b\mu} = e^{b\mu}\delta g_{\mu\nu}$$

Now, $$e^{b\mu} = e^{b}_{\rho} g^{\mu\rho}$$

so $$\delta e^{b\mu} = g^{\mu\rho}\delta e^{b}_{\rho} + e^{b}_{\rho}\delta g^{\mu\rho}$$

and hence

$$\delta e^{b}_{\nu} = g^{\rho}_{\nu}\delta e^{b}_{\rho} - g_{\mu\nu}\delta g^{\mu\rho} e^{b}_{\rho} = e^{b\mu}\delta g_{\mu\nu}$$

or

$$e^{b\mu}\delta g_{\mu\nu} + \delta g_{\mu\nu} e^{b}_{\rho}\delta g^{\mu\rho} = 0$$

$$\delta g^{\mu\rho} = -\,g^{\mu\alpha} g^{\rho\beta}\delta g_{\alpha\beta}$$. So the above reduces to

$$e^{b\mu}\delta\, g_{\mu\nu} - g_{\mu\nu} g^{\mu\alpha} g^{\rho\beta}\, e^{b}_{\rho}\delta\, g_{\alpha\beta} = 0$$

or

$$e^{b\mu}\,\delta g_{\mu\nu} - \delta g_{\nu\beta}\, e^{b\beta} = 0$$

Which is a trivial identity. This shows consistency.

Let $$\Delta\underline{\underline{G}} = \left(\left(\delta\, g_{\mu\nu}\right)\right), \Delta\underline{\underline{E}} = \left(\left(\delta e^{a}_{\mu}\right)\right)_{a,\mu}$$

$$\underline{\underline{E}} = \left(\left(e_{a\mu}\right)\right)_{a,\mu}$$

The equation for $\underline{\underline{\Delta E}}$ be solved is

$$\underline{\underline{E}}^T, \underline{\underline{\Delta E}} + \left(\underline{\underline{\Delta E}}\right)^T . \underline{\underline{E}} = \underline{\underline{\Delta G}}$$

❖❖❖❖❖

[18] Geodesic Equation in Tetra Formalism:

$$\begin{aligned}
v^a(\tau) &= v^a(\tau) e^a_\mu(x(\tau)) \\
v^\mu &= v^a e^\mu_a \frac{dv^\mu}{d\tau} = \frac{dv^\mu}{d\tau} e^\mu_a + v^a e^\mu_{a.v} v^v \\
&= \frac{dv^a}{d\tau} e^\mu_a + v^a v^b e^\mu_{a.v} e^v_b \\
\Gamma^\mu_{v\sigma} &= g^{\mu\alpha}\Gamma_{\alpha v\sigma} = \frac{1}{2} g^{\mu\alpha}\left(g_{\alpha v,\sigma} + g_{\alpha\sigma,v} - g_{v\sigma,\alpha}\right)
\end{aligned}$$

Express $\Gamma^\mu_{v\sigma}$ is terms of (e^a_μ) and simplify.

Alternate Method: The geodesic equation can be expressed as $v^v \mu^\mu_{:v} = 0$

Thus $$v^v \left(v^a e^\mu_a\right)_{:v} = 0$$

or $$v^v \left(e^\mu_a v^a_{,v} + v^a e^\mu_{a:v}\right) = 0$$

or $$e^\mu_a \frac{dv^a}{d\tau} + v^a v^b e^v_b e^\mu_{a:v} = 0$$

or $$\frac{dv^a}{d\tau} + v^c v^b e^v_b e^a_\mu e^\mu_{c:v} = 0$$

or $$\frac{dv^a}{d\tau} + v^b v^c \tau^a_{cb} = 0$$

Riemann tensor in tetrad basis:

$$\begin{aligned}
e^a_{\mu:v:\rho} - e^a_{\mu:\rho:v} &= R^\beta_{\mu v\rho} e^a_\beta \\
R^a_{bcd} &= R^\beta_{\mu v\rho} e^\alpha_\beta e^\mu_b e^v_c e^\rho_d \\
&= e^\mu_b e^v_c e^\rho_d \left(e^a_{\mu:v:\rho} - e^a_{\mu:\rho:v}\right)
\end{aligned}$$

Now,

$$\begin{aligned}
O &= g_{\mu v:\sigma}\left(\eta_{ab} e^a_\mu e^b_v\right)_{:\sigma} \\
&= \eta_{ab}\left(e^a_{\mu:\sigma} e^b_v + e^a_\mu e^b_{v:\sigma}\right) \\
&= e_{av} e^a_{\mu:\sigma} + e_{b\mu} e^b_{v:\sigma}
\end{aligned}$$

So

$$O = e^{\nu a}_{\mu:\sigma} + e^{a\nu}\, e_{b\mu}\, e^{b}_{\nu:\sigma}$$

Now

$$e^{\mu}_{b}\, e^{\nu}_{c}\, e^{a}_{\mu:\nu:\rho} = \left(e^{\mu}_{b}\, e^{\nu}_{c}\, e^{a}_{\mu:\nu}\right)_{:\rho} - e^{a}_{\mu:\nu}\left(e^{\mu}_{b:\rho}\, e^{\nu}_{c} + e^{\mu}_{b}\, e^{\nu}_{c:\rho}\right)$$

$$= \left(e^{\mu}_{b} e^{\nu}_{c} e^{a}_{\mu:\nu}\right)_{,\rho} - e^{a}_{\mu:\nu}\left(e^{\mu}_{b:\rho} e^{\nu}_{c} + e^{\mu}_{b} e^{\nu}_{c:\rho}\right)$$

$$= \tau^{a}_{bc,\rho} - e^{a}_{\mu:\nu}\left(e^{\mu}_{b:\rho}\, e^{\nu}_{c} + e^{\mu}_{b}\, e^{\nu}_{c:\rho}\right)$$

$$= \eta^{af}\,\tau_{bfc,\rho} - e^{a}_{\mu:\nu}\left(e^{\mu}_{b:\rho}\, e^{\nu}_{c} + e^{\mu}_{b}\, e^{\nu}_{c:\rho}\right)$$

Now, $$e^{\mu}_{b}\, e^{a}_{\mu:\nu}\, e^{\nu}_{c} = \tau^{a}_{bc} = \eta^{af}\,\tau_{bfc}$$

So $$e^{a}_{\mu:\nu}\, e^{\nu}_{c} = \eta^{af}\, e^{b}_{\mu}\,\tau_{bfc}$$

$$e^{a}_{\mu:\nu}\, e^{\nu}_{c}\, e^{\mu}_{b:\rho} = \eta^{af}\, e^{g}_{\mu}\, e^{\mu}_{b:\rho}\,\tau_{gfc}$$

$$= \eta^{af}\,\tau^{g}_{bh}\, e^{h}_{\rho}\,\tau_{gfc} = e^{h}_{\rho}\,\tau^{g}_{bh}\,\tau^{a}_{fc}$$

Likewise the other terms are simplified.

Exercise: Express R^{a}_{acd} in terms of the spin coefficients τ_{abc} and their first order partial derivatires.

❖❖❖❖❖

[19] Quantum General Relativistic Scattering:

The 3-metric of spaces times has the form $\tau_{rs} = -\dfrac{\tau_{rs}}{g_{oo}} + \dfrac{g_{or}g_{os}}{g_{oo}^{2}}$

This 3-metric is obtained by considering the time taken by a photon to propagate from x^r to $x^r + dx^r$ back and forth.

Let $\tau = \det(\tau_{rs})$.

The Laplace Beltamic operator in 3D curved space is them

$$\Delta_B = \tau^{-1/2}\,\partial_r\left(\tau^{1/2}\tau^{rs}\,\partial_s\right), \left(\left(\tau^{rs}\right)\right) = \left(\left(\tau_{rs}\right)\right)^{-1}$$

Summation over $r, s = 1, 2, 3$.

Approximation: Let $\tau_{rs} = \delta_{rs} + \varepsilon X_{rs}$

Then $$\tau_{rs} = \delta_{rs} + \varepsilon X_{rs} + O(\varepsilon^2).$$

Let $$\chi = \chi_{rr} = t^{r}\left(\left(\left(\chi_{rs}\right)\right)\right).$$

Then, $$\tau = 1 + \varepsilon\chi + O\left(\varepsilon^2\right).$$

$$\tau^{1/2} = 1 + \frac{1}{2}\varepsilon\xi + O(\varepsilon^2).$$

$$\tau^{-1/2} = 1-\frac{1}{2}\varepsilon\chi+O(\varepsilon^2)$$

$$\Delta_B\psi = \left(1-\frac{1}{2}\varepsilon\chi\right)\partial_r\left(\left(1+\frac{1}{2}\varepsilon\chi\right)(\delta_{rs}-\varepsilon\chi_{rs})\psi_{,s}\right)+O(\varepsilon^2)$$

$$= \left(1-\frac{1}{2}\varepsilon\chi\right)\left(\Delta_f^{\psi}+\frac{\varepsilon}{2}(\chi\psi,r)_{,r}-\varepsilon\left(\chi_{rs}\psi_{,s}\right)_{,r}\right)+O(\varepsilon^2)$$

$$= \Delta_f\psi+\varepsilon\left\{\frac{1}{2}\left(\chi\psi_{,r}\right)_{,r}-\left(\chi_{rs}\psi_{,s}\right)_{,r}-\frac{1}{2}\chi\Delta_f\psi\right\}+O(\varepsilon^2)$$

Where $\Delta f = \sum_1^3 \partial_r^2$ is the Laplacian of flat space. Schrödinger Hamiltonian of an electron board to the makes in the presence of the gravitational field. The Hamiltonian after interaction with gravitational field assumed to be concentration around the nuclear is

$$H_1 = \frac{-\hbar^2}{2m}\Delta f+\varepsilon V_1$$

Where

$$V_1 = \frac{-\hbar^2}{2m}S\left\{\frac{1}{2}\chi_{,r}\,\partial_s-\chi_{rs}\partial_r\partial_s-\chi_{rs}\,\partial_r\,\partial_s\right\}-\frac{Z^{e2}}{\left(\sum_{r=1}^{3}xr^2\right)^{1/2}}$$

i.e.

$$S\{X\} = \frac{X+X^*}{2}. \text{ We find}$$

$$\frac{-2mV_1}{\hbar^2} = \frac{1}{4}\left(\chi_{,r}\partial_r+\partial_r\chi_{,s}\right)-\frac{1}{2}\left(\chi_{rs,\,r}\partial_s-\partial_s\,\chi_{rs,r}\right)$$

$$-\frac{1}{2}\left(\chi_{rs,}\,\partial_r\partial_s+\partial_r\partial_s\chi_{rs}\right)-\frac{ze^2}{\left(\sum_1^3 x^{r2}\right)^{1/2}}$$

$$-\frac{1}{4}\chi_{,rr}+\frac{1}{2}\chi_{rs,rs}$$

$$-\chi_{rs}\partial_r\partial_s-\frac{1}{2}(\partial_r\,\chi_{rs})\partial_s-\frac{ze^2}{\left(\sum_1^3 x^{r2}\right)^{1/2}}$$

Scattering theory can be developed taking $H_0 = -\frac{\hbar^2}{2m}\Delta f$ and $H_1 = -\frac{\hbar^2}{2m}\Delta f+\varepsilon V_1$.

Time dependent quantum scattering theory:

$H_1(t) = H_0 + \varepsilon V_1(t), t \in \mathbb{R}$. $|\psi_0\rangle \simeq$ output state $|\psi_+\rangle \simeq$ scattered $|\psi_i\rangle \simeq$ input state. Let

$$U(t, s) = T\left\{\exp\left\{-i\int_s^t H_1(\tau)d\tau\right\}\right\}$$

$$= I + \sum_{n=1}^{\infty}(-i)^n \int_{s<\tau_n<..<\tau_1<t} H_1(\tau_1)H_1(\tau_2)...H_1(\tau_n)d\tau_1...d\tau_n$$

$$= U_0(t-s) + \sum_{n=1}^{\infty}(-i)^n \in^n \int_{s<\tau_n<..\tau_1<t} U_0(t-\tau_1)V_1(\tau_1)U_0(\tau_1-\tau_2)V_1(\tau_2)$$

$$U_0(\tau_{n-1}-\tau_n)V_1(\tau_n)$$

$$U_0(\tau_n - s)d\tau_1...d\tau_n$$

where $U_0(t) = \exp(-itH_0)$.

Maxwell equation in cosmology. The Robertson-Walker metric Evolution of the universe assuming homogeneity and isotropivity.

$$d\tau^2 = dt^2 - \frac{R^2(t)dr^2}{1-kr^2} - R^2(t)r^2\left(d\theta^2 + \sin^2\theta d\varphi^2\right)$$

$$X^0 = t, X^1 = r, X^2 = 0, X^3 = \varphi.$$

$$g_{00} = 1, g_{11} = -\frac{R^2(t)}{1-kr^2}, g_{22} = -R^2(t)r^2, g_{33} = -R^2(t)^{r^2}\sin^2\theta$$

$$R_{\mu\nu} - \frac{1}{2}Rg_{\mu\nu} = k(T_{\mu r}) \tag{1}$$

$$T_{\mu\nu} = (p+\rho)v_\mu v_\nu - pg_{\mu\nu}$$

The frame (t, r, θ, φ) is comoving *i.e.* Geodesic are r = constant, θ = constant, φ = constant. It can be verified that these satisfy the geodesic equation.

$$\frac{d^2x^\mu}{d\tau^2} + \Gamma^\mu_{\nu\sigma}\frac{dx^\nu}{d\tau}\frac{dx^\sigma}{d\tau} = 0$$

(Since $\Gamma^r_{00} = 0$, $r = 1, 2, 3$). Thus $(v^\mu) = (1, 0, 0, 0) = (v_\mu)$.

(1) $\Rightarrow$

$$-R = kT = k(p + \rho - 4p) = k(\rho - 3p)$$

$$R = K(3p - \rho).$$

(1). can be expressed as

$$R_{\mu\nu} = \frac{1}{2}Rg_{\mu\nu} + kT_{\mu\nu}$$

$$= \frac{K}{2}(3p-\rho)g_{\mu\nu} + K((p+\rho)v_\mu v_\nu - pg_{\mu\nu})$$

$$= K(p+\rho)v_\nu v_\nu + kg_{\mu\nu}\left(\frac{p}{2}-\frac{\rho}{2}\right)$$

i.e.

$$R_{\mu\nu} = K\left\{(p+\rho)v_\mu v_\nu + \left(\frac{p-\rho}{2}\right)g_{\mu\nu}\right\}$$

$$R_{00} = K\left\{p+\rho+\frac{p-\rho}{2}\right\}$$

$$= \frac{K}{2}\{3p+\rho\} \qquad (2a)$$

$$R_{kk} = \frac{K}{2}(p-\rho)g_{kk,}\ 1 \le k \le 3. \qquad (2b)$$

i.e.

$$R_{00} = \frac{K}{2}(3p+\rho) \qquad p : p(t),\ \rho = \rho(t)\ (\because \text{ of homogeniety})$$

$$R_{11} = -\frac{K}{2}\frac{(p-\rho)R^2(t)}{1-kr^2},$$

$$R_{22} = -\frac{K}{2}(p-\rho)R^2(t)r^2,$$

$$R_{23} = -\frac{K}{2}(p-\rho)R^2(t)r^2\sin^2\theta$$

$$R_{\mu\nu} = \Gamma^{\alpha}_{\mu\alpha,\nu} - \Gamma^{\alpha}_{\mu\nu,\alpha} - \Gamma^{\alpha}_{\mu\nu}\Gamma^{\beta}_{\alpha\beta} + \Gamma^{\alpha}_{\mu\beta}\Gamma^{\beta}_{\nu\alpha}$$

$$R_{00} = \Gamma^{\alpha}_{0\alpha,0} - \Gamma^{\alpha}_{00,\alpha} - \Gamma^{\alpha}_{00}\Gamma^{\beta}_{\alpha\beta} + \Gamma^{\alpha}_{0\beta}\Gamma^{\beta}_{0\alpha}$$

$$\Gamma^{\alpha}_{0\alpha} = \Gamma^{0}_{00} + \Gamma^{1}_{01} + \Gamma^{2}_{02} + \Gamma^{3}_{03}$$

$$\Gamma^{0}_{00} = \frac{1}{2}g^{00}g_{00,0} = 0,\ \Gamma^{1}_{01} = \frac{1}{2}g''(g_{11,0}) = \frac{1}{2}(\log g_{11})_{10} = \frac{R'(t)}{R(t)},$$

$$\Gamma^{2}_{02} = \frac{1}{2}g(\log g_{22})_{,0} = \frac{R'(t)}{R(t)},\ \Gamma^{3}_{03} = \frac{1}{2}(\log g_{33})_{,0} = \frac{R'(t)}{R(t)}.$$

$$\therefore \qquad \Gamma^{\alpha}_{0\alpha} = \frac{R'(t)}{R(t)}.$$

$$\Gamma^{\alpha}_{00} = 0,\ \Gamma^{\alpha}_{00}\Gamma^{\beta}_{\alpha\beta} = 0,\ \Gamma^{\alpha}_{0\beta}\Gamma^{\beta}_{0\alpha}$$

$$= \left(\Gamma^{1}_{01}\right)^2 + \left(\Gamma^{2}_{02}\right)^2 + \left(\Gamma^{3}_{03}\right)^2 = 3\left(\frac{R'}{R}\right)^2.$$

So,

$$R_{00} = \frac{3R'}{R} + 3\frac{R'^2}{R^2}$$

$$R_{11} = \Gamma^{\alpha}_{1\alpha,1} - \Gamma^{\alpha}_{11,\alpha} - \Gamma^{\alpha}_{11}\Gamma^{\beta}_{\alpha\beta} + \Gamma^{\alpha}_{1\beta}\Gamma^{\beta}_{1\alpha}$$

$$\Gamma^{\alpha}_{1\alpha} = \Gamma^{1}_{11} + \Gamma^{2}_{12} + \Gamma^{3}_{13}$$

$$\Gamma'_{11} = \frac{1}{2}g^{11}g_{11,1} = \frac{1}{2}(\log g_{11})_{,1} = \frac{-Kr}{(-Kr^2)}$$

$$\Gamma^{2}_{12} = \frac{1}{2}g^{22}g_{22,1} = \frac{1}{2}(\log g_{22})_{,1} = \frac{1}{r},$$

$$\Gamma^{3}_{13} = \frac{1}{2}(\log g_{33})_{,1} = \frac{1}{r}$$

So,
$$\Gamma^{1}_{1\alpha,1} = \left(\frac{-kr}{1-kr^2} + \frac{2}{r}\right)_{,1}$$

$$= \frac{-k}{1-kr^2} - \frac{2k^2r^2}{(1-kr^2)^2} - \frac{2}{r^2}$$

$$= \frac{\left(k+k^2r^2\right)}{\left(1-kr^2\right)} - \frac{2}{r^2}$$

$$\Gamma^{\alpha}_{11,\alpha} = \Gamma^{0}_{11,0} + \Gamma^{1}_{11,1}$$

$$\Gamma^{0}_{11} = \frac{1}{2}g^{00}g_{11},_0$$

$$= -\frac{1}{2}g_{11,0}\frac{R(t)R'(t)}{\left(1-kr^2\right)}, \Gamma^{1}_{11} = \frac{-kr}{1-kr^2}, \Gamma^{1}_{11,1}$$

$$= \frac{-k}{1-kr^2} - \frac{2k^2r^2}{\left(1-kr^2\right)^2}$$

$$\Gamma^{\alpha}_{11}\Gamma^{\beta}_{\alpha\beta} = \Gamma^{0}_{11}\Gamma^{\beta}_{0\beta} = \Gamma^{0}_{11}\left(\Gamma^{1}_{01} + \Gamma^{2}_{02} + \Gamma^{3}_{03}\right)$$

$$+ \Gamma^{1}_{11}\Gamma^{\beta}_{\alpha\beta} + \Gamma^{1}_{11}\left(\Gamma^{1}_{11} + \Gamma^{2}_{12} + \Gamma^{3}_{13}\right)$$

$$\Gamma_{01} = -\frac{1}{2}g^{00}g_{11,0} = -\frac{1}{2}g_{11,0} = \frac{RR'}{1-kr'^2}$$

$$\Gamma^{1}_{01} + \Gamma^{2}_{02} + \Gamma^{3}_{03} = 3R'/R.$$

$$\Gamma^{1}_{11} = \frac{-kr}{1-kr^2}, \Gamma^{1}_{11} + \Gamma^{2}_{12} + \Gamma^{3}_{13} = \frac{-kr}{1-kr^2} + \frac{2}{r}.$$

So
$$\Gamma^{\alpha}_{11}\Gamma^{\beta}_{\alpha\beta} = \frac{RR'}{\left(1-kr^2\right)}\left[\frac{3R'}{R}\right] - \frac{kr}{1-kr^2}\left[\frac{-kr}{1-kr^2} + \frac{2}{r}\right]$$

$$= \frac{3R'^2}{\left(1-kr^2\right)} + \frac{k^2r^2}{\left(1-kr^2\right)^2} - \frac{2k}{\left(1-kr^2\right)}$$

$$\Gamma^{\alpha}_{1\beta}\Gamma^{\beta}_{1\alpha} = \left(\Gamma^1_{11}\right)^2 + \left(\Gamma^2_{12}\right)^2 + \left(\Gamma^3_{13}\right)^2 + 2\Gamma^1_{10}\Gamma^0_{11}$$

$$= \frac{k^2r^2}{\left(1-kr^2\right)^2} + \frac{2}{r^2} + \left(\frac{2R'}{k}\right)\left(\frac{RR'}{1-kr^2}\right)$$

So

$$R_{11} = -\frac{\left(k+k^2r^2\right)}{\left(1-kr^2\right)^2} - \frac{2}{r^2} - \frac{(RR')}{\left(1-kr^2\right)} + \frac{k}{1-kr^2} + \frac{2k^2r^2}{\left(1-kr^2\right)^2}$$

$$= -\frac{3R/2}{\left(1-kr^2\right)} - \frac{k^2r^2}{\left(1-kr^2\right)^2} + \frac{2k}{\left(1-kr^2\right)}$$

$$= +\frac{k^2r^2}{\left(1-kr^2\right)^2} + \frac{2}{r^2} + \frac{2R/2}{\left(1-kr^2\right)}$$

$$= -\frac{\left(2R'^2+RR''\right)}{\left(1-kr^2\right)} + \frac{2k\left(1-kr^2\right)}{\left(1-kr^2\right)^2}$$

$$= \left(-(2R'+RR'')+2k/\left(1/2r^2\right)\right)$$

The R_{00} and R_{11} equation are therefore

$$3\left(\frac{R'}{R}+\frac{R'^2}{R^2}\right) = \frac{k}{2}(3p+\rho), \qquad (\alpha)$$

$$2k-\left(RR''+2R'^2\right) = -\frac{k}{2}(\rho-\rho)R^2 \qquad (\beta)$$

If we set $p = 0$ then there equation become

$$3\left(\frac{R'}{R}+\frac{R'^2}{R^2}\right) = \frac{k\rho}{2},\ -2k+\left(RR''+2R^2\right) = \frac{-K}{2}\rho R^2$$

So we get in this case,

$$-\frac{2k}{R^2}+\frac{R''}{R}+\frac{2R'^2}{R^2}$$

❖❖❖❖❖

[20] Derive Dirac Equation in Curved Space-Time Using the Newman Penrose Formalism: Special-Relativity:

$$\sigma^{\mu}_{AB}\, X^{A}_{,\mu} + im\,\eta_B = 0 \tag{1a}$$

$$\sigma^{\mu}_{AB}\,\overline{A^{1}}\,\bar{\eta}^{A}_{,\mu} + im\,\bar{\chi}_B = 0 \tag{1b}$$

$$\chi_A = \varepsilon_{AB}\chi^{B},\ \eta_A = \varepsilon_{AB}\eta^{B}, \chi^{A} = \varepsilon^{AB}\chi_B,$$

$$\eta^{A} = \varepsilon^{AB}\eta_B, ((\varepsilon_{AB}\eta)) = \begin{pmatrix} 0 & 1 \\ -1 & 0 \end{pmatrix} = \varepsilon\ ((\varepsilon^{AB}))$$

$$= ((\varepsilon_{AB}))^{-1} = \begin{pmatrix} 0 & -1 \\ 1 & 0 \end{pmatrix}$$

(1) Can be expressed as (2)

$$\sigma^{0T}\partial_0\underline{\chi} + \sum_{r=1}^{3}\sigma^{rT}\partial_r\chi + im\varepsilon\,\underline{\eta} = 0 \tag{3}$$

$$\sigma^{0} = I_2, \sigma^{1} = \begin{pmatrix} 0 & 1 \\ 1 & 0 \end{pmatrix}, \sigma^{2} = \begin{pmatrix} 0 & -i \\ i & 0 \end{pmatrix}, \sigma^{3} = \begin{pmatrix} 1 & 0 \\ 0 & -1 \end{pmatrix}$$

$$\sigma^{0T}\partial_0\bar{\eta} + \sum_{r=1}^{3}\sigma^{rT}\partial_r\bar{\eta} + im\,\varepsilon\,\bar{\chi} = 0$$

Thus

$$\partial_0\,\chi + \sum_{r=1}^{3}\sigma^{rT}\partial_r\underline{\chi} + im\,\varepsilon\,\underline{\eta} = 0 \tag{2a}$$

$$\partial_0\underline{\eta} + \sum_{r=1}^{3}\sigma^{r}\partial_r\underline{\eta} - im\,\varepsilon\,\underline{\chi} = 0 \tag{2b}$$

Where $\underline{\chi} = \left(\chi^{A}\right)^{2}_{A=1}, \underline{\eta} = \left(\eta^{A}\right)^{2}_{A=1}$

Note: $\bar{\sigma}^{rT} = \sigma^{r*} = \sigma^{r}$

(2) Gives on multiplying by i and using $p_0 = i\partial_0$, $p^{r} = -\,i\partial r$,

$$p^{0}\chi - \sum_{r=1}^{3}\sigma^{rT}p^{r}\chi - m\varepsilon\eta = 0$$

$$p^{0}\eta - \sum_{r=1}^{3}\sigma^{r}p^{r}\eta - m\varepsilon\chi = 0,$$

Or

$$\left[p^{0} - \begin{pmatrix} (\sigma^{T}, p) & 0 \\ 0 & (\sigma, p) \end{pmatrix} - m\begin{pmatrix} 0 & \varepsilon \\ -\varepsilon & 0 \end{pmatrix}\right]\begin{pmatrix} \chi \\ \eta \end{pmatrix} = 0$$

The Dirac Hamiltonian operator

$$H_0 = \begin{pmatrix} (\sigma^T, p) & 0 \\ 0 & (\sigma, p) \end{pmatrix} + m\begin{pmatrix} 0 & \varepsilon \\ -\varepsilon & 0 \end{pmatrix}$$

is clearly Hermitian a $\varepsilon^T = -\varepsilon$.

$$\begin{pmatrix} o^{rT} & 0 \\ 0 & \sigma^r \end{pmatrix}\begin{pmatrix} 0 & \varepsilon \\ -\varepsilon & 0 \end{pmatrix} + \begin{pmatrix} 0 & \varepsilon \\ -\varepsilon & 0 \end{pmatrix}\begin{pmatrix} \sigma^{rT} & 0 \\ 0 & \sigma^r \end{pmatrix}$$

$$= \begin{pmatrix} 0 & \sigma^{rT}\varepsilon \\ -\sigma^r\varepsilon & 0 \end{pmatrix} + \begin{pmatrix} 0 & \varepsilon\sigma^r \\ -\varepsilon\sigma r^T & 0 \end{pmatrix}$$

$$= \begin{pmatrix} 0 & \sigma^{rT}\varepsilon + \varepsilon\sigma^r \\ -\left(\sigma^r\varepsilon + \varepsilon\sigma^{rT}\right) & 0 \end{pmatrix}$$

$$\sigma^{1T}\varepsilon + \varepsilon\sigma^1 = \begin{pmatrix} 0 & 1 \\ 1 & 0 \end{pmatrix}\begin{pmatrix} 0 & 1 \\ -1 & 0 \end{pmatrix} + \begin{pmatrix} 0 & 1 \\ -1 & 0 \end{pmatrix}\begin{pmatrix} 0 & 1 \\ 1 & 0 \end{pmatrix}$$

$$= \begin{pmatrix} -1 & 0 \\ 0 & 1 \end{pmatrix} + \begin{pmatrix} 1 & 0 \\ 0 & -1 \end{pmatrix} = 0.$$

$$\sigma^{2T}\varepsilon + \varepsilon\sigma^2 = \begin{pmatrix} 0 & i \\ -i & 0 \end{pmatrix}\begin{pmatrix} 0 & 1 \\ -1 & 0 \end{pmatrix} + \begin{pmatrix} 0 & 1 \\ -1 & 0 \end{pmatrix}\begin{pmatrix} 0 & -i \\ i & 0 \end{pmatrix}$$

$$= \begin{pmatrix} -i & 0 \\ 0 & -i \end{pmatrix} + \begin{pmatrix} i & 0 \\ 0 & i \end{pmatrix} = 0$$

$$\sigma^{3T}\varepsilon + \varepsilon\sigma^3 = \begin{pmatrix} 1 & 0 \\ 0 & -1 \end{pmatrix}\begin{pmatrix} 0 & 1 \\ -1 & 0 \end{pmatrix} + \begin{pmatrix} 0 & 1 \\ -1 & 0 \end{pmatrix}\begin{pmatrix} 1 & 0 \\ 0 & -1 \end{pmatrix}$$

$$= \begin{pmatrix} 0 & 1 \\ 1 & 0 \end{pmatrix} + \begin{pmatrix} 0 & -1 \\ -1 & 0 \end{pmatrix} = 0$$

Thus, $\quad \sigma^{rT}\varepsilon + \varepsilon\sigma^r = 0$

and likewise premultiplying this equation by ε and postmultiplying by ε, we get

$$\varepsilon\sigma^{rT} + \sigma^r\varepsilon = 0$$

Thus

$$\left\{\begin{bmatrix} \begin{pmatrix} \sigma^{rT} & 0 \\ 0 & \sigma^r \end{pmatrix}, \begin{pmatrix} 0 & \varepsilon \\ -\varepsilon & 0 \end{pmatrix} \end{bmatrix}\right\} = 0, r = 1, 2, 3$$

(1) Thus implies on pre multiplying by

$$p^0 + \begin{pmatrix} (\sigma^T, p) & 0 \\ 0 & (\sigma, p) \end{pmatrix} + m\begin{pmatrix} 0 & \varepsilon \\ -\varepsilon & 0 \end{pmatrix}$$

and using

$$\left\{\begin{pmatrix} \sigma^{rT} & 0 \\ 0 & \sigma^r \end{pmatrix}, \begin{pmatrix} \sigma^{sT} & 0 \\ 0 & \sigma^s \end{pmatrix}\right\} = 2\delta_{rs} I$$

The Klein-Gordon equation

$$\left[p^{02} - \sum_{r=1}^{3} p^{r2} - m^2\right]\begin{pmatrix} \chi \\ \eta \end{pmatrix} = 0$$

In gtr we set

$$\sigma^{\mu}_{(ab)} = e^{\mu}_{(ab)} = \sigma^{\mu}_{AB}\zeta^A_a \bar{\zeta}^B_b$$

with ζ^A_a a dyad. We set

$$\zeta_{aA} = \varepsilon_{AB}\zeta^B_a \text{ and so } \zeta^A_a = \varepsilon^{AB}\zeta_{aB}$$

Also $$\zeta^a_A = \varepsilon^{ab}\zeta_{bA}, \zeta_{aA} = \varepsilon_{ab}\zeta^b_A,$$

When $$((\varepsilon_{ab})) = \begin{pmatrix} 0 & 1 \\ -1 & 0 \end{pmatrix}$$

We write $$e^{\mu}_{(00)} = l^{\mu}, e^{\mu}_{(01)} = m^{\mu},$$

$$e^{\mu}_{(10)} = \bar{m}^{\mu}, e^{\mu}_{(11)} = n^{\mu}$$

Just as a tetrad basis is str transforms 4 vectors into 4-scalars, the dyad basis in str transforms 2-spinors into 2-scalars.

Thus, $$\left(\left(e^{\mu}_{(ab)}\right)\right) = \left(\left(\sigma^{\mu}_{ab}\right)\right) = \begin{pmatrix} l^{\mu} & m^{\mu} \\ \bar{m}^{\mu} & n^{\mu} \end{pmatrix}$$

σ^{μ}_{AB} transform bispinors into 4-vectors:

$$\sigma^{\mu}_{AB} p^{AB} = p^{\mu}$$

and if $((\sigma^{AB}_{\mu}))$ denotes the inverse of $((\sigma^{\mu}_{AB}))$ in the trade that

$$\sigma^{AB}_{\mu}\sigma^{\nu}_{AB} = \delta^{\nu}_{\mu},$$

$$\sigma^{AB}_{\mu}\sigma^{\mu}_{CD} = \delta^A_C\delta^B_D.$$

Where $$\sigma^{AB}_{\mu} = \varepsilon^{AC}\varepsilon^{BD} y_{\mu\nu}\sigma^{\nu}_{CD}$$

We verify: $$\sigma^{AB}_{\mu}\sigma^{\nu}_{AB} = \varepsilon^{AC}\varepsilon^{BD} g_{\mu\rho}\sigma^{\rho}_{CD}\sigma^{\nu}_{AB}$$

$$\begin{aligned}
\varepsilon^{AC}\varepsilon^{BD}\sigma^{\rho}_{CD}\sigma^{\nu}_{AB} &= \varepsilon^{ac}\varepsilon^{bd}\sigma^{\rho}_{cd}\sigma^{\nu}_{ab} \\
&= \varepsilon^{01}\varepsilon^{01}\sigma^{\rho}_{11}\sigma^{\nu}_{00} + \varepsilon^{01}\varepsilon^{10}\sigma^{\rho}_{10}\sigma^{\nu}_{01} \\
&= \varepsilon^{10}\varepsilon^{01}\sigma^{\rho}_{11}\sigma^{\nu}_{10} + \varepsilon^{10}\varepsilon^{10}\sigma^{\rho}_{00}\sigma^{\nu}_{11}
\end{aligned}$$

$$= n^{\rho} l^{\nu} - \overline{m}^{\rho} m^{\nu} - m^{\rho} \overline{m}^{\nu} + l^{\rho} n^{\nu}$$

$$= g^{\nu\rho} = g^{\rho\nu}$$

Now comes Friedman's lemma.

$$\varepsilon^{kf} \zeta_a^E \overline{\zeta}_f^F \left(\zeta_{bE} \overline{\zeta}_{kF}\right)_{:\mu} = \varepsilon^{kf} \zeta_a^E \overline{\zeta}_f^F \left(\zeta_{bE:\mu} \overline{\zeta}_{kF} + \zeta_{bE} \overline{\zeta}_{kF:\mu}\right)$$

$$= \varepsilon^{kf} \varepsilon_{afk} \zeta_a^E \zeta_{bE:\mu} + \varepsilon^{kf} \varepsilon_{ab} \overline{\zeta}_f^F \overline{\zeta}_{kF:\mu}$$

$$= 2\zeta_a^E \zeta_{bE:\mu} + \varepsilon_{ab} \overline{\zeta}^{kF} \overline{\zeta}_{kF:\mu} = 2\zeta_a^E \zeta_{bE:\mu}$$

Since $\qquad \zeta^{kF} \zeta_{kF:\mu} = \frac{1}{2}\left(\zeta^{kF} \zeta_{kF}\right)_{:\mu} = 0$

Since $\qquad \zeta^{kF} \zeta_{kF} = \delta_k^k = 2$ a scalar constant field.

Note: $\qquad \zeta_a^E \zeta_{bE} = \varepsilon_{ab} \equiv \begin{pmatrix} 0 & 1 \\ -1 & 0 \end{pmatrix}$

Now consider $\sigma^{\mu}_{AB} \chi^A_{:\mu}$. We have

$$\chi^A_{:\mu} = \left(\zeta_a^A \chi^a\right)_{:\mu} = \zeta^A_{a:\mu} \chi^a + \zeta_a^A \chi^a_{,\mu}$$

Since $\qquad \chi_a = \zeta_A^a \chi^A$ is a scalar for $a = 0$.

$\therefore \qquad \sigma^{\mu}_{AB} \chi^A_{:\mu} = \sigma^{\mu}_{AB} \zeta_a^A \chi^a_{,\mu} + \sigma^{\mu}_{AB} \zeta^A_{a:\mu} \chi^c$

Now $\qquad \zeta^A_{a:\mu} = \left(\varepsilon^{AB} \zeta_{aB}\right)_{:\mu}$

$$= \varepsilon^{AB} \zeta_{aB:\mu} \text{ since } \varepsilon^{AB}_{:\mu} = 0$$

So, $\qquad \sigma^{\mu}_{AB} \zeta^A_{a:\mu} \chi^a = \sigma^{\mu}_{AB} \chi^a \varepsilon^{AC} \zeta_{ac:\mu} = \sigma^{\mu}_{Ab} \chi^a \zeta_b^A \zeta^{bc} \zeta_{ac:\mu}$

(Note that $\varepsilon^{ab} = \zeta_A^a \zeta^{bA}, \varepsilon^{AB} = \zeta_a^A \zeta^{aB} \varepsilon^{ab} = \varepsilon^{AB} \zeta_A^a \zeta_B^b = \zeta_A^a \varepsilon^{AB} \zeta_B^b = \zeta_A^a \zeta^{bA}$)

Using Friedman's lemma, we thus get

$$\sigma^{\mu}_{AB} \zeta^A_{a:\mu} \chi^a = \sigma^{\mu}_{AB} \chi^a \zeta_b^A \zeta^{bc} \zeta_{ac:\mu}$$

$$= \sigma^{\mu}_{AB} \chi^a \zeta^{bA} \zeta_b^c \zeta_{ac:\mu}$$

$$= \frac{1}{2} \sigma^{\mu}_{AB} \chi^a \zeta^{bA} \left(\varepsilon^{kf} \zeta_b^E \overline{\zeta}_f^F \left(\zeta_{aE} \overline{\zeta}_{kF}\right)_{:\mu} \right)$$

$$= \frac{1}{2} \sigma^{\mu}_{AB} \chi^a \varepsilon^{AE} \overline{\zeta}^{kF} \left(\zeta_{aE} \overline{\zeta}_{kF}\right)_{:\mu}$$

$$= \frac{1}{2}\sigma^{\mu}_{AB}\,\chi^{a}\,\bar{\zeta}^{kF}\left(\varepsilon^{AE}\,\zeta_{aE}\,\bar{\zeta}_{kF}\right)_{:\mu}$$

$$= \frac{1}{2}\sigma^{\mu}_{AB}\,\chi^{a}\,\bar{\zeta}^{kF}\left(\varepsilon^{A}_{a}\,\bar{\zeta}_{kF}\right)_{:\mu}$$

$$= \frac{1}{2}e^{\mu}_{(pq)}\,\zeta^{ap}_{A}\,\zeta^{bq}_{B}\,\chi^{a}\,\bar{\zeta}^{kF}\left(\zeta^{A}_{a}\,\bar{\zeta}_{kF}\right)_{:\mu}$$

No: $\zeta^{p}_{A}\bar{\zeta}^{kF}\left(\zeta^{A}_{a}\bar{\zeta}_{kF}\right)_{:\mu} = \zeta^{pA}\bar{\zeta}^{kF}\left(\zeta_{aA}\bar{\zeta}_{kF}\right)_{:\mu}$

Now, $\left(\zeta aA\bar{\zeta}kF\right)_{:\mu} = \left(\sigma^{\nu}_{AF}\,e_{\nu}\,(ak)\right)_{:\mu}$

$$\Big(\sigma^{\nu}_{AF}\,\zeta^{A}_{a}\,\bar{\zeta}^{F}_{k} = e^{\nu}_{(ak)}$$

So $\zeta^{A}_{a}\bar{\zeta}^{F}_{k} = \sigma^{AF}_{\nu}\,e^{\nu}_{(ak)}$

or $\zeta_{aA}\bar{\zeta}_{kF} = \sigma_{\nu AF}e^{\nu}_{(ak)} = \sigma^{\nu}_{AF}\,e_{\nu(ak)}\Big)$

$$= \sigma^{\nu}_{AF}\,e_{\nu(ak):\mu}$$

Thus $\sigma^{\mu}_{AB}\,\zeta^{A}_{a:\mu}\,\chi^{a} = \frac{1}{2}e^{\mu}_{(pq)}\,\bar{\zeta}^{q}_{B}\,\chi^{a}\,\zeta^{pA}\,\bar{\zeta}^{kF}\,\sigma^{\nu}_{AF}\,e_{\nu(ak):\mu}$

$$= \frac{1}{2}e^{\mu}_{(pq)}\,e_{\nu(ak):\mu}\sigma^{\nu(pk)}\,\bar{\zeta}^{q}_{B}\chi^{a} \qquad \text{(Note: } s^{\nu(pk)} = e^{\nu(pk)}\text{)}$$

Thus $\sigma^{\mu}_{AB}\,\chi^{A}_{:\mu} = \sigma^{\mu}_{AB}\,\zeta^{A}_{:a}\,\chi^{a}_{,\mu} + \frac{1}{2}e^{\mu}_{(pq)}\,e_{\nu(ak):\mu}e^{\nu(pk)}\,\bar{\zeta}^{q}_{B}\chi^{a}$

and thus, $\bar{\zeta}^{B}_{br}\,\sigma^{\mu}_{AB}\,\chi^{A}_{:\mu} = e^{\mu}_{(ar)}\,\chi^{a}_{,\mu} + \frac{1}{2}e^{\mu}_{(pq)}e_{\nu(ak):\mu}e^{\nu(pk)} = \bar{\zeta}^{q}_{B}\,\bar{\zeta}^{B}_{r}\,\chi^{a}$

$$= e^{\mu}_{(ar)}\chi^{a}_{,\mu} + \frac{1}{2}e^{\mu}_{(pr)}e_{\nu(ak):\mu}e^{\nu(pk)}\,\chi^{a}$$

$$(\zeta^{q}_{B}\,\zeta^{B}_{r} = \delta^{q}_{r})$$

The general relativistic Dirac equation are

$$\sigma^{\mu}_{AB}\,\chi^{A}_{:\mu} + im\,\bar{\eta}_{B} = 0,$$

$$\sigma^{\mu}_{AB}\,\bar{\eta}^{A}_{:\mu} + im\,\bar{\chi}_{B} = 0$$

and hence we get from these on premultiplying by $\bar{\zeta}^{B}_{r}$,

$$e^{\mu}_{(ar)}\chi^{a}_{,\mu} + \frac{1}{2}e^{\mu}_{(pr)}e_{\nu(ak):\mu}e^{\nu(pk)}\chi^{a} + im\,\bar{\eta}_{r} = 0 \qquad (\alpha)$$

and

$$e^{\mu}_{(ar)}\eta^{a}_{,\mu} + \frac{1}{2}e^{\mu}_{(pr)}e_{\nu(ak):\mu}e^{\nu(pk)}\eta^{a} + im\bar{\chi}_{r} = 0 \qquad (\beta)$$

(α) and (β) constitute the Dirac equation in gtr.

Note:

$$e_{\mu(pk)} = g_{\mu\nu}e^{\nu}_{(ak)}, e^{\nu(pk)} = \varepsilon^{pa}\,\varepsilon^{kb}\,e^{\nu}_{(ab)}$$

where
$$\left(\left(\varepsilon^{ab}\right)\right) = \begin{pmatrix} 0 & -1 \\ 1 & 0 \end{pmatrix},$$

$$\left(\left(e^{\mu}_{(ab)}\right)\right) = \begin{pmatrix} l^{\mu} & m^{\mu} \\ \overline{m}^{\mu} & n^{\mu} \end{pmatrix},$$

❖❖❖❖❖

[21] Proofs of Identities from S. Chandra Sekhar, " The Mathematical Theory of Blackholes". Cartain's Second Equation of Structure:

$$R(X, Y)Z = \nabla_X Z - \nabla_Y \nabla_X Z - \nabla[X, Y]\, Z$$

i.e. $$R(X, Y) = [\nabla^X \nabla_Y] - \nabla_{[X, Y]}.$$

$$\nabla_{e_\alpha} e_\beta = \Gamma^{\sigma}_{\alpha\beta}\, e_\sigma .$$

$$\nabla_X e_\alpha = \nabla_{X^\sigma e_\sigma} e_\alpha = X^\sigma \nabla_{e_\sigma} e_\alpha = X^\sigma \Gamma^{\beta}_{\sigma\alpha}\, e_\beta = \omega^{\beta}_{\alpha}(X) e_\beta, \omega^{\beta}_{\alpha}(X)$$

$$= \Gamma^{\beta}_{\alpha\alpha} X^\sigma .$$

$$\omega^{\beta}_{\alpha} = \Gamma^{\beta}_{\sigma\alpha}\, e^\sigma$$

$$d\omega^{\beta}_{\alpha} = d\Gamma^{\beta}_{\sigma\alpha} \Lambda e^\sigma + \Gamma^{\beta}_{\sigma\alpha}\, de^\sigma$$

$$R(X,Y)e_\alpha = \nabla_X \nabla_Y e_\alpha - \nabla_Y \nabla_X\, e_\alpha - \nabla_{[X,Y]}\, e_\alpha$$

$$= \nabla_X \left(\omega^{\beta}_{\alpha}(Y) e_\beta\right) - \nabla_Y \left(\omega^{\beta}_{\alpha}(X) e_\beta\right) - \omega^{\beta}_{\alpha}([X,Y]) e_\beta$$

$$= \left(X\left(\omega^{\beta}_{\alpha}(Y)\right)\right) - Y\left(\omega^{\beta}_{\alpha}(X)\right) - \omega^{\beta}_{\alpha}([X,Y]) e_\beta$$

$$+ \omega^{\beta}_{\alpha}(Y) \nabla_X\, e_\beta - \omega^{\beta}_{\alpha}(X) \nabla_Y\, e_\beta$$

$$= d\omega^{\beta}_{\alpha}([X,Y]) e_\beta + \left(\omega^{\beta}_{\alpha}(Y)\, \omega^{\sigma}_{\beta}(X) - \omega^{\sigma}_{\beta}(X)\, \omega^{\sigma}_{\beta}(Y)\right) e_\sigma$$

$$= \left(d\omega^{\sigma}_{\alpha}([X,Y]) + \omega^{\sigma}_{\beta} \Lambda\, \omega^{\beta}_{\alpha}([X,Y])\right) e_\sigma$$

Thus,

$$e^\sigma (R(X,Y) e_\alpha) = \left(d\omega^{\sigma}_{\alpha} + \omega^{\sigma}_{\beta} \Lambda\, \omega^{\beta}_{\alpha}\right)([X,Y])$$

Now,

$$\nabla_X \nabla_Y e_\alpha = \nabla_X \left(Y^\sigma\, \Gamma^{\beta}_{\sigma\alpha}\, e_\beta\right)$$

$$= X^\rho \nabla_{e_\rho}\left(Y^\sigma \Gamma^\beta_{\sigma\alpha} e_\beta\right)e_\beta \;\; X^\rho e_\rho\left(Y^\sigma \Gamma^\beta_{\sigma d}\right)e_\beta + X^\rho Y^\sigma \Gamma^\beta_{\rho\alpha} \nabla_{e_\rho} e_\beta$$

$$= X^\rho\left(e_\rho(Y^\sigma)\Gamma^\beta_{\sigma\alpha} + Y^\sigma e_\rho\left(\Gamma^\beta_{\rho\alpha}\right)\right)e_\beta + X^\rho T^\sigma \Gamma^\beta_{\rho\alpha}\Gamma^\mu_{\rho\beta} e_\mu$$

$$\therefore \quad (\nabla_X\nabla_Y - \nabla_Y\nabla_X)e_\alpha = \Gamma^\beta_{\sigma\alpha}(X^\rho e_\rho(Y^\sigma) - Y^\rho e_\rho(X^\sigma))e_\beta$$

$$+ X^\rho Y^\sigma\left(\Gamma^\beta_{\rho\alpha}\Gamma^\mu_{\rho\beta} - \Gamma^\beta_{\sigma\alpha}\Gamma^\mu_{\sigma\beta}\right)e_\mu + X^\rho Y^\sigma\left(e_\rho\left(\Gamma^\beta_{\sigma\alpha}\right) - e_\alpha\left(\Gamma^\beta_{\rho\alpha}\right)\right)e_\beta$$

$$= \Gamma^\beta_{\sigma\alpha} e^\sigma([X,Y])e_\beta + X^\rho Y^\sigma\left(\Gamma^\beta_{\sigma\alpha}\Gamma^\mu_{\rho\beta} - \Gamma^\beta_{\sigma\alpha}\Gamma^\mu_{\sigma\beta} + e_\rho\left(\Gamma^\mu_{\sigma\alpha}\right) - e_\sigma\left(\Gamma^\mu_{\rho\alpha}\right)\right)e_\mu$$

$$= \nabla_{[X,Y]}e_\alpha + X^\rho Y^\sigma\left(\Gamma^\beta_{\sigma\alpha}\Gamma^\mu_{\rho\beta} - \Gamma^\beta_{\sigma\alpha}\Gamma^\mu_{\sigma\beta} + e_\rho\left(\Gamma^\mu_{\sigma\alpha}\right) - e_\sigma\left(\Gamma^\mu_{\rho\alpha}\right)\right)e_\mu$$

Thus,

$$R(X,Y)e_\alpha = \left(\nabla_X\nabla_Y - \nabla_Y\nabla_X - \nabla_{[X,Y]}\right)e_\alpha$$

$$= X^\rho Y^\sigma\left(\Gamma^\beta_{\sigma\alpha}\Gamma^\mu_{\sigma\beta} - \Gamma^\beta_{\rho\alpha}\Gamma^\mu_{\sigma\beta} + e_\rho\left(\Gamma^\mu_{\sigma\alpha}\right) - e_\sigma\left(\Gamma^\mu_{\rho\alpha}\right)\right)e_\mu$$

$$\equiv X^\rho Y^\sigma R^\mu_{\rho\sigma\alpha} e_\mu$$

So $\quad e^\sigma(R(X,Y))e_\alpha = X^\rho Y^\mu R^\sigma_{\rho\mu\alpha} = \frac{1}{2}R^\sigma_{\rho\mu\alpha} e^\rho \Lambda e^\mu(X,Y)$

Thus we get Cartan's second equation of structure:

$$\frac{1}{2}R^\sigma_{\rho\mu\alpha} e^\rho \delta e^\mu = d\omega^\sigma_\alpha + \omega^\sigma_\beta \Lambda \omega^\beta_\alpha .$$

Verification of identities from S. Chandrasekhar's book Mathematical theory of Blackholes.

$$d(F\omega') = F d\omega' + F_{,A} dX_A \Lambda d\omega' + F_{,1} dX' \Lambda d\omega'$$

$$\omega' = e^\Psi\left(dX' - \sum q_A dX^A\right)$$

$$d\omega' = e^\Psi \Psi_{,1} dX' \Lambda\left(dX' - q_A dX^A\right) + e^\Psi \psi_{,B} dX^B \Lambda\left(dX' - q_A dX^A\right)$$

$$- e^\Psi q_{A,B} dX^B \Lambda dX^A - e^\Psi q_{A,1} dX' \Lambda dX^A$$

$$= -e^\Psi(\Psi_{,1} q_A + q_{A,1})dX' \Lambda dX^A + e^\Psi \psi, A dX^A L dX'$$

$$- e^\Psi\left(\psi_{,B} q_A dX^B \Lambda dX^A + q_{A,B}\right)$$

$$= e^\Psi(\Psi_{,1} q_A + q_{A,1})dX^A \Lambda dX' + \Psi_{,A} + e^\Psi(\Psi_{,B} q_A + q_{A,B})dX^A \Lambda dX^B$$

$$\Psi_{,1} q_A + \Psi_{,A} = \Psi_{1;A}$$

$$q_{B:A} = q_{B,A} + q_A q_{B,1}$$

$$dX^A = e^{-\mu A}w^A, dX^1 = e^{-\psi}w^1 + q_A dX^A$$
$$= e^{-\psi}w^1 + q_A e^{-\mu A}w^A.$$

So, $$dw⊠ = e^{\psi}\left(\psi_{:A} + q_{A,1}\right)e^{-\mu A}w^A \Lambda\left(e^{-\psi}w^1 + q_B e^{-\mu_B}w^B\right)$$
$$+ e^{\psi}\left(\psi_{,B} + q_A + q_{A,B}\right)e^{-(\mu_A+\mu_B)}w^A \Lambda w^B$$
$$= e^{-\mu A}\left(\psi_{:A} + q_{A,1}\right)w^A \Lambda w^1$$
$$+ e^{\psi-\mu A-\mu B}\left(q_B\left(\psi_{:A} + q_{A,1}w^A X\, w^B + \psi_{,B}q_A + q_{A,B}\right)\right.$$
$$= e^{-\mu A}\left(\psi_{:A} + q_{A,1}\right)w^A \,\Lambda\, w^1$$
$$+ e^{\psi-\mu_A-\mu_B}\left(q_{A:B} + q_B\psi_{:A} + \psi,_B\, q_A\right)w^A X\, w^B$$

Now $$e^{\psi-\mu_A-\mu_B}\left(q_B + \psi_{:A} + \psi,_B\, q_A\right)w^A \Lambda\, w^B$$
$$= e^{\psi-\mu_A-\mu_B}\left(q_B\left(\psi_{,A} + q_A\psi_{,1}\right) + \psi,_B\, q_A\right)w^A \Lambda\, w^B = 0$$

Since $q_B\psi,_A + \psi,_B\, q_A$ and $q_B q_A \psi_{,1}$

are symmetric in the indices (A, B) while $\omega^A \Lambda\, \omega^B$ is antisymmetric in (A, B).

Thus,
$$d\omega' = e^{-\mu A}(\psi_{:A} + q_{A,1})\omega^A \Lambda\omega' + e^{\psi-\mu A-\mu B}\, q_{A:B}\,\omega^A \Lambda\omega^B$$

Thus,
$$d(F\omega') = F\, e^{-\mu A}\left(\psi_{:A} + q_{A,1}\right)\omega^A \Lambda\, \omega' + Fe^{\psi-\mu A-\mu B}q_{A:B}\,\omega^A \Lambda\omega^B$$
$$+F_{,A}\, e^{-\mu A}\,\omega^A \,\Lambda\, \omega' + F_{,1}\left(e^{-\psi}\omega' + q_A\, dX^A\right)\Lambda\omega'$$

Now, $$dX^A \,\Lambda\, \omega' = e^{-\mu A}\,\omega^A \,\Lambda\, \omega'$$

Thus, $$d(F\omega') = \left(Fe^{-\mu A}\left(\psi_{:A} + q_{A,1}\right) + F_{,A}\, e^{-\mu A} + F_{,1}\, e^{-\mu A}\, q_A\right)\omega^A \Lambda\, \omega'$$
$$+ F\, e^{\psi-\mu A-\mu B}q_{A:B}\,\omega^A \,\Lambda\, \omega^B$$

Now,
$$\mathcal{D}_A(Fe^{\psi}) = (Fe^{\psi})_{:A} + q_{A,1}Fe^{\psi}$$
$$= (Fe^{\psi})_{,A} + q_A(Fe^{\psi})_{,1} + Fq_{a,1}e^{\psi}$$
$$= e^{\psi}\left(F_{,A} + F\psi_{,A} + F_{,1}q_A + Fq_A\psi_{,1} + Fq_{A,1}\right)$$
$$= Fe^{\psi}\psi_{:A} + Fe^{\psi}q_{A,1} + e^{\psi}\left(F_{,1}q_A + F_{,A}\right)$$

Thus,

$$d(Fw') = Fe^{\psi-\mu A-\mu B} q_{A:B} w^A \Lambda w^B + \rho_A(Fe^{\psi}) \cdot e^{-\psi-\mu A} w^A \Lambda w'$$

$$w^A = e^{\mu A}\, dX^4,\ A = 2, 3, 4$$

$$dw^A = e^{\mu A}\left(\mu_{A,1} dX' \Lambda\, dX^A + \mu_{A,B} dX^B \Lambda dX^4\right)$$

(Summation over B only)

$$= e^{\mu A}\mu_{A,1}\left(e^{-\psi} w' + q_B\, dX^B\right)$$

$$\Lambda\, dX^A + e^{\mu A}\mu_{A,B} dX^B \Lambda\, dX^A$$

$$= e^{\mu A-\psi}\mu_{A,1}\mu' \Lambda e^{-\mu A} w^A + \left(e^{\mu A}\mu_{A,1} q_B + e^{\mu A}\mu_{A,B}\right)$$

$$e^{-\mu A-\mu B} w^B \Lambda w^A$$

$$= e^{-\psi}\mu_{A,1} w' \Lambda w^A - e^{\mu B}\left(\mu_{A,B}\, \mu_{A,1}\, q_B\right) w^B \Lambda\, w^A$$

(Formula in eqn., (48)).

$$= -e^{-\psi}\mu_{A,1} w^A \Lambda\, w' - e^{\mu B}\mu_{A:B} w^A \Lambda w^B$$

❖❖❖❖❖

[22] Mathematical Preliminaries for Cartan's Eqn. of Structure:

$$w \in \Omega'(\mu), X, Y \in X(\mu)$$

$$w = w_\alpha\, dX^\alpha,\ X = X^\alpha \partial_\alpha,\ Y = Y^\alpha \partial_\alpha$$

Then,

$$(dW)(X, Y) = w_{\alpha\,\beta} dX^\beta \Lambda dX^\alpha (X, Y) = w_{\alpha,\beta}\left(X^\beta Y^\beta - X^\alpha Y^\beta\right)$$

$$X(w(Y)) = X^\alpha \partial_\alpha\left(w_\beta Y^\beta\right) = w_{\beta,\alpha} X^\alpha Y^\beta + w_\beta X^\alpha Y^\beta_{,\alpha}$$

$$Y(w(Y)) = w_{\beta,\alpha} Y^\alpha X^\beta + w_\beta Y^\alpha X^\beta_{,\alpha} X\left(w(Y)\right) - Y\left(w(X)\right)$$

$$= w_{\beta,\alpha}\left(X^\alpha Y^\beta + Y^\alpha X^\beta\right) + w_\beta [X, Y]^\beta$$

$$dw(X, Y) + w([X, Y])$$

or $$dw(X, Y) = X\left(w(Y)\right) - Y\left(w(X)\right) - w([X, Y])$$

Let $\{e_\alpha\}$ be a local basis for $X(\mu)$ (vector fields). Then,

$$\nabla_X e_\alpha = \nabla_X \beta e_\beta e_\alpha = X^\beta \nabla e_\beta e_\alpha \equiv X^\beta \Gamma^\sigma_{\beta\alpha} e \sigma$$ By definition,

$$\nabla e_\beta e_\alpha = \Gamma^\sigma_{\beta\alpha} e_\sigma.$$

Torsion: $T(X, Y) = \nabla_X Y - \nabla_Y X - [X, Y]$

Define $w^\sigma_\beta(X) = \Gamma^\sigma_{\beta\alpha} X^\alpha = \Gamma^\sigma_{\beta\alpha} e^\alpha(X)$

$\{e^\alpha\}$ is the dual of $\{e_\alpha\}$

i.e. $w^\sigma_\beta = \Gamma^\sigma_{\beta\alpha} e^\alpha.$

$$T^\alpha(X, Y) = e^\alpha(T(X, Y)) = e^\alpha(\nabla_X Y) - e^\alpha(\nabla_Y X) - e^\alpha([X, Y])$$

$$\begin{aligned}\nabla_X Y &= \nabla_X(Y^\alpha e_\alpha) = X(Y^\alpha) e_\alpha + Y^\alpha \nabla_X e_\alpha \\ &= X(Y^\alpha) e_\alpha + Y^\alpha X^\beta \nabla e_\beta e_\alpha = X(Y^\alpha) e^\alpha + Y^\alpha X^\beta \Gamma^\sigma_{\beta\alpha} e_\alpha \\ &= (X(Y^\sigma) + w^\sigma_\beta(Y) X^\beta) e_\sigma\end{aligned}$$

So $e^\alpha(\nabla_X Y - \nabla_Y X) = (X(Y^\sigma) - Y(X^\sigma) + \sigma^\sigma_\beta(Y) X^\beta - w^\sigma_\beta(X) Y^\beta) e_\sigma$

$$e^\alpha([X, Y]) = e^\alpha(XY - YX) = e^\alpha(X(Y^\beta e_\beta) - Y(X^\beta e_\beta))$$

$$\begin{aligned}e^\sigma(T(X, Y)) &= e^\sigma(\nabla_X Y - \nabla_Y X) - e^\sigma([X, Y]) \\ &= X(Y^\sigma) - Y(X^\sigma) + w^\sigma_\beta(Y) X^\beta - w^\sigma_\beta(X) Y^\beta - e^\sigma([X, Y])\end{aligned}$$

$$\begin{aligned}de^\alpha(X, Y) &= X(e^\alpha(Y)) - Y(e^\alpha(X)) - e^\sigma([X, Y]) \\ &= X(Y^\alpha) - Y(X^\alpha) - e^\alpha([X, Y])\end{aligned}$$

So
$$\begin{aligned}e^\alpha(T(X, Y)) &= de^\alpha(X, Y) = w^\alpha_\beta(Y) X^\beta - w^\alpha_\beta(X) Y^\beta \\ &= w^\alpha_\beta(Y) e^\beta(X) - w^\alpha_\beta(X) e^\beta(Y) \\ &= (e^\beta \Lambda w^\alpha_\beta)(X, Y)\end{aligned}$$

or $T^\alpha = de^\alpha + e^\beta \Lambda w^\alpha_\beta$

This is Cartan's first eqn. of structure.
so in the limit $N \to \infty$ (scaling limit), we get

$$\frac{d}{d\tau}\int_o^1 J(\theta)\tilde{\rho}_\tau(\theta)\,d\theta = \int_o^1 J''(\theta)\,\widehat{\psi}(\tilde{\rho}_\tau(\theta))\,d\theta$$

$$\frac{\partial \tilde{\rho}_\tau(\theta)}{\partial \tau} = \frac{\partial^2}{\partial \theta^2}\widehat{\psi}(\tilde{\rho}_\tau(\theta))$$

which is the generalized Burgen's equation.

❖❖❖❖❖

[23] Reflection and Refraction at a Metamaterial Surface:

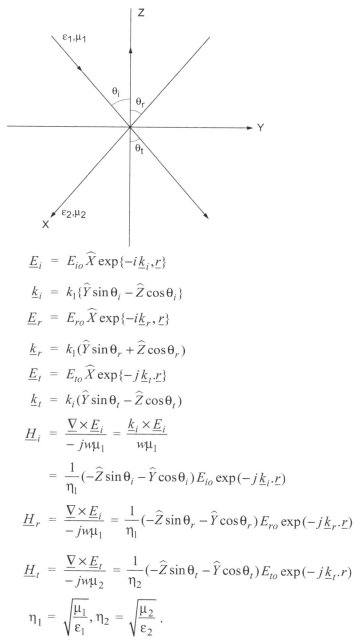

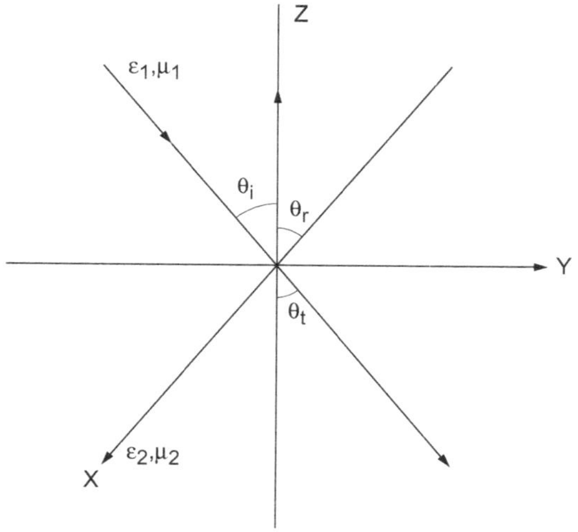

$$\underline{E}_i = E_{io}\widehat{X}\exp\{-i\underline{k}_i,\underline{r}\}$$

$$\underline{k}_i = k_1\{\widehat{Y}\sin\theta_i - \widehat{Z}\cos\theta_i\}$$

$$\underline{E}_r = E_{ro}\widehat{X}\exp\{-i\underline{k}_r,\underline{r}\}$$

$$\underline{k}_r = k_1(\widehat{Y}\sin\theta_r + \widehat{Z}\cos\theta_r)$$

$$\underline{E}_t = E_{to}\widehat{X}\exp\{-j\underline{k}_t.\underline{r}\}$$

$$\underline{k}_t = k_i(\widehat{Y}\sin\theta_t - \widehat{Z}\cos\theta_t)$$

$$\underline{H}_i = \frac{\underline{\nabla}\times\underline{E}_i}{-jw\mu_1} = \frac{\underline{k}_i\times\underline{E}_i}{w\mu_1}$$

$$= \frac{1}{\eta_1}(-\widehat{Z}\sin\theta_i - \widehat{Y}\cos\theta_i)E_{io}\exp(-j\underline{k}_i.\underline{r})$$

$$\underline{H}_r = \frac{\underline{\nabla}\times\underline{E}_i}{-jw\mu_1} = \frac{1}{\eta_1}(-\widehat{Z}\sin\theta_r - \widehat{Y}\cos\theta_r)E_{ro}\exp(-j\underline{k}_r.\underline{r})$$

$$\underline{H}_t = \frac{\underline{\nabla}\times\underline{E}_t}{-jw\mu_2} = \frac{1}{\eta_2}(-\widehat{Z}\sin\theta_t - \widehat{Y}\cos\theta_t)E_{to}\exp(-j\underline{k}_t.r)$$

$$\eta_1 = \sqrt{\frac{\mu_1}{\varepsilon_1}}, \eta_2 = \sqrt{\frac{\mu_2}{\varepsilon_2}}.$$

Surface charge density $= \sigma_s(Y)$

Surface charge density $= J_s(Y)\widehat{X}$

Continuity of E tan gives

$$F_{-io}\exp(-jk_1Y\sin\theta_i)+E_{ro}\exp(-jk_1Y\sin\theta_r)$$
$$=E_{to}\exp(-jk_2Y\sin\theta_t) \quad (1)$$

$\therefore$ $$E_{io}+E_{ro} = E_{to},\ \theta_r=\theta_i,\ k_1\sin\theta_i=k_2\sin\theta_t.$$

Discontinuity of normal component of $\underline{D}$: No equation.

Continuity of normal component of $\mu\,\underline{H}=\underline{B}$:

$$-\frac{\mu_1}{\eta_1}\sin\theta_i\,E_{io}-\frac{\mu_1}{\eta_1}\sin\theta_r\,E_{ro} = -\frac{\mu_2}{\eta_2}\sin\theta_i E_{to}$$

Or $$k_1\sin\theta_i E_{io}+k_i\sin\theta_r E_{ro} = k_2\sin\theta_t E_{to}$$

So $$E_{io}+E_{ro} = E_{to}$$

no new equation. Discontinuity of tangential component of $\underline{H}$:

$$\frac{\cos\theta_i}{\eta_1}E_{io}-\frac{\cos\theta_r}{\eta_1}E_{ro}-\frac{\cos\theta_t}{\eta_2}=J_{so}.$$

Provided

$$J_s(Y)=J_{so}\exp(-jk_1\sin\theta_i Y)$$

(By Fourier analysis, we Pick the Fourier Componenets of $J_s(Y)$ at the spatial frequency $k_1\sin\theta_i$.

Let $$\frac{E_{ro}}{E_{io}} = R,\ \frac{E_{to}}{E_{to}}=T.\text{ Then,}$$

$$1+R = T_1\cos\theta_i-R.\cos\theta_r-\frac{\eta_1}{\eta_2}\cos\theta_t T=\eta_1 T_{so}$$

Suppose $\mu_2>0$, $\varepsilon_2=-j|\varepsilon_2|$ (perfect conductor)

Then $$\eta_1>0,\ \eta_2=\sqrt{\frac{\mu_2}{\varepsilon_2}}=\sqrt{j}\,|\eta_2|=\frac{(i+j)}{\sqrt{2}}|\eta_1|.$$

$$k_1\sin\theta_i = k_2\sin\theta_t$$

$\Rightarrow$ $$\sqrt{\varepsilon_1\mu_1}\sin\theta_i = \sqrt{\varepsilon_2\mu_2}\sin\theta_t$$

$\Rightarrow$ $$\sin\theta_t = \frac{C_2}{C_1}\sqrt{j}\sin\theta_i$$

$$= \frac{C_2}{C_1}\sin\theta_i\frac{(i+j)}{\sqrt{2}}\text{ where }C_2=\sqrt{|\varepsilon_2|\mu_2}$$

$$C_1 = \sqrt{\varepsilon_1\mu_1}$$

$$\cos\theta_i-R\cos\theta_i-\frac{\eta_1}{|\eta_2|}\frac{(1-j)}{\sqrt{2}}\cos\theta_t(1+R) = \eta_1 J_{so}$$

$$\cos\theta_t = \sqrt{1-\sin^2\theta_t} = \left[1-\frac{k_1^2}{k_2^2}\sin^2\theta_i\right]^{1/2}$$

$$\left[1+\alpha\sin^2\theta_i\right]^{1/2} \text{ where } \alpha = -\frac{k_1^2}{k_2^2} = \frac{\varepsilon_1\mu_1}{|\varepsilon_2|\mu_2} > 0$$

So,
$$R\left(\cos\theta_i + \frac{\chi(1-j)}{\sqrt{2}}(1+\alpha\sin^2\theta_i)^{1/2}\right)$$

$$= \cos\theta_i - \frac{\chi(1-j)}{\sqrt{2}}(1+\alpha\sin^2\theta_i)^{1/2} - \eta_1 J_{so}$$

Or
$$R = \frac{\left\{\cos\theta_i - \frac{\chi(1-j)}{\sqrt{2}}(1+\alpha\sin^2\theta_i)^{1/2} - \eta_1 J_{so}\right\}}{\left\{\cos\theta_i - \frac{\chi(1-j)}{\sqrt{2}}(1+\alpha\sin^2\theta_i)^{1/2}\right\}}$$

Where $X = \frac{\eta_1}{|\eta_2|} > 0$. $\simeq$ Reflectivity for a metamaterial with

$\varepsilon_2 < 0, \mu_2 > 0,$

When the interface carries a surface current density

$$S_t = \max_{0\le S\le t} B_s .$$

$$X_t = (S_t - B_t + \alpha^{-1})\exp(-\alpha S_t)$$

is a Martingale (local)

Proof:
$$`dX_t = (dS_t - dB_t)\exp(-\alpha S_t) - \alpha(S_t - B_t + \alpha^{-1})\, ex(-\alpha S_t)\, dS_t$$
$$((dS_t)^2 = 0 \text{ since } S_t \uparrow \Rightarrow S_t \text{ is of bounded variation}).$$
$$= -\exp(-\alpha S_t)\, dB_t$$

Since $(S_t - B_t)dS_t = 0 \therefore dS_t > 0 \Rightarrow S_t = B_t$

Corollary: $(|B_t| + \alpha^{-1})\exp(-\alpha L_t)$ is a local Martingale where $L_t = \int_0^t \delta(B_s)\, ds$

Proof: From Skorchod's theorem the bivariate processes $\{(S_t - B_t, S_t)\}_{t\ge 0}$ and $\{(|B_t|, |L_t|)\}_{t\ge 0}$ have the same law.

$\times\{(B_t, L_t^a)\}_{t\ge 0}$ and $\left\{\left(B_{Ct}, L_{Ct}^{a\sqrt{C}}\right)\Big/\sqrt{C}\right\}_{t\ge 0}$ have to same law.

Proof:
$$L_t^a = \int_0^t \delta(B_s - a)\, dS$$

$$\frac{1}{\sqrt{C}}\Big|_{C_t}^{a\sqrt{C}} = \frac{1}{\sqrt{C}}\int_0^{C_t} \delta(B_s - a\sqrt{C})\, dS$$

Charge variables: $S = C\alpha$, $dS = Cd\alpha$. Then

$$\frac{1}{\sqrt{C}} L_{Ct}^{a\sqrt{C}} = \frac{1}{\sqrt{C}} \int_0^t \delta\left(B_{C\alpha} - a\sqrt{C}\right) C d\alpha$$

$$= \sqrt{C} \int_0^t \delta\left(B_{C\alpha} - a\sqrt{C}\right) d\alpha$$

$$= \int_0^t \delta\left(\frac{1}{\sqrt{C}} B_{C\alpha} - a\right) d\alpha$$

The proof is completed by noting that $\frac{1}{\sqrt{C}} B_{Ct}, t \geq 0$ is Brownian motion, say W_t, $t \geq 0$, and thus

$$\frac{1}{\sqrt{C}} L_{Ct}^{a\sqrt{C}} = \int_0^t \delta(W_\alpha - a)\, d\alpha \text{ is the local time of } W.$$

❖❖❖❖❖

[24] Maxwell's Equations in an Unhomogeneous Medium with Background Gravitation Taken into Account:

$\chi_{\rho\sigma}^{\mu\nu}(x)$ is the permittivity permeability tensor. Maxwell's eqns: Wave propagation equation:

$$\left(\chi_{\rho\sigma}^{\mu\nu}(x)\, F^{\rho\sigma} \sqrt{-g}\right)_{\nu} = 0$$

Thus, $\chi_{\rho\sigma}^{\mu\nu} = \chi_{\sigma\rho}^{\nu\mu}$.

$$F_{\mu\nu} = A_{\nu,\mu} - A_{\mu,\nu}$$

So

$$\left(\chi_{\rho\sigma}^{\mu\nu}\, g^{\rho\sigma}\, g^{\sigma\beta} \sqrt{-g}\, F_{\alpha\beta}\right)_{\nu} = 0$$

or

$$\left(\chi^{\mu\nu\alpha\beta} \sqrt{-g}\, (A_{\beta,\alpha} - A_{\alpha,\beta})\right)_{,\nu} = 0$$

Where $\chi^{\mu\nu\alpha\beta} = \chi_{\rho\sigma}^{\mu\nu}\, g^{\alpha\rho}\, g^{\beta\sigma}$

Example: In special relativity, $\{-F^{or}, r = 1, 2, 3\}$ is the electric field $\underline{E}$ in a Cartesian system and $\{-F^{23}, -F^{31}, -F^{12}\}$ is the magnetic field $\underline{B}$.

The electric displacement vector $\underline{D} = \underline{\underline{\varepsilon}} \cdot \underline{E}$ Where $\underline{\underline{\varepsilon}}$ is a 3×3 matrix. Thus,

$$D_1 = D_X = \varepsilon_{11}E_1 + \varepsilon_{12}E_2 + \varepsilon_{13}E_3$$

$$D_2 = D_Y = \varepsilon_{21}E_1 + \varepsilon_{22}E_2 + \varepsilon_{23}E_3$$

$$D_3 = D_Z = \varepsilon_{31}E_1 + \varepsilon_{32}E_2 + \varepsilon_{33}E_3$$

Thus, $D_r = -\varepsilon_{rs} F^{os}$ (summation over s = 1, 2, 3) Likewise the magnetic field intensity is

$$H_1 = H_X = (\mu^{-1})_{11} B_1 + (\mu^{-1})_{12} B_2 + (\mu^{-1})_{13} B_3 \text{ etc.}$$

$$H_r = (\mu^{-1})_{rs} B_s = -(\mu^{-1})_{r1} F^{23} - (\mu^{-1})_{r2} F^{31} - (\mu^{-1})_{r3} F^{12}$$

Let $$\chi^{\mu\nu\alpha\beta} = g^{\alpha\rho} g^{\sigma\beta} \left(\delta^{\mu}_{\rho} \delta^{\mu}_{\sigma} + \delta \cdot \eta^{\mu\nu}_{\rho\sigma}\right) = g^{\mu\rho} g^{\nu\sigma} + \delta^{\eta\mu\nu\rho\sigma}$$

δ being a small perturbation parameter.

In our case,

$$\underline{D}_b = \left(\left(\chi^{o\nu}_{\rho\sigma} F^{\rho\sigma}\right)\right)^3_{r=1}$$

$$= -\varepsilon_{r1} F^{01} - \varepsilon_{r2} F^{02} - \varepsilon_{r3} F^{03}$$

provided

$$\varepsilon_{r1} = \chi^{or}_{o1} - \chi^{or}_{1o},$$

$$\varepsilon_{r2} = \chi^{or}_{o2} - \chi^{or}_{3o},$$

$$\varepsilon_{r3} = \chi^{or}_{o3} - \chi^{or}_{2o}$$

and $\chi^{or}_{\alpha\beta} = 0$ if $\alpha \neq 0$, $\beta \neq 0$, *i.e.* $\chi^{or}_{km} = 0$ $\quad 1 \leq k, m \leq 3.$

$$\underline{H} = \left(-(\mu^{-1})_{r1} F^{23} - (\mu^{-1})_{r2} F^{31} - (\mu^{-1})_{r3} F^{12}\right)^3_{r=1}$$

$$= \left(-\chi^{23}_{\rho\sigma} F^{\rho\sigma}, -\chi^{31}_{\rho\sigma} F^{\rho\sigma}, -\chi^{12}_{\rho\sigma} F^{\rho\sigma}\right)$$

so
$$\chi^{23}_{23} - \chi^{23}_{32} = (\mu^{-1})_{11},\ \chi^{23}_{12} - \chi^{23}_{21} = (\mu^{-1})_{13},\ \chi^{23}_{31} - \chi^{23}_{13} = (\mu^{-1})_{12},$$

$$\chi^{31}_{23} - \chi^{31}_{32} = (\mu^{-1})_{21},\ \chi^{31}_{12} - \chi^{31}_{21} = (\mu^{-1})_{23},\ \chi^{31}_{31} - \chi^{31}_{13} = (\mu^{-1})_{22},$$

$$\chi^{12}_{23} - \chi^{12}_{32} = (\mu^{-1})_{31},\ \chi^{12}_{31} - \chi^{12}_{13} = (\mu^{-1})_{32},\ \chi^{12}_{12} - \chi^{12}_{21} = (\mu^{-1})_{33},$$

and $$\chi^{rs}_{ok} = \chi^{rs}_{ko} = 0 \qquad 1 \leq k \leq 3 \ \forall\ 1 \leq r, s \leq 3.$$

The Maxwell eqns. are

$$\left[\left(g^{\mu\rho} g^{\nu\sigma} + \delta \cdot \eta^{\mu\nu\rho\sigma}\right) F_{\rho\sigma} \sqrt{-g}\right]_{,\nu} = 0.$$

or
$$\left(F^{\mu\nu}\sqrt{-g}\right)_{,\nu} + \delta\left(\eta^{\mu\nu\rho\sigma}\sqrt{-g}\, F_{\rho\sigma}\right)_{,\nu} = 0.$$

Propagation of em waves in a inhomogeneous metamaterial.

$$\left(\nabla^2 + \varepsilon\,\mu \frac{\partial^2}{\partial t^2}\right)\psi(t, r) = 0 \qquad \varepsilon\,\mu = \frac{1}{C^2} > 0$$

Solution:
$$\psi(t, \underline{r}) = \int f(w, \hat{n}) \exp\left(iw\left(t + \frac{i(\hat{n}, r)}{c}\right)\right) dw\, d\Omega(\hat{n})$$

$$\equiv \int F\left(t + \frac{i(\hat{n}, \underline{r})}{c}, \hat{n}\right) d\Omega(\hat{n})$$

Inhomogeneous metamaterial

$$\begin{aligned} \nabla \times E &= -jw\mu \underline{H} \\ \nabla \times \underline{H} &= -jw\varepsilon(X,Y,Z)\underline{E} \qquad \varepsilon, \mu > 0. \\ \text{div}\,(\varepsilon\, \underline{E}) &= 0, \\ \text{div}\,(\underline{H}) &= 0. \end{aligned}$$

So $\qquad \nabla\,(\text{div}\,\underline{E}) - \Delta \underline{E} = -jw\mu(-jw\,\varepsilon)\underline{E}$

or $\qquad (\Delta - w^2\mu\,\varepsilon)\underline{E} = \underline{\nabla}\,(\text{div}\,\underline{E})$

$$\begin{aligned} (\nabla\,\varepsilon, \underline{E}) + \varepsilon\,\text{div}\,\underline{E} &= 0. \\ \text{div}\,\underline{E} &= -(\nabla \log \varepsilon, E) \end{aligned}$$

$$\left(\Delta - k_o^2(1+\delta\cdot\chi(\underline{r}))\right)\underline{E}(\underline{r}) + \delta(\underline{\nabla}\chi(\underline{r}), \underline{E}(\underline{r})) = 0$$

$$k_o^2 = w^2\mu\,\varepsilon_o$$

❖❖❖❖❖

[25] Relativistic Kinematics* Motions of a Rigid Body in a Gravitational Field a General Relativistic Calculation:

$$\mathcal{L} = -\rho_o c^2 \int_B \Big[g_{oo}(t, \underline{\underline{R}}(r)) + 2g_{ok}(t, \underline{\underline{R}}(t)\underline{r}) R'_{kj}(t) X^n$$

$$+ g_{km}(t, \underline{\underline{R}}(t)\underline{r}) R'_{kj}(t) R'_{ml}(t) X^j X^l \Big]^{1/2} dX' dX^2 dX^3$$

Where $\qquad \underline{r} = (X^j)_{j=1}^3.$

B is the volume of the body at $t = 0$. We shall approximate that integral using the binomial theorem:

$$\left(g_{oo} + 2g_{ok}\,\xi^k + g_{km}\,\xi^k\,\xi^m\right)^{1/2}$$

$$\approx g_{oo}^{1/2}\left[1 + \frac{g_{ok}\xi^k}{g_{oo}} + \frac{g_{km}\xi^k\xi^m}{2g_{oo}} - \frac{1}{8}\left(\frac{2g_{ok}\xi^k}{g_{oo}} + \frac{g_{km}\xi^k\xi^m}{g_{oo}}\right)^2\right]$$

In this approximation, g_{ok} is regarded to be $O(\|\xi\|)$ and then the expansion is valid upto $O(\|\xi\|^4)$.

Then

$$\mathcal{L} \approx -\rho_o c^2 \int_B g_{oo}(t, \underline{\underline{R}}(t)\underline{r})^{1/2} d^3r - \rho_o c^2 \int_B \frac{g_{oo}(t, \underline{\underline{R}}(t)\underline{r})}{\sqrt{g_{oo}(t, \underline{\underline{R}}(t)\underline{r})}} R'_{kj}(t) X^j d^3r$$

$$-\frac{\rho_o c^2}{2} \int_B \frac{g_{km}(t, \underline{\underline{R}}(t)\underline{r})}{\sqrt{g_{oo}(t, \underline{\underline{R}}(t)\underline{r})}} R'_{kj}(t) R'_{ml}(t) X^j X^l d^3r$$

$$-\frac{\rho_o c^2}{4}\int_B \frac{g_{ok}(t,\underline{\underline{R}}(t)\underline{r})}{\sqrt{g_{oo}(t,\underline{\underline{R}}(t)\underline{r})}} R'_{kj}(t)\,X^j\,d^3r$$

$$+\frac{\rho_o c^2}{8}\int_B \frac{g_{km}(t,\underline{\underline{R}}(t)\underline{r})}{\sqrt{g_{oo}(t,\underline{\underline{R}}(t)\underline{r})}} R'_{kj}(t)\,R'_{ml}(t)\,X^j\,X^l\,d^3r$$

Example: Spherically symmetric state metrics like the Schwarzchild metric.

$$d\tau^2 = A(r)\,dt^2 - B(r)\,dr^2 - C(r)r^2(d\theta^2 + \sin^2\theta\,d\varphi^2)$$

In terms of a Cartesian system.

$$r^2 = x^2 + y^2 + z^2$$

$$rdr = xdx + ydy + zdz$$

$$r^2(d\theta^2 + \sin^2\theta\,d\varphi^2)$$

$$= dx^2 + dy^2 + dz^2 - dr^2$$

Robots perturbed by quantum noise in a gravitational field.

$H_o \simeq$ Hamiltonian of the Robot

Derivation of $H_o = \quad R(t) = R(\varphi(t), \theta(t), \psi(t))$

$$= R_z(\varphi(t))\,R_X(\theta(t)\,R_Z(\psi(t)))$$

Lagrangian

$$\pounds(\varphi, \theta, \psi, \dot\varphi =, \dot\theta =, \dot\psi =) = \int_B (\underline{\underline{R}}(t)\underline{\xi}, \underline{\underline{R'}}(t)\underline{\xi})\,d^3\xi$$

$$= \int_B L(R(\varphi,\theta,\psi)\underline{\xi}, (\dot\varphi R_{,\varphi}(\varphi,\theta,\psi) + \dot\theta R_{,\theta}(\varphi,\theta,\psi)$$

$$+\dot\psi R_{,\psi}(\varphi,\theta,\psi))\,\underline{\xi})\,d^3\xi$$

Where $L(\underline{x}, \underline{v}) \simeq$ Lagrangian for a single point particle Canonical momenta:

$$p_\varphi = \frac{\partial \mathcal{L}}{\partial \dot\varphi} = \int (R_{,\varphi}\underline{\xi})^T \frac{\partial L}{\partial \underline{v}}\,d^3\xi$$

$$p_\theta = \frac{\partial \mathcal{L}}{\partial \dot\theta} = \int (R_{,\theta}\underline{\xi})^T \frac{\partial L}{\partial \underline{v}}\,d^3\xi$$

$$p_\psi = \frac{\partial \mathcal{L}}{\partial \dot\psi} = \int (R_{,\psi}\underline{\xi})^T \frac{\partial L}{\partial \underline{v}}\,d^3\xi$$

Chapter 4 (Continued) Relativistic kinematics of rigid bodies.
Rigid body motion in the Schwarzchild metric.

$$g_{oo}(\underline{r}) = \alpha(r) = 1 - \frac{2m}{r},\ m = \frac{GM}{c^2}.$$

$$g_{11}(\underline{r}) = \alpha(r)^{-1},$$

$$g_{22}(r) = r^2,\ g_{33} = r^2\sin^2\theta.$$

At time $t = 0$ r is ϕ the position of a point on the body. At time t, this point moves to $\underline{\underline{R}}(t)\underline{r}$.

So
$$\frac{d\tau}{dt}(\underline{r}) = \Big[\alpha(\| \underline{\underline{R}}(t)\underline{r} \|) - \alpha^{-1}(\| \underline{\underline{R}}(t)\underline{r} \|)\| R'(t)r \|^2$$
$$- \| \underline{\underline{R}}(t)\underline{r} \|^2 (\theta'(t,r)^2 + \sin^2(\theta'(t,r))\varphi'^2(t,\underline{r}))\Big]^{1/2}$$

where
$$\theta'(t,\underline{r}) = \frac{\partial\theta(t,r)}{\partial t}, \varphi'(t,\underline{r}) = \frac{\partial\varphi(t,r)}{\partial t}$$

Where $\theta(t,\underline{r})$ and $\varphi(t,\underline{r})$ are respectively the elevation angle and agimute angle relative to the centre of the blockhole. Here we are assuming that the top's fixed point is at the centre of the blackhole. This is too unrealised.
Assume therefore that the fixed point of the top is at $\underline{r}_o > 2m$. Let $p : \underline{\xi}$ be a point in the top relative to $\underline{r}_o$. Then the position of P after time t is

$$\underline{r}_o + \underline{\underline{R}}(t)\underline{\xi} \equiv \underline{r}(t)$$

Then for any metric,

$$g_{\mu\nu}(t,\underline{r}(t)) = g_{\mu\nu}(t,\underline{r}_o + \underline{\underline{R}}(t)\underline{\xi}) \approx g_{\mu\nu}(t,r_o) + g_{\mu\nu,n}(t,\underline{r}_o)(R'(t)\xi)^n$$

$$g_{\mu\nu}(t,\underline{r}(t))\frac{dX^\mu}{dt}\frac{dX^\nu}{dt} \approx \left[g_{\mu\nu}(t,\underline{r}_o) + g_{\mu\nu,n}(t,\underline{r}_o)(R'(t)\xi)^n\right]\frac{dX^\mu}{dt}\frac{dX^\nu}{dt}$$

$$\frac{dX^o}{dt} = 1, \frac{dX^k}{dt} = (R'(t)\xi)^n, k = 1, 2, 3.$$

So
$$g_{\mu\nu}\,dX^\mu dX^\nu \approx \left[g_{oo}(t,\underline{r}_o) + g_{oo,n}(t,\underline{r}_o)R'(t)^n_m\xi^m\right]$$
$$+ 2\left[g_{ok}(t,\underline{r}_o) + g_{ok,n}(t,\underline{r}_o)R'(t)^n_m\xi^m\right] \times R'(t)^k_s\xi^s$$
$$+ \left[g_{kl}(t,r_o) + g_{kl,n}(t,r_o)R'(t)^n_m\xi^m\right] \times R'(t)^k_s R'(t)^l_p \xi^s\xi^p$$

Upto $O(\|\xi\|^2)$, we have

$$g_{\mu\nu}\frac{dX^\mu}{dt}\frac{dX^\nu}{dt} = g_{oo}(t,\underline{r}_o) + g_{oo,n}(t,\underline{r}_o)R'(t)^n_m\xi^m$$
$$+ 2g_{ok}(t,\underline{r}_o)R'(t)^k_s\xi^s + g_{oo,nm}(t,\underline{r}_o)R'(t)^n_p R'(t)^n_p R'(r)^m_q \xi^p\xi^q$$
$$+ 2g_{ok,n}(t,\underline{r}_o)R'(t)^n_m R'(t)^k_s \xi^m\xi^s + g_{kl}(t,r_o)R'(t)^k_s R'(t)^l_p \xi^s\xi^p$$
$$= g_{oo}(t,\underline{r}_o) + \left(g_{oo,n}(t,\underline{r}_o)R'(t)^n_m + 2g_{on}(t,r_o)R'(t)^n_m\right)\xi^m$$
$$+ \Big[\left(g_{oo,nm}(t,\underline{r}_o) + R'(t)^n_p R'(t)^m_q\right) + 2g_{om,n}(t,r_o)$$
$$+ R'(t)^n_p R'(t)^m_q + g_{nm}(t,\underline{r}_o) + R'(t)^n_p R'(t)^m_q\Big]\xi^s\xi^b$$

❖❖❖❖❖

[26] Random Fluctuation in the Gravitational Field Produced by Random Fluctuation in the Energy-Momentum Tensor of the Matter Plus Radiation Field:

$$g_{\mu\nu}(x) = g_{\mu\nu}^{(a)}(x) + \delta \cdot h_{\mu\nu}(x)$$

$$g_{\mu\nu} = g^{\mu\nu(0)} - \delta h^{\mu\nu} + 0(\delta^2)$$

Where

$$\left(\left(g^{\nu\mu(0)}\right)\right) = \left(\left(g_{\mu\nu}^{(a)}\right)\right)^{-1} \cdot h^{\mu\nu} + g^{\mu\alpha(a)} g^{\nu(\mu 0)} h_{\alpha\beta}$$

$$T^{\mu\nu} = P\nu^{\mu}\mu^{\nu} \quad \nu^{\mu} = V^{\mu} + \delta \cdot \mu^{\mu}$$

$$g_{\mu\nu}\,\nu^{\mu}\nu^{\nu} = 1 \Rightarrow g_{\mu\nu} V^{\mu} V^{\nu} = 0$$

$$g_{\mu\nu} V^{\mu}\mu^{\nu} = 0 \Leftrightarrow V_{\mu}\mu^{\mu} = 0$$

$$\delta\Gamma_{\mu d}{}^{\alpha} = g^{\alpha\beta}\Gamma_{\beta\mu\alpha} = (0)\Gamma_{\mu\alpha}{}^{\alpha(0)}$$

$$+ \delta \cdot \left\{ - h^{\alpha\beta}\Gamma_{\beta\mu d}{}^{(0)} + \frac{1}{2} g^{\alpha\beta(0)}\, h_{\beta\mu,\alpha} + h_{\beta\alpha,\mu} - h_{\mu\alpha,\beta} \right\}$$

$$+ 0(\delta^2)$$

$$R_{\mu\nu} = R_{\mu\nu}^{(0)} + \delta \cdot R_{\mu\nu}^{(1)} + O(\delta^2)$$

$$R_{\mu\nu}^{(1)} = k_{\mu\nu}^{(1)\alpha\beta\rho\sigma}(x)\, h_{\alpha\beta}^{(x)}, \rho_{\sigma} + K_{\mu\nu}^{(2)\alpha\beta\rho}(x)\, h_{\alpha\beta,\rho}(x) + K_{\mu\nu}^{(3)}\,\alpha\beta h_{\alpha\beta}(x)$$

$K^{(1)}$, $K^{(2)}$, $K^{(3)}$ are expressible completely in term of $g_{\mu\nu}^{(0)}$ and its first and second order partial derivatives.

The linearized Einstein field equation is then

$$K_{(\mu\nu)}^{(1)\alpha\beta\rho\sigma}(x)\, h_{\alpha\beta,\rho\sigma}(x) + K_{\mu\nu}^{(2)\alpha\beta\rho}(x) h_{\alpha\beta,\rho}(x) + K_{\mu\nu}^{(3)\alpha\beta}(x) h_{\alpha\beta}(x)$$

$$= C \cdot \left\{ \rho_1(x)(V_{\mu}(x)) V_{\nu}(x) - \frac{1}{2} g_{\mu\nu}^{(0)}(x) \right.$$

$$\left(R_{\mu\nu} - \frac{1}{2} R_{\delta\mu\nu} = C \cdot \rho\nu_{\mu}\nu_{\nu} + \rho_0(x)(V_{\mu}(x)\mu_{\nu}(x)) + V_{\nu}(x)\mu_{\mu}(x) \right.$$

$$\Rightarrow \quad - R = C\rho$$

$$\left. \left. - \frac{1}{2} h_{\mu\nu}(x) \right) \right\}$$

$$\Rightarrow \quad R_{\mu\nu} = C \cdot \rho \left(\nu_{\mu}\nu_{\nu q} - \frac{1}{2} g_{\mu\nu} \right)$$

Where
$$\rho = \rho_0 + \delta\rho_1 + O(\delta^2),$$
$$\delta.\mu_\mu(x) = \nu_\nu(x) - V_\mu(x) = g_{\mu\nu}\nu^\nu - g_{\mu\nu}^{(0)}V^\nu$$
$$= \left(g_{\mu\nu}^{(0)} + \delta.h_{\mu\nu}\right)\left(V^\nu + \delta.u^\nu\right) - g_{\mu\nu}^{(0)}V^\nu$$
$$= \delta.\left(g_{\mu\nu}^{(0)}u^\nu + h_{\mu\nu}V^\nu\right) + O(\delta^2)$$
or
$$u_\mu = g_{\mu\nu}^{(0)}u^\nu + h_{\mu\nu}V^\nu$$
Consider the equation
$$\left(\rho\nu^\mu\nu^\nu\right)_{:\nu} = 0$$
implied by the Einstein field equation. We get
$$\left(\rho\nu^\nu\right)_{i\nu}\nu^\mu + \rho^{\mu\nu}\nu^\mu_{:\nu} = 0$$
so
$$\left(\rho\nu^\nu\right)_{:\nu} = 0 \text{ and } \nu^\nu\nu^\mu_{:\nu} = 0$$
or
$$\left(\rho\nu^\mu\sqrt{-g}\right)_{,\mu} = 0,\ \nu^\nu\left(\nu^\mu_{:\nu} + \Gamma^\mu_{\nu\rho}\nu^\rho\right) = 0$$
$$g = \det\left(\left(g_{\mu\nu}^{(0)} + \delta.h_{\mu\nu}\right)\right)$$
$$= g^{(0)}\left(1 + \delta.h^\mu_\mu\right) + 0(\delta^2)$$
$$\triangleq g^{(0)}(1 + \delta.h) + 0(\delta^2)$$
Where
$$h^\mu_\nu = g^{\mu\alpha(0)}h_{\nu\alpha},\ h = h^\mu_\mu.$$
The perturbed fluid dynamical eqn. (without viscosity) are
$$V^\nu\left(\mu^\mu_{,\nu\rho} + \Gamma^{\mu(0)}_{\nu}u^\rho + \Gamma^{\mu(1)}_{\nu\rho}V^\rho\right) + \mu^\nu\left(V^\mu_\nu + \Gamma^{\mu(0)}_{\nu\rho}V^\rho\right) = 0$$
Or since $V^\mu_{,\nu} + \Gamma^{\mu(0)}_{\nu\rho}V^\rho = 0,$

It follows that
$$V^\nu\mu^\mu_{,\nu} + \Gamma^{\mu(0)}_{\nu\rho}V^\nu\mu^\rho + \Gamma^{\mu(1)}_{\nu\rho}V^\nu V^\rho = 0,$$
Note that
$$\Gamma^\mu_{\alpha\beta} = g^{\mu\rho}\Gamma_{\rho\alpha\beta}$$
$$= \left(g^{\mu\rho(0)} - \delta\cdot h^{\mu\rho}\right)\left(\Gamma^{(0)}_{\rho\alpha\beta} + \delta\cdot\Gamma^{(1)}_{\rho\alpha\beta} + O\left(\delta^2\right)\right)\Gamma^{\mu\rho(0)}_{\alpha\beta}$$
$$+ \delta\left(g^{\mu\rho(0)}\Gamma^{(1)}_{\alpha\rho\beta} - h^{\mu\rho}\Gamma^{(0)}_{\alpha\rho\beta}\right) + O(\delta^2)$$
Where
$$\Gamma^{(0)}_{\alpha\rho\beta} = \frac{1}{2}\left(h_{\rho\alpha,\beta} + h_{\rho\beta,\alpha} - h_{\alpha,\beta\rho}\right),$$

$$\Gamma^{\mu(0)}_{\alpha\beta} = g^{\mu\rho(0)}\Gamma^{(0)}_{\alpha\rho\beta}$$

Explicit form for $K^{(1)}$, $K^{(2)}$, $K^{(3)}$:

$$R^{(1)}_{\mu\nu} = \Gamma^{\alpha(1)}_{\mu\alpha,\nu} - \Gamma^{\alpha(1)}_{\mu\nu,\alpha} - \left(\Gamma^{\alpha(0)}_{\mu\nu}\Gamma^{\beta(1)}_{\alpha\beta} + \Gamma^{\alpha(1)}_{\mu\nu}\Gamma^{\beta(0)}_{\alpha\beta}\right.$$

$$+\left(\Gamma^{\alpha(0)}_{\mu\beta}\Gamma^{\beta(1)}_{\nu\alpha} + \Gamma^{\alpha(1)}_{\mu\beta}\Gamma^{\beta(0)}_{\nu\alpha}\right)$$

Where
$$\Gamma^{\mu(1)}_{\alpha\beta} = g^{\mu\rho(0)}\Gamma^{(1)}_{\rho\alpha\beta} - h^{\mu\rho}\Gamma^{(1)}_{\rho\alpha\beta}$$

$$= \frac{1}{2}g^{\mu\rho(0)}\left(h_{\rho\alpha,g} + h_{\rho\beta,\alpha} - h_{\alpha,\beta\rho}\right)$$

$$-\frac{1}{2}h^{\mu\rho}\left(g^{(0)}_{\rho\alpha,\beta} + g^{(0)}_{\rho\beta,\alpha} - g^{(0)}_{\alpha,\beta,\rho}\right)$$

The equation to be solved for $h_{\alpha\beta}(x)$, and $u^{\nu}(x)$, $\rho_1(x)$ are the following pde's

$$K^{(1)\alpha\beta\rho\sigma}_{\mu\nu}(x)\,h_{\alpha\beta,\rho\sigma}(x) + K^{(2)\alpha\beta\rho}_{\mu\nu}(x)\,h_{\alpha\beta,\rho}(x) + K^{(3)\alpha\beta}_{\mu\nu}(x)\,h_{\alpha\beta}(x)$$

$$= C\cdot\left\{\rho_1(\nu)\left(V_\mu V_\nu - \frac{1}{2}g^{(0)}_{\mu\nu}\right) + \rho_0(x)\left(V_\mu u_\nu + V_\nu u_\mu - \frac{1}{2}h_{\mu\nu}(x)\right)\right\} \quad (1),$$

$$V^{\nu}(x)u^{\mu}_{,\nu}(x) + \Gamma^{\mu(0)}_{\nu\rho}V^{\nu}u^{\rho}$$

$$+V^{\nu}V^{\rho}\left(\frac{1}{2}g^{\mu\sigma(0)}\left(h_{\sigma\nu,\rho} + h_{\sigma\rho,\nu} - h_{\nu\rho,\sigma}\right) - h^{\mu\sigma}\Gamma^{(0)}_{\sigma\nu\rho}(x)\right) = 0 \quad (2)$$

And finally the man conservation equation

$$\left(\rho_1 V^{\mu}\sqrt{-g^{(0)}} + \rho_{(0)}\sqrt{-g}^{(0)u^{\mu}} - \frac{\rho_0 V^{\mu}}{2\sqrt{-g^{(0)}}}h\right)_{,\mu} = 0 \quad (3)$$

Where
$$h = h^{\mu}_{\mu} = g^{\mu\nu(0)}h_{\mu\nu}$$

Random fluctuation in the energy-momentum tensor of the *em* field

$$\mathcal{L}_{EM} = F_{\mu\nu}F^{\mu\nu}\sqrt{-g}$$

$$\delta_g\mathcal{L}_{EM} = -F_{\mu\nu}F^{\mu\nu}\delta g/2\sqrt{-g} + F_{\mu\nu}\sqrt{-g}\,F_{\alpha\beta}\,\delta\left(g^{\mu\alpha}g^{\nu\beta}\right)$$

$$g = -F_{\mu\nu}F^{\mu\nu}gg^{\alpha\beta}\delta g_{\alpha\beta}/2\sqrt{-g}$$

$$\pm 2\sqrt{-g}\,F_{\mu\nu}F_{\alpha\beta}\,g^{\mu\alpha}g^{\nu\rho}g^{\beta\sigma}\delta g_{\rho\sigma}$$

Coeff. of $\delta g^{\rho\sigma}$ is

$$\sqrt{-g}\left\{\frac{1}{2}F_{\mu\nu}F^{\mu\nu}g^{\rho\sigma} - 2F^{\alpha\rho}F^{\sigma}_{\alpha}\right\}$$

$\alpha S^{\rho\sigma}\sqrt{-g}$ the energy momentum tensor of the *em* field.

$$S^{\rho\sigma} = \frac{1}{4}F_{\mu\nu}F^{\mu\nu}g^{\rho\sigma} - F^{\alpha\rho}F^{\sigma}_{\alpha}$$

Motion of charged matter in an *em* field and *em* grav. Field-linearized version: Exact equation is

$$\left(\rho v^{\mu}v^{\nu}\right)_{:\nu} + K_1 S^{\mu\nu}_{:\nu} = 0$$

$$S^{\mu\nu}_{:\nu} = \frac{1}{4}\left(F_{\alpha\beta}F^{\alpha\beta}\right)^{g^{\mu\nu}}_{:\nu} - \left(F^{\alpha\mu}F^{\nu}_{a}\right)_{:\nu}$$

$$= \frac{1}{2}F^{\alpha\beta}F_{\alpha\beta:\nu}g^{\mu\nu} - F^{\alpha\mu}_{:\nu}F^{\nu}_{\alpha} - F^{\alpha\mu}F^{\nu}_{\alpha:\nu}$$

In the absence of charged matter. In the presence of charged matter, the extra term is comes from the interaction Lagrangian between the *em* field and four caurraso density.

Rough:

$$\delta_g\left(J_\mu A^\mu\sqrt{-g}\right) = \delta_g\left(\rho_q v_\mu A^\mu\sqrt{-g}\right)$$

$$= \delta_g\left(\rho_q v^\mu A_\mu\sqrt{-g}\right) = -\rho_q v^\nu A_\mu \delta g / 2\sqrt{-g}$$

$$= -\rho_q v^\mu A_\mu g g^{\alpha\beta}\delta g\alpha\beta / 2\sqrt{-g}$$

$$= \rho_q v^\mu A_\mu \sqrt{-gg}\,\delta g_{\alpha\beta} / 2$$

$\rho_q \simeq \rho_q$ where $\rho \simeq$ matter density and $q \simeq$ charge per unit man. Thus the eqn. get modified to

$$\left(\rho v^{\mu}v^{\nu}\right)_{:\nu} + K_1 S^{\mu\nu}_{:\nu} + K_2\left(\rho v^{\alpha}A_{\alpha}g^{\mu\nu}\right)_{:\nu} = 0$$

or

$$\left(\rho v^{\nu}\right)^{\nu\mu}_{:\nu} + \left(\rho v^{\nu}\right)v^{\mu}_{:\nu} + K_1 S^{\mu\nu}_{:\nu} + K_2\left(\rho v^{\alpha}A_{\alpha}\right)_{:\nu}g^{\mu\nu} = 0. \quad (1)$$

The coefficient of δA_μ gives the Maxwell eqn.

$$F^{\nu}_{:\alpha\nu} = K_3' J_\alpha = K_3\rho v_\alpha$$

So

$$v\, S^{\mu\nu}_{:\nu} = \frac{1}{2}F^{\alpha\beta}F_{\alpha\beta:\nu}g^{\mu\nu} - F^{\alpha\mu}_{:\nu}F^{\nu}_{\alpha} - \rho K_3 F^{\alpha\mu}v_\alpha$$

$$v_\mu S^{\mu\nu}_{;\nu} = \frac{1}{2}F^{\alpha\beta}F_{\alpha\beta:\nu}v^{\nu} - F^{\alpha\mu}_{:\nu}F^{\nu}_{\alpha}v_\mu - \rho K_3 F^{\alpha\mu}v_\mu v_\alpha$$

$$= \frac{1}{2}F^{\alpha\beta}F_{\alpha\beta:\nu}v^{\nu} + F^{\mu\alpha}_{:\nu}F^{\nu}_{\alpha}u^{\mu} \quad (2)$$

$\left(\because F^{\alpha\mu}v_\mu v_\alpha = 0\right)$. So we get from (1) and (2),

$$\left(\rho v^{\nu}\right)_{:\nu} + \frac{K_1}{2} F^{\alpha\beta} F_{\alpha\beta:\nu} v^{\nu} + K_1 F_{\nu}^{\mu\alpha} F_{\alpha}^{\nu} v_{\mu} + K_2 \left(\rho v^{\alpha} A_{\alpha}\right)_{:\nu} v^{\nu}$$

$$= 0 \tag{3}$$

Now,

$$\left(\rho v^{\alpha} A_{\alpha}\right)_{:\nu} v^{\nu} = \rho v^{\alpha} v^{\nu} A_{\alpha:\nu} + \left(\rho v^{\alpha}\right)_{:\nu} A_{\alpha} v^{\nu}$$

$$= \rho v^{\alpha} v^{\nu} A_{\nu:\alpha} + \left(\rho v^{\alpha}\right)_{:\nu} v^{\nu} A_{\alpha}. \tag{4}$$

Taking (1), (2), (3) into account, (1) can be expressed as

$$\rho v^{\nu} v_{:\nu}^{\nu\mu} - v^{\mu} \left(\frac{K_1}{2} F^{\alpha\beta} F_{\alpha\beta:\nu} v^{\nu} K_2 \left(\rho_{\nu}^{\alpha} A_{\alpha}\right)_{\nu} v^{\nu} \right.$$

$$+ K_1 F_{:\nu}^{\rho\alpha} F_{\alpha}^{\nu} v_{\rho} + K_2 \left(\rho v^{\alpha} A_{\alpha}\right)_{\nu} v^{\nu} + K_2 g^{\mu\nu} \left(\rho v^{\alpha} A_{\alpha}\right)_{:\nu}$$

$$+ K_1 \left(\frac{1}{2} F^{\alpha\beta} F_{\alpha\beta:\nu} g^{\mu\nu} + F_{:\nu}^{\mu\alpha} F_{\alpha}^{\nu} + \rho K_3 F^{\mu\alpha} \mathrm{v}_{\alpha} \right) = 0 \tag{5}$$

This can be rearranged as

$$\rho v^{\nu} v_{:\nu}^{\mu} + K_1 K_3 \rho F^{\mu\alpha} v_{\alpha} + \frac{K_1}{2} F^{\alpha\beta} F_{\alpha\beta:\nu} \left(g^{\mu\nu} - v^{\mu} v^{\nu}\right)$$

$$+ K_1 \left(F_{:\nu}^{\mu\alpha} F_{\alpha}^{\nu} - \frac{1}{2} F_{:\nu}^{\rho\alpha} F_{\alpha}^{\nu} v_{\rho} v^{\mu} \right) + K_2 \left(\rho v^{\alpha} A_{\alpha}\right)_{:\nu} - \left(g^{\mu\nu} - v^{\mu} v^{\nu}\right) = 0$$

Note that

$$F^{\alpha\beta} F_{\alpha\beta:\nu} = \frac{1}{2}\left(F^{\alpha\beta} F_{\alpha\beta}\right)_{,\nu} = \frac{1}{2}\left(F^2\right)_{,\nu}$$

We thus express that above equation as

$$\rho v^{\nu} v_{:\nu}^{\mu} + K_1 K_3 \rho F^{\mu d} v_{\alpha} + \left[\frac{K_1}{4} (F^2)_{,\nu} + K_2 (\rho v^{\alpha} A_{\alpha})_{,\nu} \right] \left[g^{\mu\nu} - v^{\mu} v^{\nu} \right]$$

$$+ K_1 \left[F_{:\nu}^{\mu a} F_{\alpha}^{\nu} - \frac{1}{2} F_{:\nu}^{\rho\alpha} F_{\alpha}^{\nu} v_{\rho}^{\nu} \right] = 0$$

❖❖❖❖❖

[27] Special Relativistic Plasma Physics:

(*a*) Dynamical problem, how does an initial random velocity and field evolve with time?

(*b*) In a non-relativistic plasma problem, how does an initial random velocity, density and *e-m* field evolve with time?

The special relativistic plasma (MHD)

$$T^{\mu\nu} = (\rho + p) v^{\mu} v^{\nu} - p \eta^{\mu\nu}$$

In the rest frame,

$$\left(\left(T^{\mu\nu}\right)\right) = \begin{pmatrix} \rho & & & O \\ & -p & & \\ & & -p & \\ O & & & -p \end{pmatrix}$$

Which is correct since $\rho (= \rho c^2)$ is the rest frame energy density and $-p$ is the force per unit area along any spatial direction which by Newton's second law is the momentum flux (momentum flowing per unit area per unit time, the momentum being, in the rest frame, $\perp$ to the area element)

$$T^{\mu\nu} + S^{\mu\nu} = (\rho + p)v^{\mu}v^{\nu} - pg^{\mu\nu} + K\left(\frac{1}{4}F_{\alpha\beta}\, F_{\alpha\beta}\, g^{\mu\nu} - F^{\alpha\mu}\, F_{\alpha}^{\nu}\right)$$

is the combined energy momentum tensor of matter plus radiation. Here $g^{\mu\nu} = \eta^{\mu\nu}$

$\left(T^{\mu\nu} + S^{\mu\nu}\right)_{,\nu} = 0 \;\simeq$ Conservation on the total energy-momentum tensor of the matter plus radiation field.

Thus, $$\left((\rho + p)v^{\mu}v^{\nu}\right)_{,\nu} - p^{,\mu} + K\left(\frac{1}{2}\eta^{\mu\nu}\, F^{\alpha\beta} F_{\alpha\beta,\nu} - F_{,\nu}^{\alpha\mu} F_{\alpha}^{\nu} - F^{\alpha\mu} F_{\alpha,\nu}^{\nu}\right) = 0$$

Now, $$F_{\alpha,\nu}^{\nu} = K_1 J_{\alpha}$$

and so $$\left((\rho + p)v^{\mu}v^{\nu}\right)_{,\nu} - p^{,\mu} + \frac{K}{2}F^{\alpha\beta} F_{\alpha\beta}^{,\mu}$$

$$+ K\, F_{,\nu}^{\mu\alpha} F_{\alpha}^{\nu} - KK_1 J_{\alpha} F^{\alpha\mu} = 0$$

For *mhd* analysis, we must find the 4-vector-tensor equivalent od Ohm's law

$$\underline{\underline{J}} = \sigma(E + \underline{v} \times \underline{B})$$

The natural candidate appears to be $J^{\mu} = \sigma F^{\mu\nu} v_{\nu}$ and the above *mhd* equation assume the form

$$\left((\rho + p)v^{\mu}v^{\nu}\right)_{,\nu} - p^{,\mu} + \frac{K}{2}F^{\alpha\beta} F_{\alpha\beta}^{,\mu} + K\, F_{,\nu}^{\mu\alpha} F_{\alpha}^{\nu} + K_2 F^{\mu\alpha} F_{\alpha}^{\nu} v_{\nu}$$
$$= 0 \tag{1}$$

where $$k_2 = KK_1\sigma$$

These are coupled to the maxwell equation in the form

$$F_{,\nu}^{\mu\nu} = K_1 \sigma F^{\mu\nu} v^{\nu} \tag{2}$$

(1) and (2) are the basic equation of a special-realtivistic plasma.

(1) gives

$$\left((\rho + p)_{,\nu}\, v^{\mu}\right)_{,\nu} v^{\mu} + (\rho + p)v^{\nu} v_{,\nu}^{\mu} - p^{,\mu}$$

$$+\frac{K}{2}F^{\alpha\beta}F^{,\mu}_{\alpha\beta}+K\,F^{\mu\alpha}_{,\nu}F^{,\nu}_{\alpha}+K_2F^{\mu\alpha}F^{\nu}_{\alpha}v_{\nu}=0 \quad (3a)$$

So, $$\left((\rho+p)v^{\nu}\right)_{,\nu}-v_{\mu}p^{\mu}+\frac{k}{2}F^{\alpha\beta}F^{,\mu}_{\alpha\beta}v_{\mu}$$

$$+K\,F^{\mu\alpha}_{,\nu}F^{\nu}_{\alpha}v_{\mu}+K_2F^{,\mu\alpha}F^{\nu}_{\alpha}v_{\nu}v_{\mu}=0 \quad (3b)$$

Non-relativistic case: MHD equation are

$$\left(\rho(v,\underline{\nabla})\underline{v}+\underline{v},t\right) = -\nabla p+\eta\Delta\underline{v}+\sigma(\underline{v}+\underline{B})\times\underline{B}$$

Linearization $$\underline{v}(t,\underline{r}) = \underline{V_0}(t,\underline{r})+\delta\underline{v}(t,\underline{r})$$

$$\underline{B}(t,\underline{r}) = \underline{B_0}(t,\underline{r})+\delta\underline{B}(t,\underline{r})$$

Assume $\rho=\text{constant}$, $\frac{\eta}{\rho}=v,\ \frac{\sigma}{\rho}=\alpha,$

Then, $$\left(\underline{V_0},\underline{\nabla}\right)\delta\underline{v}+(\delta\underline{v},\underline{\nabla})\underline{V_0}+\delta\underline{v}_{,t}$$

$$= -\frac{1}{\rho}\underline{\nabla}\delta p+\eta\Delta\underline{v}+\alpha\left(\delta\underline{v}+\underline{B}_0\right)\times\underline{B}_0$$

$$+\alpha\left(\underline{V}_0+\delta\underline{B}\right)\times\underline{B}_0+\alpha\left(\underline{V}_0+\underline{B}_0\right)\times\delta\underline{B}$$

$$\text{div}\,\underline{v} = 0$$

$\Rightarrow \text{div}\,\underline{V}_0=0,\ \text{div}\,(\delta,\underline{v})=0.$

Fourier expansion:

$$\underline{V}_0(t,\underline{r}) = \int\hat{V}_0(t,\underline{k})\exp(i\underline{k}.\underline{r})\,d^3\underline{k}$$

$$\delta\underline{v}(t,r) = \int\delta\hat{\underline{v}}(t,\underline{k})\exp(i\underline{k}.\underline{r})\,d^3\underline{k}$$

$$\left(\underline{V}_0,\underline{\nabla}\right)\delta\underline{v} = \int i\left(\underline{k}',\hat{\underline{V}}_0(t,\underline{k})\right)\delta\hat{\underline{v}}(t,\underline{k}')$$

$$\exp\left(i(\underline{k}+\underline{k}',\underline{r})\right)d^3k\,d^3\underline{k}'$$

Coff.' of $\exp(i(\underline{k},\underline{r}))$ in $\left(\underline{V}_0,\underline{\nabla}\right)\delta\underline{v}$ is given by

$$i\int\left(\underline{k}'-\underline{k}',\hat{\underline{V}}_0(t,\underline{k}')\right)\delta\hat{\underline{v}}(t,\underline{k}-\underline{k}')\,d^3\underline{k}'$$

$$= i\int\left(\underline{k}',\hat{\underline{V}}_0(t,\underline{k}-\underline{k}')\right)\delta\hat{\underline{v}}(t,\underline{k}')\,d^3\underline{k}'$$

$$(\delta\underline{v},\underline{\nabla})\underline{V}_0 = \int i\left(\underline{k}'\delta\hat{\underline{v}}(t,\ \underline{k})\right)\hat{\underline{V}}_0(t,\underline{k}')\ \exp\left(i(\underline{k}+\underline{k}'.\underline{r})\right)d^3k\,d^3k'$$

Coff. of $\exp(i(\underline{k}.\underline{r}))$ in $(\delta\underline{v},\underline{\nabla})\underline{V}_0$ is

$$i\int\left(\underline{k},\underline{k}',\ \delta\hat{\underline{v}}(t,\underline{k}')\right)\hat{V}_0(t,\underline{k}-\underline{k}')\,d^3\underline{k}'$$

$$= i\int\left(\underline{k}, \delta\hat{\underline{v}}(t, \underline{k}' - \underline{k}')\right)\hat{V}_0(t, \underline{k}')d^3\underline{k}'$$

$$\delta\underline{v},t = \int\delta\hat{v},t(t, \underline{k})\exp(i(\underline{k}, \underline{r}))d^3\underline{k}$$

$$\underline{\nabla}\delta p = \int i\underline{k}\delta\hat{p},(t, \underline{k})\exp(i(\underline{k}, \underline{r}))d^3\underline{k},$$

$$\underline{\Delta}\delta\underline{v} = -\int k^2\delta\hat{\underline{v}}(t, \underline{k})\exp(i(\underline{k}, \underline{r}))d^3\underline{k}$$

$$(\delta\underline{v}\times\underline{B}_0)\times\underline{B}_0 = \int\left(\delta\hat{\underline{v}}(t, \underline{k})\times\hat{B}_0(t, \underline{k})\right)\times\hat{B}_0(t, \underline{k}'')$$

$$\exp(i(\underline{k} + \underline{k}' + \underline{k}'', \underline{r}))d^3kd^3k'd^3k''$$

$$\left(\underline{V}_0\times\delta\underline{B}\right)\times\underline{B}_0 = \int\left(\hat{V}_0(t, k)\times\delta\hat{\underline{B}}(t, \underline{k}')\right)\times\hat{\underline{B}}_0(t, \underline{k}'')$$

$$\exp\left\{i(\underline{k} + \underline{k}' + \underline{k}'', \underline{r})\right\}\ d^3kd^3k'd^3k''$$

Coff. of $\exp(i(\underline{k}, \underline{r}))$ in

(a) $\delta\underline{v}_{,t}$ is $\delta\hat{v}_{,t}(t, \underline{k})$,

(b) $\underline{\nabla}\delta p$ is $i\underline{k}\delta\hat{p}(t, \underline{k})$,

(c) $(\underline{V}_0\times\delta\underline{B})\times\underline{B}_0$ is $\int\left(\hat{\underline{V}}_0(t, k - k' - k'')\times\delta\hat{\underline{B}}(t, \underline{k}')\right)\times\hat{\underline{B}}_0(t, \underline{k}'')d^3k'd^3k''$

(d) $(\delta\underline{v}\times\underline{B}_0)\times\underline{B}_0$ is $\int\left(\delta\hat{\underline{v}}(t, \underline{k}')\times\hat{\underline{B}}_0(t, k'')\right)\times\hat{\underline{B}}_0(t, \underline{k} - \underline{k}' - k'')\ d^3k'd^3k''$

The linearized MHD eqn. in the spatial frequency domain thus assume the form

$$\frac{\partial}{\partial t}\delta\hat{\underline{v}}(t, \underline{k}) + i\int\left(\underline{k}'\hat{\underline{V}}_0(t, \underline{k} - \underline{k}')\right)\delta\hat{\underline{v}}(t, k')d^3k'$$

$$+ i\int(\underline{k} - \underline{k}', \delta\hat{\underline{v}}(t, k'))\hat{V}_0(t, \underline{k} - \underline{k}')d^3k' + \frac{i}{\rho}\underline{k}\delta\hat{p}(t, \underline{k}) + vk^2\delta\hat{\underline{v}}(t, \underline{k})$$

$$= \alpha\left\{\int\left(\delta\hat{\underline{v}}(t, \underline{k}')\times\hat{B}_0(t, k'')\right)\times\hat{\underline{B}}_0(t, k - k' - k'')\ d^3k'd^3k''\right.$$

$$+ \int\left(\hat{\underline{V}}_0(t, \underline{k} - \underline{k}' - \underline{k}'')\times\delta\hat{\underline{B}}(t, \underline{k}')\right)\times\hat{B}_0(t, k'')d^3k'd^3k''$$

$$\left. + \int\left(\hat{\underline{V}}_0(t, k - k' - k'')\times\widehat{\underline{B}_0}(t, k')\right)\times\delta\hat{B}(t, k'')d^3k'd^3k''\right\} \qquad (\alpha)$$

The equation of continuity

$$\text{div}\ (\delta\underline{v}) = 0$$

becomes $\left(\underline{k}, \delta\hat{\underline{v}}(t, \underline{k})\right) = 0$ $\qquad (\beta)$

At each $\underline{k}$, we construct an orthogonal set of 3 victors

$\left(e_1(\underline{k}), e_2(\underline{k}), \underline{k}\right)$ so that $\left(e_1(\underline{k}), e_2(k)\right) = 0,$

$$\left(e_1(\underline{k}), \underline{k}\right) = \left(e_2(\underline{k}), \underline{k}\right) = 0$$

and $$\left(e_1(k), e_1(k)\right) = \left(e_2(k), e_2(k)\right) = 1.$$

$$(\beta) \Rightarrow \quad \delta\hat{\underline{v}}(t, \underline{k}) = \delta\hat{v}_1(t, \underline{k})\underline{e}_1(\underline{k}) + \delta\hat{v}_2(t, \underline{k})\underline{e}_2(k)$$

for some complex scalar fields $\delta\hat{v}_1$ and $\delta\hat{v}_2$. Substituting into (α) this expression, after taking $\hat{\underline{k}}\times$ gives using

$$\hat{\underline{k}} \times \underline{e}_1(\underline{k}) = \underline{e}_2(\underline{k}), \hat{\underline{k}} \times \underline{e}_2(\underline{k}) = \underline{e}_1(k) \text{ and so}$$

$$\frac{\partial}{\partial t}\delta\hat{\underline{v}}_1(t, \underline{k}) + i\int\left(\underline{k}', \hat{\underline{V}}_0(t, \underline{k} - \underline{k}')\right)\delta\hat{v}_1(t, \underline{k}')d^3k'$$
$$+ i\int\left(\left(\underline{k} - \underline{k}', \hat{\underline{e}}_1(\underline{k}')\right)\delta\hat{v}_1(t, k')\hat{V}_{02}(t, \underline{k} - \underline{k}')d^3k'\right.$$
$$\left. + \left(\underline{k} - \underline{k}', \hat{e}_2(\underline{k}')\right)\delta\hat{v}_2(t, \underline{k}')\right) + \nu k^2\delta\hat{v}_1(t, \underline{k})$$
$$= \alpha\left\{T_1 + T_2 + T_3\right\}$$

When $$T_1 = \int\left[\hat{k} \times \left(\left(\delta\hat{\underline{v}}(t, \underline{k}') \times \hat{\underline{B}}_0(t, k'')\right) \times \hat{\underline{B}}_0(t, k - k' - k'')\right)\right]_2 d^3k'd^3k''$$
$$= \int\hat{\underline{k}} \times \left[\left(\hat{B}_{01}(t, \underline{k} - \underline{k}' - \underline{k}'')\delta\hat{v}_1(t, \underline{k}')\right.\right.$$
$$\left. + \hat{B}_{02}(t, \underline{k} - \underline{k}' - \underline{k}''), \delta\hat{v}_2(t, \underline{k}')\right)\hat{\underline{B}}_0(t, k')$$
$$\left. - \left(\hat{B}_0(t, \underline{k} - \underline{k}' - \underline{k}''), \hat{\underline{B}}_0(t, k'')\right)\delta\hat{\underline{v}}_1(t, \underline{k}')\right]_2 d^3k'd^3k''$$
$$= \int\left\{\left(\hat{B}_{01}(t, k - k' - k''), \delta\hat{\underline{v}}_1(t, k')\right.\right.$$
$$\left. + \hat{B}_{02}(t, k - k' - k''), \delta\hat{\underline{v}}_2(t, k')\right\}\hat{\underline{B}}_{01}(t, k'')$$
$$- \left(\hat{\underline{B}}_0(t, k - k' - k''), \hat{\underline{B}}_0(t, \underline{k}'')\right)\delta\hat{v}_1(t, k')\ d^3k'd^3k''$$

where $$\left(e_1(\underline{k}), \hat{\underline{B}}_0\right), \hat{\underline{B}}_{02} = \left(e_2(\underline{k}), \hat{\underline{B}}_0\right),$$

$$T_2 = \int\left[\underline{k} \times \left(\left(\hat{V}_0(t, k - k' - k'') \times \delta\hat{\underline{B}}(t, \underline{k}')\right) \times \hat{\underline{B}}_0(t, \underline{k}'')\right)\right]_2 d^3k'd^3k''$$
$$= \int\left(\hat{\underline{k}} \times \left(\hat{\underline{B}}_0(t, \underline{k}''), \hat{\underline{V}}_0(t, k - k' - k'')\right)\delta\hat{\underline{B}}(t, \underline{k}')\right.$$
$$\left. - \left(\hat{\underline{B}}(t, \underline{k}''), \delta\hat{\underline{B}}(t, k')\right)\hat{V}_0(t, k - k' - k'')\right)_2 d^3k'd^3k''$$

$$= \int\Big\{\Big(\hat{\underline{B}}_0(t,\underline{k}''), \hat{\underline{V}}_0(t,k-k'-k'')\Big)\delta\hat{\underline{B}}(t,\underline{k}')$$

$$-\Big(\hat{\underline{B}}_0(t,\underline{k}''), \delta\hat{\underline{B}}(t,k')\Big)\hat{V}_0(t,k-k'-k'')\Big\}\, d^3k'd^3k''$$

and

$$T_3 = \int\Big[\hat{\underline{k}}\times\Big(\Big(\hat{\underline{V}}_0(t,k-k'-k'')\times\hat{\underline{B}}_0(t,k')\Big)\times\delta\hat{\underline{B}}(t,\underline{k}'')\Big)\Big]_2$$

$$d^3k'd^3k''$$

$$= \int\Big\{\Big(\hat{\underline{V}}_0(t,k-k'-k''), \delta\hat{\underline{B}}_1(t,k'')\Big)\hat{B}_1(t,k')$$

$$-\Big(\hat{\underline{B}}_0(t,\underline{k}'), \delta\hat{\underline{B}}_0(t,k'')\Big)\hat{V}_{01}(t,k-k'-k'')\Big\}$$

$$d^3k'd^3k''$$

Likewise by connecting the $(1)^{\text{th}}$ component, We get another involving $\frac{\partial}{\partial t}\delta\hat{v}_2(t,k)$ on the L.H.S. These form a pair of linear intergo-differential equations for $\left\{\delta\hat{v}(t,\underline{k}), m = 1,2, \delta\hat{B}(t,\underline{k})\right\}$.

Consider the special can when $\underline{V}_0$ and $\underline{B}_0$ are constant vectors. Then the mhd equation (linearized) for $\delta v(t.\underline{r})$ for becomes

$$(\underline{V}_0, \underline{\nabla})\,\delta\underline{v}(t,\underline{r}) = -\frac{\nabla\delta p}{\rho} + v\Delta\delta\underline{v}(t,\underline{r})\,\delta v_{,t}(t,\underline{r}) + \alpha(\delta\underline{v}(t,\underline{r})\times\underline{B}_0)\times\underline{B}_0$$

$$+\Big(\underline{V}_0\times\delta\underline{B}(t,\underline{r})\Big)\times\underline{B}_0 + (\underline{V}_0\times\underline{B}_0)\times\delta\underline{B}(t,\underline{r}))$$

The moncomeater equation (linearized) is

$$\underline{\nabla}(\delta\underline{v}(t,\underline{r})) = 0$$

and finally the linearized Maxwell eqn. are

$$\underline{\nabla}\times\delta\underline{E} = -\frac{\partial}{\partial t}\delta\underline{B},$$

$$\underline{\nabla}\times\delta\underline{B} = -\mu\sigma(\delta\underline{E} + \delta\underline{v}\times\underline{B}_0 + \underline{V}_0\times\delta\underline{B}) + \mu\varepsilon\delta E_{,t}$$

This latter equation gives on taking $\underline{\nabla}\times$ and using the previous one along with $\underline{\nabla}.\delta\underline{B} = 0$,

$$\nabla^2\delta\underline{B} - \mu\sigma\delta\underline{B}_{,t} - \mu\varepsilon\delta\underline{B}_{,tt} + \mu\sigma\delta\underline{v}\times\underline{B}_0 + \mu\sigma\underline{V}_0\times\delta\underline{B} = 0$$

In the spatial frequency domain, there read

$$\Big(\underline{V}_0, \underline{k}\Big)\delta\hat{\underline{v}}(t,\underline{k}) + \delta\hat{\underline{v}}_{,t}(t,\underline{k})$$

$$-\frac{k}{\rho}\delta\hat{p}(t,\underline{k}) - vk^2\delta\hat{\underline{v}}(t,\underline{k})$$

$$+\alpha\Big(\big(\delta\hat{\underline{v}}(t,k)\times\underline{B}_0\big)\times\underline{B}_0+\big(\underline{V}_0\times\delta\hat{\underline{B}}(t,k)\big)\times\underline{B}_0$$

$$+\big(\underline{V}_0\times\underline{B}_0\big)\times\delta\hat{\underline{B}}(t,k)\Big), \qquad (\alpha)$$

$$\big(\underline{k},\delta\hat{\underline{v}}(t,\underline{k})\big)=0 \qquad (\beta)$$

$$-k^2\delta\hat{\underline{B}}(t,\underline{k})-\mu\sigma\delta\hat{\underline{B}}_{,t}(t,\underline{k})-\mu\varepsilon\delta\hat{\underline{B}}_{,tt}(t,\underline{k})$$

$$+\mu\sigma\delta\hat{\underline{v}}(t,\underline{k})\times\hat{\underline{B}}_0+\mu\sigma\underline{V}_0\times\delta\hat{\underline{B}}(t,\underline{k})=0 \qquad (\gamma)$$

Assuming a time dependence of e^{jwt} for $\delta\hat{\underline{v}}(t,\underline{k})$ and $\delta\hat{\underline{B}}(t,\underline{k})$, we get

$$\left[\big(\underline{k},\underline{V}_0\big)+jw+vk^2-\alpha\underline{B}_0\underline{B}_0^T+\alpha\beta_o^2\right]\delta\hat{\underline{v}}$$

$$=\alpha\Big(\big(\underline{V}_0,\underline{B}_0\big)I-\underline{V}_0\underline{B}_0^T+\underline{B}_0\underline{V}_0^T-\underline{V}_0\underline{B}_0^T\Big)\delta\hat{\underline{B}}$$

$$-\frac{k}{\rho}\delta\hat{p}=0 \qquad (\alpha)$$

(γ) becomes

$$\left[k^2+jw\mu\sigma-\mu tw^2+\mu\sigma\Omega_{v_0}\right]\delta\hat{\underline{B}}-\mu\sigma\Omega_{B_0}\delta\hat{\underline{v}}=0 \qquad (\gamma')$$

When for any vector $X=\big(X_1,X_2,X_{3x}^T\in\mathbb{R}^3$

$$\Omega_X=\begin{pmatrix}0&-X_3&X_2\\X_3&0&-X_1\\-X_2&X_1&0\end{pmatrix}$$

This matrix has the defining property

$$\underline{\underline{\Omega}}_x\underline{\xi}=\underline{X}\times\underline{\xi}\forall\underline{\xi}\in\mathbb{R}^3$$

$(\alpha')(\beta),(\gamma')$ can be cast in matrix form (block) as

$$\begin{bmatrix}\Big(\big(\underline{k},\underline{V}_0\big)+jw+vk^2+\alpha B_0^2\Big)\underline{\underline{I}}_3-\alpha\underline{B}_0\underline{B}_0^T & \alpha\Big(\big(\underline{V}_0,\underline{B}_0\big)I_3-2\underline{V}_0\underline{B}_0^T+\underline{B}_0\underline{V}_0^T\Big) & -\dfrac{k}{\underline{p}}\\ \underline{k}^T & O^T & \underline{0}\\ -\mu\sigma\underline{\underline{\Omega}}B_0 & \big(k^2+jw\mu\sigma-\mu\varepsilon w^2\big)\underline{\underline{I}}_3+\mu\sigma\underline{\underline{\Omega}}_{V_0} & 0\end{bmatrix}\begin{bmatrix}\delta\hat{\underline{v}}\\ \delta\hat{\underline{B}}\\ \delta\hat{\underline{p}}\end{bmatrix}=0$$

For a non-zero solution to exist, we require that the determinant of the above 7 × 7 matrix vanish. This gives us for given $\underline{B}_0, \underline{V}_0$ a dispersion relation between w and $\underline{k}$.

Now we analyze the special-relativistic mhd equation (3*a*) and (3*b*). Substituting for $\left((\rho+p)v^{\nu}\right)_{,\nu}$ from (3*b*) into (3*a*) gives

$$v^{\nu}\left\{v^{\alpha}p'^{\alpha}-\frac{k}{2}F^{\rho\sigma}F^{,\alpha}_{\rho\sigma}v_{\alpha}-kF^{\rho\alpha}_{,\nu}F^{\nu}_{\alpha}v_{\rho}-k_2F^{\rho\alpha}F^{\nu}_{\alpha}v_{\nu}v_{\rho}\right\}$$

$$+(\rho+p)v^{\nu}v^{\mu}_{,\nu}-p^{,\mu}+\frac{k}{2}F^{\alpha\beta}F^{,\mu}_{\alpha\beta}+kF^{\mu\alpha}_{,\nu}F^{\nu}_{\alpha}+k_2F^{\mu\alpha}F^{\nu}_{\alpha}v_{\nu}=0 \tag{4}$$

We note from (4) that when there is no *em* field, *i.e.* $F^{\mu\nu}=0$, we get the standard special relativistic fluid dynamical equation without viscosity.

$$(\rho+p)v^{\nu}v^{\mu}_{,\nu}-p^{,\mu}+v^{\mu}v_{\alpha}p^{\alpha}=0 \tag{5}$$

and the man conservation equation (3*b*) becomes

$$\left((\rho+p)v^{\mu}\right)_{,\mu}-v_{\mu}p^{,\mu}=0 \tag{6}$$

Suppose we combine (5) and (6) with the equation of state

$$p=p(\rho). \tag{7}$$

These are to be combined with

$$\eta_{\mu\nu}v^{\mu}v^{\nu}=1 \tag{8}$$

or

$$v_{\mu}v^{\mu}=1.$$

(6) can be exprenal as

$$\left(\rho v^{\mu}\right)_{,\mu}+pv^{\mu}_{,\mu}=0 \tag{9}$$

Let $\left(v^{r}\right)_{r=1}^{3}$ be the spatial component of v^{μ}. We note that (8) is the same as

$$v^{02}-\sum_{r=1}^{3}v^{r^2}=1 \tag{10}$$

Define

$$\gamma(u)=\left(1-u^2\right)^{-1/2}$$

when

$$u^r=\frac{dx^r}{d\tau}=\frac{d\tau^r}{d\zeta}=v^r/v^0$$

Then (10) implies

$$v^{02}-v^{02}u^2=1$$

or

$$v^0=\left(1-u^2\right)^{-1/2}=\gamma(u), v^r=\gamma u^r_1 1\le r\le 3.$$

So (9) can be exprinced as

$$(\gamma\rho)_{,0} + \left(\gamma\rho u^{r}\right)_{,r} + p\left(\gamma_{,0} + (\gamma u^{r}),r\right) = 0 \qquad (10)$$

and (5) as

$$\gamma(\rho + p)\gamma_{,0} + \gamma(\rho + p)u^{rs}(\gamma u^{r})_{,s} - p^{r} + \gamma u^{r}\left(\gamma p^{s0} + \gamma u^{s} p,s\right) = 0$$

or

$$\gamma\gamma_{,0} + \gamma u^{s}\left(\gamma u^{r}\right)_{,s} + \frac{p,r}{(p+\rho)} + \frac{\gamma^{2}u^{r}p_{,0}}{(p+\rho)} = 0 \qquad (11)$$

$$+\gamma^{2}u^{r}u^{s}p,s/(p+\rho)$$

Perturbed version of (10) and (11).

$$\rho = \rho_0 + \delta\rho(t, \underline{r}), u^{r} = U_0^{r} + \delta\underline{u}^{r}(t, \underline{r})$$

$$p = p(\rho_0) + p'(\rho_0)\delta P(t, \underline{r})$$

The unperturbed versions of (10) and (11) are consistent with $p(\rho_0), U_0^r, \rho_0$ as constants. We look at the perturbed versions of (10) and (11):

$$\delta\gamma_{,0}\rho_0 + \gamma_0\delta\rho_{,0} + \rho_0 U_0^{r}\delta\gamma_{,r} + \gamma_0 U_0^{r}\delta\rho_{,r}$$

$$+\gamma_0\rho_0\delta u^{r}_{,r} + p_0(\delta\gamma_{,0} + U_0^{r}\delta\gamma_{,r} + \gamma_0\delta u^{r}_{,r}) = 0 \qquad (12)$$

❑❑❑

7

Quantum Signal Processing

[1] Quantum Scattering of a Rigid Body:

$$\left[\frac{\partial^2}{\partial X^2}+\frac{\partial^2}{\partial Y^2}+\frac{m}{I}\frac{\partial^2}{\partial \theta^2}+k^2\right]\psi_s(X,Y,\theta)$$

$$=\frac{2m}{£^2}\chi\psi_i(X,Y,\theta)$$

where $$k^2=\frac{2mE}{\hbar^2}=k_X^2+k_Y^2+\frac{m}{I}n^2.$$

By the Green's function for the Helmholtz equation, this has the solution

$$\Psi_s(X, Y, \theta)$$

$$=\sum_{r\in\mathbb{Z}}\frac{-m}{2\pi\hbar^2}$$

$$\int_{\substack{X',Y'\in\mathbb{R},\\ \theta'\in[0,2\pi]}}\frac{\exp\left(ik\left((X-X')^2+(Y-Y')^2+\left(\sqrt{\frac{I}{m}}(\theta-\theta'+2r\pi)\right)^2\right)^{\frac{1}{2}}\right)}{\left((X-X')^2+(Y-Y')^2+\left(\sqrt{\frac{I}{m}}(\theta-\theta'+2r\pi)\right)^2\right)^{\frac{1}{2}}}$$

$$\times(\chi\psi_i)(X',Y',\theta')dX'dY'\times\sqrt{\frac{I}{m}}d\theta'$$

❖❖❖❖❖

[2] Optimal Control of a Field:

The filed φ: $\mathbb{R}^n \to \mathbb{R}^p$

$\varphi = (\varphi_1, .., \varphi_p)$ satisfies the field equation

$$\mathcal{L}\underline{\varphi}(X) = \underline{J}(X)$$

where $\mathcal{L}$ is a linear partial differential operator.

$\underline{J}: \mathbb{R}^n \to \mathbb{R}^q$ is a source (input) signal field.

We wish to select $\underline{J}$ so that the cost functional

$$\int C(\varphi_k(X), \varphi_{k,m}(X), \underline{J}(X))d^n X$$

is a minimum. Incorporating the equations as constraints using a Lagrange multiplier field

$\underline{\Lambda}: \mathbb{R}^n \to \mathbb{R}^q$ (Note $\mathcal{L}: C^\infty(\mathbb{R}^n, \mathbb{R}^p) \to C^\infty(\mathbb{R}^n, \mathbb{R}^q)$)

the functional to be minimized is

$$S[\varphi, J, \Lambda] = \int C(\varphi_k, \varphi_{k,m}, \underline{J})d^n X - \int (\underline{\Lambda}, \mathcal{L}\varphi - J)d^n X$$

$$\delta_{\varphi_k} S = 0 \Rightarrow \frac{\partial C}{\partial \varphi_k} - \sum_m \frac{\partial}{\partial X_m} \frac{\partial C}{\partial \varphi_{k,m}} - (\mathcal{L}^* \Lambda)_k = 0$$

$$\delta_{J_k} S = 0 \Rightarrow \frac{\partial C}{\partial J_k} + \Lambda_k = 0$$

$$\delta_\Lambda S = 0 \Rightarrow \mathcal{L}\varphi - \underline{J} = 0$$

Eliminating Λ gives

$$\frac{\partial C}{\partial \varphi_k} - \sum_m \frac{\partial}{\partial X_m} \frac{\partial C}{\partial \varphi_{k,m}} + \left(\mathcal{L}^* \frac{\partial C}{\partial J}\right)_k = 0,$$

$$\mathcal{L}\varphi - \underline{J} = 0$$

❖❖❖❖❖

[3] Example from Quantum Mechanics:

Calculate $\{f(t), 0 \le t \le T\}$ so that

$$\int_0^T \|\psi_d(t) - \psi(t)\|^2 dt \text{ is a minimum}$$

where $\psi(.)$ satisfies

$$-\frac{1}{2m}\Delta\psi(t) + f(t)V\psi(t) = i\psi'(t) \tag{1}$$

Here $\psi(t) = \psi(t, \underline{r})$ and $\psi_d(t) = \psi_d(t, \underline{r})$,

$$\|\psi_d(t) - \psi(t)\|^2 = \int |\psi_d(t,\underline{r}) - \psi(t,\underline{r})|^2 d^3 r.$$

and $V = V(\underline{r})$ (multiplication operator). Incorporating the eqns. of motion (Schrödinger's equation) using the complex Lagrange multiplier $\Lambda(t, \underline{r})$, we have to minimize

$$S\{\psi, f, \Lambda] = \int |\psi_d(t,\underline{r}) - \psi(t,\underline{r})|^2 dt d^3 r$$

$$+ \int \mathrm{Re}\left(\bar{\Lambda}(t,\underline{r})\left(i\psi_{,t}(t,\underline{r}) + \frac{1}{2m}\Delta\psi(t,\underline{r}) - f(t)V(\underline{r})\psi(t,\underline{r})\right)\right) dt d^3 r$$

ψ, $\bar{\psi}$, f, Λ, $\bar{\Lambda}$ are independent functions.

$$\delta_{\bar{\psi}} S = 0$$

$$\Rightarrow -\left(\psi_d(t,r) - \psi(t,\underline{r})\right) + \frac{i}{2}\Lambda_{,t}(t,\underline{r}) + \frac{1}{4m}\Delta\Lambda(t,\underline{r})$$

$$-\frac{1}{2} f(t) V(\underline{r}) \Lambda(t,\underline{r}) = 0 \quad (2)$$

$$\Rightarrow \qquad \delta_f S = 0$$

$$\int \text{Re}\left(\bar{\Lambda}(t,\underline{r}) V(\underline{r}) \psi(t,\underline{r})\right) d^3 r = 0 \quad (3)$$

Equations (1), (2) and (3) are to be solved for ψ, Λ and f.

Note that $f(t)$ is real.

❖❖❖❖❖

[4] Example from Quantum Field Theory:

The field $\phi(t, \underline{r}) = \phi(X)$ is coupled to an external source $J(t, \underline{r}) = J(X)$ via the interaction action $\int J(X)\phi(X) d^4 X$. The Lagrangian density is thus

$$£_J\left(\phi, \phi_{,\mu}^{*}\right) = \frac{1}{2}\partial_\mu \phi \partial^\mu \phi - \frac{1}{2} m^2 \phi^2 + J\phi$$

The propagator is

$$\langle Vac \,|T\{\phi(X)\, \phi(Y)\}|\, Vac \rangle = \int \exp\left(i \int £_J\left(\phi(\xi), \phi, \mu(\xi), \xi\right) d^4 \xi\right)$$

$$\phi(X)\phi(Y) \prod_{\eta \in \mathbb{R}^4} d\phi(\eta)$$

$$= \frac{-\delta^2}{\delta J(X) \delta J(Y)} \int \exp\left(i \int £J\left(\phi(\xi), \phi_{,\mu}(\xi), \xi\right) df \xi\right)$$

$$\prod_{\eta \in \mathbb{R}^4} d\phi(\eta)$$

We wish to make this as close to a given function $K(X, Y)$ on $\mathbb{R}^4 \times \mathbb{R}^4$ as possible.

The propagator $\langle Vac|T\{\phi(X).\ \phi(Y)\}|Vac\rangle$ controls the scattering probability amplitude between times t_X and t_Y when $X = (t_X, \underline{r}_X)$, $Y = (t_Y, \underline{r}_Y)$.

Thus, we must select the function $\{J(X)\}_{X \in \mathbb{R}^4}$ so that

$$\int \left| \langle Vac | T\{\phi(X)\phi(Y)\} | Vac \rangle - K(X,Y) \right|^2 d^4 X d^4 Y$$

is a minimum. The optimal minimizing equations for the current density $J(.)$ are (J is real and ϕ is also real)

$$\int \left\{ \frac{\delta}{\delta J(Z)} \langle Vac | T\{\phi(X).\phi(Y)\} | Vac \rangle \right.$$

$$\left. \left(\langle Vac | T\{\phi(X)\phi(Y)\} | Vac \rangle - K(X,Y) \right) d^4 X d^4 Y \right. = 0$$

or equivalently,

$$\int \left\{ \left(\frac{\delta^3}{\delta J(X)\delta J(Y)\delta J(Z)} \int \exp\left(-iS_J(\phi)\right) \prod_{\xi} d\phi(\xi) \right) \right.$$

$$\left. \cdot \left(\frac{\delta^2}{\delta J(X)\delta J(Y)} \int \exp(iS_J(\phi)) \prod_{\xi'} d\phi(\xi') - K(X,Y) \right) \right\} d^4 X d^4 Y = 0$$

❖❖❖❖❖

[5] Quantum Signal Processing (Contd.):

Typical subspace: Let $\rho_n \in \mathbb{C}^{d^n \times d^n}$ be a density matrix. Assume that $\rho_n = \rho^{\otimes n}$ where $\rho \in \mathbb{C}^{d \times d}$ is a density matrix.

Let $$\rho = \sum_{x \in A} p(x) |x><x| \quad A = \{1, 2, .., d\}.$$

$p(x) \geq 0$ $\sum_{x \in A} p(x) = 1$, $\{(x): x \in A\}$ an ONB for $\mathbb{C}^d$.

Then $$\rho_n = \rho^{\otimes n} = \sum_{x_1,..,x_n \in A} p(x_1)...p(x_n) |x_1...x_n><x_1...x_n| .$$

Let $$H(\rho) = -T_r(\rho \log \rho) \equiv -\sum_{x \in A} p(x) \log p(x) .$$

Let $$T_\delta^n = \left\{ (x_1,..,x_n) \in A^n \left| \left| \frac{1}{n} \sum_{i=1}^{n} \log p(x_i) + H(\rho) \right| < \delta \right. \right\}$$

By Chebrshev's inequality

$$P(T_\delta^n) \equiv \sum_{(x_1,..,x_n) \in T_\delta^n} p(x_1)...p(x_n) \geq 1 - \frac{Var(\log p(x))}{n\delta^2}$$

where $$Var(\log p(x)) = \sum_{x \in A} p(x) (\log p(x))^2 + H(\rho)^2$$

$$= \sum_{x \in A} p(x) (\log p(x) + H(\rho))^2$$

Thus,

$$\lim_{n \to \infty} P(T_\delta^n) = 1, \forall \delta > 0.$$

Now for $(x_1, .., x_n) \in T\delta$,

$$-\delta - H(\rho) \leq \frac{1}{n}\sum_{i=1}^{n} \log p(x_i) \leq \delta - H(\rho)$$

and so

$$\exp(-h(\delta + H(\rho))) \leq p(x_1) \ldots p(x_n) \leq \exp(n(\delta - H(p)))$$

Thus $$1 \geq P(T_\delta^n) \geq \# T(_\delta^n). \exp(-n(\delta + H(\rho)))$$

or $$\#(T_\delta^n) \leq \exp(n(\delta + H(\rho)))$$

Also, $$1 - \frac{Var(\log p(x))}{n\delta^2} \leq P(T_n^\delta)$$
$$\leq \exp(n(\delta - H(\rho))) \# (T_\delta^n)$$

Thus,

$$\left(1 - \frac{v}{n\delta^2}\right)\exp(n(H(\rho) - \delta))$$
$$\leq \#(T_\delta^n) \leq \exp(n(\delta + H(\rho))) \tag{1}$$

where $$v = Var(\log p(x)).$$

Choose any $\delta > 0$ and encode the elements of T_δ^n (n-length sequences) into k length binary sequences where $2^k \leq \exp(n(\delta + H(\rho)))$.

This encoding can be performed in a 1 – 1 way.

The original product source consisted of d^n sequences from the alphabet A.

Encode all the elements of $T_d^{nc} \equiv A^n \setminus T^c$

into a single k length binary sequence.

The compression achieved in this system of encoding is smaller than

$$d^n \to \exp(n(\delta + H(\rho)))$$

or in terms of bits

$$n \log_2 d \to n(\delta + H(\rho))\log_2 e$$

i.e. the bits rate reduction is by the fraction

$$\frac{(\delta + H(\rho))\log_2 e}{\log_2 d} = \frac{(\delta + H(\rho))}{\log_e d}$$

The error of decoding probability is

$$P(T_\delta^{nc}) = 1 - P(T_\delta^n) \leq \frac{v}{n\delta^2}$$

By choosing n sufficiently large, the decoding error probability can be made arbitrarily small while retaining the bit compression rate as $\leq \dfrac{(\delta + H(\rho))}{\log_e d}$.

Thus choosing δ sufficiently small, we can achieve a compression rate arbitrarily clone to $\frac{H(\rho)}{\log_2 d}$ and making n large enough the corresponding error probability can be made arbitrarily small. This is the Shannon nóiselen coding theorem.

❖❖❖❖❖

[6] em Field Theory Based Image Processing:

$\widehat{\underline{E}}_i(w, \underline{r})$ is the electric field incident on the object $H_0(w, \underline{r})$ describes the filter structure of the object at $\underline{r}$. The e-field signal at this pixel is $H(w, \underline{r})\ \widehat{\underline{E}}_i(w, \underline{r})$. The surface magnetic current density at $\underline{r}$ ($\underline{r} \in S$, S is the object surface) is

$$\underline{M}_S(w, \underline{r}) = H_0(w, \underline{r})\hat{n}(\underline{r}) \times \widehat{\underline{E}}\upsilon(w, \underline{r})$$

where $\hat{n}(\underline{r})$ is the unit normal to S at $\underline{r}$.

The electric vector potented at $\underline{r}$ is then

$$\underline{F}(w, \underline{r}) = \frac{\varepsilon}{4\pi}\int_S \frac{\underline{M}_S(w, \underline{r}')\exp\left(-\frac{jw}{c}|\underline{r} - \underline{r}'|\right)}{|\underline{r} - \underline{r}'|} dS(\underline{r}')$$

$$= \frac{\varepsilon}{4\pi}\int_S \frac{H_0(w, \underline{r}')\hat{\underline{n}}(\underline{r}') \times \widehat{\underline{E}}\upsilon(w, \underline{r}')\exp\left(-\frac{jw}{c}|\underline{r} - \underline{r}'|\right)}{|\underline{r} - \underline{r}'|} dS(\underline{r}')$$

$\{H_0(w, \underline{r}'), \underline{r}' \in S\}$ describes the object pattern and $\{\underline{E}(w, \underline{r}), \underline{r} \in \mathbb{R}^3 \backslash S\}$ the recorded image.

The above equation tells us how to recover the function $\{H_0(w, \underline{r}'), \underline{r}' \in S\}$ from em field measurements. The magnetic field $\widehat{H}_i(w, \underline{r})$ in the incident wave is given by

$$\widehat{H}_i(w, \underline{r}) = \frac{i}{w\mu}\nabla\widehat{\underline{E}}_i(w, \underline{r})$$

and the surface electric current density is then

$$\underline{J}_S(w, \underline{r}) = -\frac{j}{w\mu}\hat{n}(\underline{r}) \times \left(\underline{\nabla} \times \widehat{\underline{E}}_i(w, \underline{r})\right)$$

Note: Assuming

$$\widehat{\underline{E}}_i(w, \underline{r}) = \int \tilde{\underline{E}}_i(w, \hat{m})\exp\left(-j\frac{w}{c}(\hat{m}, \underline{r})\right) d\Omega(\hat{m})$$

so that $\widehat{\underline{E}}_i$ satisfies the wave equation

$$\left(\nabla^2 + \frac{w^2}{c^2}\right)\widehat{\underline{E}}_i(w, \underline{r}) = 0$$

we get

$$\widehat{H}_i(w,\underline{r}) = \frac{1}{\mu c}\int\left(\widehat{m}\times\underline{\tilde{E}}_i\left(w,\widehat{\underline{m}}\right)\right)\exp\left(-\frac{jw}{c}\left(\widehat{m},\underline{r}\right)\right)d\Omega(\widehat{m})$$

and thus
$$\underline{J}_S(w,\underline{r}) = -\frac{1}{\mu c}\int\left\{\widehat{\underline{n}}(\underline{r})\times\left(\widehat{\underline{m}}\times\underline{\tilde{E}}_i\left(w,\widehat{\underline{m}}\right)\right)\right\}$$
$$\exp\left(-j\frac{w}{c}\left(\widehat{m},\underline{r}\right)\right)d\Omega(\widehat{m})$$

Problem from Revuz and Yor "Continuous Martingales and Brownian Motion" P.185

$$\underline{Z}_t = \underline{B}_t + \underline{f}(t)\in\mathbb{R}^2.$$

$$\int_{\frac{1}{n}}^{t}\left|Z_s\right|^{-1}\left|df(s)\right| = \int_{1}^{nt}\left|Z_{s/n}\right|^{-1}\left|df\left(\frac{s}{n}\right)\right|$$

$$\mathbb{E}\left\{\left|Z_{s/n}\right|^{-1}\right\} = \mathbb{E}\left\{\left|\underline{B}_{s/n}+\underline{f}\left(s/n\right)\right|^{-1}\right\}$$

$$= \mathbb{E}\left\{\left|\sqrt{\frac{s}{n}}Z_0+f\left(s/n\right)\right|^{-1}\right\}$$

where
$$Z_0 = B_{s/n}\Big/\sqrt{\frac{s}{n}} \sim N(0,\underline{I}_2).$$

Thus
$$\mathbb{E}\left\{\left|Z_{s/n}\right|^{-1}\right\} \le \sqrt{\frac{n}{s}}K$$

where
$$K = \sup_{\underline{m}\in\mathbb{R}^2}\mathbb{E}\left\{\left|\underline{m}+\underline{G}\right|^{-1}\right\} < \infty\left(\underline{G}\sim N\left(0,\underline{I}_2\right)\right)$$

Thus,
$$\int_{\frac{1}{n}}^{t}\mathbb{E}\left\{\left|Z_s\right|^{-1}\left|df(s)\right|\right\} \le K\sqrt{n}\int_{1}^{nt}\frac{1}{\sqrt{s}}\left|df\left(\frac{s}{n}\right)\right|$$

$$\le K\sqrt{n}\sup_{1\le s\le nt}\left\{\frac{\left|df\left(s/n\right)\right|}{ds}\right\}\int_{1}^{nt}\frac{1}{\sqrt{s}}ds$$

$$= \frac{2K}{\sqrt{n}}\sup_{\frac{1}{n}\le s\le t}\left\{\frac{\left|df(s)\right|}{ds}\right\}s^{\frac{1}{2}}\Big|_{1}^{nt}\left(\frac{\left|df\left(s/n\right)\right|}{ds}=\frac{1}{n}\frac{\left|df\left(s/n\right)\right|}{d\left(s/n\right)}\right)$$

$$= 2K\left(\sqrt{t}-\frac{1}{\sqrt{n}}\right)\sup_{\frac{1}{n}\le s\le t}\left\{\frac{|df(s)|}{ds}\right\}$$

$$\underset{n\to\infty}{\to} 2K\sqrt{t}\sup_{0\le s\le t}\left\{\frac{|df(s)|}{ds}\right\} < \infty$$

Let $$T = \inf\{t | \underline{Z}_t = 0\}.$$

❖❖❖❖❖

[7] Derivation of the Kallianpur-Striebel Formula:

$$dX_t = v(X_t)dt + \sigma(X_t)dW_t^{proc}$$

$$dY_t = h(X_t)dt + \sigma_Y dW_t^{obs}$$

$$p\left(X_t + dt | \mathcal{F}_{t+dt}^Y\right) = p\left(X_{t+dt} | \mathcal{F}_t^Y, dy_t\right)$$

$$= \int p\left(X_{t+dt} | X_t\right) p\left(X_t | \mathcal{F}_t^Y, dy_t\right) dX_t$$

$$\mathbb{E}\left[f\left(X_{t+dt}\right) | F_{t+dt}^Y\right]$$

$$= \int p\left(X_t | \mathcal{F}_t^Y, dy_t\right) dX_t \left(\int f(X_{t+dt}) p(X_{t+dt} | X_t) dX_{t+dt}\right)$$

$$= \int (f(X_t) + (\mathcal{L}f)(X_t)dt) p\left(X_t | \mathcal{F}_t^Y, dy_t\right) dX_t$$

$$= \frac{\int (f(X_t) + (\mathcal{L}f)(X_t)dt) p\left(dy_t | X_t\right) p\left(X_t | \mathcal{F}_t^Y\right) dX_t}{\int p\left(dy_t | X_t\right) p\left(X_t | \mathcal{F}_t^Y\right) dX_t}$$

$$= \frac{\int\left[(f(X_t) + (\mathcal{L}f)(X_t)dt)\exp\left\{-(dy_t - h(X_t)dt)^T \frac{(dy_t - h(X_t)dt)}{2\sigma Y^2 dt}\right\} \times p\left(X_t \mathcal{F}_t^Y\right) dX_t\right]}{\int \exp\left\{-(dy_t - h(X_t(dt)^T \frac{(dy_t - h(X_t)dt)}{2\sigma Y^2 dt} \times p\left(X_t | \mathcal{F}_t^Y\right) dX_t\right\}}$$

$$= \frac{\int (f(X_t) + (\mathcal{L}f)(X_t)dt)\exp\left\{h(X_t)^T dy_t - \frac{1}{2}h(X_t)^T h(X_t)dt\right\} p\left(X_t | \mathcal{F}_t^Y\right) dX_t}{\int \exp\left\{h(X_t)^T dy_t - \frac{1}{2}h(X_t)^T h(X_t)dt\right\} p\left(X_t | \mathcal{F}_t^Y\right) dX_t} \quad (1)$$

(assuming $\sigma_Y = 1$)

Consider now

$$L_t(X|Y) = \exp\left\{\int_0^t \underline{h}(\underline{X}_s)^T d\underline{y}_s - \frac{1}{2}\underline{h}(\underline{X}_s)^T \underline{h}(\underline{X}_s)ds\right\}$$

We have $$dL_t(X|Y) = \left(\underline{h}(X_t)^T d\underline{y}_t\right)L_t(X|Y)$$

Now define $$\sigma_t(f) = \int_{C_X[0,t]} f(X_t)L_t(X \mid Y)P[dX]$$

Then $$d\sigma_t(f) = \int\left[\mathcal{L}f(X_t)dt + f(X_t)dL_t(X \mid Y)\right]P[dX]$$

$$\left(\text{since } df(X_t).dL_t(X|Y) = 0 \because dW_t^{proc}.dW_t^{obs}\right)$$

$$= \int\left[\mathcal{L}f(X_t)dt + fh(X_t)^T dy_t\right]L_t(X \mid Y)P[dX] \tag{2}$$

From (1), $\mathbb{E}\left\{f(X_{t+dt})\middle|\mathcal{F}_{t+dt}^Y\right\}$

$$= \frac{\int\left(f(X_t) + \mathcal{L}f(X_t)dt\right)\left(1 + h(X_t)^T d\underline{y}_t\right)p\left(X_t|\mathcal{F}_t^Y\right)dX_t}{\int\left(1 + h(X_t)^T dy_t\right)p\left(X_t|\mathcal{F}_t^Y\right)dX_t}$$

$$= \frac{\left\{\int f(X_t)p\left(X_t|\mathcal{F}_t^Y\right)dX_t + \left(\int\left(\mathcal{L}f(X_t)\right)p\left(X_t|\mathcal{F}_t^Y\right)dX_t\right)dt + \int f(X_t)\left(h(X_t)^T dy_t\right)p\left(X_t|\mathcal{F}_t^Y\right)dX_t\right\}}{\left(1 + \int\left(h(X_t)^T dy_t\right)p\left(X_t|\mathcal{F}_t^Y\right)dx_t\right)} \tag{3}$$

We have from (2),

$$\sigma_{t+dt}(f) = \sigma_t(f) + d\sigma_t(f)$$

$$= \int\left[f(X_t) + \mathcal{L}f(X_t)dt + f(X_t)h(X_t)^T dy_t\right]L_t(X \mid Y)P[dX] \tag{4}$$

Comparing (3) and (4) shows that

$$\frac{\sigma_{t+dt}(f)}{\sigma_{t+dt}(1)} = \mathbb{E}\left[f(X_{t+dt})\middle|\mathcal{F}_{t+dt}^Y\right]$$

provided that

$$\frac{\sigma_t(f)}{\sigma_t(1)} = \mathbb{E}\left[f(X_t)\middle|\mathcal{F}_t^Y\right].$$

Thus by inductions

$$\mathbb{E}\left\{f(X_t)\middle|\mathcal{F}_t^Y\right\} = \frac{\sigma_t(t)}{\sigma_t(1)} \forall t \geq 0.$$

Let $$\mathbb{E}\left\{\xi\middle|\mathcal{F}_t^Y\right\} = \pi_t(\xi)$$

for any $r.v.\,\xi$. Then we've shown that

$$d\sigma_t(f) = \sigma_t(\mathcal{L}f)dt + \sigma_t(hf)dy_t \tag{5}$$

Thus,

$$d\sigma_t(1) = \sigma_t(h)dy_t.$$

$$d\pi_t(f) = d\left(\frac{\sigma_t(f)}{\sigma_t(1)}\right)$$

$$= \frac{d\sigma_t(f)}{\sigma_t(1)} - \frac{\sigma_t(f)}{\sigma_t(1)^2}d\sigma_t(1) + \frac{\sigma_t(f)}{\sigma_t(1)^3}(d\sigma_t(1))^2 - \frac{d\sigma_t(f).d\sigma_t(1)}{\sigma_t(1)^2}$$

$$= \frac{\sigma_t(\mathcal{L}f)}{\sigma_t(1)}dt + \frac{\sigma_t(hf)}{\sigma_t(1)}dy_t - \frac{\sigma_t(f)}{\sigma_t(1)}\frac{\sigma_t(h)}{\sigma_t(1)}dy_t$$

$$+\frac{\sigma_t(f)}{\sigma_t(1)^3}\sigma_t(h)^2 dt - \frac{\sigma_t(hf)}{\sigma_t(1)^2}\sigma_t(h)dt$$

$$= \left[\pi_t(\mathcal{L}f) + \pi_t(f)\pi_t(h)^2 - \pi_t(hf)\pi_t(h)\right]dt$$

$$+\left[\pi_t(hf) - \pi_t(f)\pi_t(h)\right]dy_t$$

$$= \pi_t(\mathcal{L}f)dt + \left[\pi_t(hf) - \pi_t(f)\pi_t(h)\right]\left[dy_t - \pi_t(h)dt\right]$$

This is Kushner's stochastic filtering equation.

❖❖❖❖❖

[8] Slave Dynamics with Parametric Uncertainties Using Adaptive Controllers:

$$M(q)\ddot{q} + C(q,\dot{q})\dot{q} + G(\underline{q},t) = \underline{\tau}(t)$$

$$\underline{\tau}(t) = Y_d\hat{\theta}_d - J^T(\underline{q},\hat{\underline{\theta}}_K)K(\Delta\hat{\dot{X}} + \alpha\Delta X)$$

θ_d = dynamics parameter uncertainties,

θ_K = kinematic parameter uncertainties.

$$\Delta\underline{X} = \underline{X}(t) - \underline{X}_d(t)$$

$\underline{X}_d(t) \simeq$ desired position.

$$\underline{q} \to \underline{\underline{J}}(\underline{q},\underline{\theta}_K)$$

is the Jacobian matrix of $\underline{q} \to \underline{X}$ (angular variables Cartesian position of the tip).

$$\dot{\underline{q}}_r = \underline{\underline{J}}^{-1}(\underline{q},\hat{\theta}_K)\dot{\underline{X}}_r \qquad (r \text{ for reference})$$

$$\dot{\underline{X}}_r = \dot{\underline{X}}_d - \alpha(X - X_d)$$

$$\therefore \quad \dot{X} - \dot{X}_r = \dot{X} - \dot{X}_d + \alpha(X - X_d)$$

$\simeq p_d$ controller.

$$\underline{S} = \dot{\underline{q}} - \dot{\underline{q}}_r = J^{-1}(\underline{q},\underline{\theta}_K)\dot{\underline{X}} - J^{-1}(\underline{q},\hat{\underline{\theta}}_K)\dot{\underline{X}}_r$$

$$\dot{\underline{X}} = \underline{\underline{J}}(\underline{q},\underline{\theta}_K)\dot{\underline{q}}$$

$$M(\underline{q})\underline{S} + C(\underline{q},\underline{\dot{q}})\underline{S} + Y_d\underline{\theta}_d = \underline{\tau}$$

$$\Leftrightarrow M(\underline{q})(\ddot{q} - \ddot{q}_r) + C(q,\dot{q})(\dot{q} - \dot{q}_r) + Y_d\theta_d = \underline{\tau}$$

$$\Leftrightarrow M(\underline{q})\underline{\ddot{q}} + C(\underline{q},\underline{\dot{q}})\dot{q} + G(\underline{q},t) - \underline{\tau} = 0 = M(\underline{q})\ddot{q}_r + C(q,\dot{q})\dot{q}_r + G(\underline{q},t) - Y_d\theta_d$$

$$\Leftrightarrow \quad Y_d\theta_d = M(q)\ddot{q}_r + C(q,\dot{q})\dot{q}_r + G(\underline{q},t)$$

$$\begin{aligned}
\Delta\widehat{\dot{X}} &= \widehat{\dot{X}} - \dot{X}_d \\
\widehat{\dot{X}} &= J\left(\underline{q},\underline{\hat{\theta}}_K\right)\underline{\dot{q}} \\
\dot{X} &= J\left(\underline{q},\underline{\theta}_K\right)\underline{\dot{q}} \\
\dot{\hat{\underline{\theta}}}_d &= -F_1(q,\dot{q},q_r,\dot{q}_r,\ddot{q}_r)\underline{S} \\
\dot{\hat{\underline{\theta}}}_K &= F_2(\underline{q},\underline{\dot{q}},q_r,\dot{q}_r,\ddot{q}_r)K(\Delta\dot{X} + \alpha\Delta X) \\
\underline{G}(\underline{q},t) &= \underline{G}_0(\underline{q}) - \underline{\underline{J}}^T(\underline{q},\underline{\theta}_K)\underline{f}_h^*(t)
\end{aligned}$$

$\underline{f}_h^*(t) \simeq$ White Gaussian noise in $\mathbb{R}^2$.

Final set of dynamical eqns:

Variables: $\underline{q}, \hat{\theta}_K, \hat{\underline{\theta}}_d, \underline{q}_r$

eqns. governing these:

$$\begin{aligned}
&\underline{\underline{M}}(\underline{q})\underline{\ddot{q}} + C(\underline{q},\underline{\dot{q}})\underline{\dot{q}} + G_0(\underline{q}) \\
&\quad = \underline{\underline{J}}^T(\underline{q},\underline{\theta}_K)\underline{f}_h^* + Y_d\hat{\theta}_d \\
&\qquad -J^T(\underline{q},\hat{\theta}_K)K\left(\underline{\underline{J}}(\underline{q},\underline{\theta}_K)\underline{\dot{q}} - \underline{\dot{X}}_d + \alpha\left(h(\underline{q},\underline{\theta}_K) - \underline{X}_d\right)\right) \qquad (1)
\end{aligned}$$

$$Y_d = Y_d(\underline{q},\underline{\dot{q}},\underline{\ddot{q}})$$

$$\dot{\hat{\theta}}_d = -F_1(\underline{q},\underline{\dot{q}},\underline{q}_r,\underline{\dot{q}}_r,\underline{\ddot{q}}_r)(\underline{\dot{q}} - \underline{\dot{q}}_r) \qquad (2)$$

$$\begin{aligned}
\dot{\hat{\underline{\theta}}}_K &= -F_2(\underline{q},\underline{\dot{q}},\underline{q}_r,\underline{\dot{q}}_r,\underline{\ddot{q}}_r) \\
&\quad K\left(\times J(\underline{q},\underline{\theta}_K)\underline{\dot{q}} - \underline{\dot{X}}_d + \alpha\left(h(\underline{q},\underline{\theta}_K) - \underline{X}_d\right)\right) \qquad (3)
\end{aligned}$$

$$\underline{\dot{q}}_r = \underline{\underline{J}}^{-1}(\underline{q},\hat{\underline{\theta}}_K)\left(\underline{\dot{X}}_d - \alpha\left(h(\underline{q},\underline{\theta}_K) - \underline{X}_d\right)\right) \qquad (4)$$

Linearized model for parametric uncertainties:

$$\begin{aligned}\hat{\underline{\theta}}_K &= \underline{\theta}_K + \varepsilon\delta\underline{\theta}_K\,,\ \underline{f}_h^* = \frac{\varepsilon d\underline{B}(t)}{dt}\\ \hat{\underline{\theta}}_d &= \underline{\theta}_d + \varepsilon\delta\underline{\theta}_d\,,\\ h\left(\underline{q},\underline{\theta}_K\right) &= h(\underline{q}) = \underline{X}_d + O(\varepsilon)\end{aligned}$$

Then $\underline{\underline{J}}\left(\underline{q},\underline{\theta}_K\right)\dot{\underline{q}} - \dot{\underline{X}}_d = \frac{d}{dt}\left(h\left(\underline{q},\underline{\theta}_K\right) - \underline{X}_d\right) = O(\varepsilon).$

We write $\underline{\underline{J}}(\underline{q})$ for $\underline{\underline{J}}(\underline{q},\underline{\theta}_K)$.

(1) approximates to

$$\begin{aligned}\underline{\underline{M}}\left(\underline{q}\right)\ddot{\underline{q}} + \underline{\underline{C}}\left(\underline{q},\dot{\underline{q}}\right)\dot{\underline{q}} + \underline{G}_0(\underline{q}) &= \varepsilon\underline{\underline{J}}^T\left(\underline{q}\right)\frac{d\underline{B}(t)}{dt} + Y_d\left(\underline{q},\dot{\underline{q}},\ddot{\underline{q}}\right)\underline{\theta}_d\\ &\quad + \varepsilon Y_d\left(\underline{q},\dot{\underline{q}},\ddot{\underline{q}}\right)\delta\underline{\theta}_d - \varepsilon\underline{\underline{J}}^T(\underline{q})K\left(\delta\dot{\underline{X}} + \alpha\delta\underline{X}\right)\end{aligned}$$

where

$$\begin{aligned}\underline{\underline{J}}\left(\underline{q},\underline{\theta}_K\right)\dot{\underline{q}} - \dot{\underline{X}}_d &= \delta\dot{\underline{X}}\,,\\ \underline{h}\left(\underline{q},\underline{\theta}_K\right) - \underline{X}_d &= \delta\underline{X}\,.\end{aligned}$$

We consider a linearized model for $\underline{q}$ too:

$$\underline{q} = \underline{q}_0 + \varepsilon\delta\underline{q}$$

Zeroth order terms:

$$\underline{\underline{M}}\left(\underline{q}_0\right)\ddot{\underline{q}}_0 + \underline{\underline{C}}\left(\underline{q}_0,\dot{\underline{q}}_0\right)\dot{\underline{q}}_0 + \underline{G}_0\left(\underline{q}_0\right) = Y_d\left(q_0,\dot{q}_0,\ddot{q}_0\right)\underline{\theta}_d$$

First order terms:

$$\begin{aligned}&M\left(\underline{q}_0\right)\delta\ddot{\underline{q}} + \underline{\underline{M}}'\left(\underline{q}_0\right)\left(\delta\dot{q}\otimes\dot{q}_0\right) + C\left(q_0,\dot{q}_0\right)\delta\dot{q}\\ &+ C_{,1}\left(q_0,\dot{q}_0\right)\delta\underline{q}\otimes\dot{q}_0 + C_{,1}\left(q_0,\dot{q}_0\right)\delta\dot{\underline{q}}\otimes\dot{q}_0 + G_0'(\underline{q}_0)\delta q\\ &= \underline{\underline{J}}^T\left(q_0\right)\frac{d\underline{B}(t)}{dt} + Y_d\left(\underline{q}_0,\dot{\underline{q}}_0,\ddot{\underline{q}}_0\right)\delta\underline{\theta}_d\\ &\quad + \left[Y_{d,1}\left(\underline{q}_0,\dot{\underline{q}}_0,\ddot{\underline{q}}_0\right)\left(\delta q\otimes I\right) + Y_{d,2}\left(q_0,\dot{q}_0,\ddot{q}_0\right)\left(\delta\dot{q}\otimes I\right)\right]\\ &\quad + Y_{d,3}\left(q_0,\dot{q}_0,\ddot{q}_0\right)\left(\delta\ddot{q}\otimes I\right)\Big]\underline{\theta}_d\\ &\quad - J^T(\underline{q}_0)K\left(\left(J(\underline{q}_0)\dot{q}_0 - \dot{\underline{X}}_d\right) + \alpha\left(\underline{h}(\underline{q}_0) - \underline{X}_d\right)\right)\end{aligned}$$

(2) approximates to

$$\begin{aligned}\delta\dot{\underline{\theta}}_d &= -F_1\left(\underline{q}_0,\underline{q}_0\right)\left(\dot{\underline{q}}_0 - \dot{\underline{q}}_{0r}\right)\\ \delta\dot{\underline{\theta}}_K &= -F_2\left(\underline{q}_0,\dot{\underline{q}}_0\right)K\left(J(\underline{q}_0)\dot{q}_0 - \dot{\underline{X}}_d + \alpha\left(h\left(\underline{q}_0\right) - \underline{X}_d\right)\right)\\ \dot{q}_{0r} &= J^{-1}(\underline{q}_0)(\dot{\underline{X}}_d)\end{aligned}$$

$$\delta \dot{\underline{q}}_r = -\alpha J^{-1}(q_0)\left(h(\underline{q}_0) - X_d\right) - \left(J^{-1}\right)_{,1}(\underline{q}_0)\left(\delta \underline{q}_0 \otimes \dot{\underline{X}}_d\right)$$

Parametric uncertainties-their origin.

$$\underline{\underline{M}}(\underline{q}) = \begin{pmatrix} \alpha(\underline{\theta}) & \beta(\underline{\theta})\cos(q_1 - q_2) \\ \beta(\underline{\theta})\cos(q_1 - q_2) & \gamma(\underline{\theta}) \end{pmatrix}$$

$$= \underline{\underline{M}}\left(\underline{q}, \underline{\theta}\right).$$

$$\underline{\theta} = \underline{\theta}_0 + \delta\underline{\theta}.$$

Then $$\underline{\underline{M}}\left(\underline{q}, \underline{\theta}\right) \approx \underline{\underline{M}}_0(\underline{q}) + \sum_{k=1}^{4} M_k(\underline{q})\delta\theta_k$$

$$M_k(\underline{q}) = \left.\frac{\partial \underline{\underline{M}}\left(\underline{q}, \underline{\theta}\right)}{\partial \theta_k}\right|_{\underline{\theta}=\underline{\theta}_0}, \; k = 1, 2, 3, 4.$$

$$\underline{\theta} = (m_1, m_2, l_1, l_2).$$

Note that $\underline{\theta}_d$ containts machine torque terms.

The equation for $\delta \underline{q}$ can be expressed as

$$\begin{aligned}
&\left[M\left(\underline{q}_0\right) - Y_{d,3}\left(q_0, \dot{q}_0, \ddot{q}_0\right)\left(I \otimes \underline{\theta}_d\right)\right]\delta\ddot{\underline{q}} \\
&+\left[M'(\underline{q}_0)\left(\underline{\underline{I}} \otimes \dot{\underline{q}}_0\right) + C\left(q_0, \dot{q}_0\right) + C_{,1}\left(q_0, \dot{q}_0\right)\left(I \otimes \dot{\underline{q}}_0\right)\right. \\
&\left. - Y_{d,2}\left(q_0, \dot{q}_0, \ddot{q}_0\right)\left(I \otimes \theta_d\right) + J^T\left(\underline{q}_0\right)\underline{\underline{KJ}}(\underline{q}_0)\right]\delta\dot{\underline{q}} \\
&+\left[G_0'\left(\underline{q}_0\right) - Y_{d,1}\left(\underline{q}_0, \dot{\underline{q}}_0, \ddot{\underline{q}}_0\right)\left(\underline{\underline{I}} \otimes \underline{\theta}_d\right)\right]\delta\underline{q}
\end{aligned}$$

$$= \underline{\underline{J}}^T(\underline{q}_0)\frac{d\underline{B}(t)}{dt} - \alpha\underline{\underline{J}}^T(\underline{q}_0)K\left(\underline{h}(\underline{q}_0) - \underline{X}_d\right)$$

$$+Y_d\left(\underline{q}_0, \dot{\underline{q}}_0, \ddot{\underline{q}}_0\right)\delta\underline{\theta}_d \qquad (\alpha)$$

$$\delta\dot{\underline{\theta}}_d = -F_1\left(q_0, \dot{q}_0, q_{0r}, \dot{q}_{0r}, \ddot{q}_{0r}\right)\left(\dot{\underline{q}}_0 - \dot{\underline{q}}_{0r}\right) \qquad (\beta)$$

$$\dot{\underline{q}}_{0r} = J^{-1}(\underline{q}_0)\dot{\underline{X}}_d \qquad (\tau)$$

$\delta\underline{q}, \delta\underline{\theta}_d$

$$\delta\dot{\underline{\theta}}_K = K\left(J(\underline{q}_0)\dot{\underline{q}}_0 - \dot{\underline{X}}_d + \alpha\left(\hbar(q_0) - \underline{X}_d\right)\right)$$

$$\alpha = 0 \; \mathbb{E}\int_0^T \left\|\underline{h}\left(\underline{q}, \underline{\theta}_K\right) - \underline{X}_d\right\|^2 dt$$

$$q \; \mathbb{E}\int_0^T \left\|\underline{h}\left(\underline{q}, \underline{\theta}_K\right) - \underline{X}_d\right\|^2 dt + b\mathbb{E}\int_0^T \left\|J\left(\underline{q}, \underline{\theta}_K\right)\dot{\underline{q}} - \dot{\underline{X}}_d\right\|^2 dt$$

$$\underline{q} = \underline{q}_0 + \delta\underline{q}(t) - \varphi(t)$$

$$
\begin{aligned}
& q\left\|h(q_0,\underline{\theta}_K)-X_d+J(q_0,\underline{\theta}_K)\delta\underline{q}\right\|^2 \\
& +b\left\|J(\underline{q}_0,\underline{\theta}_K)\underline{\dot{q}}_0-\underline{\dot{X}}_d+J_{,1}(\underline{q}_0,\underline{\theta}_K)(\delta\underline{q}\otimes\underline{\dot{q}}_0)\right. \\
& \left.+J(\underline{q}_0,\underline{\theta}_K)\delta\dot{q}\right\|^2 M(\underline{q},\underline{\theta}_d,\underline{\theta}_K)
\end{aligned}
$$

$$
\begin{aligned}
h(\underline{q}_0,\underline{\theta}_K)-\underline{X}_d &= \varphi(t). \\
\underline{\underline{J}}(q_0,\theta_K) &= F_1(t). \\
J(\underline{q}_0,\underline{\theta}_K)\underline{\dot{q}}_0-\underline{\dot{X}}_d &= -\dot{\varphi}(t) \\
J_{,1}(q_0,\underline{\theta}_K)(I\otimes\underline{\dot{q}}_0) &= F_2(t)
\end{aligned}
$$

$$
\int_0^T\left\{q\mathbb{E}\left[\left\|\underline{\varphi}(t)-\underline{\underline{F}}(t)\delta\underline{q}(t)\right\|^2\right]\right.
$$

$$
\left.+b\mathbb{E}\left[\left\|\underline{\dot{\varphi}}(t)-F_2(t)\delta\underline{q}(t)-F_1(t)\delta\dot{q}(t)\right\|^2\right]\right\}dt \quad \delta\underline{\ddot{q}}+P_1(t)\delta\underline{\dot{q}}+P_2(t)\delta\underline{q}(t)
$$

$$
\begin{aligned}
&= Q_1(t)\frac{d\underline{B}(t)}{dt}+\delta\underline{\dot{q}}Q_2(t)K(t)Q_3(t)\delta\underline{\dot{q}} \\
&\quad +Q_4(t)K(t)q_5(t)+q_6(t)
\end{aligned}
$$

Simultaneous tracking and disturbance analyzes using RLS.

$$
\begin{aligned}
\dot{Z}_t &= L(q_t,\dot{q}_t)\left(G(q_t,\dot{q}_t)-\tilde{T}(t)-p(\dot{q}_t)-Z_t\right) \\
\underline{\hat{d}}_t &= \underline{Z}_t+p(\underline{\dot{q}}_t) \\
J(q_t)\ddot{q}_t+G(q_t,\dot{q}_t) &= T(t)+\underline{\underline{J}}(q_{dt})\left(K_p(q_{dt}-q_t)+K_d(\dot{q}_{dt}-\dot{q}_t)\right)+d_t \\
T(t) &= J(q_{dt})\ddot{q}_{dt}+G(q_{dt},\dot{q}_{dt})-\hat{d}_t \\
&= T_0(t)-\hat{d}_t \\
p(\underline{q}_t) &= \sum_{k=1}^{m}C_k p_k(\dot{q}_t),\ L(q_t,\dot{q}_t) = p'(\dot{q}_t)J(q_t)^{-1} \\
&= \sum_k c_k p_k'(\dot{q}_t)J(q_t)^{-1} = \sum_k c_k L_k(q_t,\dot{q}_t) \\
\tilde{T}(t) &= T(t)+\underline{\underline{J}}(\underline{q}_{dt})\left(K_p(q_{dt}-q_t)+K_d(\dot{q}_{dt}-\dot{q}_t)\right) \\
&= T_0(t)-\hat{d}_t+J(t)\left(K_p(q_{dt}-q_t)+K_d(\dot{q}_{dt}-\dot{q}_t)\right)
\end{aligned}
$$

Discretization with time step Δ:

$$
q_t-2q_{t-1}+q_{t-2} = -\Delta^2 J(q_{t-1})^{-1}G(q_{t-1},\dot{q}_{t-1})
$$

$$+J(q_{t-1})^{-1}\left\{\Delta^2 J(q_{dt-1})\left((q_{dt-1}-q_{t-1})K_p+(\dot{q}_{dt-1}-\dot{q}_{t-1})K_d\right)\right.$$
$$\left.+\Delta^2 d_{t-1}+\Delta^2 T_0(t-1)-\Delta^2\left(Z_{t-1}+p(\dot{q}_{t-1})\right)\right\}$$

or
$$q_t = 2q_{t-1}-q_{t-2}+K_pF_1(q_{t-1},t-1)+K_dF_2(q_{t-1},q_{t-2},t-1)$$
$$+\sum_{k=1}^{m} C_k\psi_k(q_{t-1},q_{t-2})+\tilde{f}_3(t-1) \quad (1)$$

where
$$F_1(q_{t-1},t-1) = \Delta^2 J(q_{t-1})^{-1}J(q_{dt-1})(q_{dt-1}-q_{t-1})$$
$$F_2(q_{t-1},q_{t-2},t-1) = \Delta^2 J(q_{t-1})^{-1}(\dot{q}_{dt-1}-\dot{q}_{t-1})$$
$$= \Delta^2 J(q_{t-1})^{-1}\left(\dot{q}_{dt-1}-\Delta^{-1}(q_{t-1}-q_{t-2})\right)$$

Let
$$F_1(q_{t-1},t-1) \equiv F_1(t-1),$$
$$F_2(q_{t-1},q_{t-2},t-1) \equiv F_2(t-1)$$
$$\psi_k(q_{t-1},q_{t-2}) = -\Delta^2 J(q_{t-1})^{-1}p_k(\dot{q}_{t-1})$$
$$= -\Delta^2 J(q_{t-1})^{-1}p_k\left(\frac{(q_{t-1}-q_{t-2})}{\Delta}\right)$$
$$\tilde{f}_3(t-1) = -\Delta^2 J(q_{t-1})^{-1}G(q_{t-1},\dot{q}_{t-1})$$
$$+\Delta^2 \underline{\underline{J}}(\underline{q}_{t-1})^{-1}\left(\underline{d}_{t-1}+\underline{T}_0(t-1)-Z_{t-1}\right)$$
$$= \tilde{f}_3(q_{t-1},Z_{t-1},t-1)$$

Let
$$f_3(t-1) = 2q_{t-1}-q_{t-2}+\tilde{f}_3(t-1),$$
$$\psi_k(t-1) = \psi_k(q_{t-1},q_{t-2})$$

Then
$$q_t = f_3(t-1)+K_pF_1(t-1)+K_dF_2(t-1)+\sum_k C_k\psi_k(t-1) \quad (1)$$

$$Z_t = Z_{t-1}+\Delta L(q_{t-1},\dot{q}_{t-1})\left(G(q_{t-1},\dot{q}_{t-1})-T_0(t-1)-\hat{d}_{t-1}\right.$$
$$+J(q_{dt-1})\left(K_p(q_{dt-1}-q_{t-1})+K_d(\dot{q}_{dt-1}-\dot{q}_{t-1})\right)$$
$$\left.-p(\dot{q}_{t-1})-Z_{t-1}\right)$$
$$= Z_{t-1}+\Delta\sum_k C_kL_k(q_{t-1},\dot{q}_{t-1})\left(G(q_{t-1},\dot{q}_{t-1})\right.$$
$$\left.-T_0(t-1)-\hat{d}_{t-1}-Z_{t-1}\right)$$
$$-\Delta\sum_{k,m} C_kC_mL_k(q_{t-1},\dot{q}_{t-1})p_m(\dot{q}_{t-1})$$

$$+\Delta\sum_k C_k K_p L_k\left(q_{t-1},\dot{q}_{t-1}\right)J(q_{dt-1})\left(q_{dt-1}-q_{t-1}\right)$$

$$+\Delta\sum_k C_k K_d L_k\left(q_{t-1},\dot{q}_{t-1}\right)(\dot{q}_{dt-1}-\dot{q}_{t-1})$$

$$= Z_{t-1}+\sum_k C_k\underline{\Phi}_k(t-1)+\sum_{k,m} f(k,m)\underline{\chi}_{km}^{(1)}(t-1)$$

$$+\sum_k \alpha_k\underline{\chi}_k^{(2)}(t-1)+\sum_k \beta_k\underline{\chi}_k^{(3)}(t-1) \tag{2}$$

where $\Phi_k(t-1) = \Delta\cdot L_k\left(q_{t-1},\dot{q}_{t-1}\right)$

$$\left(G\left(q_{t-1},\dot{q}_{t-1}\right)-T_0(t-1)-\hat{d}_{t-1}-Z_{t-1}\right)$$

$$f(k,m) = C_k C_m,\ \chi_{km}^{(1)}(t-1)$$

$$= -\Delta.L_k\left(q_{t-1},\dot{q}_{t-1}\right)p_m\left(\dot{q}_{t-1}\right),$$

$$\mathcal{L}_k = C_k K_p,\ \beta_k = C_k K_d,$$

$$\chi_k^{(2)}(t-1) = \Delta.L_k\left(\underline{q}_{t-1},\underline{\dot{q}}_{t-1}\right)J\left(q_{dt-1}\right)\left(q_{dt-1}-q_{t-1}\right),$$

$$\chi_k^{(3)}(t-1) = \Delta.L_k\left(q_{t-1},\dot{q}_{t-1}\right)\left(\dot{q}_{dt-1}-\dot{q}_{t-1}\right)$$

$$\hat{d}_t = Z_t+\underline{p}(\underline{\dot{q}}_t)$$

For combined tracken and disturbance observer,

$$\left\{\widehat{C}_k(N),\widehat{f}(k,m,N),\widehat{\alpha}_k(N),\hat{\beta}_k(N),\widehat{K}_p(N),\widehat{K}_d(N)\right\}$$

$$= \arg\min\lambda.\sum_{t=0}^{N}\left\|Z_{t-1}+T\underline{\underline{\Phi}}_0(t-1)\underline{c}+T\underline{\underline{\chi}}^{(1)}(t-1)\underline{f}\right.$$

$$\left.+\underline{\underline{\chi}}^{(2)}(t-1)\underline{\alpha}+\chi^{(3)}(t-1)\underline{\beta}T(t)-dt\right\|^2$$

$$+\mu.\sum_{t=0}^{N}\left\|\underline{f}_3(t-1)+\underline{F}(t-1)\underline{K}+T\underline{\underline{\psi}}(t-1)\underline{c}-q_{dt}\right\|^2$$

where

$$\underline{\Phi}_0(t-1) = \underline{\Phi}(t-1)+\underline{p}\left(\dot{q}_{t-1}\right)$$

$$p(\underline{\dot{q}}_t) = \sum_k C_k p_k(\underline{\dot{q}}_t) = \underline{\underline{p}}(\underline{\dot{q}}_t)\underline{C}$$

$$\underline{K} = [K_p, K_d]^T,\ \underline{F}(t-1) = \left[\underline{F}_1(t-1)_1^1\underline{F}_2(t-1)\right]$$

$\underline{d}_t$ is obtained from $\widehat{C}(t-1)$, $\widehat{\underline{f}}(t-1)$, $\widehat{\underline{K}}(t-1)$ as

$$d_t \approx J\left(\underline{q}_t\right)\underline{\ddot{q}}_t+G\left(\underline{q}_t,\underline{\dot{q}}_t\right)-T(t)$$

$$-\underline{\underline{J}}\left(q_{dt}\right)\widehat{K}_p(t-1)\left(\underline{q}_{dt}-\underline{q}_t\right)+\widehat{K}_d(t-1)\left(\underline{\dot{q}}_{dt}-\underline{\dot{q}}_t\right)$$

where $$T(t) = J(\underline{q}_{dt})\ddot{\underline{q}}_{dt} + G(\underline{q}_{dt}, \dot{\underline{q}}_{dt}) - \hat{d}_{t|t-1}$$

where
$$\begin{aligned}\hat{d}_{t|t-1} &= Z_{t-1} + \hat{\underline{C}}^T(t-1)\underline{\Phi}_0(t-1) + \hat{\underline{f}}^T(t-1)\underline{\chi}^{(1)}(t-1) \\ &\quad + \hat{\underline{\alpha}}^T(t-1)\underline{\chi}^{(2)}(t-1) + \hat{\underline{\beta}}^T(t-1)\underline{\chi}^{(3)}(t-1) \\ \ddot{q}_{dt} &= \left(\underline{q}_{dt} - 2\underline{q}_{dt-1} + \underline{q}_{dt-2}\right)\Big/\Delta^2, \\ \dot{q}_{dt} &= \left(\underline{q}_{dt} - \underline{q}_{dt-1}\right)\Big/\Delta\end{aligned}$$

Let $$\underline{\theta} = \begin{bmatrix}\underline{K} \\ \underline{C} \\ \underline{f} \\ \underline{\alpha} \\ \underline{\beta}\end{bmatrix}$$ and

We get $\underline{F}(t-1)\underline{K} + \underline{\underline{\psi}}(t-1)\underline{C}$
$$\begin{aligned}&= \left[\underline{F}(t-1)_1^1\,\underline{\underline{\psi}}(t-1)_1^1\,\underline{\underline{O}}\right]\underline{\theta} \\ &\equiv \underline{\underline{X}}_1(t-1)\underline{\theta},\end{aligned}$$

$$\underline{\underline{\Phi}}_0(t-1)\underline{C} + \underline{\underline{\chi}}^{(1)}(t-1)\underline{f} + \underline{\underline{\chi}}^{(2)}(t-1)\underline{\alpha} + \underline{\underline{\chi}}^{(3)}(t-1)\underline{\beta}$$
$$\begin{aligned}&= \left[\underline{O}_1^1\underline{\underline{\Phi}}_0(t-1)_1^1\underline{\underline{\chi}}^{(1)}(t-1)_1^1\underline{\underline{\chi}}^{(2)}(t-1)_1^1\underline{\underline{\chi}}^{(3)}(t-1)\right]\underline{\theta} \\ &\quad \underline{\underline{X}}_2(t-1)\underline{\theta}.\end{aligned}$$

The energy function to be minimized is

$$\begin{aligned}\varepsilon(\underline{\theta}|t) &= \lambda.\sum_{t=0}^{N}\left\|\underline{Z}_{t-1} - \underline{d}_t + \underline{\underline{X}}_2(t-1)\underline{\theta}\right\|^2 \\ &\quad + \mu.\sum_{t=0}^{N}\left\|\underline{f}_3(t-1) - \underline{q}_{dt} + \underline{\underline{X}}_1(t-1)\underline{\theta}\right\|^2 \\ \underline{\nabla}_\theta\varepsilon(\underline{\theta}|t) &= 0 \Rightarrow\end{aligned}$$

$$\begin{aligned}&\lambda.\sum_{t=0}^{N}\underline{\underline{X}}_2(t-1)^T\left(\underline{\underline{X}}_2(t-1)\hat{\underline{\theta}} + \underline{Z}_{t-1} - \underline{d}_t\right) \\ &\quad + \mu.\sum_{t=0}^{N}\underline{\underline{X}}_1(t-1)^T\left(\underline{\underline{X}}_1(t-1)\hat{\underline{\theta}} + \underline{f}_3(t-1) - \underline{q}_{dt}\right) = 0\end{aligned}$$

$$\underline{\hat{\theta}}(N) = \left\{\sum_{t=1}^{N}\left[\lambda \underline{\underline{X}}_2(t-1)^T \underline{\underline{X}}_2(t-1) + \mu.\underline{\underline{X}}_1(t-1)^T \underline{\underline{X}}_1(t-1)\right]\right\}^{-1}$$

$$\left\{\sum_{t=0}^{N}\left\{\lambda \underline{\underline{X}}_2(t-1)^T\left(\underline{Z}_{t-1} - \underline{d}_t\right) + \mu \underline{\underline{X}}_1(t-1)^T\left(\underline{f}_3(t-1) - \underline{q}_{dt}\right)\right\}\right\}$$

$$\underline{p}\left[\underline{\dot{q}}_t\right] = C\begin{bmatrix}\dot{q}_{1t} \\ \dot{q}_{2t} + \dot{q}_{1t}\end{bmatrix}$$

$$p\left[\underline{\dot{q}}_t\right] = \begin{bmatrix}C_1(\dot{q}_{1t}) \\ C_2(\dot{q}_{1t} + \dot{q}_{2t})\end{bmatrix}$$

$$= C_1\begin{bmatrix}\dot{q}_{1t} \\ 0\end{bmatrix} + C_2\begin{bmatrix}0 \\ \dot{q}_{1t} + \dot{q}_{2t}\end{bmatrix}$$

$$= C_1 p_1\left(\underline{\dot{q}}_t\right) + C_2 p_2\left(\underline{\dot{q}}_t\right)$$

$$\varepsilon(\underline{C}, \underline{f}) = \lambda\sum_{t=0}^{N}\left\|\underline{\eta}_1(t) + \underline{\underline{\Phi}}_1(t-1)\underline{C} + \underline{\underline{\chi}}^{(1)}(t-1)\underline{f}\right\|^2$$

$$+\mu\sum_{t=0}^{N}\left\|\underline{\eta}_2(t) + \underline{\underline{\psi}}(t-1)\underline{C}\right\|^2$$

$$\begin{bmatrix}\lambda\underline{\eta}_1(t) \\ \lambda\underline{\eta}_2(t)\end{bmatrix} = \underline{\eta}(t),$$

$$\begin{bmatrix}\lambda\underline{\underline{\Phi}}_1(t-1) & \lambda\underline{\underline{\chi}}^{(1)}(t-1) \\ \mu\underline{\underline{\psi}}(t-1) & \underline{\underline{0}}\end{bmatrix}$$

$$= \underline{\underline{X}}(t-1).$$

$$\begin{pmatrix}\underline{C} \\ \underline{f}\end{pmatrix} = \underline{\theta}.$$

$$\varepsilon_N(\underline{\theta}) = \sum_{t=0}^{N}\left\|\underline{\eta}(t) + \underline{\underline{X}}(t-1)\underline{\theta}\right\|^2$$

$$\underline{\hat{\theta}}(N) = \left(\sum_{t=1}^{N}\underline{\underline{X}}(t-1)^T\underline{\underline{X}}(t-1)\right)^{-1}\left(\sum_{t=0}^{N}\underline{\underline{X}}(t-1)^T\underline{\eta}(t)\right)$$

$$\underline{q}_t = \underline{f}_3(t-1) + \left[F_1(t-1)_1^1 F_2(t-1)\right]\begin{bmatrix}K_p \\ K_d\end{bmatrix}$$

$$+\left[\underline{\psi}_1(t-1)_1^1\underline{\psi}_2(t-1)\right]\begin{bmatrix}C_1 \\ C_2\end{bmatrix}$$

$$\underline{q}(:,0) = \text{zeros}(2, 1);$$

First assume K_p, K_d known. Then

$$\varepsilon(\underline{C}, \underline{f}) = \lambda \sum_{t=0}^{N} \left\| \underline{\underline{Z}}_{t-1} - \underline{d}_t + \left(\underline{\underline{\Phi}}_0 (t-1) + \left(K_p \underline{\underline{\chi}}^{(2)} (t-1) + K_d \underline{\underline{\chi}}^{(3)} (t-1) \right) \right) \underline{C} \right.$$
$$\left. + \underline{\underline{\chi}}^{(1)} (t-1) \underline{f} \right\|^2$$
$$+ \mu \sum_{t=0}^{N} \left\| \underline{f}_3 (t-1) + F_1 (t-1) K_p + F_2 (t-r) K_d - \underline{q}_{dt} + \underline{\underline{\psi}}(t-1) \underline{C} \right\|^2$$

$$\underline{\underline{\Phi}}_0 (t-1) + K_p \underline{\underline{\chi}}^{(2)} (t-1) + K_d \underline{\underline{\chi}}^{(3)} (t-1)$$
$$\equiv \underline{\underline{\Phi}}_1 (t-1)$$

$$\underline{f}_3 (t-1) + KpF_1 (t-1) + K_d F_2 (t-1) - q_{dt}$$
$$= \underline{\eta}_2 (t)$$

$$\underline{\underline{Z}}_{t-1} - \underline{d}_t = \eta_1 (t)$$

$$\underline{\underline{R}}_X [N] = \sum_{t=1}^{N} \underline{\underline{X}} (t-1)^T \underline{\underline{X}} (t-1)$$

$$\underline{r}_X [N] = \sum_{t=1}^{N} \underline{\underline{X}} (t-1)^T \underline{\eta} (t)$$

$$\underline{\hat{\theta}}(N) = \underline{\underline{R}}_X [N]^{-1} \underline{r}_X [N]$$

$$\underline{\underline{R}}_X [N+1] = \underline{\underline{R}}_X [N] + \underline{\underline{X}} [N]^T \underline{\underline{X}} [N]$$

$$\underline{r}_X [N+1] = \underline{r}_X [N] + \underline{\underline{X}} [N-1]^T \underline{\eta} [N+1]$$

$$\underline{\underline{R}}_X [N+1]^{-1} = \underline{\underline{R}}_X [N]^{-1} - \underline{\underline{R}}_X [N]^{-1} \underline{\underline{X}} [N]^T$$
$$\times \left(I + \underline{\underline{X}} [N] \underline{\underline{R}}_X [N]^{-1} \underline{\underline{X}} [N]^T \right)^{-1} \underline{\underline{X}} [N] \underline{\underline{R}}_X [N]^{-1}$$

$$\underline{\hat{\theta}}[N+1] = \left\{ R_X [N]^{-1} - R_X [N]^{-1} \underline{\underline{X}} [N]^T Q[N]^{-1} \underline{\underline{X}} [N] \underline{\underline{R}}_X [N]^{-1} \right\}$$
$$\times \left\{ \underline{r}_X [N] + \underline{\underline{X}} [N-1]^T \underline{\eta} [N+1] \right\}$$
$$= \underline{\hat{\theta}}[N] + R_X [N]^{-1} \underline{\underline{X}} [N]^T \underline{\eta} [N+1]$$
$$- R_X [N]^{-1} X[N]^T Q[N]^{-1} \underline{\underline{X}} [N] \underline{\hat{\theta}}[N]$$
$$- R_X [N]^{-1} X[n]^T Q[N]^{-1} X[N] R_X [N]^{-1} X[N]^T \underline{\eta} [N+1]$$

$$= \underline{\hat{\theta}}[N] + R_X[N]^{-1} X[N]^T Q[N]^{-1}\left(\underline{\eta}(N+1) - \underline{\underline{X}}[N]\underline{\hat{\theta}}[N]\right)$$

$$\underline{\underline{Q}}[N] = \underline{\underline{I}} + \underline{\underline{X}}[N]\underline{\underline{R}}_X[N]^{-1}\underline{\underline{X}}[N]^T$$

$$\underline{\hat{\theta}}[N] = \begin{bmatrix} \underline{\widehat{C}}[N] \\ \underline{\widehat{f}}[N] \end{bmatrix}$$

$$\underline{q}[N] = \underline{f}_3[N-1] + K_p F_1[N-1] + K_d F_2[N-1]$$

$$\widehat{C}_1[N] + \psi_1[N-1] + \widehat{C}_2[N]\psi_2[N-1]$$

$$f_3[N-1] = 2q[N-1] - q[N-2] + \tilde{f}_3[N-1]$$

Convergence analysis

$$J(\underline{q}_t)\underline{\ddot{q}}_t + \underline{G}(\underline{q}_t, \underline{\dot{q}}_t) = T_0(t) + J(q_{dt})\left(K_p\left(\underline{q}_{dt} - \underline{q}_t\right) + K_d\left(\underline{\dot{q}}_{dt} - \underline{\dot{q}}_t\right)\right) + \underline{d}_t - \hat{d}_t$$

$$\overline{\lim_{t\to\infty}}\, \mathbb{E}\left\|\underline{q}_t - \underline{q}_{dt}\right\|^2 < \varepsilon.$$

$$\hat{d}_t = Z_t + p(\underline{\dot{q}}_t)$$

$$\dot{Z}_t = L\left(\underline{q}_t, \underline{\dot{q}}_t\right)\left(G\left(\underline{q}_t, \underline{\dot{q}}_t\right) - p(\underline{\dot{q}}_t) - T_0(t) + \underline{\hat{d}}_t\right.$$
$$\left. -J(q_{dt})\left(K_p\left(q_{dt} - q_t\right) + K_d\left(\dot{q}_{dt} - \dot{q}_t\right)\right)\right)$$

$$q_t, Z_t, d_t \;\; J(\underline{q}_t)\delta\underline{\ddot{q}}_t + J'(\underline{q}_t)\left(I \otimes \underline{\ddot{q}}_t\right)\delta\underline{q}_t$$
$$+ G_{,1}\left(\underline{q}_t, \underline{\dot{q}}_t\right)\delta\underline{q}_t + G_{,2}\left(\underline{q}_t, \underline{\dot{q}}_t\right)\delta\underline{\dot{q}}_t$$
$$= -K_p\delta\underline{q}_t - K_d\delta\dot{q}_t + \delta d_t - \delta\hat{d}_t$$

$$d_t = -\gamma\underline{\dot{q}}_t + \varepsilon\delta d_t - \gamma\delta\dot{q}_t$$

$$\delta\dot{\hat{d}}_t = -L_{,1}(t)\left(I \otimes \left(\gamma\dot{q}_t + \underline{\hat{\mu}}_t\right)\right)\delta\underline{q}_t$$
$$-L_{,2}(t)\left(I \otimes \left(\gamma\dot{q}_t + \underline{\hat{\mu}}_t\right)\right)\delta\underline{\dot{q}}_t$$
$$+L(t)\left(\delta dt - \delta\hat{d}t\right)$$

$$\delta\underline{\ddot{q}}_t = \left[G_{,1}(t) - J(t)^{-1}J'(t)\left(I \otimes \underline{\ddot{q}}_t\right) - K_p\right]\delta\underline{q}_t$$
$$-\left(J(t)^{-1}G_{,2}(t) + K_d\right)\delta\underline{\dot{q}}_t + \delta d_t - \delta\hat{d}_t$$

$$A(t) = G_{,1}(t) - J(t)^{-1}J'(t)\left(I \otimes \underline{\ddot{q}}_t\right) - \underline{K}_p I_2$$

$$B(t) = -J(t)^{-1}G_{,2}(t) + K_d I_2$$

$$\delta\underline{\ddot{q}}_t = \underline{\underline{A}}(t)\delta\underline{q}_t + \underline{\underline{B}}(t)\delta\underline{\dot{q}}_t + \delta\underline{d}_t - \delta\hat{d}_t$$

$$\delta\dot{\hat{d}}_t = C(t)\delta\underline{q}_t + \underline{\underline{D}}(t)\delta\dot{\underline{q}}_t + L(t)\left(\delta\underline{d}_t - \delta\hat{\underline{d}}_t\right)$$

$$\delta\hat{d}_t = \delta Z_t + p'(\dot{\underline{q}}_t)\delta\dot{\underline{q}}_t$$

$$\dot{Z}_t = L\left(\underline{q}_t, \dot{\underline{q}}_t\right)\left(d_t - p(\dot{q}_t) - J(q)\ddot{q}\right)$$

$$\dot{\hat{d}}_t = \dot{Z}_t + LJ\ddot{q}_t$$

$$\dot{\hat{d}}_t = L\left(d_t - \hat{d}_t\right)\ p'(\dot{q}_t)J^{-1} = L$$

$$\hat{d}_t = Z + p$$

$$\delta\dot{\hat{d}}_t = L_{,1}(t)\left(I\otimes\left(\mu_t - \hat{\mu}_t\right)\right)\delta\underline{q}_t + L_{,2}(t)\left(I\otimes\left(\mu_t - \hat{\mu}_t\right)\right)\delta\dot{\underline{q}}_t$$
$$+L(t)\left(\delta d_t - \delta\hat{d}_t\right)$$

$$d_t = \underline{\underline{\mu}}_t + \varepsilon\delta d_t - \gamma\dot{\underline{q}}_t + \varepsilon\delta d_t\,.$$

$$\underline{\underline{C}}(t) = -L_{,1}(t)\left(I\otimes\left(\gamma\dot{\underline{q}}_t + \hat{\underline{\mu}}_t\right)\right)$$

$$\underline{\underline{D}}(t) = -L_{,2}(t)\left(I\otimes\left(\gamma\dot{\underline{q}}_t + \hat{\underline{\mu}}_t\right)\right)$$

$$\delta\dot{\underline{q}}_t = \delta\underline{w}_t$$

$$\delta\dot{\underline{w}}_t = A(t)\delta\underline{q}_t + \underline{\underline{B}}(t)\delta\underline{w}_t + \delta\underline{d}_t - \delta\hat{\underline{d}}_t$$

$$\delta\dot{\hat{d}}_t = C(t)\delta\underline{q}_t + \underline{\underline{D}}(t)\delta\underline{w}_t + L(t)\left(\delta\underline{d}_t - \delta\hat{\underline{d}}_t\right)$$

$$\frac{d}{dt}\begin{pmatrix}\delta\underline{q}_t\\ \delta\underline{w}_t\\ \delta\breve{\underline{d}}_t\end{pmatrix} = \underbrace{\begin{pmatrix}0 & I & 0\\ \underline{\underline{A}}(t) & \underline{\underline{B}}(t) & -\underline{\underline{I}}\\ \underline{\underline{C}}(t) & D(t) & -L(t)\end{pmatrix}}_{F(t)}\begin{pmatrix}\delta\underline{q}_t\\ \delta\underline{w}_t\\ \delta\breve{\underline{d}}_t\end{pmatrix} + \underbrace{\begin{pmatrix}\underline{\underline{0}}\\ \underline{\underline{I}}\\ \underline{\underline{L}}(t)\end{pmatrix}}_{H(t)}\delta\underline{d}_t$$

$$\frac{d\underline{\xi}_t}{dt} = \underline{\underline{F}}(t)\underline{\xi}_t + \underline{\underline{H}}(t)\delta\underline{d}_t$$

$$\underline{\xi}_t = \int_0 \Phi(t,\tau)\underline{\underline{H}}(\tau)\delta\underline{d}\tau d\tau$$

$$\underline{\underline{\Phi}}(t,\tau) = \underline{\underline{I}} + \sum_{n=1}^{\infty} F(t_1)\ldots F(t_n)\ \tau < t_n < \ldots < t, < t\ dt_1\ldots dt_n$$

$$\mathbb{E}\left[\left\|\underline{\xi}_t\right\|^2\right] = \sigma_d^2\int_0^t T_r\left(\Phi(t,\tau)H(\tau)H(\tau)^T\Phi(t,\tau)^T\right)d\tau$$

$$\lim_{t\to\infty}\mathbb{E}\left[\left\|\underline{\xi}_t\right\|^2\right] = \sigma_d^2\int_0^\infty T_r\left(\Phi(\infty,\tau)H(\tau)H(\tau)^T\Phi(\infty,\tau)^T\right)d\tau$$

For stability

$$\sigma_d^2 \int_0^{\infty} \|\Phi(\infty,\tau)H(\tau)\|^2 d\tau \ < \infty.$$

$$\begin{aligned}
\dot{\hat{\mu}}_t + L\hat{\mu}_t &= L\mu_r \\
\dot{\hat{d}}_t &= L\left(d_t - \hat{d}_t\right) \\
\dot{\hat{\mu}}_t &= \sum C_k L_k\left(\underline{q}_t, \dot{\underline{q}}_t\right)\left(\mu_t - \hat{\mu}_t\right) + \beta\int_0^T \left\|\underline{\mu}_t - \hat{\underline{\mu}}_t\right\|^2 dt \\
&\quad -\int \underline{\lambda}(t)^T\left(\dot{\hat{\underline{\mu}}}t - \Sigma C_k L_k\right) \\
\hat{d}_{t+1} &= \hat{d}_t + \delta.\, L\left(d_t - \hat{d}_t\right)
\end{aligned}$$

$F(t)$, $0 \le t \le T$.

$$\underline{\underline{I}} + \sum_{n=1}^{\infty} \int F(t_1)F(t_2)\ldots F(t_n)\, dt_1 \ldots dt_n\, 0 \le t_n < \ldots < t_1 < T$$

$$e^{TF} = \frac{1}{2|\lambda_{\max}|}$$

$$\int_0^T 2_e^{\lambda_{nm}} \tau F(t_1) \ \propto \lambda_{\max}$$

$$\frac{\left(e^{2\lambda_{\max} T} - 1\right)}{2\lambda_{\max}} \sum_{n=0}^{\infty} \frac{\lambda_{\max}^n t^n T^n}{\lfloor n} \Rightarrow e^{\lambda_{\max} T}$$

$$\sigma_d^2 \int_0^T T_r\left(\Phi(T,\tau)H(\tau).H(\tau)^T \Phi(T,\tau)^T\right) - \sigma_d^2 \Big/ 2\lambda_{\max}$$

$$\begin{aligned}
\|\Phi(t,\tau)\| &\le 1 + \sum_{n=1}^{\infty} \frac{(t-\tau)^n}{\lfloor n} \Big(\sup_{\tau \le S \le t} \|F(S)\|\Big)^n \\
&= \exp\Big((t-\tau) \sup_{\tau \le S \le t} \|\underline{E}(S)\|\Big)
\end{aligned}$$

eigenvalues of $F(\tau)$, $0 \le \tau \le T \to \infty$.

Should all have – ve real part.

Then $\dfrac{1}{\sup\limits_{\tau \le t \le T} \|\underline{\underline{E}}(t)\|}$

$\Phi(T, \tau)$

$\underset{0\le t\le T}{\{F(t)\}}\ \lambda_{\max}.<\infty.$

$$\frac{1}{2|\lambda_{\max}|}$$

Stock-share price fluctuations and control.

$$dP_0(t) = r_0(t)\, P_0(t)dt,$$

$$dP_\alpha(t) = \left[b_\alpha\left(\underline{N}(t),t\right)dt + \sum \sigma_{\alpha\beta}\left(\underline{N}(t),t\right)dB_\beta(t)\right]P_\alpha(t)$$

$$X(T) = \int_0^T \underline{N}(t)^T\, d\underline{P}(t) - \int_0^T C\left(\underline{N}(t),\underline{P}(t),t\right)dt$$

choose $\left\{N(t)\middle|0\le t\ge T\right\}$ so that $\mathbb{E}\{X(T\}$is a maximum subject to the constraint

$$\mathbb{E}\int_0^T f_\alpha\left(\underline{N}(t),\underline{P}(t),t\right)dt = 1\le\alpha\le R.$$

This is a stochastic optimal control problem.

$$\mathbb{E}\int_0^T N(t)^T\, d\underline{P}(t) = \mathbb{E}\int_0^T N(t)^T\, \underline{\underline{B}}\left(\underline{N}(t),t\right)\underline{P}(t)dt$$

where $$\underline{\underline{B}}\left(\underline{N}(t),t\right) = diag\left[\left\{b_\alpha\left(\underline{N}(t),t\right)\right\}_{\alpha=0}^{d}\right]$$

After incorporating the constraints using Lagrange multipliers λ_α, the functional to be maximized is given by

$$\mathbb{E}\int_0^T \mathcal{L}\left(\underline{N}(t),\underline{P}(t),\underline{\lambda},t\right)dt$$

where

$$\mathcal{L}(\underline{N},\ \underline{P},\ \underline{\lambda},\ t) = \underline{N}^T\underline{\underline{B}}\left(\underline{N},t\right)\underline{P}(t) - C\left(\underline{N},\underline{P},t\right) - \underline{\lambda}^T\left(\underline{f}\left(\underline{N},\underline{P},t\right)-\underline{\xi}\right)$$

$N_\alpha(t) \simeq$ number of shares purchased at limit of αth types.

$P_\alpha(t) \simeq \alpha^{th}$ type share price, $\alpha \ge 1$,

$P_0(t) \simeq$ bond price at time t.

$\underline{N}(t)$ is to be choice of the form.

$\underline{N}(t) = \underline{\psi}\ (\underline{P}(t),\ t)$

(feedback)

Let $\max\limits_{\substack{\underline{N}(s),\\ t\le s\le T}} \mathbb{E}\left(\int_t^T \mathcal{L}\left(\underline{N}(s),\underline{P}(s),s,\underline{\lambda}\right)ds|\underline{P}(t)\right)$

$$= V(\underline{P}(t), t, \underline{\lambda})$$

Then $V(\underline{P}(t), t, \underline{\lambda}) = \max_{\underline{N}(t)} \left\{ \mathcal{L}\left(\underline{N}(t), \underline{P}(t), t, \underline{\lambda}\right) dt \quad + \mathbb{E}\left[V\left(\underline{P}(t+dt), t+dt, \underline{\lambda}\right) \middle| \underline{P}(t) \right] \right\}$

$$= \max_{\underline{N}(t)} \left\{ \mathcal{L}\left(\underline{N}(t), \underline{P}(t), t, \underline{\lambda}\right) dt + V\left(\underline{P}(t), t, \underline{\lambda}\right) \right.$$

$$+ \frac{\partial V\left(\underline{P}(t), t, \underline{\lambda}\right)}{\partial t} dt + dt \left(\underline{\nabla}_P V\left(\underline{P}(t), t, \underline{\lambda}\right) \right)^T \underline{\underline{B}}\left(\underline{N}(t), t\right) \underline{P}(t)$$

$$\left. + dt \frac{1}{2} \sum_{\alpha,\beta} T_r \left\{ \left(\sigma\sigma^T\right)_{\alpha\beta} \left(\underline{N}(t), t\right) P_\alpha(t) P_\beta(t) \frac{\partial^2 V\left(\underline{P}(t), t\right)}{\partial P_\alpha \partial P_\beta} \right\} \right\}$$

Thus $V(P, t, \underline{\lambda})$ satisfies the partial differential equation

$$\max_{\underline{N}} \left\{ \mathcal{L}\left(\underline{N}, \underline{P}, t, \underline{\lambda}\right) + \underline{\nabla}_P V\left(\underline{P}, t, \lambda\right)^T \underline{\underline{B}}(\underline{N}, t) \underline{P} \right.$$

$$\left. + \frac{1}{2} \sum_{\alpha,\beta} T_r \left\{ (\sigma\sigma^T)_{\alpha\beta} (\underline{N}, T) P_\alpha P_\beta \frac{\partial^2 V\left(\underline{P}, t\right)}{\partial P_\alpha \partial P_\beta} \right\} \right\}$$

$$= -\partial V\left(\underline{P}, t, \underline{\lambda}\right) \Big/ \partial t$$

Thiemann P.54

Show that

$$X_{\mu\nu} \triangleq q_\mu^\rho q_\nu^\sigma n^\lambda \nabla_\lambda \nabla_\rho n_\sigma$$

$$= -KK_{\mu\nu} + \nabla_\rho\left(n^\rho K_{\mu\nu}\right) + \left(\nabla_n n^\rho\right) n_\nu K_{\mu\rho} + \nabla_n (n_\mu n^\rho) \nabla_\rho n_\nu$$

$$K_{\mu\nu} = q_\mu^\rho q_\nu^\sigma \nabla_\rho n_\nu .$$

$$g_{\mu\nu} = q_{\mu\nu} + n_\mu n_\nu$$

$$q_{\mu\nu} n^\nu = 0$$

$$n_\mu n^\mu = 1.$$

$$K = g^{\mu\nu} K_{\mu\nu} = q^{\mu\nu} K_{\mu\nu}$$

$$n^\sigma \Delta_\rho n^\sigma = 0,\ n^\mu K_{\mu\nu} = n^\nu K_{\mu\nu} = 0 \qquad \therefore q_\nu^\mu n^\nu = 0.$$

$$q_\mu^\rho q_\nu^\sigma n^\lambda \nabla_\lambda \nabla_\rho n\sigma$$

$$= q_\mu^\rho q_\nu^\sigma n^\lambda \nabla_n \nabla_\rho n\sigma = \nabla_n K_{uv} - \left(\nabla_\rho n_\sigma\right)\left(\nabla_n \left(q_\mu^\rho q_\nu^\sigma\right)\right)$$

$$K_{\mu\nu} = \left(\delta_\mu^\rho - n^\rho n_\mu\right)\left(\delta_\nu^\sigma - n^\sigma n_\nu\right) \nabla_\rho n_\sigma$$

$$= \nabla_\mu n_\nu - n_\mu \nabla_n n\sigma_\nu \qquad \because n^\sigma \nabla_\rho n_\sigma = \frac{1}{2} \nabla_\rho \left(n^\sigma n_\sigma\right) = 0.$$

$$\nabla_\mu n_\nu = K_{\mu\nu} + n_\mu \nabla_n n_\nu$$

So
$$X_{\mu\nu} = \nabla_n K_{\mu\nu} - \left(K_{\rho\sigma} + n_\rho \nabla_n n_\sigma\right)\nabla_n\left(q^\rho_\mu q^\sigma_\nu\right)$$

$$\nabla_n\left(q^\rho_\mu q^\sigma_\nu\right) = -\delta^\rho_\mu \nabla_n\left(n^\sigma n_\nu\right) - \delta^\sigma_\nu \nabla_n\left(n^\rho n_\mu\right) + \nabla_n\left(n^\rho n^\sigma n_\mu n_\nu\right)$$

$$\begin{aligned} K_{\rho\sigma}\nabla_n\left(q^\rho_\mu q^\sigma_\nu\right) &= -K_{\mu\sigma}\nabla_n\left(n^\sigma n_\nu\right) - K_{\rho\nu}\nabla_n\left(n^\rho n_\mu\right) + K_{\rho\sigma}\nabla_n\left(n^\rho n^\sigma n_\mu n_\nu\right) \\ &= -K_{\mu\sigma} n_\nu \nabla_n n^\sigma - K_{p\nu} n_\mu \nabla_n n^\rho + K_{\rho\sigma} n_\mu n_\nu \nabla_n\left(n^\rho n^\sigma\right) \end{aligned}$$

$$\nabla_n K_{\mu\nu} = n^\rho \nabla_\rho K_{\mu\nu} = \nabla_\rho\left(n^\rho K_{\mu\nu}\right) - K_{\mu\nu}\nabla_\rho n^\rho$$

$$K_{\mu\nu} = \nabla_\mu n_\nu - n_\mu n^\rho \nabla_\rho n_\nu$$

$\Rightarrow$
$$\begin{aligned} &\nabla_\mu n^\mu - n_\mu n^\rho \nabla_\rho n^\mu \\ &= K^\mu_\mu = K \end{aligned}$$

$\Rightarrow$
$$K = \Delta_\mu n^\mu.$$

$$\nabla_n K_{\mu\nu} = \nabla_\rho\left(n^\rho K_{\mu\nu}\right) - K K_{\mu\nu}$$

So we've shown that

$$\begin{aligned} X_{\mu\nu} = {} & \nabla_\rho\left(n^\rho K_{\mu\nu}\right) - K K_{\mu\nu} \\ & + \left(K_{\mu\sigma} n_\nu \nabla_n n^\sigma + K_{\rho\nu} n_\mu \nabla_n n^\rho\right) - n_\rho\left(\nabla_n n_\sigma\right) \\ & \left(\nabla_n\left(q^\rho_\mu q^\sigma_\nu\right)\right) \end{aligned}$$

Note that

$$K_{\rho\sigma}\nabla_n\left(n^\rho n^\sigma\right) = n^\rho K_{\rho\sigma}\nabla_n n^\sigma + n^\sigma K_{\rho\sigma}\nabla_n n^\rho = 0.$$

Let
$$Y_{\mu\nu} = \left(K_{\mu\sigma} n_\nu \nabla_n n^\sigma + K_{\rho\nu} n_\mu \nabla_n n^\rho\right) - n_\rho\left(\nabla_n n^\sigma\right)\nabla_n\left(q^\rho_\mu q^\sigma_\nu\right)$$

Then
$$Y_{\mu\nu} = K_{\mu\rho} n_\nu \nabla_n n^\rho + K_{\rho\nu} n_\mu \nabla_n n^\rho - n_\rho\left(\nabla_n n\rho\right) q^\sigma_\nu \nabla_n q^\rho_\mu$$

Now,
$$\begin{aligned} K_{\rho\nu} n_\mu \nabla_n n^\rho &= \left(\nabla_\rho n_\nu - n_\rho \nabla_n n_\nu\right) n_\mu \nabla_n n^\rho \\ &= \left(\nabla_\rho n_\nu\right) n_\mu \nabla_n n^\rho \qquad \because n_\rho \nabla_n n^\rho = 0. \end{aligned}$$

Further, $n_\rho\left(\nabla_n n_\sigma\right) q^\sigma_\nu \nabla_n q^\rho_\mu$

$$\begin{aligned} &= -n_\rho\left(\nabla_n n_\sigma\right) q^\sigma_\nu \nabla_n\left(n^\rho n_\mu\right) \left(\nabla_n \delta^\mu_\nu = 0\right) \\ &= -\left(\nabla_n n_\sigma\right) q^\sigma_\nu \nabla_n n_\mu \\ &= -\left(\nabla_n n_\nu\right)\left(\nabla_n n_\mu\right) \end{aligned}$$

$$\left(\because q_\nu^\sigma = \delta_\nu^\sigma - n^\sigma n_\nu \text{ and } n^\sigma \nabla_n n_\sigma = 0\right)$$

Thus,

$$\begin{aligned} Y_{\mu\nu} &= K_{\mu\rho} n_\nu \nabla_n n^\rho + \left(\nabla_\rho n_\nu\right) n_\mu \nabla_n n^\rho + \left(\nabla_n n_\nu\right)\left(\nabla_n n_\mu\right) \\ &= K_{\mu\rho} n_\nu \nabla_n n^\rho + \left(\nabla_\rho n_\nu\right)\left(n_\mu \nabla_n n^\rho + n^\rho \nabla_n n_\mu\right) \\ &= K_{\mu\rho} n_\nu \nabla_n n^\rho + \left(\nabla_\rho n_\nu\right) \nabla_n \left(n_\mu n^\rho\right) \end{aligned}$$

Quantum Belavkin Filtering

State $j_t(X)$, $X \in \hbar$.

State satisfies Evans-Hudson Flow eqns:

$$\begin{aligned} dj_t(X) &= j_t\left(\theta_\beta^\alpha(X)\right) d\Lambda_\alpha^\beta(t) \\ d\Lambda_\alpha^\beta d\Lambda_\nu^\mu &= \varepsilon_\nu^\beta d\Lambda_\alpha^\mu, \ \varepsilon_\nu^\beta = \delta_\nu^\beta, \ \beta, \nu \geq 1, \\ d\Lambda_0^\beta d\Lambda_\nu^0 &= d\Lambda_0^0 \delta_\nu^\beta, \ \beta, \nu \geq 1. \end{aligned}$$

All other product differentials are zero.

$\{Y_t\}$ = non-demolition measurement process.

$$dY_t = j_t\left(L_\beta^\alpha\right) d\Lambda_\alpha^\beta(t)$$

Let $\eta_{t]} = \sigma\{Y_s : s \leq t\}$, $\simeq$ Abelian Von-Neumann algebra.

$$dZ_t = f(t) Z_t dY_t$$

Then $\sigma\{Z_s : s \leq t\} \in \eta_{t]}$.

$$\mathbb{E}\left\{j_t(X) \middle| \eta_{t]}\right\} = \widehat{X}(t).$$

$$\mathbb{E}\left[\left(j_t(X) - \widehat{X}(t)\right) Z_t\right] = 0$$

Let

$$d\widehat{X}(t) = F_t(X) dt + G_t(X) dY_t, \ \left(dt = d\Lambda_0^0(t)\right),$$

Where $F_t(X)$, $G_t(X) \in \eta_{t]}$.

$$\mathbb{E}\langle\xi\rangle = \langle g\varphi(u)|\xi|g\varphi(u)\rangle, \ ||g|| = 1, \ \varphi(u) = \exp\left(\frac{-\|u\|^2}{2}\right) e(u).$$

$$\mathbb{E}\left\{\left(dj_t(X) - d\widehat{X}(t)\right) Z_t + \left(j_t(X) - \widehat{X}(t)\right) dZ_t + \left(dj_t(X) - d\widehat{X}(t) dZ_t\right)\right\} = 0$$

$$\begin{aligned} \mathbb{E}[dj_t(X) Z_t] &= \mathbb{E}\left[j_t\left(\left(\theta_\beta^\alpha(X)\right)\right) Z_t d\Lambda_\alpha^\beta(t)\right] \\ &= \overline{u_\alpha(t)} u_\beta(t) dt . \mathbb{E}\left[j_t\left(\theta_\beta^\alpha(X)\right) Z_t\right] \\ \mathbb{E}\left[d\widehat{X}(t) . Z_t\right] &= dt . \mathbb{E}\left[F_t(X) Z_t\right] + \mathbb{E}\left[G_t(X) dY_t \delta . Z_t\right] \end{aligned}$$

$$= dt.\mathbb{E}[F_t(X)Z_t] + \mathbb{E}\left[G_t(X)j_t\left(L^{\alpha}_{\beta}\right)Z_t\right]\overline{u_\alpha(t)}u_\beta(t)dt$$

$$\mathbb{E}\left[\left(j_t(X) - \widehat{X}(t)dZ_t\right)\right] = f(t)dt.\mathbb{E}\left[\left(j_t(X) - \widehat{X}(t)\right)Z_t dY_t\right]$$

$$= f(t)dt\, u_\beta(t)\overline{u_\alpha(t)}\mathbb{E}\left[\left(j_t(X) - \check{X}(t)\right)j_t\left(L^{\beta}_{\alpha}\right)Z_t\right]$$

$$\left(dj_t(X) - d\widehat{X}(t)\right)dZ_t = \left[J_t\left(\theta^{\alpha}_{\beta}(X)\right)d\Lambda^{\beta}_{\alpha}(t) - F_t dt - G_t j_t\left(L^{\alpha}_{\beta}\right)d\Lambda^{\beta}_{\alpha}(t)\right]$$

$$f(t)Z_t j_t\left(L^{\mu}_{\nu}\right)d\Lambda^{\nu}_{\mu}(t)$$

$$= f(t)\left\{j_t\left(\theta^{\alpha}_{\beta}(X)L^{\mu}_{\nu}\right)Z_t\varepsilon^{\beta}_{\mu}d\Lambda^{\nu}_{\alpha}(t)\right.$$

$$\left. - G_t j_t\left(L^{\alpha}_{\beta}L^{\mu}_{\nu}\right)Z_t\varepsilon^{\beta}_{\mu}d\Lambda^{\nu}_{\alpha}(t)\right\}$$

(**Note**: $Z_t \in \eta_{t]}$, $\left[\eta_{t]}, j_t(\xi)\right] = 0 \,\forall\, \xi \in \mathcal{L}(h) \Rightarrow \left[\eta_{t]}, j_t\left(L^{\mu}_{\nu}\right)\right] = 0 \Rightarrow \left[Z_t, j_t\left(L^{\mu}_{\nu}\right)\right] = 0$)

$$\mathbb{E}\left[\left(dj_t(X) - d\widehat{X}(t)\right)dZ_t\right]$$

$$= f(t)\overline{u_\alpha}(t)u_\nu(t)dt\mathbb{E}\left\{\left[j_t\left(\varepsilon^{\beta}_{\mu}\theta^{\alpha}_{\beta}(X)L^{\mu}_{\nu}\right)\right.\right.$$

$$\left.\left. + G_t j_t\left(\varepsilon^{\beta}_{\mu}L^{\alpha}_{\beta}L^{\mu}_{\nu}\right)\right]Z_t\right\}$$

$$= f(t)\overline{u_\alpha(t)}u_\nu(t)dt\mathbb{E}\left\{\left(\pi_t\left(\varepsilon^{\beta}_{\mu}\theta^{\alpha}_{\beta}(X)L^{\mu}_{\nu}\right)\right.\right.$$

$$\left.\left. + G_t(X)\pi_t\left(\varepsilon^{\beta}_{\mu}L^{\mu}_{\beta}L^{\mu}_{\nu}\right)\right)Z_t\right\}$$

where $\pi_t(X) = \mathbb{E}\left\{j_t(X)\middle|\eta_{t]}\right\}$.

We thus get on equality coeff's of $f(t)Z_t$ and Z_t separately,

$$\overline{u_\alpha(t)}u_\beta(t).\pi_t\left(\theta^{\alpha}_{\beta}(X)\right) - F_t(X) - G_t(X)\pi_t\left(L^{\alpha}_{\beta}\right)\overline{u_\alpha(t)}u_\beta(t) = 0, \tag{1}$$

$$\overline{u_\alpha(t)}u_\beta(t)\left[\pi_t\left(X.L^{\beta}_{\alpha}\right) - \pi_t(X)\pi_t\left(L^{\beta}_{\alpha}\right)\right] + \overline{u_\alpha(t)}u_\nu(t)\left[G_t(X)\pi_t\left(\varepsilon^{\beta}L^{\alpha}_{\beta}L^{\mu}_{\nu}\right) + \pi_t\left(\varepsilon^{\beta}_{\mu}\theta^{\alpha}_{\beta}(X)L^{\mu}_{\nu}\right)\right] = 0 \tag{2}$$

Solving (2) gives $G_t(X)$ and the $F_t(X)$ is obtained from (1).

This provides us with a linear filter. For general non-linear filters, we assume that

$$d\widehat{X}(t) = F_t(X)dt + \sum_{k=1}^{\infty} G_{kt}(X)(dY_t)^k$$

where $F_t(X)$, $G_{kt}(X) \in \eta_{t]}$. $\forall k, t$.

We write

$$\widehat{X}(t) = \pi_t(X) = \mathbb{E}\left[j_t(X)\middle|\eta_{t]}\right].$$

Then with $Z_t \in \eta_{t]} = \sigma\{Y_s : s \le t\}$ defined by the *qsde*

$$dZ_t = \sum_{k=1}^{\infty} f_k(t)(dY_t)^k Z_t$$

$$= \sum_{k=1}^{\infty} f_k(t) Z_t (dY_t)^k$$

We choose $F_t(X)$ and $\{G_{kt}(X)\}^k$ so that

$$\mathbb{E}\left[\left(X(t) - \widehat{X}(t)\right)Z_t\right] = 0 \;\forall t, \{f_k(.)\}_k.$$

i.e. $\mathbb{E}\left[\left(X(t) - \widehat{X}(t)\right)dZ_t\right] + \mathbb{E}\left[\left(dX(t) - d\widehat{X}(t)\right)Z_t\right]$

$$+\mathbb{E}\left[\left(dX(t) - d\widehat{X}(t)\right)dZ_t\right] = 0$$

which gives $\quad \mathbb{E}\left[\left(dX(t) - d\widehat{X}(t)\right)Z_t\right] = 0,$

$$\mathbb{E}\left[\left(X(t) - \widehat{X}(t)\right)(dY_t)^k Z_t\right] + \mathbb{E}\left[\left(dX(t) - d\widehat{X}(t)\right)(dY_t)^k Z_t\right] = 0$$

where $\quad X(t) = j_t(X).$

We note that $\quad \mathbb{E}[\xi] = \langle g\varphi(u)|\xi|g\varphi(u)\rangle,$

$$dY_t = j_t\left(L_\beta^\alpha\right)d\Lambda_\alpha^\beta(t)$$

implies

$$(dY_t)^k = j_t\left(L_{\beta_1}^{\alpha_1}\ldots L_{\beta_k}^{\alpha_k}\right)d\Lambda_{\alpha_1}^{\beta_1}\ldots d\Lambda_{\alpha_k}^{\beta_k}$$

$$= j_t\left(L_{\beta_1}^{\alpha_1}\ldots L_{\beta_k}^{\alpha_k}\right)\varepsilon_{\alpha_2}^{\beta_1}\varepsilon_{\alpha_3}^{\beta_2}\ldots\varepsilon_{\alpha_k}^{\beta_{k-1}}d\Lambda_\alpha^{\beta_k},$$

We leave the reader to work out the details.

Band theory of semiconductors

Electronics in a crystal

$V(X+d) = V(X)\;\forall X \in \mathbb{R}$. Periodic potential.

$$\left[-\frac{1}{2m}\frac{d^2}{dX^2} + V(X)\right]\psi(X) = E\psi(X) \tag{1}$$

$\psi(X + Nd) = \psi(X)$ (periodic boundary condition) is assumed. N = Number of nuclei.

d = nuclei-interspacing. If $\psi(X)$ satisfies (1) then $\psi(X+d)$ also satisfies (1). Hence assuming non-degeneracy of the eigentials,

$$\psi(X+d) = C\psi(X), X\in\mathbb{R}$$

C is some complex constant.

$$\psi(X+Nd) = \psi(X) \Rightarrow C^N = 1 \Rightarrow$$

$$C = \exp\left(\frac{2\pi in}{N}\right) \text{ for some } n\in\mathbb{Z}.$$

Let $\psi(X) = \exp(ikX)\,\varphi k(X)$

Then $\exp(ikd)\,\varphi_k(X+d) = C\varphi_k(X)$

So we can assume

$$\varphi_k(X+d) = \varphi_k(X) \text{ and } \exp(ikd) = C = \exp\left(\frac{2\pi in}{N}\right)$$

i.e. $$k = \frac{2\pi n}{Nd}$$

Now,

$$\psi'(X) = \exp(ikX)\left(ik\varphi_k(X) + \varphi_k'(X)\right)$$

$$\psi''(X) = \exp(ikX)\left(2ik\varphi_k'(X) - k^2\varphi_k(X) + \varphi_k''(X)\right)$$

So from Schrödinger's equation,

$$\varphi_k''(X) - k^2\varphi_k(X) + 2ik\varphi_k'(X) + 2m\left(E - V(X)\right)\varphi_k(X) = 0.$$

k can take N values *i.e.* $k = \frac{2\pi n}{Nd}$, $n = 0, 1, ..., N-1$.

Since both $\varphi_k(.)$ and $V(.)$ are periodic with period it, they have a Fourier series expansion

$$\varphi_k(X) = \sum_{n\in\mathbb{Z}} C_k[n]\exp\left(\frac{2\pi inX}{d}\right)$$

$$V(X) = \sum_{n\in\mathbb{Z}} V_n \exp\left(\frac{2\pi inX}{d}\right).$$

❖❖❖❖❖

[9] Belavkin's Theory of Quantum Non-Linear Filtering:

Let $X(t)$ (the state/signal process) satisfy the *qsde*.

$$dX_t = (Z_t^b F_t Z_t - X_t \otimes \delta)_\nu^\mu \, dA_\mu^\nu(t) \qquad ...(1)$$

where $Z_t \cdot Z^b{}_t \cdot F_t$, are operators in $h_0\Gamma s(\mathcal{H}_{t]})$ with $\mathcal{H}_t] = V\otimes L^2[0,t].t \geq 0 V$. V is a p dimensional complex Hilbert space and letting $\{e_\alpha \cdot 1 \leq \alpha \leq p\}$ be an ONB for V, we define

$$A_u^\nu = \Lambda_{|e_\mu\rangle\langle e_\nu|}(t), \text{ for } \mu, \nu = 1, 2, ... p$$

$\Lambda_H(t)$ being the conservation associated with any $H \in \mathcal{L}(\mathcal{H})$ that commutes with $\chi_{[0,t]} \forall t \geq 0$.

Here $\mathcal{H} = \mathcal{H}_{\infty)} = Cl\left(\underset{t \geq 0}{U} \mathcal{H}_{t]}\right) = V \otimes L^2(\mathbb{R}_+)$

We may start with the Hilbert space $\mathcal{H}$ and define $\mathcal{H}_{t]} = \chi_{[0,t]} \mathcal{H}$

Where $\chi_{[0,t]}$ acts by multiplication on $\mathcal{H}$ to become the orthogonal projection of $\mathcal{H}$ onto $\mathcal{H}_{t]} = V \otimes L^2 [0, t]$. The operator $\Lambda_H(t)$ is defined by

$$\exp(i\alpha\Lambda_H(t)) = W(0, \exp(i\alpha H_t)), \text{ where}$$

$$H_t = \chi_{[0,t]} H = H\chi_{[0,t]} \text{ and } W(u, \cup)$$

is the Weyl operator defined on $\Gamma s(V \otimes L^2(\mathbb{R}_+)) = \Gamma s(H)$ by

$$W(u, \cup)e(v) = \exp\left(-\|u\|^2/2 - \langle u.\cup v\rangle\right) e(\cup v + u)$$

$\cup \in \mathcal{L}(\mathcal{H})$, $u, v \in \mathcal{L}\,\mathcal{H}$ and

$$e(u) = \overset{\infty}{\underset{n=0}{\otimes}} \frac{u^{\otimes n}}{\sqrt{\lfloor n}}$$

We note that $|e_u\rangle\langle e_v|$ commutes with $\chi_{[0,t]}$ in $V \otimes L^2(\mathbb{R}_+)$ and hence $\Lambda_{|e_\mu\rangle\langle e_v|}(t)$ is well defined.

We also define $A_0^0(t) = t . A_\mu^0(t)$

$$= a^*(e_\mu \chi_{[0,t]}), \mu \geq 1 \text{ and}$$

$$A_0^\mu(t) = a(e_\mu \chi_{[0,t]}), \mu \geq 1$$

Quantum $It_o's$ formula implies

$$dA_0^\mu(t)\, dA_\nu^0(t) = \delta_\nu^\mu dt = \delta_\mu^\mu A_0^0(t),$$

$$dA_u^\nu(t)\, dA_\sigma^\rho(t) = \delta_\sigma^\nu\, dA_\mu^\rho(t),\ \mu, \nu, \rho, \sigma \geq 1,$$

$$dA_0^\nu(t)\, dA_\sigma^\rho(t) = \delta_\sigma^\rho\, dA_0^\rho(t),\ \nu, \rho, \sigma \geq 1,$$

$$dA_\nu^\mu(t)\, dA_\sigma^0(t) = \delta_\sigma^\rho\, dA_\mu^0(t),\ \mu, \nu, \rho, \geq 1$$

These differential rules be summarized by defining

$$\varepsilon_\nu^\mu = \begin{cases} \delta_\nu^\mu, & \mu, \nu \geq 1 \\ 0, & \mu = 0 \text{ or } \nu = 0 \end{cases}$$

as $$dA_\nu^\mu(t)\, dA_\sigma^\rho(t) = \varepsilon_\nu^\mu\, dA_\nu^\rho(t)$$

$$p \geq \mu, \nu, \rho, \sigma \geq 0.$$

Also,

$$A_\nu^\mu(t)^* = A^\nu{}_\mu(t), \quad \mu, \nu \geq 0.$$

Let $\psi \in h_0$ and consider $\xi = \psi \otimes |e(0)\rangle$

$\in h_0 \otimes \Gamma s(\mathcal{H})$.

Define the measurement process, $Y_i(t)$, $1 \le i \le n$ by the *qsde*

$$dY_i(t) = (Z_t^b D_{it} Z_t)^{\mu}_{\nu} dA^{\nu}_{\ \mu}(t) \quad 1 \le i \le n \tag{2}$$

We define the process $U(t)$ by

$$dU(t) = U(t)\,[Z^{\mu}_{\nu} - \delta^{\mu}_{\nu}]\, dA^{\nu}_{\ \mu}(t) \tag{3}$$

The condition for $U(t)$ to be unitary is that of $d(UU^*) = 0$

or

$$dU.U^* + U.dU^* + dU.dU^* = 0 \tag{4}$$

or substituting from (3),

$$U[Z^{\mu}_{\nu} - \delta^{\mu}_{\nu}]U^*\, dA^{\nu}_{\ \mu} + U\,[Z^{\mu *}_{\nu} - \delta^{\mu}_{\nu}]U^*\, dA^{\nu}_{\ \mu}$$
$$U[Z^{\mu *}_{\nu} - \delta^{\mu}_{\nu}]\,[Z^{\rho}_{\sigma} * - \delta^{\rho}_{\sigma}]\, U^* dA^{\rho}_{\sigma} = 0 \tag{5}$$

or using the quantum It_0 to formula,

$$[z^{\mu}_{\nu} - \delta^{\mu}_{\nu} + z^{\nu\ *}_{\ \mu} - \delta^{\nu}_{\ \mu}]\, dA^{\nu}_{\mu}$$
$$+ [z^{\mu}_{\nu} - \delta^{\mu}_{\nu}]\,[z^{\rho}_{\sigma} * - \delta^{\rho}_{\sigma}]\, \varepsilon^{\nu}_{\sigma}\, dA^{\rho}_{\mu} = 0$$

or

$$z^{\mu}_{\nu} + z^{\nu}_{\ \mu} * - Z\delta^{\mu}_{\nu}$$
$$+ [z^{\mu}_{\rho} - \delta^{\mu}_{\rho}]\,[z^{\nu\ *}_{\ \sigma} - \delta^{\nu}_{\ \sigma}]\, \varepsilon^{\nu}_{\sigma} = 0 \tag{6}$$

or

$$z^{\mu}_{\nu} + z^{\nu *}_{\mu} - Z\delta^{\mu}_{\nu} + \varepsilon^{\rho}_{\nu} z^{\mu}_{\rho} + z^{\nu *}_{\sigma} + \varepsilon^{\mu}_{\nu}$$
$$- \varepsilon^{\rho}_{\nu} z^{\mu}_{\rho} - \varepsilon^{\mu}_{\sigma} z^{\nu *}_{\sigma} = 0 \tag{7}$$

Now,

$$\varepsilon^{\rho}_{\nu} z^{\mu}_{\rho} = \begin{cases} z^{\mu}_{\nu}, & \nu \ge 1 \\ 0, & \nu = 0 \end{cases}$$

$$\varepsilon^{\mu}_{\upsilon} z^{\nu}_{\sigma *} = \begin{cases} z^{\nu}_{\mu}, & \mu \ge 1 \\ 0, & \mu = 0 \end{cases}$$

$$\varepsilon^{\rho}_{\sigma} z^{\mu}_{\rho} z^{\nu}_{\sigma} *$$
$$= z^{\nu}_{k} z^{\nu}_{k} *$$

where k runs over 0, 1, 2, ., p (Greek indices like μ, ν, ρ, σ run over 0, 1, 2, ., p while Roman indices like k, l, m, n, i, j *etc.* run over 0, 1, 2, ., p.

This is in argeement with Einstein's general relativity). So (7) can be rewritten as

$$z_k^{\mu} z_k^{\nu *} = \delta_{\nu}^{\mu}, \mu, \nu \geq 1,$$
$$z_k^{\mu} z_k^{0*} + z_0^{\nu} = 0, \mu \geq 1,$$
$$z_k^{0} z_k^{\nu *} + z_0^{\nu *} = 0, \nu \geq 1,$$

The last two equns. are the same. Hence, the condition for $U(t)$ defined by (3) to be unitary $\forall t$ is that

$$z_k^m z_k^{r*} = \delta^{mr}, \tag{8a}$$
$$z_k^m z_k^{0*} = z_0^m = 0 \tag{8b}$$
$$z_0^0 + z_0^{0*} = z \tag{8c}$$

and the non-demolition conditions states that

$$[Y_i(t), X(s)] = 0, \forall s > t \forall i \tag{9}$$

i.e. measurement of Y_i up to time t do not interfere with the state at times $> t$.

We wish that (8) implies $z^b.z = I$,

i.e. $$(z^b)_{\nu}^{\mu} z_{\rho}^{\nu} = \delta_{\rho}^{\mu}$$

Equivalently $z^b = z^{-1}$ and so

$$zz^b = I \tag{10}$$

i.e. $$z_{\nu}^{\mu} (z^b)_{\rho}^{\nu} = \delta_{\rho}^{\mu} \tag{11}$$

or $$z_0^{\nu} (z^b)_{\rho}^{0} + z_k^{\mu} (z^b)_{\rho}^{k} = \delta_{\rho}^{\mu} \tag{12}$$

Let $\mu = m, \rho = r\ (m, r \geq 1)$. Then

We require

$$z_k^m (z^b)_r^k + z_0^m (z^b)_r^0 = \delta_r^m \tag{13}$$

Let $P = ((z_k^m))_{1 \leq m \leq p,\ 1 \leq k \leq p}$

Then $P* = ((z_m^{k*}))_{1 \leq m \leq p,\ 1 \leq k \leq p}$

Also let $q = ((z_0^m))_{1 \leq m \leq p},\ S = ((z_m^{0*}))$

Then (8) can be expanded as

$$PP^* = I, \tag{14a}$$
$$PS + q = 0 \tag{14b}$$

i.e. $$P* = P^{-1}.\ S = -P^{-1}q \tag{15}$$

Let $$P^b = ((z_k^{bm}))_{m,k}, \tag{16}$$
$$q^b = ((z_m^{b0}))$$

Then (13) becomes

$$PP^b + qq^{bT} = I \tag{17}$$

or in view of (14b),

$$P^b + Sq^{bT} = P^{-1} \tag{18}$$

Let $\mu = m$, $\rho = 0$ in (12). Then

$$z_0^m (z^b)_0^0 + z_k^m (z^b)_0^k = 0$$

i.e.

$$q\left(z^b\right)_0^0 + P\tilde{q}^b = 0$$

where $$\tilde{q}^b = ((z_0^b\, m))$$

Thus $$P^{-1} q (z^b)_0^0 + \tilde{q}^b = 0$$

or $$-S (z^b)_0^0 + \tilde{q}^b = 0$$

or $$\tilde{q}^b = S (z^b)_0^0 \tag{19}$$

Finally, let $\mu = 0$, $\rho = r$ in (12). we get

$$z_0^0 (z^b)_r^0 + z_k^0 (z^b)_r^k = 0$$

or $$z_0^0 z^{bT} + S^* P^b = 0 \tag{20}$$

or in view of (18),

$$z_0^0 z^{bT} + S^*(P^{-1} + Sq^{bT}) = 0$$

or

$$(z_0^0 + S^* S)\, q^{bT} = -S^* P^{-1}$$

This gives

$$\begin{aligned} q^{bT} &= (z_0^0 + S^*S)^{-1} S^*P^{-1} \\ P^b &= P^{-1} - S(z_0^0 + S^*S)^{-1} S^*P^{-1} \\ &= [I - S\,(z_0^0 + S^*S)^{-1} S]P^{-1} \end{aligned} \tag{21}$$

(12) with $\mu = \rho = 0$ gives

$$z_0^0 (z^b)_0^0 + z_k^0 (z^b)_0^k = 1$$

i.e.

$$z_0^0 (z^b)_0^0 + S^T \tilde{q}^b = 1 \tag{22}$$

From (19) and (22),

$$z_0^0 (z^b)_0^0 + S^T S\, (z^b)_0^0 = 1$$

i.e. $$(z^b)^0_0 = (z^0_0 + S^T S)^{-1} \tag{23}$$

Let $$\mathcal{B}_{t]} = \hbar_0 \oplus \Gamma s(V \otimes L^2[0, t]) = \hbar_0 \otimes \Gamma s(H_{t]})$$

and let $$\chi_{t]} \subset \mathbb{B}_{t]}$$

be an algebra that commutes with $\mathbb{B}_{t]}$

algebra, *i.e.* $\chi_{t]} \subset \mathbb{B}_{t]} \cap \mathbb{B}'_{t]}$ and $\chi_{t]} \uparrow$

In particular, $\chi_{t]}$ is a as $t\uparrow$

Commutative algebra.

Since $\chi_{t]}$ commutes with $\mathbb{B}_{t]}$ it follows that if $Y_i(s) \in \chi_{s]}\ \forall S$, we can in any state, talk of the joint probability law of $\chi(t) \in \mathbb{B}_{t]}$ and $\{Y_i(s), S \leq t\}_{i=1}^n$.

In particular, we let $\mathcal{F}^y_{t]}$ be the algebra generated by $\{Y_i(s), S \leq t\}_{i=1}^n$ and then

$$\mathcal{F}^Y_{t]} \subset \chi_{t]} \subset \mathbb{B}_{t]} \cap \mathbb{B}'_{t]}.$$

Let $\varepsilon_t : \mathbb{B}_{t]} \to \mathcal{F}^Y_{t]}$ be the orthogonal projection *w.r.t* the inner product

$$\langle W, V\rangle = T_r(WV^*)$$

Then $\varepsilon_t(X_{(t)})$ is the estimate of $X(t)$ based on $\mathcal{F}^Y_{t]}$. In any state,

$$\varepsilon_t(X(t)) = \mathbb{E}[X(t) \mid \mathcal{F}^Y_{t]}]$$

i.e.

$$\langle X(t) - \varepsilon_t(X(t)), Y_i(s)\rangle = 0$$

$$1 \leq i \leq n, 0 \leq s \leq t.$$

Let $\xi \in \hbar_0 \otimes \Gamma s(\mathcal{H})$ be arbitrary and let

$$E^\xi_t \in \mathbb{B}_{t]} = \mathcal{L}(\hbar_0 \otimes \Gamma s(V \otimes L^2[0, t]))$$
$$= \mathcal{L}(\hbar_0 \otimes \Gamma s(V \otimes \mathcal{H}_{t]}))$$

be the orthogonal projection of $\hbar_0 \otimes \Gamma s(V \otimes H_{t]})$ Onto $\pounds^Y_t \xi$. Then for any $X \in \mathbb{B}_{t]}$.

$\varepsilon_t(X) \in \mathcal{F}^Y_t$ commutes with any $A \in \mathbb{B}_{t]}$ and so

$$\varepsilon_t(X)A\xi = A\varepsilon_t(X)\xi$$
$$= AE^\xi_t X\xi\ ??$$

All the algebras are assumed to be * unital.

$$X - \varepsilon_t(X) \perp \mathcal{F}^Y_t$$

$$\Rightarrow \quad \langle X - \varepsilon_t(X), Y\rangle = 0\ \forall Y \in \mathcal{F}^Y_t$$

$$X\xi - E^\xi_t X\xi \perp Y\xi\ \forall Y \in \mathcal{F}^Y_t$$

$$\forall X \in B_{t]}$$

If for some ξ

$$Cl\,\{X\xi\,|X \in \mathbb{B}_{t]}\} = \hbar_0 \otimes \Gamma s\,(H_{t]})$$

Then $X\xi - E_t^{\xi} X\xi \perp Y\xi\ \forall Y \in \mathbb{B}_{t}]$

and Y over $\mathcal{F}_t^{Y}$

Implies by varying X over $\mathbb{B}_{t]}$

$\eta - E_t^{\xi}\eta \perp \mathcal{F}_t^{Y}\xi \ \forall\ \eta \in \hbar_0 \otimes \Gamma s\,(\mathcal{H}_{t]})$

If further $\mathcal{F}_t^{Y}\xi$ is independent of ξ for nonzero ξ, then $E_t^{\xi} = E_t$ is independent of ξ and we get

$\eta - E_t\eta \perp \mathcal{F}_t^{Y}(\hbar_0 \otimes \Gamma s\,(\mathcal{H}_{t]}))$

$\forall \eta \in h_0 \otimes \Gamma s\,(\hbar_t])$

Alternately, we can choose and fix a $\xi \in \hbar_0 \otimes \Gamma s\,(\mathcal{H}_b])$ and define $\varepsilon_t\,(X)\ (X \in B_{t]})$ as a positive projection onto $\mathcal{F}_t^{Y}$ such that

$X\xi - \varepsilon_t\,(X)\xi \perp \mathcal{F}_t^{Y}\xi$

This is equivalent to requiring that

$\langle\xi|\ Y_i\,(s)\,(X - \varepsilon_t\,(X))\,|\xi\rangle = 0 \qquad 1 \le i \le n,\ 0 \le s \le t$

or equivalently in view of the * nature of the algebras involved,

$\langle\xi|\,(X - \varepsilon_t\,(X))\ Y_j{*}\,(s)\,|\xi\rangle \ = \qquad 0,\ S \le t,\ 1 \le i \le n,$

Note that

$$\varepsilon_{t+dt} \circ \varepsilon_t = \varepsilon_t = \varepsilon_t \circ \varepsilon_{t+dt}$$

and so $\varepsilon_t \circ |\varepsilon_{t+dt}\,(X\,(t + dt)) - \varepsilon_t\,(X\,(t))]$

$$= \varepsilon_t \circ \varepsilon_{t+dt}\,(X\,(t + dt) - X\,(t))$$

$$= \varepsilon_t\,(dX\,(t))$$

i.e. $\quad \varepsilon_t \circ d\varepsilon_t\,(X\,(t)) = \varepsilon_t\,(dX(t))$

Let $\quad \widehat{Y}_i(t) = U\,(t)\,Y_i\,(t)\,U{*}\,(t)$

Then
$$d\widehat{Y}_i(t) = dUY_iU{*} + UdY_iU{*} + UY_idU{*}$$
$$+ dUdY_iU{*} + UdY_idU{*} + dUY_idU{*}$$
$$= U\left\{\left(z_\nu^\mu - \delta_\nu^\mu\right)Y_i + \left(z^b D_i z\right)_\nu^\mu + Y_i z_\mu^{\nu*}\right\}U{*}\,dA_\mu^\nu - \delta_\nu^\mu$$
$$+ U\left(z_\nu^\mu - \delta_\nu^\mu\right)\left(z^b D_i z\right)_\sigma^\rho U{*}\,dA_\mu^\nu \cdot dA_\rho^\sigma$$

$$
\begin{aligned}
&+U\left(z^b D_i \cdot z\right)^{\mu}_{\nu}\left(z^{\rho *}_{\sigma}-\delta^{\rho}_{\sigma}\right)U * dA^{\nu}_{\mu} dA^{\rho}_{\sigma} \\
&+U\left(z^{\mu}_{\nu}-\delta^{\mu}_{\nu}\right)Y_i\left(z^{\rho *}_{\sigma}-\delta^{\rho}_{\sigma}\right)U * dA^{\nu}_{\mu} dA^{\rho}_{\sigma} \\
= \; & U\left\{\left(\left(z^{\mu}_{\nu}-\delta^{\mu}_{\nu}\right)Y_i+Y_i\left(z^{\nu *}_{\mu}-\delta^{\mu}_{\nu}\right)+\left(z^b D_i z\right)^{\mu}_{\nu}\right)dA^{\nu}_{\mu}\right\} \\
&+\left(z^{\mu}_{\nu}-\delta^{\mu}_{\nu}\right)\left(z^b D_i z\right)^{\rho}_{\sigma}\varepsilon^{\nu}_{\rho}\, dA^{\sigma}_{\mu} \\
&+\left(z^b D_i z\right)^{\mu}_{\nu}\left(z^{\rho *}_{\sigma}-\delta^{\rho}_{\sigma}\right)\varepsilon^{\nu}_{\rho}\, dA^{\rho}_{\mu} \\
&+\left(z^{\mu}_{\nu}-\delta^{\mu}_{\nu}\right)Y_i\left(z^{\rho *}_{\sigma}-\delta^{\rho}_{\sigma}\right)\varepsilon^{\nu}_{\sigma}\, dA^{\rho}_{\mu}\Big\} U * \\
= \; & U\Big\{z^{\mu}_{\nu}Y_i+Y_i z^{\nu *}_{\mu}-zY_i\,\delta^{\mu}_{\nu}+(z^b D_i z)^{\mu}_{\nu} \\
&+\left(z^{\mu}_{\sigma}-\delta^{\mu}_{\sigma}\right)\left(z^b D_i z\right)^{\rho}_{\nu}\varepsilon^{\sigma}_{\rho}+\left(z^b D_i z\right)^{\mu}_{\rho}\left(z^{\mu *}_{\sigma}\,\delta^{\mu}_{\sigma}\right)^{\rho}_{\nu}\varepsilon^{\rho}_{\sigma} \\
&+\left(z^{\mu}_{\rho}-\delta^{\mu}_{\rho}\right)Y_i\left(z^{\nu *}_{\sigma}-\delta^{\nu}_{\sigma}\right)\varepsilon^{\rho}_{\sigma}\Big\} U * dA^{\nu}_{\mu} \\
= \; & UG^{\mu}_{i\nu}U * dA^{\nu}_{\mu}
\end{aligned}
$$

where

$$
\begin{aligned}
G^{\mu}_{i\nu} = \; & z^{\mu}_{\nu}Y_i+Y_i z^{\nu *}_{\mu}+\left(\varepsilon^{\mu}_{\nu}-2\delta^{\mu}_{\nu}\right)Y_i \\
&+\left(z^b D_i z\right)^{\mu}_{\nu}+\left(\varepsilon^{\sigma}_{\rho}\, z^{\mu}_{\sigma}-\varepsilon^{\mu}_{\rho}\right)\left(z^b D_i z\right)^{\rho}_{\nu} \\
&+\left(z^b D_i z\right)^{\mu}_{\rho}\left(\varepsilon^{\rho}_{\sigma}\, z^{\nu *}_{\sigma}-\varepsilon^{\rho}_{\nu}\right)+\varepsilon^{\rho}_{\sigma}\, z^{\mu}_{\rho}\, Y_i z^{\nu *}_{\sigma} \\
&-z^{\mu}_{\rho}\, Y_i\, \varepsilon^{\rho}_{\nu}-\varepsilon^{\mu}_{\sigma}\, Y_i\, z^{\nu *}_{\sigma}
\end{aligned}
$$

Thus,

$$
\begin{aligned}
G^{m}_{ik} = \; & z^{m}_{k}Y_i+Y_i\, z^{k *}_{m}-\delta^{m}_{k}Y_i \\
&+\left(z^b D_i z\right)^{m}_{k}+z^{m}_{r}\left(z^b D_i z\right)^{r}_{k} \\
&-\left(z^b D_i z\right)^{m}_{k}+\left(z^b D_i z\right)^{m}_{r} z^{k *}_{r}-\left(z^b D_i z\right)^{m}_{k}+z^{m}_{r}\, Y_i\, z^{k *}_{r} \\
&-z^{m}_{k}\, Y_i-Y_i\, z^{k *}_{r} \\
= \; & -\delta^{m}_{k}Y_i-\left(z^b D_i z\right)^{m}_{k} \\
&+z^{m}_{r}\left(z^b D_i z\right)^{r}_{k}+\left(z^b D_i z\right)^{m}_{r} z^{k *}_{r}+z^{m}_{r}\, Y_i z^{k *}_{r}
\end{aligned}
$$

For general μ, $\nu \geq 0$, we can write

$$\begin{aligned} G^{\mu}_{i\nu} &= z^{\mu}_{\nu} Y_i + Y_i z^{\nu *}_{\mu} + \left(\varepsilon^{\nu}_{\mu} - 2\delta^{\mu}_{\nu}\right) Y_i \\ &+ \left(z^b D_i z\right)^{\mu}_{\nu} + \left(z \varepsilon z^b D_i z\right)^{\mu}_{\nu} + \left(z^b D_i z \varepsilon z^*\right)^{\mu}_{\nu} \\ &- \left(\varepsilon z^b D_i z\right)^{\mu}_{\nu} - \left(z^b D_i z \varepsilon\right)^{\mu}_{\nu} \\ &+ \left(z \varepsilon Y_i z^*\right)^{\mu}_{\nu} - (z\varepsilon)^{\mu}_{\nu} Y_i - Y_i \left(\varepsilon z^*\right)^{\mu}_{\nu} \end{aligned} \qquad (\alpha_0)$$

$$(A\varepsilon)^{\mu}_{\nu} = A^{\mu}_{\rho}\, \varepsilon^{\rho}_{\nu}, (\varepsilon A)^{\rho}_{\nu} = \varepsilon^{\mu}_{\rho}\, A^{\rho}_{\nu}$$

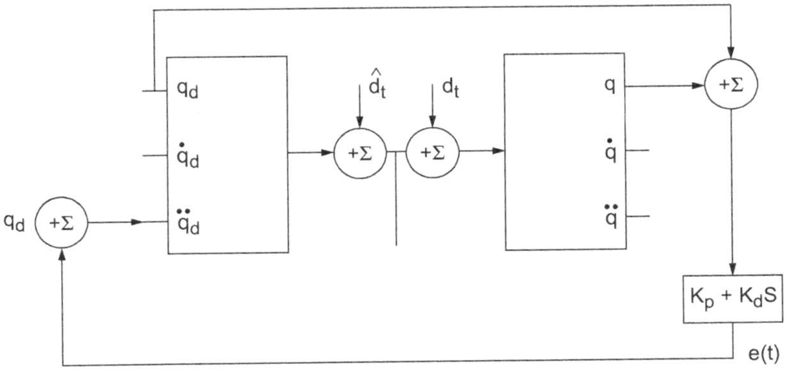

where ε^{μ}_{ν} are scalars and A^{μ}_{ν} are operators). The condition (7) for?? of $U(t)$ can be expressed as

$$z + z^* - 2I + \varepsilon + z\,\varepsilon\, z^* - z\,\varepsilon - \varepsilon\, z^* = 0 \qquad (\alpha_1)$$

i.e. or equivalently,

$$z\,(I - \varepsilon) + (I - \varepsilon)\, z^* + \varepsilon - zI + z\,\varepsilon\, z^* = 0 \qquad (\alpha_2)$$

We note that the $(p + 1) \times (p \times 1)$ matrix $\varepsilon = \left(\left(\varepsilon^{\mu}_{\nu}\right)\right)_{0 \leq \mu, \nu \leq \rho}$ has rank p since its first row and first column are zero. (α_0) can be expressed as

$$\begin{aligned} G_i &= zY_i + Y_i z^* + (\varepsilon - 2I) Y_i + z^b D_i z + z\varepsilon z^b D_i z \\ &+ z^b D_i z\, \varepsilon z^* - \varepsilon z^b D_i z \\ &- z^b D_i z\, \varepsilon + z\,\varepsilon Y_i z^* - z\,\varepsilon Y_i - Y_i\, \varepsilon z^* \\ &= z^b (I - \varepsilon) Y_i + Y_i (I - \varepsilon) z^* \\ &+ (\varepsilon - 2I) Y_i + (I - \varepsilon) z^b D_i z \\ &- z^b D_i z\, \varepsilon + z\,\varepsilon z^b D_i z \end{aligned}$$

$$+ z^b D_i z\, \varepsilon + z^* + z\, \varepsilon\, Y_i (z^* - I) - Y_i\, \varepsilon\, z^* \qquad (\alpha_3)$$

Note that

$$Y_i \varepsilon = \varepsilon\, Y_i$$

$$I - \underline{\underline{\varepsilon}} = \left(\left(\delta^\mu_\nu - \varepsilon^\mu_\nu\right)\right) = \left[\begin{array}{c|c} 1 & \underline{O}^T \\ \hline \underline{O} & \underline{\underline{O}} \end{array}\right]$$

$$\underline{\underline{\varepsilon}} = \left(\left(\varepsilon^\mu_\nu\right)\right) = \left[\begin{array}{c|c} O & \underline{O}^T \\ \hline \underline{O} & \underline{\underline{I}} \end{array}\right]$$

Suppose we do not assume that $z^b = z^{-1}$, but rather $z^b = z^*$. Then (α_3) becomes

$$\begin{aligned} G = {} & z^*(I-\varepsilon)Y_i + Y_i(I-\varepsilon)z^* \\ & + (\varepsilon - 2I)Y_i + (I-\varepsilon)z^* D_i z \\ & - z^* D_i z\, \varepsilon + z\, \varepsilon\, z^* D_i z \\ & + z^* D_i z\, \varepsilon\, z^* + z\, \varepsilon\, Y_i (z^* - I) - Y_i\, \varepsilon\, z^* \end{aligned} \qquad (\alpha_4)$$

Substituting for $z\, \varepsilon\, z^*$ from (α_2) into (α_4) gives

$$\begin{aligned} G = {} & z^*(I-\varepsilon)Y_i + (I-\varepsilon)z^* + (\varepsilon - 2I)Y_i \\ & + (I-\varepsilon)z^* D_i z - z^* D_i z\, \varepsilon \\ & + \left(2I - \varepsilon + (\varepsilon - I)z^* z(\varepsilon - I)D_i z\right) \\ & + z^* D_i \left(2I - \varepsilon + (\varepsilon - I)z^* + z(\varepsilon - I)\right) \\ & + z\, \varepsilon\, Y_i (z^* - I) z^* - Y_i\, \varepsilon\, z^* \\ = {} & \left\{z^*, (I-\varepsilon)Y_i\right\} + (\varepsilon - 2I)Y_i \\ & + z^* D_i z - \left\{\varepsilon, z^* D_i z\right\} \\ & + (2I - \varepsilon)D_i z + \left\{(\varepsilon - I), z^* D_i z\right\} \\ & z^* D_i (\varepsilon - I) z^* + z(\varepsilon - I) D_i z \\ = {} & \left\{z^*, (I-\varepsilon)Y_i\right\} + (\varepsilon - 2I)Y_i \\ & + (2I - \varepsilon)D_i z + z^* D_i (\varepsilon - I) z^* \\ & + z(\varepsilon - I) D_i z. \end{aligned}$$

Now,

$$\begin{aligned} \left(z(I-\varepsilon)D_i z\right)^\mu_\nu &= z^\mu_0 \left(D_i z\right)^0_\nu \\ &= z^\mu_0 D^0_{i\alpha} z^\alpha_\nu \text{ etc.} \end{aligned} \qquad (\alpha_5)$$

Here {.,.} denotes anticommutator. For non-demolition, we require in particular

$$\left[Y_i(t), X(t)\right] = 0$$

i.e, we require

$$[dY_i(t), X(t)] + [Y_i(t), dX(t)] + [dY_i(t), dX(t)] = 0$$

or equivalently,

$$\left[\left(z_t^b D_{it}\, zt\right)_\nu^\mu, X(t)\right] dA_\mu^\nu(t) + \left[Y_i(t), \left(z_t^b F_t z_t\right)_\nu^\mu - X(t)\delta_\nu^\mu\right] dA_\nu^\mu(t)$$
$$+\left[\left(z^b D_i z\right)_\nu^\mu, \left(z^b Fz\right)_\sigma^\rho - X\delta_\sigma^\rho\right] dA_\mu^\nu dA_\rho^\sigma = 0$$

or

$$\left[\left(z^b D_i z\right)_\nu^\mu, X\right] + \left[Y_i, \left(z^b Fz\right)_\nu^\mu - X\delta_\nu^\mu\right]$$
$$+\,\varepsilon_\rho^\sigma\left[\left(z^b D_i z\right)_\sigma^\mu, \left(z^b Fz\right)_\nu^\rho - X\delta_\nu^\rho\right] = 0 \qquad (\alpha_6)$$

or using $\left[Y_i^{(t)}, X(t)\right] = 0$, this is the same as

$$\left[\left(z^b D_i z\right)_\nu^\mu, X\right] + \left[Y_i, \left(z^b Fz\right)_\nu^\mu\right] + \varepsilon_\rho^\sigma\left[\left(z^b D_i z\right)_\sigma^\mu, \left(z^b Fz\right)_\nu^\rho\right]$$
$$-\,\varepsilon_\nu^\sigma\left[\left(z^b D_i z\right)_\sigma^\mu, X\right] = 0 \qquad (\alpha_7)$$

Taking $\mu = m,\ \nu = 0$ gives

$$\left[\left(z^b D_i z\right)_0^m, X\right] + \left[Y_i, \left(z^b Fz\right)_0^m\right] + \left[\left(z^b D_i z\right)_k^m, \left(z^b Fz\right)_0^k\right] = 0 \qquad (\alpha_8)$$

Taking $\mu = 0,\ \nu = m$ gives

$$\left[Y_i\left(z^b Fz\right)_m^0\right] + \left[\left(z^b D_i z\right)_k^0, \left(z^b Fz\right)_m^k\right] = 0 \qquad (\alpha_9)$$

Taking $\mu = m,\ \nu = r$ gives

$$\left[Y_i, \left(z^b Fz\right)_r^m\right] + \left[\left(z^b D_i z\right)_k^m, \left(z^b Fz\right)_r^k\right] = 0 \qquad (\alpha_{10})$$

Taking $\mu = \nu = 0$ gives

$$\left[\left(z^b D_i z\right)_0^0, X\right] + \left[Y_i, \left(z^b Fz\right)_0^0\right] + \left[\left(z^b D_i z\right)_k^0, \left(z^b Fz\right)_0^k\right] = 0 \qquad (\alpha_{11})$$

Martingales:

Suppose $\varepsilon_t(X(s)) = X(t)$ for $s \geq t$

w.r.t. ξ, *i,e.*

$$\varepsilon_t(X(s))\xi = X(t)\xi,\ s \geq t$$

Then we call X an ε m-Martingale.

Let
$$\xi = \psi \otimes |e(0)\rangle.$$

Then if
$$dX(t) = \left[\left(z^b(t)F(t)z^b(t)\right)^{\mu}_{v} - X(t)\delta^{\mu}_{v}\right]dA^v_{\mu}(t)$$

we get $\quad \langle \xi, \varepsilon_t (dX(t))\xi\rangle = \left\langle \xi, \left(\left(z^bFz\right)^0_0 - X\right)\xi\right\rangle dt$

and more generally, for any $P, Q \in \mathbb{B}_{t]}$
We have,

$$\langle A\xi, \varepsilon_t (dX(t))\mathbb{B}\,\xi\rangle = \langle \xi, A*B\varepsilon_t (dX(t))\xi\rangle = \langle \xi, \varepsilon_t (dX(t))A*B\xi\rangle$$

Since $\varepsilon_t(P) \in \mathcal{F}_t^Y \subset \mathbb{B}'_{t]}$

and $A*B \in \mathbb{B}_{t]}$. Here $P \in \mathbb{B} = \mathcal{L}\left(h_0 \otimes \Gamma s(\mathcal{H})\right)$
is arbitrary.

Now,

$$\left\langle \xi, \varepsilon_t \left[\left(\left(z^bFz - X\right)^{\mu}_{tv}\right)dA^v_{\mu}(t)\right]\xi\right\rangle \left\langle \xi, E_t\left[\left(\left(z^bFz - X\right)^{\mu}_{tv}\right)dA^v_{\mu}(t)\xi\right]\right\rangle$$

$$= \left\langle \xi, \left(z^bFz - X\right)^{\mu}_{tv} dA^v_{\mu}(t)\xi\right\rangle$$

Since
$$\xi = \psi \otimes |e(0)\rangle.$$

We have

$$dA^j_0(t)\xi = \psi \otimes dA^j_0(t)|e(0)\rangle = 0.$$
$$\langle \xi | dA^0_j(t) = 0$$

since

$$dA^0_j(t) = dA^{j*}_0(t)$$
$$\langle \xi, dA^m_k(t)\xi\rangle = \langle e(0)|dA^m_k(t)|e(0)\rangle = 0 \text{ and } G^{\mu}_v(t) \in \mathbb{B}_{t]}.$$
More generally for any $A, B \in \mathbb{B}_{t]}$, we have
$$\left\langle A\xi, \varepsilon_t\left[G^{\mu}_v(t)dA^v_{\mu}(t)\right]B\xi\right\rangle$$

$$= \left\langle \xi, \varepsilon_t\left(A*G^{\mu}_v(t)BdA^v_{\mu}(t)\right)\xi\right\rangle$$
$$= \left\langle \xi, \left(A*G^{\mu}_v B)dA^v_{\mu}(t)\right)\xi\right\rangle$$
$$= \left\langle \xi, \left(A*G^0_0 B)\xi\right\rangle dt$$

Since

$$\left\langle \xi \mid dA_j^0(t) \mid \xi \right\rangle = \left\langle \xi \mid dA_j^0(t) \mid \xi \right\rangle$$
$$= \left\langle \xi \mid dA_k^m(t) \mid \xi \right\rangle = 0 \, \forall j, m, k \geq 1.$$

Note:

$$\langle \xi \mid \Lambda_H(t) \mid \xi \rangle = \langle e(0) \mid \Lambda_H(t) \mid e(0) \rangle = \langle 0 \mid H_t \mid 0 \rangle = 0$$

$$\left\langle \xi \mid A_j^0(t) \mid \xi \right\rangle = \left\langle A_0'^{j}(t) e(0), e(0) \right\rangle = \langle e(0) \mid e(0) \rangle$$

$$\left\langle e_j X_{[0,t]} \mid 0 \right\rangle = 0, j = 1$$

and likewise

$$\left\langle \xi \mid A_0^j(t) \mid \xi \right\rangle = 0, j \geq 1$$

This writing

$$G_\nu^\mu(t) = \left(z^b F z\right)_\nu^\mu(t) - X(t)\,\delta_\nu^\mu,$$

it follows that

$$dX(t) = G_\nu^\mu(t)\, dA_\mu^\nu(t)$$

and

$$\varepsilon_t[dX(t)] = \varepsilon_t\left[G_\nu^\mu(t)\, dA_\mu^\nu(t)\right]$$
$$= \varepsilon_t\left(G_0^0(t)\right) dt$$

We have for $s \geq t$

$$Y_i(s) = Y_i(t) + \int_t^s dY_i(\tau)$$

and if ε_t projects onto $\mathcal{F}_t^Y$, then

$$\varepsilon_t\left(Y_i(s)\right) = Y_i(t) + \int_t^s \varepsilon_t\left(dY_i(\tau)\right)$$

Now,

$$\varepsilon_t\left(dY(\tau)\right) = \varepsilon_t \circ \varepsilon_\tau\left(dY(\tau)\right)$$
$$= \varepsilon_t \circ \varepsilon_\tau\left(\left(z^b D_i z\right)_0^0(\tau)\right) d\tau$$
$$= \varepsilon_t\left(\left(z^b D_i z\right)_0^0(\tau)\right) d\tau$$

Thus,

$$= \varepsilon_t\left(Y_i(s)\right) = Y_i(t) + \varepsilon_t \circ \int_t^s \varepsilon_t\left(\left(z^b D_i z\right)_0^0(\tau)\right) d\tau$$
$$= Y_i(t) + \varepsilon_t \circ \int_t^s \varepsilon_t\left(\left(z^b D_i z\right)_0^0(\tau)\right) d\tau$$

Hence Y is a Martingale iff

$$\varepsilon_\tau\left(\left(z^b D_i z\right)_0^0(\tau)\right) = 0 \forall \tau \qquad (\alpha_{12})$$

Now consider

$$Q(s,t) = \int_t^S (\varepsilon_s - \varepsilon_\tau)\left(\left(z^b F z\right)_v^\mu(\tau)\, dA_\mu^v(\tau)\right)$$

We have for $t < \alpha < S$,

$$\varepsilon_\alpha(Q(s,t)\xi) = \int_t^S (\varepsilon_\alpha - \varepsilon_\alpha \circ \varepsilon_\tau)\left(\left(z^b F z\right)_v^\mu(\tau)\, dA_\mu^v(\tau)\right)\xi$$

and in first, more generally, for $A, B \in \mathbb{B}_{t]}$

$$\begin{aligned}
\langle A\xi, \varepsilon_\alpha(Q(s,t))B\xi\rangle &= \int_t^S \left\langle \xi, (\varepsilon_\alpha - \varepsilon_\alpha \circ \varepsilon_\tau)\left(A * z^b F z\right)_v^\mu(\tau) B\, dA_\mu^v(\tau)\xi \right\rangle \\
&= \int_t^\alpha \left\langle \xi, (\varepsilon_\alpha - \varepsilon_\alpha \circ \varepsilon_\tau)\left(A * \left(z^b F z\right)_v^\mu(\tau) B\, dA_\mu^v(\tau)\xi \right\rangle \\
&\quad + \int_\alpha^S \left\langle \xi, (\varepsilon_\alpha - \varepsilon_\alpha \circ \varepsilon_\tau)\left(A * \left(z^b F z\right)_v^\mu(\tau) B\right) dA_\mu^v(\tau)\xi \right\rangle \\
&= \int_t^\alpha \left\langle \xi, (\varepsilon_\alpha - \varepsilon_\tau)\left(A * \left(z^b F z\right)_v^\mu(\tau) B\right) dA_\mu^v(\tau)\xi \right\rangle
\end{aligned}$$

since

$$\varepsilon_\alpha \circ \varepsilon_\tau = \begin{cases} \varepsilon_\tau \tau < \alpha \\ \varepsilon_\alpha \tau > \alpha. \end{cases}$$

Thus,

$$\langle A\xi, \varepsilon_\alpha(Q(s,t))B\xi\rangle = \langle A\xi, Q(\alpha,t)B\xi\rangle$$

Hence

$$\varepsilon_\alpha(Q(s,t)) = Q(\alpha,t), \quad t < \alpha < S.$$

In particular for $\alpha < S$,

$$\begin{aligned}
\varepsilon_\alpha \circ \int_0^S (\varepsilon_s - \varepsilon_\tau)&\left(\left(z^b F z\right)_v^\mu(\tau)\, dA_\mu^v(\tau)\right) \\
&= \int_0^\alpha (\varepsilon_\alpha - \varepsilon_\tau)\left(\left(z^b F z\right)_v^\mu(\tau)\, dA_\mu^v(\tau)\right)
\end{aligned}$$

i.e.

$$M_s = \int_0^S (\varepsilon_\alpha - \varepsilon_\tau)\left(\left(z^b F z\right)_v^\mu(\tau)\, dA_\mu^v(\tau)\right)$$

is a Martingale *w.r.t.*

$$\left\{\mathcal{F}_t^y\right\}$$

Note that

$$\begin{aligned} M_s &= \int_0^S (\varepsilon_s - \varepsilon_\tau)(dX(\tau)) \\ &= \varepsilon_s(X(s)) - \int_0^S \varepsilon_\tau(dX(\tau)) \end{aligned}$$

or equivalently,

$$\varepsilon_t(X(t)) = \int_0^t \varepsilon_\tau(dX(\tau)) + M_t$$

and since

$$\begin{aligned} \left\langle A\xi, \varepsilon_\tau(dX(\tau))B\xi\right\rangle &= \langle A\xi, dX(\tau)B\xi\rangle \\ &= \langle \xi, A^* dX(\tau)B\xi\rangle \\ &= \left\langle \xi, A^* G_\nu^\mu(\tau)Bd\, A_\mu^\nu(\tau)\xi\right\rangle \\ &= \left\langle \xi, A^* G_0^0(\tau)B\xi\right\rangle d\tau \\ &= \left\langle A\xi, \varepsilon_\tau\left(G_0^0(\tau)\right)B\xi\right\rangle d\tau \\ & \quad \forall A, B \in \mathbb{B}_{\tau]} \end{aligned}$$

Hence,

$$\varepsilon_t(dX(\tau)) = \varepsilon_\tau\left(G_0^0(\tau)\right)d\tau$$

i.e.

$$\varepsilon_t(X(t)) = \int_0^t \varepsilon_\tau\left(G_0^0(\tau)\right)d\tau + M_t$$

We write

$$H_{i\nu}^\mu(t) = \left(z^b D_i\, z\right)_\nu^\mu(t)$$

so $\quad dY_i(t) = H_{i\nu}^\mu(t)\, dA_\mu^\nu(t)$

and assume that

$$dM_t = K^i(t)(dY_i(t)) - \varepsilon_t\left(dY_i(t)\right)$$

i.e.

where $\quad K^i(t) \in \mathcal{F}_{t]}^Y \subset \mathbb{B}'_{t]}$

Then

$$\varepsilon_t(dM_t) = K^i(t)\left(\varepsilon_t\left(dY_i(t)\right)\right) - \varepsilon_t\left(dY_i(t)\right) = 0$$

Remark: Let $A \in \mathcal{F}_{t]}^{Y}$ and $B \in \mathbb{B}_{t]}$. Then $\langle \xi, (AB - \varepsilon_t(AB))C\xi \rangle = 0$

$$\forall C \in \mathcal{F}_{t]}^{Y}$$

and also

$$\begin{aligned}
\langle A^*\xi, (B - \varepsilon_t(B))C\xi \rangle &= \langle \xi, (AB - A\varepsilon_t(B))C\xi \rangle \\
&= \langle \xi, (BA - \varepsilon_t(B)A)C\xi \rangle \\
&= \langle \xi, (B - \varepsilon_t(B))AC\xi \rangle = 0
\end{aligned}$$

$$\therefore \qquad AC \in \mathcal{F}_t^Y$$

Hence

$$\varepsilon_t(AB) = A\varepsilon_t(B)$$

and likewise

$$\varepsilon_t(BA) = \varepsilon_t(B)A$$

So,

$$\varepsilon_t(dM_t) = 0.$$

Also,

$$\begin{aligned}
M_t &= \int_0^t K^i(s)\left(dY_i(s) - \varepsilon_s\left(dY_i(s)\right)\right) \\
&\in \pounds_{t]}^{Y}
\end{aligned}$$

since $\mathcal{F}_{t]}^{Y}$ is an algebra and $K^i(s) \in \mathcal{F}_{t]}^{Y} \subset \pounds_t^Y$, $dY_i(s)^{\in} \mathcal{F}_{t]}^{Y}$, $\varepsilon_s(dY_i(s)) \in \mathcal{F}_t^Y \ \forall \tau > S$.

Thus,

$$\begin{aligned}
d\varepsilon_t(X(t)) &= \varepsilon_t\left(G_0^0(t)\right)dt \\
&\quad + K^i(t)\left(dY_i(t)\right) - \varepsilon_t(dY_i(t)).
\end{aligned}$$

Now,

$$\varepsilon_t\left(dY_i(t)\right) = \varepsilon_t\left(H_{i0}^0(t)\right)dt$$

so our filtering eqn., becomes

$$\begin{aligned}
d\varepsilon_t(X(t)) &= \varepsilon_t\left(G_0^0(t)\right)dt \\
&\quad + K^i(t)\left(dY_i(t)\right) - \varepsilon_t\left(H_{i0}^0(t)\right)dt \qquad (\beta_1)
\end{aligned}$$

Note that

$$\begin{aligned}
dX(t) &= G_\nu^\mu(t)\, dA_\mu^\nu(t), \\
dY_i(t) &= H_{i\nu}^\mu(t)\, dA_\mu^\nu(t)
\end{aligned}$$

From (β_1),

$$\varepsilon_t\left\{d\varepsilon_t(X(t))\left(dY_j(t)-\varepsilon_t\left(H_0^0(t)\right)dt\right)\right\}$$

$$= \varepsilon_t\left\{d\varepsilon_t(X(t))dY_j(t)\right\}$$

$$= K^i(t)\varepsilon_t\left(dY_i(t)\,dY_j(t)\right)$$

or

$$\varepsilon_t\left\{d\varepsilon_t(X(t))H_\upsilon^\mu(t)dA_\mu^\upsilon(t)\right\}$$

$$= K^i(t)\varepsilon_t\left(H_{iv}^\mu(t)H_{j\sigma}^\rho(t)\varepsilon_\rho^v\, dA_\mu^\sigma(t)\right)$$

$$= K^i(t)\varepsilon_t\left\{\varepsilon_\rho^\sigma H_{i\sigma}^\mu(t)H_{jv}^\rho(t)\,dA_\mu^v(t)\right\}$$

$$= K^i(t)\varepsilon_t\left\{H_{i\sigma}^0(t)\varepsilon_\rho^\sigma H_{j0}^\rho(t)\right\}dt$$

$$= K^i(t)\varepsilon_t\left\{H_i(t)\in H_j(t)_0^0\right\}dt \qquad (\beta_2)$$

If we assume that

$$H_{iv}^\mu(t)\in\mathcal{F}_{t]}^Y,$$

Then

$$\varepsilon_{t+dt}\left(X(t+dt)\right)H_{iv}^\mu(t)$$

$$= \varepsilon_{t+dt}\left(X(t+dt)H_{iv}^\mu(t)\right)$$

and hence

$$\varepsilon_t\left\{d\varepsilon_t\left(X(t)\right)H_{iv}^\mu(t)\,dA_\mu^v(t)\right\}$$

$$= \varepsilon_t\left\{\varepsilon_{t+dt}\left(dX(t)\,H_{iv}^\mu(t)\,dA_\mu^v(t)\right)\right\}$$

$$= \varepsilon_t\left\{dX(t)\,H_{iv}^\mu(t)\,dA_\mu^v(t)\right\}$$

$$= \varepsilon_t\left\{G_\sigma^\rho(t)\,H_{iv}^\mu(t)\,dA_\rho^\sigma(t)\,dA_\mu^v(t)\right\}$$

$$= \varepsilon_t\left\{G_\sigma^\rho(t)\,H_{iv}^\mu(t)\,\varepsilon_\mu^\sigma\, dA_\rho^v(t)\right\}$$

$$= \varepsilon_t\left\{G_\sigma^\mu(t)\,H_{iv}^\rho(t)\,\varepsilon_\rho^\sigma\, dA_\mu^v(t)\right\}$$

$$= \varepsilon_t\left\{\left[G(t)\in H_i(t)\right]_v^\mu\, dA_\mu^v(t)\right\} = \varepsilon_t\left\{\left[G(t)\in H_i(t)\right]_0^0\right\}dt$$

Thus,

$$\varepsilon_t\left\{\left[H_i(t)\in H_i(t)\right]_0^0\right\}K^i(t)$$

$$= \varepsilon_t\left\{\left[G(t)\in H_i(t)\right]_0^0\right\},1\le i\le n \qquad (\beta_3)$$

This is the equation that determines

$$K^i(t) \in \mathcal{F}_{t]}^Y$$

Let $\rho(t)$ be the state at time t of the system plus bath. $\rho(t) \in \mathbb{B}_{t]}$.

We assume that

$$T_r\left(\rho(0)X(t)\right) = T_r\left(\rho(t)X(0)\right)$$

Assume that

$$dX(t) = L_\rho^\mu(t)X(t)\tilde{L}_\nu^\rho(t)dA_\mu^\nu(t)$$

where

$$L_\rho^\mu(t), \tilde{L}_\nu^\rho(t) \in \mathbb{B}_{t]}.$$

Thus,

$$\begin{aligned} G_\nu^\mu(t) &= L_\rho^\mu(t)X\,\tilde{L}_\nu^\rho(t) \\ G_0^0(t) &= L_\rho^0(t)X(t)\tilde{L}_0^\rho(t) \end{aligned}$$

Suppose we assume that

$$L_\rho^0(t), \tilde{L}_0^\rho(t), \varepsilon\mathcal{F}_{t]}^Y.$$

Then

$$\varepsilon_t\left(G_0^0(t)\right) = L_\rho^0(t)\varepsilon_t(X(t))\tilde{L}_0^\rho(t)$$

Suppose we further assume

$$H_{i0}^0(t) = \sum_k H_{ik}^{(1)}(t)X(t)H_{ik}^{(2)}(t) + R_i(t)$$

where

$$H_k^{(1)}(t), H_k^{(2)}(t), \underset{\Lambda}{\overset{R_i(t)}{\in}} \mathcal{F}_{t]}^Y$$

Then

$$\varepsilon_t\left(H_0^0(t)\right) = \sum_K H_k^{(1)}(t)\varepsilon_t(X(t))H_k^{(2)}(t) + R(t)$$

So our filtering eqn. assume the form

$$\begin{aligned} d\varepsilon_t(X(t)) &= L_\rho^0(t)\varepsilon_t(X(t))\tilde{L}_0^\rho(t)\,dt \\ &\quad + K^i(t)\left(dY_i(t) - \left(H_{ik}^{(1)}(t)\varepsilon_t(x(t))H_{ik}^{(2)}(t) - R_i(t)\,dt\right)\right. \end{aligned}$$

Summation over i and k is implied.

$$\begin{aligned} T_r\left(\rho(0)\varepsilon_t(X(t))\right) &= T_r\left(\varepsilon_t{}^*(\rho(0))X(t)\right) \\ &= T_r\left(\varepsilon_t{}^*(\rho(0))\varepsilon_t(X(t)\right) \end{aligned}$$

If we assume $\varepsilon_t^*(\rho(0)) = \rho(0)$, then

$$T_r\left(\rho(0)\varepsilon_t(X(t))\right) = T_r\left(\rho(0)X(t)\right)$$

We define

$$\tilde{\rho}(t) = \varepsilon_t^*(\rho(t))$$

Assume $\varepsilon_t^* = \varepsilon_t$. Then,

$$\begin{aligned} T_r\left(\tilde{\rho}(t)X(0)\right) &= T_r\left(\rho(t)\varepsilon_t(X(0))\right) \\ &= T_r\left(\varepsilon_t(\rho(t)X(0))\right) \end{aligned}$$

since $P(t) \in \text{Range}(\varepsilon_t) \subset \mathcal{F}_t^Y$

write

$$X(t) = T_t(X(0))$$

Then

$$\begin{aligned} \rho(t) &= T_t^*(\rho(0)) \\ \tilde{\rho}(t) &= \varepsilon_t^*(\rho(t)) = \varepsilon_t^* T_t^*(\rho(0)) \\ &= \left(T_t \circ \varepsilon_t\right)^*(\rho(0)) \end{aligned}$$

Thus,

$$T_r\left(\tilde{\rho}(t)X(0)\right) = T_r\left(\rho(0)T_t \circ \varepsilon_t(X(0)\right)$$

However, we prefer to define $\tilde{\rho}(t)$ by $\tilde{\rho}(t) \in \mathcal{F}_{t]}^Y$,

$$T_r\left(\tilde{\rho}(t)X(0)\right) = T_r(P(0)\varepsilon_t(X(t)))\,X(0) \in \mathcal{F}_t^Y,$$

i.e.

$$\begin{aligned} \tilde{\rho}(t) &= \varepsilon_t\left(\varepsilon_t \circ T_t\right)^*(\rho(0)) = \varepsilon_t T_t^* \varepsilon_t(\rho(0)) \\ &= \varepsilon_t T_t^*\left(\varepsilon_t^*(\rho(0))\right) \end{aligned}$$

Then,

$$T_r\left[d\tilde{\rho}(t).X(0)\right] = T_r\left[\rho(0).d\varepsilon_t(X(t))\right]$$

Assume $F(t) = 0$

$$\begin{aligned} &= T_r[\rho(0)\left\{L_\rho^0(t)\varepsilon_t(X(t))\tilde{L}_0^\rho(t)\,dt\right. \\ &\quad + K^i(t)(dY_i(t))\Big) - \Big(H_{ik}^{(1)}(t)\varepsilon_t(X(t))H_{ik}^{(2)}(t) \\ &\quad \left. + R_i(t)\Big)dt\right\}\Big] \\ &= T_r\left\{\tilde{L}_0^\rho(t)\rho(0)L_\rho^0(t)\varepsilon_t(X(t))\right\}dt \end{aligned}$$

$$+T_r\left[\rho(0)K^i(t)dY_i(t)\right]$$

$$-T_r\left\{H_{ik}^{(2)}(t)\rho(0)H_{ik}^{(1)}(t)\varepsilon_t(X(t))\right\}dt$$

$$+T_r\left[\rho(0)K^i(t)R_i(t)\right]dt$$

$$= T_r\left[\left(T_t*\varepsilon_t*\tilde{L}_0^{\rho}(t)\rho(0)L_{\rho}^0(t)\right)X(0)\right]dt$$

$$+T_r\left[\rho(0)K^i(t)dY_i(t)\right]$$

$$-T_r\left\{T_t*\varepsilon_t*\left(H_{ik}^{(2)}(t)\rho(0)H_{ik}^{(1)}(t)\right)X(0)\right\}$$

$$+T_r\left\{\rho(0)K^i(t)R_i(t)\right\}dt$$

Assume that $X(0)\in\mathcal{F}_{0]}^{Y}$ and $\varepsilon_t^* = \varepsilon_t$.

Then,

$$T_r\left[\left(T_t*\varepsilon_t*\tilde{L}_0^{\rho}(t)\rho(0)L_{\rho}^0(t)\right)X(0)\right]$$

$$= T_r\left[\varepsilon_t\circ T_t*\circ\varepsilon t\left(\tilde{L}_0^{\rho}(t)\rho(0)L_{\rho}^0(t)\right)\right]$$

$$X(0)]$$

We write $$\tilde{T}_t = \varepsilon_t\circ T_t*\circ\varepsilon_t$$

and assume that

$$\varepsilon_t T_t*\left(\tilde{L}_0^{\rho}(t)\right) = \hat{L}_0^{\rho}(t)$$

We assume

T_t* is an algebra homomorphism. Then

$$\varepsilon_t T_t*\varepsilon_t\left(\tilde{L}_0^{\rho}(t)\rho(0)L_{\rho}^0(t)\right)$$

$$= \varepsilon_t T_t*\left(\tilde{L}_0^{\rho}(t)\varepsilon_t(\rho(0))L_{\rho}^0(t)\right)$$

$$= \hat{L}_0^{\rho}(t)\tilde{T}_t(\rho(0))\hat{L}_{\rho}^0(t)$$

Assume $G_{\iota\upsilon}^{\mu}$ an linear functions of X (t) and $H_k^0(t), H_0^k(t)$ do not involve X (t), it follows that K^i (t) is a linear function of X (t). We write is as $K^i(t).K^i(X(t))$

$$= \varepsilon_t\{K_j^i(t)X(t)Q_j(t)\}$$

(sum over j is implies).

The map $X(0)\xrightarrow{T_t}X(t)$ is assumed to be a homomorphism from $\mathbb{B}_{0]}\to\mathbb{B}_{t]}$

where

$$\mathbb{B}_{oj]} = \mathbb{B}(\hbar_0), \mathbb{B}_{t]} = \mathbb{B}(\hbar_0) \otimes \mathbb{B}\left(\Gamma s(L^2[0,t])\right).$$

We have the filtering eqn.

$$d\varepsilon_t(X(t)) = \varepsilon_t\left(G_0^0(t)dt + K^i(t)(dY_i(t)) - \varepsilon_t(H_{i0}{}^0(t))\,dt\right)$$

where

$$dX(t) = G_\nu^\mu(t)\,dA_\mu^\nu(t),$$

$$dY_i(t) = H_{i\nu}^\mu(t)\,dA_\mu^\nu(t)$$

$G_\nu^\mu(t)$ are linear functions of $X(t)$ *i.e.* of the form $L_\rho^\mu(t)X(t)\tilde{L}_\nu^\rho(t)$. We write this as

$$G_\nu^\mu(t) = \tau_{\nu t}^\mu(X(t))$$

$$\tau_0^0 = \tau, \text{ so, } G_0^0(t) = \tau_t(X(t))$$

where $\tau_t : \mathbb{B}_{t]} \to \mathbb{B}_{t]}$ is linear.

Also,

$$dY_i(t) - \varepsilon_t\left(H_{i0}^0(t)\right)dt = H_{i\nu}^\mu(t)\,dA_\mu^\nu(t) + \left(H_{i0}^0(t)dt - \varepsilon_t\left(H_{i0}^0(t)dt\right)\right) - H_{i0}^0(t)\,dt$$

We have for $\rho(0) \in \mathbb{B}_0, \left(T_{rt} \equiv T_{r\mathbb{B}t]}\right)$

$$\begin{aligned}
T_{r\mathbb{B}t]}\{\rho(0)\,d\varepsilon_t(X(t))\} &\equiv T_{rt}\{\rho(0)(\varepsilon_{t+dt}(X(t+dt)) - \varepsilon_t(X(t)))\} \\
&= T_{rt}\{\varepsilon_{t+dt}{}^*(\rho(0))X(t+dt)\} \\
&\quad - T_{rt}\{\varepsilon_t{}^*\phi(\rho(0))X(t)\} \\
&= T_r\{T_{t+dt}{}^*\varepsilon_{t+dt}{}^*(\rho(0))X(0)\} \\
&\quad - T_{rt}\{T_t{}^*\varepsilon_t{}^*(\rho(0))X(0)\} \\
&= T_{rt}\{dT_t{}^*\circ\varepsilon_t{}^*(\rho(0)).X(0)\} \\
T_{rt}\left\{\rho(0)\varepsilon_t\left(G_0^0(t)\right)\right\} &= T_{rt}\left[T_t^*\tau_t{}^*\varepsilon_t{}^*(\rho(0))X(0)\right] \\
&= T_{rt}\left[T_t^*\tau_t{}^*\varepsilon_t{}^*(\rho(0))X(0)\right] \\
T_{rt}\{\rho(0)K^i(t)\,dY_i(t)\} &= T_{rt}\{\rho(0)K^i(t)\,dY_i(t)\} \\
&= T_{rt}\{\rho(0)K^i(t)H_{i\nu}^\mu(t)\}\,dA_\mu^\nu(t)
\end{aligned}$$

Equivalently,

$$dY_i(t) - H_{i0}^\circ(t)\,dt = \tilde{H}_{i\nu}^\mu(t)\,dA_\mu^\nu(t)$$

where

$$\tilde{H}^{\mu}_{i\nu}(t) = \begin{cases} H^{\mu}_{i\nu}(t) & \mu \neq 0 \text{ or } \nu \neq 0 \\ 0 & \mu = 0, \quad \nu = 0. \end{cases}$$

Thus,

$$\begin{aligned} T_{rt}\{\rho(0)K^i(t)\,dY_i(t)\} &= T_{rt}\{\rho(0)K^i(t)H^0_{i0}(t)\}dt \\ &\quad + T_{rt}\left\{\rho(0)K^i(t)\tilde{H}^{\mu}_{i\nu}(t)\right\}dA^{\nu}_{\mu}(t) \end{aligned}$$

Since $dY_i(t)$ commutes with $\mathbb{B}_{t]}$, we can regard $dY_i(t)$ as an element of $\mathbb{B}_{[t,t+dt]}$ *i.e.*

$$\mathbb{B}_{[0,t+dt]} \cong \mathbb{B}_{t]} \otimes B_{[t,t+dt]}$$

and then

$$\begin{aligned} T_{rt}\{\rho(0)K^i(t)\,dY_i(t)\} &= T_{rt}\{\rho(0)K^i(t)\}dY_i(t) \\ &= T_{rt}\{\rho(0)K^i(X(t))\}dY_i(t) \\ &= T_{rt}\{\varepsilon_t*(\rho(0))K^i(X(t))\}dY_i(t) \\ &= T_{rt}\{T_t*K_i*\varepsilon_t*(\rho(0))X(0)\}dY_i(t) \end{aligned}$$

Further,

$$\begin{aligned} & T_{rt}\{\rho(0)K^i(X(t))\varepsilon_t\left(H^0_{i0}(t)\right)\} \\ &= T_{rt}\{\varepsilon_t*(\rho(0))K^i(X(t))H^0_{i0}(t)\} \\ &\quad (\because\ K^i(X(t)) \in \mathcal{F}^Y_{t]}) \\ &= T_{rt}\{K^{i*}(\varepsilon_t*(\rho(0))H^0_{i0}(t))X(t)\} \\ &= T_{rt}\{T_t*K^{i*}(\varepsilon_t*(\rho(0))H^0_{i0}(t))X(0)\} \end{aligned}$$

Thus, we get

$$\begin{aligned} &= T_{rt}\{dT_t*\circ\varepsilon_t*(\rho(0))X(0)\} \\ &= T_{rt}\{T_t*\tau_t*\varepsilon_t(\rho(0))X(0)\}dt \\ &\quad + T_{rt}\{T_t*K_t*\varepsilon_t*(\rho(0))X(0)\}dY_i(t) \\ &\quad - T_{rt}\{T_t*K_t*(\varepsilon_t*(\rho(0))H^0_{i0}(t))X(0)\}dt \qquad (\gamma_1) \end{aligned}$$

Thus if

$$\tilde{\mathbb{B}}_{t]} = \Gamma s\left(L^2[0,t]\right)$$

so that

$$\mathbb{B}_{t]} = \mathbb{B}(\hbar_0) \otimes \tilde{\mathbb{B}}_{t]}$$

then writing

$$\tilde{T}_{rt} = T_r \tilde{\mathbb{B}}_{t]}$$

we get

$$\tilde{T}_{rt}\left\{d\left(T_t^* \circ \varepsilon_t^*\right)(\rho(0))\right\} = \tilde{T}_{rt}\{T_t * \tau_t * \varepsilon_t * (\rho(0))\}dt$$
$$+ \tilde{T}_{rt}\{T_t * K_i * \varepsilon_t * (\rho(0))\}dY_i(t)$$
$$- \tilde{T}_{rt}\{T_t * K_i * (\varepsilon_t * (\rho(0)) H_{i0}^0(t))\}dt \qquad (\gamma_2)$$

The partial trace $\tilde{T}_{rt}$ above acts on an operator in

$$\mathbb{B}(\hbar / 0) \otimes \tilde{\mathbb{B}}_{t]} \otimes \mathbb{B}_{(t,\, t+dt)}$$

to produce an operate in

$$\mathbb{B}(\hbar_0) \otimes \mathbb{B}_{[t,\, t+dt]}$$

In $(\gamma 2), \rho(0)$ could have been replaced by $\rho(0)Q$ for any $Q \in \mathcal{F}_{t]}^Y \subset \mathbb{B}_{t]}$. The result would still be valid. Consider

$$\tilde{T}_{rt}\{d(T_t * \circ \varepsilon_t *)(\rho(0)Q)X(0)\}$$

This equals

$$\tilde{T}_{rt}\{T_{t+dt} * (\varepsilon_{t+dt} * (\rho(0))Q)X(0)\}$$
$$- \tilde{T}_{rt}\{T_t * (\varepsilon_t * (\rho(0))\,Q)X(0)\}$$
$$= \tilde{T}_{rt}\{T_{t+dt} * \circ \varepsilon_{t+dt} * (\rho(0))\,QX(0)\}$$
$$= -\tilde{T}_{rt}\{T_t * \circ \varepsilon_t * (\rho(0))\,QX(0)\} \qquad (\delta_1)$$

(since by the non-demolition assumption, Q is not objected by $T_{t+dt}*$ or T_t* which given the $X(t)$ dynamics $\left(\mathcal{F}_t^Y \subset \mathbb{B}'_{t]}\, \mathcal{F}_{t]}^Y \subset \mathcal{F}_{t+dt}^Y\right)$

$$\subset \mathbb{B}'_{t+dt]}$$

The first term in (δ_1) is

$$= \tilde{T}_{rt}\{\varepsilon_{t+dt} * \circ T_{t+dt} * \circ \varepsilon_{t+dt} * (\rho(0))Q\}X(0)$$
$$= \tilde{T}_{rt}\{\tilde{\rho}(t+dt)Q\}X(0)$$

and the second term is

$$\tilde{T}_{rt}\{\varepsilon_t * \circ T_t * \circ \varepsilon_t * (\rho(0))Q\}X(0)\,.$$
$$= \tilde{T}_{rt}(\tilde{\rho}(t)Q)X(0)$$

Thus,

$$= \tilde{T}_{rt}\{d(T_t * \circ \varepsilon_t *)(\rho(0)Q)X(0)\}$$
$$= \tilde{T}_{rt}\{\tilde{\rho}(t+dt)Q\}X(0) \qquad (\delta_1)$$

$$= \tilde{T}_{rt}\{T_t * \tau_t * \varepsilon_t * (\rho(0)Q)\}$$

$$= \tilde{T}_{rt}\{T_t * \tau_t * (\varepsilon_t * (\rho(0))Q)\}$$

$$= \tilde{T}_{rt}\{\varepsilon_t * T_t * \tau_t * \varepsilon_t * (\rho(0))Q\} \qquad (\delta_3)$$

(assuming that τ_t* has no effect on $Q(\tau_t$* is built of operators in $\mathbb{B}_{t]}$ which commute with $Q)) = \tilde{T}_{rt}\{\varepsilon_t * T_t * \tau_t * T_t(\tilde{\rho}(t))Q\}$

assuming T_t to be unitary, *i.e.* $T_t* = T_t^{-1}$.

Let $\qquad T_t * \tau_t * T_t = \tilde{\tau}_t *$

Thus (δ_3) equals

$$\tilde{T}_{rt}\{\varepsilon_t * \tau_t * (\tilde{\rho}(t))Q\}$$

Likewise defining

$$\tilde{K}_{it} * = T_t * K_i * T_t, \text{ we get}$$

$$\tilde{T}_{rt}\{T_t * K_i * \varepsilon_t * (\rho(0)Q)\} dY_i(t)$$

$$= \tilde{T}_{rt}\{\varepsilon_t * \tilde{K}_{it} * (\tilde{\rho}(t))Q\} dY_i(t)$$

and finally, $\tilde{T}_{rt}\{T_t * \tilde{K}_i * \varepsilon_t * (\rho(0)Q)H_{i0}^{\circ}(t)\}$

$$= \tilde{T}_{rt}\{\varepsilon_t * \tilde{K}_{it} * (\tilde{\rho}(t))H_{i0}^{\circ}(t)Q\}$$

Thus the quantum filtering for the state $\tilde{\rho}(t)$ is

$$d\varepsilon_t * T_t * \varepsilon_t * (\rho(0))$$

$$\equiv d\tilde{\rho}(t) = d(\varepsilon_t * \tilde{\rho}(t))$$

$$= \varepsilon_t * \tilde{\tau}_t * (\tilde{\rho}(t))dt + \varepsilon_t * \tilde{K}_{it} * (\tilde{\rho}(t))dY_i(t)$$

$$- \varepsilon_t * K_{it} * (\tilde{\rho}(t))H_{i0}^{\circ}(t)dt \qquad (\rho)$$

Note:

$$\tilde{\rho}(t) = T_t * \circ \varepsilon_t * (\rho(0))$$

$$\tilde{\rho}(t) = \varepsilon_t * (\tilde{\rho}(t)) = \varepsilon_t * T_t * \varepsilon_t * (\rho(0)).$$

New notation for quantum stochastic calculus

$$\Lambda_0^i(t) = A_i(t), i \geq 1,$$

$$\Lambda_i^0(t) = A_i * (t), i \geq 1, \; \Lambda_j^i(t) = \Lambda_{|e_j\rangle\langle e_i|}(t)$$

$$\Lambda_0^{\circ}(t) = t,$$

$$d\Lambda_j^{\circ} \cdot d\Lambda_j^0 = \delta_j^i dt, \quad i, j \geq 1,$$

$$d\Lambda_k^i \cdot d\Lambda_m^j = \delta_m^i d\Lambda_k^i, k \geq 0, \quad i, m \geq 1.$$

$$d\Lambda_k^0 \cdot d\Lambda_m^j = 0, j, k, m \geq 0,$$

$$d\Lambda_j^i \cdot d\Lambda_0^k = 0, i, j, k \geq 0.$$

Consider for $X \in \mathbb{B}(\hbar_0)$, the *qsde*

$$dj_t(X) = j_t\left(Q_j^i(x)\right)d\Lambda_i^j(t)$$

(sum over $i, j \geq 0$).

Define $$\varepsilon_j^i = \begin{cases} \delta_j^i & i, j \geq 1, \\ 0, & i = 0 \text{ or } j = 0 \end{cases}.$$

Then $d\Lambda_j^i d\Lambda_m^k = \varepsilon_m^i d\Lambda_j^k, \quad i, j, k, m \geq 0.$

Suppose j_t is a homomorphism from $\mathbb{B}_0 = \mathbb{B}\hbar_0$

$$\to \mathbb{B}_{t]} = \mathbb{B}(\hbar_0) \otimes \Gamma s(\mathcal{H}_{t]}).$$

Quantum I to's formula implies

$$\begin{aligned} dj_t(XY) &= dj_t(X).dj_t(Y) \\ &= j_t\left(\theta_j^i(X)\right) j_t\left(\theta_m^k(Y)\right) d\Lambda_i^j d\Lambda_k^m \\ &= j_t\left(\theta_j^i(X)\theta_m^k(Y)\right)\varepsilon_k^j\, d\Lambda_i^m \end{aligned}$$

On the one hand and on the other,

$$dj_t(XY) = j_t\left(\theta_m^i(XY)\right)d\Lambda_i^m$$

Thus, $\quad \varepsilon_k^j\, \theta_j^i(X)\theta_m^k(Y) = \theta_m^i(XY)$

These are called the structure equations and are necessary for $j_t.\mathbb{B}_0 \to \mathbb{B}_{t]}$ to be a homomorphism $\forall t$.

$$j_t(X) = X + \int_0^t j_s\left(\theta_j^i(X)\right)d\Lambda_i^j(s)$$

$$\langle fe(u), j_t(X)ge(v)\rangle = \langle f, Xg\rangle\langle e(u), e(v)\rangle + \int_0^1\left\langle fe(u), j_s\left(\theta_j^i(X)\right)ge(v)\right\rangle$$

$$d\mu_i^j(u, v, s)$$

Note that

$$\begin{aligned} \left\langle e(u), d\Lambda_i^j(s)e(v)\right\rangle &= \left\langle e(u), d\Lambda_{|e_i\rangle\langle e_j|}^{(s)}\, e(v)\right\rangle \\ &= \bar{u}_i(s)v_j(s)ds\langle e(u), e(v)\rangle, i, j \geq 1, \end{aligned}$$

$$\left\langle e(u), d\Lambda_0^i(s)e(v)\right\rangle = \langle e(u), dA_i(s)e(v)\rangle$$
$$= v_i(s)\,ds\,\langle e(u), e(v)\rangle$$
$$\left\langle e(u), d\Lambda_i^\circ(s)e(v)\right\rangle = \langle dA_i(s)e(u), e(v)\rangle = \bar{u}_i(s)\,ds.\langle e(u), e(v)\rangle .$$

so

$$d\mu_i^j(u,v,s) = \begin{cases} \bar{u}_i(s)\,ds & \text{if } j=0,\ i\geq 1 \\ v_i(s)\,ds & \text{if } i=0,\ j\geq 1 \\ \bar{u}_i(s)v_j(s)\,ds & \text{if } i,j\geq 1 \\ ds & \text{if } i=j=0. \end{cases}$$

$$\frac{1}{T}\int_0^T \langle fe(u), j_t(X)ge(v)\rangle\,dt$$
$$= \langle f, Xg\rangle\langle e(u), e(v)\rangle$$
$$+\frac{1}{T}\int_0^T (T-S)\left\langle fe(u), j_s\left(\theta_j^i(X)\right)ge(v)\right\rangle\, d\mu_i^j(u,v,s)$$
$$= \xrightarrow[T\to\infty]{} \langle f, Xg\rangle\langle e(u), e(v)\rangle$$
$$+\int_0^\infty \left\langle fe(u), j_s\left(\theta_j^i(X)\right)ge(v)\right\rangle\, d\mu_i^j(u,v,s)$$

provided.

$$\int_0^\infty \left|\left\langle fe(u), j_s\left(\theta_j^i(X)\right)ge(v)\right\rangle\right| \left|d\mu_i^j(u,v,s)\right| < \infty .$$

Other identities. For $\tau \geq 0$

$$j_t(X)j_{t+\tau}(Y) = \left[X + \int_0^t j_s\left(\theta_j^i(X)\right)d\Lambda_i^j(s)\right]$$
$$\left[Y + \int_0^{t+\tau} j_s\left(\theta_j^i(Y)\right)d\Lambda_i^j(s)\right]$$
$$= XY + \int_0^{t+\tau} Xj_s\left(\theta_j^i(Y)\right)d\Lambda_i^j(s)$$
$$+\int_0^t j_s\left(\theta_j^i(X)\right)Y\,d\Lambda_i^j(s)$$
$$+\int_0^t\int_0^{t+\tau} j_s\left(\theta_j^i(X)\right)j_{s2}\left(\theta_m^k(Y)\right)d\Lambda_i^j(s_1)\,d\Lambda_k^m(s_2)$$

Now, $\langle fe(u), j_t(X)j_{t+\tau}(Y)ge(v)\rangle$

$$= \langle f, XYg\rangle\langle e(u), e(v)\rangle$$
$$+\int_0^{t+\tau}\left\langle fe(u), Xj_s\left(\theta_j^i(Y)\right)ge(v)\right\rangle\, d\mu_i^j(u,v,s)$$

$$+\int_0^t \left\langle fe(u), j_s\left(\theta_j^i(X)\right)Yge(v)\right\rangle \, d\mu_i^j(u,v,s)$$

KRP Quantum Stochastic Calculus.

$h_0 \otimes \Gamma s\left(L^2(\mathbb{R}_+)\right)$ is the Fock space.

$$dU = \left[L\,dA^* - L^*\,dA - \left(iH + \frac{1}{2}L^*L\right)dt\right]U\,,$$

$$dW_f = \left[f\,dA^* - \bar{f}\,dA - \frac{1}{2}(f)^2\,dt\right]W_f$$

$$T_r(\rho_t) = T_r\{(\rho_0 \otimes |e(0)\rangle\langle e(0)|)U(t)^*W_f(t)U(t)\}$$

$$\rho_t = T_{r2}\{(\rho_0 \otimes |e(0)\rangle\langle e(0)|)U^*(t)W_f(t)U(t)\}$$

$$\begin{aligned}
d(U^*W_fU) &= dU^*W_fU + U^*W_f dU \\
&\quad + U^*dW_fU + U + U^*dW_f dU \; + dU^*W_f dU \\
&= U^*\left\{\left(L^*\,dA - L\,dA^* + \left(iH - \frac{1}{2}L^*L\right)dt\right)W_f\right. \\
&\quad + W_f\left(L\,dA^* - L^*\,dA - \left(iH + \frac{1}{2}L^*L\right)dt\right) \\
&\quad \left.\left(f\,dA^* - \bar{f}\,dA - \frac{1}{2}|f|^2\,dt\right)W_f\right\}U \\
&\quad + U^*(L^*\,dA - L\,dA^*) - \left(f\,dA^* - \bar{f}\,dA\right)W_fU \\
&\quad + U^*\left(f\,dA^* - \bar{f}\,dA\right)\, W_f\,(L\,dA^* - L^*\,dA)U \\
&\quad + U^*(L^*\,dA - L\,dA^*)\,W_f\,(L\,dA^* - L\,dA^*)U \\
&= U^*\{L^*W_f - W_fL^* - \bar{f}W_f\}U\,dA \\
&\quad + U^*\{W_fL - LW_f + f\,W_f\}U\,dA^* \\
&\quad + U^*\left\{-\frac{1}{2}\left(L^*LW_f + W_fL^*L\right) + i\left[H, W_f\right]\right. \\
&\quad - \frac{1}{2}|f|^2\,W_f + L^*f\,W_f \\
&\quad \left. - \bar{f}W_fL + L^*W_fL\right\}U\,dt
\end{aligned}$$

Since

$$dA|0\rangle = 0, \; \langle 0|\, dA^* = 0, \text{ we get}$$

$$\frac{d\rho_t}{dt} = T_{r2}\left\{\left(\rho_0 \otimes |e(0)\rangle\langle e(0)|\right)\right\}$$

$$U^*\{i(HW_f - W_fH)$$

$$-\frac{1}{2}\left(L^*LW_f + W_fL^*L\right) + L^*W_fL$$

$$-\frac{1}{2}|f|^2 W_f + L^*fW_f - \bar{f}W_f\Big\}U \; (f \in L^2(\mathbb{R}_+))$$

$$= \frac{d}{dt}T_r(\rho_t) = T_r\left(\frac{d\rho_t}{dt}\right)$$

$$= T_r\left\{U\left(\rho_0 \otimes |e(0)\rangle\langle e(0)|\right)U^*\right.$$

$$\times\left(i\left[H, W_f\right] - \frac{1}{2}\left(L^*LW_f + W_fL^*L - 2L^*W_fL\right)\right)$$

$$\left.-\frac{1}{2}|f|^2 W_f + L^*fW_f - \bar{f}W_fL\right)\Big\}$$

Note that W_f acts completely is $\Gamma s\left(L^2(\mathbb{R}_+)\right)$ *i.e.* on $\hbar_0$ it is the identity. Hence

$$\left[H, W_f\right] = 0, \left(L, W_f\right) = 0, \left[L^*, W_f\right] = 0$$

Hence,

$$T_r\left(\frac{d\rho_t}{dt}\right) = T_r\left(\rho_0 \otimes |e(0)\rangle\langle e(0)|\right)$$

$$U^*\left(-\frac{1}{2}|f|^2 W_f + \left(L^*f - \bar{f}L\right)W_f\right)U$$

$$= T_r\left\{U\left(\rho_0 \otimes |e(0)\rangle\langle e(0)|\right)U^*\left(fL^* - \bar{f}L - \frac{1}{2}|f|^2\right)W_f\right\}$$

($[f, L] = 0$, since $f, \bar{f}$ act in $\Gamma s\left(L^2(\mathbb{R}_+)\right)$ $[\bar{f}, L] = 0$, which L^*L act in $\hbar_0$)

Equivalently,

$$\frac{d}{dt}T_r\left\{U\left(\rho_0 \otimes |e(0)\rangle\langle e(0)|\right)U^*W_f\right\}$$

$$= T_r\left\{U\left(\rho_0 \otimes |e(0)\rangle\langle e(0)|\right)U^*\left(fL^* - \bar{f}L - \frac{1}{2}|f|^2\right)W_f\right\}$$

$$\left(fL^* - \bar{f}L - \frac{1}{2}|f|^2\right)W_f\Big\}$$

$$fL^* - \bar{f}L - \frac{1}{2}|f|^2$$

acts in $\hbar_0$ ($f \equiv f(t)$ is simply multiplying. Hence it can be visualized as acts in $\hbar_0$ (as multiplication by the scalar $f(t)$ or in $\Gamma s\left(L^2(\mathbb{R}_+)\right)$. Thus it commutes with W_f. Hence we have equivalently,

$$\frac{d}{dt}T_r\left\{(\rho_0 \otimes |e(0)\rangle\langle e(0)|)U^*W_f U\right\} = T_r\left((\rho_0 \otimes |e(0)\rangle\langle e(0)|)\right.$$

$$\left.\left(fL^* - \bar{f}L - \frac{1}{2}|f|^2\right)U^*W_f U\right\} = T_r\left\{\left(fL^* - \bar{f}L - \frac{1}{2}|f|^2\right)\rho_0\right.$$

$$\left.\times T_{\gamma 2}\left(|e(0)\rangle\langle e(0)|U^*W_f U\right)\right\}$$

$$= T_{\gamma 1}\left\{\rho_0\left(T_{r2}\left(|e(0)\rangle\langle e(0)|U^*W_f U\right)\right.\right.$$

$$\left.\left.\left(fL^* - \bar{f}L - \frac{1}{2}|f|^2\right)\right)\right\}$$

or

$$\frac{d}{dt}T_{r1}\left\{\rho_0 T_{r2}\left(|e(0)\rangle\langle e(0)|U^*W_f U\right)\right\} = T_{r1}\left\{\rho_0\left(T_{r2}\left(|e(0)\rangle\langle e(0)|U^*W_f U\right)\right.\right.$$

$$\left.\left.\left(fL^* - \bar{f}L - \frac{1}{2}\right)|f|^2\right\}\right\}$$

Thus writing

$$X(t) = T_{r2}\left\{|e(0)\rangle\langle e(0)|U^*W_f(t)U(t)\right\}$$

we get

$$\frac{dX(t)}{dt} = X(t)\left(f(t)L^* - \bar{f}(t)L - \frac{1}{2}|f(t)|^2\right)$$

$$= f(t)X(t)\left(L^* - \bar{f}(t)X(t)L - \frac{1}{2}|f(t)|^2 X(t)\right).$$

$$T_r\left[\frac{d\rho_t}{dt}\right] = \frac{d}{dt}T_r\left[(\rho_0 \otimes |e(0)\rangle\langle e(0)|)U^*W_f(t)U(t)\right]$$

$$= \frac{d}{dt}T_r\left[U(t)(\rho_0 \otimes |e(0)\rangle\langle e(0)|)U^*(t)W_f(t)\right]$$

$$= T_r\left[U\left(\rho_0 \otimes |e(0)\rangle\langle e(0)|\right)U^*\right.$$

$$\left.\left(\theta\left(W_f\right) + fL^*W_f - \bar{f}W_f L - \frac{1}{2}|f|^2 W_f\right)\right]$$

$$= T_r\left\{\left[\theta'\left(U\left(\rho_0 \otimes |e(0)\rangle\langle e(0)|\right)U^*\right)\right.\right.$$

$$+\bar{f}L\left(U\left(\rho_0 \otimes |e(0)\rangle\langle e(0)|\right)U^*\right)$$
$$+f\left(U\left(\rho_0 \otimes |e(0)\rangle\langle e(0)|\right)U^*\right)L^*$$
$$\left.-\frac{1}{2}|f|^2\, U\left(\rho_0 \otimes |e(0)\rangle\langle e(0)|\right)U^*\right]W_f\Big\}$$
$$= T_r\left[\theta'(\rho_t)+\bar{f}L\rho_t + f\rho_t L^* - \frac{1}{2}|f|^2\rho_t\right]$$

since $\qquad [L^* W_f] = 0, [L, W_f] = 0$

where $\qquad \theta'(\rho) = -i[H.\rho] - \frac{1}{2}(\rho L^* L + L^* L\rho - 2L\rho L^*)$

The Dirac δ-function and its applications to quantum mechanics and signal analysis.

*The δ-function is a generalised function *i.e.* a distribution defined by

$$\langle \delta, \psi \rangle = \psi(0)$$

i.e.

$$\int_{\mathbb{R}} \delta(x)\psi(x)\,dx = \psi(0)$$

where ψ is any Rapidly decreasing infinitely differential function:

$$\psi \in C^\infty(\mathbb{R}),\ x^m \frac{d^m\psi(x)}{dx^n} \to 0$$

as $\qquad = |x| \to \infty, m, n = 0, 1, 2, \ldots$

*Derivatives of the δ-function are defined by

$$\left\langle \delta^{(n)}, \psi \right\rangle = (-1)^n \psi^{(n)}_{(0)}, n = 0, 1, 2, \ldots$$

$\psi \in \xi(\mathbb{R}), \xi(\mathbb{R})$ is the space of rapidly decreasing function on $\mathbb{R}$.

*Some useful properties:

$$(a) \int_{\mathbb{R}} \delta^{(n)}(x)\psi(y-x)\,dx = \psi^{(n)}(y)$$

Follows by the definition of $\delta^{(n)}$ and the fact that $X \to \psi(y-x)$ is a rapidly decreasing function if ψ is $\forall y \in \mathbb{R}$.

The formula for $\left\langle \delta^{(n)}, \psi \right\rangle$ can be visualized using the integration by parts formula (formally)

$$\int_{\mathbb{R}} \delta^{(n)}(x)\psi(x)\,dx = \psi(x)\delta^{(n-1)}(x)\Big|_{-\infty}^{+\infty} - \int_{\mathbb{R}} \psi'(x)\delta^{(n-1)}(x)\,dx$$

$$= -\left\langle \delta^{(n-1)}, \psi^{(n)} \right\rangle$$

$$= \left\langle \delta^{(n-2)}, \psi^{(2)} \right\rangle = ... = (-1)^n \left\langle \delta, \psi^{(n)} \right\rangle$$

$$= (-1)^n \psi^{(n)}(0) .$$

(*b*)

$$\int_{\mathbb{R}} \psi(x)\delta^{(n)}(x)\varphi(x)\,dx = (-1)^n \int_{\mathbb{R}} \delta(x)\frac{d^n}{dx^n}(\varphi(x)\psi(x))\,dx$$

$$= (-1)^n \sum_{k=0}^{n} \binom{n}{k} \varphi^{(k)}(0)\psi^{(n-k)}(0)$$

$$i = \left\langle \delta^{(n)}\psi, \varphi \right\rangle = \left\langle \delta^{(n)}, \psi\varphi \right\rangle$$

$$= (-1)^n \sum_{k=0}^{n} \binom{n}{k} \psi^{(n-k)}(0) \left\langle \delta^{(n)}, \varphi \right\rangle (-1)^k$$

So
$$\psi\delta^{(n)} = \sum_{k=0}^{n} \binom{n}{k} (-1)^{n-k} \psi^{(n-k)}(0)\,\delta(k)$$

*Grains function and δ-functions. Define the N-dimensional δ-function.

$$\delta(\underline{x}) = \prod_{\alpha=1}^{N} \delta(x\alpha) = \underline{x}(x\alpha)_{\alpha=1}^{N}$$

Let $\mathcal{L}$ be a linear partial differentiator on $C^\infty(\mathbb{R}^N)$. Let $S(\underline{x})$ be a function on $\mathbb{R}^N$ (the source function). We wish to solve for $\psi(\underline{x})$ the equation

$$\mathcal{L}\psi(x) = S(\underline{x}), \ \underline{x} \in B \tag{1}$$

with boundary condition $\psi(\underline{x}) = 0, \underline{x} \in \partial B$.

Suppose we solve

$$L_x G(\underline{x}, \underline{x}') = \delta(\underline{x} - \underline{x}'), \quad \underline{xx}' \in B$$

With the boundary condition

$$G(\underline{x}, \underline{x}') = 0, \qquad \underline{x} \in \partial B, \underline{x}' \in B$$

Thus,

$$\mathcal{L}_x \int_B G(\underline{x}, \underline{x}')\, S(\underline{x}')\, d^N x' = \int_B (\delta_x G(\underline{x}, \underline{x}'))\, S(\underline{x}')\, d^N x'$$

$$= \int_B (\delta(\underline{x}, \underline{x}'))\, S(\underline{x}')\, d^N x'$$

$$= S(\underline{x}), \ x \in B.$$

Further, the boundary condition

$$\int_B G(\underline{x},\underline{x}')\,S(\underline{x}')\,d^N x' = 0,\ \underline{x}\in\partial B$$

is satisfied since

$$G(\underline{x},\underline{x}') = 0,\ \underline{x}\in\partial B, \underline{x}'\in B\,.$$

Thus the solution to (1) with the vanishing boundary condition is

$$\psi(\underline{x}) = \int_B G(\underline{x},\underline{x}')S(\underline{x}')d^N x'\,,\ \underline{x}\in B$$

Here B is an open subset of $\mathbb{R}^N$ with boundary ∂B.

Example:

$$\Delta\frac{1}{|\underline{r}|} = -4\pi\delta(\underline{r}) \quad \underline{r}\in\mathbb{R}^3\,.$$

$$\Delta = \nabla^2 = \frac{\partial^2}{\partial x^2}+\frac{\partial^2}{\partial y^2}+\frac{\partial^2}{\partial z^2},\ \delta(r) = \delta(x)\delta(y)\delta(z).$$

To verify this identity, we take a test function $\psi(r,\theta,\varphi)$. Then

$$\int_{\mathbb{R}^3}\psi(r,\theta,\varphi)\Delta\frac{1}{r}d^3r = \int_{\mathbb{R}^3}\frac{1}{r}\Delta\psi(r)d^3r \qquad \text{(integration by parts)}$$

$$= \int_{\substack{0\le r\le\infty,\\ 0\le\theta\le\pi,\\ 0\le\varphi\le 2\pi}}\frac{1}{r}(\Delta\psi(\underline{r}))r^2\sin\theta\,dr\,d\theta\,d\varphi$$

$$= \int(\Delta\psi(\underline{r}))r\sin\theta\,dr\,d\theta\,d\varphi$$

$$= \int\left[\frac{1}{r}\frac{\partial^2}{\partial r^2}(r\psi)+\frac{1}{r^2\sin\theta}\frac{\partial}{\partial\theta}\left(\sin\theta\frac{\partial\psi}{\partial\theta}\right)\right.$$
$$\left.+\frac{1}{r^2\sin^2\theta}\frac{\partial^2\psi}{\partial\varphi^2}\right]r\sin\theta\,dr\,d\theta\,d\varphi$$

$$= \int\sin\theta\frac{\partial^2}{\partial r^2}(r\psi)\,dr\,d\theta\,d\varphi$$
$$+\int\frac{1}{r}\frac{\partial}{\partial\theta}\left(\sin\theta\frac{\partial\psi}{\partial\theta}\right)dr\,d\theta\,d\varphi$$
$$+\int\frac{1}{r\sin\theta}\frac{\partial^2\psi}{\partial\varphi^2}dr\,d\theta\,d\varphi$$

The second and third integrals vanish since

$$\int_0^{\pi} \frac{\partial}{\partial\theta}\left(\sin\theta\frac{\partial\psi}{\partial\theta}\right)d\theta = \sin\theta\frac{\partial\psi}{\partial\theta}\bigg|_{\theta=0}^{\theta=\pi} = 0,$$

and $$\int_0^{2\pi}\frac{\partial^2\psi}{\partial\varphi^2}d\varphi = \frac{\partial\psi}{\partial\varphi}\bigg|_{\varphi=0}^{\varphi=2\pi} = 0$$

Thus, $$\int_{\mathbb{R}^3}\frac{1}{r}(\Delta\psi(\underline{r}))d^3r = \int\sin\theta\frac{\partial^2}{\partial r^2}(r\psi(r,\theta,\varphi))\,dr\,d\theta\,d\varphi$$

$$= \int\sin\theta\,d\theta\,d\varphi\int_0^{\infty}\frac{\partial^2}{\partial r^2}(r\psi)\,dr$$

$$= \int\sin\theta\,d\theta\,d\varphi\frac{\partial}{\partial r}(r\psi(r,\theta\varphi))\bigg|_{r=0}^{r=\infty}$$

$$= \int\sin\theta\,d\theta\,d\varphi\left(r\frac{\partial\psi}{\partial r}+\psi\right)\bigg|_{r=0}^{r=\infty}$$

$$= -4\pi\psi(\theta)\text{ since } r\frac{\partial\psi}{\partial r}\bigg|_{r=0} = 0$$

and $$\int\sin\theta\,d\theta\,d\varphi = 4\pi.$$

Thus, $$\Delta\frac{1}{r} = -4\pi\delta(\underline{r})$$

$$\psi\to 0,\ r\to\infty.$$

Hence if $$\Delta\psi(\underline{r}) = S(\underline{r}),\qquad \underline{r}\in\mathbb{R}^3$$

and $$\psi(\underline{r})\to 0 \text{ as } |\underline{r}|\to\infty\text{, then}$$

$$\psi(\underline{r}) = -\int_{\mathbb{R}^3}\frac{S(\underline{r}')}{4\pi|\underline{r}-\underline{r}'|}d^3r'$$

which is Coulomb's law.

Another example:

Likewise

$$(\Delta+k^2)\frac{\exp(-ikr)}{4\pi r} = -\delta(\underline{r})$$

To see this, we compute

$$I = \int\psi(\underline{r})(\Delta+k^2)G(r)\,d^3r$$

where $$G(\underline{r}) = \frac{\exp(-ikr)}{4\pi r}$$

Integration by parts gives

$$\int \psi(\underline{r})\Delta G(\underline{r})d^3r = \int (\Delta\psi(\underline{r}''))G(\underline{r})d^3r$$

$$= \int G(\underline{r})\left(\frac{1}{r}\frac{\partial^2}{\partial r^2}(r\psi) + \frac{1}{r^2\sin\theta}\frac{\partial}{\partial\theta}\left(\sin\theta\frac{\partial\psi}{\partial\theta}\right)\right.$$

$$\left.+\frac{1}{r^2\sin^2\theta}\frac{\partial^2\psi}{\partial\varphi^2}\right)r^2\sin\theta\, dr\, d\theta\, d\varphi$$

$$= \int rG(\underline{r})\frac{\partial^2}{\partial r^2}(r\psi)\sin\theta\, dr\, d\theta\, d\varphi$$

(by the previous argument, the angular part is zero)

Now,

$$\int_0^\infty rG(\underline{r})\frac{\partial^2}{\partial r^2}(r\psi)\,dr = \int_0^\infty \frac{\exp(-jkr)}{4\pi}\frac{\partial^2}{\partial r^2}(r\psi)\,dr$$

$$= \left.\frac{\exp(-ikr)}{4\pi}\frac{\partial}{\partial r}(r\psi)\right|_{r=0}^{r=\infty}$$

$$+\frac{ik}{4\pi}\int_0^\infty \exp(-ikr)\frac{\partial}{\partial r}(r\psi)\,dr$$

$$= -\frac{1}{4\pi}\psi(0) + \left.\frac{ik}{4\psi}\exp(-ikr)(r\psi)\right|_{r=0}^{r=\infty}$$

$$= -\frac{k^2}{4\pi}\int_0^\infty \exp(-ikr)r\psi\, dr$$

Generalized Cartan matrices. The homomorphism theorem from V. Kac,

$$\tilde{\alpha}_i^{VT}\tilde{\alpha}_j = \alpha_i^{VT}\alpha_j \quad 1 \le i,\ j \le n.$$

Infinite dimensional lie algebras.

$$\sum_{j=1}^{2n-l}\tilde{\alpha}_i^{VT}\tilde{\alpha}_j S_{jk} = \alpha_i^{VT}\alpha_k \quad 1 \le i \le n,\ 1 \le k \le 2n-l$$

$$\tilde{\alpha}_j = \sum_{k=1}^{2n-l} T_{kj}\alpha_k = T(\alpha_j),\ 1 \le j \le 2n-l$$

$$\sum_{j=1}^{2n-l}\tilde{\alpha}_i^{VT}\,T(\alpha_j)\,S_{jk} = \alpha_i^{VT}\alpha_k$$

$$1 \le i \le n,\ 1 \le k \le 2n - l.$$

$$\tilde{\alpha}_i^{V^T} TS(\alpha_k) = \alpha_i^{V^T} \alpha_k, 1 \le i \le n$$

$$1 \le k \le 2n - l$$

$$\{\alpha_k\}_{k=1}^{2n-1}$$

$\simeq$ standard basis for $\mathbb{C}^{2n-l}$.

$$\left[(TS)^T(\tilde{\alpha}_i^V)\right]^T \alpha_k = \alpha_i^{V^T} \alpha_{k, \substack{1 \le i \le n \\ 1 \le k \le 2n-l}}$$

$$(TS)^T\left(\tilde{\alpha}_i^V\right) = \alpha_i^V 1 \le i \le n$$

$$T(\alpha_j) = \tilde{\alpha}_j, 1 \le j \le 2n - l$$

$$T^T(\tilde{\alpha}_i^V) = S^{-T}\left(\alpha_i^V\right), 1 \le i \le n.$$

$$S = \begin{bmatrix} I_n & \underline{\underline{X}} \\ \underline{\underline{O}} & \underline{\underline{Y}} \end{bmatrix} \in \mathbb{C}^{2n-l \times 2n-l}.$$

$$\begin{bmatrix} A_1 & B \\ A_2 & D \end{bmatrix}\begin{bmatrix} I_n & X \\ O & \underline{\underline{Y}} \end{bmatrix} = \begin{bmatrix} A_1 & A_1X + BY \\ A_2 & A_2X + DY \end{bmatrix}$$

$$= \begin{bmatrix} A_1 & O \\ A_2 & I \end{bmatrix} \in \mathbb{C}^{n \times 2n-l}.$$

$$\Rightarrow \qquad Y = D^{-1}(A_2X - I)$$

$$A_1X - BD^{-1}(A_2X - I) = 0$$

$$\Rightarrow \qquad X = -(A_1 - BD^{-1}A_2)^{-1}BD^{-1}.$$

$$\therefore \qquad S(\alpha_j) = \alpha_j,\ 1 \le j \le n.$$

$$S^T = \begin{bmatrix} I_n & O \\ X^T & Y^T \end{bmatrix} \cdot S^{-T} = \begin{bmatrix} P & Q \\ R & S_0 \end{bmatrix}.$$

$$\begin{bmatrix} I_n & O \\ X^T & Y^T \end{bmatrix}\begin{bmatrix} P & Q \\ R & S_0 \end{bmatrix} = \begin{bmatrix} I_n & O \\ O & I_{n-1} \end{bmatrix}$$

$$\Rightarrow \qquad P = I_n,\ Q = 0,\ X^T + Y^TR = 0,$$

$$Y^TS_0 = I_{n-l} \Rightarrow P = I_n,\ Q = 0,$$

$$S_0 = Y^{-T},\ R = -Y^{-T}X^T = -(XY^{-1})^T.$$

So

$$S^{-T} = \begin{bmatrix} I_n & O \\ R & S_0 \end{bmatrix}$$

$$S^{-T}\left(\alpha_i^V\right) = \begin{bmatrix} I_n & O \\ R & S_0 \end{bmatrix}\left(\left(\alpha_{ij}^V\right)\right)_{j=1}^{2n-l} = \begin{pmatrix} \left(\left(\alpha_{ij}^V\right)\right)_{j=1}^{n} \\ R\left(\left(\alpha_{ij}^V\right)\right)_{j=1}^{n} \end{pmatrix}$$

$\because$ $$\alpha_{ij}^V = 0, n+1 \le j \le 2n-l$$

$$\left(\alpha_{ij}^V = \underline{\alpha}_i^{V^T} \underline{\alpha}_j\right)$$

$\therefore$ $$\underline{\alpha}_j^T S^{-T} \alpha_i^V = \left(S^{-T}\alpha_j^V\right)_j = \alpha_{ij}^V,\ 1 \le i,\ j \le n.$$

So we get

$$T(\alpha_i) = \tilde{\alpha}_i, 1 \le i \le n,$$

$$T^T\left(\tilde{\alpha}_i^V\right)_j = T^T\left(\tilde{\alpha}_i^V\right)_j,\ 1 \le i \le n, 1 \le j \le n.$$

KRP Quantum Stchastic calculus.

$$\tilde{a}(u) f(\sigma) = \int_0^\infty \bar{u}(s)(-1)^{\#(\sigma\cap[0,s])} f(\sigma \cup \{s\}) ds$$

$$\tilde{a}^*(u) f(\sigma) = \sum_{s\in\sigma} (-1)^{\#(\sigma\cap[0,s])} u(s) f(\sigma \setminus \{s\}).$$

$$\tilde{a}^*(u)\tilde{a}(u) f(\sigma) = \sum_{s\in\sigma} (-1)^{\#(\sigma\cap[0,s])} v(s)\,(\tilde{a}(u)f)(\sigma\setminus\{s\})$$

$$= \sum_{s\in\sigma} (-1)^{\#(\sigma\cap[0,s])} v(s)$$

$$\int_0^\infty \bar{u}(s')(-1)^{\#(\sigma\setminus\{s\}\cap[0,s'])\setminus\{s\}} f(\sigma\setminus\{s\}\cup\{s'\}) ds'$$

$$\tilde{a}(u)\tilde{a}^*(v) f(\sigma) = \int_0^\infty \bar{u}(s)(-1)^{\#(\sigma\cap[0,s])}\left(\tilde{a}^*(v)f\right)(\sigma\cup\{s\}) ds$$

$$= \int_0^\infty \bar{u}(s)(-1)^{\#(\sigma\cap[0,s])}$$

$$\sum_{s'\in\sigma\cup\{s\}} (-1)^{\#((\sigma\cup\{s\})\cap[0,s'])} v(s') f(\sigma\cup\{s\}/\{s'\}) ds$$

$$= \int_0^\infty \bar{u}(s)(-1)^{\#(\sigma\cap[0,s])}$$

$$\sum_{s' \in \sigma} (-1)^{\#(\sigma \cap [0, s']) \cup \{s\} \cap [0,s']} v(s') f\left((\sigma/\{s'\}) \cup \{s\}\right) ds$$

$$+ \int_0^\infty \bar{u}(s)(-1)^{\#(\sigma \cap [0, s])} (-1)^{\#\left((\sigma \cup \{s\}) \cap [0, s]\right)} v(s) f(\sigma) ds$$

(**Note**: σ is a finite set and hence, its Lebesgue measure is zero.)

Now,

$$\begin{aligned}
&\#(\sigma \cap [0, s]) + \#\left((\sigma \cup \{s\}) \cap [0, s]\right) \\
&\qquad = \#(\sigma \cap [0, s]) + \#\left((\sigma \cap \{0, s\}) \cup [s]\right) \\
&\qquad = 2\#(\sigma \cap [0, s]) \text{ if } s \notin \sigma \\
&\qquad = 2\#^{+1}(\sigma \cap [0, s]) \text{ if } s \in \sigma
\end{aligned}$$

Thus, $$\tilde{\alpha}(u)\tilde{\alpha}^*(v) f(\sigma) = \left[\sum_{s' \in \sigma} \int_0^\infty \bar{u}(s)(-1)^{\#(\sigma \cap [0,s]) + \#\left((\sigma \cap [0,s']) \cup (\{s\} \cap [0,s'])\right)} f\left((\sigma/\{s'\}) \cup \{s\}\right) ds\right] v(s') - f(\sigma)\left\{\int_0^\infty \bar{u}(s) v(s) ds\right\}$$

On the other hand,

$$\tilde{a}^*(v)\tilde{a}(u) f(\sigma) =$$

$$\sum_{s' \in \sigma} (-1)^{\#(\sigma \cap [0, s'])} v(s') \int_0^\infty \bar{u}(s)(-1)^{\#\left((\sigma \setminus \{s'\}) \cap [0, s]\right)} f\left((\sigma/\{s'\}) \cup \{s\}\right)$$

$$= \sum_{s' \in \sigma} \left(\int \bar{u}(s)(-1)^{\#\left((\sigma \cap [0, s'])/\{s'\}\right) + \#(\sigma \cap [0,s'])} f\left((\sigma/\{s'\}) \cup \{s\}\right) ds\right) v(s')$$

Now, if $s \in [0, s'], S \notin \sigma$,

Then $\#\left((\sigma \cap [0, s]) \cup (\{s\} \cap [0, s'])\right) = \#(\sigma \cap [0, s']) + 1$

While if $s \in [0, s'], s \notin \sigma$, then

$$\#\left((\sigma \cap [0, s']) \cup (\{s\} \cap [0, s'])\right) = \#(\sigma \cap [0, s'])$$

On the other hand, if

$s \in [0, s'], s \notin \sigma$ then,

$$\#\left((\sigma\cap[0,s])\setminus\{s'\}\right) = \#(\sigma\cap[0,s])$$

While if $s\in[0,s'], s\notin\sigma$, then

$$\#\left((\sigma\cap[0,s])\setminus\{s'\}\right) = \#\left(\sigma\cap[0,s]\right)-1$$

Provided $s'\in\sigma$. These facts show that

$$\left(\tilde{a}(u)\tilde{a}^*(v)+\tilde{a}^*(v)\tilde{a}^*(u)\right)f(\sigma) = -\langle u,v\rangle f(\sigma)$$

which is the CAR.

❖❖❖❖❖

[10] A Problem Related to the Generator of Brownian Motion:

(Revuz and Yor) [p 192] TPT

Ex. 3.13

$$f(B_t) = P_t f(0)+\int_0^t\left(P_{t-s}f\right)'\left(B_s\right)dB_s \tag{1}$$

$$P_t = \exp\left(\frac{t}{2}\Delta\right)\Delta = D^2, D=\frac{d}{dx}.$$

$$\left(P_{t-s}f\right)'\left(B_s\right) = \exp\left(\frac{t-s}{2},\Delta\right)f'\left(B_s\right)$$

$$= \exp\left(\left(\frac{t-s}{2}\right)D^2\right)Df\left(B_s\right)$$

$$P_t f(0) = \exp\left(t\frac{D^2}{2}\right)f(0)$$

$$d\,(\text{L.H.S. of 1}) = f'\left(B_t\right)dB_t+\frac{1}{2}f''\left(B_t\right)dt$$

$$d\,(\text{R.H.S. of 1}) = \frac{1}{2}\Delta\exp\left(\frac{t}{2}\Delta\right)f(0)\,dt$$

$$+f'\left(B_t\right)dB_t+\frac{dt}{2}\int_0^t\left(\Delta P_{t-s}f\right)'\left(B_s\right)dB_s$$

So we've to prove that

$$\Delta\exp\left(\frac{t\Delta}{2}\right)f(0)+\int_0^t\left(\Delta P_{t-s}f\right)'\left(B_s\right)dB_s$$

$$= f''(B_t)\equiv\Delta f(B_t) \tag{2}$$

$$\int_0^t\left(\Delta P_{t-s}f\right)'\left(B_s\right)dB_s = \int_0^t\left(D\Delta e^{\frac{(t-s)\Delta}{2}}f\left(B_s\right)dB_s\right.$$

Now,

$$d_s\left(\Delta e^{\frac{(t-s)\Delta}{2}} f\right)(B_s) = \left(D\Delta e^{\frac{(t-s)\Delta}{2}} f\right)(B_s)dB_s$$

$$+\frac{1}{2}\Delta^2\left(e^{\frac{(t-s)\Delta}{2}} f\right)(B_s)ds - \frac{\Delta^2}{2}e^{\frac{(t-s)\Delta}{2}} f(B_s)ds$$

$$= D\Delta e^{\frac{(t-s)\Delta}{2}} f(B_s)ds$$

Thus,

$$\int_0^t\left(D\Delta e^{\frac{(t-s)}{2}} f(B_s)\right)dB_s = \int_0^t ds\Delta e^{\frac{(t-s)\Delta}{2}} f(B_s)$$

$$= \Delta f(B_t) - \Delta e^{t\Delta/2} f(0)$$

❖❖❖❖❖

[11] A Problem Related to Poisson Random Fields:

(Revuz and Yor, Ex. 1.19)

$$M_\varphi \equiv \left|\mathbb{E}\exp\left(\int_E \varphi(x)\, N^*(dx)\right)\right.$$

$$= \int_0^\infty e^{-s}ds\left\{\mathbb{E}\left[\exp\left(\int_E \varphi(x) N(dx)\right)\middle| S = s\right]\right.$$

$$\left.\times\mathbb{E}\left[\exp(\varphi(x))\,|\, S = s\right]\right\}$$

$$= \int_0^\infty e^{-s}ds\left(\exp\left(\int\left(e^{\varphi(x)}-1\right)g_s\lambda(dx)\right)\right.$$

$$\left.\times\int\exp(\varphi(x))\, f_s\,\lambda(dx)\right)$$

$$\left(g_s\lambda(dx) = \left(\int_s^\infty f_t(x)\,dt\right)\lambda(dx)\right)$$

$$= \int_{[0,\,\infty]}^\infty e^{-s}ds\,\psi(s)e^{\varphi(x)} f_s(x)\lambda(dx)$$

(where $\psi(s) = \exp\left(\int\left(e^{\varphi(x)}-1\right)g_s\lambda(dx)\right)$

Now,

$$\frac{d}{ds}\psi(s) = \frac{d}{ds}\exp\left(\int_E\left(e^{\varphi(x)}-1\right)g_s(x)\,d\lambda(x)\right)$$

$$= \left(-\int\left(e^{\varphi(x)}-1\right)f_s(x)\,d\lambda(x)\right)\psi(s)$$

$$= \left(1-\int e^{\varphi(x)}f_s(x)\,d\lambda(x)\right)\psi(s)$$

so

$$= \left(\int e^{\varphi(x)}f_s(x)\,d\lambda(x)\right)\psi(s) = \psi(s)-\psi'(s)$$

and hence,

$$M_\varphi = \int_0^\infty e^{-s}(\psi(s)-\psi'(s''))\,ds$$

$$= -\int_0^\infty \frac{d}{ds}\left(e^{-s}\psi(s)\right)ds = \psi(0)$$

$$= \exp\left(\int\left(e^{\varphi(x)}-1\right)g_0(x)\,d\lambda(x)\right)$$

QED

Remark from V. Kac

$$\left\langle \beta, \alpha_i^V \right\rangle \geq 0 \ \forall i \in R$$

$$\beta = \sum_{i\in R} m_i\alpha_i + \sum_{i\notin R} m_i\alpha_i = \beta' + \sum_{i\notin R} m_i\alpha_i$$

$i \notin \text{Supp}\,(\alpha)\, j \in \text{Supp}\,(\alpha) \Rightarrow \left\langle \alpha_i, \alpha_j \right\rangle$

$$\left\langle \beta', \alpha_j^v \right\rangle \geq -\left\langle \sum_{i\notin R} m_i\alpha_i, \alpha_j^v \right\rangle, \ j \in R$$

$TPT \quad \left\langle \sum_{i\notin R} m_i\alpha_i, \sigma_j^v \right\rangle \leq 0 \forall j \in R$

$$i \notin R,\ j \in R \Rightarrow i \neq j \Rightarrow \left\langle \alpha_i, \alpha_s^v \right\rangle \leq 0$$

Thus, $\left\langle \sum_{i\notin R} m_i\alpha_i, \alpha_j^v \right\rangle \geq 0$. Hence

$$\left\langle \beta', \alpha_j^v \right\rangle \geq 0,\ j \in R.$$

$TPT \left\langle \beta', \alpha_j^v \right\rangle > 0$ for some $j \in R$.

Suppose $\left\langle \beta', \alpha_j^v \right\rangle = 0 \ \forall j \in R$.

Then $\left\langle \beta, \alpha_j^v \right\rangle = \left\langle \sum_{i\notin R} m_i\alpha_i, \alpha_j^v \right\rangle \leq 0 \quad \forall j \in R$.

Now if $i \in \text{Supp}(\beta) \cap R^c$ then $m_i > 0$ and $\left\langle \alpha_i, \alpha_j^v \right\rangle < 0, \ \forall j \in R$ (since $i \neq j$). Thus if supp $(\beta) \cap R^c \neq \phi$,

Then $\left\langle \sum_{i \notin R} m_i \alpha_i, \alpha_j^v \right\rangle < 0$

and we get $\left\langle \beta, \alpha_j^v \right\rangle < 0, \ \forall j \in R$

which conditions $(\hat{s}.34)$. Thus,

$\left\langle \beta', \alpha_j^V \right\rangle > 0$ for some $j \in R$.

It remains to prove that

$$\text{Supp}\ (\beta) \cap R^c \neq \phi,$$

But supp (β) = supp (α). So we must show that

$$\text{supp}\ (\alpha) = \cap R^c \neq \phi$$

i.e.

$$\text{Supp}\ (\alpha) \cap P \neq \phi$$

i.e $P \neq \phi$ which is true.

(Revuz and Yor)

$$A_t = t - g_t.$$

$$\mathbb{E}\left[f(|B_t|) \mid \check{\mathcal{F}}_t \right] = A_t^{-1} \int_0^\infty \exp\left(\frac{-y^2}{2A_t} \right) y\, f(y)\, dy$$

$$\check{\mathcal{F}}_t = \mathcal{F}g_t.\ \text{Let}\ f(y) = y^2.\ \check{\mathcal{F}}_t \subset \mathcal{F}_t$$

$$\mathbb{E}\left[B_t^2 \middle| \mathcal{F}_s \right] = t - s + B^2{}_s$$

$$\begin{aligned}
\mathbb{E}\left[B_t^2 \middle| \check{\mathcal{F}}_s \right] &= \mathbb{E}\left[t - s + B_s^2 \middle| \check{\mathcal{F}}_s \right] \\
&= t - s + \mathbb{E}\left[B_s^2 \middle| \check{\mathcal{F}}_s \right] \\
&= t - s + A_s^{-1} \int_0^\infty \exp\left(\frac{-y^2}{2A_s} \right) y^3\, dy \\
&= t - s + \int_0^\infty \exp(-\xi)\, d\xi (2A_s \xi) \\
&= t - s + 2\, A_s
\end{aligned}$$

So, $\mathbb{E}\left[t - B_t^2 \middle| \overset{\vee}{\mathcal{F}}_s\right] = s - 2\,A_s,\ t > s$. It follows that $\{t - 2\,A_t,\ t \geq 0\}$ is a $\overset{\vee}{\mathcal{F}}_t$ – Martingale

Now take $f(y) = |y|$. Then, for $t \geq s, \mathbb{E}\,[|\,B_r\,||\,\mathcal{F}_s] = \mathbb{E}\,[L_t \,|\, \mathcal{F}_s] - L_s + |\,B_s\,|$

Since

$$d\,|B_t| = \text{Sqn}\,(B_t)\,dB_t + dL_{t]}$$

and hence

$$|\,B_t\,| - |\,B_s\,| = \int_s^t \text{Sqn}\,(B_\tau)\,dB_\tau + L_t - L_s$$

Thus $\quad \mathbb{E}\,[|\,B_t] - L_t \,|\, \mathcal{F}_s] = |\,B_s\,|\,L_s$

$$\therefore \mathbb{E}\left[|\,B_t\,| - L_t \,|\, \overset{\vee}{\mathcal{F}}_s\right] - L_s = \mathbb{E}\left[|\,B_s\,|\, \overset{\vee}{\mathcal{F}}_s\right] - L_s$$

since L_s is $\overset{\vee}{\mathcal{F}}_s$ measurable. It $\{u - g_u, u \leq s\}$ is known, then $\{g_u, u \leq s\}$ is known, *i.e.* all the zeros of B before s are known and hence L_s is known. Now,

$$|\,E\left[|\,B_s\,|\,\widehat{\mathcal{F}}_s\right] = A_s^{-1}\int_0^\infty \exp\left(-y^2 / 2A_s\right)y^2 dy$$

$$= \left(\frac{1}{2A_s}\right)\int_{-\infty}^\infty \exp\left(\frac{-y^2}{2A_s}\right)y^2 dy$$

$$= \frac{1}{2A_s} \times \sqrt{2\pi\,A_s} \times A_s = \sqrt{\frac{\pi}{2}A_s}$$

Then,

$$\mathbb{E}\left[|\,B_t\,| - L_t \,|\, \widehat{\mathcal{F}}_s\right] = \sqrt{\frac{\pi}{2}A_s - L_s} = \sqrt{\frac{\pi}{2}(s - g_s)} - L_s$$

and hence $\sqrt{\frac{\pi}{2}(t - g_t)} - L_t, t \geq 0$ is an $\widehat{\mathcal{F}}_t$ Martingale.

Let $z \sim N(0, \sigma^2)$.

$$|\,\mathbb{E}\,f(|\,z\,|) = \int_{\mathbb{R}} \frac{1}{\sigma\sqrt{2\pi}} \exp\left(\frac{-z^2}{2\sigma^2}\right) f(|\,z\,|)\,dz$$

$$= \frac{2}{\sigma\sqrt{2\pi}}\int_0^\infty \exp\left(-\frac{z^2}{2\sigma^2}\right) f(z)\,dz$$

Let $B(t)$, $t \geq 0$ be B.M. Calculate

$$|\,\mathbb{E}\,[f(B(t))\,|\,T >]$$

where $T = \min\{s \geq 0 | B(s) = 0\}$. Assume $B(0) = x > 0$.

$$\{T > t\} = \{\min_{0 \le s \le t} B(s) > 0\}$$

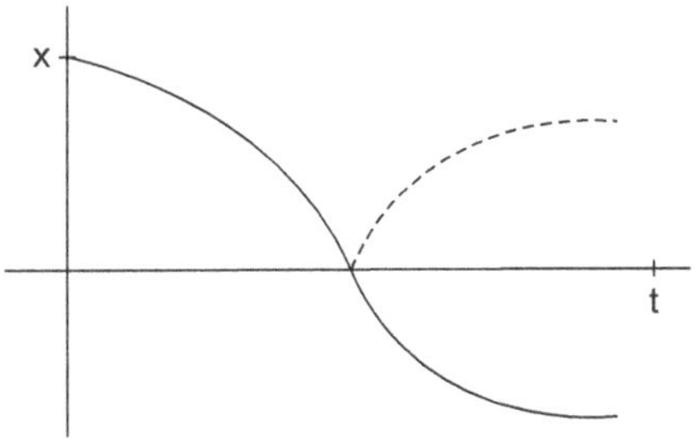

By the reflection principle,

$$\{B(t) < \xi, T < t\} = \{\tilde{B}(t) > -\xi\}$$

where $\tilde{B}$ is B reflected at T.

So,

$$P_x\{B(t) < \xi, T > t\} = P_x\{B(t) < \xi\} - P_x\{B(t) > -\xi\}$$

$$\frac{1}{\sqrt{2\pi t}}\int_{-\infty}^{\xi} \exp\left(-\frac{(y-x)^2}{2t}\right) dy$$

$$-\frac{1}{\sqrt{2\pi t}}\int_{-\xi}^{\infty} \exp\left(-\frac{(y-x)^2}{2t}\right) dy$$

So $\quad P_x\{B(t) \in d\xi, T > t\}$

$$= \frac{1}{\sqrt{2\pi t}}\left(\exp\left(-\frac{(\xi-x)^2}{2t}\right) + \exp\left(-\frac{(\xi+x)^2}{2t}\right)\right) d\xi,$$

$$\xi > 0.$$

Thus,

$$|\mathbb{E}\{f(B(t)|T > t\} = \frac{1}{\sqrt{2\pi}}\int_{-x/\sqrt{t}}^{\infty} f\left(x + \eta\sqrt{t}\right) + \exp\left(-\frac{\eta^2}{2}\right) d\eta$$

$$+\frac{1}{\sqrt{2\pi}}\int_{x/\sqrt{t}}^{\infty} f\left(-x + \eta\sqrt{t}\right)\exp\left(-\frac{\eta^2}{2}\right) d\eta$$

Time changed Brownian motion:

Let X_t, $t \ge 0$ be a continuous semi-martingale. Then $X_t = M_t + A_t$ where M_t is a Martingale (continuous) and A_t is a process of bounded variation. We can write $A_t = A_t^{(1)} - A_t^{(2)}$ where $A_t^{(1)}, A_t^{(2)}$ an non-decreasing continuous processes. We have

$$d\langle X, X\rangle_t = (dX_t)^2 = (dM_t)^2 = d\langle M, M\rangle_t$$

or

$$\langle X, X\rangle_t = \langle M, M\rangle_t \, \forall t \geq 0.$$

Let $$T_t = \min\{s \geq 0 \mid \langle M, M\rangle_s \geq t\}$$

Then $B_t = M_{T_t}, t \geq 0$ is a Brownian motion up to time $\langle M, M\rangle_\infty$. Then $M_t = B_{\langle M, M\rangle_t}$. The proof of this result depends on Levy's characterization of Brownian motion. $\{T_t\}_{t\geq 0}$ is an increasing family of stop times for the M process:

$$\{T_t \leq \alpha\} = \{\langle M, M\rangle_s = t \text{ for some } s \leq \alpha\} \in \mathcal{F}_\alpha^M.$$

Thus by Doob's optional stopping theorem, $M_{T_t}, t \geq 0$ is a martingale (continuous). Further,

$$(d M_{T_t})^2 = d\langle M, M\rangle_{T_t} = dt$$

Hence $\{M_{T_t}\}_{t\geq 0}$ is Brownian motion.

Generalised Sudershan Lindblad equation.

$$\begin{aligned}
dU(t) &= [-iH(t)dt + L_1(t)dA(t) - L_2{}^*(t)dA^*(t) \\
&\quad + S(t)d\Lambda(t)]U(t) \\
dA dA^* &= dt,\ dA^* dA = 0,\ dA d\Lambda = dA, \\
d\Lambda dA^* &= dA^*,\ d\Lambda dA = 0,\ dA^* d\Lambda = 0 \\
d\Lambda d\Lambda &= d\Lambda. \\
X(t) &= U^*(t)XU(t), X \in \mathcal{L}(\mathfrak{h}), \\
&\quad A(t), A^*(t), \Lambda(t) \in \mathcal{L}(\Gamma s(\mathcal{H})) \\
&\quad L(t), L^*(t), S(t) \in \mathcal{L}(\mathfrak{h}). \\
dX(t) &= dU^* XU + U^* XdU + dU^* XdU \\
&= U^*((iH^* dt + L_1{}^* dA^* - L_2 dA + S^* d\Lambda)X \\
&\quad + X(-iH^* dt + L_1 dA - L_2{}^* dA^* - Sd\Lambda))U \\
&\quad + U^*(iH^* dt + L_1{}^* dA^* - L_2 dA + S^* d\Lambda)X \\
&\quad (-iHdt + L_1 dA - L_2{}^* dA^* + Sd\Lambda)U \\
&= U^*\{i(H^* X - XH)\}U\,dt \\
&\quad + U^*(L_1{}^* X - XL_2{}^*)U\,dA^* \\
&\quad + U^*(XL_1 - L_2 X)UdA + U^*(S^* X + XS)Ud\Lambda \\
&\quad + U^* L_2 X L_2{}^* U\,dt - U^* L_2 X\,SU\,dA \\
&\quad - U^* S^* XL_2^* U\,dA^* + U^* S^* X\,SU\,d\Lambda
\end{aligned} \tag{1}$$

In particular, taking $X = I$ gives

$$\begin{aligned} d(U^*U) &= [iU^*(H^*-H)U + U^*L_2L_2^*U]dt \\ &\quad + U^*(L_1^*-L_2^*-S^*XL_2^*)U\,dA^* \\ &\quad + U^*(L_1-L_2-L_2S)U\,dA \\ &\quad + U^*(S^*+S+S^*S)U\,d\Lambda \end{aligned}$$

For unitarity of $U(t)\forall t$, we require $d(U^*U) = 0$

i.e.

$$\begin{aligned} L_2L_2^* &= (H-H^*), \\ S^*+S+S^*S &= 0, \\ L_1-L_2-L_2S &= 0 \end{aligned}$$

Assume that these conditions are satisfied.

Let $\rho(0) = \rho_s(0)\otimes|\varphi(u)\rangle\langle\varphi(u)|$

where $\rho_s(0)\in\mathcal{L}(\mathcal{h}), u\in\mathcal{H} = L^2(\mathbb{R}_+)$,

$$|\varphi(u)\rangle = \exp\left(-\frac{\|u\|^2}{2}\right)|e(u)\rangle.$$

Then
$$\begin{aligned} dT_r[\rho(0)X(t)] &= T_r[\rho(0)dX(t)] \\ &= T_r[U\rho(0)U^*(i(H^*X-XH)+L_2XL_2^*)]dt \\ &\quad + T_r[U\rho(0)U^*(L_1^*X-XL_2^*-S^*XL_2^*)]\bar{u}(t)dt \\ &\quad + T_r[U\rho(0)U^*(XL_1-L_2X-L_2XS)]u(t)dt \\ &\quad + T_r[U\rho(0)U^*(S^*X+XS+S^*XS)]\,|u(t)|^2\,dt \end{aligned}$$

Let
$$\begin{aligned} \rho_s(t) &= T_{r2}[U(t)\rho(0)U^*(t)] \\ &\equiv T_{r2}\left[U(t)\left(\rho_s(0)\otimes|\varphi(u)\rangle\langle\varphi(u)|\right)U^*(t)\right] \end{aligned}$$

Then,

$$\begin{aligned} \frac{d}{dt}T_r[\rho_s(t)X] &= \frac{d}{dt}T_r[\rho(0)X(t)] \\ &= T_r[\rho_s(t)(i(H^*X-XH)+L_2XL_2^*)] \\ &\quad + T_r[\rho_s(t)(L_1^*X-XL_2^*-S^*XL_2^*)]\bar{u}(t) \\ &\quad + T_r[\rho_s(t)(XL_1-L_2X-L_2XS)]u(t) \\ &\quad + T_r[\rho_s(t)(S^*X+XS+S^*XS+S^*XS)]\,|u(t)|^2 \end{aligned}$$

Since $X\in\mathcal{L}(\mathcal{h})$ is arbitrary Hermitian, it follows that

$$\begin{aligned}\rho_s'(t) &= -i(H(t)\rho_s(t)) - \rho_s(t)H^*(t)\\ &+ L_2^*(t)\rho_s(t)L_2(t)\\ &+ \bar{u}(t)[\rho_s(t)L_1^*(t) - L_2^*(t)\rho_s(t) - L_2^*(t)\rho_s(t)S^*(t)]\\ &+ u(t)[L_1(t)\rho_s(t) - \rho_s(t)L_2(t) - S(t)\rho_s(t)L_2(t)]\\ &+ \rho_s(t)S^*(t) + S(t)\rho_s(t) + S(t)\rho_s(t)S^*(t)\end{aligned}$$

Now let $H = H_0 + iP$, where H_0, P are both Hermitian. Then

$$L_2L_2^* = i(H - H^*) = -2P$$

or

$$P = -\frac{1}{2}L_2L_2^*$$

Also, $$L_1 = L_2(1+S).$$

Thus, we get

$$\begin{aligned}\rho_s'(t) &= -i[H_0, \rho_s(t)] - \frac{1}{2}(L_2L_2^*\rho_s(t) + \rho_s(t)L_2L_2^*)\\ &- 2L_2^*\rho_s(t)(L_2(t)) + \bar{u}(t)[\rho_s(t)\end{aligned}$$

❖❖❖❖❖

[12] Relativistic Kinematics of Rigid Bodies:

Let a top after time t be subject to the rotation $R(t) = R(\varphi(t), \theta(t), \psi(t))$ (Eular angles). Then its Lagrangian in a gravitational field has the form

$$\mathcal{L}(\varphi, \theta, \psi, \varphi', \theta', \psi') = \int_B F(R(t)\xi, R'(t)\xi)\, d^3\xi$$

where B is the volume of the body occupied at time $t = 0$.

Example 1: In non-relativistic kinematics,

$$F(\xi, v) = \frac{1}{2}\rho\|\underline{v}\|^2 - \rho g\xi^3$$

Then $$\mathcal{L} = \int_B \left[\frac{1}{2}\rho\|R'(t)\xi\|^2 - \rho_g[R'(t)]_{3k}\xi^k\right]d^3\xi$$

$$= \frac{1}{2}T_r\left(R'(t)\underline{\underline{I}}\,R'(t)^T\right) - [R'(t)]_{3k}\, mgd^k$$

When $$\underline{\underline{I}} = \rho\int \xi\xi^T d^3\xi$$

and $$d_k = \int_B \xi^k d^3\xi \Big/ \int_B d^3\xi^{\,1m} = \int_B \rho d^3\xi$$

Example 2: For special relativistic approximation,

$$F(\xi, \underline{v}) = -\rho c^2\sqrt{1 - v^2/c^2} - \rho g\xi^3$$

and then

$$\mathcal{L} = -\rho c^2 \int_B \left(1 - \|R'(t)\xi\|^2 / c^2\right)^{1/2} d^3\xi - mgR(t)_{3k} d^k$$

Example 3: For a general relativistic approximation,

$$F(\xi, v) = -\rho_0 \left[g_{00}(\xi) + 2g_{or}(\xi)v^r + g_{rs}(\xi)v^r v^s \right]^{1/2}$$

$$\approx -\rho_0 g_{00}(\xi)^{1/2} \left[1 + \frac{g_{or}(\xi)v^r}{g_{00}(\xi)} \right.$$

$$\left. + \frac{1}{2}\left[-\frac{g_{or}(\xi)g_{os}(\xi)}{g_{00}{}^2} + \frac{g_{rs}(\xi)}{g_{00}(\xi)} \right] v^r v^s \right]$$

Writing

$$\alpha_0(\xi) = -\rho_0 \sqrt{g_{00}(\xi)},$$

$$\beta_r(\xi) = -\rho_0 g_{or}(\xi) / \sqrt{g_{00}(\xi)},$$

$$\gamma_{rs}(\xi) = \rho_0 \sqrt{g_{00}(\xi)} \frac{(g_{or}(\xi)g_{os}(\xi) - g_{00}(\xi)g_{rs}(\xi))}{g_{00}^2(\xi)}$$

we get

$$F(\xi, v) \approx \alpha_0(\underline{\xi}) + \beta_r(\xi)v^r + \frac{1}{2}\gamma_{rs}(\xi)v^r v^s$$

and

$$\mathcal{L} = \int_B \alpha_0\left(\underline{\underline{R}}(t)\underline{\xi}\right) d^3\xi + \int_B \underline{\beta}^T\left(\underline{\underline{R}}(t)\underline{\xi}\right) \underline{\underline{R}}'(t)\xi d^3\underline{\xi}$$

$$+ \frac{1}{2}\int_B \underline{\xi}^T \underline{\underline{R}}'(t)^T \Gamma(\underline{\underline{R}}(t)\underline{\xi}) \underline{\underline{R}}'(t) \underline{\xi} d^3\xi$$

In general, suppose we have a Taylor expansion

$$F(\underline{\xi}, \underline{v}) = \sum_{r,s=0}^{\infty} \left\langle \underline{a}(r,s),\ \underline{\xi}^{\otimes r} \otimes \underline{v}^{\otimes s} \right\rangle$$

where

$$\underline{a}(r,s) \in \mathbb{R}^{3r+s}$$

Then

$$\mathcal{L} = \sum_{r,s=0}^{\infty} \left\langle \underline{a}(r,s), \left(\underline{\underline{R}}(t)^{\otimes r} \otimes \underline{\underline{R}}'(t)^{\otimes s}\right) \int_B \underline{\xi}^{\otimes(r+s)} d^3\xi \right\rangle$$

$$= \sum_{r,s=0}^{\infty} \left\langle \underline{a}(r,s), \left(R(t)^{\otimes r} \otimes R'(t)^{\otimes s}\right) \underline{I}(r,s) \right\rangle$$

where

$$\underline{I}(r,s) = \int_B \underline{\xi}^{\otimes(r+s)} d^3\xi \qquad \in \qquad \mathbb{R}^{3r+s}$$

Problem from Revuz of Continuous Martingale and Brownian motion.

$\{B(t),\, t \geq 0\}$ standard Brownian motion.

$$L(t) \simeq \text{local time of } B(.) \text{ at } 0.$$

$$T_a = \min\{t \geq 0 | B(t) = a\},\ a > 0.$$

$$L(t) = \int_0^t \delta(B(s))\, ds$$

Calculate the law of $L(T_a)$. Let

$$u_\lambda(a, x) = \mathbb{E}[\exp(-\lambda L(T_a)) | B(0) = x]$$

$$= \mathbb{E}\left[\exp(-\lambda \int_0^{T_a} \delta(\beta(s))\, ds \mid B(0) = x\right]$$

Then

$$u_\lambda(a, x) = (1 - \lambda\delta(x)h) \cdot E\left[\exp\left(-\lambda \int_h^{T_a} \delta(B(s))\, ds\right)\right.$$

$$|B(0) = x] + O(h^2)$$

$$= (1 - \lambda\delta(x)h) \,|\, \mathbb{E}[u_\lambda(a, B(h)) \,|\, B(0) = x] + O(h^2),$$
$$(h \to 0+)$$

$$= (1 - \lambda h\delta(x))\left(u_\lambda(a, x) + \frac{h}{2} u_{\lambda, xx}(a, x)\right) + 0(h)$$

Thus,

$$\frac{1}{2} u_{\lambda, xx}(a, x) - \lambda u_\lambda(a, 0)\delta(x) = 0 \tag{1}$$

So

$$u_\lambda(a, x) = \begin{cases} \alpha_1 x + \beta_1, & x > 0 \\ \alpha_2 x + \beta_2, & x < 0. \end{cases}$$

Also integrating (1) from 0 – to 0 + gives

$$\frac{1}{2}(u_{\lambda, x}(a, 0+) - u_{\lambda, x}(a, 0-1) - \lambda u_\lambda(a, 0) = 0 \tag{2}$$

Further integrating (1) from $x = -\infty$ to x gives

$$\frac{1}{2} u_{\lambda, x}(a, x) - \lambda u_x(a, 0) O(x) = 0 \tag{3}$$

Integrating (3) from 0 – to 0 + gives

$$u_\lambda(a, 0+) = u_\lambda(a, 0-) \equiv u_\lambda(a, 0) \tag{4}$$

From (2),

$$\frac{1}{2}(\alpha_1 - \alpha_2) - \lambda u_\lambda(a, 0) = 0 \tag{5}$$

From (4), $\beta_1 = \beta_2 \equiv \beta$ say.

$$u_\lambda(a,a) = 0 \Rightarrow \alpha_1 a + \beta_1 = 0.$$

Thus, $\quad b = -\alpha_1 a,$

$$u_\lambda(a,0) = \beta$$

Thus $\quad \alpha_1 - \alpha_2 = 2\lambda\beta$

or $\quad \alpha_2 = \alpha_1 - 2\lambda\beta$

or $\quad \alpha_2 = \alpha_1 + 2\lambda\alpha_1\alpha$

So, $\quad \alpha_1 = \dfrac{\alpha_2}{1+2\lambda a}$

Now, $\quad u_{\lambda,x}(a,x) = \alpha_2, x<0$

$$\beta = -\alpha_1 a = \frac{-\alpha_2 a}{1+2\lambda a} \equiv u\lambda(a,0)$$

$$\alpha_2 = \alpha_2(\lambda,a)$$

$$\alpha_2 = \sum_{n=0}^{\infty} \alpha_{2n}(a)\lambda^n$$

Than since $\lim_{x\to\infty} u_\lambda(a,0) = 0$, we must have

$$\lim_{\lambda\to\infty} \frac{-a\sum_{n=0}^{\infty}\alpha_{2n}(a)\lambda^n}{1+2\lambda a} = 0$$

and hence, $\alpha_{2n}(\alpha) = 0 \forall n \geq 1$.

Thus, $\quad u_\lambda(a,0) = \dfrac{-\alpha_{20}(a)a}{1+2\lambda a}$

$$\lim_{x\to\infty} u_\lambda(a,0) = 1 \Rightarrow -\alpha_{20}(a)a = 1$$

$\Rightarrow \quad \alpha_{20}(a) = -\dfrac{1}{a}$, so

$$u_\lambda(a,0) = \frac{1}{1+2\lambda a}$$

Hence $\quad P_r\{L(T_a)\in dx\} = \dfrac{1}{2a}\exp\left(\dfrac{-x}{\alpha a}\right), x \geq 0$

So, $\quad P_r\{L(T_a) > x\} = \exp\left(\dfrac{-x}{\alpha a}\right)$

$$\left(\Delta - K_0^2\right)\underline{E}(\underline{r}) = \delta\{k_0^2 X(\underline{r})\underline{E}(\underline{r}) - (\underline{\nabla}X(r), \underline{E}(\underline{r}))\}$$

Solution:

Quantum strings theory in a background metric:

$g_{\mu\nu}(x)$ is the ambient metric,

$\mu_1, \nu = 0, 1, \ldots, d$, We may for example have

$X = \tau, X = \sigma.$

String: $(X^{\mu}(\tau, \sigma))_{\mu=0}^{d}, \tau \geq 0, 0 \leq \sigma \leq L$.

Lagrangian: $\mathcal{L}\left(x^{\mu} X^{\mu}\tau, X^{\mu}\sigma\right)$

$$= \left[g_{\mu\nu}(X)\left(\frac{\partial X^{\mu}}{\partial\tau}\frac{\partial X^{\nu}}{\partial\tau} - \frac{\partial X^{\mu}}{\partial\sigma}\frac{\partial X^{\nu}}{\partial\sigma}\right)\right]^{1/2}$$

Action: $\int \mathcal{L}\, d\tau\, d\sigma$

$$\frac{\partial\mathcal{L}}{\partial X^{\mu}\tau} = g_{\mu\nu} X,^{\nu}\tau/\mathcal{L}$$

$$\frac{\partial\mathcal{L}}{\partial X,^{\mu}\sigma} = g_{\mu\nu} X,^{\nu}\sigma/\mathcal{L}$$

Equations of motion: $\frac{\partial}{\partial\tau}\frac{\partial\mathcal{L}}{\partial X,^{\nu}\tau} + \frac{\partial}{\partial\sigma}\frac{\partial \pounds}{\partial X^{\mu}\sigma} = \frac{\partial\mathcal{L}}{\partial X,^{\mu}_{\sigma}}$

$$\frac{\partial\mathcal{L}}{\partial X^{\mu}} = g_{\rho'\sigma',\mu}(X)\left(X,^{\rho'}_{\tau} X,^{\sigma'}_{\tau}\right) - X^{\rho'},_{\sigma} X^{\sigma'}_{,\sigma}\Big)/2\mathcal{L}$$

So $\frac{\partial}{\partial\tau}\frac{\left(g_{\mu\nu}X,^{\nu}\tau\right)}{\mathcal{L}} - \frac{\partial}{\partial\sigma}\frac{\left(g_{\mu\nu}X,^{\nu}{}_{\sigma}\right)}{\mathcal{L}} = g_{\alpha\beta,\mu}(X,^{\alpha}_{\tau} X,^{\beta}_{\tau} - X,^{\alpha}_{\sigma} X,^{\beta}_{\sigma})/2\mathcal{L}$

Let $\mathcal{L}d\tau \to d\tau, \mathcal{L}d\sigma \to d\sigma$

Then the above equations can be expressed as

$$\frac{\partial}{\partial\tau}(g_{\mu\nu}X,^{\nu}_{\tau}) - \frac{\partial}{\partial\sigma}(g_{\mu\nu}X,^{\nu}_{\sigma}) = \frac{1}{2}g_{\alpha\beta,\mu}(X,^{\alpha}_{\tau} X,^{\beta}_{\tau} - X,^{\alpha}_{\sigma} X,^{\beta}_{\sigma})$$

or

$$g_{\mu\nu}(X,^{\nu}_{\tau\tau} - X,^{\nu}_{\sigma\sigma}) + g_{\mu\nu},_{\rho} X,^{\tau}_{\tau} X,^{\nu}_{\tau} - g_{\mu\nu,\rho}X,^{\rho}_{\sigma} X^{\nu}_{i\sigma}$$

$$-\frac{1}{2}g_{\nu},_{\rho\mu}(X,^{\nu}_{\tau} X,^{\rho}_{\tau} - X,^{\nu}_{\sigma} X,^{\rho}_{\sigma}) = 0$$

or

$$g_{\mu\nu}(X,^{\nu}_{\tau\tau} - X,^{\nu}_{\sigma\sigma}) + \Gamma_{\mu\nu\rho}(X,^{\rho}_{\tau} X,^{\nu}_{\tau} - X,^{\rho}_{\sigma} X,^{\nu}_{\sigma})$$

or

$$X,^{\mu}_{\tau\tau} - X,^{\mu}_{\sigma\sigma} + \Gamma^{\mu}_{\nu\rho}(X,^{\nu}_{\tau} X,^{\rho}_{\tau} - X,^{\nu}_{\sigma} X,^{\rho}_{\sigma}) = 0$$

This is the string Geodesic equation.

More generally, we can have a 2 × 2 matrix metric.

$$\underline{\underline{G}}_{\mu\nu} = \begin{bmatrix} g_{\mu\nu}^{(11)} & (X) & g_{\mu\nu}^{(12)} & (X) \\ g_{\mu\nu}^{(12)} & (X) & g_{\mu\nu}^{(22)} & (X) \end{bmatrix}$$

The String Lagrangian Corresponding to this background matrix is

$$S(X) = \left\{ \left[\frac{\partial X^{\mu}}{\partial \tau} \; \frac{\partial X^{\mu}}{\partial \sigma} \right] G_{\mu\nu}(X) \begin{bmatrix} \frac{\partial X^{\nu}}{\partial \tau} \\ \frac{\partial X^{\nu}}{\partial \sigma} \end{bmatrix} \right\}^{1/2}$$

$$= \left\{ g_{\mu\nu}^{(11)} \frac{\partial X^{\mu}}{\partial \tau} \frac{\partial X^{\nu}}{\partial \tau} + g_{\mu\nu}^{(22)} \frac{\partial X^{\mu}}{\partial \sigma} \frac{\partial X^{\nu}}{\partial \sigma} + 2 g_{\mu\nu}^{(12)} \frac{\partial X^{\mu}}{\partial \tau} \frac{\partial X^{\nu}}{\partial \sigma} \right\}^{1/2}$$

The previous is a special case of this obtained by taking

$$g_{\mu\nu}^{(12)} = 0, \; g_{\mu\nu}^{(22)} = -g_{\mu\nu}^{(11)}$$

❖❖❖❖❖

[13] Quantum Robotics Disturbance Observer:

Clamial Lagrangian

$$\mathcal{L}(\underline{\theta}, \underline{\theta}', t) = \frac{1}{2} \underline{\theta}'^{T} \underline{\underline{J}}(\tilde{\underline{\theta}}) \underline{\theta}' - V(\underline{\theta}) + \underline{T}(t)^{T} \underline{\theta} + \underline{d}(t)^{T} \underline{\theta}$$

$\underline{d}(t) \simeq$ disturbance

$T(t) \simeq$ torque

$V \simeq$ potential.

$$\underline{p} = \frac{\partial \alpha}{\partial \theta} = J(\underline{\theta}) \underline{\theta}' .$$

Hamiltonian:

$$H(\underline{\theta}, p, t) = \underline{p}^{T} \underline{\theta}' - \mathcal{L} = \frac{1}{2} \underline{p}^{T} \underline{\underline{J}}(\theta)^{-1} \underline{p} + V(\underline{\theta}) - T(t)^{T} \underline{\theta} - \underline{d}(t)^{T} \underline{\theta}$$

Schrodinger equation

$$\frac{i \partial \psi_t(\underline{\theta})}{\partial t} = H\left(\underline{\theta}, -i \frac{\partial}{\partial \theta}, t \right) \psi_t(\underline{\theta})$$

$$= -\frac{1}{2} \left(\frac{\partial}{\partial \underline{\theta}} \right)^{T} \underline{\underline{J}}(\underline{\theta})^{-1}, \frac{\partial}{\partial \underline{\theta}} \psi_t(\underline{\theta}) + V(\underline{\theta}) \psi_t(\underline{\theta})$$

$$- T(t)^{T} \underline{\theta} \, \psi_t(\underline{\theta}) - \underline{d}(t)^{T} \underline{\theta} \, \psi_t(\underline{\theta})$$

To take damping into account, we introduce the Sudarshan. Lindblad operator

$$L = L(\underline{\theta}, \underline{p})$$

$$= \left.\sum_{r,s=0}^{\infty} L_{rs}^T(\underline{\theta}^{\otimes r} \otimes \underline{p}^{\otimes s}\right), L_{rs} \in\in d^{r+s}$$

Noisy, Schrödinger equation for the density operator kernel $\rho_t(\underline{\theta}, \underline{\theta}')$ becomes

$$i\frac{\partial \rho_t(\underline{\theta}, \underline{\theta}')}{\partial t} = H\left(\underline{\theta}, -i\frac{\partial}{\partial \underline{\theta}}, t\right)\rho_t(\underline{\theta}, \underline{\theta}')$$

$$-\rho_t(\underline{\theta}, \underline{\theta}')\, H\left(\underline{\theta}' - i\frac{\partial}{\partial \underline{\theta}'}, t\right)$$

$$-\frac{i}{2}[L^* L\rho_t + \rho_t L^* L - 2L\rho_t L^*]$$

where
$$L^* L\rho_t \equiv (L^* L)\left(\underline{\theta}, -i\frac{\partial}{\partial \underline{\theta}}\right)\rho_t(\underline{\theta}, \underline{\theta}')$$

$$\rho_t L^* L \equiv \rho_t(\underline{\theta}, \underline{\theta}')\,(L^* L)\left(\underline{\theta}', -i\frac{\partial}{\partial \underline{\theta}'}\right)$$

and
$$L\rho_t L^* \equiv L\left(\underline{\theta}', -i\frac{\partial}{\partial \underline{\theta}}\right)\rho_t(\underline{\theta}, \theta')\, L^*\left(\underline{\theta}', -i\frac{\partial}{\partial \underline{\theta}'}\right)$$

Note that the kernel of the operator $\underline{F}(\underline{\theta}, \theta')^T \dfrac{\partial}{\partial \theta'}$ is $K(\underline{\theta}, \underline{\theta}', \underline{\theta}'')$

where $\int K(\underline{\theta}, \underline{\theta}', \underline{\theta}'') f(\underline{\theta}'')\, d\theta''$

$$= \underline{F}(\underline{\theta}, \underline{\theta}')^T \frac{\partial f(\underline{\theta}')}{\partial \underline{\theta}'}$$

$$= \underline{F}(\underline{\theta}, \underline{\theta}')^T [\underline{\nabla}\theta' \delta(\theta' - \theta'')]\; f(\underline{\theta}'')\, \underline{d\theta}''$$

i.e.
$$K(\underline{\theta}, \underline{\theta}', \underline{\theta}'') = \underline{F}(\theta, \theta')^T\; \underline{\nabla}_{\theta'} \delta(\underline{\theta}' - \theta'')$$

The kernel of the operator

$$\underline{L}_{rs}^T(\underline{\theta}^{\otimes r} \otimes \underline{\nabla}_{\theta}^{\otimes s})\, F(\underline{\theta}, \underline{\theta}')$$

is $K(\underline{\theta}, \underline{\theta}', \underline{\theta}'')$ where

$$\int K(\underline{\theta}, \underline{\theta}', \underline{\theta}'') f(\underline{\theta}'')\, d\underline{\theta}'' = \underline{L}_{rs}^T(\underline{\theta}^{\otimes r} \otimes \underline{\nabla}_{\theta}^{\otimes s})\, F(\underline{\theta}, \underline{\theta}') f(\underline{\theta}')$$

Problem: Let K be an operator represented by the kernel $K(\underline{\theta}, \underline{\theta}')$. Then what is the kernel of the operator

$$K = \underline{L}_{rs}^T(\underline{\theta}^{\otimes r} \otimes \underline{\nabla}_{\theta}^{\otimes s})$$

Let it be $K(\underline{\theta}, \underline{\theta}')$. Then

$$\int K(\underline{\theta},\underline{\theta}')\,f(\underline{\theta}')d\underline{\theta}' = \underline{L}_{rs}^{T}(\underline{\theta}^{\otimes r}\otimes\underline{\nabla}_{\theta}^{\otimes s})\,f(\underline{\theta})$$
$$= \int \underline{L}_{rs}^{T}(\underline{\theta}^{\otimes r}\otimes\underline{\nabla}_{\theta}^{\otimes s})\,\delta(\underline{\theta}-\underline{\theta}')\,f(\theta')d\underline{\theta}'$$

Then,
$$K(\underline{\theta},\underline{\theta}') = \underline{L}_{rs}^{T}\,(\underline{\theta}^{\otimes r}\otimes\nabla_{\theta}^{\otimes s})\,\delta(\underline{\theta}-\underline{\theta}')$$
$$\equiv \sum_{\substack{i_1,\dots,i_r,\\ j_1,\dots,j_s}} L_{rs}[i_1,i_2,\dots,i_r,j_1,j_2,\dots,j_s]$$
$$\theta_{i1},\theta_{i2}\dots\theta_{ir}\,\frac{\partial^{s}\delta(\underline{\theta}-\underline{\theta}')}{\partial\theta_{j1}\dots\partial\theta_{js}}.$$

Likewise, the kernel of the operator.
$\underline{F}(\underline{\theta},\underline{\theta}')^{T}\underline{\nabla}_{\theta'}$ is $K(\underline{\theta},\underline{\theta}',\underline{\theta}'')$, where $K(\underline{\theta},\underline{\theta}',\underline{\theta}'') = \underline{F}(\underline{\theta},\underline{\theta}')^{T}\,\underline{\nabla}_{\theta'}\delta(\underline{\theta}'-\underline{\theta}'')$

In the context of the above Sudarshan-Lindblad equation, the Kernel of an operator of the form $\underline{F}(\underline{\theta},\underline{\theta}')^{T}\dfrac{\partial}{\partial\underline{\theta}'}$ should be a matrix *i.e.* a function of two variables. This kernel is thus
i.e.
$$K(\underline{\theta}',\underline{\theta}'') = \int \underline{F}(\underline{\theta},\underline{\theta}')^{T}\,(\underline{\nabla}_{\theta'}\delta(\underline{\theta}'-\theta''))\,d\theta'$$
$$= -\int \underline{\nabla}_{\theta'}^{T}\,\underline{F}(\underline{\theta},\underline{\theta}')\delta(\underline{\theta}',\underline{\theta}')d\underline{\theta}'$$
$$= -\,\underline{\nabla}_{\theta''}^{T}\,\underline{F}(\underline{\theta},\underline{\theta}'')$$

i.e.
$$K(\underline{\theta},\underline{\theta}') = -\,\mathrm{div}_{\underline{\theta}'}\underline{F}\,(\underline{\theta},\underline{\theta}')$$

Likewise, the kernel of
$$\underline{L}_{rs}^{T}\left(\underline{\theta}^{\otimes s}\otimes\underline{\nabla}_{\theta}^{\otimes s}\right)F(\underline{\theta},\theta')$$
is $\int \underline{L}_{rs}^{T}\underline{\theta}^{\otimes r}\otimes(\underline{\nabla}_{\theta}^{\otimes s}(\delta(\underline{\theta}-\underline{\theta}')F(\underline{\theta}',\theta'')))d\underline{\theta}' \equiv K(\underline{\theta},\theta'')$

i.e.
$$K(\underline{\theta},\underline{\theta}'') = \underline{L}_{rs}^{T}\underline{\theta}^{\otimes r}\otimes\underline{\nabla}_{\theta}^{\otimes s}F(\underline{\theta},\underline{\theta}'')$$
or
$$K(\underline{\theta},\underline{\theta}') = \underline{L}_{rs}^{T}\left(\underline{\theta}^{\otimes r}\otimes\underline{\nabla}_{\theta}^{\otimes s}\right)F(\underline{\theta},\underline{\theta}')$$

Finally, the kernel of
$$F(\underline{\theta},\underline{\theta}')\underline{L}_{rs}^{T}\left(\underline{\theta}'^{\otimes r}\otimes\underline{\nabla}_{\theta'}^{\otimes s}\right)$$
is
$$K(\underline{\theta}-\underline{\theta}'') = \int F(\underline{\theta},\underline{\theta}')\underline{L}_{rs}^{T}\left(\underline{\theta}'^{\otimes r}\otimes\underline{\nabla}_{\theta'}^{\otimes s}\right)\delta(\underline{\theta}'-\underline{\theta}'')d\underline{\theta}'$$
$$= \int F(\underline{\theta},\underline{\theta}')\underline{L}_{rs}^{T}\left(\underline{\theta}'^{\otimes r}\otimes\underline{\nabla}_{\theta''}^{\otimes s}\right)\delta(\underline{\theta}'-\underline{\theta}'')(-1)^{|s|}\,d\underline{\theta}'$$
$$= (-1)^{|s|}\,\underline{L}_{rs}^{T}\left(\underline{\theta}''^{\otimes r}\otimes\underline{\nabla}_{\theta''}^{\otimes s}\right)F(\underline{\theta},\underline{\theta}'')$$

i.e. $$K(\underline{\theta}-\underline{\theta}') = (-)^{|s|}\underline{L}_{rs}^{T}\left(\underline{\theta}'^{\otimes r}\otimes\underline{\nabla}_{\theta t}^{\otimes s}\right)F(\underline{\theta}-\underline{\theta}')$$

Quantum disturbance observer:

$$\hat{d} = L(t)(\underline{d}(t)-\underline{\hat{d}}(t))$$

If $L(t)\geq 0$, then in the scalar case, for $d(t)=d$

Constant, we get

$$\underline{\hat{d}}(t) = \underline{d} = \exp\left(-\int_0^t L(\tau)\,d\tau\right)$$

$$(\hat{d}(0)-d) \to 0,\ t\to\infty$$

So we can use this formula to create a disturbance observer in the quantum case.

❖❖❖❖❖

[14] Quantum Field Theory in the Presence of Stochastic Disturbance:

Real Valued Klein-Gordon field interacting with a random vector potential. Lagrangian density:

$$\mathcal{L}(\varphi,\varphi,_{\mu}) = \frac{1}{2}(\partial_\mu+ieA_\mu)\varphi(\partial^\mu-ieA^\mu)\varphi-\frac{1}{2}m^2\varphi^2$$

$\bar{\varphi}=\varphi$. Classical field equation:

$$\begin{aligned}\delta_\varphi\mathcal{L} &= \frac{1}{2}(\delta\varphi,_{\mu}+ieA_\mu\delta\varphi)(\partial^\mu-ieA^\mu)\varphi\\ &\quad+\frac{1}{2}(\partial_\mu+ieA_\mu)\varphi(\delta\partial^\mu\varphi-ieA^\mu\delta\varphi)-m^2\varphi\delta\varphi\\ &\approx\frac{1}{2}\delta\varphi(-\partial_\mu+ieA_\mu)(\partial^\mu-ieA^\mu)\varphi\\ &\quad+\frac{\delta\varphi}{2}(-\partial^\mu-ieA^\mu)(\partial_\mu+ieA_\mu)\varphi\\ &= -\frac{\delta\varphi}{2}\{(\partial_\mu-ieA_\mu)(\partial^\mu-ieA^\mu)\varphi\\ &\quad+(\partial^\mu+ieA^\mu)(\partial_\mu-ieA_\mu)\varphi\}\\ &= -\delta\varphi\,\mathrm{Re}\{(\partial^\mu+ieA^\mu)(\partial_\mu+ieA_\mu)\varphi\}\end{aligned}$$

where $\approx$ means that a total divergence has been removed since it does not contribute to the action $\int\mathcal{L}(\varphi,\varphi,_{\mu})\,d^4x$. The classical field equations are thus

$$\mathrm{Re}\{(\partial_\mu+ieA_\mu)(\partial^\mu+ieA^\mu)\varphi\}+m^2\varphi = 0$$

i.e. $$\partial_\mu\partial^\mu\varphi-e^2A_\mu A^\mu\varphi+m^2\varphi = 0$$

or $$\partial_\mu\partial^\mu\varphi+\left(m^2-e^2A_\mu A^\mu\right)\varphi = 0$$

Thus for a real KG field, the four potential of the external *em* field contributes a man term to it.

Let $m^2-e^2A_\mu A^\mu = \psi(x). \quad x=(t, r)$. Then

$$(\partial_\mu\partial^\mu+\psi)\varphi = 0.$$

Suppose A_μ is a random field. Then so is ψ and we get from

$$\left(\partial_t^2-\Delta+\psi\right)\varphi = 0$$

or $$(\Delta-\partial_t^2)\varphi = -\psi\varphi,$$

$$\varphi(t,\underline{r}) = \frac{1}{4\pi}\int\frac{\delta(t'-t+|\underline{r}-\underline{r}'|)}{|\underline{r}-\underline{r}'|}\psi(t',\underline{r}')\,\varphi(t',\underline{r}')\;dt'd^3r'$$

This equation can be iterated to obtain a series expansion for $\varphi(t,\underline{r})$. But $\varphi_0(t,\underline{r})$ be the initial field satisfying

$$\left(\Delta-\partial_t^2\right)\varphi_0 = 0.$$

Then

$$\varphi(t,\underline{r}) = \varphi_0(t,\underline{r})+\sum_{n=1} G_\psi^m\varphi_0(t,\underline{r})$$

where $$G\psi f(t,\underline{r}) = \frac{1}{4\pi}\int\frac{\delta(t'-t+|\underline{r}-\underline{r}'|)}{|\underline{r}-\underline{r}'|}\psi(t',\underline{r}')\,f(t',\underline{r}')\;dt'd^3r'$$

Problem: Use this expression to calculate the moments $\mathbb{E}\left\{\prod_{k=1}^{p}\varphi(t_k,\underline{r}_k)\right\}$ of the K.G. field in terms of the moments $\mathbb{E}\left\{\prod_{k}^{q}\psi(t_k,\underline{r}_k)\right\}$, $p,q\geq 1$.

Generalization of *e-m* fields for stiring charges: Interaction Lagrangian between field and charges:

$$A_\mu\frac{dx^\mu}{d\tau}\to B_{\mu\nu}^{(1)}\frac{\partial X^\mu}{\partial\tau}\frac{\partial X^\nu}{\partial\tau}+2B_{\mu\nu}^{(2)}\frac{dX^\mu}{d\tau}\frac{\partial X^\nu}{\partial\sigma}+B_{\mu\nu}^{(3)}\frac{\partial X^\mu}{\partial\sigma}\frac{\partial X^\nu}{\partial\sigma}$$

$B_{\mu\nu}^{(1)}$ and $B_{\mu\nu}^{(3)}$ are symmetric while $B_{\mu\nu}^{(2)}$ is skew-symmetric.

$$d(A_\mu\xi^\mu) = dA_\mu\Lambda\xi^\mu+A_\mu d\xi^\mu$$

If $d\xi^\mu = 0\left(e_j\xi^\mu=dx^\mu\right)$, then

$$d(A_\mu\xi^\mu) = dA_\mu\Lambda\xi^\mu$$

$$
\begin{aligned}
dX^{\mu} &= \frac{\partial X^{\mu}}{\partial \tau} d\tau + \frac{\partial X^{\mu}}{\partial \sigma} d\sigma \\
dX^{\mu} \otimes dX^{\nu} &= \frac{\partial X^{\mu}}{\partial \tau}\frac{\partial X^{\nu}}{\partial \tau} d\tau \otimes d\tau \\
&= \frac{\partial X^{\mu}}{\partial \sigma}\frac{\partial X^{\nu}}{\partial \sigma} d\sigma \otimes d\sigma + \frac{\partial X^{\mu}}{\partial \tau}\frac{\partial X^{\nu}}{\partial \sigma} d\tau \otimes d\sigma \\
&+ \frac{\partial X^{\mu}}{\partial \sigma}\frac{\partial X^{\nu}}{\partial \tau} d\sigma \otimes d\tau \\
B_{\mu\nu} &\equiv B^{(1)}_{\mu\nu} \frac{\partial}{\partial \tau} \otimes \frac{\partial}{\partial \tau} + B^{(1)}_{\mu\nu} \frac{\partial}{\partial \tau} \otimes \frac{\partial}{\partial \sigma} \\
&+ B^{(2)}_{\nu\mu} \frac{\partial}{\partial \sigma} \otimes \frac{\partial}{\partial \tau} + B^{(3)}_{\mu\nu} \frac{\partial}{\partial \sigma} \otimes \frac{\partial}{\partial \sigma}
\end{aligned}
$$

❖❖❖❖❖

[15] Simultaneous Tracking, Parametric Uncertainties Estimation and Disturbance Observer in a Single Robot:

$$
\begin{aligned}
&\underline{\underline{\widehat{M}}}(\underline{q})\left(\ddot{\underline{q}} + \ddot{\underline{e}} + K_p \underline{e} + K_d \dot{\underline{e}}\right) + \widehat{N}(\underline{q} \cdot \dot{\underline{q}}) \\
&\qquad = \text{Computed torque} = \tau(t) \\
&\qquad = \underline{\underline{M}}(\underline{q})\ddot{q} + \underline{N}(\underline{q} \cdot \dot{\underline{q}})
\end{aligned}
$$

or

$$
\begin{aligned}
\ddot{\underline{e}} + K_p \underline{e} + \underline{\underline{K}}_d \dot{e} &= -M(\underline{q})^{-1}\left(\delta M(\underline{q})\ddot{\underline{q}} + \delta \underline{N}(\underline{q} \cdot \dot{\underline{q}})\right) \\
&= W\left(\underline{q}, \dot{\underline{q}}, \ddot{\underline{q}}\right)\delta\underline{\theta}
\end{aligned}
$$

Where $0\left(\| \delta(\underline{\theta}) \|^2\right)$ terms have been neglected. Approximation

$$
\begin{aligned}
\ddot{\underline{e}} + \underline{\underline{K}}_p e + \underline{\underline{K}}_d \dot{e} &= -M(q)\left(\delta M(\underline{q})\ddot{\underline{q}}_d + \delta N(\underline{q}, \dot{\underline{q}})\right) \\
&= W_1(t, \underline{q}, \dot{\underline{q}})\delta\underline{\theta} \\
&= W_2(t, \underline{e}, \dot{e})\delta\underline{\theta}
\end{aligned}
$$

In the presence of disturbance $d(t)$,

$$
M(\underline{q})\ddot{\underline{q}} + \underline{N}(\underline{q}, \dot{\underline{q}}) = \underline{\tau}(t) + \underline{d}(t)
$$

Disturbance observer:

$$
\begin{aligned}
\dot{\underline{z}} &= -L(\underline{q}, \dot{\underline{q}})\underline{z} + L(\underline{q}, \dot{\underline{q}})\left(N(\underline{q}, \dot{\underline{q}}) - \underline{\tau}(t) - \underline{p}(\dot{\underline{q}})\right) \\
&= L(\underline{q}, \dot{\underline{q}})\left(N(\underline{q}, \dot{\underline{q}}) - \underline{\tau}(t) - \underline{p}(\dot{\underline{q}}) - z\right) \\
\hat{d} &= p(\dot{q}) + z \\
\dot{\hat{d}} &= p'(\dot{\underline{q}})\ddot{\underline{q}} + \dot{z} = L\left(N - \tau - \hat{d}\right) + p'(\dot{q})\ddot{\underline{q}}
\end{aligned}
$$

$$= L\left(-M\ddot{q} + d - \hat{d}\right) + p'(\dot{q})\ddot{q}$$

$$= L(d - \hat{d}) + (p'(\dot{q}) - LM)\ddot{q}$$

$$\left(M\ddot{q} + N = \tau + d\right)$$

Taking $p'(\underline{\dot{q}}) = L(\underline{q}, \underline{\dot{q}})M(q)$, *i.e.*,

$$\underline{\underline{L}}(q, \underline{\dot{q}}) = \underline{p}'(\underline{\dot{q}})\underline{\underline{M}}^{-1}(\underline{q}), \text{ then,}$$

$$\dot{\hat{d}} = L(d - \hat{d})$$

Showing that $\left|\hat{d}(t) - d(t)\right| \to 0$ if $L(\underline{q}, \underline{\dot{q}}) > 0$

$t \to \infty$ (positive definite matrix)

Let $\underline{\tilde{e}} = \left[\begin{smallmatrix} e \\ \dot{e} \end{smallmatrix}\right]$. Then,

$$\dot{\underline{\tilde{e}}} = \begin{bmatrix} \dot{\underline{e}} \\ -K_p\underline{e} - K_d\underline{\dot{e}} \end{bmatrix} + \begin{bmatrix} 0 \\ W_2 \end{bmatrix}\delta\underline{\theta} + \begin{bmatrix} 0 \\ -M^{-1}(\underline{q})\underline{d}(t) \end{bmatrix}$$

$$W_2(t, \underline{\tilde{e}}) = -M\left(\underline{q}_d(t) - \underline{e}(t)\right)^{-1}\left(\frac{\partial \underline{M}}{\partial \underline{\theta}}(\underline{q}, \theta_0)(I_p \otimes \underline{\ddot{q}}_d) + \frac{\partial N}{\partial \underline{\theta}}(\underline{q}, \underline{\dot{q}}, \underline{\theta}_0)\right)$$

$$= -M(\underline{q}_d - \underline{e})^{-1}\left[\frac{\partial \underline{\underline{M}}}{\partial \underline{\theta}}(\underline{q}_d - \underline{e}, \theta_0)\,(\underline{\underline{I}}_p \otimes \underline{\ddot{q}}_d) + \frac{\partial N}{\partial \underline{\theta}}(\underline{q}_d - e, \underline{\dot{q}}_d - \underline{\dot{e}}, \underline{\theta}_0)\right]$$

Update formula for $\delta\underline{\theta}$:

Lyapunov function

Let $V(\underline{\tilde{e}}, \delta\underline{\theta}) = \dfrac{1}{2}\underline{\tilde{e}}^T\underline{\underline{Q}}\underline{\tilde{e}} + \dfrac{1}{2}\delta\underline{\tilde{\theta}}^T\underline{\underline{P}}\delta\underline{\theta}$

where $\underline{\underline{Q}}, P \geq 0$ (Positive definite matrics). We write the eqn. for $\underline{\tilde{e}}$ as

$$\frac{d\tilde{e}(t)}{dt} = \begin{pmatrix} \underline{\underline{O}} & \underline{\underline{I}}_d \\ -\underline{\underline{K}}_p & -\underline{\underline{K}}_d \end{pmatrix}\underline{\tilde{e}}(t) + \begin{pmatrix} \underline{\underline{O}} \\ \underline{\underline{W}}_2 \end{pmatrix}\delta\underline{\theta}(t) + \begin{pmatrix} O \\ -M^{-1}(\underline{q}) \end{pmatrix}\underline{d}(t)$$

Then writing

$$\underline{\underline{A}} = \begin{pmatrix} \underline{\underline{O}} & \underline{\underline{I}}_d \\ -\underline{\underline{K}}_p & -\underline{\underline{K}}_d \end{pmatrix}, \underline{\underline{B}} = \begin{pmatrix} \underline{\underline{O}} \\ \underline{\underline{W}}_2 \end{pmatrix},$$

$$\underline{\underline{C}} = \begin{pmatrix} O \\ -M^{-1}(\underline{q}) \end{pmatrix}$$

We have
$$\underline{\underline{A}} = \underline{\underline{A}}\left(\underline{\underline{K}}_p, \underline{\underline{K}}_d\right), \underline{\underline{B}} = \underline{\underline{B}}(t, \tilde{\underline{e}}(t)),$$

$$\underline{\underline{C}} = \underline{\underline{C}}(t, \underline{e}(t)) \equiv \underline{\underline{C}}(t, \tilde{\underline{e}}(t))$$

and
$$\frac{d\tilde{\underline{e}}}{dt} = \underline{\underline{A}}\tilde{\underline{e}} + \underline{\underline{B}} \cdot \delta\underline{\theta} + \underline{\underline{C}}\, d(t)$$

$$\begin{aligned}\frac{dV}{dt} &= \tilde{\underline{e}}^T \underline{\underline{Q}} \frac{d\tilde{\underline{e}}}{dt} + \delta\underline{\theta}^T \underline{\underline{P}} \frac{d\delta\underline{\theta}}{dt} \\ &= \tilde{\underline{e}}^T \underline{\underline{Q}}\left(\underline{\underline{A}}\,\tilde{\underline{e}} + \underline{\underline{B}}\delta\underline{\theta} + \underline{\underline{C}}\underline{d}(t)\right) \\ &\quad + \delta\underline{\theta}^T \underline{\underline{P}} \frac{d\delta\theta}{dt} \\ &= \frac{1}{2}\tilde{\underline{e}}^T (QA + A^T Q)\tilde{\underline{e}} \\ &\quad + \delta\underline{\theta}^T \left[B^T Q\tilde{e} + P\frac{d\delta\theta}{dt}\right] + \tilde{\underline{e}}^T \underline{\underline{Q}}\,\underline{\underline{C}}\underline{d}(t)\end{aligned}$$

If $\underline{d}(t)$ is white noise then

by taking

$$P\frac{d\delta\underline{\theta}(t)}{dt} = -\underline{\underline{B}}^T \underline{\underline{Q}}\tilde{\underline{e}} + \underline{\underline{R}}\delta\underline{\theta}$$

or equivalently,

$$\frac{d\delta\underline{\theta}}{dt} = -\underline{\underline{P}}^{-1}\underline{\underline{B}}^T \underline{\underline{Q}}\tilde{\underline{e}} + \underline{\underline{P}}^{-1}\underline{\underline{R}}\,\delta\underline{\theta}$$

When $R \le 0$, we get

$$\frac{d}{dt}\mathbb{E}(V) = \mathbb{E}\left\{\frac{1}{2}\tilde{\underline{e}}^T \left(QA + A^T Q\right)\tilde{\underline{e}}\right\} \pm \mathbb{E}\left\{\delta\underline{\theta}^T R\delta\underline{\theta}\right\} \le 0$$

provided $QA + A^T Q \le 0$

We may even have white noise $d_\theta(t)$ in the $\delta\underline{\theta}$-evolution equation:

$$\frac{d\delta\underline{\theta}}{dt} = -P^{-1}B^T Q\tilde{e} + P^{-1}R\delta\theta + \underline{\underline{C}}_1(t, \tilde{\underline{e}}, \delta\underline{\theta})\, d_\theta(t)$$

Then $$\frac{d}{dt}\mathbb{E}(V) = \mathbb{E}\left\{\frac{1}{2}\tilde{e}^T(QA + A^TQ)\tilde{e}\right\} + \mathbb{E}\left\{\delta\underline{\theta}^T \underline{\underline{R}}\delta\underline{\theta}\right\} \le 0$$

Then $\begin{pmatrix}\tilde{e}\\ \delta\underline{\theta}\end{pmatrix}$ satisfy the side

$$d\begin{pmatrix}\tilde{e}\\ \delta\underline{\theta}\end{pmatrix} = \begin{bmatrix}0 & I_d & 0\\ -K_p & -K_d & \underline{\underline{W}}_2\\ (-p^{-1} & B^TQ) & p^{-1}R\end{bmatrix}\begin{bmatrix}\underline{\tilde{e}}\\ \delta\underline{\theta}\end{bmatrix}dt$$

$$+\begin{pmatrix}C(t,\underline{\tilde{e}}) & 0\\ 0 & C_1(t,\underline{\tilde{e}}\delta\underline{\theta})\end{pmatrix} d\begin{pmatrix}\underline{B}_1(t)\\ \underline{B}_2(t)\end{pmatrix}$$

where $\underline{B}_1(t)$ and $\underline{B}_2(t)$ are respectively

$\mathbb{R}^d$ and $\mathbb{R}^p$-valued independent Brownian motions where P is the number of parameters $\underline{\theta}$ and d is the number of links

Write $\underline{\underline{Q}} = \begin{bmatrix}\underline{\underline{Q}}_1\\ \underline{\underline{Q}}_2\end{bmatrix}$ in partitioned from. Then

$$P^{-1}B^TQ = [P^{-1}W_2{}^TQ_2]$$

Note $P \in \mathbb{R}^{p\times p}W_2 \in R^{d\times p}$, $Q_2 \in \mathbb{R}^{d\times 2d}$. So

$$P^{-1}B^TQ = P^{-1}W_2{}^TQ_2 \in \mathbb{R}^{p\times 2d}$$

We write $$Q_2 = \left[\underline{\underline{Q}}_{21} \mid \underline{\underline{Q}}_{22}\right]$$

where $\underline{\underline{Q}}_{21} \in \mathbb{R}^{d\times d}, \underline{\underline{Q}}_{22} \in \mathbb{R}^{d\times d}$. Then, the joint dynamics of $\underline{\tilde{e}}$ and $\delta\underline{\theta}$ is

$$d\begin{pmatrix}\tilde{e}\\ \delta\underline{\theta}\end{pmatrix} = \begin{bmatrix}0 & I_d & 0\\ -K_p & -K_d & W_2\\ -P^{-1}W_2^TQ_{21} & -P^{-1}W_2^TQ_{22} & P^{-1}R\end{bmatrix}\begin{bmatrix}\underline{\tilde{e}}\\ \delta\underline{\theta}\end{bmatrix}$$

$$+\begin{bmatrix}C(t,\underline{\tilde{e}}) & 0\\ 0 & C_1(t,\tilde{e},\delta(\underline{\theta})\end{bmatrix} d\underline{B}(t)$$

Quantization: Consider a state variable system (nonlinear)

$$\frac{d\underline{X}(t)}{dt} = F(t,\underline{X}(t)), t \ge 0$$

$$\underline{X}(t) \in \mathbb{R}^n, F : \mathbb{R}_+ \times \mathbb{R}^n \to \mathbb{R}^n$$

Define the Lagrangian

$$\mathcal{L}(\underline{X}, \underline{X}', \underline{Y}', t) = \underline{Y}(t)^T \underline{X}'(t) - \underline{Y}(t)^T \underline{F}(t, \underline{X}(t))$$
$$\cong \frac{\left(Y^T \underline{X}' - \underline{X}^T \underline{Y}'\right)}{2} - \underline{Y}^T \underline{F}(t, \underline{X})$$
$$= \mathcal{L}_1(\underline{X}, \underline{X}', \underline{Y}, \underline{Y}', t)$$

The Euler-Lagrange equation

$$\frac{d}{dt}\frac{\partial \mathcal{L}}{\partial \underline{X}'} = \frac{\partial \mathcal{L}}{\partial \underline{X}}, \frac{d}{dt}\frac{\partial \mathcal{L}}{\partial Y'} = \frac{\partial \mathcal{L}}{\partial Y}$$

give $Y'(t) + \left(\frac{\partial \underline{E}}{\partial \underline{x}}\right)^T \underline{Y} = 0$

and $X'(t) = F(t, \underline{X}(t'))$

Hamiltonian: $\underline{\pi}_x = \frac{\partial \mathcal{L}_L}{\partial \underline{X}'} = \frac{1}{2}\underline{Y}$

$$\underline{\pi}_y = \frac{\partial \mathcal{L}_L}{\partial \underline{Y}'} = -\frac{1}{2}\underline{X}'$$

$$H = \pi_x{}^T X + \pi_y{}^T Y - £_1$$

Cannot be expressed in terms of $\underline{X}, \underline{Y}, \pi_{\underline{x}}, \pi_y, t$ alone.

Consider the Lagragian

$$\mathcal{L} = Y^T X' - Y^T F(t, \underline{X})$$

Add to this a term $\frac{\varepsilon_1 \| \underline{X}' \|^2}{2} + \frac{\varepsilon_2 \| \underline{Y}' \|^2}{2}$ to get

$$\tilde{\mathcal{L}}(\underline{X}, \underline{X}', \underline{Y}, \underline{Y}') = \frac{\varepsilon_1}{2} \| \underline{X}' \|^2 + \frac{\varepsilon_2}{2} \| \underline{Y}' \|^2 + \underline{Y}^T \underline{X}' - \underline{Y}^T F(t, \underline{X})$$

$$\pi_x = \frac{\partial \tilde{\mathcal{L}}}{\partial \underline{X}'} = \varepsilon_1 \underline{X}' + \underline{Y}$$

$$\pi_y = \frac{\partial \tilde{\mathcal{L}}}{\partial \underline{Y}'} = \varepsilon_2 Y'$$

$$\tilde{H} = \pi_x{}^T \underline{X}' + \pi_y{}^T \underline{Y}' - \tilde{\mathcal{L}}$$
$$= \varepsilon_1 \| X' \|^2 + \varepsilon_2 Y'^T Y'$$
$$+ Y^T F(t, X) - \frac{\varepsilon_1}{2} \| \underline{X}' \|^2 - \frac{\varepsilon_2}{2} \| \underline{Y}' \|^2$$
$$= \frac{1}{2\varepsilon_1} \| \pi_x - Y \|^2 + \frac{\| \pi_y \|^2}{2\varepsilon_2} + Y^T F(t, \underline{X})$$

$$= \frac{1}{2\varepsilon_1}\| \pi_x - \underline{Y} \|^2 + \frac{\| \pi_y \|}{2\varepsilon_2} + \underline{Y}^T \underline{F}(t, \underline{X})$$

$$= \tilde{H}(\underline{X}, \underline{Y}, \underline{\pi}_x, \underline{\pi}_y, t)$$

We take $\varepsilon_1 = \varepsilon_2 = \varepsilon$. Thus,

$$\tilde{H} = \frac{1}{2\varepsilon}\| \pi_x - Y \|^2 + \frac{1}{2\varepsilon}\| \pi_y \|^2 + \underline{Y}^T \underline{F}(t, \underline{X})$$

Hamilton equation of motion:

$$\frac{\partial \tilde{H}}{\partial \pi_x} = \frac{1}{\varepsilon}(\underline{\pi}_x - \underline{Y})$$

$$\frac{\partial \tilde{H}}{\partial \pi_y} = \frac{1}{\varepsilon}\underline{\pi}_y$$

$$\frac{d\underline{X}}{dt} = \frac{\partial \tilde{H}}{\partial \pi_x} = \frac{1}{\varepsilon}(\underline{\pi}_x - Y)\frac{d\pi_x}{dt} = -\frac{\partial \tilde{H}}{\partial X}$$

$$\frac{dY}{dt} = \frac{\partial \tilde{H}}{\partial \pi_y} = \frac{1}{\varepsilon}\pi_y = \left(\frac{\partial F}{\partial X}\right)^T \underline{Y}$$

$$\frac{d\pi_y}{dt} = -\frac{\partial \tilde{H}}{\partial y} = \frac{1}{\varepsilon}(\pi_x - Y)\underline{F}(t, \underline{X})$$

Thus, $$\varepsilon \frac{d^2 y}{dt^2} = \frac{d\underline{X}}{dt} - \underline{F}(t, \underline{X})$$

As $\varepsilon \to 0$, this becomes the original dynamical system

$$\frac{d\underline{X}}{dt} = \underline{F}(t, \underline{X}) .$$

QSDE

Let $\underline{X}, \underline{Y} \in \mathbb{R}'$. Then

$$H(x, y, \pi_x, \pi_y, t) = \frac{1}{2\varepsilon}(\pi_x - y)^2 + \frac{1}{2\varepsilon}\pi_y{}^2 + y f(t, x)$$

$$dU(t) = \left[\left(-i\tilde{H} + \frac{1}{2}LL^*\right)dt + L dA_t - L^* dA_t^*\right]U(t)$$

Let $$L = \alpha x + \beta y + \gamma \pi_x + \delta \pi_y$$

$\alpha, \beta, \gamma, \delta \in \mathbb{C}$.

$$X(t) = U^*(t) X U(t) \qquad X \equiv X \otimes I$$

$$Y(t) = U^*(t) Y U(t)$$

$$\pi_x(t) = U^*(t) \pi_x U(t)$$

$$\pi_y(t) = U^*(t)\pi_y U(t)$$

In general for any system observable ξ,

$$\begin{aligned}
\xi(t) &= U^*(t)\xi U(t) \\
d\xi(t) &= dU^*\xi U + U^*\xi dU + dU^*\xi dU \\
&= U^*\left\{L^*L\xi + \frac{1}{2}\xi L^*L\right. \\
&\quad \left. - i\left[\tilde{H},\xi\right] - L\xi L^*\right\}U\,dt + dA, dA \text{ terms.}
\end{aligned}$$

KRP paperon quantum probability and strong quantum Markov processes.

$$\begin{aligned}
J\lambda(X, u) &= \lambda(X, u) \\
J\lambda(Y, v) &= \lambda(Y, v) \\
\langle \lambda(Y,v), J^*J\lambda(X,u)\rangle &= \langle \lambda(Y,v), \lambda_1(X,u)\rangle \\
&= \left\langle T_1(Y)^v \otimes \delta_1(Y)v, T(X)u \otimes \delta_1(X)u\right\rangle \\
&= \langle v, T_1(Y)_T{}^*(X)u\rangle + \langle v, \delta_1(Y)^*\delta_1(X)u\rangle \\
J^*J_{jT}(X)\lambda(Y,u) &= J^*J\begin{bmatrix} T(X) & \delta^*(X^*) \\ \delta(X) & \delta(X) \end{bmatrix}\begin{bmatrix} T(Y)u \\ \delta(Y)u \end{bmatrix} \\
&= J^*J\begin{bmatrix} T(X)^T((Y) + \delta^*(X^*)\delta(Y))u \\ (\delta(X)T(Y) + \delta(X)\delta(Y))u \end{bmatrix} \\
&= J^*J\begin{bmatrix} T(XY)u \\ \delta(XY)u \end{bmatrix} \\
&= J^*J\lambda(XY, u) = J^*\lambda_1(XY, u) \\
&\quad \left\langle \lambda(z,v), J^*J^j{}_T(X)\lambda(Y,u)\right\rangle \\
&\quad \langle \lambda_1(z,v), \lambda_1(XY,u)\rangle \\
&= \left\langle \begin{bmatrix} T_1(z)v \\ \delta_1(z)v \end{bmatrix}, \begin{bmatrix} T_1(XY)u \\ \delta_1(XY)u \end{bmatrix} \right\rangle \\
&= \langle v, T_1(z)^*T_1(XY)u\rangle + \langle v, \delta_1(z)^*\delta_1(XY)u\rangle \\
&= \langle v, (T_1(z)^*T_1(XY) + \delta_1(z)^*\delta_1(XY) \,|\, u\rangle \\
&= \langle v, (T_1(z^*)T_1(XY) + \delta_1{}^*(z^*)\delta_1(XY) \,|\, u\rangle \\
&= \langle v, T_1(z^*XY)u\rangle
\end{aligned}$$

On the other hand,

$$J^*J\lambda(Y,u) = J^*\lambda_1(Y,u).$$

So, $\langle \lambda(z,v), j_T(X)J^*J\lambda(Y,u)\rangle$

$$= \langle J_T(X^*)\lambda(z,v), J^*\lambda_1(Y,u)\rangle$$

$$= \langle J\, j_T(X^*)\lambda(z,v), \lambda_1(Y,u)\rangle$$

Now, $$j_T(X^*)\lambda(z,v) = \begin{bmatrix} T(X^*) & \delta(X)^* \\ \delta(X^*) & S(X^*) \end{bmatrix}\begin{bmatrix} T(z)v \\ \delta(z)v \end{bmatrix}$$

$$= \begin{bmatrix} (T(x^*)T(z)+\delta(X)^*\delta(z))v \\ (\delta(x^*)T(z)+\delta(X^*)\delta(z))v \end{bmatrix}$$

$$= \begin{bmatrix} T(X^*z)v \\ \delta(X^*z)v \end{bmatrix} (T(X)^* = T(X^*))$$

$$= \lambda(X^*z, v)$$

So,

$\langle \lambda(z,v), j_T(X)J^*J\lambda(Y,u)\rangle$

$$= \langle \lambda_1(X^*z,v), \lambda_1(Y,u)\rangle$$

$$= \left\langle \begin{bmatrix} T_1(X^*z)v \\ \delta_1(X^*z)v \end{bmatrix}, \begin{bmatrix} T_1(Y)u \\ \delta_1(Y)u \end{bmatrix} \right\rangle$$

$$= \langle v, (T_1(\underline{X}^*z)^*T_1(Y)+\delta_1(X^*z)^*\delta_1(Y))u\rangle$$

$$= \left\langle v, (T_1(z^*\underline{X})^*T_1(Y)+\delta_1^+(z^*X)^*\delta_1(Y))u \right\rangle$$

$$= \langle v, T_1(z^*XY)u\rangle$$

Hence $[J^*J, j_T(X)] = 0 \ \forall X$

Let $$\rho = \begin{bmatrix} \gamma & \alpha^* \\ \alpha & \beta \end{bmatrix}, \rho = J^*J.$$

$[\rho, j_T(X)] = 0$ has been proved.

Then, $$\begin{bmatrix} \gamma & \alpha^* \\ \alpha & \beta \end{bmatrix}\begin{bmatrix} T(X) & \delta^+(X) \\ \delta(X) & \delta(X) \end{bmatrix}$$

$$= \begin{bmatrix} T(X) & \delta^+(X) \\ \delta(X) & \delta(X) \end{bmatrix}\begin{bmatrix} \gamma & \alpha^* \\ \alpha & \beta \end{bmatrix}$$

Equivalently,

$$\rho\, j_T(X)\lambda(Y,u) = j_T(X)\rho\lambda(Y,u) \qquad \forall XY, u$$

Thus,

$$\begin{bmatrix} \gamma & \alpha^* \\ \alpha & \beta \end{bmatrix}\begin{bmatrix} T(X) & \delta^+(X) \\ \delta(X) & \delta(X) \end{bmatrix}\begin{bmatrix} T(Y) & u \\ \delta(Y) & u \end{bmatrix}$$

$$= \begin{bmatrix} T(X) & \delta^+(X) \\ \delta(X) & \delta(X) \end{bmatrix} J^* \begin{bmatrix} T_1(Y) & u \\ \delta_1(Y) & u \end{bmatrix}$$

Since $\rho\lambda(Y, u) = J^* J\lambda(Y, u) = J^*\lambda_1(Y, u)$

Taking $X = 1$ gives noting that

$$T(1) = 1$$

$$\delta(1) = 1$$

$$\delta(1) = 0$$

$$\delta^+(1) = 0$$

i.e. $$j_T(1) = 1,$$

$$\begin{bmatrix} \gamma T(Y)u + \alpha^* \delta(Y)u \\ \alpha T(Y)u + \beta\delta(Y)u \end{bmatrix} = J^* \begin{bmatrix} T_1(Y)u \\ \delta_1(Y)u \end{bmatrix}$$

Take inner product on both sides with $\lambda(1, v) = \begin{bmatrix} v \\ 0 \end{bmatrix}$ to set,

$$\langle v, (\gamma T(Y) + \alpha^* \delta(Y))u \rangle = \left\langle J\lambda(1, v), \begin{bmatrix} T_1(Y)u \\ \delta_1(Y)u \end{bmatrix} \right\rangle$$

$$= \left\langle \lambda_1(1, v), \begin{bmatrix} T_1(Y)u \\ \delta_1(Y)u \end{bmatrix} \right\rangle$$

$$= \left\langle \begin{bmatrix} v \\ 0 \end{bmatrix}, \begin{bmatrix} T_1(Y)u \\ \delta_1(Y)u \end{bmatrix} \right\rangle$$

$$= \langle v, T_1(Y)u \rangle$$

Thus,

$$T_1(Y) = \gamma T(Y) + \alpha^* \delta(Y) \forall Y.$$

❖❖❖❖❖

[16] Quantum Filtering to Signal Estimation and Image Processing:

The pixel at (X, Y) has an complex, amplitude $\psi_t(X, Y)$ at time t.$|\psi_t(X, Y)|^2 = I_t(X, Y)$ describes the intensity at (X, Y) while $\text{Arg}(\psi_t(X, Y)) = \varphi_t(X, Y)$ describes the phase at (X, Y) at time t.

ψ_t satisfies the 2-D wave equation.

$$\frac{\partial^2 \psi_t(X,Y)}{\partial t^2} - c^2 \Delta \psi_t(X,Y) = 0.$$

$$\Delta = \frac{\partial^2}{\partial X^2} + \frac{\partial^2}{\partial Y^2}.$$

Let $\psi_t = I_t e^{i\varphi_t}$

$$\frac{\partial \psi_t}{\partial t} = \left(\frac{\partial I_t}{\partial t} + i I_t \frac{\partial \varphi_t}{\partial t}\right) e^{i\varphi_t},$$

$$\frac{\partial^2 \psi_t}{\partial t^2} = \left[\frac{\partial^2 I_t}{\partial t^2} + 2i \frac{\partial I_t}{\partial t}\frac{\partial \varphi_t}{\partial t} - I_t \left(\frac{\partial \varphi_t}{\partial t}\right)^2 + i I_t \frac{\partial^2 \varphi_t}{\partial t^2}\right] e^{i\varphi_t}$$

$$\nabla \psi_t = (\underline{\nabla} I_t + i I_t \underline{\nabla} \varphi_t) e^{i\varphi_t}$$

$$\nabla^2 \psi_t = \Delta \psi_t = [\nabla^2 I_t + 2i(\underline{\nabla} I_t, \underline{\nabla} \varphi_t) + i I_t \nabla^2 \varphi_t - I_t |\underline{\nabla} \varphi_t|^2] e^{i\varphi_t}$$

Equating real and imaginary parts in the wave eqn. then gives

$$\frac{\partial^2 I_t}{\partial t^2} - I_t \left(\frac{\partial \varphi_t}{\partial t}\right)^2 - c^2 \nabla^2 I_t + c^2 I_t |\underline{\nabla} \varphi_t|^2 = 0,$$

$$I_t \frac{\partial^2 \varphi_t}{\partial t^2} - 2c^2 (\underline{\nabla} I_t, \underline{\nabla} \varphi_t) - c^2 I_t \nabla^2 \varphi_t = 0$$

or

$$\ddot{I}_t - c^2 \Delta I_t + I_t \left(c^2 |\underline{\nabla} \varphi_t|^2 - \dot{\varphi}_t^2\right) = 0$$

$$I_t \left(\ddot{\varphi}_t c^2 \Delta \varphi_t\right) - 2c^2 \left(\underline{\nabla} I_t, \underline{\nabla} \varphi_t\right) = 0$$

❖❖❖❖❖

[17] Image Transmission Through Quantum Channels:

Let $\delta(X, Y)$ be the complex amplitude of the image at pixel (X, Y) at the frequency w.

Each pixel has a rgb value. Let $w_1 \simeq$ red frequency $w_2 \simeq$ green frequency, $w_3 \simeq$ blue frequency. Then Image amplitude at (X, Y) is

$$S(X, Y, t) = \sum_{k=1}^{3} S_k(X,Y) \cos\left(w_k t + \phi_k(X,Y)\right)$$

The near field electromagnetic four potential produced by $S(X, Y, t)$ is taking

$$J_S^t(X, Y) = \sum_{k=1}^{3} \left(S_{k1}(X,Y)\widehat{X} + S_{k2}(X,Y)\widehat{Y}\right)$$

$$\cos(w_k t + \phi_k(X, Y))$$

or more generally, with polarization accounted for,

$$\underline{J}_s(t, X, Y) = \sum_{k=1}^{3} \{S_{k1}(X, Y)\cos(w_k t + \phi_{k1}(X, Y))\widehat{X}$$

$$+ S_{k2}(X, Y)\cos(w_k t + \phi_{k2}(X, Y_1))\,\widehat{y}\}$$

and

$$\underline{A}(t, X, Y, Z)$$

$$= \frac{\mu}{4\pi}\sum_{n, m=1}^{N} \frac{\underline{J}_s\left[t - \left((X - n\Delta)^2 + (y - m\Delta)^2 + z^2\right)^{1/2}\Big/c, n\Delta, m\Delta\right]}{\left((x - n\Delta)^2 + (Y - m\Delta)^2 + z^2\right)^{1/2}}$$

$$= \frac{\mu}{4\pi}\widehat{X}\sum_{k=1}^{3} \frac{S_{k1}(n, m)\cos\left\{\begin{matrix} w_k\left(t - \left((x - n\Delta)^2 + (y - m\Delta)^2 + z^2\right)^{\frac{1}{2}}\Big/c\right) \\ + \phi_{k1}(n, m)\end{matrix}\right\}}{\left((x - n\Delta)^2 + (y - m\Delta)^2 + z^2\right)^{1/2}}$$

The electromagnetic four potential $\underline{A}(t, \underline{r})$ and $\phi(t, \underline{r})$ an linear functions of the image intesite and phases. Denote by $f_t[m, n], 1, \leq m, n \leq N, t \geq 0$ the image amplitude values. Then,

$$\underline{A}(t, \underline{r}) = \sum_{m, n=1}^{N} \int_0^t \underline{\underline{K}}(t - \tau, \underline{r}, m, n)\, \underline{f_\tau}[m, n]\, d\tau$$

where $\underline{f}\tau[m, n]$ is a 2 × 1 electric field vector corresponding to the amplitude at the $(m, n)^{th}$ pixel. Where $\underline{\underline{K}}$ is a 3 × 2 matrix valued kernal defined on $\mathbb{R} + X\mathbb{R}^3 x\{1, 2,., N\}^2$. Let ρ_0 denote the initial density operator of a quantum system. Assume that the system is a 3-D harmonic oscillates with Hamiltonian

$$H_0 = \frac{1}{2}\sum_{k=1}^{3}(p_k^2 + \alpha_k^2 q_k^2) = \frac{1}{2}\left(\underline{p}^T \underline{p} + \underline{q}^T \underline{\underline{D}} q\right)$$

After interaction with the image *em* field, it becomes

$$H(t) = \frac{1}{2}\left(\underline{p} + e\underline{A}\right)^2 + \frac{1}{2}\underline{q}^T \underline{\underline{D}} q - e\phi$$

or upto linear orders in $\underline{A}$, assumes the coulomb gauge condition $\phi = 0$, we get

$$H(t) \approx \frac{1}{2}\left(p^2 + \underline{q}^T \underline{\underline{D}}\, \underline{q}\right) - \frac{ie}{2}[div\underline{A} + 2(\underline{A}, \underline{\nabla})]$$

where $O(A^2)$ terms have been neglected. The state of the system evolves as

$$\rho'(t) = -\frac{i}{2}\left[p^2 + \underline{q}^T \underline{\underline{D}}\, \underline{q}, \rho(t)\right]$$

$$+\frac{e}{2}\left[-div\,A-2i(\underline{A},-\underline{p}),\rho(t)\right]$$

$\rho(0) \simeq$ known. Let

$$\rho(t) = \rho_0(t)+e\rho_1(t)+O(e^2)$$

Then,

$$\rho_0'(t) = -i[H_0,\rho_0(t)],$$

$$\rho_1'(t) = -i[H_0,\rho_1(t)]$$

$$+\frac{1}{2}[-i\,div\,\underline{A}+2(\underline{A},\underline{p}),\rho_0(t)]$$

So,

$$\rho_0(t) = T_t(\rho(0)) \qquad .T_t = \exp(-itadH_0)$$

and

$$\rho_1(t) = -\frac{1}{2}\int_0^t T_t-\tau(ad\,div\,\underline{A}\tau+2iad(\underline{A},\underline{p}))(\rho(\tau))\,d\tau$$

Let

$$H_{0k} = \frac{1}{2}\left(p_k^2+\alpha_k^2q_k^2\right), k = 1, 2, 3.$$

The eignstates of

$$H_0 = \sum_{k=1}^{3} H_{0k}$$

are

$$|n_1\,n_2\,n_3\rangle, n_1, n_2, n_3 = 0, 1, 2, ...$$

with

$$H_0|n_1\,n_1\,n_3\rangle = \left(\sum_{j=1}^{3}\left(n_j+\frac{1}{2}\right)\alpha_j\right)|n_1\,n_2\,n_3\rangle.$$

Let

$$a_k = (p_k-i\alpha_kq_k)/\sqrt{2}, 1\le k\le 3.$$

Then

$$\left[a_k,a_j^+\right] = \delta_{kj}\alpha_k$$

$$H_0 =$$

❖❖❖❖❖

[18] Entanglement Theory in Quantum Mechanics:

Let $\{|e_\alpha\rangle : 1\le\alpha\le N\}$ and $\{|f_\alpha\rangle : 1\le\alpha\le N\}$ be respectively $ONB'S$ for Hilbert spaces $\mathcal{H}_1$ and $\mathcal{H}_2$.

Consider the following pure state in $\mathcal{H}_1\otimes\mathcal{H}_2$:

$$|\psi\rangle = \sum_{\alpha=1}^{N}\sqrt{p(\alpha)}|e_\alpha\rangle\otimes|f_\alpha\rangle$$

where $$p(\alpha) \geq 0, \sum_{\alpha=1}^{N} p(\alpha) = 1.$$

There if A, B are respectively observables in $\mathcal{H}_1$ and $\mathcal{H}_2$, we have

$$\langle \psi | A \otimes B | \psi \rangle = \sum_{\alpha, \beta=1}^{N} \sqrt{p(\alpha) p(\beta)} \langle e_\alpha | A | e_\beta \rangle \langle f_\alpha | B | f_\beta \rangle$$

In particular if $\{|e_\alpha\rangle\}$ is an eigenbasis for A, there $\langle e_\alpha | A | e_\beta \rangle = \lambda(\alpha) \delta_{\alpha, \beta}$ and we get

$$\langle \psi | A \otimes B | \psi \rangle = \sum_{\alpha=1}^{N} \lambda(\alpha) p(\alpha) \langle f_\alpha | B | f_\alpha \rangle$$

Stochastic optimal control in financial economics

$$dP_k(t) = (r_k(t)\, dt + s_k(t)\, dB_k(t))\, P_k(t)$$
$$1 \leq k \leq d.$$
$$P_k(t) \simeq \text{Price of the } k^{\text{th}} \text{ share.}$$

$r_k(.)$ and $\sigma_k(.)$ are adopted process which are of the form

$$r_k(t) = \tilde{r}_k(t) \cdot \underline{P}(t')$$
$$\sigma_k(t) = \tilde{\sigma}_k(t) \underline{P}(t).$$

Number of shares of type k held in time $(t, t + dt)$ is $N_k(t)$.

Consumption process per unit time = $C(t)$.

Total wealth at time T is

$$X(T) = \sum_{k=1}^{d} \int_0^T N_k(t)\, d\, P_k(t) - \int_0^T C(t)\, dt$$

$$\mathbb{E} X(T) = \sum_{k=1}^{d} \int_0^T \mathbb{E}\{N_k(t) r_k(t) P_k(t)\}\, dt - \int_0^T \mathbb{E}\{C(t)\}\, dt$$

must be maximized subject to constraints

$$\mathbb{E}\{\varphi_m(\underline{r}(t), \underline{\sigma}(t), \underline{N}(t), C(t), \underline{P}(t))\} = 0$$
$$m = 1, 2, .., N$$
$$\underline{N}(t) = \tilde{\underline{N}}(t, \underline{P}(t)), C(t) = \tilde{C}(t, \underline{P}(t))$$

are allowed. $\tilde{\underline{r}}, \tilde{\underline{\sigma}}, \tilde{\underline{N}}, \mathcal{C}$ functions must be determined. Solution based on dynamic programming method. Let $\lambda_m(t)$ be (non-random) Lagrange multiplies. Maximize

$$\sum_{k=1}^{d} \int_0^T \mathbb{E}\{N_k(Ar_k(t) P_k(t))\}\, dt - \int_0^T \mathbb{E}\{C(t)\}\, dt$$

$$-\int_0^T \lambda(t)^T \mathbb{E}\{\underline{\varphi}(\underline{r}(t), \underline{\sigma}(t), \underline{N}(t), C(t), \underline{P}(t))\}\,dt$$

Let

$$= \underline{N}(t)\max, \underline{r}(t), C(t), \sum_{k=1}^{d} \int_S^T \mathbb{E}\{N_k(t) r_k(t) P_k(t) \mid \underline{P}(s)\}\,dt\; \sigma(t)$$

$$s \le t \le T$$

$$-\int_S^T \underline{\lambda}(t)^T \mathbb{E}\{\underline{\varphi}(\underline{r}(t), \underline{\sigma}, (t), N(t), C(t), \underline{P}(t)) \mid \underline{P}(s)\}\,dt$$

$$\equiv V(S, \underline{P}(s))$$

A Hamiltonian-Jacoli equation $V(S, P, \lambda)$ with an I to correction kevm can be claimed.

Remarks from Helgason.

$$G = KAN \text{ Iwasawa decomposition.}$$

$$g = k(g)\exp(H(g))\,n(g).$$

$$m \in M' \cdot M' \simeq \text{ normalizer of } d \text{ in } K, ak(\tilde{r}) = A.$$

$$b = m^{-1}a^{-1}m, a \in A.$$

$$b\bar{n}b^{-1} = m^{-1}a^{-1}m\bar{n}m^{-1}am$$

$$\bar{n} = k(\bar{n})h(\bar{n})n(\bar{n}),\; h(\bar{n}) = \exp(H(\bar{n})) \in A.$$

$$b\bar{n} = bk(\bar{n})h(\bar{n})n(\bar{n})$$

$$= bk(\bar{n})b^{-1}\,bh(\bar{n})\,n(\bar{n}).$$

Since $bk(\bar{n})b^{-1} \in K$, we get

$$H(b\bar{n}) = \log(bh(\bar{n})) = \log b + H(\bar{n}).$$

$$b\bar{n}b^{-1} = bk(\bar{n})\,b^{-1}bh(\bar{n})\,n(\bar{n})b^{-1}$$

$$= bk(\bar{n})b^{-1}\,bh(\bar{n})b^{-1}\,bn(\bar{n})b^{-1}.$$

Since $bn(\bar{n})\,b^{-1} \in N$, we get

$$H(b\bar{n}\,b^{-1}) = H(bh(\bar{n})b^{-1})$$

$$= H(\bar{n})$$

$$bk(\bar{n}) = bk(\bar{n})b^{-1}b$$

so $\quad H(bk(\bar{n})) = \log b$, so $H(bk(\bar{n})) = \log b + H(b\bar{n}b^{-1}) - H(\bar{n})$.

Suppose instead we use the decomposition

$$G = NAK$$

$$g = n(g)\,h(g)\,k(g), \quad h(g) = \exp(H(g)),$$

Then

$$\bar{n} = n(\bar{n})h(\bar{n})k(\bar{n})$$

$$b\bar{n}\,b^{-1} = bn(\bar{n})b^{-1}bh(\bar{n})b^{-1}bk(\bar{n})b^{-1}$$

and so,

$$H(b\bar{n}b^{-1})\log bh(\bar{n})b^{-1} = bh(\bar{n}) = H(\bar{n})$$

so $\quad H(b\bar{n}b^{-1}) = H(\bar{n})$

Further, $\quad H(bk(\bar{n})) = H(b) = \log b.$

Hence $\quad H(bk(\bar{n})) = \log b + H(b\bar{n}b^{-1}) - H(\bar{n})$

(P. 149) →

P. 136

$$S'_{\lambda,s} = \int_{\bar{N}_s} \delta_{m_s k(\bar{n}_s)M} \exp\left(-(i\lambda+\rho)\left(H(\bar{n}_s)\right)\right)d\bar{n}_s$$

So

$$S'_{\lambda,s}(F) = \int_{K/M} F\,dS'_{\lambda,S}$$

$$= \int F(m_s k(\bar{n}_s)M)\,\exp\left(-(i\lambda+\rho)\,(H(\bar{n}_s 1))\right)\,d\bar{n}_s$$

Robot in the presence of quantum noise.

$$H = \frac{1}{2}\underline{p}^T \underline{\underline{J}}(\underline{q})^{-1}\underline{p} - V(\underline{q})$$

$$dU(t) = \left[-\left(iH + \frac{\varepsilon^2}{2}LL^*\right)dt + \varepsilon\,LdA - \varepsilon\,L^*dA^*\right]U(t)$$

$\mathfrak{h}_0$ = system Hilbert space = $L^2(\mathbb{R}^2)$

Bath space $= \Gamma s(L^2(\mathbb{R}_+))$

$L \in \delta(\mathfrak{h}_0)$. Let $\quad L = \underline{\alpha}^T\underline{q} + \underline{\beta}^T\underline{p}$

$$U(t) = U_0(t) + \varepsilon\,U_1(t) + \varepsilon^2 U_2(t) + O(\varepsilon^3)$$

$$dU_0 = -iHU_0 dt$$

$$dU_1 = -iHU_1 dt + (LdA - L^*dA^*)U_0$$

$$dU_2 = -iHU_2 dt - \frac{1}{2}LL^*U_0 dt + (LdA - L^*dA^*)U_1$$

Solution: $\quad U_0(t) = \exp(-itH)$

$$U_1(t) = \int_0^t U_0(t-\tau)(LdA(\tau) - L^*dA^*(\tau))U_0(\tau)\,d\tau$$

$$U_2(t) = \int_0^t U_0(t-\tau)\Big(-\frac{1}{2}LL^*U_0(\tau)d\tau$$

$$+ LdA(\tau) - L^*dA^*(\tau))U_1(\tau)d\tau$$

Let $\{B(t), t \geq 0\}$ be standard $BM, B(0) = 0$. Calculate the law of

$$\tilde{T}_a = \min\{t \geq 0 \,||\, B(t)| = a\}, a > 0.$$

Solution:

$$\tilde{T}_a = T_a \Lambda T_{-a} T_x = \min\{t \geq 0\}$$

$$B(t) = a\}, a \in \mathbb{R}.$$

Now by Doob's optional stopping theorem.

$$\mathbb{E}\exp\left(\lambda B(T_a) - \lambda^2 T_a / 2\right) = 1, \quad (1)$$

$$\mathbb{E}\left[\exp(-\lambda B(\tilde{T}_a) - \lambda^2 \tilde{T}_a / 2)\right] = 1, \quad (2)$$

$$\mathbb{E}\left[\exp(-\lambda B(\tilde{T}_a) - \lambda^2 \tilde{T}_a / 2)\right] = 1 \quad (3)$$

From (2) and (3), on adding,

$$\mathbb{E}\left[\exp(-\lambda^2 \tilde{T}_a / 2)\cosh(\lambda B(\tilde{T}_a)\right] = 1$$

But $B(\tilde{T}_a) \in \{\pm a\}$. So cosh $(\lambda B(\tilde{T}_a))$ = cosh (λ_a).

Thus, $\mathbb{E}\left[\exp(-\theta\tilde{T}_a)\right] = \operatorname{sech}\left(a\sqrt{2\theta}\right), \theta \geq 0$

We replace $\underline{d}(t)$ in the disturbance observer equation by = $\tilde{\underline{d}}(t)$ where

$$\tilde{\underline{d}}(t) = \underset{\underline{d}}{\arg\min} \int | i\psi(\underline{\theta}) + \frac{1}{2}\underline{\nabla}_\theta^T \underline{\underline{J}}(\underline{\theta})^{-1}\underline{\nabla}_\theta \psi_t(\theta)$$

$$- V(\underline{\theta})\psi_t(\underline{\theta}) + \underline{T}(t)^T \underline{\theta}\psi_t(\theta) + \underline{d}^T \underline{\theta}\psi_t(\underline{\theta})|^2 \, d\underline{\theta}$$

We then get

$$\operatorname{Re}\left\{\int \underline{\theta}\psi_t(\underline{\theta})\left[-\overline{\dot{\psi}_t(\underline{\theta})} + \frac{1}{2}\underline{\nabla}_\theta^T \underline{\underline{J}}(\underline{\theta})^{-1}\underline{\nabla}_\theta \overline{\psi_t(\underline{\theta})} - V(\underline{\theta})\overline{\psi_t(\underline{\theta})} + \underline{T}(t)^T \underline{\theta}\overline{\psi_t(\underline{\theta})}\right]d\underline{\theta}\right\}$$

$$= -\left[\int \underline{\theta}\,\underline{\theta}^T |\psi_t(\underline{\theta})|^2 \, d\underline{\theta}\right]\tilde{\underline{d}}(t)$$

i.e.

$$\tilde{\underline{d}}(t) = -\left[\int \underline{\theta}\,\underline{\theta}^T |\psi_t(\underline{\theta})|^2 \, d\underline{\theta}\right]^{-1}$$

$$\times \operatorname{Re}\left\{\int \underline{\theta}\psi_t(\underline{\theta})\left[-i\dot{\psi}(\underline{\theta}) + \frac{1}{2}\underline{\nabla}_\theta^T \underline{\underline{J}}(\underline{\theta})^{-1}\underline{\nabla}_\theta \overline{\psi_t(\underline{\theta})}\right.\right.$$

$$\left.\left. - V(\underline{\theta})\overline{\psi_t(\underline{\theta})} + \underline{T}(t)^T \underline{\theta}\overline{\psi_t(\underline{\theta})}\right]d\underline{\theta}\right\}$$

A damped quantum system that generalizes the Classical Robot equation (the damping arising from feedback)

$$\underline{\underline{J}}(\underline{\theta})\underline{\theta}'' + G(\underline{\theta}, \underline{\theta}') = T(t) - k_p(\underline{\theta}'\ \underline{\theta}_d) - k_d(\dot{\underline{\theta}} - \dot{\underline{\theta}}_d) + \underline{d}(t)$$

can be modelled using the Sudarshan Lindblad equation

$$i\frac{\partial \rho_t}{\partial t} = [H(t), \rho_t] - \frac{i}{2}(L^* L\rho_t + \rho_t L^* L - 2L\rho_t L^*)$$

The dual of this equation for observables is

$$\frac{dX(t)}{dt} = i[H(t), X(t)]$$

$$-\frac{1}{2}(X(t)L^* L + L^* LX(t) - 2L^* X(t)L)$$

Let

$$H(t) \equiv H(\underline{\theta}, \underline{p}, t)$$

$$= \frac{1}{2}\underline{p}^T \underline{\underline{J}}(\underline{\theta})\underline{p} + V(\underline{\theta}) - T(t)^T \underline{\theta}$$

$$= +\frac{1}{2}k_p \| \underline{\theta} - \underline{\theta}_d(t) \|^2 \text{ and}$$

Quantum Robotics

$$H_0 = \frac{1}{2}\underline{p}^T M(\underline{q})^{-1} \underline{p} - V(\underline{q})$$

Without external torque.

$$\frac{d\underline{q}}{dt} = \frac{\partial H_0}{\partial \underline{p}} = M(\underline{q})^{-1}\underline{p}$$

$$\frac{d\underline{p}}{dt} = \frac{\partial H_0}{\partial \underline{q}} = -\frac{1}{2}\frac{\partial}{\partial \underline{q}}\underline{p}^T \underline{\underline{M}}(\underline{q})^{-1}\underline{p} - V'(\underline{q})$$

$$\frac{\partial M(\underline{q})^{-1}}{\partial \underline{q}} = \left[\frac{\partial M^{-1}}{\partial q_1}, \ldots, \frac{\partial M^{-1}}{\partial q_d}\right] \in \mathbb{R}^{d\times 2d}$$

So,

$$\left(\frac{\partial M(q)^{-1}}{\partial \underline{q}}\right)^T = \begin{bmatrix} \frac{\partial M^{-1}}{\partial q_1} \\ \vdots \\ \frac{\partial M^{-1}}{\partial q_d} \end{bmatrix} \in \mathbb{R}^{2d\times d}$$

$$M^T = M \cdot \frac{\partial}{\partial \underline{q}}\left(\underline{p}^T \underline{\underline{M}}(\underline{q})^{-1}\underline{p}\right) = \left(\frac{\partial M(q)^{-1}}{\partial \underline{q}}\right)^T \underline{p}$$

$$\neq \left[p^T \cdot\cdot\, p^T\right] = \left(p^T \otimes \underline{\underline{Id}}\right)\left(\frac{\partial M^{-1}}{\partial \underline{q}}\right)^T \underline{p}$$

Desired equation of motion with external torque and $p-d$ controller:

$$\frac{d\underline{p}}{dt} = -\frac{1}{2}\frac{\partial}{\partial \underline{q}}\left(\underline{p}^T \underline{\underline{M}}(\underline{q})^{-1}\underline{p}\right) - V'(\underline{q})$$

$$+F(\underline{q})\left(k_p(\underline{q}_d(t)-\underline{q})\right)$$

$$+k_d(\underline{p}_d(t)-\underline{p})+\underline{\tau}(t)$$

Try realizing this Sudarshan-Lindblad equation:

$$\frac{dX}{dt} = i[-H(t), X(t)] - \frac{1}{2}(L^*LX + XL^*L - 2L^*XL)$$

$$H(t) = \frac{1}{2}\underline{p}^T \underline{\underline{M}}(\underline{q})^{-1}\underline{p} + V(\underline{q}) - \underline{\tau}(t)^T \underline{q}$$

Then,
$$\frac{dp_\alpha}{dt} = i\left[\frac{1}{2}\underline{p}^T \underline{\underline{M}}(q)^{-1}\underline{p} + V(\underline{q}), p_\alpha\right] + \tau_\alpha(t)$$

$$-\frac{1}{2}\left(L^*Lp_\alpha + p_\alpha L^*L - 2L^*p_\alpha L\right)$$

$$L^*Lp_\alpha + p_\alpha L^*L - 2L^*p_\alpha L$$

$$= L^*[L_1 p_\alpha] + [p_\alpha, L^*]L$$

Let
$$L = \psi(\underline{t}, \underline{q}, \underline{p})$$

Then,
$$[L, p_\alpha] = i\frac{\partial\psi}{\partial q_\alpha}, [p_\alpha, L^*] = -i\frac{\partial\psi^*}{\partial q_\alpha}$$

We require that

$$-\frac{i}{2}\left(\psi^*\frac{\partial\psi}{\partial q_\alpha} - \frac{\partial\psi^*}{\partial q_\alpha}\psi\right)$$

$$= k_p X_{1\alpha}(\underline{q}, p) + k_d X_{2d}(\underline{q}, \underline{p})$$

$$1 \le d \le d.$$

We consider a slightly more general case with the Lindblad term

$$-\frac{1}{2}(\psi^*\psi X + X\psi^*\psi - 2\psi^* X\psi)$$

$$-\frac{1}{2}(\varphi^*\varphi X + X\varphi^*\varphi - 2\varphi^* X\varphi)$$

so that

$$-\frac{i}{2}\left(\psi^*\frac{\partial\psi}{\partial q_\alpha} - \frac{\partial\psi^*}{\partial q_\alpha}\psi\right) = \chi_{1\alpha}(\underline{q}, p)$$

and

$$-\frac{i}{2}\left(\varphi^*\frac{\partial\varphi}{\partial q_\alpha}-\frac{\partial\varphi^*}{\partial q_\alpha}\varphi\right) = \chi_{2\alpha}(\underline{q},\underline{p})$$

Let us for the moment assume commutativity so that

$$\mathrm{Im}\left(\psi^*\frac{\partial\psi}{\partial q_\alpha}\right) = \chi_{1\alpha}$$

$$\mathrm{Im}\left(\varphi^*\frac{\partial\varphi}{\partial q_\alpha}\right) = \chi_{2\alpha}$$

Then,

Let

$$\psi(t,\underline{q},\underline{p}) = \sum_n C_n(t)\psi_n(\underline{q},\underline{p}) \qquad C_n(t),$$

$$\varphi(t,\underline{q},\underline{p}) = \sum_n d_n(t)\varphi_n(\underline{q},\underline{p}) \qquad d_n(t)\in\mathbb{R}$$

where $\{\psi_n\}$, are $\{\varphi_n\}$ basis function. We get

$$\sum_{n,m} C_n(t)C_m(t)\,\mathrm{Im}\left(\bar{\psi}_n{}^*(\underline{q},\underline{p})\frac{\partial\psi_m(\underline{q},\underline{p})}{\partial q_\alpha}\right) = \chi_{1\alpha}(\underline{q},\underline{p}) \tag{1}$$

$$\sum_{n,m} d_n(t)d_m(t)\,\mathrm{Im}\left(\bar{\varphi}_n(\underline{q},\underline{p})\frac{\partial\varphi_m(\underline{q},\underline{p})}{\partial q_\alpha}\right) = \chi_{2\alpha}(\underline{q},\underline{p}) \tag{2}$$

We assume that the function

$$Q_{nm\alpha}(\underline{q},\underline{p}) = \mathrm{Im}\left(\bar{\psi}_n(q,p)\frac{\partial\psi_m}{\partial q_\alpha}(\underline{q},\underline{p})\right)$$

$$\chi_{nm\alpha}(\underline{q},\underline{p}) = \mathrm{Im}\left(\bar{\varphi}_n(\underline{q},\underline{p})\frac{\partial\varphi_m}{\partial q_\alpha}(\underline{q},\underline{p})\right)$$

$n, m \geq 1$, $\alpha = 1, 2, ., d$ are linearly independent. We assume $\{C_n(t), d_n(t)\}$ to be classical random variables and average over than so that the Sudarshan Lindblad equation is actually

$$\begin{aligned}\frac{dX}{dt} &= i[H,X]-\frac{i}{2}(\mathbb{E}\,(\psi^*\psi X+X\psi^*\psi-2\psi^*X\psi)\\&\quad+\mathbb{E}\,(\varphi^*\varphi X+X\varphi^*\varphi-2\varphi^*X\varphi))\\&= i[H,X]-\frac{i}{2}\sum_{n,m}\mathbb{E}\,\{C_n(t)C_m(t)\}\left(\psi_n{}^*\psi_m X\right.\\&\quad\left.+X\psi_n{}^*\psi_m-2\psi_n{}^*X\psi_m\right)\end{aligned}$$

$$-\frac{i}{2}\sum_{n,m}\mathbb{E}\left\{d_n(t)d_m(t)\right\}\left(\varphi_n{}^*\varphi_m X + X\varphi_n{}^*\varphi_m - 2\varphi_n{}^* X\varphi_m\right)\Big)$$

Let $\quad \rho_{nm}(t) = \mathbb{E}\left\{C_n(t)C_m(t)\right\}$

$$\sigma_{nm}(t) = \mathbb{E}\left\{d_n(t)d_m(t)\right\}$$

Then (1) and (2) get replaced by

$$\sum_{nm}\rho_{nm}(t)\theta_{nm\alpha}\left(\underline{q},\underline{p}\right) = \chi_{1\alpha}(\underline{q},\underline{p}),$$

$$\sum_{nm}\sigma_{nm}(t)\chi_{nm\alpha}(\underline{q},\underline{p}) = \chi_{2\alpha}(\underline{q},\underline{p}),$$

(**Note** that we can assume that $\varphi_n = \psi_n \forall_n$).

Note: Suppose

$L = \sum_{k=1}^{p} C_k L_k$ where L_k an operators and $C_k' s$ Jointly distributed real random variables.

Then

$$T(X) \equiv \mathbb{E}\left\{L^* LX + XL^* L - 2L^* LXL\right\}$$

$$= \sum_{k,m=1}^{p}\mathbb{E}\left\{C_k C_m\right\}\left(L_k{}^* L_m X + XL_k{}^* L_m - 2L_k{}^* XL_m\right)$$

Consider the Hermitian matrix

$$\underline{\underline{C}} = ((\mathbb{E}\{C_k\, C_m\}))_{1\le k,\, m\le p}$$

Let $\underline{\underline{C}} = \sum_{k=1}^{p}\underline{\xi}_k\underline{\xi}_k{}^*$ (spectral decomposition) where the ξ_k, s are orthogonal.

Then

$$T(X) = \sum_{k=1}^{p}\left(\sum_r \xi_k[r]L_r{}^*\right)\left(\sum_r \overline{\xi_k[t]}\,L_r\right)X$$

$$+X\left(\sum_r \xi_k[r]L_r{}^*\right)\left(\sum_r \overline{\xi_k[r]}L_r\right)$$

$$-2\left(\sum_r \xi_k[r]L_r{}^*\right)X\left(\sum_r \overline{\xi_k[r]}\,L_r\right)$$

$$= \sum_{k=1}^{p}\left(M_k{}^* M_k X + XM_k{}^* M_k - 2M_k{}^* XM_k\right)$$

where

$$M_k = \sum_r \overline{\xi_k[r]}\, L_r$$

The Sudarshan-Lindblad eqn. this becomes

$$\frac{dX}{dt} = i[H, X] - \frac{1}{2}\sum_{k=1}^{p}\left(M_k{}^*[M_k, X] + [X, M_k{}^*]M_k\right)$$

Writing $M_k = M_k\left(\underline{q}, \underline{p}\right)$, we get

$$[M_k, q_\alpha] = -i\frac{\partial M_k}{\partial p_\alpha},$$

$$[M_k, p_\alpha] = i\frac{\partial M_k}{\partial q_\alpha},$$

$$[q_\alpha, M_k{}^*] = i\frac{\partial M_k{}^*}{\partial p_\alpha},$$

$$[p_\alpha, M_k{}^*] = -i\frac{\partial M_k{}^*}{\partial q_\alpha}$$

We with therefore (ignoring non-commutativity) to choose $\{M_k\}$ so that

$$-\frac{i}{2}\sum_{k=1}^{p}\left[M_k{}^*\frac{\partial M_k}{\partial p_a} - \frac{\partial M_k{}^*}{\partial p_\alpha}M_k\right] = F_\alpha(\underline{q}, \underline{p})$$

and

$$-\frac{i}{2}\sum_{k=1}^{p}\left[M_k{}^*\frac{\partial M_k{}^*}{\partial p_\alpha} - \frac{\partial M_k{}^*}{\partial q_\alpha}M_k\right] = G_\alpha(\underline{q}, \underline{p})$$

i.e.

$$\sum_{k=1}^{p}\mathrm{Im}\left(\frac{\partial M_k^*}{\partial p_\alpha}M_k\right) = F_\alpha,$$

$$\sum_{k=1}^{p}\mathrm{Im}\left(\frac{\partial M_k^*}{\partial q_\alpha}M_k\right) = -G_\alpha,\ 1 \le \alpha \le d$$

Helgason Volume III, P. 571

$$\eta(X)f = Xf - \frac{(n-2)}{2n}(div\,X)f$$

$$\eta(Y)f = (X)f = YXf - \frac{(n-2)}{2n}(div\,Y)Xf$$

$$-\left(\frac{(n-2)}{2n}\right)(Y(div\,X)f + (div\,X)Yf)$$

$$+\left(\frac{n-2}{2n}\right)^2(div\,Y)(div\,X)f$$

$$\therefore \qquad [\eta(Y), \eta(X)]f = [Y, X]f - \left(\frac{n-2}{2n}\right)((div\,Y)X - (div\,X)Y)f$$

$$-\left(\frac{n-2}{2n}\right)((Y(div\,X) - X(div\,Y))f$$

$$+((div\,X)Y - (div\,Y)X)f)$$

$$= [Y, X]f - \left(\frac{n-2}{2n}\right)(Y(divX) - X(div\,Y))f$$

$$= \eta([Y, X])f = [Y, X]f - \left(\frac{n-2}{2n}\right)div([Y, X])f$$

$$div([Y, X]) = \left(Y^j X,^i_j - X^j Y,^i_j\right)_{,i}$$

$$= Y^j X,^i_{ji} - X^j Y,^i_{ji}$$

$$= Y(div\,X) - X(div\,X)$$

So, $$[\eta(Y), \eta(X)] = [Y, X]f - \left(\frac{n-2}{2n}\right)div([Y, X])f$$

$$= \eta([Y, X])f$$

Suppose $f_{,ii} = 0$. Then

$$(\eta(X)f)_{,ii} = (Xf)_{,ii} - \left(\frac{n-2}{2n}\right)((div\,X)f)_{,ii}$$

$$= \left(X^j f_{,j}\right)_{,ii} - \left(\frac{n-2}{2n}\right)\left(X^j_{,j} f\right)_{,ii}$$

$$= X^j_{,ii} f_{,j} + X^j f_{,jii} + 2X^j_{,i} f_{,ij}$$

$$-\left(\frac{n-2}{2n}\right)\left(X^j_{,jii} f + X,^i_j f_{,ii} + 2X^j_{,ji} f_{,i}\right)$$

$$= \left(X^i_{,ij} - \left(\frac{n-2}{n}\right)X^j_{,ij}\right)f_{,i} + 2X^j_{,i} f_{ij} - \left(\frac{n-2}{2n}\right)X^j_{,jii} f$$

Helgason Volume III, P. 138

$$\sum\nolimits^+_S(1) \cup \{\beta\} = \left(\sum\nolimits^+_0 \cap S^{a1-1} \sum\nolimits^-_0\right) \cup \left\{\left(S_{\alpha 2} \ldots S_{\alpha p}\right)^{-1} \alpha_1\right\}$$

$$\sum\nolimits^+_S = \sum\nolimits^+_0 \cap S^{-1} \sum\nolimits^-_0 = \sum\nolimits^+_0 \cap \left(S_{\alpha 2} \ldots S_{\alpha p}\right)^{-1} S^{-1}_{\alpha 1} \sum\nolimits^-_0$$

$$\sum\nolimits^+_{S(1)} \cup \{\beta\} = \left(\sum\nolimits^+_0 \cap \left(S_{\alpha 2} \ldots S_{\alpha p}\right)^{-1} \sum\nolimits^-_0\right) \cup$$

$$\left\{\left(S_{\alpha 2}\ldots S_{\alpha p}\right)^{-1}\alpha_1\right\}$$

$$S^{(1)} = S_{\alpha 2}\ldots S_{\alpha p}$$

$$S = S_{\alpha 1}S^{(1)} = S_{\alpha 1}\ldots S_{\alpha p}.$$

TPT

$$\sum\nolimits_0^+ \cap \left(S_{\alpha 2}\ldots S_{\alpha p}\right)^{-1} S_{\alpha 1}^{-1} \sum\nolimits_0^-$$

$$= \left(\sum\nolimits_0^+ \cap S_{\alpha 2}\ldots S_{\alpha p}\right)^{-1} \sum\nolimits_0^- \cup \left\{\left(S_{\alpha 2}\ldots S_{\alpha p}\right)^{-1}\alpha_1\right\}$$

i.e.

$$\sum\nolimits_0^+ \cap S_{\alpha p}\ldots S_{\alpha 2}S_{\alpha 1}\sum\nolimits_0^- = \left(\sum\nolimits_0^+ \cap S_{\alpha p}\ldots S_{\alpha 2}\sum\nolimits_0^-\right) \cup \left\{\left(S_{\alpha p}\ldots S_{\alpha 2}\right)\alpha_1\right\}$$

$$= \left(\sum\nolimits_0^+ \cup S_{\alpha p}\ldots S_{\alpha 2}\,\alpha_1\right) \cap \left\{S_{\alpha p}\ldots S_{\alpha 2}\right\} \left(\sum\nolimits_0^- \cup \{\alpha_1\}\right)$$

$$= \left(\sum\nolimits_0^+ \cup S_{\alpha p}\ldots S_{\alpha 2}\alpha_1\right) \cap S_{\alpha p}\ldots S_{\alpha 2}S_{\alpha 1}\left(S_{\alpha 1}\sum\nolimits_0^- \cup \{-\alpha_1\}\right)$$

Consider $-\alpha_1 \in \sum\nolimits_0^-$

Then $$S_{\alpha 2}(-\alpha_1) = -\alpha_1 + 2\frac{\langle \alpha_1, \alpha_2\rangle}{\langle \alpha_2, \alpha_2\rangle}\alpha_2 = -\alpha_1 + \alpha_{12}\alpha_2$$

is a negative root since $\langle \alpha_1, \alpha_2\rangle < 0$.

$$S_{\alpha 3}\ldots S_{\alpha 2}(-\alpha_1) = -\alpha_1 + \alpha_{12}\alpha_2 + \alpha_{13}\alpha_3 - \alpha_{12}\alpha_{23} + \alpha_3$$

is again a negative root since

$$\alpha_{12} < 0,\ \alpha_{13} - \alpha_{12}\alpha_{23} < 0\ (\alpha_{13} < 0,\ \alpha_{12},\ \alpha_{23} < 0).$$

$S_{\alpha 3}\ldots S_{\alpha 2}(-\alpha_1)$ is a negative root.

So $\sum\nolimits_0^+ \cap S_{\alpha p}\ldots S_{\alpha 2}\sum\nolimits_0^- = \sum\nolimits_0^+ \cap S_{\alpha p}\ldots S_{\alpha 2}\left(\sum\nolimits_0^- / \{-\alpha_1\}\right)$

$$S_{\alpha 2}(-\alpha_2) = \alpha_2,\ S_{\alpha 3}S_{\alpha 2}(-\alpha_2) = \alpha_2 - \alpha_{23}\alpha_3 \text{ is } a + ve \text{ root etc.}$$

so

$S_{\alpha p}\ldots S_{\alpha 2}(-\alpha_2)$ is a $+$ *ve* root

$S_{\alpha 2}(-\alpha_3)$ is a $-$ *ve* root etc.

$S_{\alpha p}\ldots S_{\alpha 2}(-\alpha_3)$ is a $-$ *ve* root.

Thus $\sum\nolimits_0^+ \cap S_{\alpha p}\ldots S_{\alpha 2}\sum\nolimits_0^-$

Also, $\sum\nolimits_0^+ \cap S_{\alpha p}\ldots S_{\alpha 3}\{\alpha_2\}$

$$S_{\alpha 2}\alpha_1 = \alpha_1 - \alpha_{12}\alpha_2 \text{ is a positive root continuing}$$

$S_{\alpha p}\ldots S_{\alpha 2}\,\alpha_1$ is a + *ve* root.

Now the + *ve* roots in

$S_{\alpha p}S_{\alpha p-1}S_{\alpha 1}\sum_0^-$

are $\quad S_{\alpha p}\ldots S_{\alpha 2}\,\{-\alpha_1\} = S_{\alpha p}\ldots S_{\alpha 2}\,\{\alpha_1\}$

Thus, $\sum_0^+ \cap S_{\alpha p}\ldots S_{\alpha 1}\sum_0^-$

$$= \sum_0^+ \cap S_{\alpha p}\ldots S_{\alpha 2}\{\alpha_1\}.$$

So we must prove that

$$\sum_0^+ \cap S_{\alpha p}\ldots S_{\alpha 2}\{\alpha_1\} = \left(\sum_0^+ \cap S_{\alpha p}\ldots S_{\alpha 3}\{\alpha_2\}\right)\cup\left(S_{\alpha p}\ldots S_{\alpha 2}\{\alpha_1\}\right)$$

Since $S_{\alpha p}\ldots S_{\alpha 3}\,\{\alpha_2\}$

If $S_{\alpha p}\ldots S_{\alpha 2}\,\{\alpha_1\}$ is indivisible and $S_{\alpha p}\ldots S_{\alpha 3}\,\{\alpha_2\}$ is divisible, we are through.

Helgason Vol. 3. Geometric analysis on symmetric spaces.

$$B = K/M \quad g \in G.$$

$$KM \xrightarrow{g} gkM, \qquad k \in K$$

$$\begin{aligned} A(Kk, kM) &= A(Mk^{-1}gK) \\ &= A\left(Mn(k^{-1}g)e^{A(k^{-1}g)}k(k^{-1}g)K\right) \end{aligned}$$

Let $m \in M$. Then $m \in K$ and

$$\begin{aligned} &A\left(Mn(k^{-1}g)e^{A(k^{-1}g)}k(k^{-1}g)K\right) \\ &\qquad = A\left(Mn(k^{-1}g)m^{-1}\,e^{A(k^{-1}g)}\,mk(k^{-1}g)K\right) \\ &\qquad = A\left(n_1e^{A(k^{-1}g)}K\right) \end{aligned}$$

(where $mn\left(k^{-1}g\right)m^{-1} \in N$)

$$= A\left(k^{-1}g\right)$$

$$A(gK, kM) \equiv A(g, k) = A\left(k^{-1}g\right).$$

$$\frac{dg^{-1}(b)}{db} = \frac{d(g^{-1}kM)}{d(kM)}$$

K/M is parametrized by the root vectors $\{X_\alpha\}$ in it, the Lie algebra of K.

$$g^{-1} k = n(g^{-1}k)e^{A(g^{-1}k)}$$

$$g^{-1}kM = n(g^{-1}k)e^{A(g^{-1}k)}k(g^{-1}k)M$$

$$ne^A k_0 M = e^A e^{-A} ne^A k_0 M$$

Writing $dn = X_\alpha$, we get

$$e^{-A}dne^A = e^{-ad(A)}(X_\alpha)$$

$$= e^{-\alpha(A)}X_\alpha$$

Thus, the Jacobian determinant of

$$n \to e^{-A}ne^A$$

is $\prod_{\alpha \in \Delta t} e^{-\alpha(A)} = e^{-2\rho(A)}$

Thus the Jacobian determinant of

$$n \to ne^A k_0 M$$

is also the same.

Now we compute the Jacobian determinant of

$$A \to ne^A k_0 M$$

We note that

$$nAk_0 M = nAn^{-1}k_0 M$$

and $nAn^{-1} = Ad(n)A$

For $n = e^{tX\alpha} \quad Ad(n)A = e^{tad(X\alpha)}(A)$

$$= A - t\alpha(A)X_\alpha$$

and hence $A \to ne^A k_0 M$ has Jacobian determinant 1.

Example of a Martingale

[From Revuz and Yor (continuous Martingale and Brownian motion)].

Let $\{X_t, t \geq 0\}$ be a continuous Martingale and define $X_t{}^* = \sup_{0\leq s\leq t} |X_s|$. $X_t{}^*$ is an increasing process and hence has bounded variation. Thus $(dX^*{}_t)^2 = 0$. Let

$dL_t = \delta(X_t)d\langle X\rangle_t$,

$$d\langle X\rangle_t = (dX_t)^2 . L_t = \int_0^t \delta(X_s)d\langle X\rangle_s,$$

is the local time of X at 0.

Let $M_t = |X_t|\log^+ X_t{}^* - X_t{}^* V| - \int_0^t \log^+ X_s{}^* dL_s$

where $\log^+ x = 0v \log x = \log(xvi), x > 0.$

$$X_t^* \, vi = \exp(\log^+ X_t^*).$$

Hence $\quad d(X_t^* vi) = (X_t^* vi) d\log^+ X_t^*$

and $\quad dM_t = (\log^+ X_t^*)(d|X_t| - dL_t) + (|X_t| - X_t^* vi)\, d\log+ X_t^*$

Now, $(X_t^* vi)\, d\log^+ X_t^* = (|X_t|\, vi)\log + X_t^*$

Since X_t^* changes only if $X_t^* = |X_t|$.

Also, $\quad d|X_t| = dL_t + \delta gn(X_t)\, dX_t$

$$\equiv dL_t + dW_t$$

where $\quad W_t = \int_0^t Sgn(X_s)\, dX_s$ is a Martingale.

Thus $\quad dM_t = (\log^+ X_t^*) dW_t + (|X_t| - |X_t| vi) d\log^+ X_t^*$

Now,

$$(|X_t| - |X_t| vi)\, d(\log^+ X_t^*)$$

$$= (X_t^* - X_t^* vi) d\log^+ X_t^* = 0$$

Since if $X_t^* > 1$, then $X_t^* - X_t^* vi = 0$

and if $X_t^* < 1$, then $d\log^+ - X_t^* = 0$

$$\begin{aligned} d\log^+ X_t^* &= d[\log(X_t^*)\theta(X_t^* - 1)] \\ &= \left[\frac{1}{X_t^*}\theta(X_t^* - 1) + \log(X_t^*)\delta(X_t^* - 1)\right] dX_t^* \\ &= \frac{\theta(X_t^* - 1)}{X_t^*} dX_t^* = 0 \text{ if } X_t^* < 1. \end{aligned}$$

Thus, $\quad M_t = \int_0^t \log^+(X_s)\, dW_s$

$$= \int_0^t \log^+(X_s)\, Sgn(X_s)\, dX_s \text{ is a Martingale.}$$

❖❖❖❖❖

[19] Some Other Remarks Related to The Above Problem:

$$\begin{aligned} d(|X_t| \log^+ |X_t|) &= |X_t| d\log^+ |X_t| + (\log^+ |X_t|) d|X_t| \\ &\quad + d|X_t| d\log^+ |X_t| \end{aligned}$$

Now,

$$\frac{d}{dx}\log^+ x = \frac{\theta(x-1)}{x}$$

$$\frac{d^2}{dx^2}\log^+ x = -\frac{\theta(x-1)}{x^2} + \delta(x-1)$$

So,
$$d\log^+|X_t| = \frac{\theta(|X_t|-1)}{|X_t|}d|X_t|$$
$$+\frac{(\delta(|X_t|-1)-\theta(|X_t|-1)}{d|X_t|^2}(d|X_t|)^2$$
$$= d(\text{Martingale}) + \frac{\theta(|X_t|-1)}{|X_t|}dL_t$$
$$+\frac{(\delta(|X_t|-1)-\theta(|X_t|-1))}{d|X_t|^2}d\langle X\rangle_t$$
$$d|X_t|.d\log^+|X_t| \; \frac{\theta(|X_t|-1)}{|X_t|}d\langle X\rangle_t$$

Thus,
$$d(|X_t|\log^+|X_t|) = d(\text{Martingale}) + \frac{(\delta(|X_t|-1)-\theta(|X_t|-1))}{2|X_t|}d\langle X\rangle_t$$
$$+(\log^+|X_t|dL_t) + \frac{\theta(|X_t|-1)}{|X_t|}d\langle X\rangle_t$$

$\frac{\theta(|X_t|-1)}{|X_t|+t}dL_t = 0$ since $dL_t > 0$ iff $|X_t| = 0)$

Also $\log^+|X_t|dL_t = 0$ since $dL_t > 0$ iff $|X_t| < 0$

Thus, $d(|X_t|\log^+|X_t|) = d\,\text{Martingale})$
$$+\frac{1}{2|X_t|}(\delta|X_t|-1)+\theta(|X_t|-1)d\langle X\rangle_t$$
$$= \left\{\frac{\delta(|X_t|-1)}{2}+\frac{\theta(|X_t|-1)}{2|X_t|}\right\}d\langle X\rangle_t$$

Quantization of a string:
$$\mathcal{L} = \frac{1}{2}\eta_{\mu\nu}\left(\dot{X}_\mu\dot{X}^\nu - \dot{X}^{\mu'}X^{\nu'}\right)$$
$$X^\mu = X^\mu(\tau,\sigma), \dot{X}^\mu = \frac{\partial X^\mu}{\partial\tau}, X^{\mu'} = \frac{\partial X^\mu}{\partial\sigma}$$
$$\pi_\mu = \frac{\partial\mathcal{L}}{\partial\dot{X}^\mu} = \eta_{\mu\nu}\dot{X}^\nu = \dot{X}_\mu$$
$$\mathcal{H} = \pi_\mu\dot{X}^\mu - \mathcal{L} = \frac{1}{2}(\pi,\pi)+\frac{1}{2}(X,X)$$
$$= \mathcal{H}\left(\{\pi_\mu\},\{X^\mu\}\right)$$

$$(\pi, \pi) = \eta_{\mu\nu}\pi^{\mu}\pi^{\nu}, (X, X) = \eta_{\mu\nu}X^{\mu}X^{\nu}$$

Let

$$X^{\mu}(\tau, \sigma) = \sum_{n \in Z} X_n^{\mu}(\tau)\exp\left(j\frac{2\pi n\sigma}{L_0}\right) \quad 0 \le \sigma \le L_0$$

$\mathcal{H}$ = Hamiltonian density (Linear),

$\mathcal{L}$ = Lagrangian density (Linear). Then

$$\begin{aligned}
L\left(\{X_n^{\mu}, \dot{X}_n^{\mu}\}_{\mu, n}\right) &= \int_0^{L_0} \pounds d\sigma \\
&= \sum_{n, m \in Z} \frac{1}{2}\left(\dot{X}_n, \dot{X}_m\right)\int_0^{L_0} \exp(j2\pi(n+m)\sigma / L_0)d\sigma \\
&\quad + \frac{1}{2}\sum\left(X_n, X_m\right)\frac{(2\pi)^2(nm)}{L_0^2}\cdot\int_0^{L_0}\exp(j2\pi(n+m)\sigma / L_0)d\sigma \\
&= \frac{L_0}{2}\sum_{n \in Z}\left(\dot{X}_n, \dot{X}_{-n}\right) - \frac{2\pi^2}{L_0}\sum_{n \in Z} n^2\left(X_n, X_{-n}\right)
\end{aligned}$$

After normalization of time and length,

$$L\left(\{X_n^{\mu}, X_n^{\mu}\}\right) = \frac{1}{2}\sum_{n \in Z}\left\{\dot{X}_n, \dot{X}_{-n} - n^2(X_n, X_{-n})\right\}$$

$$X_{-n}^{\mu} = \bar{X}_n^{\mu} \text{ since } X^{\mu}(\tau, \sigma) \text{ is real.}$$

$$\therefore \quad L = \frac{1}{2}(\dot{X}_0, \dot{X}_0) + \text{Re}\sum_{n \ge 1}\left\{\left(\dot{X}_n, \dot{X}_{-n}\right) - n^2(X_n, X_{-n})\right\}$$

Let $X_n^{\mu} = X_{Rn}^{\mu} + jX_{In}^{\mu}$

Then $X_{-n}^{\mu} = X_{Rn}^{\mu} - jX_{In}^{\mu}$

So,

$$\text{Re}\left(\dot{X}_n, \dot{X}_{-n}\right) = \left(\dot{X}_{Rn}, \dot{X}_{Rn}\right) + \left(\dot{X}_{In}, \dot{X}_{In}\right)$$

$$\text{Re}\left(X_n, X_{-n}\right) = \left(X_{Rn}, X_{Rn}\right) + \left(X_{In}, X_{In}\right)$$

$$L = \frac{1}{2}\left(\dot{X}_0, \dot{X}_0\right) + \sum_{n \ge 1}\left(\dot{X}_{Rn}, \dot{X}_{Rn}\right) - n^2\left(X_{Rn}, X_{Rn}\right) + \left(\dot{X}_{In}, \dot{X}_{In}\right) - n^2\left(X_{In}, X_{In}\right)\Big\}$$

We may then assume that

$$L = \frac{1}{2}\left(\dot{X}_0, \dot{X}_0\right) + \frac{1}{2}\sum_{n \ge 1}\left\{\left(\dot{X}_n, \dot{X}_n\right) - n^2\left(X_n, X_n\right)\right\}$$

where $X^{\mu}_{-n}(\tau)$, $0 \le \mu \le d$, $n \ge 0$ are real function.

This is our String Lagrangian in the fourier series representation. String Hamiltonian in the fourier series representation:

$$\pi_{\mu n} = \frac{\partial L}{\dot{X}^{\mu}_{n}} = \dot{X}_{\mu n} = \eta_{\mu\nu}\dot{X}^{\nu}_{n}$$

$$H = \sum_{\mu, n} \pi_{\mu n} X^{\mu}_{n} - L = \sum_{n \ge 0} \left(\dot{X}_n, \dot{X}_n\right) - L$$

$$= \frac{1}{2}\left(\dot{X}_0, \dot{X}_0\right) + \frac{1}{2}\sum_{n \ge 1}\left\{(\dot{X}_n, \dot{X}_n)\right.$$

$$\left. + n^2 (X_n, X_n)\right\}$$

$$= \frac{1}{2}(\pi_0, \pi_0) + \frac{1}{2}\sum_{n \ge 1}\left\{(\pi_n, \pi_n) + n^2 (X_n, X_n)\right\}$$

Commutation relations for quantization:

$$\left[X^{\mu}_{n}, \pi^{\nu}_{m}\right] = i\eta^{\mu\nu}\delta[n-m]$$

The corresponding Lagrangian will then have to be taken as

$$L = \frac{1}{2}(\dot{X}_0, \dot{X}_0) + \frac{1}{2}\sum_{n \ge 1}\left\{(\dot{X}_n, \dot{X}_n) - n^2 (X_n, X_n)\right\}$$

Equation of motion:

$$\frac{d}{dt}\frac{\partial L}{\partial X^{\mu}_{0}} = 0 \Rightarrow \ddot{X}_0 = 0$$

$$X^{\mu}_{0}(\tau) = \alpha^{\mu}_{0}\tau + \beta^{\mu}_{0}$$

$$\frac{d}{dt}\frac{\partial L}{\partial \dot{X}^{\mu}_{n}} = \frac{\partial \mathcal{L}}{\partial X^{\mu}_{n}} \Rightarrow \ddot{X}^{\mu}_{n} = -n^2 X^{\mu}_{n}$$

$$X^{\mu}_{n}(\tau) = \alpha^{\mu}_{n} e^{-jn\tau} + \beta^{\mu}_{n} e^{jn\tau}$$

$$X^{\mu}(\tau, \sigma) = \sum_{n \in Z} X^{\mu}_{n}(\tau)\exp(jn\sigma)$$

$$= \alpha^{\mu}_{0}\tau + \beta^{\mu}_{0} + 2\,\mathrm{Re}\sum_{n \ge 1} X^{\mu}_{n}(\tau) e^{jn\sigma}$$

$$= \alpha^{\mu}_{0}\tau + \beta^{\mu}_{0} + \sum_{n \ge 1}\left\{\alpha^{\mu}_{n}\exp(-jn(\tau - \sigma))\right.$$

$$+ \beta^{\mu}_{n}\exp(jn(\tau + \sigma))$$

$$\left. + \bar{\alpha}^{\mu}_{n}\exp(jn(\tau - \sigma)) + \bar{\beta}^{\mu}_{n}\exp(-jn(\tau + \sigma))\right\}$$

Creation and annihilation operators:

$$a_n^\mu = \frac{1}{\sqrt{2}}\left(\pi_n^\mu - inX_n^\mu\right)$$

$$a_n^{\mu *} = \frac{1}{\sqrt{2}}\left(\pi_n^\mu + in\,X_n^\mu\right)\cdot n \geq 1$$

$$a_0^\mu = \frac{1}{\sqrt{2}}\pi_0^\mu .$$

$$\begin{aligned}(a_n, a_n{}^*) &= \eta_{\mu\nu} a_n^\mu a_n^{\nu *} \\ &= \frac{1}{2}\{(\pi_n, \pi_n) + n^2(X_n, X_n) + in\{(\pi_n, X_n) - (X_n, \pi_n)\} \\ (\pi_n, X_n) - (X_n, \pi_n) &= \eta_{\mu\nu}\left[\pi_n^\mu, X_n^\nu\right] = -i\eta_{\mu\nu}\eta^{\mu\nu} = -i\delta_\mu^\nu \\ &= i(d+1)\end{aligned}$$

So,

$$\begin{aligned}(a_n, a_n^+) &= \frac{1}{2}\left\{(\pi_n, \pi_n) + n^2(X_n, X_n)\right\} \\ &\quad + \frac{n(d+1)}{2}\end{aligned}$$

Thus,

$$\begin{aligned}H &= \frac{1}{2}(\pi_0, \pi_0) + \frac{1}{2}\sum_{n \geq 1}\left\{(\pi_n, \pi_n) + n^2(X_n, X_n)\right\} \\ &= (a_0, a_0{}^*) + \sum_{n \geq 1}(a_n, a_n{}^*)\end{aligned}$$

$$a_0 = \pi_0 = a_0{}^* = a_0^2 + \sum_{n \geq 1}(a_n, a_n{}^*)$$

With negled of an infinite constant.

$$\begin{aligned}\left[a_n^\mu, a_m^{\nu *}\right] &= \frac{in}{2}\left[\pi_n^\mu, X_m^\nu\right] - \frac{in}{2}\left[X_n^\mu, \pi_m^\nu\right] \\ &\left(\frac{n}{2}\delta[n-m] + \frac{n}{2}\delta[n-m]\right)\eta^{\mu\nu} \\ &= n\eta^{\mu\nu}\delta[n-m].\end{aligned}$$

We have the following Hesinberg equations of motion:

$$\begin{aligned}\frac{da_n^\mu(t)}{dt} &= j\left[H, a_n^\mu\right] \\ &= i\sum_{m \geq 1}\left[(a_m, a_m{}^*), a_n^\mu\right] \\ &= i\sum_{m \geq 1} m\eta_{\alpha\beta} a_m^\beta\left[a_m^{\alpha *}, a_n^\mu\right]\end{aligned}$$

$$= -i \sum_{m \geq 1} \eta_{\alpha\beta} m a_m^{\beta} \delta[m-n] \eta^{\alpha\mu}$$

$$= -i \sum_{m \geq 1} n a_m^{\mu} \delta[m-n] = -ina_n^{\mu}, n \geq 1$$

Thus, $$a_n^{\mu}(t) = a_n^{\mu}(0) \cdot \exp(-i^n t), n \geq 1,$$

$$0 \leq \mu \leq d$$

$$\pi_n^{\mu} = \frac{1}{\sqrt{2}}\left(a_n^{\mu} + a_n^{\mu *}\right) = \dot{X}_n^{\mu}$$

$$X_n^{\mu} = \frac{\left(a_n^{\mu *} + a_n^{\mu}\right)}{\sqrt{2} in}$$

So,

$$X_n^{\mu}(t) = \frac{\left\{a_n^{\mu}(0) * \exp(i^n t) - a_n^{\mu}(0) \exp(-i^n t)\right\}}{in\sqrt{2}},$$

$$X_n^{\mu}(t) = \frac{\left\{a_n^{\mu}(0) \exp(-i^n t) - d_n^{\mu}(0) * \exp(i^n t)\right\}}{\sqrt{2}}$$

$\therefore$
$$L = \frac{1}{2}\left(\dot{X}_0, \dot{X}_0\right) + \frac{1}{4} \sum_{n \geq 1} (a_n(0) \exp(-i^n t) + a_n(0) * \exp(i^n t),$$

$$a_n(0) \exp(-i^n t) + (0) * \exp(i^n t))$$

$$+ \frac{1}{4} \sum_{n \geq 1} (a_n(0) * \exp(i^n t) - a_n(0) \exp(-i^n t),$$

$$a_n(0) * \exp(int) - a_n(0) \exp(-i^n t))$$

$$= \frac{1}{2}(\dot{X}_0, \dot{X}_0) + \frac{1}{2} \sum_{n \geq 1} (a_n(0), a_n(0)) \exp(-2int)$$

$$+ (a_n(0)^*, a_n(0)^*) \exp(2int)$$

Current problems in image processing

1. Principal component based image comparision

2. Histogram based optimization.

3. Diffusion of images for the removed of sharp edge.

4. Noise removed from images defined on a C^{∞}-mainfold.

5. Diffusion of edges on a C^{∞}-manifold.

$$\frac{\partial u(t, \underline{X})}{\partial t} = \frac{1}{2}\Delta u(t, \underline{X})\left(\frac{g}{2}\right)^{-1/2} \frac{\partial}{\partial X^{\alpha\beta}}\left(g^{\alpha\beta}\sqrt{g}\,\frac{\partial u}{\partial X^{\alpha}}\right)$$

Solution to the diffusion equation of a manifold using perturbation theory.

$$g_{\alpha\beta} = \delta_{\alpha\beta} + \varepsilon\, h_{\alpha\beta}\,.$$

$$g = 1 + \varepsilon h + 0(\varepsilon^2)$$

$$g^{\alpha\beta} = \delta_{\alpha\beta} - \varepsilon h_{\alpha\beta} + 0(\varepsilon^2)$$

6. Recovery of images from noisy Radon transform.

$$\frac{1}{2}\Delta u = \frac{1}{2} g^{\alpha\beta} \frac{\partial^2 u}{\partial X^{\alpha}\partial X^{\beta}} + \frac{1}{2} g^{-1/2} \frac{\partial (g^{\alpha\beta}\sqrt{g})}{\partial X^{\beta}} \frac{\partial u}{\partial x^{\alpha}}$$

Laplacian on a semi simple Lie algebra

$$g = \hbar \otimes \bigoplus_{\alpha\varepsilon\Delta} g_{\alpha}$$

$$= \hbar \otimes \bigoplus_{\alpha\varepsilon\Delta+} (g_{\alpha} \oplus g_{\alpha})$$

$$\{H_1, H_2, \ldots, H_t\} \simeq ONB \text{ for } \hbar$$

$\{X_{\alpha}, X_{-\alpha}, \alpha = 1, 2, \ldots, n\}$ root vectors.

$$B(H_{\alpha}, H_{\beta}) = \delta_{\alpha\beta}, 1 \le \alpha, \beta \le l$$

$$B(H_{\alpha}, H_{\beta}) = 0$$

Since $\forall H \in \hbar, 0 = B([H_{\alpha}, H], X_{\beta})$

$$= B(H_{\alpha}, [H, X_{\beta}]) = \beta(H) B(H_{\alpha}, X_{\beta})$$

Choosing $H \in \hbar\, 1\, s.t.\, \beta(H) \neq 0$ gives.

$$B(H_{\alpha}, H_{\beta}) = 0$$

$$-\alpha(H) B(X_{\alpha}, X_{\beta}) = B([X_{\alpha}, H], X_{\beta}) = B(X_{\alpha}, [H, X_{\beta}])$$

$$= \beta(H) B(X_{\alpha}, X_{\beta})$$

$$\Rightarrow (\alpha(H) + \beta(H)) B(X_{\alpha}, X_{\beta}) = 0 \forall H \in \hbar$$

$$\Rightarrow \quad B(X_{\alpha}, X_{\beta}) = 0 \text{ if } \alpha + \beta \neq 0$$

Let $\lambda_{\alpha} = B(X_{\alpha}, X_{-\alpha})\ \alpha \in \Delta +$

Using these identify and taking as our metric $(X, Y) \to B(X, Y)$, we can write down the Laplace-Beltramic operator on g in terms of $H_{\alpha}, X_{\alpha}, X_{-\alpha}$.

❖❖❖❖❖

[20] Image Processing from the Em Field View Point:

There *e-m* waves satisfy the wave equation but are randomly polarized.

Let $((F_{\mu\nu}))$ denote the *em* field tensor comparing from an object. The $(m, n)^{th}$ pixel located at $(m\Delta, n\Delta, 0)$ $(1 \le m, n \le N)$ has an amplitude $S_{mn}(t)$ and assuming that spherical waves travel from it, the total signal amplitude at $\underline{r}$ due to this image field is

$$\psi(t, \underline{r}) = \sum_{m, n=1}^{N} \frac{S_{mn}\left(t - \frac{\left((X - m\Delta)^2 + (Y - n\Delta)^2 + z^2\right)^{1/2}}{C}\right)}{\left((X - m\Delta)^2 + (Y - n\Delta)^2 + z^2\right)^{1/2}}$$

Assume that a light ware described by an electric field $\underline{E}_i(t, \underline{r})$ falls on the image. At the site of the $(m, n)^{th}$ pixel at $(m\Delta, n\Delta, 0)$.

This incident field gets filtered by the black white characteristic of pixel. We denote this filter by $h_{mn}(t)$. In the frequency domain, the resulting electric field that is presents at the $(m, n)^{th}$ pixel site so $\underline{\underline{H}}_{m, n}(w)\hat{\underline{E}}(w, r)$

where $\underline{r} = \underline{r}_{mn} = (m\Delta, n\Delta, 0)$.

The induced surface magnetic current density at this pixel site is

$$\begin{aligned}\underline{M}(w, m, n) &= \left(\hat{z} X \hat{\underline{E}}(w, \underline{r}_{mn})\right) H_{mn}(w) \\ &= \left[-\hat{E}_y(w, r_{mn})\hat{x} + \hat{E}_x(w, \underline{r}_{mn})\,\hat{y}\right] H_{mn}(w)\end{aligned}$$

and the result electric vector potential at $\underline{r}$ can be computed using retarded potentials,

Application of spherical function transform to image processing

G = KAN, Iwasawa over decomposition.

$\varphi \simeq$ spherical finction *i.e.*

$$\int_k \varphi(xky)\, dk = \varphi(X)\varphi(Y), X, Y \in G$$

$$\begin{aligned}\therefore \quad &\left(\int_G f(x)\varphi(x^{-1})\,dx\right)\left(\int_G g(y)\varphi(y^{-1})\,dy\right) \\ &= \int_{K\times G\times G} f(x)g(y)\varphi\left(x^{-1}\, ky^{-1}\right) dk\, dx\, dy \\ &= \int f(x)\, g\left(z^{-1}x^{-1}k\right)\varphi(z)\, dk\, dx\, dz \\ &= \int f(x)\, g\left(z^{-1}x^{-1}\right)\varphi(z)\, dx\, dz\end{aligned}$$

$$= \int f(x) g\left(zx^{-1}\right) \varphi\left(z^{-1}\right) dx\, dz$$

$$= \int \varphi\left(z^{-1}\right)\left(\int f(y^{-1}z) g(y)\, dy\right) dz$$

So if $\quad C_{\varphi}(f) = \int f(x) \varphi\left(x^{-1}\right) dx$

Then $\quad C_{\varphi}(f {*t}\, g) = (\varphi(f)\, C_{\varphi}(g))$

where $f, g \in C^{\#}(G)$ the space of k-binvariat function on G.

❖❖❖❖❖

[21] CAR:

$$\left\langle F^{+}_{m'}(t) e(u), F^{+}_{m}(t) e(v)\right\rangle \qquad J^{+}(t) = J(t)$$

$$t = \left\langle e(u), F_{m'}(t)\, F^{+}_{m}(t) e(v)\right\rangle$$

$$= \left\langle \int_0^t J(t_1)\, dA^{+}_{m'}(t_1) e(u), \int_0^t J(t_2)\, dA^{+}_{m'}(t_2) e(u)\right\rangle$$

$$d\left(F_{m'}(t)\, F^{+}_{m}(t)\right) = J(t) F^{+}_{m}(t)\, dA_{m'}(t)$$

$$+ F_{m'}(t) J(\theta\, dA^{+}_{m}(t)) + d\langle\langle m', m\rangle\rangle(t)$$

$$d\left(F^{+}_{m}(t) F_{m'}(t)\right) = J(t) F_{m'}(t)\, dA^{+}_{m}(t) + F^{+}_{m}(t) J(t)\, dA_{m'}(t)$$

$$d\left\{F_{m'}(t), F^{+}_{m}(t)\right\} = d\langle\langle m', m\rangle\rangle(t) + \left\{J(t), F^{+}_{m}(t)\right\} dAm'(t)$$

$$+ \{J(t), F_{m'}(t)\}\, dA^{+}_{m}(t) \left\langle e(u), \{J(t), F_{m}{}^{+}(t)\} e(v)\right\rangle$$

$$= \left\langle J(t) e(u), \int_0^t J(t_1)\, dA^{+}_{m}(t_1) e(v)\right\rangle d\left\{J(t) F^{+}_{m}(t)\right\}$$

$$= dA^{+}_{m}(t) - 2J(t) F^{+}_{m}(t)\, d\Lambda_1(t) - 2dA^{+}_{m}(t)$$

$$= -2J(t) F^{+}_{m}(t)\, d\Lambda_1(t) - dA^{+}_{m}(t)\,.$$

$$d(F^{+}_{m}(t) J(t)) = 2F^{+}_{m}(t) J(t)\, d\Lambda_1(t) + dA^{+}_{m}(t)$$

$$d\left\{J(t), F^{+}_{m}(t)\right\} = -2\left\{J(t), F^{+}_{m}(t)\right\} d\Lambda_1(t)$$

$$\left\{J(0), F^{+}_{m}(0)\right\} = 0 \Rightarrow \left\{J(t), F^{+}_{m}(t)\right\} = 0 \forall t$$

[22] An Identity Concerning Positive Definite Functions on a Group:

Let $\varphi \in P(G)(+ve$ definite function on $G)$.

Let $$f = \sum_i \alpha_i \varphi_0 \bar{x}_i^1, g = \sum_i \beta_i \varphi_0 y^{-1} i$$

$$x_i, y_i \in G, \alpha_i, \beta_i \in \mathbb{C}.$$

Suppose also $f = \sum_i \alpha_i' \varphi_0 x_i^{-1}$, ($\{x_i\}$ are definite and $\{y_i\}$ are also distinct)

$$g = \sum_i \beta_i' \varphi_0 y_2^{-1}.$$ Then,

$$\sum_{i,j} \alpha_i \bar{\beta}_j \varphi_0 \left(x_i^{-1} y_j\right) - \sum_{i,j} \alpha_i' \bar{\beta}_j' \varphi\left(x_i^{-1} y_j\right)$$

$$= \sum_{i,j} \left(\alpha_i \bar{\beta}_j - \alpha_i' \bar{\beta}_j'\right) \varphi\left(x_i^{-1} y_j\right)$$

$$= \sum_{i,j} \left((\alpha_i - \alpha_i')\bar{\beta}_j - \alpha_i'\left(\bar{\beta}_j - \bar{\beta}_j'\right)\right) \varphi\left(x_i^{-1} y_j\right)$$

$$\left(\sum_j \bar{\beta}_j, \left(\sum_i (\alpha_i - \alpha_i') \varphi_0 x_i^{-1}\right) 0 y_j = 0\right),$$

$$\sum_{i,j} \alpha_i' \left(\bar{\beta}_j - \bar{\beta}_j'\right) \varphi\left(x_i^{-1} y_j\right)$$

$$= \sum_{i,j} \alpha_i' \left(\bar{\beta}_j - \bar{\beta}_0'\right) \bar{\varphi}\left(y_i^{-1} x_i\right)$$

$$= \sum_i \alpha_i' \left(\overline{\sum_j \left(\beta_j - \beta_j'\right) \varphi_0 y_j^{-1}}\right) 0 x_i = 0$$

❖❖❖❖❖

[23] Example of Construction of a Spherical Function:

Let $K \subset G$ be a subgroup and $\pi : G \to U(\mathcal{H})$ a unitary representation of G in $\mathcal{H}$. Let

$$\mathcal{H}_k^\pi = \{\xi \in \mathcal{H} \mid \pi(k)\xi = \xi \forall k \in K\}$$

For $\xi, \eta \in \mathcal{H}_k^\pi$, put

$$\varphi(x) = \langle \xi, \pi(x)\eta \rangle, x \in G$$

Then $$\varphi(k_1 x k_2) = \left\langle \pi(k^{-1})\xi, \pi(x)\pi(k_2)\eta \right\rangle = \langle \xi, \pi(x)\eta \rangle = \varphi(x).$$

Let $$\int_k \pi(k)\, dk = P.$$

Then $$\int \varphi(xky)\, dk = \langle \xi, \pi(x) P \pi(y)\eta \rangle = \sum_\alpha \langle \xi, \pi(x) e_\alpha \rangle \langle e_\alpha, \pi(y)\eta \rangle$$

where $\{e_\alpha\}$ is an ONB for $\mathcal{H}_k^\pi$.

Writing $\varphi_{\xi,\eta}$ for φ, we get

$$\int_k \varphi_{\xi,\eta}(xky)\,dk = \sum_\alpha \varphi_{\xi}, e_\alpha(x)\varphi_{e\alpha,\eta}(y)$$

In particular, suppose

$$\lim \mathcal{H}_k^\pi = 1.$$

Let $\quad \xi \in \mathcal{H}_k^\pi, \quad \|\xi\| = 1$, *i.e.*

$$\mathcal{H}_k^\pi = \mathbb{C}\cdot\xi$$

Then $\quad \int_k \varphi_{\xi,\xi}(xky)\,dk = \varphi_{\xi,\xi}(x)\varphi_{\xi,\xi}(y)$

and $P_{\xi,\xi}$ its therefore a spherical function. Matric valued spherical functions: G is a group, $K \subseteq G$ is a compact subgroup, $\pi : G \to \mathbb{B}(\mathcal{H})$ is a representation of G is $\mathcal{H}, \pi / k$ is a unitary.

Let $$P = \int_k \pi(k)\,dk$$

Then $P^2 = P = P^*$. Let $\{e_\alpha\}_{\alpha=1}^\infty$ be an ONB for $P(\mathcal{H}) = \mathbb{R}(P)$. Define

$$\varphi_{\alpha\beta}(x) = \langle e_\alpha | \pi(x) | e_\beta \rangle$$

Then
$$\begin{aligned}\int_k \varphi_{\alpha\beta}(xky)\,dk &= \langle e_\alpha | \pi(x)P\pi(y) | e_\beta \rangle \\ &= \sum_\gamma \langle e_\alpha | \pi(x)e_\gamma \rangle \langle e_\gamma | \pi(y)e_\beta \rangle \\ &= \sum_\gamma \varphi_{\alpha\gamma}(x)\varphi_{\gamma\beta}(y)\end{aligned}$$

Or equivalently, defining

$$\Phi(x) = ((\varphi_{\alpha\beta}(x)),$$

We have

$$\int_k \Phi(xky)\,dk = \Phi(x)\cdot\Phi(y),\ x, y \in G.$$

$$\left[-\frac{1}{2}\frac{\partial}{\partial \underline{q}} + \underline{\underline{J}}(\underline{q})^{-1}\frac{\partial}{\partial \underline{q}}V(\underline{q})\right]\psi(\underline{q}) = E\psi(\underline{q}).$$

$$\underline{\underline{J}}(\underline{q}) \approx \underline{\underline{J}}_0 + J_0'(\delta\underline{q} \otimes \underline{\underline{I}})$$

$$\underline{\underline{J}}(\underline{q})^{-1}\underline{\underline{J}}_0$$

❖❖❖❖❖

[24] Spherical Functions in Image Processing:

$$\exp\left(i\langle \underline{\lambda}, \underline{x}\rangle\right) = f(x) \text{ say.}$$

$$G = O(n) \, s \, \mathbb{R}^n \text{ act on } \mathbb{R}^n$$

$$\underline{\lambda}, x \in \mathbb{R}^n .$$

$$k = O(n), X = G/K = \mathbb{R}^n.$$

$D \simeq$ Invariant differential operator on $C^\infty(X) \equiv \xi(X)$:

$$D = \sum_{k_1,\dots k_n} \underline{\underline{C}}(k_1 \dots k_n) \frac{\partial^{k_1+\dots+k_n}}{\partial x_1^{k_1} \dots \partial x_n^{k_0}}$$

Since $\underline{\underline{C}}(k_1,..,k_n)$ are constants, D is invariant under translation $= \mathbb{R}^n$.

If D is to be invariant under $O(n)$ to, then

$$D = \psi(\Delta), \Delta = \sum_{\alpha=1}^{n} \frac{\partial^2}{\partial x_\alpha^2}$$

When ψ is a polynomial.

Then $$Df(x) = \psi\left(-\|\lambda\|^2\right) f(x) \text{ . Let } g \in G .$$

$$\int_k f(gk.x)\,dk = \int_K \exp\left(i\langle \underline{\lambda}, gk.x\rangle\right) dk$$

$$= \int_k \exp(i\langle g^* \underline{\lambda}, k.\underline{x}\rangle dk)$$

Consider $$g = (k_0, \underline{\xi}), k_0 \in O(n) = K,$$

$$\underline{\xi} \in \mathbb{R}^n = X.$$

i.e. $$g.X = k_0 X + \xi, \quad X \in X.$$

Then $$\langle \lambda, gkx\rangle = \langle \lambda, k_0 kx + \xi\rangle$$

$$= \langle \lambda, \xi\rangle + \langle \lambda, k_0 kx\rangle .$$

So $$\int_K f(gkx)\,dk = \exp\left(i\langle \lambda, \xi\rangle\right) \int_K \exp\left(i\langle \lambda, kx\rangle\right) dk$$

$$= f(\xi)\varphi_\lambda(x).$$

Note that $$g \cdot 0 = k_0 \cdot 0 + \xi = \xi$$

So the above result can be expressed as

$$\int_k f(gkx)\,dk = f(g \cdot 0)\varphi_\lambda(x)$$

where $$\varphi_\lambda(x) = \int_k \exp\left(i\langle \lambda, kx\rangle\right) dk$$

Note that $\varphi_{\lambda}(kx) = \varphi_{\lambda}(k) \forall k \in k, x \in X.$

Also,

$$\psi(\Delta)\varphi_{\lambda}(x) = \int \psi\left(-\left\|k^T \underline{\lambda}\right\|^2 \exp\left(i\left\langle k^T \lambda, x\right\rangle\right) dk\right)$$

$$= \psi\left(-\|\underline{\lambda}\|^2\right) \int \exp\left(i\langle \lambda, kx\rangle\right) dk$$

$$= \psi\left(-\|\underline{\lambda}\|^2\right) \varphi_{\lambda}(x).$$

❖❖❖❖❖

[25] CAR from CCR (KRP Quantum Stochastic Calculus):

$$J(t) = W(0, U(t)),$$

$$U(t) = -\xi[0,t] + \xi(t,\infty)$$

ξ is a spectral measure on $\mathbb{R}^+$ with values in $\mathcal{H}$. $U(t)$ is a unitary operator in $\mathcal{H}$

Since $U^*(t) = -\xi[0,t] + \xi(t,\infty) = U(t),$

$$U^*(t)U(t) = -\xi[0,t] + \xi(t,\infty) = \xi(\mathbb{R}_+) = 1.$$

Note $\xi(A)\xi(B) = \xi(A \cap B).$

qsde for $J(t)$.

$$\langle e(u), J(t)e(v)\rangle = \langle e(u), e(U(t)v)\rangle$$

$$= \exp\left(\langle u, U(t)v\rangle\right)$$

Assume $\mathcal{H} = L^2(\mathbb{R}_+)$ and $\xi(B) = X_B$ (indicator). Then

$$\frac{d}{dt}\langle e(u), J(t)e(v)\rangle = \left(\frac{d}{dt}\langle u, U(t)v\rangle\right) J(t)$$

$$\langle u, U(t)v\rangle = -\int_0^t \overline{u(s)}v(s) + \int_t^{\infty} \overline{u(s)}v(s)$$

So $\frac{d}{dt}\langle u, U(t)v\rangle = -2\overline{u(t)}v(t).$

Thus, $\langle e(u), J(t)e(v)\rangle = -\int_0^t 2J(s)\bar{u}(s)v(s)\,ds$

More generally, for arbitrary $\mathcal{H}$,

$$\langle e(u), J(t)d\Lambda_1(t)e(v)\rangle = \langle e(u), J(t)e(v)\rangle d\left(\langle u, \xi v\rangle\right)(t)$$

$$\equiv \langle e(u), J(t)e(v)\rangle\, d\langle u, \xi[0,t]v\rangle$$

On the other hand.

$$\langle e(u), dJ(t)e(v)\rangle = d\langle e(u), J(t)e(v)\rangle$$

$$= J(t)\,d\langle u, U(t)v\rangle$$

$$\begin{aligned}\langle u, U(t)v\rangle &= -\langle u, (\xi[0,t]-\xi(t,\infty))v\rangle \\ &= -2\langle u, \xi[0,t]v\rangle + \langle u, v\rangle\end{aligned}$$

So, $$dJ(t) = -2J(t)\,d\Lambda_1(t)\,.$$

Now let m be a ξ-martingale and let

$$F_m(t) = \int_0^t J(s)\,dA_m(s)$$

Then $$F_m{}^*(t) = \int_0^t J(s)\,dA_m{}^*(s)$$

Then, $$\begin{aligned}J^2(t) &= W^2(0, U(t)) \\ &= W\left(0, U(t)^2\right) = W(0, I) = I.\end{aligned}$$

Thus,

$$\begin{aligned}dA_m(t) &= J(t)\,dF_m(t)\,, \\ dA_m{}^*(t) &= J(t)\,dF_m{}^*(t).\end{aligned}$$

We have by the quantum Ito formula,

$$\begin{aligned}d(F_{m1}{}^*(t)F_{m2}(t)) &= dF_{m1}{}^*(t)\,F_{m2}(t) + F_{m1}{}^*(t)\,dF_{m2}(t) \\ &\quad + dF_{m1}{}^*(t).dF_{m2}(t) \\ &= J(t)^2\,dA_{m1}{}^*(t)\,dA_{m2}(t) \\ &\quad + J(t)\,F_{m2}(t)\,dA_{m1}{}^*(t)\,F_{m1}{}^*(t)\,J(t)\,dA_{m2}(t) \\ &= J(t)\,F_{m2}(t)\,dA_{m1}{}^*(t) + F_{m1}{}^*(t)\,J(t)\,dA_{m2}(t)\end{aligned}$$

Thus,

$$\begin{aligned}&d\left\langle F_{m1}(t)e(u), F_{m2}(t)e(v)\right\rangle \\ &\qquad = d\left\langle\langle u, m_1\rangle\right\rangle(t)\left\langle e(u), J(t)\,F_{m2}(t)\,e(v)\right\rangle \\ &\qquad\quad + d\left\langle\langle m_2, v\rangle\right\rangle(t)\left\langle e(u), F_{m1}{}^*(t)J(t)\,e(v)\right\rangle\end{aligned}$$

Likewise

$$\begin{aligned}d(F_{m2}(t)\,F_{m1}{}^*(t)) &= F_{m2}(t)\,J(t)\,dA_{m1}{}^*(t) + J(t)F_{m1}{}^*(t)\,dA_{m2}(t) \\ &\quad + d\left\langle\langle m_2, m_1\rangle\right\rangle(t)\end{aligned}$$

so

$$\begin{aligned}&d\left\langle F_{m2}{}^*(t)\,e(u), F_{m1}{}^*(t)\,e(v)\right\rangle \\ &\qquad = d\left\langle\langle u, m_1\rangle\right\rangle(t)\left\langle e(u)\,F_{m2}(t)\,J(t)\,e(v)\right\rangle \\ &\qquad\quad + d\left\langle\langle m_2, v\rangle\right\rangle(t)\left\langle e(u), J(t)\,F_{m1}{}^*(t)\,e(v)\right\rangle \\ &\qquad\quad + d\left\langle\langle m_2, m_1\rangle\right\rangle(t)\left\langle e(u), e(v)\right\rangle.\end{aligned}$$

Thus,

$$d\langle e(u),(F_{m1}*(t)F_{m2}(t)+F_{m2}(t)F_{m1}*(t))e(v)\rangle$$
$$= d\langle\langle m_2, m_1\rangle\rangle(t)\langle e(u),e(v)\rangle.$$
$$d\langle\langle u, m_1\rangle\rangle(t)\langle e(u),\{J(t),F_{m2}*(t)\}e(v)\rangle.$$
$$+\,d\langle\langle m_2, v\rangle\rangle(t)\langle e(u),\{J(t),F_{m1}*(t)\}e(v)\rangle.$$

We shall now that $\{J(t),F_m(t)\} = 0$

and hence $\{F_{m1}*(t),F_{m2}(t)\} = \langle\langle m_2,m_1\rangle\rangle(t)$

$$\begin{aligned}+\big(J(t).dF_m(t)\big) &= dJ(t).dF_m(t)\\ &\quad + J(t).dF_m(t)+dJ(t).dF_m(t)\\ &= -2JF_m d\Lambda_1 + dA_m - d\Lambda_1 dA_m\\ &= -2JF_m d\Lambda_1 + dA_m.\end{aligned}$$

and secondly,

$$\begin{aligned}d(F_m(t)J(t)) &= dF_m.J+F_m.dJ+dF_m.dJ\\ &= dA_m - 2F_m J d\Lambda_1 - 2dA_m\cdot d\Lambda_1\\ &= -dA_m - 2F_m J d\Lambda_1\end{aligned}$$

Hence $\quad d\{J,F_m\} = -2\{J,F_m\}d\Lambda_1$

So $\quad d\langle e(u),\{J,F_m\}(t)e(v)\rangle$

$$= -2\langle e(u),\{J,F_m\}(t)e(v)\rangle\; d\langle\langle u,\xi v\rangle\rangle(t)$$

where $\langle e(u),\{J,F_m\}(t)e(v)\rangle = 0$

$\Rightarrow \langle e(u),\{J,F_m\}(t+dt)e(v)\rangle = 0$

Since $\quad \{J,F_m\}(0) = 0(F_m(0)=0).$

It follow, that $\{J,F_m\}(t) = 0\forall t.$

Hence we have proved the $CAR:\{F_{m2}(t),F_{m1}*(t)\} = \langle\langle m_2,m_1\rangle\rangle(t)$

❖❖❖❖❖

[26] Spherical Function (Helgason Vol. II):

$$\int_{G/MN}\exp((-i\lambda+\rho)(A(x,b)))\,db$$
$$= \int_{k\times A}\exp((-i\lambda+\rho)(A(x,kaMN)))\,dkda$$

$$= \int_{k \times A} \exp\left((-i\lambda + \rho) A\left(a^{-1}k^{-1}x\right)\right) dkda$$

$$= \varphi - \lambda(k_1 x k_1)$$

$$= \int_{k \times A} \exp\left((-i\lambda + \rho)(A(a^{-2}k^{-1}k_1 x k_2))\right) dkda$$

So $\varphi - \lambda(k_1 x k_2)$

$$= \int_{K \times A} \exp\left((-i\lambda + \rho) A(a^{-1}kx)\right) dkda$$

$$(G = NAK$$

$$g = n(g)\, a(g)\, k(g)$$

$$\Rightarrow gk_2 = n(g)\, a(g)\, k(g)\, k_2)$$

$$\Rightarrow a(gk_2) = a(g)$$

$$n(gk_2) = n(g))$$

$$= \varphi - \lambda(x).$$

$$A\,(gk,\, kaMN)$$

$$\underline{\underline{dif}}\;\; A\left(a^{-1}k^{-1}g\right) \qquad A(g_1 k, g_2 MN)$$

$$\underline{\underline{dif}}\;\; A\left(g_2^{-1} g_1\right)$$

To check that this definition is correct. we note that

$$A\left((g_2 mn)^{-1} g_1 k\right) = A\left(n^{-1}m^{-1}g_2^{-1}g_1 k\right)$$

$$n^{-1}m^{-1}g_2^{-1}g_1 k = n^{-1}m^{-1}n\left(g_2^{-1} g_1\right) a\left(g_2^{-1} g_1\right) k\left(g_2^{-1} g_1\right) k$$

$$= n^{-1}m^{-1}n\left(g_2^{-1} g_1\right) mm^{-1} a\left(g_2^{-1} g_1\right) k\left(g_2^{-1} g_1\right) k$$

$$= n^{-1}m^{-1}n\left(g_2^{-1} g_1\right) m\, a\left(g_2^{-1} g_1\right) m^{-1}\, k\left(g_2^{-1} g_1\right) k$$

Now, $m^{-1}n\left(g_2^{-1}g_1\right) m \in N$

$\therefore$ $n^{-1}m^{-1}n\left(g_2^{-1} g_1\right) m \in N$

$$m \in k \;\Rightarrow\; m^{-1}k\left(g_2^{-1} g_1\right) k \in k$$

So, $A\left(n^{-1}m^{-1}g_2^{-1}g_1 k\right) = a\left(g_2^{-1} g_1\right)$

$$(g = n(g)a(g)k(g) \in NAK = G).$$

❖❖❖❖❖

[27] Clarifications of Problems Related to the I to measure of Brownian motion:

$$Y(t) = \sum_{k=1}^{N(t)} X_k \quad N(t) \simeq \text{Pounion process, } \{X_k\}_{k=1}^{\infty}$$

are *iid* r.v's independent of $N(.)$.

Then $$Y(t) = \int_0^t X_{N(s)} dN(s)$$

$$\mathbb{E}\{Y(t)\} = \int_0^t \mathbb{E}\{X_{N(s)}\}\mathbb{E}\{dN(s)\}$$

$$= \mathbb{E}(X_1)\mathbb{E}\{N(t)\} = \lambda t \mathbb{E}(X_1)$$

$Pr\{Y(.)$ Makes a jump in time $[t, t+dt]$ into the set $B\}$

$$= \lambda dt F(B)$$

where $B \in \mathbb{B}$ and F is the probability distribution of X_1. Here X takes values in the measurable space $(E, \mathbb{B})$.

I to measure: $[\tau_{s-}, \tau_s]$ is an excursion interval for $\{B(t)\}$. Here

$$\tau_{s-} = \text{int}\{t \geq 0\} | L_t \geq s\},$$

$$\tau_s = \text{int}\{t \geq 0\} | L_t > s\},$$

The process $\{B(\tau_{s-} + t): 0 \leq t \leq \tau_s - \tau_s\} \equiv e_s$. is the excursion process of $B.M.$ at S.

For each $s \in G$ (G is the set of zeroes of $B.M.$) e_s is an excursion.

Master formula:

$$E \sum_{s \in G} f(s, e_s) = \int f(s, u) dn(u) ds$$

where $n(.)$ is the characteristic measure of an excursion $n(.)$ is concentrated over paths of the form $\{\xi(t): 0 \leq t \leq R\}$

where $R < \infty$. Equivalently, we may assume that $n(.)$ is concentrated over paths of the form $\{\xi : \xi(t) = 0, t > R\}$

where $R = R(w)$ is a finite random variable. For the excursion e_s, $R = \tau_s - \tau_{s-}$. At each $s \in G$, the point process $\{e_s\}$ makes a jump into an excursion path *i.e.* a path that hits zero only at its end points.

The master formula should be compared with the formula

$$\mathbb{E} \sum_{k=1}^{\infty} f(t_k, X_R) = \mathbb{E} \int_0^{\infty} f(t, X_{N(t)}) dN(t)$$

$$= \int_0^{\infty} \mathbb{E}\{f(t, X_1)\} \lambda dt$$

$$= \int_{E\times[0,\infty]} f(t,x)\lambda dF(x)dt$$

where $\{X_k\}_{k=1}^{\infty}$ are *iid* random variables having distribution F and independent of $N(.)$, a Poisson and process with rate λ.

[28] Problem from Revuz and Yor:

Continuous Martingale and Brownian Motion

Let n denote the I to measure of B.M. show that

$$\int_0^\infty P_0^{\tau s} dS \circ \int_0^\infty n^n (Y \cap \{u < R\})\, du = \int_0^\infty P_0^t\, dt(Y)$$

$$\text{L.H.S.} = \sum_{S\in G_w} P_0^{\tau s} \Delta S_0 \int_0^\infty \left(\sum_{s=0} \int_{-0}^{r} \mathbb{E}\{Y \circ \theta_{\tau s-} \mid \tau_s - \tau_{s-} = u, \tau_{s-}\}\, du \right)$$

$$n_R(r)dr$$

where $n_R(\tau) = \dfrac{1}{2\sqrt{2\pi r^3}}$ is the It_0 measure of R.

Note. $P_0^{\tau s} \int_0^\infty n^u (Y \cap \{u < R\})\, du$

$$\mathbb{E}_n\left\{\int_{\tau_{s-}}^{\tau_s} \mathbb{E}\left[Y \mid \tau_s - \tau_s \equiv {}^{u_{s-}}_{\tau_s} \right] du\right.$$

where $\tau_s = \tau_{s-} + R$

(where $|F_n$ denotes integral *w.r.t.* $n_R(\tau)\, dr$ the I to measure of R).

$$= \int_0^\infty n_R(r)\, dr \int_0^{\tau_s - \tau_s = r} \mathbb{E}\left[Y \circ \theta_{\tau s-} \mid \tau_s - \tau_{s-} = u, t_{s-} \right] du$$

Thus,

$$\sum_s P^{\tau_s} \Delta s \circ \int_0^\infty n^u (Y \cap \{u < R\})\, du$$

$$= \mathbb{E} \sum_{s\in W} \Delta S \int_0^\infty n_R(r)\, dr \int_0^r \mathbb{E}\,[Y \circ \theta_{\tau s} \mid \tau_{s-}, \tau_s - \tau_s - = u]\, du$$

$$= \mathbb{E} \sum_{s\in W} \Delta s \int_0^\infty du \left(\int_u^\infty n_R(r)\, dr \right) n[z|R = u, \tau_{s-}]$$

(where $z = Y \circ \theta_{\tau s-}$ is defined on $[0, \tau_s - \tau_s - = R]$ *i.e.* on the excursion path space.)

$$= \mathbb{E} \sum_S \Delta s \int_0^\infty \frac{du}{\sqrt{2\pi u}} n[z \mid R = u, \tau_{s-}]$$

$$= \mathbb{E} \int_0^\infty ds \int_0^\infty \frac{du}{\sqrt{2\pi u}} n[z \mid R = u, \tau_{s-}]$$

Note: $n(B)\,ds$ is the probability that in time $[s,\ s+ds]$ an excursion $\{B_{\tau_{s-}}+u \mid 0 \le u \le \tau_{s-}\tau_{s-}\}$ will occur taking values in the let B.

Thus, $\sum_{S} P^{\tau s}\Delta s\int_0^\infty n^u(Y\cap\{u<R\})\,du = \mathbb{E}\int_0^\infty \frac{du}{\sqrt{2\pi u}}\int_0^\infty ds\, n[z \mid R=u, \tau_{s-}]$

On the other hand,

$$\begin{aligned}\int_0^\infty P_0^t\,dt(Y) &= \mathbb{E}\int P_0^{\tau_s}\,d\tau_s(Y)\\ &= \mathbb{E}\sum_{s\in G_w} P^{\tau_s}(\tau_{s-}\tau_{s-})(Y)\\ &= \mathbb{E}\sum_{s\in G_w}\mathbb{E}\,[(\tau_{s-}\tau_{s-})Y \mid \tau_{s-},\tau_s]\\ &= \mathbb{E}\int_{\substack{r\ge 0\\ r\ge 0}} n_R(r)\,dr\,ds\,\mathbb{E}\,[Y\circ\theta\tau_{s-} \mid \tau_{s-},\tau_s-\tau_{s-}=r]\end{aligned}$$

Now $\qquad rn_R(r) = \dfrac{1}{\sqrt{2\pi r}}$. So

$$\int_0^\infty P_0^t\,dt(Y) = \mathbb{E}\int_{R_+^2}\frac{dr\,ds}{\sqrt{2\pi r}}n[z \mid \tau_{s-}, R=r]$$

Thus, we've proved

$$\int_0^\infty P_0^t\,dt = \int_0^\infty P_0^{\tau s}\,ds\int_0^\infty n_u(\cap\{u<R\})\,du$$

❖❖❖❖❖

[29] Group Theory and Robotics:

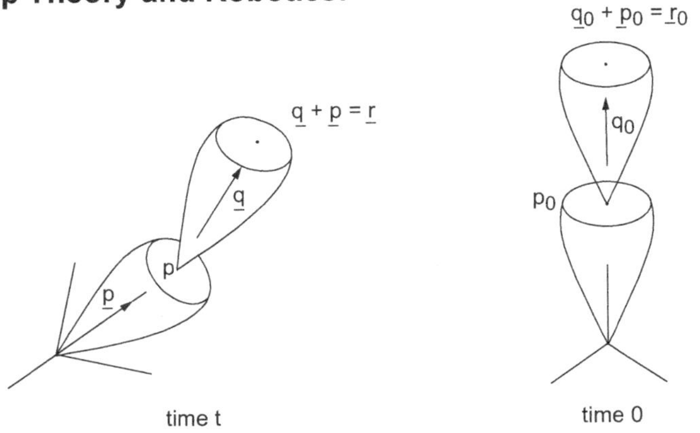

We apply a rotation R_1 first to the 2-link system taking $p_0 \to R_1p_0, q_0 \to R_1q_0$ and hence $\underline{r}_0 = \underline{p}_0+\underline{q}_0 \to \underline{R}_1\underline{p}_0+\underline{R}_1\underline{q}_0 = \underline{R}_1\underline{r}_0$. Then a rotation R_2 is applied to R_1q_0 *i.e.* to the second top relation to the first. The find a point $\underline{r}$ is thus

$$\begin{aligned}\underline{r} &= R_1\underline{p}_0 + R_2R_1\underline{q}_0 \\ &= R_1\underline{p}_0 + R_2R_1(\underline{r}_0 - \underline{p}_0) \\ &= R_2R_1\underline{r}_0 + (R_1 - R_2R_1)\underline{p}_0\end{aligned}$$

and $p_0 \to R_1 p_0$.

Thus the group action is

$$\begin{bmatrix}\underline{r}_0 \\ \underline{p}_0\end{bmatrix} \to \begin{bmatrix}\underline{r} \\ \underline{p}\end{bmatrix} = \begin{bmatrix}R_2R_1 & R_1 - R_2R_1 \\ 0 & R_1\end{bmatrix}\begin{bmatrix}\underline{r}_0 \\ \underline{p}_0\end{bmatrix}$$

With $R_1, R_2 \in SO(3)$.

The Lie algebra g of G:

$$G = \left\{\begin{bmatrix}R_2R_1 & R_1 - R_2R_1 \\ 0 & R_1\end{bmatrix} : R_1R_2 \in SO(3)\right\}$$

Let $$T(R_1, R_2) = \begin{bmatrix}R_2R_1 & R_1 - R_2R_1 \\ 0 & R_1\end{bmatrix}$$

Then, $$T(S_1, S_2)\,T(R_1, R_2) = \begin{bmatrix}S_2S_1 & S_1 - S_2S_1 \\ 0 & S_1\end{bmatrix}.\begin{bmatrix}R_2R_1 & R_1 - R_2R_1 \\ 0 & R_1\end{bmatrix}$$

$$= \begin{bmatrix}S_2S_1R_2R_1 & S_1R_1 - S_2S_1R_2R_1 \\ 0 & S_1R_1\end{bmatrix}$$

$$= T(S_1R_1, S_2S_1R_2S_1^{-1})$$

The Lie algebra of G:

Let $$R_1 = I + \delta.X_1,\ R_2 = I + \delta.X_2,$$
$$S_1 = I + \delta.Y_1,\ R_2 = I + \delta.Y_2,$$

Then

$$T(R_1, R_2) = T(I + \delta.X_1, I + \delta.X_2)$$

$$\begin{bmatrix}I_3 + \delta(X_1 + X_2) & -\delta X_2 \\ 0 & I_3 + \delta X_1\end{bmatrix} + O(\delta^2) = I_6 + \delta\begin{bmatrix}X_1 + X_2 & -X_2 \\ 0 & X_1\end{bmatrix} + O(\delta^2)$$

Thus, $$[T(R_1, R_2), T(S_1, S_2)] = \delta^2\left\{\begin{bmatrix}X_1 + X_2 & -X_2 \\ 0 & X_1\end{bmatrix}\begin{bmatrix}Y_1 + Y_2 & -Y_2 \\ 0 & Y_1\end{bmatrix} - \begin{bmatrix}Y_1 + Y_2 & -Y_2 \\ 0 & Y_1\end{bmatrix}\begin{bmatrix}X_1 + X_2 & -X_2 \\ 0 & X_1\end{bmatrix}\right\}$$

$$= \delta^2\begin{bmatrix}[X_1 + X_2, Y_1 + Y_2] & [Y_2, X_2] - X_1Y_2 - X_2Y_1 + Y_1X_2 + Y_2X_1 \\ 0 & [X_1, Y_1]\end{bmatrix}$$

$$= \delta^2 \begin{bmatrix} [X_1 + X_2, Y_1 + Y_2] & [Y_2, X_2] + [Y_1, X_2] + [Y_2, X_1] \\ 0 & [X_1, Y_1] \end{bmatrix}$$

$$= \delta^2 \begin{bmatrix} [X_1 + X_2, Y_1 + Y_2] & [Y_1 + Y_2, X_2] + [Y_2, X_1] \\ 0 & [X_1, Y_1] \end{bmatrix}$$

One parameter subgroups:

$$\Phi(t) = \exp\left\{ t \begin{bmatrix} X_1 + X_2 & -X_2 \\ 0 & X_1 \end{bmatrix} \right\}$$

$$X_1, X_2 \in SO(3) = g(SO(3)) \qquad \text{(zero trace skew-symmetric } 3 \times 3 \text{ matrics)}$$

We can identify two basic one parametric subgroups, namely those defined by

$$\tilde{X}_1 = \begin{bmatrix} X_1 & 0 \\ 0 & X_1 \end{bmatrix} = I_2 \otimes X_1$$

and $$\begin{bmatrix} X_2 & -X_2 \\ 0 & 0 \end{bmatrix} = \begin{bmatrix} 1 & -1 \\ 0 & 0 \end{bmatrix} \otimes X_2 - \widehat{X}_2$$

We have $$\exp(t\tilde{X}_1) = \begin{bmatrix} \exp(tX_1) & 0 \\ 0 & \exp(tX_1) \end{bmatrix} = I_2 \otimes \exp(tX_1),$$

$$\exp(t\widehat{X}) = I_6 + \sum_{n=1}^{\infty} \frac{1}{\underline{|n}} \left(\begin{bmatrix} 1 & -1 \\ 0 & 0 \end{bmatrix}^n \otimes X_0^n \right)$$

Let $f \mathbb{R}^n \Rightarrow \mathbb{R}^n$, Let $y \in \mathbb{R}^n$ be fixed.

$$\int_{O(n)} f(x + ky)\, dk = g(x) \equiv (Tf)(x)$$

Then, $$g(k_1 x) = \int_{O(n)} f(k_1 x + ky)\, dk = \int f(k_1(x + k^{-1}ky))\, dk$$

$$= \int f(k_1(x + ky))dk$$

$$= (T(fok_1))(x)$$

Thus, $$(Tf)ok_1 = T(fok_1) \tag{1}$$

Let $$L(k) f(x) = f(k^{-} x),\ k \in O(n),\ x \in \mathbb{R}^n.$$

Then L is a representation of $O(n)$ in $\mathcal{L}^2(\mathbb{R}^n)$ and we have from (1) that

$$L(k)oT = ToL(k),\ k \in O(n)$$

i.e. $$[L(k), T] = 0 \forall \in O(n).$$

It follows that X, T

If $X \in g(O(n))$ = any $n \times n$ real show symmetric matrix,

Then $\quad [L(e^{tx}), T] = 0 \; \forall t \in \mathbb{R}$
and hence if

$$\pi_L(x) = \frac{d}{dt} L(e^{tx})\Big|_{t=0}$$

Then $\quad [\pi_L(x), T] = 0$

From which we reduce that

$$\square TF(x) = 0$$

where $\square = \infty\left\{\sum_{\alpha=1}^{n} \frac{\partial^2}{\partial \times \alpha^2}\right\}\infty(.)$ Denotes angular part of a differential

($\square$ is in the universal enveloping algebra operator of the lie algebra $\{\pi_L(X) : X \in O(n)\}$

$$O(n) = g(O(n)).$$

$$(L(e^{tX})f)(x) = f(e^{-tX}.x)$$

so

$$\pi_L(X) f(x) = \frac{d}{dt} f(e^{-tX} x)\Big|_{t=0}$$

$$= -(Xx, \underline{\nabla}) f(x)$$

$$= -\underline{x}^T X^T \underline{\nabla} f(x)$$

$$= \underline{x}^T X^T \underline{\nabla} f(x)$$

We can also deduce this result using fourier analysis:

$$f(x) = \int_{\mathbb{R}^n} \hat{f}(\xi) \exp(i < \xi, x >) d\xi$$

$$(Tf)(x) = \int_{\mathbb{R}^n \times O(n)} \hat{f}(\xi) \exp(i < \xi, x >) \exp(i < \xi, ky >) d\xi\, dk$$

$$= \int_{\mathbb{R}^n} \hat{f}(\xi) \exp(i < \xi, x >) \psi_y(\xi) d\xi$$

where

$$\psi_y(\xi) = \int_{O(n)} \exp\left(i\langle \xi, ky\rangle\right) dk$$

$$= \int_{O(n)} \exp\left(i\langle \xi, y\rangle\right) dk$$

(ψ_y depends only on $||y||$ not its direction Likewise (Tf) depends only on $||y||$)

Now let ∇^2 denote the Laplacian in $\mathbb{R}^x$. Then

$$\nabla^2 Tf(x) = -\int_{\mathbb{R}^n} ||\xi||^2 \hat{f}(\xi) \exp\left(i\langle \xi, x\rangle\right) \psi_y(\xi) d\xi$$

$$-\|\xi\|^2 \psi_y(\xi) = \nabla_y^2 \psi_y(\xi) \text{ since } \|k\xi\|^2 = \|\xi\|^2 \ \forall k \in O(n)$$

Since $\psi_y(\xi)$ depends only on $\|y\|$, it follows that $\nabla_y^2 \psi_y(\xi) = R_{\|y\|}\psi_y(\xi)$ where $R_{\|y\|}$ denotes the radial part of the Laplacian ∇_y^2. Thus,

$$\begin{aligned}\nabla^2 Tf(x) &= \int (R_{\|y\|}\psi_y(\xi))\widehat{f}(\xi)\exp(i\langle\xi, x\rangle)d\xi \\ &= R_{\|y\|} = \int \psi_y(\xi)\widehat{f}(\xi)\exp(i\langle\xi, x\rangle)d\xi \\ &= R_{\|y\|}Tf(x)\end{aligned}$$

Writing $T = T_y$ (to be precise), we have $Ty \equiv T_{\|y\|}$ and $\nabla^2 T_y f(x) = R_{\|y\|}T_y f(x)$

i.e. $$\nabla^2 T_y = R_{\|y\|}T_y$$

❖❖❖❖❖

[30] Suppose $\nabla^2 f(x) = 0$, $x \in D$

where D is a connected open subject of $\mathbb{R}^n$. Let $B = B(x, \delta) \subset D$.

Then $$f(x) = \int_{O(n)} f(x + \delta k\underline{e}_1)dk$$

where $\left\|\underline{e}_1\right\| = 1, \underline{e}_1 \in \mathbb{R}^n$ is arbitrary

Equivalently,

$$f(x) = \int_{O(n)} f(x + ke_1)dk$$

Where $\|e_1\| = 1$ and $\nabla^2 f(x') = 0$ for $x' \in B(x, 1)$.

Let π be any representation of $O(n)$ in $\mathcal{H}$. Consider

$$P_\pi = \int_{O(n)} \pi(k)dk$$

Then P_π is the orthogonal projection onto

$$\mathcal{H}_\pi = \{\xi \in \mathcal{H} \mid \pi(k)\xi = \xi \forall k \in O(n)\}$$

Thus $\int_{O(n)} k^{\otimes m}dk$ is the orthogonal projection onto

$$\mathcal{H}_m = \{\xi \in (\mathbb{R}^n)^{\otimes m} \mid k^{\otimes m}\xi = \xi \forall k \in O(n)\}$$

Now, $$\begin{aligned}f(x + ke_1) &= f(x) + \sum_{m=1}^{\infty} \frac{1}{\underline{|m}}\left\{\langle k_{e1}, \nabla\rangle^m\right\}f(x) \\ &= f(x) + \sum_{m=1}^{\infty} \frac{1}{\underline{|m}}\left\langle k^{\otimes m} e_1^{\otimes m}, \nabla^{\otimes m}\right\rangle f(x)\end{aligned}$$

Clearly, $\int_{O(n)} k^{\otimes m} e_1^{\otimes m} dk = 0$ for m odd ≥ 1.

That follows by replacing k by kk_1, where $k_1 \in O(n)$, $k_1 e_1 = -e_1$.
Hence,

$$\int_{O(n)} f(x+ke_1)dk = f(x) + \sum_{m=1}^{\infty} \frac{1}{\lfloor 2m} \int \left\langle k^{\otimes 2m} e_1^{\otimes 2m}, \nabla^{\otimes 2m} \right\rangle dk\, f(x)$$

Now, $\int_{O(n)} k^{\otimes 2m} e_1^{\otimes 2m} dk = \underline{\xi}_m$

is independent of the unit vector $e_1 \in \mathbb{R}^n$ since if $e_2 \in \mathbb{R}^n$ is another unit vector, then $\exists k_1 \in O(n)$ such that $e_2 = k_1 e_1$.

$$\text{Thus,} \int_{O(n)} k^{\otimes 2m} e_2^{\otimes 2m} dk = \int_{O(n)} (kk_1)^{\otimes 2m} e_1^{\otimes 2m} dk$$

$$= \int_{O(n)} k^{\otimes 2m} e_1^{\otimes 2m} dk \; (d(kk_1) + dk).$$

Now since $d(k_1 k) + dk$ also, it follows that

$$k^{\otimes 2m} \xi_m = \xi \forall k \in O(n)$$

So writing $k = e^{tX}$, $X \in O(n) = g(O(n))$,
we get

$$(e^{tX})^{\otimes 2m} \xi_m = \xi_m \forall t \in \mathbb{R}.$$

Whence taking $\frac{d}{dt}$ at $t = 0$ gives

$$\left(\sum_{k=1}^{2m} \otimes X_k \right) \xi_m = 0$$

Where

$$\begin{aligned} \sum_{k=1}^{2m} \otimes X_k &= X \otimes I \otimes .. \otimes I \\ &\quad + I \otimes X \otimes . \otimes I \\ &\quad \vdots \\ &\quad + I \otimes I \otimes . \otimes X \\ &= X^{\otimes 2m} (X_k = X) \end{aligned}$$

$$i.e. \qquad X^{\otimes 2m} \xi_m = 0.$$

we get

$$\int_{O(n)} F(x+ke_1)dk = f(x) + \sum_{m=1}^{\infty} \frac{1}{\lfloor 2m} \left\langle \underline{\xi}_m, \nabla^{\otimes 2m} \right\rangle f(x)$$

Now, $$\underline{\xi}_2 = \left(\int_{O(n)} k \otimes k dk\right)(e_1 \otimes e_1)$$

By the Peter-Weyl theorem (orthogonality of matrix elements of irreducible, rep'ns),

$$\int_{O(n)} k_{\alpha\beta} k_{\rho\sigma} dk = c \cdot \delta_{\alpha\rho} \delta_{\beta\sigma}$$

so $$\left\langle u \otimes v, \underline{\xi}_2 \right\rangle = \int_{O(u)} \langle u, ke_1 \rangle \langle v, ke \rangle dk$$

$$c\delta_{\rho\sigma} u_\rho v_\sigma = c\langle u, v \rangle$$

Thus, $\left\langle \underline{\xi}_2, \nabla^{\otimes 2} \right\rangle = c\langle \underline{\nabla}, \underline{\nabla} \rangle = c.\nabla^2$ and since $\nabla^2 f(x) = 0$, it follows that

$$\left\langle \underline{\xi}_2, \nabla^{\otimes 2} \right\rangle f(x) = 0.$$

Decompose the rep'n $k \to k^{\otimes 2m} (k \in O(n))$ into irrep's:

$$k^{\otimes 2m} \approx \overset{\infty}{\underset{r=0}{\otimes}} m_r \pi_r(k)$$

$\pi_0(k) = 1$

$m_r \geq 0$

$\downarrow$

Multiply of π_r in $k^{\otimes 2m}$

Then $\int_{O(n)} k^{\otimes 2m} dk \approx m_0 1$

Specifically there is for each $m \geq 1$, a linear operator (non-singular) $T_m (\mathbb{R}^n)^{\otimes m} \to (\mathbb{R}^n)^{\otimes m}$ to that

$$k^{\otimes 2m} = T_m^{-1} \left(\overset{p}{\underset{r=0}{\otimes}} m_r \pi_r(k) \right) T_m$$

where $\pi_r(k)$ is an $p_r \times p_r$ matrix and

$$\sum_{r=0}^{p} m_r p_r = nm$$

Thus since $\int_{O(n)} \pi_r(k) dk = 0$ for $r \geq 1$, we get

$$\int_{O(n)} k^{\otimes 2m} dk = T_m^{-1} [I_{m0} \otimes O] T_m$$

$$\left(p_0 = 1, \int_{O(n)} \pi_0(k)\,dk = 1 \text{ since } \pi_0(k) = 1 \right)$$

Thus,

$$\int_{O(n)} f(x + ke_1)\,dk = f(x) + \sum_{m=1}^{\infty} \frac{1}{\lfloor 2m} \left\langle T_m^{-1}[I_{m_0} \otimes 0] T_m e_1^{\otimes 2m} \nabla \otimes 2m \right\rangle f(x)$$

(Note m_0 varies with m).

❖❖❖❖❖

[31] Entropy Evolution in Sudarshan-Lindblad Equation:

$$\frac{d\rho(t)}{dt} = T(\rho(t))$$

where $$T(\rho) = -i[H, \rho] - \frac{1}{2}(L^* L\rho + \rho L^* L - 2L\rho L^*)$$

$$\rho(t) + d\rho(t) = \rho(t + dt) = \rho(t) + T(\rho(t))\,dt$$

Let $\{|v_\alpha(t)\rangle\}_{\alpha=1}^{d}$, $\{\lambda_\alpha(t)\}_{\alpha=1}^{d}$ be respectively orthogonal set of eigenvectors of $\rho(t)$ and the corresponding eigenvalues,

Thus $$\rho(t) = \sum_{\alpha=1}^{d} \lambda_\alpha(t) |v_\alpha(t)\rangle\langle v_\alpha(t)|$$

Entropy $$S(\rho(t)) = -T_r(\rho(t)\log\rho(t)) = -\sum_{\alpha=1}^{d} \lambda_\alpha(t)\log\lambda_\alpha(t)$$

$$(\rho + dr)(v_\alpha + dv_\alpha) = (\lambda_\alpha + d\lambda_\alpha)(v_\alpha + dv_a)$$

$$\Rightarrow \quad (\rho - \lambda_\alpha)\,dv_\alpha = (d\lambda_\alpha - d\rho)v_\alpha.$$

$$\Rightarrow \quad (\lambda_\beta - \lambda_\alpha)\langle v_\beta | dv_\alpha \rangle = d\lambda_\alpha \delta_{\alpha\beta} - \langle v_\beta | d\rho | v_\alpha \rangle$$

Thus,

$$d\lambda_\alpha = \langle v_{\beta\alpha} | d\rho | v_\alpha \rangle$$

and for $\beta \neq \alpha$,

$$\langle v_\beta | dv_\alpha \rangle = \frac{\langle v_\beta | d\rho | v_\alpha \rangle}{\lambda_\alpha - \lambda_\beta}$$

$$= \frac{\langle v_\beta | \rho + dtT(\rho) | v_\alpha \rangle}{(\lambda_\alpha - \lambda_\beta)}$$

$$= \frac{\{\lambda_\alpha \delta_{\alpha\beta} + dt\langle v_\beta | T(\rho) | v_\alpha \rangle\}}{\lambda_\alpha - \lambda_\beta}$$

$$= \frac{dt\left\langle v_\beta \mid T(\rho) \mid v_\alpha \right\rangle}{(\lambda_\alpha - \lambda_\beta)}$$

Thus,

$$\left| dv_\beta \right\rangle = dt \sum_{\beta \neq \alpha} \frac{\left| v_\beta \right\rangle \left\langle v_\beta \right| T(\rho) \left| v_\alpha \right\rangle}{(\lambda_\alpha - \lambda_\beta)}$$

or equivalently,

$$\frac{d}{dt}\left| v_\alpha(t) \right\rangle = \sum_{\beta \neq \alpha} \frac{\left| v_\beta(t) \right\rangle \left\langle v_\beta(t) \right| T(\rho(t)) \left| v_\alpha(t) \right\rangle}{(\lambda_\alpha(t) - \lambda_\beta(t))}$$

$$\frac{d\lambda_\alpha(t)}{dt} = \left\langle v_\alpha \right| T(\rho(t)) \left| v_\alpha \right\rangle$$

$$\left\langle v_\alpha \mid T(\rho) \mid v_\alpha \right\rangle = \left\langle v_\beta \mid -i[H, \rho] - \frac{1}{2}(L^* L\rho + \rho L^* L - 2L\rho L^*) \mid v_\alpha \right\rangle$$

$$= -i\lambda_\alpha \left\langle v_\beta \mid H \mid v_\alpha \right\rangle + i\lambda_\beta \left\langle v_\beta \mid H \mid v_\alpha \right\rangle$$

$$-1/2\left((\lambda_\alpha + \lambda_\beta)\left\langle v_\beta \mid L^* L \mid v_\alpha \right\rangle - 2\left\langle v_\beta \mid L\rho L^* \mid v_\alpha \right\rangle\right)$$

This equation can be expressed as

$$\frac{d}{dt}\left| v_\alpha \right\rangle = \sum_\beta \left| v_\beta \right\rangle (\lambda_\alpha - \lambda_\beta)^{-1}$$

$$\beta \neq \alpha\left(i(\lambda_\beta - \lambda_\alpha)\left\langle v_\beta \mid H \mid v_\alpha \right\rangle - \frac{1}{2}(\lambda_\beta + \lambda_\alpha)\left\langle v_\beta \mid L^* L \mid v_\alpha \right\rangle - \left\langle v_\beta \mid L\rho L^* \mid v_\alpha \right\rangle \right)$$

and

$$\frac{d\lambda_\alpha}{dt} = \left\langle v_\alpha \mid T(\rho) \mid v_\alpha \right\rangle - \lambda_\alpha \left\langle v_\alpha \mid L^* L \mid v_\alpha \right\rangle + \left\langle v_\alpha \mid L\rho L^* \mid v_\alpha \right\rangle$$

Note that

$$\left\langle v_\alpha \mid L\rho L^* \mid v_\alpha \right\rangle = \sum_\beta \left| \left\langle v_\alpha \mid L \mid v_\beta \right\rangle \right|^2 \lambda_\beta$$

So,

$$\frac{d\lambda_\alpha}{dt} = \left(\sum_\beta \lambda_\beta \left| \left\langle v_\alpha \mid L \mid v_\beta \right\rangle \right|^2 \right) - \lambda_\alpha \left\langle v_\alpha \mid L^* L \mid v_\alpha \right\rangle$$

$$= \sum_\beta \left\{ \lambda_\beta \left| \left\langle v_\alpha \mid L \mid v_\beta \right\rangle \right|^2 - \lambda_\alpha \left| \left\langle v_\beta \mid L \mid v_\alpha \right\rangle \right|^2 \right\}$$

Finally,

$$\frac{d}{dt} S(\rho(t)) = -\frac{d}{dt} \sum_\alpha \lambda_\alpha \log \lambda_\alpha = -\sum_\alpha \frac{d\lambda_\alpha}{dt} \log \lambda_\alpha$$

Since $$\sum \lambda_\alpha = 1 \Rightarrow \sum_\alpha \frac{d}{dt}\lambda_\alpha = 0.$$

Thus,

$$\frac{d}{dt}S(\rho(t)) = \sum_{\alpha,\beta}\left(\lambda_\alpha\left|\langle v_\beta | L | v_\alpha\rangle\right|^2 - \lambda_\beta\left|\langle v_\alpha | L | v_\beta\rangle\right|^2\right)(\log \lambda_\alpha)$$

Construction of quantum non-demolition processes.

$$dU(t) = \left\{-iHdt + Gdt + \sum_j L_j^{(1)}dA_j(t) + \sum_j L_j^{(2)}dA_j^*(t) + \sum_{kj} S_k^j dA_j^k(t)\right\}$$

$$U(t)dA_j(t)dA_k^*(t)$$
$$= \delta_{jk}dt$$

$$d\Lambda_j^k(t)\,dA_q^*(t) = \delta_{pq}dA_j^*(t)$$

$$d\Lambda_k^j(t)dA_m^*(t) = \delta_m dA_k^*(t)$$

$$dA_m(t)\,d\Lambda_j^k(t) = \delta_{jm}dA_k(t)$$

All other quantum I to products are zero.

$H, G, L_j^{(1)}, L^{(2)}, S_k^j$ are system operators selected to that $U(t)$ is unitary. $\forall t$

Let $$dY_{in}(t) = \sum_j \lambda_j^{(t)}dA_j(t) + \overline{\lambda_j(t)}\,dA_j^*(t) + \sum_{j,k} C_k^j(t)d\Lambda_j^k(t)$$

where $\lambda_j, C_k^j \in L^2(\mathbb{R}_+)$.

Assume $Y_{in}(0) \in \mathcal{L}\,(\Gamma s\,(L^2(\mathbb{R}_+)))$

Then $Y_{in}(t) \in \mathcal{L}\left(\Gamma s\left(L^2[0,t]\right)\right)$

Let

$$Y_{out}(t) = U^*(t)(I \otimes Y_{in}(t))U(t)$$

Claim:

$$Y_{out}(t) = U^*(T)(I \otimes Y_{in}(t))U(t)$$

$\forall T > t$

$Pf\ d_T\{U^*(T)(I \otimes Y_{in}(t))U(T)\}$

$$= dU^*(T)(I \otimes Y_{in}(t))U(T)$$
$$+ U^*(T)(I \otimes Y_{in}(t))\,dU(T)$$
$$+ dU^*(T)(I \otimes Y_{in}(t))\,dU(T) = 0,\ T > t$$

Since $$dU(T) = \sum_k P_k dW_k(T)$$

Where P_k is a system operator, *i.e.*, $P_k \in \mathcal{L}(L^2(\mathfrak{h}))$ and $W_k(T) \in L^2\left(\Gamma s(L^2(0,T))\right)$

So

$$d_T\{U^*(T)(I \otimes Y_{in}(t))U(t)\}$$

$$= \sum_k U^*(T)\left\{P_j^* \otimes Y_{in}(t)\right\}U(t)dW_k^*(T)$$

$$+ \sum_k U^*(T)\left\{P_k \otimes Y_{in}(t)\right\}U(t)dW_k(T)$$

$$+ \sum_{k,m} U^*(T)\left\{P_k^* P_m \otimes Y_{in}(t)\right\}U(t)dW_k^*(T)\,dW_m(T)$$

and since $dT(U^*(T)\,U(T)) = 0$, we have

$$\sum_k \left(P_k^* dW_k^*(T) + P_k dW_k(T)\right) + \sum_{k,m} P_k^* P_m\, dW_k^*(T)\,dW_m(T) = 0$$

Note: $dW_k(T)$, $dW_k^*(T)$

Commute with $U(T)$, $U^*(t)$, $Y_{in}(t)$

Now, put $X \in L^2(\mathfrak{h})$ (system observable) and $X(t) = U^*(t)(X \otimes I)U(t), t \geq 0$

Then for $t \geq s$,

$$[X(t) = Y_{out}(s)]$$

$$= \left[U^*(t)(X \otimes I)U(t), U^*(t)(I \otimes Y_{in}(s))U(t)\right]$$

$$= U^*(t)[X \otimes I, I \otimes Y_{in}(s)]U(t) = 0$$

i.e.

Hence $\{Y_{out}(t), t \geq 0\}$ is a non-demolition measurement for $\{X(t), t \geq 0\}$ in the sense of Belavkin,

Now,

$$dY_{out}(t) = dY_{in}(t) + dU^*(t)\,dY_{in}(t)U(t) + U^*(t)\,dY_{in}(t)\,dU(t)$$

$$= \sum \lambda_j dA_j + \bar{\lambda}_j d\bar{A}_j + \sum C_k^j d\Lambda_j^k$$

$$+ U^*\left(\sum_j \left(L_j^{(1)*} dA_j^* + L_j^{(2)*} dA_j\right) + \sum S_k^{j*} d\Lambda_k^j\right)$$

$$\left(\sum_r \lambda_r dA_r + \bar{\lambda}_r dA_r^*\right) + C.C.\Big)U$$

Problem:

(a) Find

$$T_r\{(A + \varepsilon B)\log(A + \varepsilon B)\} - T_r(A \log A)$$

Up to $O(\varepsilon)$ when A, $B > 0$.

(b) Find $T_r\{A(\log(A+\varepsilon B)-\log A)\}$

Up to $O(\varepsilon)$ where A, $B > 0$.

Remark: The first formula can be used in finding the rate of change of the entropy (Von-Neumann) of a quantum system whose density matrix satisfies the Sudarshan-Lindblad equation.

$$\frac{d\rho_t}{dt} = -i[H,\rho_t]-\frac{1}{2}\{L^*L\rho_t+\rho_t L^*L-2L\rho_t L^*\}$$

and the second can be used to find the relative entropy of ρ_t per unit time.

❖❖❖❖❖

[32] Problem on Rigid Body Motion:

Consider the two link Robot with each link being a 3-D top. We have seen that the group of motions is defines by

$$G = \left\{\begin{bmatrix} R_2R_1 & R_1-R_2R_1 \\ 0 & R_1 \end{bmatrix} : R_1, R_2 \in SO(3)\right\}$$

$$= \left\{\begin{bmatrix} R & S-R \\ 0 & S \end{bmatrix} : R, S \in SO(3)\right\}$$

Show that if we define

$$T(R, S) = \begin{bmatrix} R & S-R \\ 0 & S \end{bmatrix}$$

Then

$$T(R_2, S_2).T(R_1, S_1) = \begin{bmatrix} R_2 & S_2-R_2 \\ 0 & S_2 \end{bmatrix}\begin{bmatrix} R_1 & S_1-R_1 \\ 0 & S_1 \end{bmatrix}$$

$$= \begin{bmatrix} R_2R_1 & S_2S_1-R_2R_1 \\ 0 & S_2S_1 \end{bmatrix} = T(R_2R_1, S_2S_1)$$

❖❖❖❖❖

[33] Problem from Revuz and Yor:

$$\lim_{x\to\infty}\frac{\sqrt{n}}{2}\int_0^t\left(\delta\left(B_s-\frac{x+a}{n}\right)-\delta\left(B_s-\frac{x}{n}\right)\right)ds$$

$$= \gamma(aL_t), t \geq 0$$

in distribution, *i.e.*
$$= \lim_{x\to\infty}\frac{\sqrt{n}}{2}\left(L_t^{\frac{(x+a)}{n}} - L_t^{x/n}\right) = \gamma(aL_t), t\geq 0$$

In distribution, where γ is a Brownian motion.

❑❑❑